Ecology

The Experimental Analysis of Distribution and Abundance

FOURTH EDITION

Charles J. Krebs

The University of British Columbia

HarperCollins*College*Publishers

To Joe Connell and Bob Paine
Ecologists extraordinaire

Editor-in-Chief: Glyn Davies
Developmental Editors: Meryl R. G. Muskin, Dick Morel
Project Editor: Brigitte Pelner
Design Supervisor and Cover Design: Jill Yutkowitz
Text Design: Judy Allan
Cover Photo: Jim Brandenburg/Minden Pictures
Production Administrator: Jeffrey Taub
Compositor: Ruttle, Shaw & Wetherill, Inc.
Printer and Binder: R. R. Donnelley & Sons Company
Cover Printer: The Lehigh Press, Inc.

For permission to use copyrighted material, grateful acknowledgment is made to the copyright holders on pp. 776–786, which are hereby made part of this copyright page.

Ecology: The Experimental Analysis of Distribution and Abundance, Fourth Edition
Copyright © 1994 by HarperCollins College Publishers

Library of Congress Cataloging-in-Publication Data

Krebs, Charles J.
   Ecology : the experimental analysis of distribution and abundance
   Charles J. Krebs. —4th ed.
       p.   cm.
   Includes bibliographical references (p.   ) and index.
   ISBN 0-06-500410-8
   1. Ecology.   2. Population biology.   3. Biogeography.   I. Title.
QH541.K67   1994   93-29542
574.5′24—dc20   CIP

93   94   95   96   9   8   7   6   5   4   3   2   1

Contents in Brief

Contents in Detail

Foreword

Twenty years ago, in 1973, I was a sophomore at Yale University, a biology major with an interest in ecology despite the fact that I really only had a vague notion as to what the field actually encompassed. That year, I took my first ecology course (as it turns out, the first of many), and the field came into clear focus for me. That I subsequently decided to become an ecologist was in part due to the exemplary efforts of the excellent instructors I was fortunate to encounter in that course; it was, I believe, equally due to the fact that those same instructors had the wisdom to select an excellent textbook for the course. It was the brand-new book *Ecology: The Experimental Analysis of Distribution and Abundance,* by Charles J. Krebs. I still have my 20-year-old copy of that first edition; it has actually never been far from my desk. Through the remainder of my undergraduate days, through graduate school at Cornell, and through a dozen years as an insect ecologist at the University of Illinois at Urbana-Champaign, the book has been an indispensable reference and guide for research and teaching—a source for clear and concise definitions, for apt and absorbing examples, and for background information in preparation for ventures into less familiar areas of ecology.

I happily now sever my sentimental ties to that well-worn book, to welcome this new, fourth edition. To say that much has changed in ecology over the past two decades is understatement of the greatest magnitude; the discipline has matured, embracing new technologies, incorporating new paradigms, and greatly expanded its theoretical as well as applied boundaries. Charles Krebs has not lost his touch in 20 years; this fourth edition sparkles with the same insightful, crisp prose and thorough, careful coverage that so impressed me as an undergraduate. I am no less impressed today; if anything, I am more impressed, because the discipline, so much broader than ever before, demands even greater expository skills in order to teach it. I have every expectation that, 20 years from now, many of you readers will feel much the same way about this copy of the fourth edition of *Ecology* as I did about the first edition; Charles Krebs is an eloquent and seductive spokesperson for the intellectual attraction of ecology.

May Berenbaum

Preface

Y ou are living in the Age of Ecology. As citizens, it becomes more and more important with each passing day that you understand your world and the ramifications of this age. A revolution has taken place over the last 30 years that has centered on the relationships between humans and their environment. The broad policy problems this revolution has illuminated have become the focus of the environmental movement. The applied scientific problems stemming from this are the focus of environmental science, and the basic science behind it all is the science of ecology.

Just as it is important to know something about physics if you aspire to be an engineer, it is necessary to learn something about ecology if you wish to understand the environmental problems we humans face today. *Ecology: The Experimental Analysis of Distribution and Abundance* presents a scientific overview of this science. By understanding how the natural world works, you will be in a better position to understand the Age of Ecology as it unfolds.

Textbook writers face two dilemmas. First, they must plot a course that places the book between the mistakes of the past and the trends of the present. We can recognize past mistakes, and lest history repeat itself we are wise to point them out through our educational system and in our texts. Unfortunately we are not as adept at recognizing future errors—at determining which of the current bandwagons in ecology are enduring and which are only ephemeral. Like most subjects of human endeavor, science is subject to fads, only some of which will serve a useful purpose.

Second, the writer must create a textbook that pleases not only the student but the instructor as well. In addressing both these issues, determining where to tread while satisfying the needs of students and instructor alike, I have tried to make this book both readable and informative.

In this fourth edition of *Ecology,* I have made two additions to the content. First, I have added material on conservation biology and community organization. Conservation biology is a focus for practical problems that cry out for ecological understanding. Many attempts to conserve biodiversity hinge on concepts of community organization that need careful thought and analysis. Second, I have strengthened the integration of evolutionary and functional ecology. Many chapters deal with ecological attributes and their evolutionary background. Ecologists can benefit by stepping back and looking at ecological systems from an evolutionary perspective, and students of evolution, in order to understand natural selection, can benefit from understanding how ecological systems function.

This book is not an encyclopedia of ecology and does not focus on descriptive ecology. Underlying the numerous approaches that characterize the study of ecology are a few basic issues. This text represents my attempt to present modern ecology as an interesting and dynamic subject by approaching these issues as a series of problem (or controversies, if you will) to be solved. These problems are illustrated diversity of examples chosen from both the plant and animal kingdoms and are sufficiently universal to be studied in a number of areas.

To understand the basic issues facing ecologists, students must have some background in biology and mathematics. While they will find that they can understand ecology without knowing any math, students who wish to proceed beyond the simplest level of analysis need this training. In this respect, ecology is no different from chemistry and physics. Statistics and calculus are useful but not essential to understand the material in this book. I present mathematical analyses step by step and illustrate them with graphs. Students who have difficulty following the mathematics should be able to understand the essence of the arguments from the graphs.

While mathematics plays a key role in the study of ecology, ecological controversies are *biological* in nature and will be resolved by biologists rather than mathematicians. An important part of ecological training is appreciating those controversies and understanding why people may reach opposing conclusions after examining the same data. While students may not have immediate answers, they can learn much about this discipline by analyzing a few of these controversies, exploring their scope, and determining the kinds of observations needed to solve them. Controversy is a sign of good science.

Each chapter in this book poses a problem and then gives you enough information to think through that problem intelligently. If you need to delve deeper, I've provided a list of selected references at the end of each chapter and a bibliography at the end of the book to use as your starting point. Each chapter ends with a series of questions and problems devised to stimulate thought, because no one can appreciate the quantitative aspects of ecology without going through some of the calculations. Most of the calculations are simple, but I have tried to leave some of them open-ended so that interested students can pursue them further through their own initiative. I have not provided answers to these problems, because for many of these the answer has yet to be found.

New to this edition are overview questions, one at the end of each chapter, which take the thought process a step further through their more general nature. These questions can easily serve as the focus of class discussion. Many overview questions are action oriented, with the goal of addressing the question, "What are the practical consequences of this idea?"

Technical terms are kept to a minimum; labeling with words should not be confused with understanding. The glossary of technical words, together with the indexes, should be adequate to cover technical definitions.

If there is a message in this book, it is a simple one: Progress in answering ecological questions comes when experimental techniques are used. The habit of asking, "What experiment could answer this question?" is the most basic aspect of the scientific method that students should learn to cultivate.

ACKNOWLEDGMENTS

I thank many friends and colleagues who have contributed to formulating and clarifying the material presented here. In particular I thank my colleagues Dennis Chitty, Judy Myers, Jamie Smith, and Tony Sinclair for their assistance and Brian Walker and the many ecologists at the CSIRO Division of Wildlife and Ecology in Canberra who answered endless queries during this revision. Dick Morel provided excellent advice and support to keep this edition on track, and Glyn Davies, Meryl

R. G. Muskin, and Brigitte Pelner at HarperCollins did more than their share of editorial work to help improve this edition. To all of these I am most grateful.

For a detailed critique of this revision I am indebted to the following reviewers: John T. Baccus, Southwest Texas State University; Robert M. R. Barclay, University of Calgary; May Berenbaum, University of Illinois/Urbana; Naomi Cappuccino, Université du Québec à Montréal; George Dale, Fordham University; Kenneth K. Ebel, Christ College Irvine; Aaron Ellison, Mount Holyoke College; Ron Etter, University of Massachusetts at Boston; Dale Hyde, Bloomsburg University; Ralph J. Larson, San Francisco State University; Sandra J. Newell, Indiana University of Pennsylvania; Frank A. Pitelka, University of California at Berkeley; Leslie Real, North Carolina State University; Stephen G. Tilley, Smith College; Charles F. Thompson, Illinois State University; and Mary K. Wicksten, Texas A&M University.

Finally, I want to thank the real authors of this book, the hundreds of ecologists who have toiled in the field and laboratory to extract from the study of organisms the concepts discussed here. A person's life work may be boiled down to a few sentences in this book, and we ecologists owe a debt that we cannot pay to our intellectual ancestors.

Charles J. Krebs

 *What Is Ecology?*

1 Introduction to the Science of Ecology

DEFINITION

The word *ecology* came into use in the last half of the nineteenth century. Ernst Haeckel in 1869 defined *ecology* as the total relations of the animal to both its organic and its inorganic environment. This very broad definition has provoked some authors to point out that if this is ecology, there is very little that is *not* ecology. Since four biological disciplines are closely related to ecology—genetics, evolution, physiology, and behavior—the problem of defining ecology may be viewed schematically in the following way:

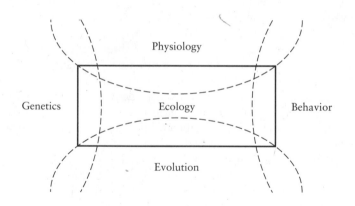

Broadly interpreted, ecology overlaps each of these subjects; hence we need a more restrictive definition.

Charles Elton (1927) in his pioneering book *Animal Ecology* defined ecology as *scientific natural history*. Although this definition does point out the origin of many of our ecological problems, it is again uncomfortably vague. Eugene Odum (1963) defined ecology as *the study of the structure and function of nature*. This statement has the merit of emphasizing the form-and-function idea that permeates biology, but it is still not a completely clear definition. A clear and restrictive definition of ecology is this: *Ecology is the scientific study of the distribution and abundance of organisms* (Andrewartha 1961). Nevertheless, this definition is static and leaves out the important idea of *relationships*. Ecology is about relationships, and we can modify Andrewartha's definition as follows: *Ecology is the scientific study of the interactions that determine the distribution and abundance of organisms*. This definition of ecology restricts the scope of our quest to a manageable level and forms the starting point for this book. We are interested, then, in *where* organisms are found, *how many* occur there, and *why*.

HISTORY OF ECOLOGY

The roots of ecology lie in natural history, which is as old as humans. Primitive tribes, which depended on hunting, fishing and food gathering, needed detailed knowledge of where and when their quarry might be found. The establishment of agriculture increased the need to learn about the practical ecology of plants and domestic animals.

Spectacular plagues of animals attracted the attention of the earliest writers. The Egyptians and Babylonians feared locust plagues, and supernatural powers were often believed to cause these outbreaks. Exodus (7:14–12:30) describes the plagues that God called down upon the Egyptians. In the fourth century B.C. Aristotle tried to explain these plagues of field mice and locusts in his *Historia Animalium*. He pointed out that the high reproductive rate of field mice could produce more mice than could be reduced by their natural predators, such as foxes and ferrets, or by the control efforts of man. Nothing succeeded in reducing these mouse plagues, Aristotle stated, except the rain, and after heavy rains the mice disappeared rapidly.

Ecological harmony was a guiding principle basic to the Greeks' understanding of nature, and Egerton (1968a) has traced this concept from ancient times to the modern term *balance of nature*. This concept of "providential ecology," in which nature is designed to benefit and preserve each species, was implicit in the writings of Herodotus and Plato. The assumptions of this world view were that the numbers of every species remain essentially constant. Outbreaks of some populations might occur, but these could usually be traced to divine intervention for the punishment of evil-doers. Each species had a special place in nature, and extinction did not occur because it would disrupt this balance and harmony in nature.

Little conceptual advance occurred until students of natural history and human ecology began to focus the ideas of population ecology and to provide an analytical framework. Graunt (1662), who described human population change in quantitative terms, can be called the father of demography* (Cole 1958). He recognized the importance of measuring in a quantitative way the birth rate, the death rate, sex ratio, and age structure of human populations, and he complained about the inadequate census data available in England in the seventeenth century. Graunt estimated the potential rate of population growth for London and concluded that even without immigration, London could double its population in 64 years.

Leeuwenhoek studied the reproductive rate of grain beetles, carrion flies, and human lice. In 1687 he counted the number of eggs laid by female carrion flies and calculated that one pair of flies could produce 746,496 flies in three months. Leeuwenhoek made one of the first attempts to calculate theoretical rates of increase for an animal species (Egerton 1968b).

Buffon in his *Natural History* (1756) touched on many of our modern ecological problems and recognized that populations of men, other animals, and plants are subjected to the same processes. Buffon discussed, for example, how the great fertility of every species was counterbalanced by innumerable agents of destruction. He believed that plague populations of field mice were checked partly by diseases and scarcity of food. Buffon did not accept Aristotle's idea that heavy rains caused the decline of dense mouse populations but thought that control was achieved by bio-

---

* Demography originated as the study of human population growth and decline. It is now used as a more general term to cover plant and animal population changes.

logical agents. Rabbits, he stated, would reduce the countryside to a desert if it were not for their predators. Buffon dealt with problems of population regulation that are still unsolved today.

Malthus published one of the earliest controversial books on demography. In his *Essay on Population* (1798), he calculated that although the numbers of organisms can increase geometrically (1, 2, 4, 8, 16, . . .), their food supply may never increase faster than arithmetically (1, 2, 3, 4, . . .). The arithmetic rate of increase in food production seems to be somewhat arbitrary, and Malthus may have presented this rate as a reasonable maximum (Flew 1957). The great disproportion between these two powers of increase led Malthus to infer that reproduction must eventually be checked by food production. The thrust of Malthus' ideas was negative: What prevents populations from reaching the point where they deplete their food supply? What checks operate against the tendency toward a geometric rate of increase? Two centuries later we still ask these questions. These ideas were not new, since Machiavelli had said much the same thing in about 1525, and Buffon in 1751, and several others had anticipated Malthus. It was Malthus, however, who brought these ideas to general attention. Darwin used the reasoning of Malthus as one of the bases for his theory of natural selection.

Other workers questioned the ideas of Malthus and made different predictions for human populations. For example, in 1841 Doubleday brought out his *True Law of Population.* He believed that whenever a species was threatened, nature made a corresponding effort to preserve it by increasing the fertility of its members. Human populations that were undernourished had the highest fertility; those that were well fed had the lowest fertility. Doubleday explained these effects by the oversupply of mineral nutrients in well-fed populations. Doubleday observed a basic fact that we recognize today, low birth rates in wealthy countries, although his explanations were completely wrong.

Interest in the mathematical aspects of demography increased after Malthus. Quetelet, a Belgian statistician, suggested in 1835 that the growth of a population was checked by factors opposing population growth. In 1838 his student Pierre-François Verhulst derived an equation describing the initial rapid growth and eventual levelling off of a population over time. This S-shaped curve he called the *logistic curve.* This work was overlooked until modern times but it is of fundamental importance, and we shall return to it later in detail.

Farr (1843) was one of the earliest demographers concerned with mortality. He discovered that in England there was a relation between the local density of the human population and the death rate, such that mortality increased with population. Farr returned in 1875 to further consideration of the human population of England. He pointed out that even though the death rate had been steadily declining in England during the 1800s, this event did not automatically lead to a population increase, since the birth rate might fall an equivalent amount. Farr pointed out that Malthus' postulate that food supply increases arithmetically was not necessarily true, at least in the United States, where food production had increased geometrically at a rate even greater than that of the human population.

During most of this time, philosophical thinking had not changed from the idea of Plato's day that there was harmony in nature. Providential design was still the guiding light. In the late eighteenth and early nineteenth centuries two ideas that undermined the idea of balance of nature gradually gained support: (1) that many species had become extinct and (2) that resources are limited and competition caused by population pressure is important in nature. The consequences of these two ideas became clear with the work of Malthus, Lyell, Spencer, and Darwin in the nineteenth century. Providential ecology and the balance of nature were replaced by natural selection and the struggle for existence (Egerton 1968c).

Yet, the balance-of-nature idea, redefined after Darwin, has continued to persist in modern ecology (Pimm 1991). The idea that natural systems are stable and in equilibrium with their environments unless humans disturb them is still believed by many ecologists and theoreticians. We discuss the equilibrium view in Chapters 24 and 25.

Many of the early developments in ecology came from the applied fields of agriculture, fisheries, and medicine. Work on the insect pests of crops has been one important source of ideas. The regulation of population size in insect pests is a basic problem that has long been under study. In 1762 the mynah bird was introduced from India to the island of Mauritius to control the red locust. By 1770 the locust threat was a negligible problem (Moutia and Mamet 1946). Forskål wrote in 1775 about the introduction of predatory ants from nearby mountains into date-palm orchards to control other species of ants feeding on the palms in southwestern Arabia. In subsequent years an increasing knowledge of insect parasitism and predation led to many such introductions all over the world in the hope of controlling introduced and native agricultural pests (Doutt 1964). We discuss this problem of *biological control* in Chapter 18.

Medical work on infectious diseases such as malaria around the 1890s gave rise to the study of epidemiology and interest in the spread of disease through a population. Before malaria could be controlled adequately, it was necessary to know in detail the ecology of mosquitoes. The pioneering work of Robert Ross (1908, 1911) attempted to describe in mathematical terms the propagation of malaria, which is transmitted by mosquitoes. In an infected area, the propagation of malaria is determined by two continuous and simultaneous processes: (1) The number of new infections among people depends on the number and infectivity of mosquitoes, and (2) the infectivity of mosquitoes depends on the number of people in the locality and the frequency of malaria among them. Ross could write these two processes as two simultaneous differential equations:

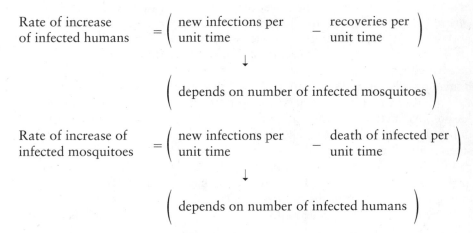

Ross had described an ecological process with a mathematical model, and his work represents a pioneering attempt at *systems analysis* (see Chapter 28). Such models can help us to clarify the problem—we can analyze the components of the model—and predict the spread of the disease.

Production ecology, the study of the harvestable yields of plants and animals, had its beginnings in agriculture, and Egerton (1969) traced this back to the eighteenth-century botanist Richard Bradley. Bradley recognized the fundamental similarities of animal and plant production, and he proposed methods of maximizing agricultural yields (and hence profits) for vineyards, trees, poultry, rabbits, and fish. The conceptual framework that Bradley used—monetary investment versus profit— could be applied to any organism. This *optimum-yield problem* is an important part of applied ecology (see Chapter 17).

Recognition of communities of living organisms in nature is very old, but specific recognition of the interrelations of the organisms in a community is relatively recent. Edward Forbes in 1844 described the distribution of animals in British coastal waters and part of the Mediterranean Sea, and he wrote of zones of differing depths that were distinguished by the associations of species they contained. Forbes noted that some species are found only in one zone and that other species have a maximum of development in one zone but occur sparsely in other adjacent zones. Mingled in are stragglers that do not fit the zonation pattern. Forbes recognized the dynamic aspect of the interrelations between these organisms and their environment. As the environment changed, one species might die out, and another might increase in abundance. Similar ideas were expressed by Karl Möbius in 1877 in a classic essay on the oyster-bed community as a unified collection of species. Möbius coined the word *biocoenosis* to describe such a community.

In a classic paper, "The Lake as a Microcosm," S. A. Forbes (1887) suggested that the species assemblage in a lake was an organic complex and that by affecting one species we exerted some influence on the whole assemblage. Each species maintains a "community of interest" with the other species, and we cannot limit our studies to a single species. Forbes believed that there is a steady balance of nature, which holds each species within limits year after year, even though each species is always trying to increase its numbers. The balance of nature kept community structure constant.

Studies of communities were greatly influenced by the Danish botanist J. E. B. Warming (1895, 1909). Warming raised questions about the structure of plant communities and the associations of species in these communities. The dynamics of vegetation change was emphasized first by North American plant ecologists. In 1899 H. C. Cowles described *plant succession* on the sand dunes at the southern end of Lake Michigan. This aspect of the development of vegetation was analyzed by Clements (1916) in a classic book that began a long controversy about the nature of the community (see Chapter 20).

With the recognition of the broad problems of populations and communities, ecology was by 1900 on the road to becoming a science. Its roots lay in natural history, human demography, biometry (statistical approach), and applied problems of agriculture and medicine.

The development of ecology during this century has followed the lines developed by naturalists during the last century. The struggle to understand how nature works has been carried on by a collection of colorful characters quite unlike the mythical stereotypes of scientists. From Alfred Lotka, who worked for the Metropolitan Life Insurance Company in New York while laying the groundwork of mathematical ecology (Kingsland 1985), to Charles Elton, the British ecologist who wrote the first animal ecology textbook in 1927 and founded the Bureau of Animal Population at Oxford (Crowcroft 1991), ecology has blossomed with an increasing understanding of our world and how we humans affect its ecological systems (McIntosh 1985).

Until the 1960s ecology was not considered an important science. The continuing increase of the human population and the associated destruction of natural environments with pesticides and pollutants awakened the public to the world of ecology. Much of this recent interest centers on the human environment and human ecology. Unfortunately, the word *ecology* became identified in the public mind with the much narrower problems of the human environment, and "ecology" came to mean everything and anything about the environment and especially human impact on the environment and its social ramifications. The science of ecology is not solely concerned with human impact on the environment, but with the interrelations of all plants and animals. As such, ecology has much to contribute to some of the broad questions about humans and their environment. Ecology should be to environmental science as physics is to engineering. Just as we humans are constrained by the laws of physics when we build airplanes and bridges, so also we are constrained by the principles of ecology when altering the environment.

BASIC PROBLEMS AND APPROACHES

We can approach the study of ecology from three points of view—*descriptive*, *functional*, or *evolutionary*. The descriptive point of view is mainly natural history and proceeds by describing the vegetation groups of the world, such as the temperate deciduous forests, tropical rain forests, grasslands, and tundra, and by describing the animals and plants within each of these ecosystems. The functional point of view, on the other hand, is oriented more toward dynamics and relationships and seeks to identify and analyze general problems common to most or all of the different areas. Functional studies deal with populations and communities as they exist and can be measured now. Functional ecology studies *proximate* causes—the dynamic responses of populations and communities to immediate factors of the environment. Evolutionary ecology studies *ultimate* causes—the historical reasons why natural selection has favored the particular adaptations we now see. The evolutionary point of view considers organisms and relationships between organisms as historical products of evolution. Functional ecologists ask *how*: How does the system operate? Evolutionary ecologists ask *why*: Why does natural selection favor this particular ecological solution? Since evolution has not only occurred in the past but is also going on at the present time, the evolutionary ecologist must work closely with the functional ecologist to understand ecological systems (Pianka 1988). The environment of an organism contains all the selective forces that shape its evolution, so ecology and evolution are two viewpoints of the same reality.

All three approaches to ecology can have shortcomings. The primary difficulty with the descriptive approach is that one can get entirely lost in it. We could use all the space in this book just to describe the temperate deciduous forest of North America. With the functional approach, there is a tendency to get far removed from reality, in the absence of detailed biological knowledge. The evolutionary approach can degenerate into undisciplined speculation about past events and provide hypotheses that can never be tested in the real world. No single approach can encompass all ecological questions. In this book I shall use a mixture of the functional and evolutionary approaches and emphasize the general problems faced by ecologists today.

The basic problem of ecology is to determine the causes of the distribution and abundance of organisms. Every organism lives in a matrix of space and time. Consequently, the two ideas of distribution and abundance are closely related, although at first glance they may seem quite distinct. What we observe for many species is that the numbers in an area vary in space.

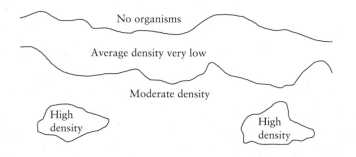

Figure 1.1 illustrates this idea for the blue jay of eastern North America. Blue jays are common in southern United States. They are rare in western Kansas and Nebraska and absent altogether in Montana. Why should these patterns of abundance occur? What limits the western and northern extension of the blue jay's range? These are examples of the fundamental questions an ecologist brings to nature.

The red kangaroo occurs throughout the arid zone of Australia (Figure 1.2). It is absent from tropical areas of northern Australia and most common in western New South Wales and central Queensland. Why are there no red kangaroos in tropical Australia above 14° south latitude? Why is this species absent from Victoria in southern Australia and from Tasmania?

We can view the average density of any species as a contour map, with the provision that the contour map may change with time. Throughout the area of distribution, the abundance of an organism must be greater than zero, and the limit of distribution equals the contour of zero abundance. Distribution may be considered a facet of abundance, and distribution and abundance may be said to be reverse sides of the same coin (Andrewartha and Birch 1954). The factors that affect the distribution of a species may also affect its abundance.

The problems of distribution and abundance can be analyzed at the level of the single species population or at the level of the community, which contains many species. The complexity of the analysis may increase as more and more species are considered in a community; consequently, in this book we shall consider first the simpler problems involving single species populations.

There is considerable overlap between ecology and its related disciplines which we cannot cover thoroughly in this book. *Environmental physiology* has developed with a wealth of information that is needed to analyze problems of distribution and abundance. *Population genetics* and *ecological genetics* are two additional foci of interest that we shall touch only peripherally. *Behavioral ecology* is another interdisciplinary area that has implications for the study of distribution and abundance. *Evolutionary ecology* is an important focus for problems of adaptation and studies of natural selection in populations.

Levels of Integration

In ecology we are dealing primarily with the three starred (*) levels of integration:

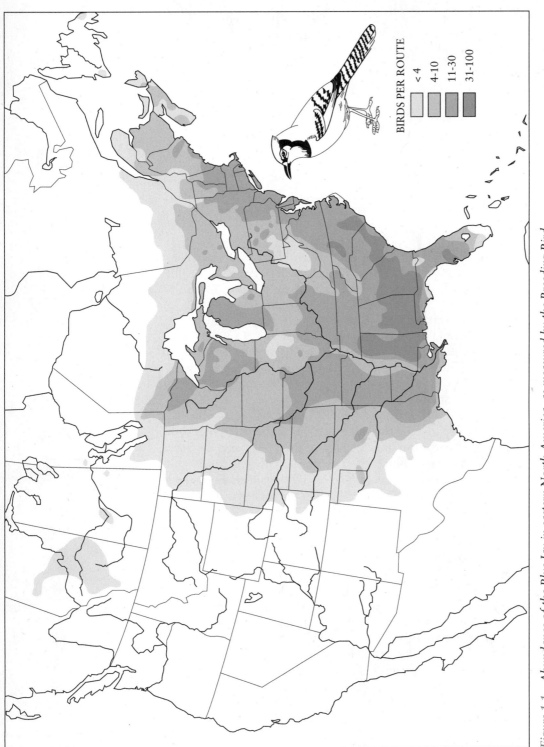

Figure 1.1   Abundance of the Blue Jay in eastern North America, as measured by the Breeding Bird Survey, 1966–1980. (From Robbins et al. 1986.)

BIRDS PER ROUTE

< 4
4-10
11-30
31-100

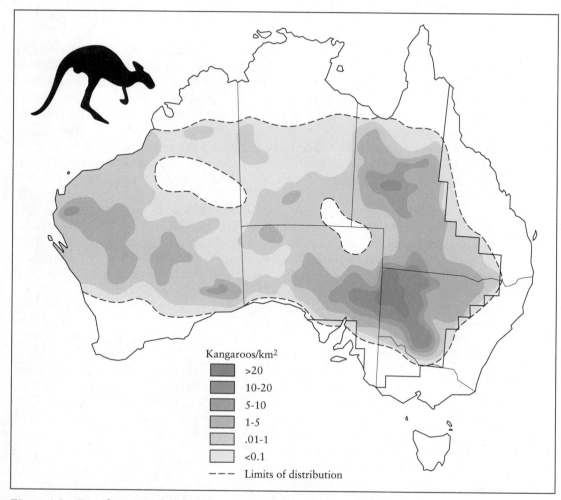

Figure 1.2   Distribution and abundance of the red kangaroo in Australia, from aerial surveys 1980–1982. (From Caughley et al. 1987a.)

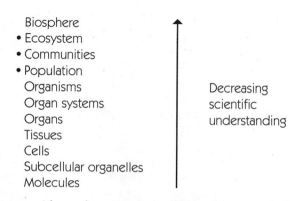

On one side, ecology overlaps with environmental physiology and behavior in studies of individual organisms, and on the other side, ecology fades into meteorology, geology, and geochemistry when we consider the biosphere, the whole-earth ecosystem. The boundaries of the sciences are not sharp but diffuse, and nature does not come in discrete packages.

Each level of integration involves a separate and distinct series of attributes and problems. For example, a population has a *density* (e.g., number of deer per square kilometer), a property that cannot be attributed to an individual organism. A community has a *species diversity,* an attribute without meaning at the population level. In general, a scientist dealing with a particular level of integration seeks explanatory mechanisms from lower levels of integration and biological significance from higher levels. For example, to understand mechanisms of changes in a population, an ecologist will study mechanisms that operate on individual organisms and will try to view the significance of these population events in a community and ecosystem framework.

Some ecologists have suggested that the ecosystem—the biotic community and its abiotic environment—is the basic unit of ecology (Tansley 1935, Rowe 1961, Evans 1956). A particular significance may be attached to the ecosystem level from the viewpoint of human ecology, but this is only one of the levels of organization at which ecologists operate. There are meaningful and important questions to be asked at each level of integration, and none of them should be neglected.

The extent of scientific understanding varies with the level of integration. We know a good deal about the molecular and cellular levels of organisms, we know something about organs and organ systems and about whole organisms, but we know relatively little about populations and even less about communities and ecosystems. This point is illustrated very nicely when you look at the levels of integration—ecology comprises about one-quarter of the levels of biology but no biology curriculum can be one-quarter ecology and do justice to *current* biological knowledge. The reasons for this situation are not hard to find—they include the increasing complexity of these higher levels and the inability to deal with them in the laboratory.

Whatever the reasons for this decrease in knowledge at the higher levels, it has serious implications for the study we are about to undertake. You will not find in ecology the strong theoretical framework that you find in physics, chemistry, molecular biology, or genetics. It is not always easy to see where the pieces fit in ecology, and we shall encounter many isolated parts of ecology that are well developed internally but are not clearly connected to anything else. This is typical of a young science. Many students unfortunately think of science as a monumental pile of facts that must be memorized. But science is more than a pile of precise facts; it is a search for systematic relations, for explanations to problems, and for unifying concepts. This is the growing end of science, so evident in a young science like ecology. It involves many unanswered questions and much more controversy.

The theoretical framework of ecology may be weak at the present time, but this must not be interpreted as a terminal condition. Eighteenth-century chemistry was perhaps in a comparable state of theoretical development as ecology is at the present time. Sciences are not static, and ecology is in a strong growth phase.

Methods of Approach

Ecology has been approached on three broad fronts: the *theoretical*, the *laboratory*, and the *field*. These three approaches are interrelated, but some problems have arisen when the results of one approach fail to verify those of another. For example, theoretical predictions may not be borne out in field data. We are primarily interested in understanding the distribution and abundance of organisms in *nature*, that is, in the *field*. Consequently, the field will always be our benchmark for comparisons, our basic standard.

Plant and animal ecology have tended to develop along separate paths. Historically, plant ecology got off to a faster start than animal ecology, despite the early interest in human demography. Because animals are highly dependent on plants, many of the concepts of animal ecology are patterned on those of plant ecology; *succession* is one example. Moreover, since plants are the ultimate source of energy for all animals, to understand animal ecology we must also know a good deal of plant ecology. This is illustrated particularly well in the study of community relationships.

Some important differences, however, separate plant and animal ecology. First, animals tend to be highly mobile, whereas plants are stationary. So a whole series of new techniques and ideas must be applied to animals, for example, to determine population density. Second, animals fulfill a greater variety of functional roles in nature—some are herbivores, some are carnivores, some are parasites. This distinction is not complete because there are carnivorous plants and parasitic plants, but the possible interactions are on the average more numerous for animals than for plants.

In the 1960s population ecology was stimulated by the experimental field approach in which natural populations were manipulated to test specific predictions arising from controversial ecological theory. During these years ecology was transformed from a static, descriptive science to a dynamic, experimental one in which theoretical predictions and field experiments were linked. At the same time, ecologists realized that populations were only parts of larger ecosystems and that we needed to study communities and ecosystems in the same experimental way as populations. To study a complex ecosystem, teams of ecologists had to be organized and integrated, and work of this scale was first attempted during the late 1960s and the 1970s.

Modern ecology is advancing particularly strongly in three major areas. First, communities and ecosystems are being studied with experimental techniques and analyzed as systems of interacting species processing nutrients and energy. Insights into ecosystems have been provided by the comparative studies of communities on different continents. Second, modern evolutionary thinking is being combined with ecological studies to provide an explanation of how evolution by natural selection has molded the ecological patterns we observe today. Behavioral ecology is a particularly strong and expanding area combining evolutionary insights with the ecology of individual animals. Third, conservation biology is becoming a dominant theme in scientific and in political arenas, and this has increased the need for ecological input in habitat management. All of these developments are providing excitement for students of ecology in the 1990s.

Application of the Scientific Method to Ecology

The essential features of the scientific method are the same in ecology as in other sciences (Figure 1.3). An ecologist begins with a problem, often based on natural history observations. For example, pine tree seedlings do not occur in mature forests on the Piedmont of North Carolina. If the problem is not based on correct observations, all subsequent stages will be useless; thus accurate natural history is a prerequisite for all ecological studies. Given a problem, an ecologist suggests a possible answer. This answer is called a *hypothesis*. In many cases, several answers

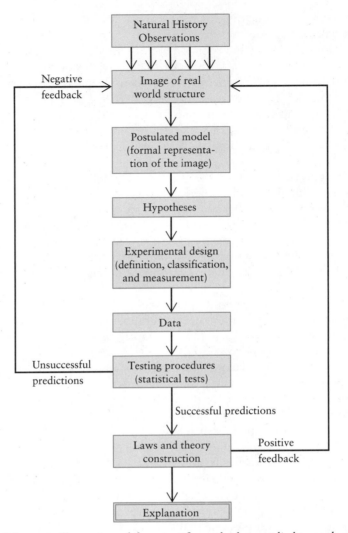

Figure 1.3   Schematic illustration of the scientific method as applied to ecological questions.

might be possible, and several different hypotheses can be proposed to explain the observations. Hypotheses arise from previous research, intuition, or inspiration. The origin of a hypothesis tells us nothing about its likelihood of being correct.

A hypothesis makes *predictions*, and the more predictions it makes the better. Predictions follow logically from the hypothesis, and mathematical reasoning is most useful here in checking on the logic of the predictions. An example of a hypothesis is that pines do not grow under hardwoods because of a shortage of light. Alternative hypotheses might be that it is a shortage of pine seeds, or a shortage of soil water. Predictions from simple hypotheses like these are often straightforward: If you provide more light, pine seedlings will grow (under the light hypothesis). A hypothesis is tested by making observations to check the predictions—an *experiment*. An experiment is defined as any set of observations that test a hypothesis. Experiments can be manipulative or natural. We could provide light artificially under the mature forest canopy, or we could look for natural light gaps in the forest. The protocol for the experiments and the data to be obtained are called the *experimental design*. From the data that result from the experiments we either accept or reject the hypothesis. And so the cycle begins again (Figure 1.3)

There are many qualifications that need to be attached to this simple scheme. Popper (1963) pointed out that we should always look for evidence that falsifies a hypothesis, and that progress in science consists of getting rid of incorrect ideas. In the real world we cannot achieve this ideal. We should also prefer simple hypotheses over complex ones, according to Popper, because we can reject simple hypotheses more quickly. This does not mean that we must be simple-minded; in ecology we must deal with complex hypotheses because the natural world is not simple. Every hypothesis must forbid something and part of the predictions of a hypothesis must be exactly what it forbids. If a hypothesis predicts everything and forbids nothing, it is quite useless in science.

Ecological systems are complex and this fact causes difficulty in applying the simple method outlined in Figure 1.3. In some cases factors operate together, so it may not be an "either light or water" situation for pine seedlings but "both light and water" together. Systems in which many factors operate together are most difficult to analyze, and ecologists must be alert for their presence (Quinn and Dunham 1983). The principle, however, remains: No matter how complex the hypothesis, it must make some predictions that we can check in the real world.

All ecological systems have an evolutionary history, and this provides another fertile source of possible explanations. There is controversy in ecology about whether one needs to invoke evolutionary history to explain present-day population and community dynamics. Evolutionary hypotheses can be tested as Darwin did by comparative methods but not by manipulative experiments (Diamond 1986).

Some ecological hypotheses are statistical in nature and do not fall into the "either A or B" category of hypotheses. Statistical hypotheses postulate quantitative relationships. For example, in North Carolina forests pine seedling abundance (per square meter) is linearly related to incident light in summer. Tests of statistical hypotheses are well understood and are discussed in all statistics textbooks. They are tested in the same way as indicated in Figure 1.3.

Some ecological hypotheses have been very fruitful in stimulating work, even though they are known to be incorrect. The progress of ecology, and science in general, occurs in many ways using mathematical models, laboratory experiments, and field studies. The following chapters will illustrate the variety of problems and approaches that characterize the young science of ecology.

Selected References

BARBOUR, M. G., J. H. BURK, and W. D. PITTS. 1980. A brief history of plant ecology. In *Terrestrial Plant Ecology*, chap. 2. Benjamin/Cummings, Menlo Park, California.

CAUGHLEY, G., J. SHORT, G. C. GRIGG, and H. NIX. 1987. Kangaroos and climate: An analysis of distribution. *Journal of Animal Ecology* 56:751–761.

CROWCROFT, P. 1991. *Elton's Ecologists: A History of the Bureau of Animal Population*. University of Chicago Press, Chicago. 177 pp.

EGERTON, F. N., III. 1973. Changing concepts of the balance of nature. *Quarterly Review of Biology* 48:322–350.

GOODLAND, R. J. 1975. The tropical origin of ecology: Eugen Warming's jubilee. *Oikos* 26:240–245.

KINGSLAND, S. E. 1985. *Modeling Nature: Episodes In the History of Population Ecology.* University of Chicago Press, Chicago. 267 pp.

McINTOSH, R. P. 1985. *The Background of Ecology: Concept and Theory.* Cambridge University Press, Cambridge. 383 pp.

MERTZ, D. B. and D. E. McCAULEY. 1980. The domain of laboratory ecology. *Synthese* 43:195–255.

PLATT, J. R. 1964. Strong inference. *Science* 146:347–353.

POPPER, K. R. 1963. *Conjectures and Refutations,* chap. 1. Routledge & Kegan Paul, London.

SHEAIL, J. 1987. *Seventy-Five Years In Ecology: The British Ecological Society.* Blackwell Scientific Publications, Oxford.

Questions and Problems

1. "The definition . . . 'ecology is the branch of biological science that deals with relations of organisms and environments' would provide the title for an encyclopedia but does not delimit a scientific discipline" (Richards 1939, p. 388). Discuss.

2. Is it necessary to define a scientific subject before one can begin to discuss it? Contrast the introductions in several textbooks of ecology with those in textbooks of some areas of physics and chemistry, as well as other biological areas, such as genetics and physiology.

3. Is it necessary to study the methodology and philosophy of science in order to understand ecology? Consider this question before and after reading the essays by Popper (1963) and Platt (1964).

4. Ask several of your nonbiologist friends how they would define ecology, and discuss the distinction between *ecology* and *environmental studies*.

5. Discuss the application of the distribution and abundance model on page 9 to the human population.

6. Quinn and Dunham (1983) argue that the conventional methods of science cannot be applied to ecological questions because many factors act together to produce ecological changes. Discuss the problem of "multiple causes" and how scientists can deal with complex systems that have multiple causes.

Overview Question

How much has the science of ecology depended upon prior advancements in other sciences such as statistics, chemistry, physics and computing? Does ecology make progress only after progress in these sciences, or do the sciences progress in tandem?

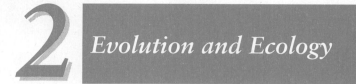

2 Evolution and Ecology

Charles Darwin was an ecologist before the term had even been coined, and he is an appropriate patron for the science of ecology because he realized the intricate connection between ecology and evolution. As we discuss ecological ideas throughout this book, we will use evolutionary concepts. This chapter provides a brief survey of the basic principles of evolution that are important in evolutionary ecology.

WHAT IS EVOLUTION?

Evolution is change, and biological evolution can be defined as changes in any attribute of a population over time. But we must be more specific than this. Evolutionary changes lead to adaptation and must involve a change in the frequency of individual genes in a population from generation to generation. What produces evolutionary changes?

Natural selection, said Charles Darwin and Alfred Wallace independently in 1858, is the mechanism that drives adaptive evolution. Natural selection operates through the following steps:

1. Variation occurs in every group of plants and animals. Individuals of the same species are not identical in any population, as was known with the breeding of domestic animals.

2. Every population of organisms produces an excess of offspring. (The high reproductive capacity of plants and animals was well known to Malthus and Buffon long before Darwin.)

3. Life is difficult, and not all individuals will survive.

4. Among all the offspring competing for limited resources, only those individuals best able to obtain and use these resources will survive and reproduce.

5. If the characteristics of these organisms are inherited, the favored traits will be more frequent in the next generation.

Natural selection will favor traits that allow individuals possessing those traits to leave more descendants. The process of natural selection is the end result of the processes of ecology in action. The environments that organisms inhabit shape the evolution that occurs. The present distribution, abundance and diversity of animals and plants are set by the evolutionary processes of the past impinging on the environment of the present.

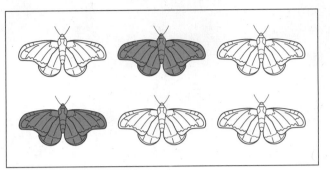

(a) Variability in wing color

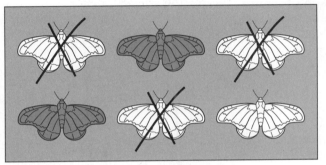

(b) Life is hard – only the "fit" will survive

(c) Heritability of wing color

Figure 2.1   *Schematic illustration of how natural selection operates using the example of industrial melanism in the moth* Biston betularia. *Air pollution changed the background color of tree bark and visual predation by birds removed the moths that did not match the background.*

A simple example of natural selection is shown in Figure 2.1. The moth *Biston betularia* shows variation in the amount of black color on the wings. When industrial pollution in central England caused lichens on tree bark to die, leaving a dark background, black-colored moths survived better because bird predators could not see them (Figure 2.1b). Black wing color is inherited in these moths, and the result has been an increase in the frequency of black moths during industrialization (Figure 2.1c). Ironically, since industrial pollution has decreased in the last 30 years, this process of natural selection is reversing (Murray et al. 1980).

Evolution through natural selection results in *adaptation* and under appropriate conditions produces new species *(speciation)*. Both these processes have ecological implications.

ADAPTATION

Natural selection acts on *phenotypes*—the observable attributes of individuals. Different genotypes give rise to different phenotypes, but not in a simple way because embryological and subsequent development is affected in many ways by environmental factors such as temperature. Therefore, it is simpler to observe the effect of natural selection directly on the phenotype and to ignore the underlying genotype. Ecologists, like plant and animal breeders, are primarily interested in phenotypic characters like seed numbers or body size.

   Three types of selection can operate on phenotypic characters (Figure 2.2). *Directional selection* is the simplest form, in which phenotypes at one extreme are selected against. Directional selection produces genotypic changes more rapidly than

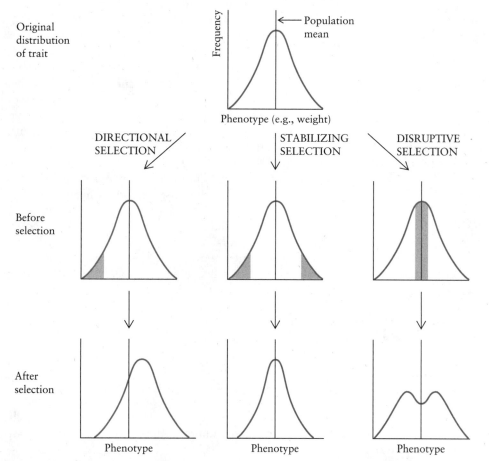

Figure 2.2   Three types of selection on phenotypic characters. Individuals in the shaded areas are selected against. (After Tamarin 1993.)

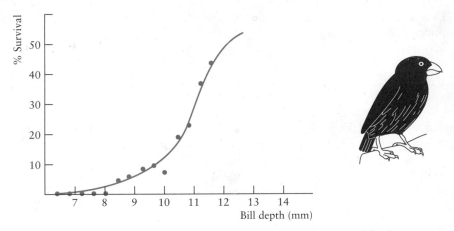

Figure 2.3   Directional selection for beak size in the Galápagos ground finch Geospiza fortis. *From 1976 to 1978 a severe drought in the Galápagos Islands caused an 85 percent drop in the population, and birds with larger beaks survived better because they could crack larger, harder seeds. (From Grant 1986, p. 213.)*

any other form, so most artificial selection is of this type. Figure 2.3 illustrates directional selection in the Galápagos ground finch *Geospiza fortis*. During a prolonged drought the birds that survived were predominantly those with large beaks that could crack large seeds (Boag and Grant 1981). Birds with large beaks can eat both large and small seeds, while birds with small beaks can eat only small seeds. Directional selection probably accounts for most of the phenotypic changes that occur during evolution. In wild populations resistance of pests to insecticides or herbicides is produced by directional selection.

Stabilizing selection is very common in present-day populations. In stabilizing selection, phenotypes near the mean of the population are more fit than those at either extreme, and thus the population mean value does not change. Figure 2.4 illustrates stabilizing selection for birth weight in humans. Early mortality is lowest for babies around 3.3 kilograms, very close to the observed mean birth weight of 3.2 kilograms for the population.

Figure 2.5 shows another example of stabilizing selection in lesser snow geese. Snow geese nest in colonies in northern Canada, and clutches hatch over a two-week period in early summer. Because predation is concentrated on whole colonies, eggs hatching synchronously confer a safety-in-numbers advantage against predators like foxes. Females whose eggs hatch synchronously on or near the mean date for the colony are more likely to raise their young successfully. Nests that hatch early suffer more predation loss, as do nests that hatch later. The result is natural selection favoring an optimum hatching time (Cooke and Findlay 1982).

Disruptive selection is a third type of selection in which the extremes are favored over the mean. But because the extreme forms breed with one another, every generation will produce many intermediate forms doomed to be eliminated. In any environment favoring the extremes, mechanisms that would prevent the opposite extremes from breeding with one another would be advantageous. Isolating mechanisms are thus an important adjunct of disruptive selection. There are no good examples known of disruptive selection.

The net result of all this selection is that organisms are adapted, and the great diversity of biological forms is a graphic essay on the power of adaptation by natural

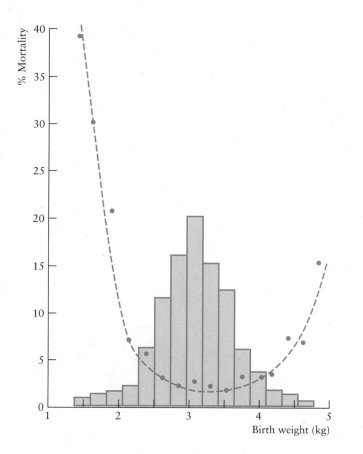

Figure 2.4   Stabilizing selection for birth weight in humans. The shaded histogram gives the distribution of birth weights in the population. The curve represents percentage of mortality at birth. (Data from Karn and Penrose 1951.)

selection. But we must be careful to note that adaptation does not produce the "best" phenotypes or "optimal" phenotypes (defined as phenotypes that are theoretically the most efficient in surviving and reproducing). The better survive, not the "best," and the biological world can never be described as "the best of all possible worlds."

Adaptation is constrained in real populations by four major forces. First, genetic forces prevent perfect adaptation because of mutation and gene flow. Mutation is always occurring, generating variation in populations, and most mutations are detrimental to organisms rather than adaptive. The immigration of individuals into an area where local environments differ will add other alleles to the gene pool and acts to smooth out local adaptations. Second, environments are continually changing, and this is the most significant short-term constraint on adaptation. Third, adaptation is always a compromise because organisms have at their disposal only a limited amount of time and energy. There are tradeoffs between adaptations, such as wing-shape in birds. A loon's wings are efficient for diving but not so efficient for flying. Fourth, historical constraints are always present because organisms have a history and change in small increments. Let us look in detail at one example of adaptation to illustrate some of these principles.

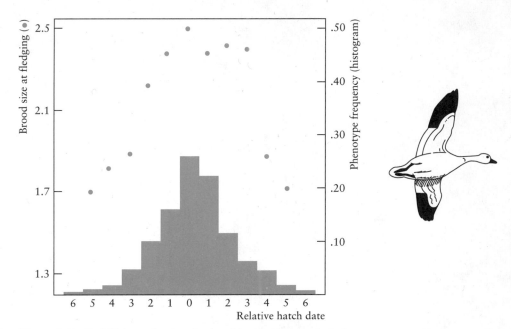

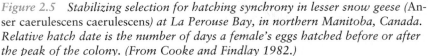

*Figure 2.5   Stabilizing selection for hatching synchrony in lesser snow geese (An-
ser caerulescens caerulescens) at La Perouse Bay, in northern Manitoba, Canada.
Relative hatch date is the number of days a female's eggs hatched before or after
the peak of the colony. (From Cooke and Findlay 1982.)*

Clutch Size in Birds

Each year emperor penguins lay 1 egg, pigeons lay 1 or 2 eggs, gulls typically
lay 3 eggs, the Canada goose 4 to 6 eggs, and the American merganser 10 or 11
eggs. What determines clutch size in birds? We must distinguish two different levels
of this question—proximate and ultimate. *Proximate factors* explain *how* a trait is
regulated by an individual. Proximate factors that determine clutch size are the
physiological factors that control ovulation and egg laying. Proximate factors involve
physiological machinery and how it works. Ultimate factors are always selective
factors, and ultimate explanations for clutch size differences always involve evolutionary arguments about adaptations. Proximate factors affecting clutch size have
to do with how an individual bird decodes its genetic information on egg laying.
Ultimate factors have to do with changes in this genetic program through time and
with the reason for these changes (Mayr 1982). Clutch size may be modified by the
age of the female, spring weather, population density, and habitat suitability. The
ultimate factors that determine clutch size are the requirements for long-term (evolutionary) survival. Clutch size is viewed as an adaptation under the control of
natural selection, and we seek the selective forces that have shaped the reproductive
rates of birds. We shall not be concerned here with the proximate factors determining
clutch size, which are reviewed by Drent and Daan (1980) and by Murphy and
Haukioja (1986).

Natural selection will favor the birds that leave the most descendants to future
generations. At first thought we might hypothesize that natural selection might favor
a clutch size that is the physiological limit the bird can lay. We can test this hypothesis
by taking eggs from nests as they are laid. When we do this, we find that some birds,

such as the common pigeon, are *determinate* layers; they lay a given number of eggs, no matter what. The pigeon lays 2 eggs; if you take away the first, it will incubate the second egg only. If you add a third, it will incubate 3 eggs. But many other birds are *indeterminate* layers; they will continue to lay eggs until the nest is "full." If eggs are removed as they are laid, these birds will continue laying. This subterfuge has been used on a mallard female, which continued to lay 1 egg per day until she had laid 100 of them. In other experiments, herring gull females laid up to 16 eggs (normal clutch 2–3), a yellow-shafted flicker female 71 eggs (normal clutch 6–8), and a house sparrow 50 eggs (normal clutch 3–5) (Klomp 1970). This evidence suggests that most birds under normal circumstances do not lay their physiological limit of eggs but that ovulation is stopped long before this limit is reached.

The next hypothesis we might suggest is that the clutch size of birds is limited by the maximum number of eggs a bird can cover with its brood patch. This may be the case for a few birds that lay many eggs. But in many cases, the brooding capacity can be shown experimentally to be larger than the actual clutch size. For example, the partridge in England typically lays 15 eggs, but up to 20 eggs can be successfully hatched (Jenkins 1961). The gannet lays 1 egg but will incubate 2 eggs successfully if one is added (Nelson 1964). Clutch size in most birds is probably not limited by brooding capacity.

One way to think about this problem of clutch size is to use a simple economic approach. Everything an organism does has some *costs* and some *benefits*. Organisms integrate these costs and benefits in evolutionary time. The benefits of laying more eggs are very clear—more descendants in the next generation. The costs are less clear. There is an energy cost to make each additional egg. There is a further cost to feeding each additional nestling. If the adult birds must work harder to feed their young, there is also a potential cost in adult survival—they may not live until the next breeding season. If adults are unable to work harder, there is a potential reduction in offspring quality. A cost-benefit model of this general type is shown in Figure 2.6.

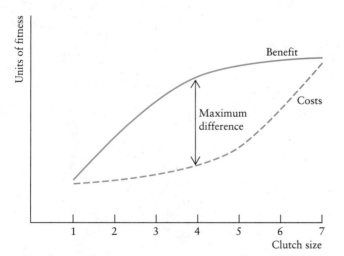

Figure 2.6   A cost-benefit model for the evolution of clutch size in birds. An individual benefits from laying more eggs because it will have more descendants, but it incurs costs because of increasing parental care in larger clutches. The point of maximal difference between benefits and costs is the optimal point for an individual (the Lack clutch size).

Models of this type are called *optimality models*. They are useful because they help us to think about what the costs and what the benefits are for a particular ecological strategy.

No organism has an infinite amount of energy to spend on its activities. The reproductive rate of birds can be viewed as one sector of a bird's energy balance, and the needs of reproduction must be maximized within the constraints of other energy requirements. The total requirements involve metabolic maintenance, growth, and energy used for predator avoidance, competitive interactions, and reproduction. In 1947 David Lack suggested that the clutch size of birds that feed their young in the nest was adapted by natural selection to correspond to the largest number of young for which the parents can provide enough food. This has been a very fertile hypothesis in evolutionary ecology because it has stimulated a variety of experiments. According to Lack's hypothesis, if additional eggs are placed in a bird's nest, the whole brood will suffer from starvation so that, in fact, fewer young birds will fledge from nests with larger numbers of eggs. In other words, clutch size is postulated to be under stabilizing selection (Figure 2.2). Let us look at a few examples to test this idea.

In England, the swift normally lays a clutch of two or three eggs. What would happen if swifts had a brood of four? Perrins (1964) artificially created broods of four by adding a chick at hatching and found that the survival of the young swifts in broods of four was poor (Figure 2.7). Swifts feed on airborne insects and apparently cannot feed four young adequately, so all the young starve. Consequently, it would not pay a swift in the evolutionary sense to lay four eggs, and the results are consistent with Lack's hypothesis.

Tropical birds usually lay small clutches, and Skutch (1967) argued that this was an adaptation against nest predators. If the intensity of nest predation increases with the number of parental feeding trips away from the nest, natural selection would favor a reduced clutch size. Hole-nesting passerine birds lay more eggs than comparable species that nest in the open (Slagsvold 1989), and predation rates are much lower for hole-nesting species (Murphy and Haukioja 1986). This suggests that a high risk of predation on the whole brood in the nest is a strong selective factor

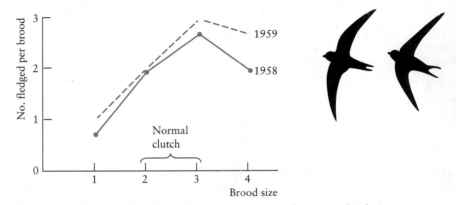

Figure 2.7   Production of young swifts (Apus apus) *in relation to clutch size in England. The normal clutch is two to three; broods were increased to four artificially. Larger broods do not produce more young, and natural selection is stabilizing. (After Perrins 1964.)*

that reduced clutch size in open-nesting birds, and also favors a shortened nesting period, independent of the ability of the parents to provide food to the nestlings.

Temperature regulation is an important component of the development of young birds and may have a bearing on clutch size. Small broods will not have the added warmth of the huddling that occurs in large broods, and consequently much energy may be used by nestlings in small broods just for thermoregulation (Royama 1969). In very large broods, the opposite problem, overheating, may occur. Thermoregulation is another component of reproduction that may place some restraint on clutch size.

Natural selection would seem to operate to maximize reproductive rate, subject to the constraints imposed by thermoregulation, feeding, and predator avoidance. This is called the *theory of maximum reproduction*, and Lack's hypothesis is part of this theory. It is a good example of how stabilizing selection can operate on a phenotypic trait such as reproductive rate. The maximum clutch size is called the *Lack clutch size* (Figure 2.6).

A major difficulty of the work done to date on the clutch-size problem is that most of the information available covers one part (*) of the life cycle:

Adults—clutch size—*fledged young—breeding adults

The critical variable is the breeding adults that emerge from a given size of clutch, and this is missing from many studies because the fledged young-to-adult stage cannot easily be studied (Murphy and Haukioja 1986). We must assume that more fledged young implies more breeding adults for the studies reported, and this assumption may not be correct for some species. A second problem is the genetic basis of clutch-size differences, on which little information is available. If clutch size is variable only because of environmental influences and not heritable, natural selection cannot operate on this trait and the studies reported here are less meaningful. If clutch size is variable in part because of genetic variation, we can reasonably ask whether adults of different "clutch-size genotypes" give rise to equal numbers of adults in the next generation.

Not all manipulation experiments confirm Lack's hypothesis. Nur (1984) manipulated clutch size in the blue tit (*Parus caeruleus*) in England to produce clutches ranging from 3 to 15. Figure 2.8 shows the resulting offspring produced. The number of surviving offspring per brood was maximized for broods of 12. Since average brood size was 9, the most common clutch size was less than the most productive clutch size. Lessells (1986) surveyed 35 bird species in which clutches had been manipulated, and found that 22 species (63 percent) were like the blue tit—the most common clutch was not the most productive. Why should this be?

The presence of tradeoffs is one explanation of why clutches should be smaller than the Lack clutch size. Clutch size may affect the chances of the adult birds surviving to breed again. Birds may become exhausted by rearing large clutches—a delayed cost of reproduction. Alternatively, laying a large clutch may postpone the next breeding attempt, leading to reduced lifetime reproduction.

The Lack clutch size is not constant for a species, and different individuals will vary in their parenting abilities and have a personal Lack clutch size. There is as yet no clear evidence for this individual optimization in birds (Godfray et al. 1991).

An alternative explanation of why the average clutch may be smaller than the Lack clutch size is that observed clutch sizes are a nonadaptive compromise. If gene flow occurs between two habitats, one good and one poor, clutches may be larger than optimal in poor habitats and smaller than optimal in good habitats. Blue tits

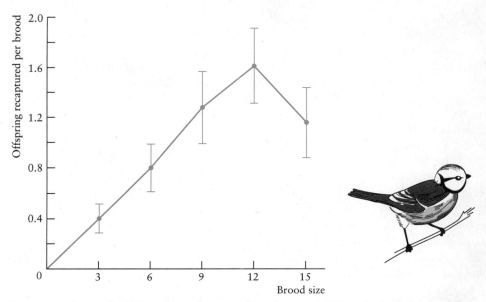

Figure 2.8   Survival of blue tits in relation to brood size in 1978 at Oxford, England. Brood size was experimentally manipulated. The most common brood size was 9, but the most productive brood size was 12 (the Lack brood size). The results are contrary to what Lack's hypothesis predicts. (From Nur 1984.)

and great tits in Belgium rarely breed in woodlands where they were born and show this nonadaptive compromise (Dhondt et al. 1990).

Recent work on bird reproduction addresses the question as to how individual parents adjust their reproductive costs in relation to environmental conditions to maximize the output of young (Godfray et al. 1991). The proximate controls of reproduction operate through the energy available to reproducing birds, and the role of female condition is critical in determining reproductive effort. Reproductive effort this year may affect the chances of surviving until next year, and parents must balance short-term and long-term costs of breeding.

COEVOLUTION

The term *coevolution* was popularized by Ehrlich and Raven (1964) to describe the reciprocal evolutionary influences that plants and plant-eating insects have had on each other. Coevolution occurs when a trait of species A has evolved in response to a trait of species B, which has in turn evolved in response to the trait in species A. Coevolution is specific and reciprocal. In the more general case, several species may be involved instead of just two, and this is called diffuse coevolution (Futuyma and Slatkin 1983).

Coevolution is simply a part of evolution, and it provides important linkages to ecology. The interactions between herbivores and their food plants have been emphasized as a critical coevolutionary interaction (Chapter 15). Predator–prey interactions (Chapter 14) can also be coevolutionary, and in some cases can lead to "arms races" between species.

Coevolution shapes the characteristics of coevolving pairs of species, while diffuse coevolution might also occur in communities of many species. There is considerable controversy about whether community-level patterns could be caused by coevolution (Orians and Paine 1983), and we shall explore this question in Chapters 24 and 25.

SPECIATION

One of the central problems of biology is to understand the origin of species. The most widely accepted hypothesis explaining the process of multiplication of species is the geographic theory of speciation (or *allopatric speciation*), which operates in four steps:

1. Reproductive isolation occurs because of physical, geographic separation of two populations.

2. The two reproductively and geographically isolated populations undergo independent evolution and become adapted to their separate environments.

3. Reproductive isolation must evolve, so that mechanisms occur to reduce interbreeding between the two populations.

4. If geographic isolation stops and the two populations come into contact, and if some reproductive isolating mechanism has evolved, the speciation process is complete, and two species now exist.

Reproductive isolating mechanisms are environmental, behavioral, mechanical, and physiological barriers that prevent two individuals of two species from producing viable offspring. A variety of isolating mechanisms have been described (Futuyma 1986, p. 112):

A. Prezygotic mechanisms: Fertilization and zygote formation are prevented.

   1. Habitat: The populations live in the same region but occupy different habitats.

   2. Seasonal or temporal: The populations exist in the same regions but are sexually mature at different times.

   3. Ethological (in animals only): The populations are isolated by different and incompatible behavior before mating.

   4. Mechanical: Cross-fertilization is prevented or restricted by incompatible differences in reproductive structures (genitalia in animals, flowers in plants).

B. Postzygotic mechanisms: Fertilization takes place and hybrid zygotes are formed but are inviable or give rise to weak or sterile hybrids.

The geographic theory of speciation is not the only way in which new species can arise (Figure 2.9). Two additional theories of speciation have been suggested as alternative models.

Parapatric speciation can occur when a population of a widespread species enters a new habitat. Although no physical barrier separates this new population from other populations, occupancy of the new habitat will result in a barrier to gene flow between the population in the new habitat and the rest of the species. Parapatric speciation may be common in organisms that move very little, such as plants, moles, and flightless insects. Insect herbivores that mate on the host plant could undergo

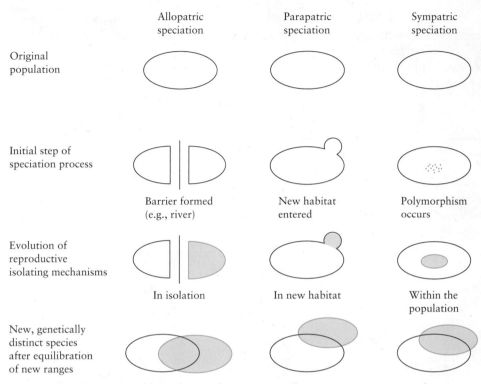

Figure 2.9   *Three general hypotheses of speciation. Allopatric speciation is probably the most common type. (After Tamarin 1993.)*

parapatric speciation if different races live on different plant species. Parapatric speciation is probably not as common as allopatric speciation.

   Sympatric speciation occurs when reproductive isolation occurs within the range of a population and before any differentiation of the two species can be detected. Sympatric speciation may be rare in nature but may be common in parasites. Parasites are usually very host-specific and often mate on the host. Consequently, if a new genotype is able to colonize a new host successfully, it may immediately become reproductively isolated from the parent population. Plants may speciate through changes in ploidy (Futuyma 1986).

   Speciation has ecological consequences, as we shall see in subsequent chapters. Closely related species may have similar resource requirements and thus compete for resources. The distribution of many species reflects the barriers that have shaped geographic isolation.

   There has been considerable controversy in evolutionary biology about the timing of evolutionary changes during speciation. On the one extreme, the gradualists think that evolution proceeds slowly and morphological changes occur because of natural selection operating gradually (Figure 2.10). At the other extreme, the punctuated equilibrium school thinks that evolution occurs in steps, with periods of rapid change when speciation occurs followed by periods of little or no change (Figure 2.10). Because the fossil record is so discontinuous, it is difficult in many cases to decide which of these two models is closer to the historical pattern of evolution. The

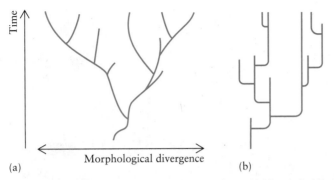

Figure 2.10   Two possible scenarios for the evolution of morphological changes in a lineage of species. Time is in millions of years. (a) An idealized version of phyletic gradualism—change is slow and steady and is not accelerated during speciation. (b) An idealized case of punctuated equilibrium—change occurs rapidly and only during speciation events. (After Futuyma 1986.)

punctuated equilibrium model postulates that stabilizing selection is the norm for most of evolutionary time, and that disruptive selection occurs rapidly during speciation. The important point in this controversy is the time scale. Fifty thousand years is only a brief moment in the fossil record and it is difficult to measure evolutionary changes on a finer time scale. Figure 2.11 shows the phylogeny of the horse family over the last 55 million years. Two points stand out in this phylogeny. First, the rate of evolution has varied greatly in geological time. During part of the Miocene (10 to 15 million years ago) there was rapid speciation of horses. Second, almost all of these horses are now extinct. In particular we have no living representatives of all the browsing horse species that lived from 10 to 55 million years ago. Our present flora and fauna is the result of both speciation and extinction, and ecologists are interested in both these processes. We will discuss the ecology of extinction in more detail in Chapter 19.

UNITS OF SELECTION

Darwin conceived of natural selection as operating through the reproduction and survival of individuals who differ in their genetic constitution. Most discussion of natural selection proceeds at this level of Darwinian selection, or individual selection.

But natural selection is not restricted to individuals. It can act on any biological units so long as these units meet the following criteria: (1) They have the ability to replicate, (2) they produce an excess number of units above replacement needs, (3) survival depends on some attribute, and (4) a mechanism exists for the transmission of these attributes. Three units of selection other than the individual can fulfill the stated criteria: *gametic, kin,* and *group* selection.

Gametic Selection

Gametes (eggs and sperm) have a genetic composition that differs from the diploid organisms that produce them. They are produced in vast excess and may

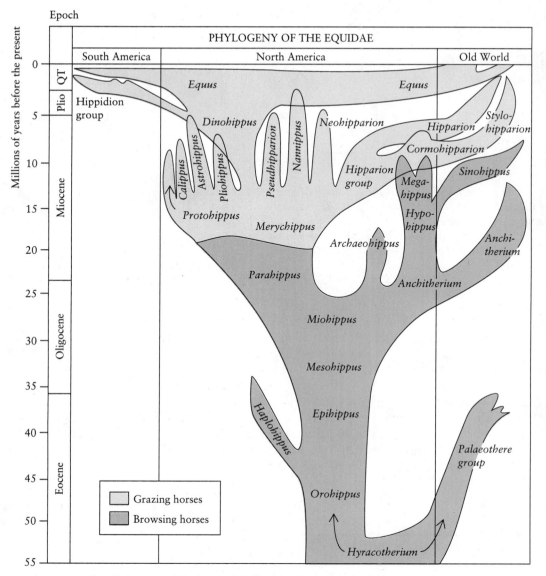

Figure 2.11   *The phylogeny of the horse family, Equidae. Note that the grazing habit evolved in the Miocene, in only one of the lineages extant at that time. (From MacFadden 1985.)*

have characteristics that they transmit through the zygote and adult organism to the next generation of gametes. Consequently, natural selection can act on a population of gametes independently from the natural selection that operates on the parent organisms. Many different characteristics of gametes could be under natural selection. Sperm mobility, for example, may be under strong selection. In plants, pollen grains that produce a faster-growing pollen tube will have a better chance of releasing their sperm nuclei and fertilizing an egg. Gametic selection is an interesting and important aspect of natural selection, but it does not directly impinge on ecological relationships.

Kin Selection

If an individual is able to increase the survival or reproduction of its relatives who carry some of the same genes, natural selection can operate through kin selection. Kin selection and individual selection may act together, and this action is described by the concept of inclusive fitness: Natural selection not only favors alleles that benefit an individual but also alleles that benefit close relatives of an individual because close relatives share many alleles in common. All relatives can help to pass copies of an individual's genes to future generations.

Kin selection was recognized as one way of explaining the existence of altruistic traits like making alarm calls. When ground squirrels sight a predator, they give an alarm call. The result of the alarm call is that the individual calling (1) draws attention to itself and thus may be attacked by the predator (detrimental to the individual) and (2) warns nearby squirrels to run for cover (beneficial to relatives nearby).

Kin selection has important consequences for ecological relationships because of its effects on social organization and population dynamics. Competition between individual organisms will be affected by the proximity of close relatives; thus it can be important for an ecologist to know the degree of kinship among members of a population.

Group Selection

Group selection can occur when populations of a species are broken up into discrete groups more or less isolated from other such groups. Groups that contain detrimental genes can become extinct, and the conditions for natural selection could occur at the level of the group as well as at the level of the individual organism.

Group selection is highly controversial, and most biologists feel that it is rare in nature. Most of the characteristics of organisms that are favorable to groups can also be explained by individual or kin selection. Controversy erupts over traits that appear to be good for the group but bad for the individual. A classic example has been the evolution of reproductive rates in birds. Group selectionists argue that many birds reproduce at less than maximal rates because populations with low reproductive rates will not overpopulate their habitats. Any populations with higher reproductive rates will overpopulate their habitats and become extinct. Restraint is selected for at the group level. But low reproductive rates are bad for the individual, and individual selection will act to favor higher reproductive rates, so group selection and individual selection are acting in opposite directions. In a group in which all members restrain themselves, a cheater will always be favored.

The alternative argument is that all reproductive rates are in fact maximal and have responded only to individual selection favoring individuals that leave the most offspring for future generations. Restraint does not exist, according to this view.

Group selection may occur, but at present it is not believed to be an important force shaping the adaptations that ecologists see while trying to understand the distribution and abundance of organisms.

SUMMARY

Evolution is the genetic adaptation of organisms to the environment. Ecology and evolution are intricately connected because evolution operates through natural selection, which is ecology in action. Organisms survive and reproduce and, because not all individuals are equally good at these activities, natural selection occurs.

Evolutionary changes involve adaptive characteristics of organisms, and these changes result from *directional selection, stabilizing selection,* and *disruptive selection.* Most of observed evolution has probably resulted from directional selection, and in existing ecological situations, stabilizing selection is probably most common.

The origin of species usually results from the geographic separation of two populations and the subsequent evolution of reproductive isolating mechanisms. Speciation may occur without geographic isolation, but probably less frequently. The distribution of many species reflects the barriers that have shaped geographic isolation.

Evolution does not occur at a constant rate over geological time. Periods of rapid speciation occur at different times in different species groups. Many species have gone extinct over evolutionary time, and the ecologist is interested in the factors that promote speciation and the processes that cause extinctions.

Natural selection may operate on four different levels. *Gametic* selection operates on characteristics of the gametes such as mobility. *Individual*, or *Darwinian*, selection is the classic form of selection on individual phenotypes, and is the level of selection responsible for most of the adaptations we see in nature. *Kin* selection introduces the idea that actions that favor the survival or reproduction of close relatives carrying the same genes will be favored by natural selection. *Group* selection occurs if whole groups or populations become extinct because of genetic characters present in some individuals in the group. Group selection is probably uncommon in nature.

Selected References

BIRCH, L. C., and P. R. EHRLICH. 1967. Evolutionary history and population biology. *Nature* 214:349–352.

COOKE, F., P. TAYLOR, C. FRANCIS, and R. ROCKWELL. 1990. Directional selection and clutch size in birds. *American Naturalist* 136:261–267.

FUTUYMA, D. J. 1986. Adaptation. In *Evolutionary Biology,* chap. 9, Sinauer Associates, Sunderland, Mass.

GODFRAY, H. C. J., L. PARTRIDGE, and P. H. HARVEY. 1991. Clutch size. *Annual Review of Ecology and Systematics* 22:409–429.

GOULD, S. J. and R. C. LEWONTIN. 1979. The spandrels of San Marco and the Panglossian paradigm: A critique of the adaptationist programme. *Proceedings of the Royal Society of London B* 205:581–598.

GRANT, P. R. 1986. *Ecology and Evolution of Darwin's Finches.* Princeton University Press, Princeton, N. J.

LACK, D. 1965. Evolutionary ecology. *Journal of Animal Ecology* 34:223–231.

PRICE, T., M. KIRKPATRICK, and S. J. ARNOLD. 1988. Directional selection and the evolution of breeding date in birds. *Science* 240:798–799.

PRIMACK, R. B., and H. KANG. 1989. Measuring fitness and natural selection in wild plant populations. *Annual Review of Ecology and Systematics* 20:367–396.

WILLIAMS, G. C. 1966. *Adaptation and Natural Selection*. Princeton University Press, Princeton, N. J.

Questions and Problems

1. Birds living on oceanic islands tend to have a smaller clutch size than the same species (or close relatives) breeding on the mainland (Klomp 1970, p. 85). Explain this on the basis of Lack's hypothesis.

2. Natural selection responds to environmental changes immediately but the resulting adaptations will not occur immediately but with a time lag, so adaptations are always slightly out of date. How could you determine whether this time lag was short or long for a particular species?

3. How can traits of cooperation evolve in a population? Does the Darwinian theory support the belief that traits of competition will evolve more readily than traits of cooperation?

4. In many ungulate species, males have horns or antlers that they use in combat with other males during the breeding season. How can natural selection limit the "arms race" in species where combat is part of the reproductive cycle? See Geist (1978) for a discussion.

5. Royama (1970, pp. 641-642) states:

   Natural selection favors those individuals in a population with the most efficient reproductive capacity (in terms of the number of offspring contributed to the next generation), which means that the present-day generations consist of those individuals with the highest level of reproduction possible in their environment.

   Is this correct? Discuss.

6. Discuss how the concept of time applies to evolutionary changes and to ecological situations. Do ecological time and evolutionary time ever overlap?

7. In many temperate zone birds, those individuals that breed earlier in the season have higher reproductive success than those that breed later in the season. Explain why this might or might not lead to directional selection for earlier breeding dates. Read Price et al. (1988) for a discussion of this problem.

8. Some birds, such as grouse and geese, have young that are mobile and able to feed themselves at hatching (precocial chicks). Discuss which factors might limit clutch size in these bird species. Winkler and Walters (1983) have reviewed studies on clutch size in precocial birds.

9. In arctic ground squirrels adult females are more likely to give alarm calls than adult males. If alarm calls are favored by kin selection, why might this difference occur? Could alarm calls be explained by group selection? Why or why not?

10. Apply the cost-benefit model in Figure 2.6 to seed production in a herbaceous plant. Discuss biological reasons for the general shape of these curves. Can you apply this model to both annual and perennial plants in the same way?

Overview Question

Humans in industrialized countries have been increasing in average body size during the last century. List several explanations for this change, and discuss how you could decide if an evolutionary explanation is needed to interpret it? How does a physiological explanation for this change differ from an evolutionary explanation?

Part 2

The Problem of Distribution: Populations

3 *Methods for Analyzing Distributions*

Why are organisms of a particular species present in some places and absent from others? This is the simplest ecological question one can ask, and hence it forms a good starting point for introducing you to ecology. This simple question about the distributions of species can be of enormous practical importance. Two examples illustrate why. Five species of Pacific salmon live in the North Pacific Ocean and spawn in the river systems of western North America, Asia, and Japan. Salmon are valuable fish both for the commercial fishermen and for sport fishermen. Why not transplant such valuable fish to other areas, for example, to the North Atlantic region or to the Southern Hemisphere? Why are five species of salmon present in the Pacific but only one species in the Atlantic? Sockeye salmon have been transplanted to Argentina but did not survive there (Foerster 1968). Why?

The African honey bee is a second example that illustrates the practical consequences of species distributions. The African honey bee *(Apis mellifera scutellata)* is a very aggressive subspecies of honey bee that was brought to Brazil in 1956 in order to develop a tropical strain with improved honey productivity. They escaped by accident, and the spread of the African bee since 1956 is shown in Figure 3.1. Because African bees are aggressive, they may drive out the established colonies of the Italian honey bee *(Apis mellifera ligustica).* In other situations, hybrids may be formed between the African and Italian subspecies. Unfortunately, the African bees are also aggressive toward humans and domestic animals, and accounts of severe stinging and even deaths have served to map the spread of the African bee. At present the African bee is spreading northward and has reached the southern United States, moving roughly 110 kilometers north per year: In 1982 it crossed the Panama Canal and reached Costa Rica in 1984. It crossed the border into Texas in 1990. United States bee keepers are understandably worried that the African bee will colonize the southern states and damage the established honey bee industry. What factors limit the distribution of the African honey bee? Will this species be able to live as far north as California and North Carolina?

TRANSPLANT EXPERIMENTS

To answer the question of distribution, we must first determine whether the limitation on distribution comes from the inaccessibility of the particular area to the species. One way to determine the source of limitation is through a transplant experiment. In a transplant experiment, we move individuals of a species to an unoccupied area and determine whether they can survive and reproduce successfully in the new environment. Some organisms can survive in areas but cannot reproduce

Figure 3.1   Spread of the African honey bee in the Americas since 1956. Southward and westward expansion in South America has been slight since 1971. (Data provided by J. G. Thomas and O. R. Taylor, personal communication.)

there, so we should follow transplant experiments through at least one complete generation. The two outcomes of the transplant experiment tell us where to go next:

Outcome	Interpretation
Transplant successful	Distribution limited either because the area is inaccessible or because the species fails to recognize the area as suitable living space
Transplant unsuccessful	Distribution limited either by other species or by physical and chemical factors

A proper transplant experiment should have a *control*, transplants done within the distribution to provide data on the effects of handling and transplanting the individual plants or animals.

If a transplant is successful, it indicates that the potential range of a species is larger than its actual range. Figure 3.2 shows this schematically for a hypothetical plant or animal. The results of transplant experiments thus direct our further investigations in one of two ways. If a species does not seem to occupy all of its potential range, we must determine if it can move into its potential range or if it lacks suitable means of transport to reach new areas. We discuss the problem of movement, or *dispersal*, in Chapter 4. Some animal species can move into new areas but do not do so. For these species, we must study their mechanisms of *habitat selection* (Chapter 5).

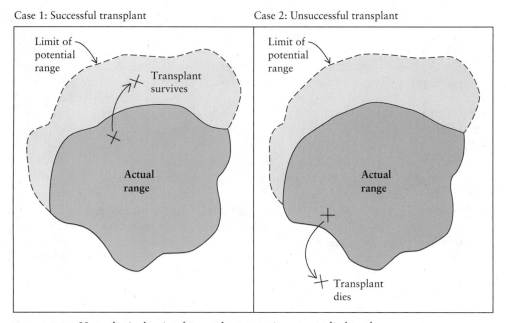

Figure 3.2   *Hypothetical pair of transplant experiments applied to the same species. The shaded area represents the actual distribution of the population. Many separate transplant experiments would be needed to define the limit of the potential range.*

If the species cannot survive and reproduce in the transplant areas, we ask whether abiotic factors or interactions with other species exclude it from these areas. Limits imposed by other species may involve either the negative effects of predators, parasites, disease organisms, or competitors, or the positive effects of interdependent species within the actual range. We can often determine if other species are restricting distribution by transplant experiments with protective devices such as cages designed to exclude the suspected predators or competitors. For example, we can transplant barnacles in mesh cages to deeper waters along the coast to see if they can survive in deep water when starfish or gastropod predators are kept away. Some examples of such experiments are described in Chapter 6.

If other species do not set limits on the actual range, we are left with the possibility that some physical or chemical factors set the range limits. For example, many tropical plant species cannot withstand freezing temperatures, and the frost line effectively limits their distributions. Limitations imposed by physical or chemical factors have been studied extensively and are the subject of a whole discipline called physiological ecology.

PHYSIOLOGICAL ECOLOGY

Physiological ecologists study the reactions of organisms to physical and chemical factors. To live in a given environment, an organism must be able to survive, grow, and reproduce; consequently, physiological ecologists must try to measure the effect of environmental factors on survival, growth, and reproduction. The major conceptual tool of physiological ecologists is *Liebig's law of the minimum*, which can be stated as follows: *The distribution of a species will be controlled by that environmental factor for which the organism has the narrowest range of adaptability or control* (Bartholomew 1958).

The job of physiological ecologists is to determine the tolerances of organisms to a range of environmental factors. This is not a simple chore. We could, for example, determine the range of temperatures over which a species can survive. Figure 3.3 shows some data of this type for two fish species. We can then repeat these studies for oxygen, pH, and salinity and build up a detailed picture of the limits of tolerance of the particular species. The stages of the life cycle may differ in their limits of tolerance. The young stages of both plants and animals are often most sensitive to environmental factors. These limits of tolerance define the fundamental niche of the species (see page 245).

Two other factors complicate the determination of tolerance limits. First, species can acclimate physiologically to some environmental factors. Figure 3.3 illustrates this concept: The lethal temperatures depend on the acclimation temperature, the temperature at which the fish have been living. Second, tolerance limits for one environmental factor will depend on the levels of other environmental factors. Thus, in many fish, pH differences will affect temperature tolerances.

Another problem arises when we try to apply these limits of tolerance to situations in the real world. Animals are particularly difficult because they are mobile and can resort to a variety of tactics that help them avoid lethal environmental conditions. Both plants and animals have evolved many types of escape mechanisms. Many birds and some insects migrate from polar to temperate or equatorial regions to avoid the polar winters. Other mammals, such as the arctic ground squirrel, hibernate during the winter and thereby avoid the necessity of feeding during the

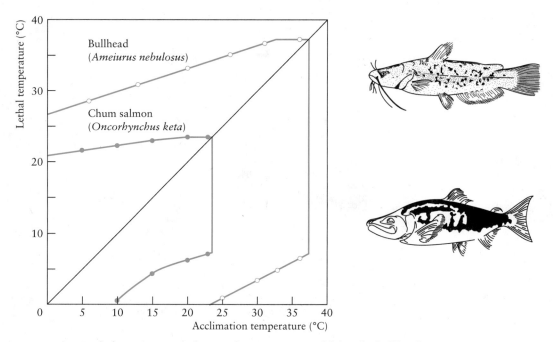

Figure 3.3 *Lethal temperature relations for two species of fish. The bullhead is a highly tolerant species in contrast to the chum salmon. The area enclosed by each trapezium is the zone of tolerance. The acclimation temperature is the temperature at which the fish had been living before the experimental tests. (After Brett 1956.)*

cold months of the year. Plants become dormant and resistant to cold temperatures in winter, while many insects enter a cold-tolerant diapause.

ADAPTATION

The organisms whose distribution we study today are products of a long history of evolution, and the physiological ecologist studies their adaptations in much the same way that we might study a single frame of a motion picture. The tolerances of species can change by the process of natural selection. In less than 50 years the grass *Agrostis tenuis* has evolved populations that live on mine wastes in Great Britain (Antonovics et al. 1971). Soils on mine wastes are contaminated with high concentrations of lead, zinc, and copper, and plants from pastures will not survive on mine soil. Normal *Agrostis tenuis* populations contain a few tolerant individuals. If you sow 2000 seeds on mine soil, only 4 or 5 grass plants will grow (Wu et al. 1975). On toxic mine soils only these tolerant genotypes survive. Figure 3.4 illustrates the differences that occur in copper tolerance along transects through a copper mine in North Wales. Genes for copper tolerance have spread downwind in this valley. Copper-tolerant populations of the grass *Agrostis* have evolved by rapid natural selection acting on very rare individual grasses that are partly tolerant to high copper levels (Wu et al. 1975). Present populations are maintained by strong disruptive selection dictated by contaminated soils. Not all plant species are able to evolve metal tolerances; many

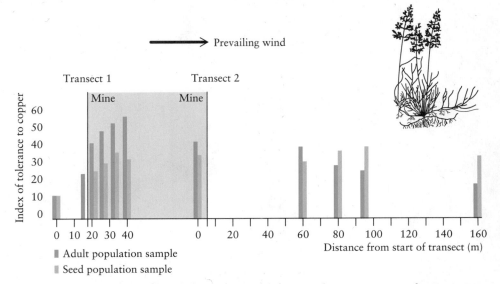

Figure 3.4 Copper tolerance of the grass Agrostis tenuis *along transects at the edge of the Drws y Coed mine in North Wales. The tolerance of both adult plants and seeds produced in situ is shown. The contaminated area is shaded. (After Macnair 1981.)*

species do not have the appropriate genetic variation in their normal populations (Bradshaw and Hardwick 1989).

There is a cost to being tolerant to heavy metal pollution. The tolerant genotypes of *Agrostis tenuis* grow poorly compared with normal genotypes when they are grown in normal soil under crowded conditions (Macnair 1981). One possible reason is that tolerant plants require more than trace amounts of heavy metals to be able to grow properly. Tolerant plants are at a selective disadvantage away from contaminated soils.

Such evolutionary changes further complicate the task of the ecologist who is trying to understand the distribution of a species. We must ask the question, What factor sets the current limitation on the geographic distribution of a species? But then we must ask further, why for many species has natural selection not been able to increase the limits of tolerance of a species and thereby to expand its geographic range? If the grass *Agrostis tenuis* has been able to increase its limits of tolerance to heavy metals, why has this not occurred in many other plant species? Many range shifts may be due to changes in the environment, but some range shifts are caused by evolutionary changes in the physiological attributes of the individuals in a population. We can also determine how much evolutionary adaptation has occurred by studying the variation in tolerance levels between different populations of the same species.

Transplant experiments can be disastrous when pests are introduced to new areas, as we shall see in the next chapter. It is critical that all transplant experiments be done safely with due regard for the ecosystem. Indiscriminate transplanting of organisms contains all the seeds of ecological disaster (Krebs 1988). Most governments have stringent rules prohibiting the importation of plants and animals from other regions.

In the next four chapters you will find many examples to illustrate the ideas given in this chapter. One cautionary note: We will begin by assuming that the factors affecting geographic distributions operate in isolation from one another. We know that this is not true from our personal experience—a spring day with a 15°C temperature will be pleasant if there is no wind but it will seem cold if a strong wind is blowing. The effects of temperature and wind, temperature and moisture, and moisture and soil nutrients are not independent but interacting. Let us begin simply, however, and see how much we can understand by treating factors as separate effects and then by adding factors together when necessary.

SUMMARY

Why are organisms of a particular species present in some places and absent from others? This simple ecological question has significant practical consequences and thus deserves careful analysis. A transplant experiment is the major technique used to analyze the factors that limit geographic ranges. This technique leads sequentially through the following steps:

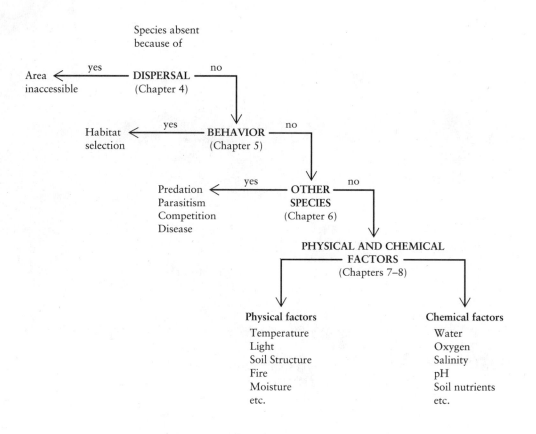

To examine any particular problem of distribution, ecologists proceed down this chain, eliminating things one by one. In the next five chapters we will see many examples in which part of this chain has been experimentally analyzed, but in no case has this chain been studied completely for a species.

The analytical question, What limits distribution now? is complementary to the evolutionary question, Why has there not been more adaptation? Thus we are led to investigate the genetic variation within populations and to look for range extensions or contractions that are associated with evolutionary shifts in the adaptations of organisms to their environment.

Selected References

BARTHOLOMEW, G. A. 1958. The role of physiology in the distribution of terrestrial vertebrates. In *Zoogeography*, ed. C. L. Hubbs, pp. 81–95. American Association for the Advancement of Science Publication No. 51, Washington, D.C.

GASTON, K. J. 1991. How large is a species' geographic range? *Oikos* 61:434–438.

HENGEVELD, R. 1989. *Dynamics of Biological Invasions*. Chapman and Hall, London.

MACNAIR, M. R. 1981. Tolerance of higher plants to toxic materials. In *Genetic Consequences of Man Made Change*, ed. J. A. Bishop and L. M. Cook, pp. 177–207. Academic Press, New York.

TAYLOR, O. R. Jr. 1985. African bees: potential impact in the United States. *Bulletin of the Entomological Society of America* 31(4): 15–28.

Questions and Problems

1. What assumptions must be made to map the potential spread of the African honey bee in North America? Read Taylor (1985) and discuss how to test these assumptions.

2. In discussing the role of physiology in studying distributions of animals, Bartholomew (1958, p. 84) states:

   It usually develops that after much laborious and frustrating effort the investigator of environmental physiology succeeds in proving that the animal in question can actually exist where it lives.

   Is this a difficulty with respect to the problem of analyzing geographic distributions?

3. Discuss the problem of defining exactly the "geographic distribution" of a plant or animal. Gaston (1991) reviews this problem.

4. Discuss the application of general methods for studying distributions to the problems of what limits the geographic range of human beings both at the present time and early in our evolutionary history.

5. In discussing Liebig's law of the minimum, Colinvaux (1973, p. 278) states:

   The idea of critically limiting physical factors may serve only to obstruct a theoretical ecologist in his quest for a true understanding of nature. . . . To say that animals live where their tolerances let them live has an uninteresting sound to it. It implies that animals have been designed by some arbitrary engineer according to some preconceived sets of tolerances, and that they then have to make do with whatever places on the face of the earth will provide enough of the required factors.

Evaluate this critique.

6. Why are the genes for copper tolerance in *Agrostis tenuis* more common in non-mine soils that are downwind of the mine site in Figure 3.4 (page 42)?

7. A hypothetical population of frogs consists of 50 individuals in each of two ponds. In one pond all of the individuals are green. In the other pond, half are green and half are brown. During a drought, the first pond dries up, and all the frogs die. In the population as a whole, the frequency of the brown phenotype has gone from 25 percent to 50 percent. Has evolution occurred? Has there been natural selection for the brown color morph?

Overview Question

Find a local field guide to birds, flowers, or mammals and discuss what the maps showing geographic ranges mean. On what scale would you map these ranges?

---

4 *Factors That Limit Distributions: Dispersal*

Some organisms do not occupy all of their potential range, and if transplanted outside their normal range, they survive, reproduce, and spread. The absence of an organism from a particular area may thus be due to the species' having failed to reach the area being studied. This simple possibility should be examined before more involved possibilities.

The transport, or *dispersal*, of organisms is a vast subject that has been of primary interest not only to ecologists but also to *biogeographers*, who wish to understand the historical changes in distributions of animals and plants. There are some very difficult problems associated with the study of dispersal. First, the detailed distribution is known for so few species that most dispersals are probably not noticed. Dispersal of individuals between different parts of the species' range may occur often. Second, an organism may disperse to a new area but not colonize it because of biotic or physical factors.

If colonization is successful, dispersal will result in *gene flow* and thus affect the genetic structure of a population. If the dispersing individuals are not a random sample of the population, dispersal will result in a founder effect, and the new population may be genetically quite distinct from the source population. Not all dispersing individuals survive to breed, so gene flow may be quite restricted in many species (Ehrlich and Raven 1969). Dispersal is thus simultaneously an ecological process affecting distributions and a genetic process affecting geographic differentiation.

The most spectacular examples of transport affecting distribution are species that are introduced by humans and explode to occupy a new area. Other examples are exploited species recolonizing their original range. Let us look into a couple of these situations.

GYPSY MOTH (Lymantria dispar)

In 1850 a French astronomer employed at Harvard University near Boston brought some eggs of the European gypsy moth *(Lymantria dispar)* to his house in Massachusetts. A few of the caterpillars escaped in 1869 and they began one of the most devastating caterpillar plagues of the New England states. Gypsy moths defoliate a great variety of deciduous and coniferous trees, from apple to alder, basswood, oaks, poplars, willows and birches. Because of severe defoliation of deciduous trees, Massachusetts in 1889 initiated a control program and by 1900 the severity of the outbreaks was reduced so the program was terminated by the state. After 1900 gypsy moths began to spread in a wave across New England (Figure 4.1). The spread slowed during the middle part of this century and then about 1980 began to accelerate. The gypsy moth defoliated 13.8 million acres of forest in 1981, and another major outbreak began in 1989 (U.S. Department of Agriculture, 1990). Timber losses associated with these outbreaks are sufficiently large that a major research program on gypsy moth control is now under way (Montgomery and Wallner 1988).

CHESTNUT BLIGHT (Cryphonectria parasitica)

The American chestnut *(Castanea dentata)* was an important component of many deciduous forests of the eastern United States from Maine to Georgia and west to Illinois (Figure 4.2). Chestnut made up more than 40 percent of the overstory trees in the climax forests of this area. In 40 years this species has been eliminated as a canopy tree from its entire range by chestnut blight.

The chestnut blight is a fungal disease that attacks chestnut trees. The disease was first noticed about 1900 in the area around New York City, where it killed all its hosts. The fungus was apparently introduced on nursery stock from Asia. Although found on other species of trees, *Cryphonectria parasitica* is lethal only to the American chestnut. The fungus enters the host tree through a wound in the bark, grows chiefly in the cambium, penetrates only short distances into the wood, and kills the tree by girdling. Once a chestnut tree is attacked, it is killed in 2 to 10 years. The fungus kills only the above-ground part of the tree, and the root systems of trunks killed 60 years ago are still sending up shoots.

Closely related native species of *Cryphonectria* are usually saprophytic on chestnuts and do not harm their host tree. *Cryphonectria parasitica* is a very weak parasite on related species of oaks and never seems to attack the closely related American beech (Shear et al. 1917).

The U.S. Department of Agriculture sponsored an expedition to China, Japan, Korea, and Taiwan from 1927 to 1930 to collect Asiatic chestnut seed. The dual purpose of importing seeds was (1) to determine if blight-resistant Oriental chestnuts could replace the vanishing American chestnut and (2) to establish Oriental chestnuts for crossbreeding with the American species. Both Chinese chestnuts *(Castanea mollissima)* and hybrid trees have been successful in areas of the central Appalachians and the Ohio valley but not in northern New York and southern New England, where the American chestnut once lived in abundance (Diller and Clapper 1969). There has also been a search for a native blight-resistant American chestnut. Large surviving native trees are found on occasion, but none has proved blight-resistant. They appear to have been lucky individuals that somehow escaped infection (Jaynes 1968) and are examples of how the fungus may be absent because of insufficient dispersal of spores. At present a breeding program has been established to produce resistant American chestnuts by backcrossing with resistant Chinese chestnuts. Two genes seem to control blight resistance and if these can be incorporated into the American chestnut genome we may once again see the chestnut become a common canopy tree in eastern United States (Burnham 1988).

Most of the chestnuts killed by the blight have been replaced by codominant trees, especially oak species, but also by beech, hickories, and red maple (Good 1968, Keever 1953). The previous oak-chestnut forests have now become oak forests or oak-hickory forests.

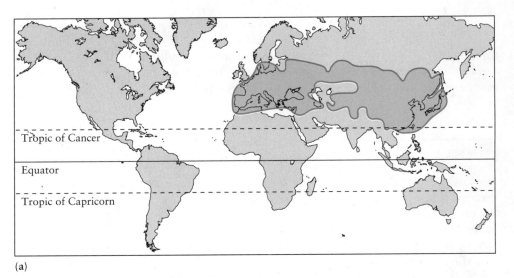

(a)

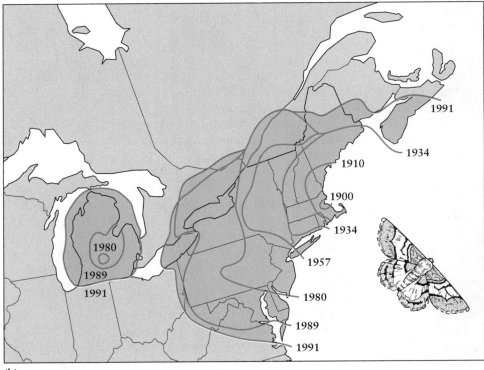

(b)

Figure 4.1   *(a) Original distribution of the gypsy moth in Eurasia. (b) Spread of the gypsy moth in the northeastern United States after its accidental introduction at the end of the nineteenth century. (After Elton 1958 and U.S. Department of Agriculture, 1990.)*

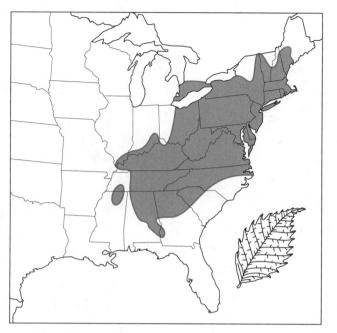

Figure 4.2   *Original distribution of the American chestnut tree* (Castanea den-tata). *Shoots from roots still occur throughout this range, but chestnut blight kills all the above-ground stems. (After Grimm 1967.)*

CALIFORNIA SEA OTTER

The sea otter was hunted by fur traders to near extinction by 1900. The few remaining populations were protected by international treaty in 1911, and the California sea otter was believed to be extinct at that time. In 1914 a small population was discovered at Point Sur in central California. Since then, otters have increased in numbers and expanded their geographic range to reoccupy areas from which they had been exterminated in the nineteenth century (Figure 4.3). The rate of spread of the sea otter is easy to estimate because it lives along the coastline in a linear habitat. Figure 4.4 shows the increase in range with time. The southern range expanded 3.1 kilometers per year between 1938 and 1972, and the northern range expanded 1.4 kilometers per year. These differences could result from the southern otters moving more as individuals, or the northern otters suffering more mortality (Lubina and Levin 1988).

THREE MODES OF DISPERSAL

The spectacular cases mentioned are undoubtedly a biased sample, but the important point they illustrate is how rapidly some organisms can spread to new areas if other conditions are favorable. Before we discuss the ecological consequences of dispersal, let us define more carefully what we mean by dispersal.

There are three ways in which species spread geographically. All are loosely labeled as *dispersal* but should be defined more precisely (Pielou 1979):

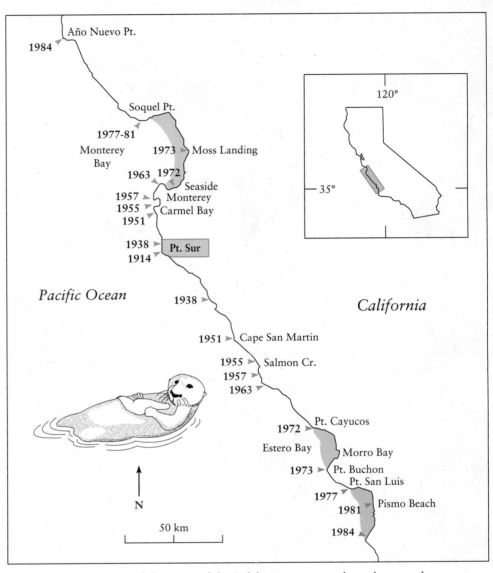

Figure 4.3 *Expansion of the range of the California sea otter along the central California coast. The current range expansion began from Point Sur where 50 sea otters were rediscovered in 1914. (After Lubina and Levin 1988.)*

1. *Diffusion:* Diffusion is the gradual movement of a population across hospitable terrain for a period of several generations. This is a common form of dispersal, illustrated by the gypsy moth (Figure 4.1) and the chestnut blight after their introduction to North America.

2. *Jump dispersal:* Jump dispersal is the movement of individual organisms across large distances followed by the successful establishment of a population in the new area. This form of dispersal occurs in a short time during the life of an individual, and the movement usually occurs across unsuitable terrain. Island colo-

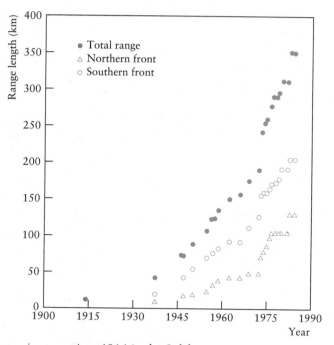

Figure 4.4   Range increase since 1914 in the California sea otter. The overall range increase is shown, along with the movement of the northern and southern fronts. The overall rate of range increase was 5.6 km per year. (After Lubina and Levin 1988.)

nization is achieved by jump dispersal, and human introductions such as the African bee (see page 38) can be viewed as an assisted form of jump dispersal.

3. *Secular dispersal:* If diffusion occurs in evolutionary time, the species that is spreading undergoes extensive evolutionary change in the process. The geographic range of a secularly dispersing species expands over geologic time, but at the same time natural selection is causing the migrants to diverge from the ancestral population. Secular dispersal is an important process in biogeography, but, since it occurs in evolutionary time, it is rarely of immediate interest for ecologists working in ecological time.

All of these forms of dispersal are affected by *barriers.* Many kinds of barriers exist for different kinds of animals and plants. Let us consider a few cases in which barriers on a local scale are important in restricting jump dispersal.

Freshwater organisms to which both land and sea are barriers might be expected to show local distributions strongly affected by jump dispersal. Water can also be a barrier to some terrestrial animals. Ruffed grouse *(Bonasa umbellus)* were originally found only on three Michigan islands in the Great Lakes, all three within 800 meters of the mainland. All other islands more than 800 meters from shore were uninhabited by this bird. Palmer (1962) suggested that lack of dispersal explained this island pattern of distribution, and he tested the flight capacity of several grouse over water. None could fly as much as 800 meters over water, and Palmer concluded that ruffed grouse are not capable of flying across 1600 meters of open water to colonize offshore islands by jump dispersal. Some of the offshore islands were artificially stocked with ruffed grouse, and populations have been very successful (Moran and Palmer 1963).

On a local or a global scale, dispersal may not limit distribution, because introduced species may be unable to survive. Long (1981) and Ebenhard (1988) list some of the early attempts to introduce foreign birds to North America. Unfortunately, failures to establish a species are rarely studied to obtain an explanation, and accidental introductions are often recorded only when they are successful. The wool trade has been responsible for the introduction of 348 plant species into England because of seeds adhering to wool fleeces shipped from overseas, but of these only 4 species have become established (Salisbury 1961). Few plant species introduced into continental areas are able to become established except in disturbed areas.

Bird introductions into continental areas are usually failures (Mayr 1964). In North America, only 4 species of introduced birds are common, although 50 species have been introduced. In Europe, only thirteen successful establishments of birds are recorded from 85 species introduced. There are about 204 species of breeding birds around Sydney, Australia, and 50 or more bird species were introduced to this area. Only 15 species got established, and only 8 species are common. Thus a rough estimate of 10 to 30 percent success is obtained in continental bird introductions.

Many species of game birds have been introduced into North America in the hope of providing food and sport. Between 1883 and 1950 there were 23 attempts to introduce four species of grouse into North America (Bump 1963). All of these attempts ended in failure; the reasons are not known.

Plants disperse primarily by means of their seeds and spores, and transport is rarely an important factor limiting distributions of plants on a local scale. Few experimental data are available to substantiate this general conclusion. *Rumex crispus* var. *littoreus* is a British plant confined to a narrow zone of the seashore at the level of the highest tides. Cavers and Harper (1967) sowed seed of this species in a variety of nonmaritime habitats and found that seedlings became established in large numbers, but most did not survive the first year. The local distribution of this plant is thus not limited by dispersal or by seed-germination requirements. Salisbury (1961, p. 100) discusses the invasion of weeds into bombed areas of London during World War II. Seeds and spores that were wind-dispersed colonized these areas very quickly.

Small animals often have a stage in their life cycles during which they can be transported by wind; such species resemble plants in that their distributions are rarely limited by lack of dispersal. Many insect species are transported by wind for long distances. Mosquitoes are a good example. The flight patterns of disease-carrying mosquitoes have been studied so that adequate control measures can be set up. The distances mosquitoes disperse determine the limits to which a given breeding location may allow contact with people and the area where control work must be done if a given human habitation is to be protected from diseases like malaria. Eyles (1944) summarized the flight ranges of several important *Anopheles* mosquitoes:

	Range (mi)	(km)
Maximum observed flight	0.11–8.0	0.18–12.9
Maximum seasonal flight	3.7–12.0	6.0–19.3

Hocking (1953) calculated the maximum flight range of four species of northern mosquitoes to be between 13.7 and 32.9 miles (22 and 53 km). These flight ranges could be increased or reduced by wind in natural populations.

COLONIZATION AND EXTINCTION

If dispersal occurs rapidly on a local scale, areas that are cleared of organisms should be recolonized rapidly. Some large-scale colonization experiments have occurred naturally. On August 26, 1883, the small volcanic island of Krakatoa in the East Indies was completely destroyed by a volcanic eruption. Six cubic miles (25 km^3) of rock was blown away, and all that remained of the island was a peak covered with ashes. This sterilized island was in effect a large natural experiment on dispersal. The nearest island not destroyed by the explosion was 40 kilometers away. Nine months after the eruption, only one species—a spider—could be found on the island. After only three years, the ground was thickly covered with blue-green algae, and 11 species of ferns and 15 species of flowering plants were found. Ten years after the explosion, coconut trees began growing on the island. After 25 years, there were 263 species of animals on the island, which was covered by a dense forest. Within 50 years, there were 47 species of vertebrates on the island, 36 bird species, 3 species of bats, a rat with 2 subspecies, 5 lizards, a crocodile, and a python (Hesse et al. 1951, pp. 68–69). There is some controversy about the methods of transport, but the majority of the plants and animals were probably transported by wind. Larger vertebrates probably arrived on driftwood rafts or in a few cases by swimming. The suggestion that emerges from these observations is that when there is vacant space, animals and plants are not long in finding it.

In 1980 the eruption of Mount St. Helens in Washington has provided a more recent opportunity to study recolonization on an area sterilized by volcanic eruption. During the first six years following the eruption, vascular plant invasion of the barren landscape has been very limited, despite the proximity of seed sources. Is colonization limited by the seed rain, or by a lack of tolerance to the volcanic ash cover of the soil? Wood and del Moral (1987) studied 22 species of grasses and herbs on Mount St. Helens. They recognized two groups—"dispersers" whose seeds were lighter and moved by the wind, and "non-dispersers" whose seeds rarely got as far as 3 meters from the parent plant. They planted 16,000 seeds in barren sites at the volcano— only 1745 seedlings emerged. Species with the largest seeds had the highest chances of growing. The slow colonization of the barren soils results from the lack of dispersal of heavy seeds onto the devastated area. Plants like *Aster* that have light seeds can reach the barren soil but because their seeds are small they die from drought stress. Recovery on Mount St. Helens will be patchy and slow because of limited seed dispersal from plants that can effectively colonize the bare soil.

The subalpine and alpine flora of Mount St. Helens consists of 95 species of vascular plants. Three nearby volcanos contain 2 to 3 times as many plants. More than half of the plant species missing from Mount St. Helens seem to be absent because of a lack of dispersal (del Moral and Wood 1988). The subalpine zone of Mount St. Helens is isolated by 50 to 80 kilometers of lowland forest, and as such resembles an oceanic island. Furthermore, volcanic eruptions have also eliminated species. About 20 species of subalpine and alpine plants became locally extinct with the 1980 eruption.

Colonization experiments can be done on a much smaller scale. Amy Schoener (1974) set out plastic-mesh "sponges" in barren, sandy parts of a lagoon in the Bahamas and measured the rate of colonization of these new habitats. Marine sponges harbor a great diversity of species within their internal chambers. Schoener recorded at least 220 species of colonists in her plastic sponges, and the same types

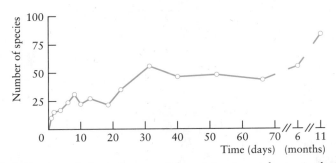

Figure 4.5 *Colonization of plastic "sponges" by marine invertebrates in the Bimini Lagoon, Bahamas. Sponges were placed in bare sandy areas within the lagoon and floated freely in the water column, anchored with rope to a bag of ballast. (After A. Schoener 1974.)*

of animals also occurred in natural sponges. Figure 4.5 shows a colonization curve for some of the sponges. Within 30 days, about 50 species had colonized the plastic sponges, and this equilibrium persisted until 70 days. As the sponges aged, more species accumulated. The important point to note is the rapidity of colonization of these new habitats. Some of the species that colonized rapidly subsequently disappeared and became locally extinct.

The examples we have analyzed suggest that dispersal may limit *local* distributions of a few plants and some animals, but in most cases empty places get filled rapidly. Let us look now at the other extreme and consider *global* distribution patterns.

Terrestrial mammals other than bats do not easily cross saltwater barriers (Darlington 1965), so whole faunas can diverge if they are isolated by ocean. Marsupials, for example, became isolated in South America and in Australia early in the Tertiary period (60 million years ago). Of the placental mammals, only rodents and bats were able to colonize Australia before the arrival of man. South America was also isolated by a water gap across Central America for most of the Tertiary and became connected to North America only during the last 2 million years. Once a land connection was established, a flood of dispersing mammals moved in both directions. The results for North America were relatively minor: We received the opossum, the porcupine, and the armadillo as additions to our mammal fauna. But in South America the results of colonization were dramatic. Many South American mammals became extinct and were replaced by North American species. Carnivores from North America have completely replaced the carnivorous marsupials that previously occupied South America. Ungulates from North America have entirely replaced the unique set of South American ungulates (Darlington 1965).

The faunas and floras of oceanic islands also show in graphic detail the limitations of distribution on a global scale. New Zealand had no native marsupials or other land mammals except for two species of bats. All of the plants and animals that colonize New Zealand or any oceanic island must do so across water. The unique combination of difficult access, limited dispersal powers of different species, and adaptive radiation has produced island floras and faunas of an unusual nature, such as the plants and animals of Hawaii and the species Charles Darwin found on the Galápagos Islands off Ecuador.

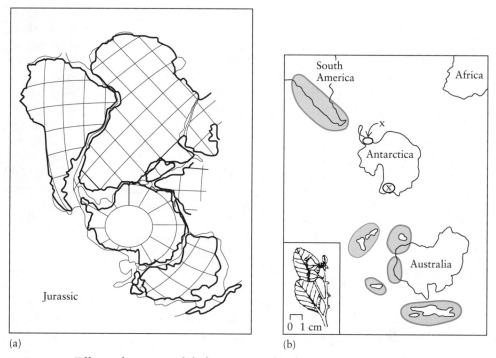

(a)   (b)

Figure 4.6 *Effects of continental drift on geographic distribution. (a) Fit of the Gondwana continents during the Jurassic period, about 135 million years ago, before breakup. (b) Modern distribution of the genus* Nothofagus *(southern beech).* Nothofagus *pollen of the Oligocene age (30 million years ago) has been found in Antarctica at the two sites indicated by x. (From Pielou 1979.)*

The southern beech (*Nothofagus* spp.) is a good example of how present geographic distributions are set by geologic events. Until 135 million years ago the southern continents were attached together in a large land mass called Gondwana (Figure 4.6). Groups present on Gondwana now have a disjunct distribution—*Nothofagus* is a good example (Figure 4.6b). *Nothofagus* seeds are heavy and not adapted for jump dispersal. Species of *Nothofagus* have probably spread slowly overland by diffusion and have been stopped by the sea, so their present distribution is a by-product of continental drift.

Continental drift not only takes certain continents farther apart, it also brings some continents closer together. As the Australian tectonic plate, for example, drifted northward after becoming detached from Antarctica, it made contact with the Asian plate about 20 million years ago. As distances over water decreased, jump dispersal of plants between Australia and Asia has been becoming steadily easier.

The Quaternary ice age is a more recent example of how geographic distributions are affected by geological events. The ice age began about 2 million years ago and has not ended. During the last 500,000 years, ice sheets in North America and Eurasia have undergone great oscillations, waxing and waning at least four times. We are now in the fourth interglacial period. At the height of the last glaciation, about 20,000 years ago, the ice volume was 77 million cubic kilometers, three times the present amount. Sea level at the height of the last glaciation was 130 meters below its present level. If all the present ice melted, sea level would rise 70 meters (Pielou 1979, 1991). The biological effects of glaciations are spectacular but slow. Dropping sea levels open up migration routes for land organisms and may restrict dispersal of marine organisms.

Disjunct distributions, such as that of *Nothofagus* (Figure 4.6) can be explained historically in two quite different ways. *Dispersal explanations* assume that the organism dispersed across preexisting barriers like mountains or rivers and then underwent speciation. *Vicariance explanations*, on the other hand, assume that the species was present on the entire area and subsequently was fragmented by the formation of barriers (Brown and Gibson 1983). If the timing of barrier formation can be determined and the phylogeny of the disjunct species is well known, one can test vicariance explanations explicitly. *Nothofagus* is a good example of a geographic pattern of distribution that is better explained by vicariance than by dispersal.

The flora and fauna of the world today have been strongly affected both by the dispersal of species and the geological formation of barriers that prevent organisms from colonizing all of their potential range. The great sweep of evolutionary history is a prolonged essay on the role of dispersal and barrier formation in limiting species distributions.

EVOLUTIONARY ADVANTAGES OF DISPERSAL

Why disperse? The answer seems obvious: to find and colonize a new area. Natural selection will clearly favor an individual that leaves a relatively crowded habitat and colonizes an empty one in which it can leave many descendants. But the evolutionary problem is this: Most dispersing organisms die, and only a few are successful. In an evolutionary sense, the individual can do one of two things: stay at home and live in a suitable place but have only a few descendants (if any), or disperse and take a chance on surviving, colonizing a new habitat, and having many descendants.

Very few species have abandoned dispersal altogether. The best-known examples are flightless birds and insects on remote islands. Darwin noted the high frequency of flightless animals on oceanic islands during the voyage of the Beagle, and Carlquist (1974) has shown that dispersal ability is reduced in many plants on islands. On subantarctic islands, for example, an average of 76 percent of the insect species are flightless (Carlquist 1974, p. 494). Flightless insects are also found on ecological islands such as the alpine zone of tropical mountains. Figure 4.7 shows a striking example of a flightless crane fly from Mount Kilimanjaro.

Other organisms devote most of their effort to dispersal. *Fugitive* species are one extreme. These are the "weeds" of the plant and animal kingdoms, which colonize temporary habitats, reproduce, and leave quickly before the temporary habitat disappears or competition with other organisms overwhelms them. Fugitive species grow entirely or predominantly in disturbed areas, and they produce large numbers of seeds adapted to long-distance dispersal by wind or by animals. The common dandelion *(Taraxacum officinale)* is an example of a fugitive or weedy species.

Strategies of dispersal in insects have been studied extensively because many pest species have high powers of dispersal (Johnson 1969, Dingle 1978). Dispersal in insects is largely a prereproductive phenomenon of the adult stage of the life cycle. Energy is first put into flight muscles and migration, and only after migration does egg formation start. The more migratory insect species are those associated with temporary habitats. Brown (1951) was one of the first to recognize this effect in

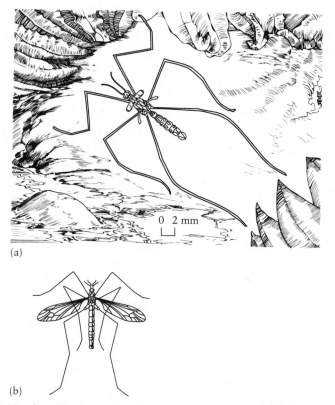

(a)

(b)

Figure 4.7   *(a)* Tipula subaptera, *a flightless crane fly from the alpine zone of Mount Kilimanjaro, Tanzania. (From Carlquist 1974.) (b)* Tipula trivittata, *a typical crane fly from eastern North America. (From Swan and Papp 1972.)*

water boatman (Corixidae). A temporary pond received many immigrant insects of species that occurred in ponds but few immigrants from species that lived in large lakes and streams:

Species	No. of Immigrants to Temporary Pond	% Total Numbers Collected in Temporary Habitats
Corixa nigrolineata	209	81
Corixa falleni	11	23

Southwood (1962) summarized an abundance of evidence that shows that dispersal occurs most often in insects that occupy temporary habitats. These insects are good examples of fugitive animal species.

Some insects alternate in the production of winged and wingless forms within the same species. Aphids are good examples. When conditions are favorable for reproduction, wingless forms are produced. When the environment deteriorates, winged forms are produced that disperse to a new habitat. In aphids, the tactile stimuli associated with high density stimulate the production of winged forms and subsequent dispersal. Water striders (gerrids) are another insect group in which both winged and wingless forms occur within the same species (Vepsalainen 1978).

Dispersal is clearly advantageous when habitats are patchy or unstable. But not all organisms live in unstable environments, and it is important to ask if dispersal can also be adaptive in uniform and predictable environments. Hamilton and May (1977) have shown with a simple model that dispersal is highly adaptive in uniform environments, as long as there are potentially empty sites away from the parent organism, even if the likelihood of successful dispersal and colonization is low. This result accords with natural history observations that nearly all organisms devote considerable resources to the dispersal of propagules.

We reach this general conclusion: Natural selection has molded the anatomy, physiology, and behavior of organisms to provide the dispersal powers needed to complete the life cycle. The many adaptations for dispersal in plants and animals illustrate in a graphic way the importance of dispersal in the lives of most species.

SUMMARY

Dispersal operates in three general ways. *Diffusion* is the slow spread of a species across hospitable terrain over several generations. *Jump dispersal* operates quickly across great distances of inhospitable terrain. *Secular dispersal* is spread in geological time, with extensive evolutionary change in the process, and may be associated with continental drift.

A species may not occur in an area because it has not been able to disperse there. This hypothesis can be tested by artificial introductions of the organism into unoccupied habitats. Some species introduced from one continent to another, such as the African bee and the gypsy moth, have spread very rapidly. Other introduced species (the majority) die out and are not successful in new areas. On a local scale, few species seem to be restricted in distribution by poor powers of dispersal.

Thus, dispersal is rarely an important factor limiting the *local* distribution of plants and animals. Organisms have many special adaptations for dispersal, and consequently rapidly colonize new areas. On a global scale, however, dispersal is a critical factor in biogeography, and barriers to dispersal help to determine distribution patterns among continents and islands.

Dispersal is adaptive if it permits individuals to colonize new areas successfully. Some species inhabit temporary habitats and exist only because of their high powers of dispersal. Other species live in more permanent habitats and disperse less.

Selected References

DINGLE, H. (ed.) 1978. *Evolution of Insect Migration and Diapause.* Springer-Verlag, New York.

EBENHARD, T. 1988. Introduced birds and mammals and their ecological effects. *Swedish Wildlife Research* 13(4):1–107.

ELTON, C. S. 1958. *The Ecology of Invasions by Animals and Plants.* Methuen, London.

HAMILTON, W. D., and R. M. MAY. 1977. Dispersal in stable habitats. *Nature* 269:578–581.

LUBINA, J. A., and S. A. LEVIN. 1988. The spread of a reinvading species: range expansion in the California sea otter. *American Naturalist* 131:526–543.

PIELOU, E. C. 1991. *After the Ice Age: The Return of Life to Glaciated North America.* Univ. Chicago Press, Chicago.

RIDLEY, H. N. 1930. *The Dispersal of Plants Throughout the World.* L. Reeve, Ashford, Kent, England.

ROFF, D. A. 1986. The evolution of wing dimorphism in insects. *Evolution* 40:1009–1020.

WOOD, D. M., and R. DEL MORAL. 1987. Mechanisms of early primary succession in subalpine habitats on Mount St. Helens. *Ecology* 68:780–790.

Questions and Problems

1. Cohen (1967, p. 17) has argued that in an environment that is predictable and constant over time, organisms can never gain any advantage by changing their location or dispersing, even if the environment is variable in space. Do you agree with this conclusion? See Hamilton and May (1977) and discuss.

2. Salisbury (1961, p. 82) states:

   The geographical distribution of weeds can usually be expressed in general terms, but for them, as indeed for most categories of plants, any attempt to map their distribution with great precision could be wholly misleading and unscientific, since the area and density of occupation is not static but dynamic.

   Explain why you agree or disagree with this.

3. One of the recurrent themes in studying introduced species, such as the starling (Lever 1987), is that several unsuccessful introductions have often preceded the successful introduction. Are there cases of successful introductions by humans on the first attempt? What might account for this pattern? How does this pattern complicate the interpretation of transplant experiments tried only one time (see page 39)?

4. Dutch elm disease is a viral disease spread by bark beetles and lethal to the American elm *(Ulmus americana).* Compare the American chestnut–chestnut blight interaction to that of the American elm–Dutch elm disease interaction. Discuss in particular the factors that promote the spread of the two diseases. Anderbrant and Schlyter (1987) and Brasier (1988) provide some literature references on Dutch elm disease.

5. In discussing dispersal and colonization in birds, Mayr (1964, p. 32) states: "The history of faunal changes on islands proves that island faunas offer far less resistance to immigrants than mainland faunas." Refer to Mayr's paper, and obtain some data on island and mainland faunas to evaluate his conclusion. List some possible reasons for these facts.

6. Review the global distributions of flightless birds and comment on the ecological situations that lead to flightlessness in birds. Carlquist (1965, pp. 224–

241) provides a general background, and Olson (1973, p. 31) gives details for a specific example.

7. Review the introduction and spread of the zebra mussel in the Great Lakes of North America (Roberts 1990, Strayer 1991). How might this invasion have been prevented? What can be done now to alleviate the problem?

8. The spread of human diseases is often more carefully documented than the spread of other diseases of plants and animals. Review the spread of cholera in the United States from 1832 to 1866, and the spread of measles in Iceland from 1945 to 1970 (references in Hengeveld 1989, p. 18–24).

Overview Question

Would you expect the same dispersal abilities in plants from tropical rain forests and boreal conifer forests? In animals? Why?

Factors That Limit Distributions: Habitat Selection

Some animals do not occupy all their potential range even though they are able to disperse into the unoccupied areas. Thus individuals "choose" not to live in certain habitats, and the distribution of a species may be limited by the behavior of individuals in selecting their habitat. We define a *habitat* as any part of the biosphere where a particular species can live, either temporarily or permanently.

Habitat selection is one of the most poorly understood ecological processes. If we assume that an animal cannot live everywhere, natural selection will favor the development of sensory systems that can recognize suitable habitats. What elements of the habitat do animals recognize as relevant? We must be careful here to define the perceptual world of the animal in question before we begin to postulate the mechanism of habitat selection. Areas that appear "similar" to the human observer may appear very different to a mosquito or a fish. Conversely, habitats we think are very different may be treated as the same by a bird.

There are two broad approaches to the study of habitat selection. The proximal approach looks at habitat choice as a result of behavioral mechanisms, and asks how, in a physiological sense, does an animal choose its habitat? The ultimate or evolutionary approach looks at the adaptive reasons for habitat choice and the evolutionary significance of the behaviors involved. In this chapter we will first consider the behavioral mechanisms of habitat selection and then discuss the evolution of habitat preference.

Plants show habitat preferences in quite different ways than animals do because they cannot actively move from one habitat to another. Seeds or spores arrive in different habitats through dispersal, and then either survive and grow or die because of biological or physical factors discussed in the next three chapters. Animals use behavioral mechanisms to choose their habitats, and individual movements are an essential component of the resulting habitat selection.

BEHAVIORAL MECHANISMS

In many invertebrates, habitat selection is accomplished in a simple manner. For example, when the isopod *Porcellio scaber* is placed in a humidity gradient, it moves at random with respect to the gradient, but it moves much more rapidly in dry air than in moist air. An individual that in the course of its random movements happens to find moist air will slow down and become motionless. The result is that most of the animals eventually come to rest at the moist end of the gradient, a very simple form of habitat selection (Fraenkel and Gunn 1940).

A more complex form of stereotyped behavior is involved in the choice of oviposition sites by many insects. In certain species of dragonflies, the males occupy territories that determine where females will oviposit (Macan 1974). The European corn borer larva will feed on a wide variety of plants. Corn borers occur mainly on corn because the ovipositing females are attracted by volatile odors produced by the corn plant (Schoonhoven 1968). Leaf beetles of the tribe Luperini (Coleoptera: Chrysomelidae: Galerucinae) use the secondary chemicals cucurbitacins as cues for host selection (Metcalf 1986). This association is present worldwide in about 1500 species of beetles feeding on plants of the family Cucurbitaceae (squash, cucumbers, melons). The southern corn rootworm (*Diabrotica undecimpunctata*) is one example from this tribe. It can feed on many species of plants but is most frequently found on cucurbits because of its attraction to cucurbitacins.

Anopheline mosquitoes are often important disease vectors, and their ecology has been studied a great deal because of the practical problems of malaria eradication. Each mosquito species is usually associated with a particular type of breeding place, and one of the striking observations that a student of malaria first makes is that large areas of water seem to be completely free of dangerous mosquitoes. Large areas of rice fields in Malaya are free of *Anopheles maculatus*, as are the majority of shallow pools in some breeding grounds of *Anopheles gambiae* (Muirhead-Thomson 1951). Why are some habitats occupied by larvae and others not? Early workers assumed that something in the water prevented the larvae from surviving, and they neglected to study the behavior of females in selecting sites in which to lay eggs. More recent work has emphasized the role of habitat selection for oviposition sites in female mosquitoes and shown that larvae can develop successfully over a much wider range of conditions than those in which eggs are laid. Thus, although we presume that the female selects a type of habitat most suitable for the larvae, many of the places she avoids are suitable for growth and development.

In southern India, the mosquito *Anopheles culicifacies* (a malaria vector) does not occur in rice fields after the plants grow to a height of 12 inches (30 cm) or more, even though these older rice fields support two other *Anopheles* species. Russell and Rao (1942) could find no eggs of *A. culicifacies* in old rice fields, yet when they transplanted this mosquito's eggs into old rice fields, the larvae survived and produced normal numbers of adults. The absence of *A. culicifacies* from this particular habitat is apparently due to the selection of oviposition sites by the females. In a series of simple experiments, Russell and Rao were able to show that mechanical obstruction of the rice plants was the main limiting factor. Glass rods placed vertically in small ponds deterred female *A. culicifacies* from laying eggs, as did barriers of vertical bamboo strips. Shade did not influence egg laying. This mosquito oviposits while flying and performing a hovering dance, never touching the water but remaining 2 to 4 inches (5 to 10 cm) above it. Mechanical obstructions seem to prevent the female mosquitoes from the free performance of this ovipositing dance and thereby restrict the species to a habitat range less than that which it could otherwise occupy.

Habitat selection is a process that operates at the level of the individual animal. Decision making or choices by mobile animals like migratory birds must occur in a hierarchical manner (Figure 5.1) from a larger spatial scale to the local microhabitat. We should recognize that the "explanation" of habitat selection is really a series of explanations appropriate to the spatial scale of interest. Most studies of habitat selection are done on the small spatial scale of microhabitats. For these studies of

Figure 5.1   An illustration of the hierarchical decision-making process involved in the choice of nonbreeding habitats by a migratory bird in Mexico. At different spatial scales the reasons for the bird's choice may differ. (From Hutto 1985.)

individual habitat selection Figure 5.2 gives a conceptual model of the variables that can influence the observed habitat choice of a species.

Habitat selection in birds has been studied in more detail than in most other groups. Two kinds of factors must be kept separate in discussing habitat selection: (1) evolutionary factors, conferring survival value on habitat selection; and (2) behavioral factors, giving the mechanism by which birds select areas. Habitat selection is the result of stimuli from (1) landscape and terrain; (2) nest, song, watch, feeding, and drinking sites; (3) food; and (4) other animals (Cody 1985).

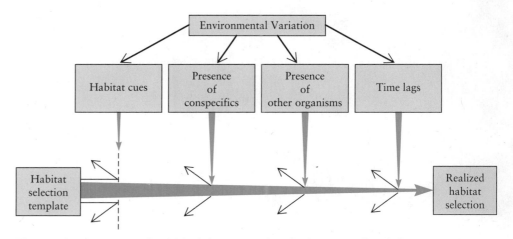

Figure 5.2   A conceptual model of the various factors that may affect habitat selection at the local level. (From Wiens 1985.)

Habitat cues for birds of prey may involve perch sites. Three species of buteos (broad winged hawks) breed in grassland and shrub-steppe areas of western North America. The red-tailed hawk selects areas with many perch trees or bluffs, while the Swainson's hawk and Ferruginous hawk select more open areas with few trees (Janes 1985). These hawks all eat much the same prey (ground squirrels, jackrabbits), and their habitat choice corresponds with their foraging behavior. Red-tailed hawks sit on perches and look for prey; their wings are less suited to soaring. Swainson's hawks are best at soaring and hunt from the air much more than from perches. Ferruginous hawks are intermediate in soaring abilities. Flapping flight is uncommon in all these hawks, and habitat selection is closely tied to their hunting methods.

In Britain the tree pipit *(Anthus trivialis)* and the meadow pipit *(Anthus pratensis)* have similar requirements except that the tree pipit breeds only in areas having one or more tall trees. For this reason, the tree pipit is absent from many treeless areas in Britain that the meadow pipit inhabits. Both pipits are ground nesters and feed on the same variety of organisms. Lack (1933) found tree pipits breeding in one treeless area close to a telegraph pole. The only use to which the tree or pole is put by the tree pipit is as a perch on which to land at the end of its aerial song. The meadow pipit has a similar song but ends it on the ground. Thus the tree pipit is excluded from colonizing heathlands only because it needs a perch on which to land after singing. Lack (1933) concluded that the distribution of bird species in the Breckland heaths of England and their associated pine plantations was largely a result of idiosyncratic behaviors and specific habitat selection restricting each bird to a habitat range less than that which it could potentially occupy.

Habitat selection in birds is partly a genetic trait, although it can be modified somewhat by learning and experience. The genetic basis of habitat selection is probably responsible for a slow response by some birds to human changes in the environment. The lack of genetic variability for habitat selection could be responsible for the lack of a response to environmental changes. The original habitat selected by a bird is often reinforced by tenacity of individuals to their sites. Many old birds return year after year to the same nesting site, even if the habitat at that site is deteriorating. Unfortunately, there has been little experimental analysis of habitat selection in birds. Klopfer (1963) has shown that chipping sparrows *(Spizella passer-*

ina) raised in the laboratory preferred to spend their time in pine rather than oak leaves, just as wild birds do. However, laboratory birds reared with oak leaves showed a decreased preference for pine as adults. Thus the innate preference for staying in pine could be modified somewhat by early experience:

Chipping Sparrows	% TIME SPENT IN	
	Pine	Oak
Wild-caught adults	71	29
Laboratory-reared, no foliage exposure	67	33
Laboratory-reared, oak foliage exposure only	46	54

The environmental cues by which gulls (*Larus* spp.) select their nesting habitat are both social and nonsocial. Laughing gulls on the coast of North Carolina nest on salt-marsh islands off the coast. Klopfer and Hailman (1965) observed two similar islands within a mile (1.6 km) of each other, but in 1959 and 1960 laughing gulls occupied only one island. This island was destroyed by a storm in 1961, and in 1962 the gulls established a thriving colony on the second island. Thus laughing gulls are selecting some general habitat features (islands free from ground predators, marsh areas, shallow-water feeding areas nearby), but within these constraints they respond to social stimulation, the presence of other breeding gulls. A "suitable" habitat, therefore, cannot be defined by its structural features alone.

Songbirds may have dialects that distinguish subpopulations, and these song dialects can influence habitat selection. White-crowned sparrows in the coastal scrub along Point Reyes National Seashore in California are subdivided into four dialects (Figure 5.3). The habitat is homogeneous, and these sparrows can clearly fly over the whole area in a few minutes, yet the dialects persist with stable boundaries (Baker et al. 1982). Dialects in bird songs are learned and are culturally transmitted from parents to offspring. Individual birds prefer to mate with birds of the same song dialect, so the socially determined dialect boundaries also become a genetic boundary that restricts gene flow. Habitat selection can thus have a cultural component independent of other biological, physical, or chemical factors.

Desert rodents have been well studied in both North America and Asia. Habitat selection is related to body size—larger species prefer open habitats and smaller species prefer bush microhabitats (Figure 5.4). These preferences have been verified experimentally. Price (1978) removed half of the bushes on half of her study grid and the kangaroo rat *Dipodomys merriami* increased in density at the more open sites within six weeks. Thompson (1982) did the opposite experiment—he added cardboard shelters to desert habitat to increase the cover in open areas (Figure 5.5). The results were as predicted: a decline in the numbers of the large kangaroo rat (*Dipodomys*) that prefers open areas, and an increase in the numbers of the smaller pocket mice (*Perognathus*). Since cardboard shelters clearly do not add food to the system, these responses of desert rodents must be to the risk of predation in open versus sheltered areas. The large kangaroo rats have adapted to living in open patches. They hop on their enlarged hind feet and this speedy locomotion is coupled with well-developed hearing to detect the approach of predators (Kotler and Brown 1988).

Other good examples of habitat selection can be found in small mammals. The

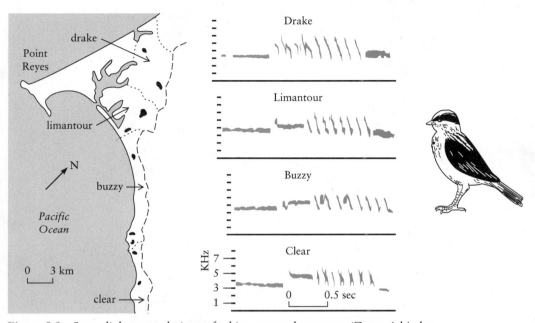

Figure 5.3   *Song dialect populations of white-crowned sparrows* (Zonotrichia leu-
cophrys nuttali) *at Point Reyes National Seashore, California. Suitable habitat is
bounded on the west by the ocean and on the east by Douglas fir forests (dashed
line). Dotted lines indicate boundaries between dialects. Representative sound
spectrograms of each of the four dialects are on the right. The sound spectrogram
displays frequency in kilohertz on the vertical axis and time on the horizontal axis.
(From Baker et al. 1982.)*

deer mouse *(Peromyscus maniculatus)* has received the most attention from this point
of view. The deer mouse is very widespread, found over most of North America,
and can be divided into two ecologically adapted types: (1) the long-tailed, long-
eared forest form and (2) the short-tailed, short-eared prairie form. The prairie form
that has been most intensively studied is the subspecies *P. m. bairdi*. This subspecies
avoids forested areas, even those with a grassy substratum, as can be readily seen
from the results of trapping lines (Harris 1952).

It was first necessary to determine whether *P. m. bairdi* could live in woody
areas, and this was done in field enclosures. Other studies with *Peromyscus* have
shown little difference in the food preferences or temperature requirements of grassland subspecies compared with forest subspecies. Harris (1952) therefore concluded
that *P. m. bairdi's* absence from forested areas is due to its behavior, not to the
unsuitability of the habitat.

Experiments in which the mice were given a choice between "woods" and "grassland" showed that *P. m. bairdi* usually chose the grassland. This behavior might be
either genetically determined or learned from early experiences, and an attempt was
made to distinguish between these causes. Early experience can play an important
part in the development of adult behavioral characteristics, and perhaps the early
experience of young mice in a particular habitat could determine the adult reactions
to different habitats.

Wecker (1963) used laboratory stock and wild mice in three types of situations:

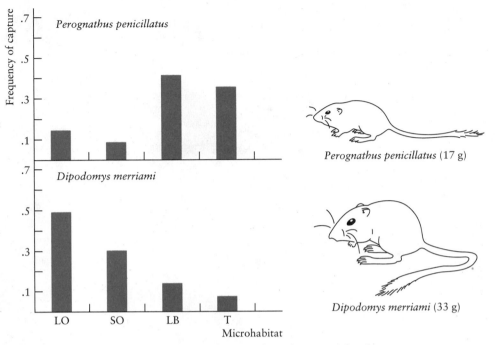

Figure 5.4   Summer microhabitat distribution of two heteromyid rodent species near Tucson, Arizona. Relative capture frequencies in live traps are given for four microhabitats: LO = large open, SO = small open, LB = large bush, T = tree. (From Price 1978.)

(1) direct testing of adults, (2) testing of offspring reared in field habitat, and (3) testing of offspring reared in woodland habitat. The results are given in Table 5.1. The wild-caught field animals showed an overwhelming preference for the field habitat. The laboratory stock, held in laboratory conditions for 12 to 20 generations, had apparently lost some of its preference for field habitat. Exposure of this stock to early experience in a field greatly increased the preference for the field habitat, but exposure to woods caused only an insignificant increase in the preference for woodland. The conclusion was that habitat selection in the prairie deer mouse is normally predetermined by heredity and results in the animal's occupying a more restricted range of habitats than is theoretically possible.

EVOLUTION OF HABITAT PREFERENCES

Why do organisms prefer some habitats and avoid others? Natural selection will favor individuals that use the habitats in which most progeny can be raised successfully. Individuals that choose the poorer, marginal habitats will not raise as many progeny and consequently will be selected against. Populations in marginal habitats may thus be sustained only by a net outflow of individuals from the preferred habitats. A variety of physical clues can be adopted by organisms as the proximate stimuli in choosing a particular type of habitat. Natural selection may act directly

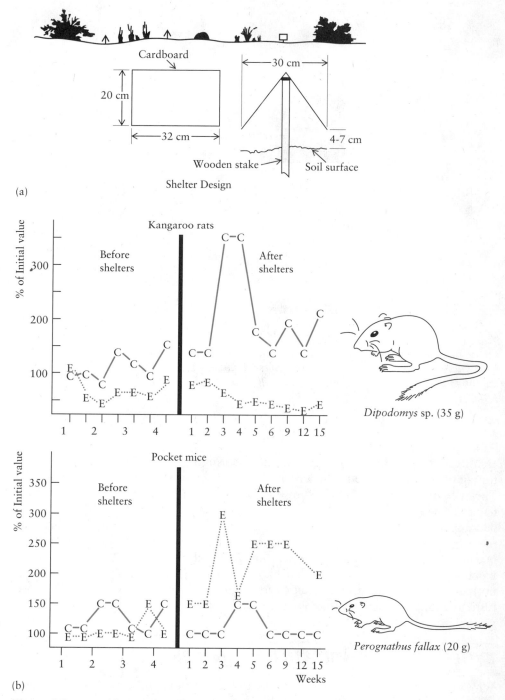

Figure 5.5   *(a) Addition of cardboard shelters to open areas, Mojave Desert, California. (b) Population changes before and after the shelters were added (E = experimental areas, C = control areas). (After Thompson 1982.)*

Table 5.1   Habitat Selection in Prairie Deer Mice (Peromyscus maniculatus bairdi)[a]

| | TIME SPENT IN FIELD PART OF PEN | |
Treatment	Active Time (%)	Inactive Time (%)
LABORATORY STOCK		
Tested directly	**48**	**40**
Offspring reared in field	85	97
Offspring reared in woods	59	44
FIELD ANIMALS		
Tested directly	**84**	**86**
Offspring reared in field	72	61
Offspring reared in woods	77	53

[a] Mice in each group were placed in a pen 100 ft × 16 ft (30m × 5m) that was half in a grassy field and half in an oak-hickory woodlot. The time spent in each half of the pen was recorded for both the part of the day in which the mice were active and the part in which they were inactive. Test results for adult controls are shown in boldface type.

Source: After Wecker (1963).

upon the behaviors that result in habitat choice, or it may select for individuals that have the capacity to learn which habitat is appropriate.

Habitat recognition may be very imprecise or very exacting. Aphids, such as the crop pest *Myzus persicae*, for example, land on host and nonhost plants with equal frequency (Emden et al. 1969). Aphids test the chemical suitability of the plant after landing by probing the leaf in several places with their mouthparts. If the plant is of the wrong species or in the wrong condition, the aphid flies to another plant. Many aphids land on suitable plants, test them, and leave anyway. Sockeye salmon, by contrast, return to spawn in the same stream in which they were hatched (Foerster 1968). After several years of feeding in the North Pacific, individual salmon are somehow capable of returning to the "correct" river system and the "correct" creek or spawning bed within that system. Olfactory clues in the water seem to be the critical stimuli, and if eggs are transferred from a natural spawning bed to a distant hatchery on a different river system, the salmon will return to the hatchery when they are adults. Some time period in the first few weeks of life is the critical period of tuning the fish's olfactory expectations (Hasler 1966).

Thus habitat recognition can be extremely specific if it benefits individuals to act accordingly. Salmon stocks adapt to the peculiar seasonality, temperature, and water-flow conditions of their own spawning grounds, and this precise adaptation can be preserved only if an excellent homing behavior has developed. At the other extreme, animals like aphids may face habitats that are favorable one month but not the next. Organisms faced with habitat unpredictability must adopt more flexible habitat selection behavior.

How can we explain the idiosyncratic behaviors of animals like the tree pipit that reject perfectly good habitats? We have seen several examples of this behavior earlier in this chapter. The important point illustrated by these examples is that evolution does not produce perfect organisms. Not all behaviors are adaptive, and environmental conditions may change such that behaviors that were formerly adaptive are now maladaptive. The genetic blueprint carried by a species cannot change overnight, and unless suitable genetic variation is present in a population, natural selection cannot operate to adjust habitat selection behaviors.

A THEORY OF HABITAT SELECTION

A simple theory of habitat selection can be used to illustrate how habitat selection may operate in a natural population (Fretwell 1972). Assume that three habitats are available to a species. For any particular species, we define a *habitat* as any part of the earth where that species can live, either temporarily or permanently. Each habitat is assumed to have a *suitability* for that species. Suitability is equivalent to *fitness* in evolutionary time, and we will assume that females produce more young in more suitable habitats than they do in less suitable habitats. Suitability is not constant but will be affected by many factors in the habitat, such as the food supply, shelter, and predators. But in addition, suitability in any habitat is usually a function of the density of other individuals, so that overcrowding reduces suitability (Figure 5.6). As a population fills up the best habitat (A in Figure 5.6), it reaches a point where the suitability of the intermediate habitat (B) is equal to that in habitat A. Individuals will now enter both habitats A and B. As these two habitats fill even more, the poor

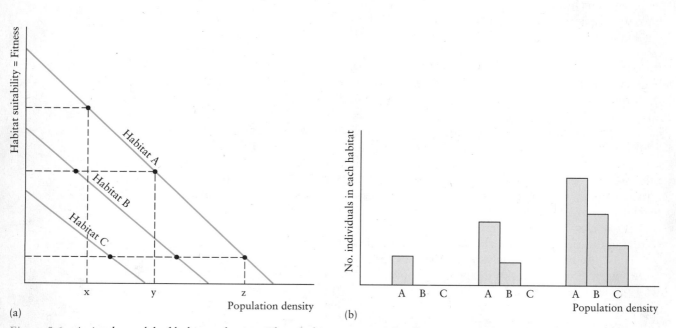

(a)   (b)

Figure 5.6   A simple model of habitat selection. Three habitats are used for illustration (A = good habitat, C = poor habitat). Habitat suitability is measured by the fitness of individuals living in that habitat. For illustrative purposes, three levels of population density are indicated (x, y, z). The model assumes that in all habitats fitness declines as population density goes up and crowding occurs. (a) At low density x, an individual can achieve the highest fitness by living in habitat A, and habitats B and C will be empty [as shown in (b)]. At high density z, an individual can choose to live in habitat A under crowded conditions, in habitat B under less crowded conditions, or in the poorest habitat, C, with the least crowding. If individuals choose their habitat as in this simple model, fitnesses of individuals will be equal in all three habitats at high density. (Modified from Fretwell 1972.)

habitat (C) finally has a suitability equal to that of habitats A and B. We assume in this simple model that all individuals are free to move into any habitat without any constraints (Fretwell 1972). The prediction that arises from this simple model of habitat selection is interesting because it is counterintuitive—we would predict from the model that, when density is high, good and poor habitats in the field would have equal suitabilities (but different densities). Individuals would be crowded in the best

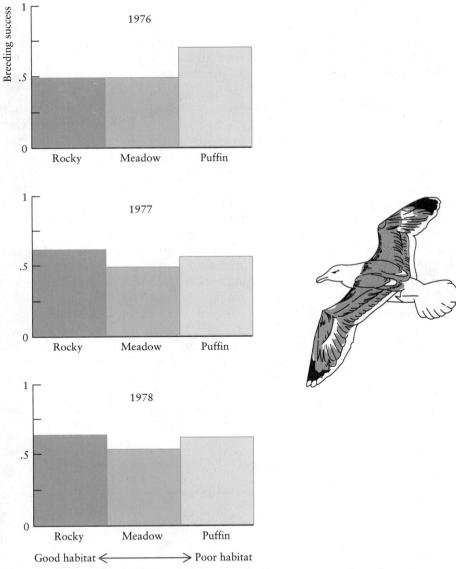

Good habitat ⟵⟶ Poor habitat

Figure 5.7 Breeding performance of herring gulls (Larus argentatus) *on Great Island, Newfoundland, 1976–1978. Breeding success is defined as the number of chicks fledged per egg laid and is a measure of fitness. Even though gulls prefer to nest in a rocky habitat, their breeding success is no higher in a rocky habitat than in the poorer puffin habitat. (Data from Pierotti 1982.)*

habitats and at low density in the poor habitats (Figure 5.6). Is there any evidence that this might in fact be the case in natural populations?

Herring gulls nest in three habitats on Great Island, Newfoundland: rocky terraces, grassy meadows, and turf slopes inhabited by Atlantic puffins (Pierotti 1982). Gulls prefer to nest in rocky habitats. Although only 10 percent of the island was rocky habitat, 22 percent of the herring gulls nested in it. Conversely, puffin slopes were least preferred by gulls. Reproductive parameters also differed among habitats. Gulls in the preferred rocky habitats laid more eggs than gulls in the puffin habitat, and gull chicks grew more rapidly in rocky habitats than they did on puffin slopes. By all the available criteria, a rocky habitat is a good habitat for nesting gulls, and puffin slopes are a poor habitat. But chick mortality is higher in the rocky habitat, and this offsets the higher egg production. Many chicks are killed by adult gulls in the crowded rocky habitat. Puffin slopes provide more hiding places for herring gull chicks, and nests are more spaced out so that few gull chicks are killed by their neighbors on puffin slopes. Puffins nest in burrows and do not compete with gulls for nesting sites.

Figure 5.7 shows the breeding success of herring gulls from the three habitats. Fitness (measured by breeding success) is nearly equal in all three habitats, exactly as predicted by Fretwell's model (Figure 5.6). Gulls crowd into the preferred rocky habitat until the expected fitness of a breeding pair falls to the level that can be reached on the poorer, but uncrowded, puffin slopes.

The proximate mechanisms by which habitats are selected are thus underlain by evolutionary expectations in fitness. We do not know how rapidly organisms can change the genetic and behavioral machinery that results in habitat selection.

Problems can arise whenever habitats change, and this has been a source of difficulty for many organisms since humans have modified the face of the earth. People provide many new habitats and destroy others. Some species of organisms, but not all, have responded by colonizing *Homo sapiens'* habitats. Other natural events, such as ice ages, cause slower habitat changes. Organisms with carefully fixed, genetically programmed habitat selection may require considerable time to evolve the necessary machinery to select a new habitat that is suitable for them. Adaptation can never be exact and instantaneous, and we must be careful not to expect perfection in organisms. We may judge the mosquito *Anopheles culicifacies* deficient for not laying eggs in all suitable habitats, but this may only reflect the fact that rice fields are a recent habitat in evolutionary time. Species evolve under a given set of environmental conditions, and populations adapt to their own particular environment. No species can truly be a jack-of-all-trades. Adapting to one type of habitat may make it impossible to live in another.

SUMMARY

If individuals are introduced into areas not occupied by the species and are able to survive, grow, and reproduce in their new habitat, the distribution of the species must be restricted either by lack of dispersal or by behavioral reactions. The behavior of individuals in selecting their habitat may thus restrict the distribution of some species of animals. Habitat selection of ovipositing insects provides some good examples in which a species can survive in a wider range of habitats than it usually occupies. Birds have also been studied from this point of view, although little experimental work has been done so far to find out why birds select some habitats for breeding and avoid others.

Behavioral limitations on distribution are usually subtle and may be the hardest to study. At present, few animal distributions are believed to be restricted by behavioral reactions but this may reflect the shortage of detailed behavioral studies on habitat preference. Plants do not have this mechanism at their disposal, and plant distributions must be limited in other ways.

Habitat selection evolves because organisms in some habitats leave more descendants than organisms in other habitats. In a predictable environment, habitat selection may be very exact. When habitats change, some species are not able to adapt quickly and therefore inhabit only a portion of their potential habitat range.

Selected References

CODY, M. L. 1985. An introduction to habitat selection in birds. In *Habitat Selection in Birds*, ed. M. L. Cody, pp. 3–56. Academic Press, New York.

JAENIKE, J., and R. D. HOLT. 1991. Genetic variation for habitat preference: Evidence and explanations. *American Naturalist* 137 (suppl.):S67–S90.

LACK, D. 1937. The psychological factor in bird distribution. *British Birds* 31:130–136.

ORIANS, G. H., and J. F. WITTENBERGER. 1991. Spatial and temporal scales in habitat selection. *American Naturalist* 137 (suppl.):S29–S49.

ROSENZWEIG, M. L. 1991. Habitat selection and population interactions: The search for mechanism. *American Naturalist* 137 (suppl.):S5–S28.

THOMPSON, S. D. 1982. Structure and species composition of desert heteromyid rodent species assemblages: Effects of a simple habitat manipulation. *Ecology* 63:1313–1321.

WECKER, S. J. 1964. Habitat selection. *Scientific American* 211(4):109–116.

WERNER, E. E., J. F. GILLIAM, D. J. HALL, and G. G. MITTELBACH. 1983. An experimental test of the effects of predation risk on habitat use in fish. *Ecology* 64: 1540–1548.

WIENS, J. A. 1985. Habitat selection in variable environments: Shrub-steppe birds. In *Habitat Selection in Birds*, ed. M. L. Cody, pp. 227–251. Academic Press, New York.

Questions and Problems

1. Morse (1980, p. 111) states:

   This is a very good time to study the mechanisms of habitat selection, in that many natural habitats are undergoing changes as a result of man's activities. These changes provide animals with potential opportunities and challenges, ones that in many instances they may never have previously experienced.

   Discuss.

2. After a theoretical analysis of the evolution of habitat selection, Holt (1985, p. 200) concluded:

    Natural selection, by favoring habitat selection by individuals, may thus have the long term evolutionary consequence of restricting the geographical range eventually occupied by a species.

    How might this occur?

3. Discuss the evolution of habitat selection and try to relate your conclusions to some of the examples presented in this chapter. How can natural selection maintain the particular ovipositing dance of *Anopheles culicifacies*, for example, if it results in suitable habitats' being left unoccupied? Does natural selection always favor the broadest possible habitat range for a species?

4. Review the factors affecting the settlement of larval barnacles (references in Hurley 1973). Discuss the evolutionary strategies needed for successful settlement of larval barnacles for species that live subtidally and those that live intertidally.

5. The aphid *Brevicoryne brassicae* avoids red cabbage varieties when selecting a host plant (Emden et al. 1969, p. 205). How would you determine whether this habitat-selection behavior was adaptive for the aphid?

6. Is it possible for a transplant experiment to be successful and yet lead to the conclusion that neither dispersal nor habitat selection is responsible for range limitation? Discuss the transplant experiment of Dayton et al. (1982) on the antarctic acorn barnacle and comment on Dayton's conclusions.

7. How are the predictions of the model given in Figure 5.6 affected if the habitat relationships are not straight lines but are curves? Can you imagine a situation in which these lines in Figure 5.6 might cross? See Rosenzweig (1985, p. 523) for a discussion.

8. Habitat selection has rarely been considered in plants. Why should this be? Read Bazzaz's (1991) review of habitat selection in plants and discuss his use of these concepts for plants.

Overview Question

Suppose you discovered a new species of lizard. Discuss how you would go about describing the habitat of this species. What would you measure and how could you decide when you had an understanding of its habitat selection?

6 Factors That Limit Distributions: Interrelations with Other Organisms

Up to now we have discussed cases in which the organism could actually live in places that it did not occupy. From now on we shall be dealing with cases in which the organism cannot complete its full life cycle if transplanted to areas it did not originally occupy. One reason for this inability to survive and reproduce could be the actions of other organisms. Predation is one of the clearest actions between species and we begin by considering the role of enemies in affecting geographical distributions.

PREDATION

Restriction of Prey by Predators

The local distribution of some species seems to be limited by predation. Work on intertidal invertebrates has provided some graphic examples of the influence of predation on distribution. Kitching and Ebling (1967) have summarized a series of studies at Lough Ine, an arm of the sea on the south coast of Ireland, and their studies are an excellent example of ecological work on distribution. Lough Ine is connected to the Atlantic Ocean by a narrow channel called the Rapids, through which the tide ebbs and flows approximately twice a day (Figure 6.1).

The common mussel *(Mytilus edulis)* is a widespread species on exposed rocky coasts in southern Ireland and throughout the world. Small mussels (less than 25 mm in length) are abundant on the exposed rocky Atlantic coast, but within Lough Ine and the more protected parts of the coast, this mussel is rare or absent (Figure 6.1). The only abundant populations are in the northern end of the lough, but these animals are typically very large (30 to 70 mm in length).

Kitching and his co-workers transferred pieces of rock with *Mytilus* attached from various parts of the lough to others. Figure 6.2 gives some typical results. Small *Mytilus* disappeared quickly from all stations to which they had been transferred within the lough, the Rapids, and the protected bays; they survived only on the open coast. The rapid loss, shown in Figure 6.2, suggested that predators were responsible. Large mussels that were transplanted around the lough also disappeared rapidly from most stations except places where they occurred naturally (C in Figure 6.1). Continuous observations on the transplanted mussels showed that three species of crabs and one starfish were the principal agents of mortality. By placing mussels of various sizes and crabs of the three species together in wire cages, Kitching and Ebling were able to show that one of the smaller species of crabs could not kill large *Mytilus* but that the other crabs could open all sizes of mussels. The areas of the lough where large *Mytilus* survive have few large crabs, and, where the large crabs are common, *Mytilus* is scarce or absent. Predatory crabs are probably restricted in

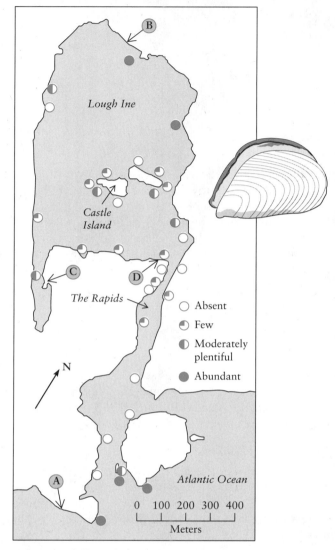

Figure 6.1 *Distribution of the mussel* Mytilus edulis *in Lough Ine and the adjacent Atlantic coast of Ireland, July 1955. For an explanation of points A to D, see Figure 6.2. (After Kitching and Ebling 1967.)*

their distribution by wave action, by strong currents, and by low salinity. Crabs also require an escape habitat in which they spend the day.

The distribution of this mussel in the intertidal zone at Lough Ine is thus controlled as follows: On the open coast, heavy wave action restricts the size of mussels and prevents predators from eliminating small mussels. In sheltered waters, predators eliminate most of the small mussels, and *Mytilus* survive only in areas safe from predators (such as steep rock faces), where they may grow to large sizes.

In another set of experiments, Kitching and Ebling (1961) studied the relation-

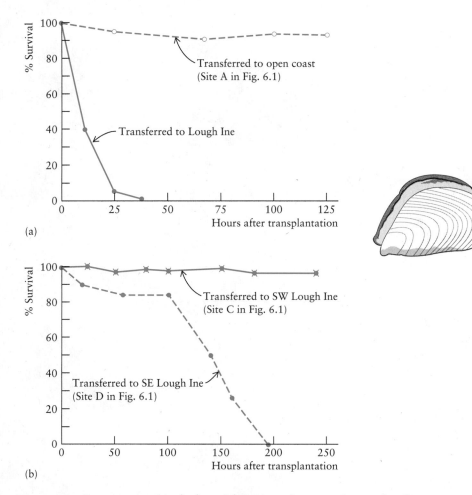

Figure 6.2   Percentage survival of mussels in transplant experiments. Small mussels (a) disappear rapidly when transplanted anywhere in Lough Ine but do not disappear if transplanted to the open coast (A in Figure 6.1). Large mussels (b) disappear if transplanted to some parts of Lough Ine such as the southeastern part (D in Figure 6.1) but do not disappear if transplanted to other parts of the lough, such as the southwestern part (C in Figure 6.1), where they occur naturally. (After Kitching and Ebling 1967.)

ship between the sea urchin *Paracentrotus lividus* and the algae on which it grazes. *Paracentrotus* lives in the shallow part of the sublittoral zone, just below the tide level, and is common in Lough Ine, although nearly absent from the open coast. One of the most extensive beds of this sea urchin occurs on the north side of Castle Island in the center of the lough, and this area is practically free from obvious algal growth. By contrast, algae are abundant in areas where this sea urchin is less common (Figure 6.3).

In July 1959 an area of 290 square meters was completely cleared of sea urchins by the removal of 1957 *Paracentrotus*. Algae immediately began to colonize the cleared area as follows:

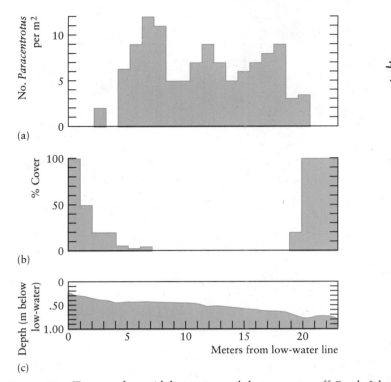

Figure 6.3   *Transect from tidal zone toward deeper water off Castle Island in*
Lough Ine, Ireland: (a) sea urchins, (b) algal cover, and (c) bottom depth profile.
Note that algal abundance is almost zero where sea urchins are common. (After
Kitching and Ebling 1961.)

	Algal Cover (%) on Experimental Area
July 7, 1959	0
July 23, 1959	10
August 10, 1959	25
September 3, 1959	50
July 1960	100

Adjacent control areas with sea urchins continued to have almost no visible algal
growth. Kitching and Ebling transferred the sea urchins removed from this area into
several areas of dense algal growth and found that they began to clear these areas
of algae as well. There is thus an inverse relationship between the occurrence of
Paracentrotus and algae.

Kitching and Ebling (1967) proposed four criteria to be fulfilled before one could
conclude that a predator restricts the distribution of its prey:

1. Prey individuals will survive when transplanted to a site where they do not
   normally occur if they are protected from predators.

2. The distributions of prey organisms and suspected predator(s) are inversely
   correlated.

3. The suspected predator is able to kill the prey, both in the field and in the laboratory.

4. The suspected predator can be shown to be responsible for the destruction of the prey in transplantation experiments.

In Australia, several species of small kangaroos have been driven to near extinction by the combined action of predation by introduced mammals like the red fox and competition for food by the introduced European rabbit. The burrowing rat kangaroo *(Bettongia lesueur)* is one example (Figure 6.4). It is extinct over the entire mainland area where it was originally so common as to be a pest, and now lives on only three islands off the western Australia coast where there are no foxes or rabbits (Ovington 1978).

Restriction of Predators by Prey

In the cases just discussed, the predator is believed to restrict the distribution of its prey; consequently, the reasons for the predator's distributional limits must be sought elsewhere. In these situations, the predator may feed on a variety of prey species, and each prey species may in turn be fed upon by many predatory species. The relationship may also operate in the other direction, and the prey may restrict the distribution of its predator. The "prey" may be a food plant and the "predator" a herbivore; alternatively, the prey may be a herbivore and the predator a carnivore. But if the prey is to restrict the predator's range, the predator must be very specialized and feed on only one or two species of prey. Such a predator is called a specialist or a monophagous predator. Many insect predators are specialists, but most vertebrate predators are not.

One example of a species that is limited in its distribution by its food source is

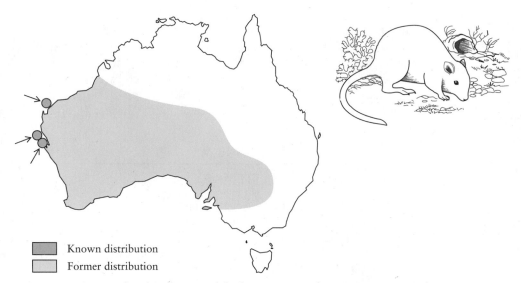

Figure 6.4   Geographic distribution of the burrowing rat kangaroo (Bettongia le-sueur) *in Australia. This small kangaroo is now extinct on the mainland and survives on only three islands. (From Ovington 1978.)*

Drosophila pachea, a rare fruit fly, which breeds only in the stems of senita cactus *(Lophocereus schottii)* throughout the Sonoran Desert of the southwestern United States and northern Mexico. This fly will not breed on the standard laboratory medium for *Drosophila* unless the medium is supplemented by a cube of fresh or autoclaved senita cactus. Conversely, medium supplemented by the cactus is toxic in varying degrees to the adults and larvae of other local species of *Drosophila*. The preliminary hypothesis is that the senita cactus contains a factor that is necessary for the development of *D. pachea* and a factor that is toxic to other species.

Heed and Kircher (1965) demonstrated that a unique sterol, schottenol (Δ^7-stigmasten-3β-ol), is the factor required by *D. pachea* for growth and reproduction. Every insect species that has been investigated requires a sterol in its diet. *Drosophila pachea* requires Δ^7-sterols, which are intermediates in the phytosterol biosynthetic pathway (Fogleman et al. 1982), and they accumulate in senita cactus. The function of sterols in the diet is probably twofold: (1) They are precursors for the molting hormone ecdysone, and (2) they affect female fertility in some way.

Senita cacti are rich in alkaloids, and 3 to 15 percent of the dry weight of the cactus consists of alkaloids. These alkaloids, particularly pilocereine, are toxic to eight species of *Drosophila* that live in the Sonoran Desert. The process of adaptation of *D. pachea* to this habitat has been possible only because of its tolerance of these potentially toxic alkaloids (Heed 1978).

The leaf-feeding beetle *Chrysolina quadrigemina* was introduced into the United States to control the Klamath weed *(Hypericum perforatum)*. In any introduction of this type to control weeds, it is important that the insect should not eat crop plants. Holloway (1964) discusses this problem for the *Chrysolina* beetle. The feeding habits of the beetle are very specific. Adult and larval beetles will not feed and will die when confined with plants other than those of the genus *Hypericum*. Adult beetles display an obligatory feeding response to the chemical hypericin in the foliage (Rees 1969). Beetles often refuse to stand on the leaf surfaces of plants which have leaves of different surface texture than that of *Hypericum*. The adults explore the leaf edges with their antennae, and if they encounter a serrated (toothed) leaf edge rather than a smooth one, they drop off the plant. The life history of this beetle is synchronized with that of its host plant and includes a summer (dry season) aestivation of the adults and subsequent response to fall rains. Thus the feeding habits, behavior, and life history of this leaf-eating beetle restrict it to a single host plant; also the range of distribution of *Chrysolina quadrigemina* is thereby restricted.

Predation is a major process affecting the distribution and the abundance of organisms. It will be discussed in more detail in Chapter 14 and in Chapters 24 and 25.

DISEASE AND PARASITISM

Enemies include predators as well as organisms that cause diseases and parasitism. Pathogens may eliminate species from areas and thereby restrict geographical distributions. We have already discussed one example of this—chestnut blight, which eliminated the American chestnut tree from eastern North America (see page 47). Another example is the native bird fauna of Hawaii.

Many of the endemic bird species of the Hawaiian Islands have become extinct in historical times, and one possible reason for these losses is introduced diseases. Warner (1968) postulated that both avian malaria and avian pox were both instrumental in causing extinctions in the Hawaiian Islands. The idea that diseases might be involved arose from the observation that native birds in Hawaii are relatively common only above 1500 meters of elevation (Figure 6.5a), while introduced birds occupy the lowland areas. The main malarial vector, the mosquito *Culex quinque-fasciatus*, is conversely most common in the lowland areas (Figure 6.5c). Because the native birds are much more susceptible to malaria than the introduced species, the malaria parasite is most common at intermediate elevations (Figure 6.5b) where vectors and hosts overlap (Van Riper et al. 1986).

The extinction of the native Hawaiian bird fauna occurred in two pulses. Before 1900 many of the low elevation bird species disappeared, coincident with extensive habitat clearing for agriculture and the introduction of rats, cats, and pigs. It is possible that other introduced diseases, such as avian pox, played a role in the early extinctions, but avian malaria did not because it was uncommon before 1900 (Van Riper et al. 1986). The second period of extinction in Hawaiian birds began in the early 1900s and was most likely the result of avian malaria. Birds that went extinct at this time lived in the midelevation forests where malaria is most prevalent (Figure 6.5c). At the same time the geographic distribution of many native birds was also reduced as they retreated to the highest elevation forests where mosquitoes are rare.

Diseases and parasites have always been a major factor in the ecology of humans (McNeill 1976, Desowitz 1991). Their role in the geographic ecology of plants and animals has been studied far less than their potential importance would warrant. We will discuss the impact of disease and parasitism on populations in Chapter 16.

ALLELOPATHY

Some organisms, plants in particular, may be limited in local distribution by poisons, antibiotics, or *allelopathic* agents. The action of penicillin among microorganisms is a classical case (Brock 1966, p. 127). Interest in toxic secretions of plants arose from a consideration of *soil sickness*. It was observed in the nineteenth century that, as one piece of ground was continuously cropped to one plant, the yields decreased and could not be improved by additional fertilizer. As early as 1832 DeCandolle suggested that the deleterious effects of continuous one-crop agriculture might be due to toxic secretions from roots. Several cases were also observed of detrimental effects of plants growing with one another: for example, grass and apple trees (Pickering 1917). Experiments of the general type shown in Figure 6.6 were performed. Apple seedlings were grown with three different sources of water: a primary source, a secondary source passing through grass and soil, and a secondary source passing through soil only. The growth of the young apple trees was apparently inhibited by something produced by the grass and carried by the water.

In the early 1900s several agronomists commented on the effect of black walnut trees (*Juglans nigra*) on nearby grass and alfalfa plants. Massey (1925) observed that the zone of dead alfalfa around a walnut tree extended over an area two to three times greater than that covered by the tree canopy and suggested that this zone was determined by the outer limits of walnut roots. The roots were suspected of secreting a toxin to which some crop plants, for example, alfalfa (lucerne) and tomatoes, were susceptible; other plants, for example, corn (maize) and beets, showed no ill effects. Schneiderhan (1927) showed that black walnut trees injured and killed apple trees up to 25 meters away. The average limit of the toxic zone was about 15 meters in radius from the walnut trunk. In every case, the toxic zone was greater than the area

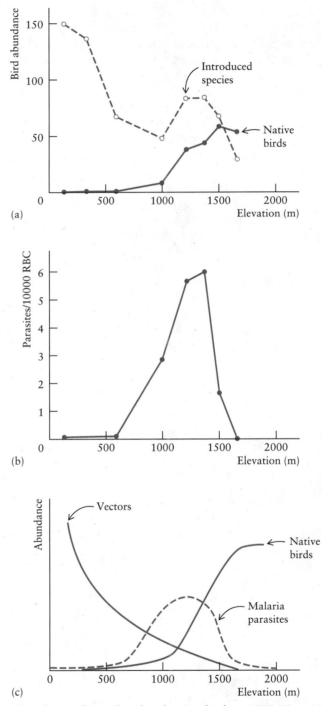

Figure 6.5   *(a) Abundance of introduced and native birds in 1978-79 at 16 sampling stations on Mauna Loa, Hawaii. (b) Avian malaria parasite loads (parasites per 10,000 red blood cells) along this same altitudinal gradient on Mauna Loa. (c) A generalized model of native bird abundance, malaria parasite incidence, and mosquito vector levels on Mauna Loa, Hawaii. (From Van Riper et al. 1986.)*

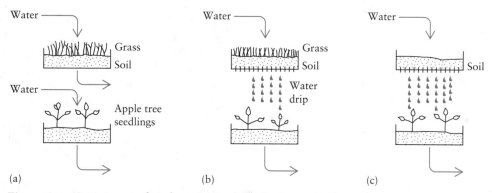

Figure 6.6   Experiments that demonstrated the detrimental effects of grass on apple tree seedlings. Grass and tree seedlings are grown in separate flats in a greenhouse. Water is provided either (a) independently to both grass and trees, (b) as a single source to the grass and soil, or (c) to the soil alone. Water drip provides moisture for the apple seedlings in (b) and (c). Apple tree seedlings do not grow properly when the water has passed through grass first (b).

covered by the walnut canopy, but larger walnut trees did not necessarily have much larger toxic zones.

Davis (1928) extracted a crystalline substance called juglone (5-hydroxy-α-naph–thaquinone) from the roots and hulls of the black walnut and showed that this chemical would kill tomato and alfalfa plants. Ponder (1987) found evidence of antagonism between walnut and some other timber tree species, like European alder, and he emphasized the selective effects of the walnut toxin: Some species, like alfalfa, are killed; others, like Kentucky bluegrass, become more abundant than usual near walnut trees. Not all walnut species secrete toxic chemicals. The closely related English walnut *(Juglans regia)* and the California walnuts *(Juglans hindsii* and *Juglans californica)* apparently do not secrete growth inhibitors (Garb 1961).

Agriculturalists have recognized the action of *smother crops* as weed suppressors. These smother crops include barley, rye, sorghum, millet, sweet clover, alfalfa, soybeans, and sunflowers. Their inhibition of weed growth was assumed to be due to competition for water, light, or nutrients. Barley, for example, is rated as a good smother crop and has extensive root growth. Overland (1966) showed that barley *(Hordeum vulgare)* inhibited the germination and growth of several weeds, even in the absence of competition for nutrients or water. Growth experiments with barley and chickweed *(Stellaria media)* gave the following results:

	AVERAGE DRY WEIGHT PER PLANT (g) AFTER 2 MONTHS OF GROWTH		No. of Chickweed Flowers
	Barley	**Chickweed**	
Controls (each grown alone)	4.15	3.20	100+
1:1 Barley/chickweed mixture	4.85	1.43	10

Extracts of living roots were more inhibitory than extracts of dead roots. The active inhibitory agent was found to be an alkaloid, but its specific chemical nature is not known. Thus the adverse effect of barley is partly due to the secretion by its roots of chemicals that reduce growth of weeds and germination of weed seeds.

The chaparral of southern California is a mixture of shrubs with a sparse understory of herbaceous vegetation. Recurrent fires at intervals of ten to forty years destroy this dense shrub cover, and in the first growing season after the fire, a luxuriant growth of herbs is produced. The shrubs regenerate principally by root sprouting. As the shrubs regenerate, seeds of herbs stop germinating, usually about five to six years after the fire, and herbs again become sparse. Even in open stands of shrubs with 50 percent of the ground bare and adequate rainfall, there is no seed germination. Fire is not the immediate cause of these changes in the herbs. Muller, Hanawalt, and McPherson (1968) cleared some chaparral stands by clipping close to the soil surface and removing the shrubs without disturbing the soil litter. In the next growing season, these clipped plots showed rapid germination of 30 herb species and looked like a typical area one year after a fire; adjoining uncleared areas again produced no germination of herbs.

In southern California, chaparral shrubs, such as the aromatic *Salvia leucophylla* and *Artemisia california*, are often separated from adjacent grassland by a bare area 1 to 2 meters wide (Figure 6.7). Apparently, volatile terpenes, which are released from the leaves of these aromatic shrubs, are able to inhibit growth in nearby grasses. To demonstrate this aerial transmission, Muller (1966) grew cucumber *(Cucumis sativus)* seedlings in a closed chamber with beakers containing 2 grams of *Salvia* leaves so that there was no physical contact between the growing seedlings and the test leaves; in one experiment, he obtained these results:

	Length of Cucumber Seedlings (mm) After 48 Hours
Control (no leaves)	28.6
Salvia apiana leaves	11.4
Salvia leucophylla leaves	2.9
Salvia mellifera leaves	2.3

Grasses were even more drastically suppressed in growth. Muller (1966) suggested that the deterioration of old *Salvia* stands might be caused by autointoxication.

An alternative explanation of the bare zone shown in Figure 6.7 was given by Bartholomew (1970). Animals concentrate their feeding activity in this zone close to the escape cover provided by the shrubs. Birds, rabbits, and mice removed seeds at 86 percent of the feeding stations in the bare zone within one day, but at only 12 percent of the feeding stations in the grassland (C in Figure 6.7). Bartholomew placed two types of one-foot-square exclosures in the bare zone of *Salvia leucophylla*. One type was complete wire mesh and excluded all herbivores, and the other type was open sided so that feeding could occur underneath it. After one year he got the following results:

	Dry Weight of Plants (g)
Open-sided exclosures	0.5
Complete exclosures	11.6

Figure 6.7 *The shrub* Salvia leucophylla *producing a bare zone in annual grass-land in California: (1) to the left of A,* Salvia *shrubs 1 to 2 meters tall; (2) between A and B, a zone 2 meters wide bare of all herbs (the root systems of the shrubs, on the average, failing to reach B); (3) between B and C, a zone of inhibited grassland consisting of small plants of* Erodium cicutarium, Bromus rubens, B. mollis, *and* Festuca megalura; *(4) to the right of C, uninhibited grassland, with large plants of* E. cicutarium, F. megalura, B. rubens, B. mollis, B. rigidus, A. fatua, *and other herbs. (After Muller 1966.)*

Bartholomew suggested that both animal grazing activities and toxins produced by shrubs could be involved in the production of a bare zone around shrubs in the annual grasslands of California.

Root competition for water or nutrients could also be contributing to the lack of herbs under chaparral shrubs. Swank and Oechel (1991) manipulated nutrients and water, reduced root competition by trenching, and eliminated herbivores by caging in stands of chamise *(Adenostoma fasciculatum)* in southern California. They found that herbs were restricted by all these factors; herbivory, root competition, inadequate water, and inadequate nutrients were all significant. Allelopathic chemicals were considered of minor importance in this system and are only one of several limiting factors in chaparral stands. Rarely in natural systems is one factor like allelopathy the complete explanation.

OTHER TYPES OF COMPETITION

The presence of other organisms may limit the distribution of some species through competition. Allelopathy is one specific type of competition for living space. But competition can occur between any two species that use the same types of resources and live in the same sorts of places. Note that two species do not need to be closely related to be involved in competition. For example, birds, rodents, and ants may compete for seeds in desert environments. Herbs and shrubs may compete for water in dry chaparral stands. Competition among animals is often over food. Plants can compete for light, water, nutrients, or even pollinators.

Competition is an important process affecting the distribution and abundance of plants and animals, and it will be discussed in more detail in Chapter 13 and in Chapters 24 and 25.

How can we determine whether competition could be restricting geographic distributions? One indication of competition may be the observation that when species A is absent, species B lives in a wider range of habitats. In extreme cases a habitat will contain only species A or species B and never both together. The principal difficulty in understanding these situations is that competition is only one of several hypotheses that will account for these facts.

The geographic distribution of closely related species has provided many examples of distributional patterns that could result from competition. Diamond (1975) described several examples of *checkerboard distributions*. Two ecologically similar species may have mutually exclusive but interdigitating distributions in an island archipelago, each island supporting only one species. Figure 6.8 gives an

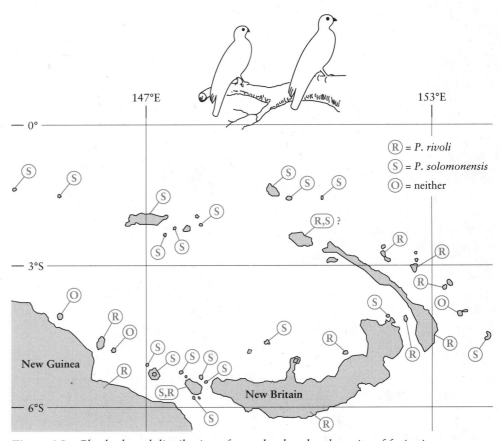

Figure 6.8   Checkerboard distribution of two closely-related species of fruit pigeons in the Bismarck Archipelago of the western Pacific Ocean. Ptilinopus rivoli *(R) or* P. solomonensis *(S) occur on most of these islands. Note that most islands have only one of these two species, and only one or two islands have both. (From Diamond 1975.)*

example of a checkerboard distribution for two fruit pigeons from the western Pacific. The identity of the successful colonist on an island may have been determined on a first-come-first-served basis, or on the basis of slight competitive advantages among the species.

Competition between species can at times be observed directly in the field. A striking case of competition for space occurs in the blackbirds of western North America. Orians and Collier (1963) describe competition for breeding space between the closely related red-winged blackbird *(Agelaius phoeniceus)* and the tricolored blackbird *(A. tricolor)*. Male redwings established territories in marshy areas in the winter before the colonial nesting tricolored blackbirds begin to establish a colony. Figure 6.9 shows one marsh completely occupied by male redwing territories in early spring. When large numbers of tricolors move into a marsh inhabited by redwings, the male redwings show strong aggression, but tricolors win out by virtue of over-whelming numbers (Figure 6.9). No successful redwing nests have been found in large tricolor colonies. A similar interaction occurs between the redwing and the yellow-headed blackbird *(Xanthocephalus xanthocephalus)* in western North America (Orians and Wilson 1964).

A particularly well studied case of competition occurs between the two terrestrial salamanders of eastern United States, *Plethodon jordani* and *Plethodon glutinosus* (Hairston 1980). These species have altitudinal distributions that overlap very little (70 to 120 meters on any one transect up the Black Mountains of North Carolina). Intensive competition between these two species was tested by removing one species from plots in the overlap zone for five years, and observing an increase in the

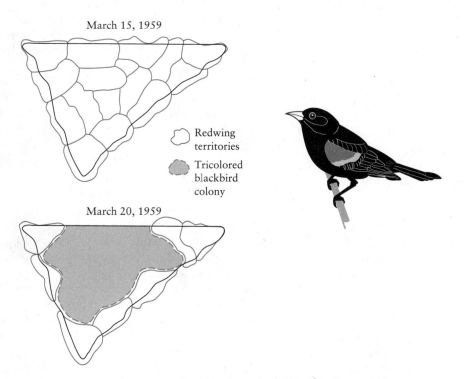

Figure 6.9   Interactions between redwing and tricolored blackbirds at Hidden Valley Marsh, Ventura County, California, in 1959. (After Orians and Collier 1963.)

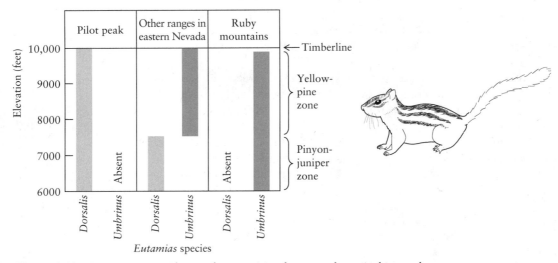

Figure 6.10 *Presumptive evidence of competition between the two chipmunks* Eutamias dorsalis *and* E. umbrinus *in the Great Basin area of Nevada. When one species is absent, the other expands its habitat range. (Data from Hall 1946.)*

abundance of the other salamander species. By proper long-term experiments like this, Hairston (1980) could demonstrate the role of competition in limiting altitudinal ranges of these salamanders.

The mountain ranges of the Great Basin of the western United States are isolated islands in a sea of desert, and many species are present in some ranges and absent from others. Brown (1971) investigated two chipmunk species that show evidence of competition (Figure 6.10). *Eutamias dorsalis* excludes *E. umbrinus* from the sparse forests of lower elevations, and *umbrinus* excludes *dorsalis* from the denser forests at higher elevations. The mechanism by which these two species compete was studied by Brown (1971) in the very narrow zone of overlap, around 2100 meters elevation, where the two species interact. *Eutamias dorsalis* prefers to live on the ground and is very aggressive toward other chipmunks of its own and of other species. *Eutamias umbrinus* spends much of its time in trees and often moves from tree to tree along interlocking branches. When trees are sparse at lower elevations, *dorsalis* is able to exclude *umbrinus* by superior aggression. Aggression becomes ineffective when trees are closely spaced because the arboreal *umbrinus* escapes through the trees. Thus the competitive success of *umbrinus* is determined by habitat structure.

The aggressive species *E. dorsalis* is eliminated from higher-elevation forests because it wastes all its time and energy on fruitless chases of the subordinate species. When Brown (1971) provided artificial feeding stations, several *umbrinus* would feed until a *dorsalis* approached. Then *dorsalis* would chase one *umbrinus* over to a tree and then return to the feeding station only to find it occupied by another *umbrinus*. By the time *dorsalis* chased the second *umbrinus* away to another tree, the first *umbrinus* was back at the feeding station. In this way, the aggressive behavior of *dorsalis* worked to its own disadvantage. In these two chipmunks, the effects of strong aggression are thus counterbalanced by the ability of the subordinate species to take refuge in some habitat from which it cannot be dislodged by the dominant species.

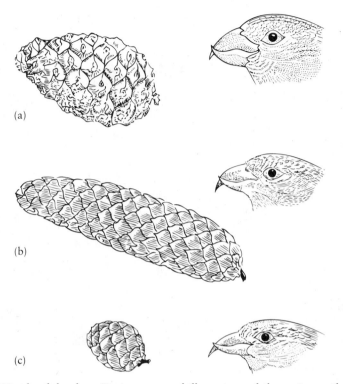

(a)

(b)

(c)

Figure 6.11 *Heads of the three European crossbill species and the main conifer cones each eats: (a) parrot crossbill and Scotch pine cone, (b) common crossbill and spruce cone, (c) white-winged crossbill and larch cone. (After Newton 1972.)*

When two species compete for resources, one species will always be better than the other in gathering or utilizing the resource that is scarce. In the long run, one species must lose out and disappear, unless it evolves some adaptation to escape from competition. There are two general evolutionary strategies a species can adopt: (1) avoid the superior competitor by selecting a different part of the habitat, or (2) avoid the superior competitor by a change in diet. The chipmunk *E. umbrinus* adopted the first solution. Let us look at an example where possible competition is avoided by a diet shift.

Crossbills are finches that have curved, crossed tips on the mandibles (Figure 6.11). Crossbills extract seeds from closed conifer cones by lateral movements of the lower jaw, and the jaw muscles are asymmetrically developed to provide the necessary leverage. Three species of crossbills live in Eurasia, and they are adapted for eating different foods (Newton 1972). The smallest crossbill is the white-winged crossbill *(Loxia leucoptera),* which has a small bill and feeds mainly on larch seeds (Figure 6.11). Larch cones are relatively soft. The medium-sized common crossbill (*L. curvirostra*) eats mainly spruce seed, and the larger parrot crossbill (*L. pytyopsittacus*) feeds on the hard cones of Scotch pine. These dietary differences are not necessarily preserved when the species live in isolation. Thus the common crossbill has evolved a Scottish subspecies, which has a large bill and feeds on pine cones, and an Asiatic subspecies, which has a small bill and feeds on larch seeds. The white-winged crossbill has an isolated subspecies on Hispaniola in the West Indies that feeds on pine seeds and has a large beak. The bill adaptations of crossbills can thus be interpreted as devices for minimizing dietary overlap in regions where all three possible competitors live.

SUMMARY

Many animals and plants are limited in their local distribution by the presence of other organisms—their food plants, predators, diseases, and competitors. Experimental transfers of organisms can test for these factors, and cages or other protective devices can be used to determine the critical interactions. Much of the evidence currently available is only circumstantial and not experimental.

Predators can affect the local distribution of their prey, and studies on intertidal organisms have illustrated this influence. The converse can also occur, whereby the prey's distribution determines the distribution of its predators, but this does not seem to be common. In some cases, an animal depends on a single food source and may have its distribution limited by the distribution of the food. Few such cases have been described but this limitation might be common for insect parasites.

Diseases and parasites may restrict geographic distributions but few such cases have been studied in natural systems. They may play a larger role than we currently suspect in species-rich tropical communities.

Some organisms poison the environment for other species, and local distributions may be affected by these chemical poisons, or *allelopathic agents*. The action of penicillin is a classic example. Chemical interactions have been described in a variety of crop plants and in native vegetation, and in most cases they are only one of several factors affecting local distributions of plants.

Competition among organisms for resources may also restrict local distributions. Some species drive others out by aggressive interactions. In other examples, the distributions of closely related species do not overlap, suggesting possible competitive interactions. Species may evolve differences in diet or habitat preferences as a result of competitive pressures.

Selected References

BULL, C. M. 1991. Ecology of parapatric distributions. *Annual Review of Ecology and Systematics* 22:19–36.

HAIRSTON, N. G. 1980. The experimental test of an analysis of field distributions: Competition in terrestrial salamanders. *Ecology* 61:817–826.

JACKSON, J. B. 1981. Interspecific competition and species distributions: The ghosts of theories and data past. *American Zoologist* 21:889–902.

KITCHING, J. A., and F. J. EBLING. 1967. Ecological studies at Lough Ine. *Advances in Ecological Research* 4:197–291.

PICKERING, S. 1917. The effect of one plant on another. *Annals of Botany* 31:181–187.

RICE, E. L. 1984. *Allelopathy*. Academic Press, New York.

VAN RIPER III, C., S. G. VAN RIPER, M. L. GOFF, and M. LAIRD. 1986. The epizo-otiology and ecological significance of malaria in Hawaiian land birds. *Ecological Monographs* 56:327–344.

Questions and Problems

1. Species interactions may be beneficial for all involved *(mutualism)*. Do mutualistic interactions ever limit the distribution of a species? (See Boucher et al. 1982 for a recent review.)

2. Norwegian lemmings do not live in lowland forests in Scandinavia even though they are regularly seen in these areas when their alpine populations are at high density. Suggest three hypotheses to explain the failure of lemmings to establish permanent populations in lowland forest, and discuss experiments to test these ideas. Oksanen (1993) discusses this question.

3. Macan (1974, p. 124), in discussing aquatic organisms, states:

   All species are probably limited to places that offer refuges from predators, unless they live in waters which, because they are temporary or offer extremes of some factor such as salinity, harbour no predators.

   Can you suggest any exceptions to this generalization? Does it apply to terrestrial organisms? To plants?

4. The sea urchin *Paracentrotus lividus* is nearly absent from the open coast of Ireland (page 76). Look up the natural history of this marine organism and suggest some possible reasons for this limitation of distribution.

5. Review the arguments about the relative importance of animals and plant toxins in the suppression of herbs in the California annual grassland (*Science* 173:462–463, 1971). Discuss an experimental program to evaluate the points of contention in this argument. Compare your experimental program with that used by Halligan (1976) and by Swank and Oechel (1991).

6. Pulliainen (1972) studied the summer diet of the three European crossbills (see Figure 6.11) in Lapland and found no differences in the food items being eaten by the three species. Reconcile these observations with the interpretation given on page 89.

7. One criterion of interspecific competition is that closely related species having mutually exclusive ranges are in competition at their zones of contact. Discuss this criterion in relation to the factors that limit distribution. Bull (1991) reviews this question.

8. Do chemical interactions between species ever affect the local distributions of marine organisms? Bakus et al. (1986) review the chemical ecology of marine organisms and provide references.

9. If sea urchins can graze algae (Figure 6.3, page 78) down to low abundance, why should there be any areas of high algal biomass, as shown in Figure 6.3? Suggest three hypotheses for these observations and indicate how you would test them experimentally.

Overview Question

Are the principles by which humans limit the distributions of species the same as those for predators? Classify and describe the various ways humans can cause effects on geographic distributions.

---

7 *Factors That Limit Distributions: Temperature and Moisture*

Temperature and moisture are the two master limiting factors to the distribution of life on earth. So it is not surprising that there is an enormous amount of literature on the effects of temperature and moisture on organisms. Before we analyze the ecological effects of these physical factors, let us look at the global temperature and moisture picture to which organisms must adapt.

CLIMATOLOGY

There are large temperature differentials over the earth, and these are a reflection of two basic variables: incoming solar radiation and the distribution of land and water. Solar radiation lands obliquely in the higher latitudes (Figure 7.1) and thus delivers less heat energy per unit area. Increased day-length in summer partially compensates for the reduced heat input at high latitudes, but total annual insolation is still lower in the polar regions. The amount of heat energy delivered to the poles is only about 40 percent of that delivered to the equator.

Land and sea absorb heat differently, and this effect produces more contrasts, even within the same latitude. Land heats quickly but cools rapidly as well, so land-controlled, or *continental*, climates have large daily and seasonal fluctuations of temperature. Water heats and cools more slowly because of vertical mixing and a high specific heat. The net result, shown in Figure 7.2, is that annual temperature ranges are greatest over the large continental land masses.

Water, alone or in conjunction with temperature, is probably the most important physical factor affecting the ecology of terrestrial organisms. Land animals and plants are affected by moisture in a variety of ways. Humidity of the air is important in controlling water loss through the skin and lungs of animals. All animals require some form of water in their food or as drink in order to operate their excretory systems. Plants are affected by the soil water levels as well as the humidity of the air around leaf surfaces. Cells are 85 to 90 percent water, and without adequate moisture there can be no life.

Moisture circulates from the ocean and the land back into clouds only to fall again as rain in a never-ending cycle. The global distribution of rainfall resulting from these processes is shown in Figure 7.3. There is a belt of high precipitation around the equator and a secondary peak in rainfall between latitudes 45° and 55°. Low precipitation around latitude 30° is associated with the distribution of deserts around the world. The distribution of continents and oceans also has a strong effect on the pattern shown in Figure 7.3. More rain falls over oceans than land. The average ocean weather station for the globe has 110 centimeters of precipitation, compared with 66 centimeters for the average land weather station. Finally, moun-

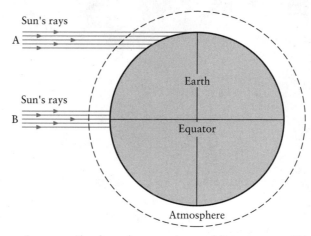

Figure 7.1   The sun's rays strike the polar areas in an oblique manner (A) and deliver less energy at the earth's surface than the vertical rays (B) for two reasons: (1) because the energy is spread over a larger surface in A and (2) because it passes through a thicker layer of absorbing, scattering, and reflecting atmosphere.

tains and highland areas intercept more rainfall and also leave a "rain shadow," or area of reduced precipitation, on their leeward side.

Water that falls on the land circulates back to the ocean as runoff or back to the air directly by evaporation or transpiration from plants. Only about 30 percent of precipitation is returned via runoff, and hence the remaining 70 percent must move directly back into the air by evaporation and transpiration. The rates of

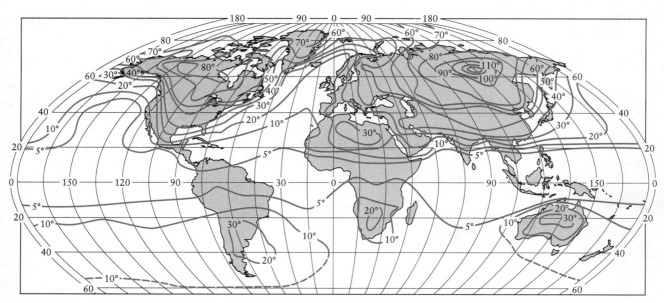

Figure 7.2   Average annual ranges of temperature (°F) for the earth. The annual range is defined as the difference between the average temperatures of the warmest and coldest months. Temperature ranges are smallest in the low latitudes and over oceans and largest over continents. (After Trewartha 1954.)

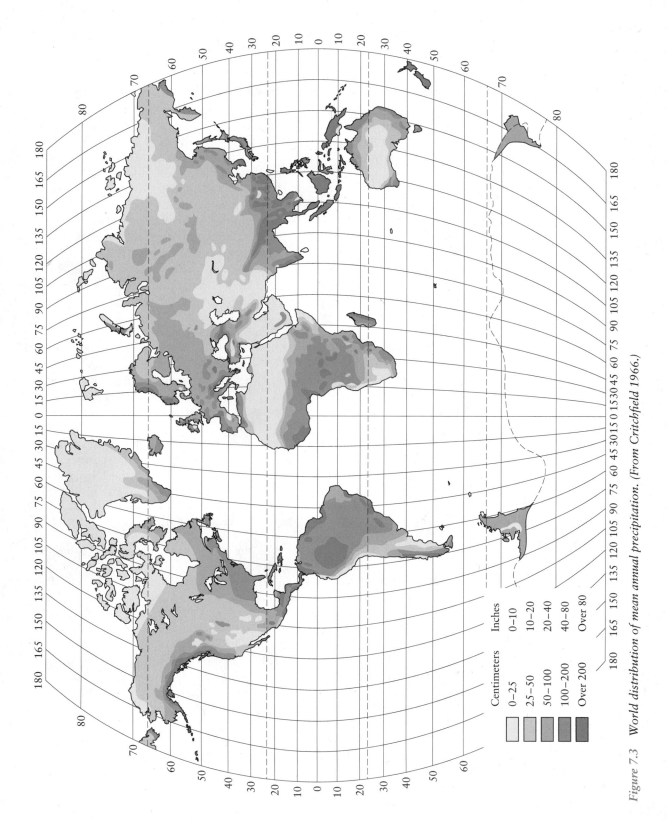

Figure 7.3 World distribution of mean annual precipitation. (From Critchfield 1966.)

evaporation and transpiration depend primarily on temperature; consequently, there is a strong interaction between temperature and moisture in affecting the water relations of animals and plants. The absolute amounts of rainfall and evaporation are less important than the relationship between the two variables. Polar areas, for example, have low precipitation but are not arid because the amount of evaporation is also low. About one-third of the land area of the globe has a rain deficit (evaporation exceeds precipitation), and about 12 percent of the land surface is extremely arid (evaporation at least twice as great as precipitation).

The vegetation of any site is usually considered a product of the climate of the area. This implies that climatic factors, temperature and moisture primarily, are the main factors controlling the distribution of vegetation. Geographers have often adopted this viewpoint and then turned it around to set up a classification of climate on the basis of vegetation. Native vegetation is assumed to be a meteorological instrument capable of measuring all the integrated climatic elements.

Some geographers have tried to set climatic boundaries independent of vegetation. One classification has been developed by Thornthwaite (1948). The basis of his climatic classification is *precipitation*, which is balanced against *potential evapotranspiration*. Potential evapotranspiration is the amount of water that would be lost from the ground by evaporation and from the vegetation by transpiration if an unlimited supply of water were available. There is no way of measuring potential evapotranspiration directly, and it is normally computed as a function of temperature. Diagrams can then be constructed; two extreme examples are shown in Figure 7.4 to illustrate the climatic regime at a desert station and a temperate deciduous forest station. Vegetation patterns can be described more accurately if we use *actual evapotranspiration*—the evaporative water loss from a site covered by a standard crop, given the precipitation. Major vegetational types such as the grassland, temperate deciduous forest, and tundra are closely associated with certain climatic types defined by the water balance (Stephenson 1990).

IMPORTANCE OF TEMPERATURE AND MOISTURE

Organisms have two options in dealing with the climatic conditions of their habitat. They can simply put up with the temperature and moisture as they are, or they can escape by some evolutionary adaptation. Let us begin our consideration of the effects of temperature and moisture by first determining how well organisms tolerate these two factors. Every organism has an upper and a lower lethal temperature, but these are not constants for each species. Organisms can acclimate physiologically to different conditions (Hoar 1975, chaps. 9–10). We illustrated this idea for two fish species in Chapter 3 (Figure 3.3, page 41). The resistance of woody plants to freezing temperatures is another example. Willow twigs (*Salix* spp.) collected in winter can survive freezing at temperatures below −150°C, while the same twigs in summer are killed by −5°C temperatures (Sakai 1970).

Temperature and moisture may act on any stage of the life cycle and can limit the distribution of a species through their effects on the following factors:

1. Survival

2. Reproduction

3. Development of young organisms

4. Interactions with other organisms near the limits of temperature or moisture tolerance (competition, predation, parasitism, diseases)

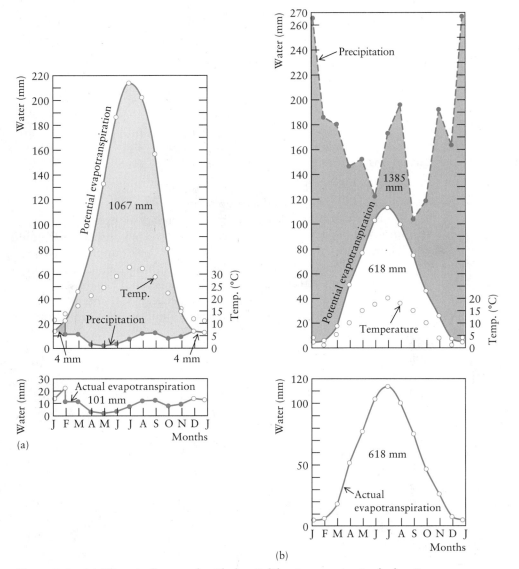

Figure 7.4   *(a) Climatic diagram for Blythe, California, a station in the hot So-noran desert. Available moisture limits plant activity. (After Major 1963.) (b) Cli-matic diagram for a station in Great Smoky Mountains National Park, Tennessee, at 1160 meters elevation, where available heat limits plant activity. The station is in a temperate deciduous forest. (After Major 1963.)*

    If temperature or moisture act to limit a distribution, what aspect of temperature or moisture is relevant—maximums, minimums, averages, or the level of variability? No overall rule can be applied here; the important measure depends on the mechanism by which temperature or moisture acts and the species involved. Plants (and animals) respond differently to the same environmental variables during different phases of their life cycle. For this reason, mean temperatures or average precipitation will not always be correlated with the limits of distributions, even if temperature or moisture is the critical variable.

To show that temperature or moisture limits the distribution of an organism, we should proceed as follows:

1. Determine the phase of the life cycle that is most sensitive to temperature or moisture.

2. Determine the physiological tolerance range of the organism for this life cycle phase.

3. Show that the temperature or moisture range in the microclimate where the organism lives is permissible for sites within the geographic range and lethal for sites outside the normal geographic range (Figure 7.5).

We will now consider a set of examples that illustrate this approach and show some of the biological complications that may occur.

The range limits of warm-blooded animals may correlate with climatic variables. Winter distributions of passerine birds in North America often correlate with minimum January temperature (Root 1988). Figure 7.6 shows one example of the Eastern phoebe, which has a northern winter range limit at −4°C. These climatic limitations in temperate zone birds are directly linked with the energetic demands associated with cold temperatures. Higher metabolic rates are needed to maintain body temperature in cold weather, and presumably there is a limit on the feeding rates of passerines that sets these limits (Root 1988).

It is less common for geographic range limits to coincide with rainfall contours. One example is the red kangaroo *(Megaleia rufa),* which occurs over most of the arid zone of inland Australia (Figure 7.7). Its distributional boundary coincides with the 400-millimeter rainfall contour, but this correlation is not a direct cause and

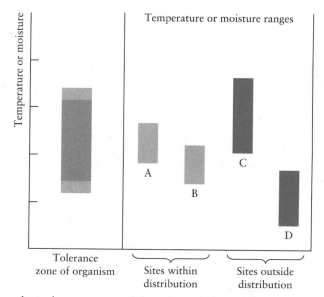

Figure 7.5   Hypothetical comparison of the tolerance zone of a species and the temperature or moisture ranges of the microclimates where the species lives. The tolerance zone is measured for the stage of the life cycle that is most sensitive to temperature or moisture and is subdivided into two zones, the optimal zone and the marginal zone.

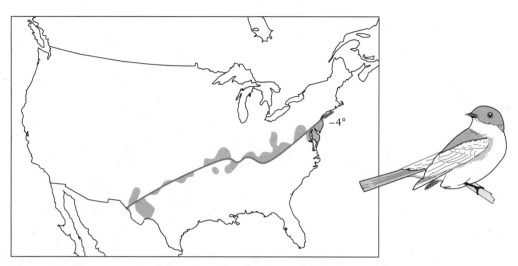

Figure 7.6 *Contour map of the winter distribution and abundance of the Eastern phoebe* (Sayornis phoebe) *from Christmas Bird Counts of 1962–1972. The bold line marks the edge of the winter range, which is associated with the −4°C isotherm of average minimum January temperatures. The stippling indicates the area of deviation between the range boundary and the isotherm. (From Root 1988.)*

effect. Red kangaroos are now more common in central Australia than they were before white settlers began grazing sheep and cattle in the arid and semiarid regions (Figure 1.2, page 11) and they have benefited from the additional watering points provided to sheep and cattle (Newsome 1975). Red kangaroos do not spread outside the 400-millimeter rainfall line because for food they depend on arid zone grasses that are restricted to the low rainfall areas. Few geographic distributions of animals are probably set directly by precipitation.

A detailed analysis of the water relations of plants is needed if we are to apply general ideas about moisture limitation to distributional studies on single plant species. The water balance of plants is difficult to measure directly, and botanists usually measure the water content of plant tissues as an index of water balance. The leaves are particularly sensitive because most evaporation occurs there. Different plants vary greatly in their ability to withstand water shortages.

Drought resistance is achieved by (1) improvement of water uptake by roots; (2) reduction of water loss by stomatal closure, prevention of cuticular respiration, and reduction of leaf surface; and (3) storage of water. Rapid root growth into deeper areas of the soil is often effective in increasing drought resistance; young plants will therefore suffer the worst from drought. Leaves of plants subject to poor water supply are often small and have thicker cuticles that reduce evaporation losses. By shedding their leaves in the drought season, plants have another very effective means of reducing their water loss. *Xerophytes* (plants that live in dry areas) show many of these special adaptations that decrease water loss. Leaves may be oriented in a vertical position which reduces the amount of radiation and evaporation. Other xerophytes, such as cacti, store water in their stems and thereby overcome drought.

While plants differ greatly in their ability to tolerate drought, they also differ in their ability to tolerate flooding. Bottomland hardwood forests occupy swamps and

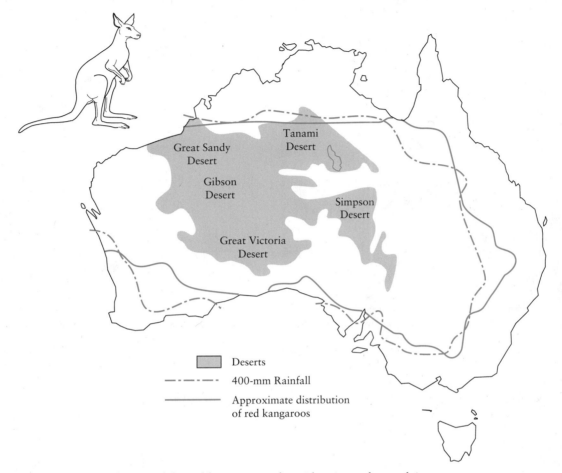

Figure 7.7   *Distribution of the red kangaroo in the arid regions of central Australia, and the 400-mm (15-in.) rainfall line. Red kangaroos are relatively rare in the large desert areas shown (see Figure 1.2). (Courtesy of G. Caughley.)*

river floodplains of the southern United States. These forests are the fastest-growing hardwood forests in the United States, and they contain a set of tree species that can survive in a flooded habitat. Some species do better than others under severe flooding. Water tupelo (*Nyssa aquatica*) is one tree species that survives and grows best under complete waterlogging; hence it occurs in areas that suffer prolonged flooding (Hosner and Boyce 1962). Other hardwood trees like sycamore (*Platanus occidentalis*) grow best on moist soil but do not grow well in waterlogged, flooded soil. Thus they are excluded from areas with prolonged flooding. Neither sycamore nor water tupelo can tolerate moisture stress from drought, and both are usually excluded from upland sites.

INTERACTION BETWEEN TEMPERATURE AND MOISTURE

In some cases, the moisture requirements of plants can restrict their geographic distributions. In other cases, moisture and temperature interact to limit geographic distributions and the ecologist must consider explanations like "both-temperature-and-moisture" rather than "either-temperature-or-moisture." Parker (1969) has reviewed drought resistance in woody plants and concluded that *frost drought* and *soil drought* are both critical in determining ranges of species. Soil drought is the usual "drought" in which soil moisture is deficient (as in the desert); it can usually be described as an *absolute* shortage of water in the soil. Frost drought or winter drought is a situation in which water is present but not available to plants (as in the tundra in winter); it can be described as a *relative* shortage of water for the plant. In both situations, water loss from the plant's leaves and stems is greater than water intake through the roots. Thus low temperatures can produce symptoms of drought. This emphasizes that water availability is the critical variable, which has led to considerable research on how to measure "available" water in the soil. Many of the distributional effects attributed to temperature may operate through the water balance of plants.

Hocker (1956) tried to describe the distribution range of the loblolly pine *(Pinus taeda)* from the meteorological data available from 207 weather stations in the southeastern United States. He included seasonal means for (1) average monthly temperature, (2) average monthly range of temperature, (3) number of days per month of measurable rainfall, (4) number of days per month with rainfall over 13 millimeters, (5) average monthly precipitation, and (6) average length of frost-free period. Weather stations were divided into two groups, one within the natural range of the pine and the other outside the range; from the difference between these two groups, Hocker mapped the climatic limits for loblolly pine (Figure 7.8). There is good agreement between observed limits of range and the limits mapped from meteorological data. The northern limit of this pine is probably set by winter tempera-

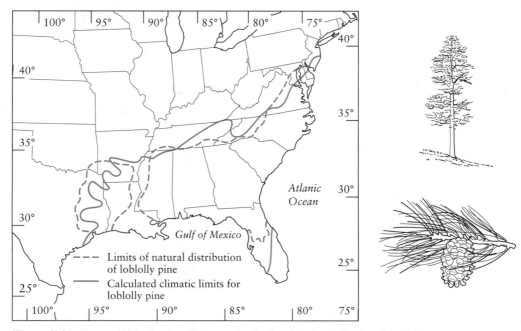

Figure 7.8 *Natural distribution limits and calculated climatic limits of loblolly pine* (Pinus taeda) *in the southeastern United States. (After Hocker 1956.)*

ture and rainfall. The rate of water uptake in loblolly pine roots decreases rapidly at lower temperatures, and this would accentuate winter drought* in more northerly areas. Hocker predicted that a northern extension of the limits of loblolly pine was not feasible because of these basic climatic limitations.

The *tree line*, or *timberline*, is a particularly graphic illustration of the limitation on plant distribution imposed by the physical environment. Not all tree lines are controlled by the same factors, and Stevens and Fox (1991) list nine factors that have been suggested to affect them:

1. Lack of soil
2. Desiccation of leaves in cold weather
3. Short growing season
4. Lack of snow, exposing plants to winter drying
5. Excessive snow lasting through the summer
6. Mechanical aspects of high winds
7. Rapid heat loss at night
8. Excessive soil temperatures during the day
9. Drought

These factors can be condensed into three primary variables: temperature, moisture, and wind. Proceeding up a mountain, temperature decreases, rainfall increases, and wind velocity increases. Because of freezing temperatures during much of the year, available soil moisture decreases. How can we determine the effects of temperature, moisture, and wind?

Daubenmire (1954) analyzed alpine timberlines in North America and reviewed the various factors that might affect them. Upper timberlines in North America decrease about 100 meters in altitude for every degree of latitude one moves north, except between the equator and 30°N, where timberlines are approximately constant at 3500 to 4000 meters. In North America, timberlines for any given latitude are lowest in the Appalachians and highest in the Rocky Mountains (Figure 7.9). The uniformity in changes of timberlines with latitude is surprising because many different tree species are involved.

Snow depth can affect the local distribution of trees near the timberline but cannot explain the existence of the timberline. In depressions where snow accumulates early and stays late, tree seedlings cannot become established. Only ridges will support trees in these circumstances, but these ridges also show a timberline; consequently, snow depth cannot be a primary factor.

Trees at the upper timberline in the Northern Hemisphere are often windblown and dwarfed, and this suggests wind as a major factor limiting trees on mountains. Within the tropics and in the Southern Hemisphere, wind effects seem to be absent. One difficulty with the wind hypothesis is that all the evidence is relevant to old trees, whereas it is the establishment of very young seedlings that is crucial to timberline formation. Daubenmire (1954) suggests that wind has secondary effects in altering timberlines in local situations, but, like snow depth, wind does not seem to be the primary cause of timberlines.

---

* Winter drought or frost drought occurs in plants when the roots are unable to take up water because of low soil temperatures but the leaves are losing water by transpiration.

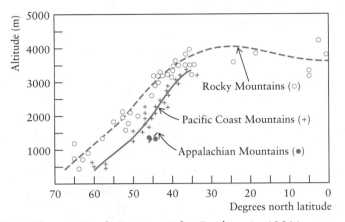

Figure 7.9   Timberlines in North America. (After Daubenmire 1954.)

Despite these temperature correlations, alpine timberlines are determined not by *summer* temperature but by *winter* desiccation, or frost drought (Tranquillini 1979). Winter desiccation is severe above the timberline because new leaves cannot grow enough in the short summer period to mature and thus become drought-resistant by laying down a thick cuticle. Figure 7.10 summarizes the causal mechanisms that produce alpine timberlines.

Almost no experimental work has been done to determine the causes for timberlines. In New Zealand, the beech *(Nothofagus)* forms an evergreen forest that stops abruptly at the timberline between 900 and 1500 meters above sea level. Wardle (1965) showed that this timberline was produced by factors that reduce seedling establishment. Good seed years near the timberline are uncommon for *Nothofagus*, and timberline seed has poor germination (0 to 3 percent). Seedling survival is poor, and the death of seedlings is associated with drying out of the tops. Wardle transplanted small seedlings above the timberline. Seedlings planted in the open all died in their first year, but shaded seedlings survived well and became established 180 meters above the timberline.

The intertidal zone of rocky coastlines is a tension zone between sea and land and as in the tree line the distributional boundaries are very clear. The upper and lower limits of dominant invertebrates and algae are often very sharp in the intertidal zone (Figure 7.11), and this zonation is a particularly graphic example of distribution on a local scale. Two barnacles dominate the British coasts; their British distributions are shown in Figure 7.12. *Chthamalus stellatus* is a southern species that is absent from the colder waters of the east British coast and is the common barnacle of the upper intertidal zone of western Britain and Ireland. Going farther north in the British Isles, one finds it restricted to a zone higher and higher on the intertidal rocks. *Chthamalus* is relatively tolerant of long periods of exposure to air, and the upper limit of its distribution on the shore is set by desiccation. This basic limitation does not seem to change over its range. The lower limit on the shore is often determined by competition for space with *Balanus balanoides*, a northern species. Connell (1961b) showed that *Balanus* grew faster than *Chthamalus* in the middle part of the intertidal zone and simply squeezed *Chthamalus* out. He also showed that *Chthamalus* could survive in the *Balanus* zone if *Balanus* were removed.

The upper limit of *B. balanoides* is also set by weather factors, but since this barnacle is less tolerant of desiccation and high temperatures than *Chthamalus*, there

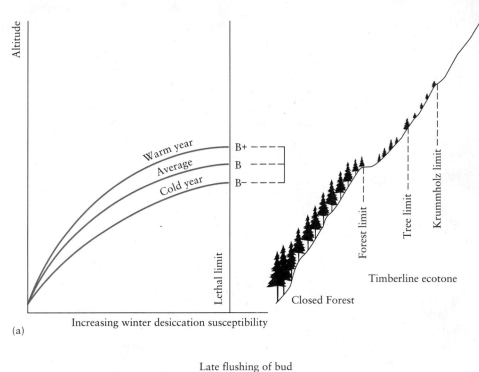

(a)

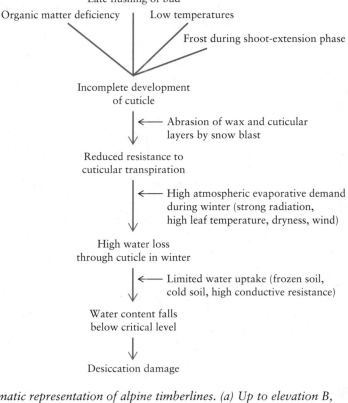

(b)

Figure 7.10   *Schematic representation of alpine timberlines. (a) Up to elevation B, natural tree regeneration can develop into a closed forest (forest limit). Above B (B + in warm years, B − in cold years), susceptibility to frost drought rapidly increases. The Krummholz limit fluctuates according to long-term climatic trends. (b) Causal chain leading to desiccation of terminal shoots in trees during the winter at the timberline. (After Tranquillini 1979.)*

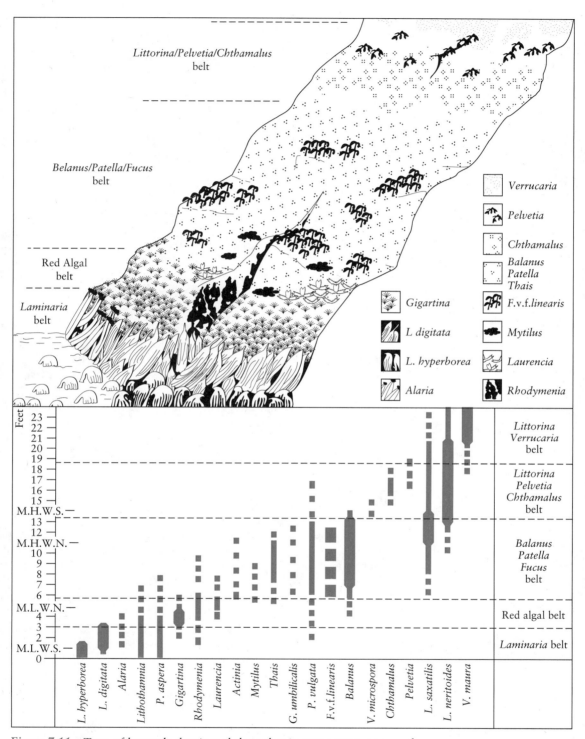

Figure 7.11 *Type of barnacle-dominated slope that is very common on moderately exposed rocky shores of northwestern Scotland and northwestern Ireland. M.H.W.S. = mean high water, spring; M.H.W.N. = mean high water, neap; M.L.W.N. = mean low water, neap; M.L.W.S. = mean low water, spring. (After Lewis 1964.)*

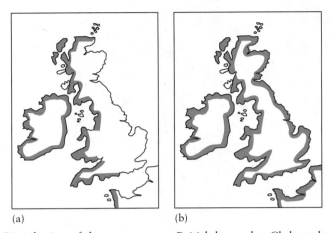

Figure 7.12 Distribution of the two common British barnacles. Chthamalus stel-latus *(a) is a southern species, and* Balanus balanoides *(b) is a northern species. (After Lewis 1972.)*

is a zone high on the shore where *Chthamalus* can survive but *Balanus* cannot (Connell 1961a). The sensitivity of young barnacles sets this upper limit. The lower limit of *Balanus* is set by competition for space with algae and by predation, particularly by a gastropod, *Thais lapillus.* Connell (1961b) summarized these results in Figure 7.13.

The distribution of these barnacles is a striking example of limitations imposed by both physical factors (temperature, desiccation) and biotic factors (competition, predation).

ADAPTATIONS TO TEMPERATURE AND MOISTURE

We have begun by assuming that there are certain physiological tolerances built into all the individuals of a particular species. But we know that local adaptation can occur and that genetic and physiological uniformity cannot be assumed throughout the range of a species. Darwin recognized that species could extend their distribution by local adaptation to limiting environmental factors, such as temperature, but the full implications of Darwin's ideas were not appreciated until the early 1900s when a Swedish botanist, Göte Turesson, began looking at adaptations to local environmental conditions in plants. Turesson (1922) coined the word *ecotype* to describe genetic varieties within a single species. He recognized that much of ecology had been pursued as if hereditary diversity within species did not exist. In a series of publications, he described some variation associated with climate and soil in a variety of plant species (Turesson 1925). The basic technique was to collect plants from a variety of areas and grow them together in field or laboratory plots at one site. The type of result he obtained in this early work can be illustrated with one example. *Plantago maritima* grows as a tall, robust plant in marshes along the coast of Sweden and also as a dwarf plant on exposed sea cliffs in the Faeroe Islands. When plants from marshes and from sea cliffs are grown side by side in an experimental garden, this height difference is not as extreme but remains significant (Turesson 1930):

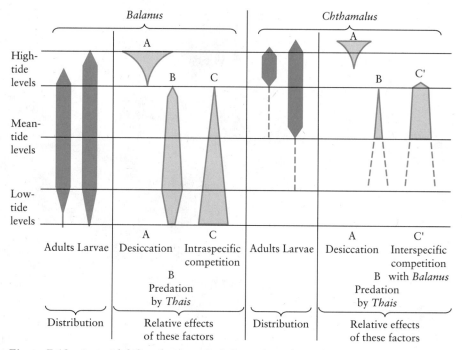

Figure 7.13   Intertidal distribution of adults and newly settled larvae of the barnacles Balanus balanoides *and* Chthamalus stellatus *at Millport, Scotland, with a diagrammatic representation of the relative effects of the principal limiting factors. (After Connell 1961b.)*

Plantago maritima Source	Mean Height (cm) in Experimental Garden
Marsh population	31.5
Cliff population	20.7

Turesson's early studies on ecotypes such as these helped to create a new research field of ecological genetics.

This common garden technique is an attempt to separate the *phenotypic* (environmental) and *genotypic* components of variation. Plants of the same species growing in such diverse environments as sea cliffs and marshes can differ in morphology and physiology in three ways: (1) All differences are phenotypic, and if seeds are transplanted from one situation to the other, they will respond exactly as the resident individuals; (2) all differences are genotypic, and if seeds are transplanted between areas, the mature plants will retain the form and the physiology typical of their original habitat; or (3) some combination of phenotypic and genotypic determination produces an intermediate result. In natural situations, the third case is most usual. Many examples are now described, particularly in plants (Heslop-Harrison 1964).

One of the most intensively studied sets of ecotypic races occurs in the perennial herb *Achillea* (yarrow), analyzed by Clausen, Keck, and Hiesey (1948) in a classic paper. Three very similar species of *Achillea* are described taxonomically, two from western North America and one from Europe. Clausen and colleagues studied the two North American species in detail. A maritime form of *Achillea borealis* lives in

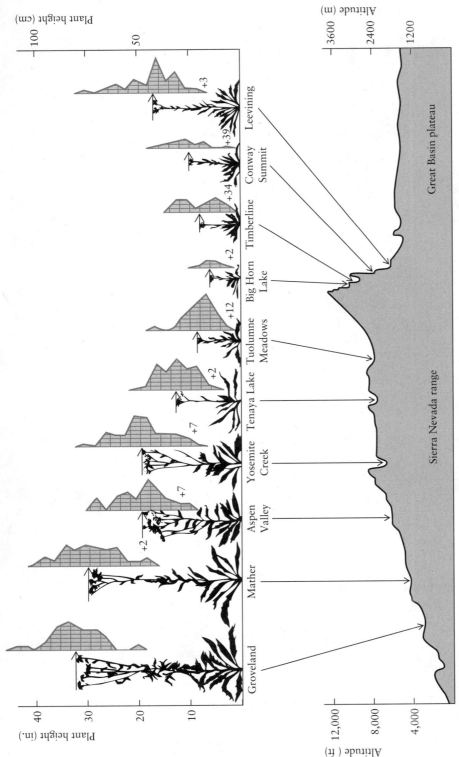

Figure 7.14 Representatives of populations of *Achillea lanulosa* as grown in a uniform garden at Stanford. These originated in the localities shown in the profile of a transect across central California at approximately 38°N latitude. Altitudes are to scale; horizontal distances are not to scale. The plants are herbarium specimens, each representing a population of approximately 60 individuals. The frequency diagrams show variation in height within each population; the horizontal lines separate class intervals of 5 cm according to the marginal scale, and the distance between vertical lines represents two individuals. The numbers to the right of some frequency diagrams indicate the number of nonflowering plants. The specimens represent plants of average height, and the arrows point to mean heights. (After Clausen et al. 1948.)

coastal areas of California as a low succulent evergreen plant that grows throughout the winter. Slightly farther inland there is an evergreen race that is similar but taller. A third race lives in the Pacific Coast range; it grows during the mild winter and flowers quickly by April, becoming dormant during the hot, dry summer. In the Central Valley of California can be found a giant race of *A. borealis* that survives under high summer temperatures, a long growing season, and ample moisture.

In the Sierra Nevada, races of *Achillea lanulosa* occur. As one proceeds up these mountains, the average winter temperature decreases below freezing, so winter dormancy is necessary and plants are smaller. On the eastern slope of the Sierra Nevada, plants of *A. lanulosa* are late-flowering and are adapted to cold, dry conditions. Clausen, Keck, and Hiesey collected seeds from a series of populations of *A. lanulosa* across California and raised plants in a greenhouse at Stanford, with the results shown in Figure 7.14. The major attributes of these races are maintained when grown under uniform conditions in the same place.

Clausen and his co-workers also raised *Achillea* in plots at Mather in the coniferous forest zone of the Sierra Nevada and at Timberline in the alpine zone. Table 7.1 gives some results of these transplant experiments for *A. lanulosa*. Under the extreme conditions at Timberline, the Groveland and Mather ecotypes survived poorly. Transplants of *A. borealis* from the coastal region all died at Timberline.

Latitudinal races of *Achillea* also occur but have not been studied in detail. Figure 7.15 illustrates the magnitude of the difference that can occur between ecotypes of the same species and shows graphically how much the gene pool of a species can be altered by local adaptations to the environment.

In many tree species at the timberline, the subalpine, erect tree form is replaced by an alpine elfinwood, a low bush form of the same species (Figure 7.16). These elfinwood forms may extend several hundred meters farther upslope in some species and thus extend the distribution locally. Clausen (1965) suggested that elfinwood forms were inherited growth forms or ecotypes that were selectively favored at the timberline. Griggs (1938) did not think that elfinwood forms of Rocky Mountain trees were genetically different from the normal growth types. The question remains unsolved.

Ecotypic variation has been analyzed in relatively few plants. Adaptations to

Table 7.1 Growth and Survival of Ecotypes of *Achillea lanulosa* Grown in Experimental Plots at Stanford (30 m elevation), Mather (1400 m), and Timberline (3050 m) in California for 3 Years.

	LONGEST STEMS (cm)			SURVIVAL[b] (%)		
Origin of Plants[a]	Stanford	Mather	Timberline	Stanford	Mather	Timberline
Groveland	83.6	58.2	15.5	100	93	40
Mather	79.6	82.4	34.3	93	90	39
Aspen Valley	47.4	56.8	25.3	93	100	73
Yosemite Creek	42.6	56.2	30.1	97	97	90
Tenaya Lake	33.9	33.7	33.4	100	97	97
Tuolumne Meadows	24.5	32.7	28.4	90	97	93
Timberline	21.2	31.6	23.7	90	67	90
Big Horn Lake	15.4	19.5	23.6	83	67	91

[a] Origin of plants may be located in Figure 7.14
[b] Based on samples of 30 plants (except Big Horn Lake, 12 plants)

Source: After Clausen et al. (1948).

Figure 7.15 *Ecotypic variation between (a) a southern and (b) a northern race of* Achillea borealis *and (c) the approximate temperature ranges of their native habitats. A race from Selma, California (a), and another from Seward, Alaska (b), growing in the Stanford garden reproduced to the same scale. (After Clausen et al. 1948.)*

moisture stress are involved in many tree ecotypes. Sugar maple *(Acer saccharum)* is an important hardwood tree in eastern North America. A study of genetic variation in sugar maple by Kriebel (1957) in Ohio used seed from 37 localities from New Brunswick and Quebec south to Florida. He recognized three ecotypes in sugar maple:

1. A *northern ecotype:* Low genetic resistance to drought, very low resistance to leaf damage from high solar radiation, and high resistance to winter injury.

2. A *central ecotype:* High resistance to drought, moderately high resistance to leaf scorch from high insolation, and high resistance to winter injury.

3. A *southern ecotype:* High drought resistance and high resistance to leaf injury by insolation, low resistance to winter injury, and poor growth form because of repeated forking of main and lateral shoots.

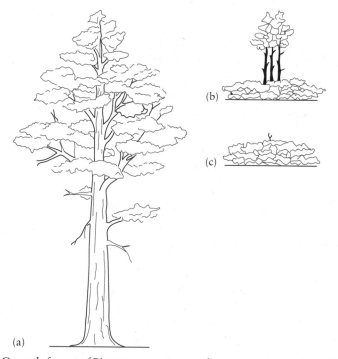

Figure 7.16   Growth forms of Pinus murrayana *on the eastern slope of the Sierra Nevada in California. (a) Single-trunk tree approximately 30 meters tall on south-facing slope at 3100 meters altitude. (b) Intermediate multitrunk tree 6 meters tall with 6-meter-wide elfinwood base on east-facing slope at 3230 meters. (c) Multi-trunk elfinwood 2 meters tall on south-facing slope at 3300 meters. (After Clausen 1965.)*

These genetic variations were ecologically important. For example, a severe summer drought in 1954 killed fewer seedlings of central and southern maples grown in a common garden:

Source of Seedlings	% Surviving Summer Drought
Northern	22.6
Central and southern	38.9

Growth patterns were also variable. Trees from northern sources stopped growing first; trees from the Gulf Coast continued growing until killed back by autumn frosts.

Many species have expanded their geographic range during this century (Hengeveld 1989); but in nearly all cases we do not know if genotypic changes accompanied these range changes. If we could study a species in the midst of a range extension, we might obtain some insight as to how organisms can extend their limits of tolerance. Crop plants selected by humans illustrate the type of changes that are possible. One of the most remarkable examples of crop-plant evolution is the case of the annual cottons. Some 600 years ago, from all four of the cultivated cotton species, forms were selected that fruited early enough to produce a sizable crop in the first growing season. These early-fruiting cottons were then planted in temperate climates with cold winters and hot summers, and the annual growth habit was imposed on them by winter frost. Selection for high productivity completed this cycle, and now almost all cultivated cottons are obligate annuals that can be grown in cold-winter areas and also in semiarid climates. Thus in a maximum of 600 generations, the cotton plant has been selected to live in environmental conditions of the temperate zone that were formerly lethal (Hutchinson 1965, p. 170).

We rarely catch a species in the process of an extension of its ecological range, yet this should be a focus of attention for ecologists interested in the problem of distribution. This opportunity may become more frequent in the future, if climatic warming occurs.

CLIMATE CHANGE AND DISTRIBUTIONS

If temperature and moisture are the master limiting factors for the geographic ranges of plants and animals, the climatic warming that is predicted for the twenty-first century will have profound effects on the earth's biota. One way to get a glimpse of the kind of changes that may occur is to look backward at the changes that have occurred in temperate regions since the end of the last Ice Age.

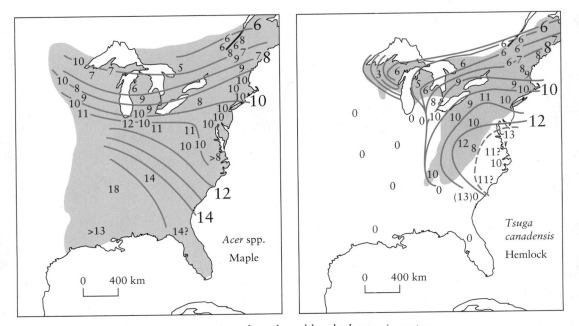

Figure 7.17 *Northward range extension of maple and hemlock trees in eastern North America since the end of the Ice Age, 16,000 years ago. Contour lines represent the advancing frontier at 1000-year intervals, numbered as thousands of years before the present. Stippled area is the present geographic range. Contours are estimated from the numbers of fossil pollen grains in lake sediments. (From Davis 1986.)*

The last continental glaciers began retreating in North America and Eurasia about 16,000 years ago. The northward expansion of tree distributions lagged behind the retreat of the ice. A detailed record of these migrations is captured in fossil pollen deposited in lakes and ponds (Davis 1986). In North America oaks and maples moved rapidly in a northeastward direction from the Mississippi Valley, while hickories advanced more slowly (Delcourt and Delcourt 1987). Hemlocks and white pines moved northwestward from refuges along the Atlantic Coast (Figure 7.17). The important finding of this paleoecological work is that the range of each species advanced individualistically. If you were sitting in New Hampshire, you would have seen sugar maple arrive 9000 years ago, hemlock 7500 years ago, and beech 6500 years ago (Davis 1986).

If we can determine the climatic limits of current geographic distributions, we can make predictions about how distributions will change with climatic warming. A major assumption of using this approach for plants is that seed dispersal is assumed to be adequate to sustain the migrations of each species. Davis (1986) suggested that hemlock was delayed nearly 2500 years in its movement north at the end of the Ice Age, partly because of slow seed dispersal. If we use climate change models to predict temperature and rainfall changes over the next 100 years, we can begin to estimate the size of the problem animals and plants will face over the next few hundred years.

Figure 7.18 shows the current and potential geographic range of American beech *(Fagus grandifolia)* under two climate change models. These models predict that the potential northern range limit of beech will move 700 to 900 kilometers north in the next century and the southern range limit will shrink to the north as well. If left to natural processes, beech must move 7 to 9 kilometers per year to the north. By contrast, since the end of the Ice Age, beech migrated into its present range at a rate of 0.2 kilometers per year. If these predictions are even approximately correct,

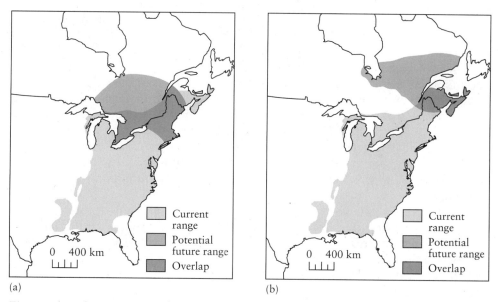

(a)   (b)

Figure 7.18   Current geographic range and predicted future range for American beech under two climatic-change scenarios: (a) the milder scenario (4.5°C warming) and (b) the more severe scenario (6.5° warming). (From Roberts 1989.)

slowly-colonizing species like trees will have to have human assistance to move into their new ranges. If this does not occur, species like beech may become extinct.

These effects of climate change will not appear immediately. Long-lived species such as trees will survive for many years as adult trees in inappropriate places. As the climate changes, they will reduce their seed production, and finally be unable to produce viable seedlings (Roberts 1989). All of these effects are complicated by ecotypic variation within tree species. If there are specific ecotypes adapted to northern or southern climatic conditions, then the problem of range shifts shown in Figure 7.18 is even more difficult to solve without the loss of ecotypic variation.

The geographic ranges of species are thus not static but dynamic, and as climate changes in the future, species will if time permits move into new areas that become climatically suitable. The major concerns of ecologists are, first, that the speed of climatic change in the next 100 years may be too great for slowly colonizing forms to move, and second, that genetic adaptations to local temperature and rainfall patterns may be lost for some species.

SUMMARY

Temperature and moisture are the major factors that limit the distributions of animals and plants. They may act on any stage of the life cycle and affect survival, reproduction, or development. Temperature and moisture may also act indirectly to limit distributions through their effects on competitive ability, disease resistance, predation, or parasitism.

From a global viewpoint, the distribution of plants can be associated with climate. Geographers have recognized this by making climatic classifications based on vegetational distributions. The tropical rain forest and tundra, for example, occupy areas with different temperature and moisture regimes. The effects of climate are less clearly seen at the level of the distribution of the individual species. In only a few cases has experimental work been done in local populations, first to pinpoint the life-cycle stage affected by climate and then to describe the physiological processes involved.

Water availability is the critical key to moisture effects on plants, and drought occurs when adequate amounts of water are not present or available to the plant. The soil may be saturated with water, but if it is all frozen, none may be taken up by plants, and they may suffer frost drought. Many of the distributional effects attributed to temperature may operate through the water balance of plants.

Species may adapt to temperature or moisture phenotypically or genotypically and thereby circumvent some of the restrictions imposed by climate. Turesson, a Swedish botanist, was one of the first to recognize the importance of *ecotypes*— genetic varieties within a single species. By transplanting individuals from a variety of habitats into a common garden, Turesson showed that many of the adaptations of plant forms were genotypic. Many ecotypes have now been described, particularly in plants, and these may involve adaptations to any environmental factor, including temperature and moisture. Ecotypic differentiation has often proceeded to the point where one ecotype cannot survive in the habitat of another ecotype of the same species.

Organisms have evolved an array of adaptations to overcome the limitations of high and low temperatures and drought. Some adaptations might allow a species to extend its geographic range. Many species are known to have extended or reduced their geographic range in historical times, but few cases have been studied in detail. Consequently, we rarely know whether a species changed its distribution because the climate changed or because the genetic makeup of the species altered.

The climatic warming that is predicted for the next century will have strong effects on geographic distributions. The major concerns are that species will not be able to migrate fast enough to keep pace with global warming, and that genetic adaptations to local environments may be lost.

Selected References

GEAR, A. J., and B. HUNTLEY. 1991. Rapid changes in the range limits of Scots Pine 4000 years ago. *Science* 251:544–547.

RICHARDSON, D. M., and W. J. BOND. 1991. Determinants of plant distribution: Evidence from pine invasions. *American Naturalist* 137:639–668.

ROOT, T. 1988. Energy constraints on avian distributions and abundances. *Ecology* 69:330–339.

STEPHENSON, N. L. 1990. Climatic control of vegetation distribution: The role of the water balance. *American Naturalist* 135:649–670.

STEVENS, G. C. and J. F. FOX. 1991. The causes of treeline. *Annual Review of Ecology and Systematics* 22:177–191.

TURESSON. G. 1930. The selective effect of climate upon the plant species. *Hereditas* 14:99–152.

WARDLE, P. 1981. Is the alpine timberline set by physiological tolerance, reproductive capacity, or biological interactions? *Proceedings of the Ecological Society of Australia* 11:53–66.

WOLCOTT, T. G. 1973. Physiological ecology and intertidal zonation in limpets *(Acmaea):* A critical look at "limiting factors." *Biological Bulletin* 145:389–422.

YOM-TOV, Y. 1979. Is air temperature limiting northern breeding distribution of birds? *Ornis Scandinavica* 11:71–72.

Questions and Problems

1. "To become widespread, a species must develop many ecological races" (Clausen, Keck, and Hiesey 1948, p. 121). Is this true in both animals and plants? Discuss with reference to colonizing species such as the starling.

2. Cain (1944) states:

   Physiological processes are multi-conditioned, and an investigation of the effects of variation of a single factor, when all others are controlled, cannot be applied directly to an interpretation of the role of that factor in nature. It is impossible, then, to speak of a single condition of a factor as being the cause of an observed effect in an organism.

Discuss the implications of this principle that the factors of the environment act collectively and simultaneously, with regard to methods for studying distributional problems.

3. Jarvis (1963, p. 310) after studying the laboratory responses of some British plants to water stress concluded:

    It is therefore difficult to apply to field conditions the results of experiments revealing physiological differences between species in their response to soil or atmospheric drought.

    Why did she come to this conclusion? How could you avoid this problem?

4. "The frost line . . . is probably the most important of all climatic demarcations in plants" (Good 1964, p. 353). Locate the frost line in a climatological atlas, and compare the distributions of some tropical and temperate species of any particular taxonomic group with respect to this boundary.

5. Are the grasses that have evolved to live on contaminated mine-waste soils (Chapter 3, page 42) an example of ecotypes? What types of selection (see Figure 2.2, page 19) would you expect to see among ecotypes?

6. The British barnacle *Elminius modestus* extends higher on the shore in the intertidal zone than does the barnacle *Balanus balanoides* when the two species occur together. However, these two species have similar tolerances to desiccation, salinity, and temperature. The range of initial settlement of young barnacles is the same for the two species. Given these facts, can you suggest an explanation for the observation that *E. modestus* extends higher on the shore than *B. balanoides*? Consult Foster (1971, p. 47), and compare his explanation with yours.

7. Adult male dark-eyed juncos *(Junco hyemalis)* remain farther north in winter than females and juveniles (Ketterson and Nolan 1982). Review the arguments of Root (1988) on energy balance and winter bird ranges, and suggest an explanation for these observations.

8. Elfinwood growth forms of trees are found near the timberline (see Figure 7.16). How would you test the suggestion that elfinwood trees are genetically different from normal trees?

9. The cane toad *(Bufo marinus)* is an introduced pest in parts of Australia. Beurden (1981) tried to predict the geographic limits of possible spread of this pest within Australia. Analyze his approach, and list the critical assumptions he uses to make these predictions.

Overview Question

Ecologists are concerned that the effects of predators, diseases, and competitors may depend upon temperature and moisture. This is called an interaction between biotic and climatic factors. Discuss how these interactions might affect our abilities to predict the impact of global warming on geographic distributions.

8 Factors That Limit Distributions: Other Physical and Chemical Factors

LIGHT

Light is important to organisms for two different reasons. Light is used as a cue for the timing of daily and seasonal rhythms in both animals and plants, and it is essential for photosynthesis. Timing is a central issue in the life cycles of organisms. Nocturnal desert animals, for example, use light as a cue for their activity cycles. The breeding seasons of many animals and plants are set by the organisms' responses to day-length changes. The seasonal impact of day length on physiological responses is called *photoperiodism*, and this has been an important focus of work in environmental physiology.

Breeding in most organisms occurs during a part of the year only, and organisms thus need a reliable cue to trigger their breeding physiology. Day length is an excellent cue because it provides a perfectly predictable pattern of change within the year. In the temperate zone in spring, temperatures fluctuate greatly from day to day, but day length increases steadily by a predictable amount. Hence many organisms use day length as a behavioral cue. The experimental verification of photoperiodism is impressive. For example, one can bring some species of birds into breeding condition in midwinter simply by increasing day length artificially (Wolfson 1964). Other examples of photoperiodism occur in plants. A *short-day plant* flowers when the day length is less than a certain critical length. A *long-day plant* flowers after a certain critical day length is exceeded. In both cases, the critical day length differs from species to species. *Day-neutral plants* flower after a period of vegetative growth, regardless of photoperiod. Experimental work has shown that flowering in plants is a response to the *dark* period rather than to the *light* period (Devlin 1969, p. 375).

Breeding seasons have evolved to occupy the part of the year in which offspring have the greatest chances of survival. Thus many temperate-zone birds use the increasing day lengths in spring as a cue to begin the nesting cycle at a point when adequate food resources will be available to the young birds both in the nest and after fledging. Before the breeding season begins, adequate food reserves must be built up to support the energy cost of reproduction. The timing of reproduction in plants and animals is strongly affected by their life cycles. Flowering plants range from *annuals*, some of which require only four to six weeks to go from seed to seed, and *biennials*, which typically flower in the year following germination, to *perennials*, which typically accumulate growth over several years and continue to flower for more than one season.

The adaptive significance of photoperiodism in plants is clear at the general level and has also been analyzed in detail for individual species. Short-day plants that flower in the spring in the temperate zone are adapted to maximizing seedling growth during the growing season. Long-day plants are adaptive for situations that require insect pollination for fertilization or a long period of seed ripening. Short-day plants that flower in the autumn are able to build up food reserves over the growing season and overwinter as seeds. Day-neutral plants will be selected when there is great uncertainty about the timing of the favorable period in relation to day length. For example, desert annuals germinate, flower, and seed whenever suitable rainfall occurs, regardless of the day length.

The breeding season of some plants can be delayed to extraordinary lengths. Bamboos are perennial grasses that remain in a vegetative state for many years and then flower, fruit, and die (Evans 1976). Every bamboo of the species *Chusquea abietifolia* on the island of Jamaica flowered, set seed, and died during 1884. The next generation of bamboo flowered and died between 1916 and 1918, which suggests a vegetative cycle of about 31 years. The climatic trigger for this flowering cycle is not yet known, but the adaptive significance is clear. The sudden production of masses of bamboo seeds (in some cases lying 12 to 15 centimeters deep on the ground) is more than all the seed-eating animals can cope with at the time so that some seeds escape being eaten and grow up to form the next generation (Evans 1976).

The second reason light is important to organisms is that it is essential for *photosynthesis*. Photosynthesis is the process by which plants convert radiant energy from the sun into chemical bond energy. Photosynthesis is remarkably inefficient. During the growing season, about 0.5 to 1 percent of the incoming radiation is captured and stored by photosynthesis. In photosynthesis, carbon is taken up from the air (or water in the case of aquatic plants) in the form of CO_2 and converted into organic compounds. We can measure the rate of photosynthesis by measuring the rate of uptake of CO_2. Figure 8.1 illustrates the great diversity of photosynthetic responses of plants to variations in light intensity. Some plants reach maximal photosynthesis at one-quarter full sunlight, and others like sugarcane never reach a maximum but continue to increase photosynthesis rate as light intensity rises.

Plants in general can be divided into two groups: *shade-tolerant* species and

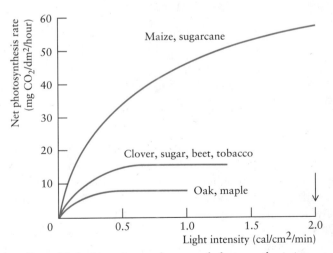

Figure 8.1   The effect of light intensity on the rate of photosynthesis in several species of plants. Photosynthesis was measured by CO_2 uptake at 30°C and 300 ppm CO_2 in air. The arrow on the light axis marks the approximate equivalent of full summer sunlight. (After Zelitch 1971.)

shade-intolerant species. This classification is commonly used in forestry and horticulture. Plant physiologists have discovered that shade tolerance is a complex of traits, and that it is not fixed for each species but varies with plant age, microclimate, and geographic area (Kozlowski et al. 1991, p. 139). Shade-tolerant plants have lower photosynthetic rates and hence would be expected to have lower growth rates than those of shade-intolerant species. The metabolic rate of shade-tolerant seedlings is apparently lower than that of shade-intolerant seedlings. Plant species become adapted to live in a certain kind of habitat and in the process evolve a series of characteristics (an "adaptive syndrome") that prevent them from occupying other habitats. Grime (1966) suggests that light may be one of the major components directing these adaptations (Table 8.1). Failure of seedlings in shaded situations is almost always associated with fungal attack, and part of adaptation to shade involves becoming resistant to fungal infections. For example, eastern hemlock seedlings are shade-tolerant and can survive in the forest understory under very low light levels. Hemlock seedlings grow slowly and have a low metabolic rate that allows them to survive low-light conditions. One consequence of these adaptations is that hemlock seedlings die easily in droughts because their roots do not grow quickly to penetrate deep into the soil.

The principle illustrated by Table 8.1 is exceedingly important in evolutionary ecology. *Individuals of a species cannot do everything in the best possible way.*

Table 8.1   Examples of Ecological Limitations in Plants and the Suggested Adaptive Reasons

Species; Geographic Location; Habitat	Ecological Limitation	Probable Basis	Suggested Consequence in Field	Causal Adaptation	Suggested Role of Adaptation in Field
Tsuga canadensis (eastern hemlock seedling phase); NE United States; floor of closed forest	Slow growth rate under conditions of ample water, nutrients, and light	Low metabolic rate	Slow root penetration leading to drought failure in dry situations	Selection for low respiration losses over long periods of inadequate light	Allows persistence of seedling under dense forest shade
Ailanthus altissima (tree-of-heaven, seedling phase); NE United States; moist, productive unshaded situations	Rapid failure in deep shade	Large respiration losses in shade	Failure in deep shade of closed forest	Selection for high growth rate on productive sites	Allows rapid exploitation of cleared ground and shade avoidance by rapid growth in height
Deschampsia flexuosa (common hair grass); Denmark; highly acidic grassland and health	Iron chlorosis on nutrient solutions and soils of high pH	Insufficient iron reaching leaves	Failure on calcareous soils	Selection for low rate of absorption or translocation of iron and other toxic metals from acid soils	Prevents accumulation of heavy metals in toxic concentrations from acid soils

Source: After Grime (1965).

Adaptations to live in one ecological habitat make it difficult or impossible to live in a different habitat. Thus life cycles have evolved as tradeoffs between contrasting habitat requirements. There can be no superanimals or superplants.

Seaweeds grow in a variety of forms and present, to ecologists, an interesting array of adaptations to light (Hay 1986). Seaweeds differ in the maximum water depth they will tolerate. Since light is attenuated very quickly with depth in the oceans, some of the morphological adaptations of seaweeds are to light levels. Two general types of seaweed occur (Figure 8.2a). Some species have flat, wide, mono-layered thalli, while others have highly dissected, narrow thalli that are multilayered. The same distinction has been applied to the leaves of trees (Horn 1971). If light intensity is important in controlling thallus shape and layering, we would predict that multilayered seaweeds would be at an advantage at high light intensity and monolayered seaweeds would be at an advantage with low light levels in deeper water (Figure 8.2b). Figure 8.3 shows data on thallus width for seven species of seaweeds from the genus *Sargassum* in the Caribbean. Species that have wider blades occur in deeper waters where light is limiting, and thus the local distribution of these seaweeds is set by light. For these seaweeds the upper or shallow limits of distribution could be determined by biotic interactions involving competition for light, or by

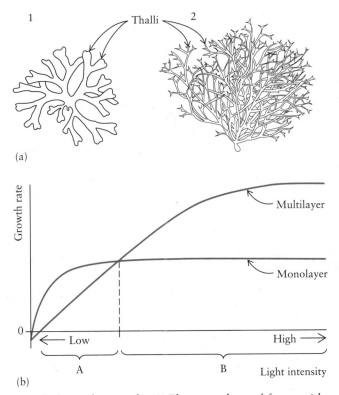

Figure 8.2 *(a) Morphology of seaweeds. (1) Flat, monolayered forms with opaque wide thalli. (2) Upright, multilayered forms with narrow thalli. (b) Proposed model for the effects of light intensity on net photosynthesis for whole plants of monolayered and multilayered seaweeds. In range A monolayers will be favored over multilayers. In range B with high light intensity multilayers will be favored. (Modified from Hay 1986.)*

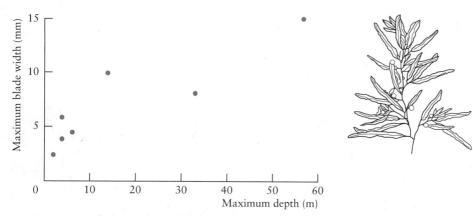

Figure 8.3 *Morphology of seven species of* Sargassum *seaweeds from the Caribbean Sea. Species with wider blades (monolayered) occur in deeper waters, while those with narrow blades (multilayered) occur in shallow waters where light intensity is highest. (From Hay 1986.)*

other biotic or abiotic factors that have yet to be studied. With plants that require light for photosynthesis, the rate of photosynthesis under given light conditions can affect whether the plant can survive in different environments.

One reason photosynthetic rate varies among plants is that there are three different biochemical pathways by which the photosynthetic reaction can occur. Most plants use the C_3 pathway first described by Calvin and often called the *Calvin cycle*. In the C_3 pathway, CO_2 from the air is first converted to 3-phosphoglyceric acid, a 3-carbon molecule (hence the name C_3). Until the mid-1960s this pathway was believed to be the only important means of fixing carbon in the initial steps of photosynthesis. In 1965, sugar cane was discovered to fix CO_2 by first producing malic and aspartic acids (4-carbon acids), and a new C_4 pathway of photosynthesis was uncovered (Bjorkman and Berry 1973). Such C_4 plants have all the biochemical elements of the C_3 pathway, so they can use either method to fix CO_2.

The ecological consequences of the C_4 pathway are profound. Figure 8.4 shows the rate of photosynthesis of a pair of closely related species of C_3 and C_4 plants. Type C_4 plants do not reach saturation light levels even under the brightest sunlight, and they always produce more photosynthate per unit area of leaf than do C_3 plants. Therefore, C_4 plants are more efficient than C_3 plants. Figure 8.5 shows the leaf anatomy of typical C_3 and C_4 plants. Chlorophyll in C_3 leaves is found throughout the leaf, but in C_4 leaves the chloroplasts are concentrated in two-layered bundles around the veins of the leaf (called *Krantz anatomy*). The bundle sheath cells in C_4 plants also have a high concentration of mitochondria. The C_4 leaf anatomy is more efficient for utilizing low CO_2 concentrations and for recycling the CO_2 produced in respiration and for rapidly translocating starches to other parts of the leaf. The biochemical reason for this is simple: The first step in fixing CO_2 in these two types of plants differs:

$$C_3: \text{Atmospheric } CO_2 + \begin{array}{c}\text{ribulose-diphosphate}\\ \text{(RuDP)}\end{array} \xrightarrow{\text{RuDP carboxylase}} \begin{array}{c}\text{phosphoglyceric}\\ \text{acid}\end{array}$$

$$C_4: \text{Atmospheric } CO_2 + \begin{array}{c}\text{phospho-enol-}\\ \text{pyruvate}\\ \text{(PEP)}\end{array} \xrightarrow{\text{PEP carboxylase}} \begin{array}{c}\text{malic acid +}\\ \text{aspartic acid}\end{array}$$

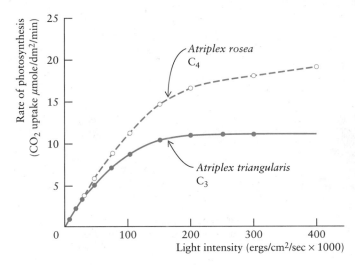

Figure 8.4   Comparative photosynthetic production of the C_3 species Atriplex triangularis *and the related C_4 species* Atriplex rosea. *The plants were grown under identical controlled conditions of 25°C day/20°C night, 16-hour days, and ample water and nutrients. (After Bjorkman 1973.)*

The enzyme RuDP carboxylase is inhibited by oxygen in the air and has a lower affinity for CO_2. The enzyme PEP carboxylase is not inhibited by oxygen and has a higher affinity for CO_2. From this biochemical information, we could predict that C_4 plants would be at an advantage when photosynthesis is limited by CO_2 concentration. This occurs under high light intensities and high temperatures and when water is in short supply (Bjorkman 1975).

Type C_4 grasses, sedges, and dicotyledons are all more common in tropical areas than in temperate or polar areas (Hattersley 1983). Figure 8.6 shows the percentage of C_4 grass species in different parts of North America and Australia and confirms the suggestion that C_4 grasses are at a selective advantage in warmer areas with high solar radiation. Figure 8.4 illustrates one possible reason for this selective advantage with respect to light. On Hawaiian mountains, which have small seasonal changes in temperature, C_3 grasses predominate at high elevations and C_4 grasses at low elevations (Rundel 1980). In Japan, C_4 plants are more prominent in areas with higher temperatures (Ueno and Takeda 1992).

Some desert succulents, such as cacti of the genus *Opuntia*, have evolved a third modification of photosynthesis, *crassulacean acid metabolism* (CAM). These plants are the opposite of typical plants because they open their stomata to take up CO_2 at night, presumably as an adaptation to minimize water loss through the stomata. This CO_2 is stored as malic acid, which is then used to complete photosynthesis during the day. The CAM plants have a very low rate of photosynthesis and can switch to the C_3 mode during daytime. They are adapted to live in very dry desert areas where little else can grow. Table 8.2 summarizes the main characteristics of C_3, C_4, and CAM plants.

The C_3 pathway is presumably the ancestral method of photosynthesis since no algae, bryophytes, ferns, gymnosperms, or more primitive angiosperms have the C_4 pathway or the capacity for CAM (Pearcy and Ehleringer 1984, Monson 1989). Almost half of the C_4 plant species are grasses (Black 1971), and this pathway has apparently increased their competitive ability.

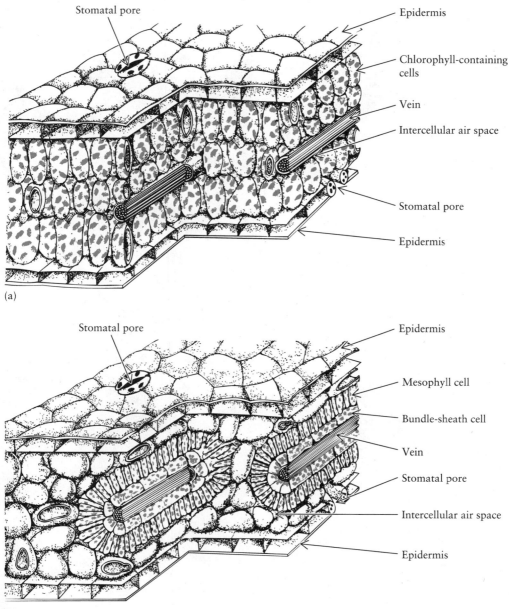

Figure 8.5   *Leaf anatomy of C_3 and C_4 plants. (a) Leaf structure of the C_3 plant*
Atriplex triangularis. *As in other typical leaves, the cells containing chlorophyll are*
of a single type, and they are found throughout the interior of the leaf. (b) Atriplex
rosea, *a C_4 plant, illustrates the modified leaf structure of C_4 species. The special-*
ized leaf of A. rosea *has nearly all its chlorophyll in two types of cells, which form*
concentric cylinders around the fine veins of the leaf. The cells of the outer cylinder
are mesophyll cells; those of the inner cylinder are bundle-sheath cells. (After
Bjorkman and Berry 1973.)

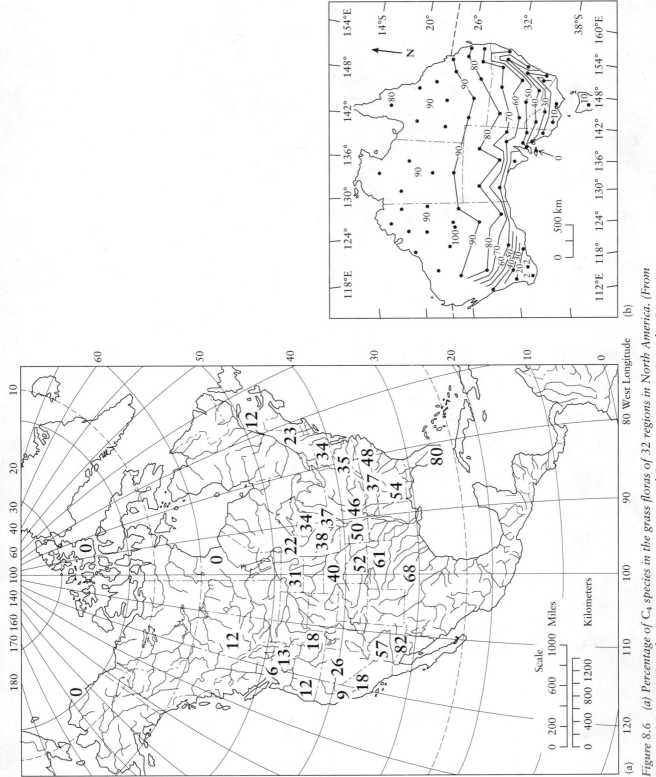

Figure 8.6   (a) Percentage of C$_4$ species in the grass floras of 32 regions in North America. (From Teeri and Stowe 1976.) (b) Approximate contour map of C$_4$ native grasses in Australia. Lines give percentages of C$_4$ species in total grass flora for 75 geographic regions. (From Hattersley 1983.)

Table 8.2   Characteristics of Photosynthesis in Three Groups of Higher Plants.

| Characteristic of Plants | TYPE OF PHOTOSYNTHESIS | | |
	C_3	C_4	CAM
Leaf anatomy (cross section)	Diffuse mesophyll	Mesophyll compact around vascular bundles	Spongy appearance, mesophyll variable
Enzymes used in CO_2 fixation in leaf	RuDP carboxylase	PEP carboxylase and then RuDP carboxylase	Both PEP and RuDP carboxylases
CO_2 compensation point[a] (ppm CO_2)	30–70	0–10	0–5 in dark, 0–200 with daily rhythm
Transpiration rate (water loss)	High	Low	Very low
Maximum rate of photosynthesis (mg CO_2/dm^2 leaf surface/hr)	15–40	40–80	1–4
Respiration in light	High rate	Apparently none	Difficult to detect
Optimum day temperature for growth	$20°-25°C$	$30°-35°C$	Approx. 35°C
Response of photosynthesis to increasing light intensity at optimum temperature	Saturation about 1/4 to 1/3 full sunlight	Saturation at full sunlight or at even higher light levels	Saturation uncertain but probably well below full sunlight
Dry matter produced (t/ha/yr)	22	39	Extremely variable

[a] The compensation point is the CO_2 concentration at which photosynthesis just balances respiration so that there is no net oxygen generated and no net CO_2 taken up.

Source: From Black (1971).

We do not know how the different photosynthetic pathways may interact with other factors to affect the geographic distribution of plant species. It is clear that the response of a plant species to temperature and moisture is strongly affected by the type of photosynthetic process it uses. Implications for animal distributions have still to be considered. Plants possessing the C_4 pathway do not seem to be either avoided or preferred by insect herbivores (Boutton et al. 1978). Further work is needed on the ecological consequences of the three different photosynthetic mechanisms, both for the plants and for the animals that depend on them.

If C_4 plants are more productive than C_3 plants, why do they not displace C_3 plants everywhere? Photosynthetic leaf productivity (Figure 8.4) cannot be directly translated into productivity in natural vegetation (Snaydon 1991). Competition in natural stands is not always for light, and mineral nutrients and water are often limiting to plants. Competition for soil resources can result in plants developing large root systems rather than large above-ground structures.

SOIL STRUCTURE AND NUTRIENTS

The structure and nutrient content of the soil are important, particularly for plants. Intricate connections among climate, soil, and vegetation make it difficult to separate cause and effect with regard to plant distribution. The soil is affected by the vegetation that grows on it and can in turn affect the nature of the vegetation. Most plant species are tolerant of a broad range of soil types; consequently, soil factors are not generally a major limitation to plant distribution. To survive and grow plants require nitrogen, potassium, phosphorus, calcium, magnesium, iron, and sulfur as well as minor amounts of boron, copper, zinc, manganese, molybdenum, and chlorine.

Some of the best examples of soil effects on the distribution of plants are from soils that develop on unusual geological formations. One example is the *serpentine soils* that occur in scattered areas all over the world. These serpentine areas have many features in common (Whittaker 1954): (1) They are sterile and unproductive for farming or forestry; (2) they possess unusual floras, characterized by narrowly endemic species found nowhere else; and (3) they support vegetation that is strikingly different from that on normal soils. Serpentine vegetation is often stunted.

Serpentine rock is basically a magnesium iron silicate, but many other minerals may be present (Walker 1954). Plants that grow well on serpentine areas must, first of all, be tolerant of low calcium levels in the soil. In addition, some serpentine soils have high concentrations of nickel and chromium, high magnesium, low nitrogen and phosphorus, and low amounts of the trace element molybdenum. These soil characteristics are often lethal to plants, but some species have become adapted to this peculiar array of soil nutrients.

Emmenanthe penduliflora and *Emmenanthe rosea* are two California herbs that grow in chaparral areas. The first species is unknown on serpentine soils, whereas the second species occurs only on such soils (Tadros 1957). The distributions of these two herbs thus present two questions: (1) Why does *E. penduliflora* not colonize serpentine areas? (2) Why does *E. rosea* not live in normal soils? Tadros showed that seedlings of *E. rosea*, the serpentine species, survived and grew on sterilized garden soil but died on unsterilized soil, and he suggested that a soil microbe was responsible for keeping this herb confined to serpentine soils. *E. penduliflora* presumably does not invade serpentine areas because of the chemistry of the soil. Wicklow (1966) questioned this interpretation; he notes that fire is necessary for these herbs to become established and indicates that *E. penduliflora* may occur in sparse numbers on serpentine areas.

Some plant species occur on both serpentine and nonserpentine soils in California, and Kruckeberg (1951) showed that several of these species had serpentine and nonserpentine races. This ecotypic variation was found in *Gilia capitata*, an annual herb of the foothills of California:

Source of Seeds	DRY WEIGHT (g) OF PLANTS AT MATURITY	
	In Serpentine Soil	In Normal Soil
Serpentine localities	4.30	4.93
Nonserpentine localities	0.37	4.48

This indicates the existence of *edaphic* races, or soil ecotypes, within a plant species. The edaphic races may be produced in a short time period. In the western United States, two introduced weeds (*Prunella vulgaris* and *Rumex acetosella*) have evolved strains tolerant to serpentine soils, probably within the last 75 years (Kruckeberg 1967).

Ecotypes adapted to serpentine soils often cannot colonize adjacent nonserpentine soils because of poor competitive ability. In a serpentine area on Jasper Ridge near Stanford, California, there is a great abundance of a serpentine ecotype of *Plantago erecta*, yet this species does not live on adjacent sandstone soils (Proctor and Woodell 1975). Proctor set up experimental plots in the sandstone area, fertilized some and cleared others, and planted seeds of *Plantago erecta*. After five months, only the cleared plots contained *Plantago erecta*, which, growing without competition from other vegetation, were able to seed normally.

Patches of yellow pines (*Pinus ponderosa* and *Pinus jeffreyi*) occur in the western Great Basin of Nevada and California, scattered in the sagebrush and pinyon-juniper vegetation. These patches contain pines but almost no herbaceous vegetation or shrubs (Figure 8.7). Billings (1950) found that these unusual stands occurred on a yellow soil that contrasts sharply with the brownish soil of the surrounding desert. The yellow soils are derived from highly weathered volcanic rocks and are strongly acid and deficient in phosphorus and nitrogen. Sagebrush and its associated plants will not grow on these yellow soils because of these mineral deficiencies and the low pH.

Bristlecone pines *(Pinus aristata)* are subalpine trees of the southwestern United States, and some of these trees are more than 4000 years old, possibly the oldest living organisms. Wright and Mooney (1965) studied the distribution of this pine in the White Mountains of California and suggested that moisture balance was an important limiting factor that interacted with the soil type. Three types of soils occurred in this zone, with the following characteristics:

*Figure 8.7   Stands of yellow pines on altered andesite (left) surrounded by pinyon-juniper vegetation (*Pinus monophylla *and* Juniperus utahensis*) on unaltered andesite. Note the lack of shrubs and herbs on the altered soils at the left compared with abundant ground cover of low shrubs, grasses, and forbs on the unaltered soils at the right. Altered soils are acid and deficient in phosphorus and nitrogen. Geiger Grade area, Virginia Mountains, 19 kilometers southeast of Reno, Nevada. (From Billings 1950.)*

	SOILS DERIVED FROM:		
	Dolomite	**Sandstone**	**Granite**
Soil Moisture	Highest	High	Lowest
pH	Alkaline	Slightly acidic	More acidic
Soil nutrients	Very low	Low	Low
Soil temperature	Low	Higher	Higher
Pinus aristata cover (%)	15.4	3.6	1.8
Artemisia tridentata (sagebrush) cover (%)	0.6	11.1	16.6

The distribution of bristlecone pine is complementary to that of sagebrush. Bristlecone pine is well developed on dolomitic soils and is favored by north-facing slopes. Sagebrush is best developed on granitic or sandstone soils, particularly on south-facing slopes, and is more drought-resistant. Bristlecone pine does best on dolomitic soils because of the greater soil moisture, its tolerance of this soil's poor nutrient availability, and the lack of competition by sagebrush and other plants. The oldest bristlecone pines known have been found growing on dolomite under very adverse conditions.

Soil structure and nutrients are very important for plant species but only rarely under extreme conditions is the soil a limiting factor for local distributions.

WATER CHEMISTRY, pH, AND SALINITY

Marine and freshwater organisms may be affected in their distribution by the chemistry of the waters in which they live. Salinity in the open ocean is not variable and consequently does not limit marine planktonic organisms, but near shores and in estuaries, the dilution of seawater by freshwater runoff may reduce salinity to critical levels.

Many freshwater ecologists have studied water chemistry in the hope of explaining distributional problems and in most cases have been unsuccessful (Macan 1974, p. 260). Some associations can be described, but they cannot always be interpreted.

Sessile rotifers live attached to a solid substrate for most of their lives, and Edmondson (1944) studied the distribution of these rotifers in 194 localities of the northeastern United States and Wisconsin. Certain rotifers occurred in only some of the lakes Edmondson studied. Of the various factors of water chemistry studied, pH and bicarbonate seemed most important (Figure 8.8). Edmondson found eight species limited with reference to pH, not bicarbonate; six species limited with reference to bicarbonate, not pH; and three species limited with reference to both pH and bicarbonate. This relationship between water chemistry and distribution may be an indirect one: Water chemistry may limit substrate plant distribution, and sessile rotifers may be substrate-specific. However, Edmondson found only three species of rotifers that were highly selective with respect to substrate.

Calcium is probably the most variable ion in most freshwater lakes and streams. Soft waters may contain less that 1 miligram per liter of calcium; hard waters may contain up to 100 milligrams per liter. Attempts have been made in many cases to relate distributions of invertebrates to calcium levels in fresh water.

Reynoldson (1958) surveyed the distribution of planarians in British lakes and showed that the species composition could be related to the calcium content of the water, which in turn was correlated with the productivity of the lake:

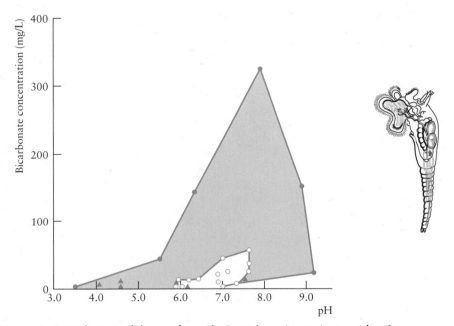

Figure 8.8 *Distribution of the sessile rotifer* Beauchampia crucigera *with reference to pH and bicarbonate concentration. Localities in which* Beauchampia *occurred are indicated by open circles. For comparison, the lakes containing* Collotheca corynetis *are represented by triangles. The shaded area represents the range of values for all lakes studied and thus the potential range of occurrence. (After Edmondson 1944.)*

Characteristic Triclad Flatworm Species	Calcium Range (mg/L)	Lake Productivity	Number of Exceptions
Phagocata vitta	≤2.4	Very low	0
Polycelis nigra alone	≤5.0	Low	4 of 31 lakes
Polycelis nigra, P. hepta, P. tenuis, and *P. felina*	>5.0	Intermediate	1 of 33 lakes
Polycelis spp. and *Dugesia polychroa* and/or *Dendrocoelum lacteum*	>10.0	High	1 of 44 lakes

Macan (1974, p. 256) emphasizes this point—that calcium may operate indirectly on invertebrates through its correlation with productivity and the amount of organic matter decomposing in a lake.

Rooted aquatic plants may be affected both by the substratum in which they grow and by the lake water in which they are immersed. Spence (1967) suggests that the water chemistry of a lake controls whether or not aquatic plant species will grow there. In Sweden, three species of water plants of the genus *Myriophyllum* inhabit waters of different ranges of chemical composition. *Myriophyllum spicatum* and *M. verticillatum* both extend into waters having more calcium than one finds in waters with *M. alterniflorum* (Figure 8.9). Hutchinson (1970) suggested that *M. spicatum*

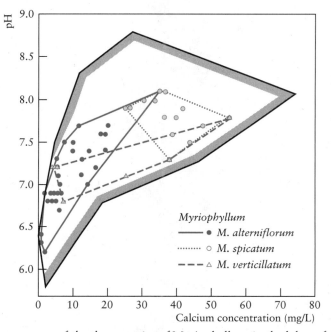

Figure 8.9   Occurrences of the three species of Myriophyllum *in the lakes of central Sweden in relation to calcium concentration and pH. The shaded envelope encloses the points for all the lakes studied in the region. (After Hutchinson 1970.)*

may be able to use the bicarbonate ion as a source of carbon in photosynthesis, whereas *M. verticillatum* cannot. In northern Sweden, where *M. verticillatum* does not occur, *M. spicatum* is found in soft-water (low calcium) lakes as well as hard-water (high calcium) lakes.

Acid rain has become an important pollution problem in North America and in Europe in the last 20 years. Fish populations are strongly affected as lakes become more acidic. Lake trout, for example, disappear from lakes in Ontario and eastern United States when the pH falls below from 5.2 to 5.6 (Mills et al. 1987). Low pH does not affect adult fish directly but it affects the survival of juvenile fish just after hatching. It also affects the survival and reproduction of many aquatic invertebrates that are the food for both juvenile and adult fish. Many lakes in southern Norway have lost all of their fish fauna because of acidification caused by air pollution.

Soil pH was believed to be a primary factor influencing plant distribution in the early days of plant ecology. Less significance is attached to it now. Some plants have strict pH requirements; others are tolerant. For plants with strict pH requirements, it may be a related soil nutrient, not pH itself, that is important.

Tansley (1917) discussed one of the classic cases of the herb *Galium sylvestre*, which grows only on limestone soils of pH > 7.0, and its close relative *Galium saxatile,* which grows only on acidic soils mostly below pH 5.0. For most species of plants soil pH may affect growth and vigor but it is rarely the most significant factor restricting distributions.

Windborne salt determines the distribution of dune plants on the North Carolina banks. Sea oats *(Uniola paniculata)* predominate on the exposed side of the foredune and at the crest of the rear dune, and *Andropogon littoralis,* another dune grass, is

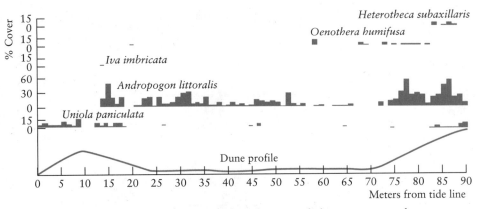

Figure 8.10 *Percent cover of the major dune plants in a belt transect over the dune profile from spring tide line to the crest of the hind dune, New Dunes, Bogue Bank, North Carolina. (After Wagner 1964.)*

best developed in protected areas (Figure 8.10). The areas dominated by sea oats match the areas exposed to high atmospheric salt, and Oosting and Billings (1942) showed that *A. littoralis* was seriously injured by daily spraying with seawater, while *U. paniculata* was only slightly affected. The distribution of these two species can therefore be explained partly by their tolerance to windborne salt.

Wagner (1964) transplanted *Uniola paniculata* to an area in the older, established dunes and showed that it did not survive there, possibly because of root competition with the established pine trees, shrubs, and other grasses. In contrast to *Uniola*, the loblolly pine is very sensitive to salt spray and will not grow near the coast in the spray zone (Wells and Shunk 1938).

Salt marshes are particularly severe environments for land plants and present another good example of zonation. Figure 8.11 shows the zonation of a Dutch salt marsh and illustrates the sequence of plant species that are associated with differing amounts of sea water flooding. By growing 16 salt marsh species in the greenhouse, and manipulating salinity and flooding regimes, Rozema et al. (1985) showed that zonation in the field (Fig. 8.11) would be predicted from an index of salt tolerance measured in the greenhouse. The physiological effects of sea water on plants operate directly by toxic ion excess, water stress due to the salinity, and anaerobic conditions (Rozema et al. 1988). These physiological effects are altered in different types of soils (sand, clay) and wave action itself may affect salt marsh plants mechanically. The apparently simple salt marsh environment is more complex ecologically than one might expect.

FIRE

Although fire is an important source of disturbance in many plant communities, it is rarely a major factor affecting geographic distributions. There is one major exception to this rule: the forest-prairie boundary in North America. The eastern side of the forest-prairie boundary in North America used to extend from Alberta to Texas in a most complex and interesting transition (Figure 8.12). This boundary was remarkably abrupt. When settlers began to colonize the Great Plains, one could pass from closed forest to grassland in a few meters. Numerous tongues of forest extended

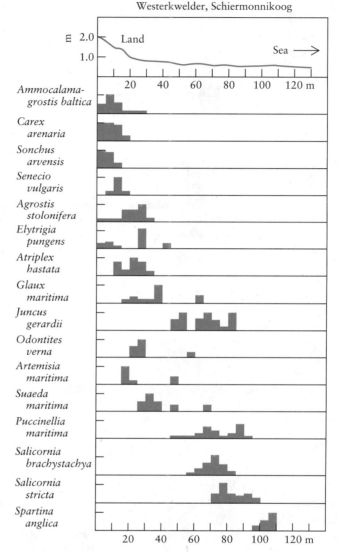

Figure 8.11   *Zonation of vegetation of the Westerkwelder salt marsh on the Wadden Island of Schiermonnikoog in the Netherlands. Histograms represent percent biomass for each species. (From Rozema et al. 1988.)*

far out into the grassland along river valleys, and isolated patches of prairie existed as far east as Indiana and Ohio. What prevented deciduous trees from colonizing the prairies?

Two hypotheses can be used to explain why the prairies of North America are treeless: (1) The areas are deficient in precipitation; or (2) fire destroyed the trees, and they are not able to return. Several points argue in favor of the fire hypothesis. Early explorers and settlers almost without exception commented on the extensive prairie fires. Woodlands occurred on escarpments along waterways and other topographic breaks throughout the grassland area in North America (Fig. 8.13). Early

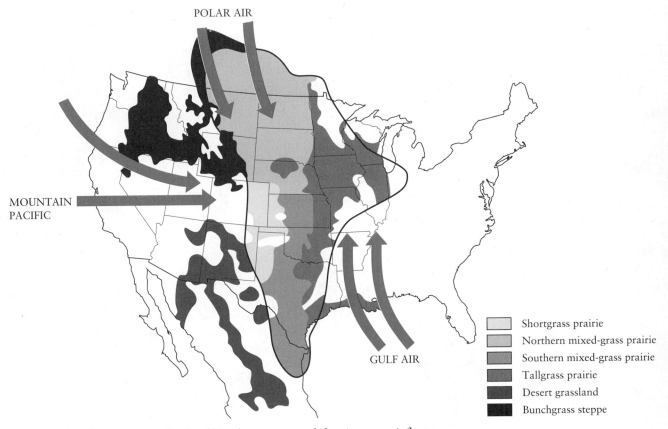

Figure 8.12 *The major grasslands of North America, and the air masses influencing the climate of the central grassland. (Modified from Anderson 1990.)*

settlers did not build houses beyond the shelter of these wooded escarpments until plowing had stopped the autumn prairie fires (Sauer 1969). These fires may have been set by Indians, but some were caused by lightning. Fire in grasslands favors grasses and represses trees.

The northern and eastern parts of the Ozark Highlands in Missouri were largely grassland with open groves of trees, according to historical records from the seventeenth century (Beilmann and Brenner 1951). Areas that are now densely forested were open habitats of mixed forest and prairie as recently as 100 to 150 years ago. Beilmann and Brenner suggest that fire was the principal agent responsible for these conditions, but they also indicate a change toward a wetter climate in this area.

Grasslands in southern Texas have apparently decreased, while woody vegetation has increased. Johnston (1963) emphasized that this involved a change in relative proportions of these plants and that mesquite *(Prosopis glandulosa)*, a woody plant, was already present in small numbers in these grasslands over 100 years ago. The change has been from mesquite prairie to mesquite brush and not an invasion of mesquite from distant areas. Control of fires was suggested to explain this shift toward woody plants.

The alternative view, that climate set the prairie-forest boundary, is difficult to

Vegetation of McLean County Presettlement (1820)

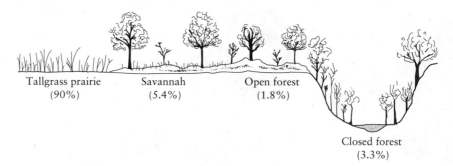

Tallgrass prairie (90%) Savannah (5.4%) Open forest (1.8%)

Closed forest (3.3%)

Figure 8.13 The relationship between presettlement vegetation patterns and to-pography in McLean County, Illinois, in 1820. In the prairie peninsula of Illinois and Indiana, fires restricted trees to areas of natural firebreaks like waterways. (From Anderson 1990.)

sustain for the eastern, moist edge of the prairie. Prairie areas in Illinois and Indiana by almost any criterion have a climate suitable for forest development (Anderson 1990, p. 14). Native grasses of the tallgrass prairie can successfully resist invasion by trees. Tree-root systems cannot compete with grasses for soil moisture especially during droughts; therefore, the tree seedlings die (Pearson 1936). This suggests that once a prairie becomes established, it might maintain itself without fire. Most of the evidence now seems to be against this idea. For example, in western North Dakota, Potter and Green (1964) showed that ponderosa pine is successfully invading grass-lands that are protected from fire. Ponderosa pine is a western pine and consequently could not be an invader on the eastern side of the prairie.

Rather extensive areas (100 km^2) of forest have been established in Nebraska by planting native trees, such as ponderosa pine, that are drought-adapted (Wells 1965). Unfortunately, these observations are equivocal because the seedlings were planted by humans, and the alternative view is that grasses prevent the *establishment* of trees from seed.

Weaver (1968, p. 112) interprets the prairie-forest boundary in climatic terms and suggests that trees survive well in the grasslands only in wet years and die back in years of drought. Weaver estimated that 50 to 60 percent of the trees in Nebraska and Kansas died during the drought of the 1930s. Weaver treated fire as a rare and largely detrimental factor in the grassland communities. If fire is detrimental, then fire suppression is an important management tool. This view of fire has now been rejected by research on prairie ecosystems (Collins and Wallace 1990), and fire is now viewed as an important natural disturbance essential to prairie grasses.

The extension of the prairie into the midwestern United States, the "prairie peninsula," has been mapped and analyzed in detail by Transeau (1935). He points out that these eastern prairies existed in both upland and lowland sites on both poorly drained and well-drained soils and that the same grasses dominated these prairies from Ohio west to Iowa. Most ecologists now believe that the prairie pen-insula was maintained by fire and that in the absence of fire the tallgrass prairie would revert to deciduous forest (Risser 1990).

The grassland climate is distinctive, principally because of its precipitation (Borchert 1950). The shortgrass steppe receives much less summer rainfall than the eastern tallgrass prairie. Winter snow and rainfall are low. Summer rainfall is very variable in the prairie regions. A hot, dry airflow in summer from the eastern base of the Rocky Mountains accompanies severe drought years and accelerates the effects of lack of moisture (Figure 8.12). During most winters and the summers of drought years, the prairie peninsula has a climate more like that of the shortgrass steppe than like that of the eastern deciduous forests. The observed boundaries of the grassland coincide with climatic gradients. As climate changes, the prairie boundary should also change.

Grasslands throughout the world respond to fire, drought, and grazing pressure, and we should not assume that all grasslands will show identical responses. In the western Great Plains the shortgrass prairie is more strongly affected by drought and grazing than it is by fire (Risser 1990). Climatic change during the next century could affect the shortgrass prairie region much more strongly than it will affect the eastern tallgrass region.

The frequency and intensity of fires may limit the geographic distribution of some tree species. Red pine (*Pinus resinosa*) occurs around the Great Lakes of North America and is well adapted to fires of low to moderate intensity. High intensity fires kill red pines and they must recolonize the burned area by seeding from trees that escaped the fire. In the northern part of its range in Quebec, red pine occurs only on islands. Bergeron and Brisson (1990) showed that the fire regime of these islands was very patchy and low intensity fires were frequent. On adjacent mainland areas red pine has been eliminated by large and intense fires that kill all seed sources. Climatic changes that alter the frequency of lightning may have important consequences for the distributions of fire-susceptible species.

SUMMARY

Many physical and chemical factors, in addition to temperature and moisture, can limit the distributions of plants and animals. Most of these cases involve details of local distributions rather than continental or worldwide distributions. Often they are concerned with the factors involved in habitat selection.

Light is used as a behavioral stimulus for animals and as a timing device to set breeding seasons and other critical events in the life cycles of plants and animals. Light is necessary for photosynthesis, and plants differ greatly in their photosynthetic rate. Some plants cannot tolerate shade, and their local distribution is affected by light requirements. Three different biochemical pathways of photosynthesis have been described for higher plants, and, depending on the photosynthetic pathway used, plants respond differently to temperature and moisture stresses.

Soil or substrate structure can be important for plants growing on extreme soil types, and the nutrient content of the soil may also affect local distributions. Water chemistry is relatively constant in the sea but highly variable in fresh waters. Numerous attempts to relate freshwater distributions to water chemistry have led to relatively few successes, however. Windblown salt from the ocean may affect plants growing in coastal areas, and the zonation of plants in coastal salt marshes is related to the salt tolerance of each species. Fire may affect plant distributions. The eastern side of the forest-prairie boundary in North America may have been set by fire, but soil drought is also a critical factor that favors grasses over trees.

It is easier to draw up a list of the physical and chemical factors affecting a plant or an animal than it is to determine if these factors are significant limitations on geographic distributions.

Selected References

COLLINS, S. L. and L. L. WALLACE. 1990. *Fire in North American Tallgrass Prairies.* University of Oklahoma Press, Norman.

FITTER, A. H. and R. K. M. HAY. 1987. *Environmental Physiology of Plants,* 2d ed. Academic Press, London.

GIVNISH, T. J. 1988. Adaptation to sun and shade: A whole-plant perspective. *Australian Journal of Plant Physiology* 15:63–92.

GRIME, J. P. 1966. Shade avoidance and shade tolerance in flowering plants. In *Light as an Ecological Factor*, ed. R. Bainbridge, G. C. Evans, and O. Rackham, pp. 187–207. Blackwell, Oxford, England.

OSMOND, C. B., K. WINTER, and H. ZIEGLER. 1982. Functional significance of different pathways of CO_2 fixation in photosynthesis. In *Encyclopedia of Plant Physiology* (new ser.), vol. 12B, ed. O. L. Lange, P. S. Nobel, C. B. Osmond, and H. Ziegler, pp. 479–547. Springer-Verlag, New York.

PEARCY, R.W. and J.R. EHLERINGER. 1984. Comparative ecophysiology of C_3 and C_4 plants. *Plant, Cell, and Environment* 7:1–13.

PROCTOR, J. and S. R. J. WOODELL. 1975. The ecology of serpentine soils. *Advances in Ecological Research* 9:255–366.

ROZEMA, J., M. C. T. SCHOLTEN, P. A. BLAAUW, and J. DIGGELEN. 1988. Distribution limits and physiological tolerances with particular reference to the salt marsh environment. In *Plant Population Ecology*, ed. A. J. Davy, M. J. Hutchings, and A. R. Watkinson, pp. 137–164. Blackwell, Oxford, England.

SNAYDON, R. W. 1991. The productivity of C_3 and C_4 plants: A reassessment. *Functional Ecology* 5:321–330.

SPENCE, D. H. N. 1982. The zonation of plants in freshwater lakes. *Advances in Ecological Research* 12:37–125.

Questions and Problems

1. Review the effects of acid rain on the local distributions of fishes and aquatic invertebrates in freshwater lakes. What are the biological reasons for the disappearance of these animals as pH falls? Schindler (1988, *Science* 239:149–157) reviews this topic.

2. Rooted aquatic plants occur around the shores of most freshwater lakes but do not occur in deeper waters offshore. List the factors that could limit the local distribution of rooted aquatics in lakes, and compare your list with that of Spence (1982).

3. Grime and Hodgson (1969, p. 68), in discussing why plants fail at the very edge of their distribution, state:

   More typically, however, fatalities are due to a complex of factors and in particular, the contribution of mineral nutritional factors remains obscure. Seedlings may persist for an indefinite period in a state of chronic nutrient deficiency and whilst it is often possible to recognize terminal phenomena, it is difficult to measure the extent to which plants may be predisposed to killing factors, by nutritional disorders.

Is this a serious problem in studying plant distributional problems?

4. Macan (1974, p. 249) points out that lakes that have a high concentration of salts in proportions unlike those in the ocean contain a fauna that consists almost exclusively of animals of freshwater origin. Why should this be?

5. Many investigators attempt to identify the mechanisms that control plant distributions and animal distributions by comparing the environments in which a species occurs with those from which it is excluded. Discuss this approach and its strengths and weaknesses.

6. Caves have very few species of animals, and often the species have special adaptations for existence in the cave environment (Culver 1970, p. 463). Discuss how light may affect the distributions of cave animals.

7. Describe an experimental design to measure the relative importance of grazing, fire, and drought on the trees and grasses at the western edge of the shortgrass prairie in western United States.

8. There is only one known C_4 tree species (Pearcy, 1983). Why should this be?

Overview Question

An insect is to be brought into your country to control a noxious weed. Describe how you would determine the factors limiting the insect's distribution and how you might predict its new geographic range in your country.

9

The Relationship Between Distribution and Abundance

We have considered in the last six chapters the ways in which ecologists answer the simple ecological questions of *who lives where*? and *what constrains these geographic distributions*? In the next section of the book we will investigate the dynamics of populations within this occupied zone, but before we embark on these questions we will consider the broad question of whether there is any relationship between distribution and abundance. Are species that have large geographic ranges any more or less abundant than species that have small geographic ranges?

DEFINITION OF GEOGRAPHIC RANGE

We began our analysis of distribution by assuming that we can easily describe the geographic range of a species. No species occurs everywhere. Figure 9.1 illustrates the problem of describing a species' geographic range. We can view the geographic range of a species on a series of spatial scales. At one extreme the range of a species is defined by the worldwide extent of occurrence, a line drawn on a map around the outermost points at which the species has been observed. This is the definition of geographic range used in bird field guides and other regional natural history guides. At the other extreme, we could map the location of individuals within the large geographic range and measure a much smaller area occupied by the species. If a habitat is not occupied by the species, this region is not included in its geographic range. We would like to know the actual area occupied by each species but this in not possible because ecologists have not collected or mapped species occurrences in this much detail for most plants and animals.

None of this is a problem if there is a good correlation between the different measures of geographic range size (Gaston 1991). For some taxonomic groups, there is a good correlation but not a perfect one. For example, Bock (1987) measured the geographic distribution of birds in the Huachuca Mountains of southern Arizona at four spatial scales:

1. Survey plots, each 35 meters in radius
2. Habitats
3. Christmas bird count circles of 24-kilometer radius
4. Map blocks of 5° latitude and longitude for western United States*

As the spatial scale increases, the correlations become weaker (Figure 9.2). For example, if we know the distribution of these birds in survey plots, we can predict

---

* A 5° block in southern Arizona is approximately 101,135 square miles (261,921 km²).

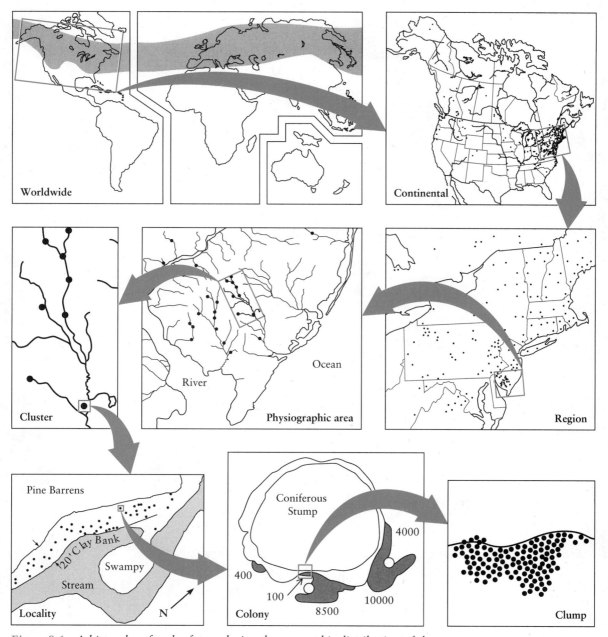

Figure 9.1   *A hierarchy of scales for analyzing the geographic distribution of the moss* Tetraphis. *The answer to the question What limits geographic distribution? may have different answers when analyzed at the continental scale versus at the scale of the individual tree stump. (After Forman 1964.)*

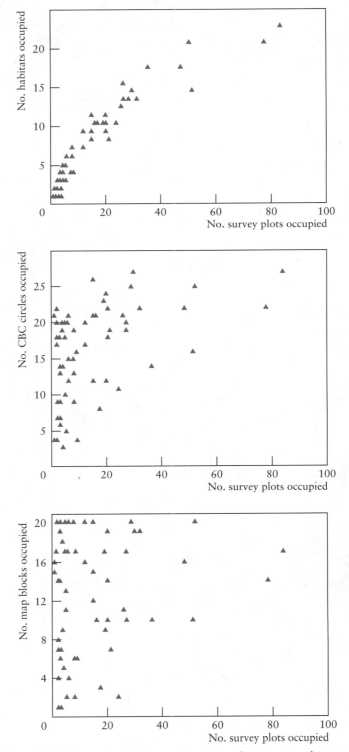

Figure 9.2 *Correlations between the geographic range of 62 species of Arizona wintering birds measured at four spatial scales: survey plots (35-m radius), habitats, Christmas Bird Count plots (24-km diameter), and 5° latitude and longitude blocks. (Data from Bock 1987.)*

very closely their distribution in habitats but we cannot predict their distribution at larger map scales. In freshwater fishes, by contrast, there seems to be a good correlation between geographic ranges measured by the number of points of collection and the area of the entire geographic range (McAllister et al. 1986). The important point is to measure geographic ranges at several spatial scales, and to remember that they will not always be perfectly correlated.

GEOGRAPHIC RANGE SIZE

After we have decided on a measure of geographic range, we can investigate the spread of range sizes of species within a taxonomic group. A general pattern has emerged in many separate groups. That is, most species within a group have small geographic ranges and only a few have very large ranges. Figure 9.3 illustrates this point for the freshwater fishes of North America and for the birds of North America. This pattern, the "hollow curve" of Figure 9.3, seems to be the rule for all groups that have been studied (Gaston 1990). Schoener (1987a) for example, showed that most Australian bird species have small geographic ranges. Anderson (1977) and Pagel and co-workers (1991) found the same pattern for mammal ranges in North America. There are other interesting patterns in these range-size data. In North America, geographic ranges of mammals are smaller toward the tropics (Figure 9.4).

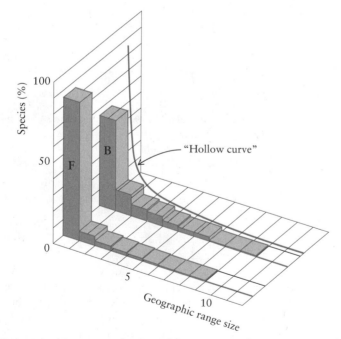

Figure 9.3 *Frequency histogram of geographic range sizes for 635 species of North American fishes (F) and 1370 species of North American birds (B). Ranges are grouped into 1 million square kilometer classes. There are many more species with small geographic ranges. (After Anderson 1985.)*

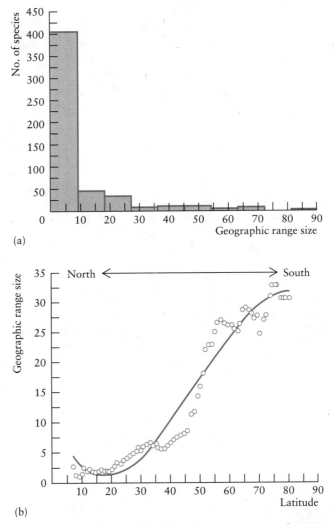

(a)

(b)

Figure 9.4   (a) Geographic range sizes for North American mammals (523 species). Range size is measured as the percentage of the total land area of North America. (b) Relationship between range size and latitude. Southern species have smaller ranges than northern species. (From Pagel et al. 1991.)

The average Canadian mammal species inhabits a range 25 times larger than the average Mexican mammal (Pagel et al. 1991).

The boundaries of geographic ranges can be reached abruptly or gradually (Caughley et al. 1988). In many cases a species' geographic distribution is a contour map with a gradual falloff in density as you approach the edge of the range. Figure 9.5 shows the distribution of the western grey kangaroo in Australia and shows how the edge of the distribution resembles a rounded hill rather than a cliff of density. Brown (1984) showed that many species have this type of geographic pattern in North America, a gradual decline in abundance from the center of the range to the boundaries.

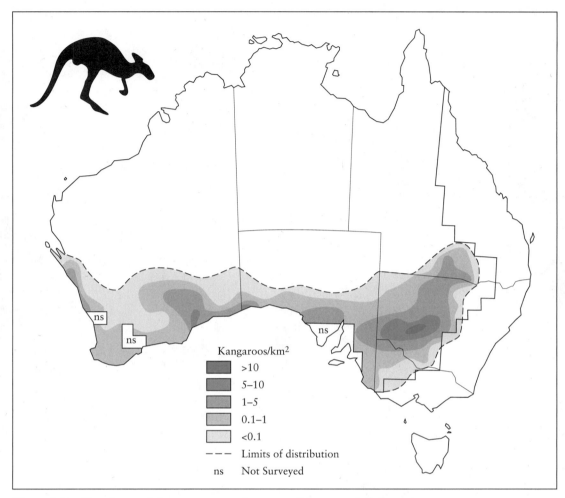

Kangaroos/km²

- >10
- 5–10
- 1–5
- 0.1–1
- <0.1
- – – –   Limits of distribution
- ns   Not Surveyed

Figure 9.5   Abundance of the western grey kangaroo (Macropus fuliginosus) *from aerial surveys, 1980–1982. (From Caughley et al. 1987a.)*

RANGE SIZE AND ABUNDANCE

Is there any relationship between geographic range size and the abundance of a species? If a species is widespread, is it always a common species? Or conversely, if a species is rare or threatened, does it have a small geographic range? Ecologists have assembled data on a wide variety of plants and animals to show that there is a correlation between distribution and abundance such that more widespread species are typically more abundant (Gaston 1990). Figure 9.6 shows data for 263 species of British moths. Moths were collected at light traps in 50 sites throughout Britain, and the geographic distribution was measured as the number of these sites occupied by a given species. Abundance data for each species were averaged over 6 to 14 years at each light trap site. Not all light traps could be operated every year. There is much variability in these moth data, but a clear trend exists for more widespread moth

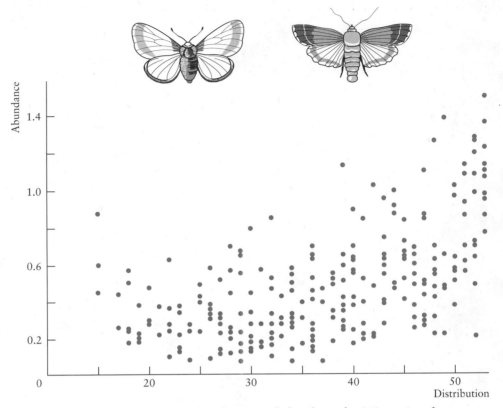

Figure 9.6   *Relationship between distribution and abundance for 263 species of British moths. Distribution is the number of trap sites throughout Britain at which the species was caught. Abundance is averaged across all sites for all years. Each dot represents one species. (From Gaston 1988.)*

species to be more abundant. Figure 9.7 shows similar data for 62 bird species in Arizona.

Plants also show a positive relationship between distribution and abundance. Gotelli and Simberloff (1987) compiled the local distribution and abundance data for 170 species of tallgrass prairie plants for the 3487-hectare Konza Prairie in Kansas. Figure 9.8 shows that the more widespread species were the more abundant species, as was the case for birds and moths.

Hanski (1982) was one of the first to call attention to the association between distribution and abundance, and he suggested the *core-satellite species hypothesis* to explain this relationship. Species tend to fall into two groups, according to Hanski (1982). *Core* species are common and widespread in distribution, while *satellite* species are rare and local in distribution. If this hypothesis is applicable, the frequency distribution of range sizes should have two peaks, one for the core species occupying large areas and one for the satellite species occupying small ranges. Figures 9.3 and 9.4 do not show any trace of a bimodal or two-peaked pattern. There are some data that do show a bimodal distribution of range sizes (Gotelli and Simberloff 1987) but most data are not consistent with the core-satellite model (Nee et al. 1991).

There are three other explanations currently available to explain why distribution and abundance may be correlated. The first explanation is the *sampling model*

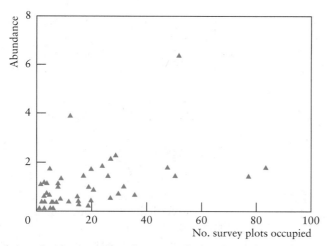

Figure 9.7 *Relationship between distribution and abundance for 62 bird species in southeastern Arizona. Abundance was averaged over all plots occupied. Distribution was measured at the local scale of 35-m radius plots. These data are from the same birds analyzed in Figure 9.2. Each triangle represents one species. (Data from Bock 1987a.)*

which argues that the observed relationship is an artifact of sampling. Rare species are more difficult to find than common species, and thus, if the biological studies are not done carefully, one will automatically observe the patterns shown in Figures 9.6 and 9.7. The second explanation is the *ecological specialization model*. This model argues that species that can exploit a wide range of resources become both widespread and common. These species are called *generalists* and are to be distinguished from *specialists* that exploit only a few resources. Provided that one can

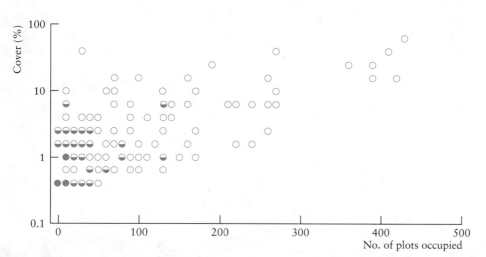

Figure 9.8 *The relationship between the distribution and abundance of 170 species of prairie plants on 433 quadrats in the Konza Prairie Research Area of central Kansas. Distribution measured by number of plots occupied, abundance measured by mean percentage cover for each species. (○ = 1 species, ◔ = 2–9 species, ● = 10 or more species) (From Gotelli and Simberloff 1987.)*

determine which species are generalists and which are specialists, one should be able to test this model. The third explanation is the *local population model*. In this model a population is subdivided into a series of discrete patches, or local population, and these patches interact because animals or plants move between the patches. Since species differ in their capacity to disperse, some will occupy more local patches than others (Hanski et al. 1992). If this model is correct, we would expect species that disperse more to be more common and more widespread, when compared with less migratory species.

Available data at present support both the sampling model and the local population model (Hanski et al. 1992). Since all studies of distribution and abundance involve sampling, in an imperfect world the sampling model will always be part of the explanation for these positive relationships between distribution and abundance. More data are needed to determine the ecological attributes of successful species that are widespread and abundant. These attributes may help us to understand the reasons for species being rare or endangered, and could assist in conservation biology (see Chapter 19). We are led in this way out of the world of geographic distributions and into the larger and more complex world of abundance. We turn now to look at what happens within the zone of distribution in which populations of animals and plants increase and decrease in size in response to many of the same environmental factors we have just considered.

SUMMARY

The geographic distribution of a species is more complex than one first suspects. The scale at which a distribution is mapped can affect the answer to the simple question of *what limits geographic distributions?* For many species we do not know the detailed geographic range because too few collections have been made. Even for larger plants and animals in developed countries we have few details about local distributions.

Geographic ranges measured on large-scale maps show a common pattern in all groups studied—most species have small geographic ranges and only a few species are very widespread. This holds for data on fishes, birds, and mammals, as well as for plant groups. Geographic range sizes may also be smaller in tropical areas than in temperate or polar areas.

There is a broad positive correlation between distribution and abundance for all kinds of animals and plants. Widespread species are more abundant than species that have small geographic ranges. There is considerable scatter in this relationship. Three explanations are currently proposed for this relationship. Ecological specialization may be one factor reducing abundance, and migration among suitable local patches may affect overall abundance and distribution. We are led to ask the question *what makes a species successful?* and we turn to consider the problem of abundance in more detail in Part Three.

Selected References

ANDERSON, S. 1984. Geographic ranges of North American birds. *American Museum Novitates* 2785:1–17.

BOCK, C. E. 1987. Distribution-abundance relationships of some Arizona landbirds: A matter of scale? *Ecology* 68:124–129.

BROWN, J. H. 1984. On the relationship between abundance and distribution of species. *American Naturalist* 124:255–279.

CARTER, R. N., and S. D. PRINCE. 1988. Distribution limits from a demographic viewpoint. In *Plant Population Ecology*, ed. A. J. Davy, M. J. Hutchings, and A. R. Watkinson, pp. 165–184. Blackwell, Oxford.

CAUGHLEY, G., D. GRICE, R. BARKER, and B. BROWN. 1988. The edge of the range. *Journal of Animal Ecology* 57:771–785.

GASTON, K. J. 1990. Patterns in the geographical ranges of species. *Biological Reviews* 65:105–129.

GASTON, K. J. 1991. How large is a species' geographic range? *Oikos* 61:434–438.

GOTELLI, N. J. and D. SIMBERLOFF. 1987. The distribution and abundance of tallgrass prairie plants: A test of the core-satellite hypothesis. *American Naturalist* 130:18–35.

Questions and Problems

1. In Neotropical forest mammals there is a *negative* relationship between geographic distribution and abundance (Arita et al. 1990). Suggest two ecological explanations for this pattern.

2. For terrestrial birds in Australia Schoener (1987a) suggested that the greater the range size, the *less* the average abundance, opposite to the trends reported in this chapter. Gaston (1990, p. 111) rejected Schoener's results. Read Gaston's comments and discuss whether or not you accept Schoener's interpretation.

3. Discuss the implications of the relationship between distribution and abundance for conservation biology.

4. The relationship between distribution and abundance is often very loose with much scatter, as shown in Figures 9.6–9.8. How does this scatter affect the interpretation of these data?

Overview Question

Discuss the classification of species or plants and animals into specialists and generalists. Choose a species and describe what you would measure to convince someone that your species is a generalist or a specialist. How does this classification assist in understanding the positive relationship between distribution and abundance that has been observed?

Part **3**

The Problem of Abundance: Populations

10 *Population Parameters*

Within their areas of distribution, animals and plants occur at varying densities. We recognize this variation when we say, for example, that sugar maples are common in one woodlot and rare in another. If we are to make these statements more precise, we must quantify density. This chapter discusses, in a preliminary way, the techniques used to estimate densities of animals and plants.

POPULATION AS A UNIT OF STUDY

A population may be defined as a *group of organisms of the same species occupying a particular space at a particular time*. Thus we may speak of the deer population of Glacier National Park, the deer population of Montana, or the human population of the United States. The ultimate constituents of the population are individual organisms that can potentially interbreed. The population may be subdivided into local populations, called demes, which are groups of interbreeding organisms, the smallest collective unit of a plant or animal population. Individuals in local populations share a common gene pool. The boundaries of a population both in space and in time are vague and in practice are usually fixed arbitrarily by the investigator.

A good deal of interest has centered on populations as units of study, from both the field of ecology and the field of genetics. One of the fundamental principles of modern evolutionary theory is that natural selection acts on the individual organism, and through natural selection, populations evolve. The fields of population ecology and population genetics have much in common.

The population has group characteristics, which are statistical measures that cannot be applied to individuals. These group characteristics are of three general types. The first and most basic characteristic of a population that we are interested in is its *density*. The second type includes four population parameters that affect density: *natality* (births), *mortality* (deaths), *immigration*, and *emigration*. In addition to these attributes, one can derive a third type of attribute to describe the secondary characteristics of a population, such as its age distribution, genetic composition, and distribution of individuals in space. Note that these population parameters result from a summation of individual characteristics.

ESTIMATION OF POPULATION PARAMETERS

The population attributes concerned with changes in abundance are interrelated as follows:

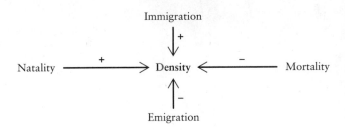

These four processes—natality, mortality, immigration, and emigration—are the *primary population parameters*. When we ask why population density has gone up or gone down in a particular species, we are asking which one or more of these parameters has changed. Let us briefly look at the methods employed in estimating these vital statistics.

Density

Density is defined as numbers per unit area, per unit volume or per unit of habitat. We can appreciate the problems involved in estimating density by considering some approximate densities of organisms in nature:

	Density in Conventional Units[a]	Density per m^2 (or m^3)
Diatoms	5,000,000/m^3	5,000,000
Soil arthropods	500,000/m^2	500,000
Barnacles (adult)	20/100cm^2	2,000
Trees	500/ha	0.0500000
Field mice	250/ha	0.0250000
Woodland mice	10/ha	0.0010000
Deer	4/km^2	0.0000040
Human beings		
Netherlands	346/km^2	0.0003460
Canada	2/km^2	0.0000020

[a] 1 hectare = 10,000 m^2 = 2.47 acres
1 sq. kilometer = 100 hectares = 0.386 sq. mile

This range of figures, covering more than a dozen orders of magnitude, gives you some idea of what we have to study. Techniques for estimating density that work nicely with deer cannot be applied to protozoa. The two fundamental attributes that affect our choice of techniques for population estimation are the *size* and *mobility* of the organism with respect to humans.

Small animals are usually more abundant than large animals. Figure 10.1 shows this trend for 212 species of animals and allows us to predict the approximate density for a species of given size. There are systematic differences between groups that are clear in Figure 10.1. Birds, for example, are less abundant than mammals of equivalent size. If you were to study a 1-kilogram bird, you should expect a density of approximately 1 per square kilometer. For a mammal of similar size, you ought to expect about 100 per square kilometer. We shall see later when we discuss energetics (Chapter 27) why this might be so.

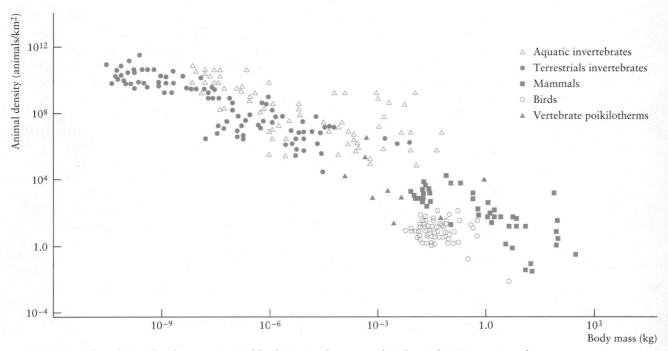

Figure 10.1   *The relationship between animal body size and average abundance for 212 species of poikilotherms and homeotherms from the temperate zone. (After Peters and Wassenberg 1983.)*

In many cases, it will be impractical to determine the *absolute density* of a population (for example, in numbers per hectare or per square meter), and we may find it adequate to know the *relative density* of the population (that, for two areas of equal size, area *x* has more organisms than area *y*). This division is reflected in the techniques developed for measuring density.

Measurement of Absolute Density

Total Counts   The most direct way to find out how many organisms are living in an area is to count them. One good example of this is the human population census. Other examples come from plant populations and from vertebrate animals. With trees one can easily count all the individuals in a given area. With territorial birds one can count all the singing males in an area. Or with bobwhite quail, one can count the number of birds in each covey. Other animals, such as the northern fur seal, may be counted when they are all gathered in breeding colonies. Few invertebrates can be counted in total, the exceptions being the barnacles and other sessile invertebrates such as some rotifers. Large animals on small areas can sometimes be counted in total, but in general this is possible for very few organisms.

Sampling Methods   Usually the investigator must be content to count only a small proportion of the population and to use this sample to estimate the total. There are two general ways of sampling, by the use of quadrats or the capture-recapture method.

Use of quadrats. The general procedure here is to count all the individuals on several quadrats of known size and to extrapolate the average to the whole area. A quadrat is a sampling area of any shape. Although the word literally describes a four-sided figure, it has been used in ecology for all shapes of areas, including circles. An example will illustrate this estimation procedure: If you counted 19 individuals of a beetle species in a soil sample of 0.1 square meter, you could extrapolate this to 190 beetles per square meter of soil surface.

The reliability of the estimates obtained by this technique depends on three things: (1) The population of each quadrat examined must be determined exactly, (2) the area of each quadrat must be known, and (3) the quadrats counted must be representative of the whole area. This last condition is usually achieved by random sampling procedures, and students acquainted with statistics will find a good discussion of this problem in Cochran (1977 chaps. 1–2). The population of each quadrat may be counted without error in some organisms but only estimated in other species. Many special techniques have been developed for applying quadrat sampling methods to different kinds of animals and plants in terrestrial and aquatic systems. I shall give just two examples of this procedure.

Lloyd (1967) sampled centipedes in 37 quadrats in a beech forest in central England and obtained the results shown in Figure 10.2. The mean density is

$$\frac{30 \text{ individuals}}{37 \text{ quadrats}} = 0.811 \text{ centipede per quadrat}$$

or, since each quadrat was 0.08 square meter, the estimated density was 10.1 centipedes per square meter. Lloyd (1967) was interested in the density of centipedes in one small patch, and he used contiguous quadrats to cover the entire patch.

Wireworms are click beetle larvae that live in the soil and feed on seeds and

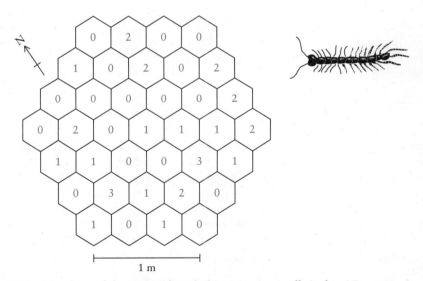

1 m

Figure 10.2   *Numbers of the centipede* Lithobius crassipes *collected in 37 contiguous hexagonal quadrats of beech litter at Wytham Woods, near Oxford, England, on October 30, 1958. The quadrats are 30 cm (1 ft) across, so that the total area sampled is about 2 m from side to side. Individual quadrats have an area of 78 cm² (0.866 ft²). (After Lloyd 1967.)*

damage the roots of agricultural crops. To estimate populations of wireworms, Salt and Hollick (1944) devised a technique of extracting all larvae from soil samples. This was done by breaking the lumps of soil, separating the very coarse and very fine material by sieves, and separating the wireworms from other organic material by benzene flotation (insects accumulate at the benzene-water interface; the plant matter stays in the water). Exhaustive tests were made at each step in this process to see if larvae were lost. They sampled soil by use of a corer that removed a cylinder of soil 10 centimeters in diameter and 15 centimeters deep. In one pasture near Cambridge, England, they obtained 240 random samples with a total of 3742 larvae of the wireworm *Agriotes*, an average of 15.6 per 10-centimeter core, or an infestation of 19.3 million per hectare. Variation among the samples can be used to estimate confidence limits for this density estimate. Salt and Hollick were able to show by this careful work that wireworm populations were about three times higher in English pastureland than people had previously supposed.

Quadrats have been used extensively in plant ecology, and this is the most common method for sampling plants. There is an immense literature on the problems of sampling plants with quadrats, dealing, for example, with the relative efficiency of round, square, and rectangular quadrat shapes. We shall not go into these detailed problems of methodology; interested students should refer to Greig-Smith (1983, chap. 2) or Krebs (1989, chap. 3).

A single example will illustrate quadrat sampling of plant populations. A transect was set out in an upland hardwood forest in southern Indiana by my ecology class. Three lines, each 106.7 meters (350 ft) long, were used, and all trees more than 25 centimeters tall within a swath of 1 meter on each side of the line were counted. Each transect line is, in effect, a very long, thin quadrat, comprising 213.4 square meters (0.02134 hectare). We obtained these results:

	NO. COUNTED			ESTIMATED NO./HA		
	Line A	Line B	Line C	Line A	Line B	Line C
Chestnut oak	20	28	18	937	1312	843
Sugar maple	5	4	7	234	187	328
American beech	13	15	16	609	703	750

By doing this type of quadrat sampling for old trees and then again for seedlings, we could determine if populations were likely to change with time. Foresters have devised a series of ingenious techniques for estimating the abundance of forest trees, and these are reviewed by Cottam and Curtis (1956) and by Krebs (1989).

Capture-recapture method. The technique of capture, marking, release, and recapture is an important one for mobile animals, because it allows not only an estimate of density but also estimates of birth rate and death rate for the population being studied.

Several models can be used for capture-recapture estimation. Basically, they all depend on the following line of reasoning: If you capture animals, mark and release them, in subsequent samples taken from this population the proportion marked should be representative of the proportion marked in the entire population. This is illustrated here for a simple example with two capture sessions:

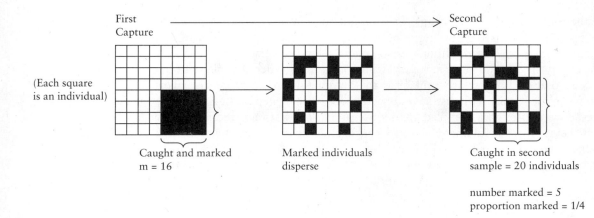

First Capture

Second Capture

(Each square is an individual)

Caught and marked
m = 16

Marked individuals
disperse

Caught in second
sample = 20 individuals

number marked = 5
proportion marked = 1/4

This simple type of population estimation is known as the Petersen method.* There are only two sampling periods: capture, mark, release (time 1) and capture and check for marks (time 2). The time interval between the two samples must be short because this method assumes a closed population with no recruitment of new animals into the population between times 1 and 2, and no losses of marked animals. We assume that, if it is random, a sample will contain the same proportion of marked animals as that in the whole population:

$$\frac{\text{Marked animals in second sample}}{\text{Total caught in second sample}} = \frac{\text{Marked animals in first sample}}{\text{Total population size}}$$

For this simple example:

$$\frac{5}{20} = \frac{16}{N}$$

or $N = 64$.

One of the first uses of the Petersen method was by Dahl (1919) who marked trout (*Salmo fario*) in small Norwegian lakes to estimate the size of the population that was subject to fishing. He marked 109 trout and in a second sample a few days later caught 177 fish, of which 57 were marked. From these data we estimate:

$$\text{Proportion of population marked} = \frac{57}{177} = 0.322$$

The number of fish marked in the first sample was 109 and therefore,

$$\text{Population estimate} = \frac{\text{size of marked population}}{\text{Proportion of population marked}} = \frac{109}{0.322} = 338$$

To estimate density two situations must be considered—closed and open populations. A population is defined as *closed* if it is not changing in size during the

---

* Also called the *Lincoln method* by wildlife ecologists.

period of capture, marking, and recapturing. A population is defined as *open* if it is changing in size during the study period. Real populations are clearly open, unless we sample them for a very brief period.

For a *closed* population, as we have just seen, we know the size of the marked population because it is equal to the number of individuals we have actually marked in the first sample. For an open population, the estimation problem is more difficult. The marked population will decrease from one sampling period to the next because of death and emigration of marked individuals and will increase during the sampling as we mark new individuals.

Consider now the marked population only (M) for an open population. This segment contains two kinds of marked animals for an arbitrary time period of year 7:

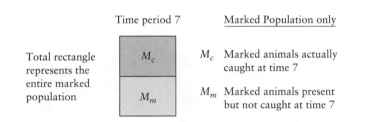

Some marked animals always miss capture, and a portion of them turn up at later sampling times. In this hypothetical example, individuals could have been marked at time 6, missed at time 7, and caught again at time 8. Such individuals are included in the lower part of the rectangle.

We need to know two things to estimate the size of the marked population: (1) the number of marked animals actually caught (M_c) and (2) the proportion of marked animals actually caught (the ratio of M_c to $M_c + M_m$). We already know the first because it is simply a count of marked animals in the catch. There are a number of ways in which the ratio of M_c to $M_c + M_m$ can be estimated, and this gives rise to several models of estimation, which are discussed in detail by Seber (1982) and Cormack (1968). Appendix II gives details for one model of estimation.

All capture-recapture models make three critical assumptions:

1. Marked and unmarked animals are captured randomly.

2. Marked animals are subject to the same mortality rate as unmarked animals. The Petersen method assumes there is no mortality during the sampling interval.

3. Marks are not lost or overlooked.

All these assumptions have caused trouble at one time or another. For example, field mice may become trap-happy or trap-shy and thus violate assumption 1. Fish tagged on the high seas may be weakened by the nets and the tagging procedure (being held out of water) so that they suffer abnormal mortality just after release. In some cases, fishermen have not returned tags from marked fish because they considered them good-luck charms. Leg rings may be lost from long-lived birds. There are numerous variations of the techniques of marking and recapture analysis, some of which may cleverly circumvent some of these problems. Fisheries scientists in particular have been very active in this field (e.g., Ricker 1975).

Two additional points may be mentioned. First, under some conditions it is possible to test from field data the crucial assumption that animals are captured randomly. Second, the reliability of these population estimates can be determined in standard statistical fashion. Both these points are reviewed and discussed by Seber (1982) and Krebs (1989). Otis, Burnham, White, and Anderson (1978) have presented an elegant capture-recapture model of population estimation for closed populations that includes a test of the assumptions.

Usually we are not interested in just a single population estimate, and we operate a mark-and-recapture scheme for several months or years, a *multiple census*. When we do this, we can begin to study the dynamics of a population. I have pointed out that it is possible to get an estimate of the birth rate and death rate of the population at the same time as we estimate its size. The gist of this procedure is as follows: Consider just two samples of the population, obtained by the general estimation technique outlined in Appendix II. We have estimates of the marked population at the end of time 7 (M_7) and at the start of time 8 (M_8), and we also know the total population size at each of these times:

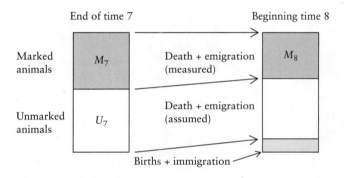

Between times 7 and 8, the marked population can only decrease, owing to death and emigration. (The marked population can *increase* only during a sampling period when we are marking animals that were formerly unmarked.) We define the survival rate as the percentage of animals that survive the time interval:

$$\text{Survival rate (\%)} = \frac{M_8}{M_7} \times 100$$

Suppose, for example, that we have estimated 500 marked animals at the end of time 7 and 400 at the start of time 8. Then

$$\text{Survival rate} = \frac{400}{500} \times 100 = 80\% \text{ per unit of time}$$

The death rate (or, more properly, *loss rate*, since it includes emigration) is defined simply as

$$\text{Loss rate} = 100\% - \text{survival rate}$$

These rates apply to the time interval between sample times 7 and 8; if this is 1 year, the estimated survival rate is 80% per year.

The birth rate, which is more commonly called the *dilution rate* because it covers both immigration and births, is obtained indirectly. We assume that the survival rate estimated for the marked animals also applies to the unmarked animals. This is shown in the diagram as a dotted line, and a numerical example will illustrate how this operates.

Using the techniques just described, we obtained a population estimate for time 7 (1500) and, repeating the whole procedure, for time 8 (1400). We also obtained an estimate for the size of the marked population at the end of time 7 (500) and at the start of time 8 (400). These values are shown in the following table:

	Time 7	Time 8	
Marked animals	500	400	Therefore, survivial rate = 0.80
Unmarked animals	1000	800	Estimated survival
Total population estimate	1500	1400	Therefore, dilution = 200 new animals

If we project the observed survival rate, we find that we can account for 1500 × 0.80 = 1200 animals at time 8 as survivors of those present at time 7. Clearly, 200 new animals have appeared through birth or immigration, and this is the *dilution*.

The principle here is summed up in the equation

$$\text{Total population size at time } t + 1 = \text{total population size at time } t + \text{dilution} - \text{losses}$$

If we know any three of the variables in the equation, we can find the fourth by subtraction.

Developments in statistics have been very important for the study of ecology. It was not until 1936 that this technique of capture-recapture analysis was worked out in detail so that it could become an important ecological tool. In the last 15 years several important developments occurred in this field of analysis (Otis et al. 1978, Seber 1982). Other statistical concepts—for example, the problems of sampling—have been worked out only in the last 35 years. This is a good example of how developments in a science are dependent on progress in both pure and applied mathematics.

The capture-recapture technique has been used mainly on larger forms such as butterflies, snails, beetles, and many vertebrates that can be readily marked.

Measurement of Relative Density

The characteristic feature of all methods for measuring relative density is that they depend on the collection of samples that represent some relatively constant but unknown relationship to the total population size. They provide an index of abundance that is more or less accurate. When an index of abundance is 4.0 on area x and 8.0 on area y, you can conclude that area y has a higher density of animals than area x. You *cannot* conclude that area y has twice the density of area x. There are a great many indices of relative density, and we shall list only a few:

1. *Traps.* Traps are often used in capture-recapture studies to estimate absolute density, as described. They may also be used to get an index of relative density by the number of individuals caught per day per trap. The traps include mousetraps spread across a field, light traps for night-flying insects, pitfall traps in the ground for beetles, suction traps for aerial insects, and plankton nets. The number of organisms trapped depends not only on the population density but also on their activity and range of movement, and the researcher's skill in placing traps, so we get only a rough idea of abundance from these techniques.

2. *Number of fecal pellets.* The technique has been used for snowshoe hares, deer, field mice, and rabbits in Australia. If you know the number of fecal pellets in an area and the average rate of defecation, you can get an index of population size.

3. *Vocalization frequency.* The number of pheasant calls heard per 15 minutes in the early morning has been used as an index of the size of the pheasant population.

4. *Pelt records.* The number of animals caught by trappers has been used to estimate population changes in several mammals; some records extend back 300 years.

5. *Catch per unit fishing effort.* The measure can be used as an index of fish abundance, for example, number of fish per 100 hours of trawling.

6. *Number of artifacts.* The count can be used for organisms that leave evidence of their activities, for example, mud chimneys for burrowing crayfish, tree-squirrel nests, and pupal cases from emerged insects.

7. *Questionnaires.* Questionnaires can be sent to sportsmen or trappers to get a subjective estimate of population changes. This is useful only when large changes in population size need to be detected among animals large enough to be noticed.

8. *Cover.* The percentage of the ground surface covered by a plant as a measure of relative density has been used by botanists and invertebrate ecologists studying the rocky intertidal zone.

9. *Feeding capacity.* The amount of bait taken by rats and mice can be measured to obtain an index of density.

10. *Roadside counts.* The number of birds of prey observed while driving a standard distance can be used as an index of abundance.

These methods for measuring relative density all need to be viewed skeptically until they have been carefully evaluated. They are most useful as a supplement to more direct census techniques and for picking up large changes in population density.

To conclude our discussion of techniques for measuring density, I should like to point out two things. First, detailed, accurate census information is obtainable on few animals. In many cases, we must be content with an order-of-magnitude estimate. Second, because of this, a disproportionate amount of work has been done on the more easily censused forms, particularly the birds and mammals. This introduces a possible bias into the discussions that follow.

Natality

Populations increase because of natality. The natality rate is equivalent to the birth rate; *natality* is simply a broader word covering the production of new individuals by birth, hatching, germination, or fission.

Two aspects of reproduction must be distinguished. The concept of *fertility* is a physiological notion that indicates that an organism is capable of breeding. *Fecundity* is an ecological concept that is based on the numbers of offspring produced during a period of time. We must distinguish between *potential fecundity* and *realized fecundity*. For example, the realized fecundity rate for an actual human population may be only one birth per 15 years per female in the childbearing ages, whereas the potential fecundity rate for humans is one birth per 10 to 11 months per female in the childbearing ages.

Natality rate may be expressed as the number of organisms produced per female per unit time. The measurement of natality or birth rate is highly dependent on the type of organism being studied. Some species breed once a year, some breed several times a year, and others breed continuously. Some produce many seeds or eggs, others few. For example, a single oyster can produce 55 million to 114 million eggs. Fish commonly lay eggs in the thousands, frogs in the hundreds. Birds usually lay between 1 and 20 eggs, and mammals rarely have litters of more than 10 offspring and often have only 1 or 2. Fecundity is inversely related to the amount of parental care given to the young. We shall not go into the details of natality measurements here because we shall discuss them with examples later.

Mortality

The biologist is interested not only in why organisms die but also why they die at a given age. Mortality, or its converse, survival, can be looked at from several perspectives. Longevity focuses on the age of death of individuals in a population. Two types of longevity can be recognized: *potential longevity* and *realized longevity*. Potential longevity is the maximum life span attainable by an individual of a particular species. This limit is set by the physiology of the plant or animal, and the organism simply dies of old age. Another way of describing potential longevity is the average longevity of individuals living under optimum conditions. Few organisms in nature live in optimum conditions. Most animals and plants die from disease, or are eaten by predators, or succumb to any one of a number of natural hazards. The actual life span of an organism is known as realized longevity. Realized longevity is the average longevity of the individuals in a population living under real environmental conditions, and this measure of longevity can be measured in the field. Two examples will illustrate this. The European robin has an average life expectancy of 1 year in the wild, whereas it can live at least 11 years in captivity (Lack 1954). In ancient Rome, the average life expectancy at birth for human females was about 21 years, and in England in the 1780s it was about 39 years (Pearl 1922). In the United States in 1990, females could expect to live 79 years on the average.

The measurement of mortality may be done directly or indirectly. The direct measure is achieved by marking a series of organisms and observing how many survive from time t to time $t + 1$. We have discussed this technique previously in the treatment of capture-recapture methods.

An indirect measure of survival may be gotten in several ways. For example, if one knows the abundance of successive age groups in the population, one can estimate the mortality between these ages. Data of this type are widely used in fisheries work, in the analysis of *catch curves*. Figure 10.3 illustrates a catch curve for bluegill sunfish in an Indiana lake. The survival rate can be estimated from the decline in relative abundance from age group to age group (except in the case of 1-year-old fish, which are usually too small to be caught in the nets used). For example,

$$\text{Survival rate between II and III years} = \frac{\text{relative abundance of III fish}}{\text{relative abundance of II fish}} = \frac{147}{292} = 0.50$$

Similarly, a drop of 147 to 54 between ages III and IV gives a survival estimate of 0.37 between these ages. In making these calculations, we are assuming that the initial number of fish in each of the two age groups was the same and that the survival rate has been constant over time for each age group. Ricker (1975, chap. 2)

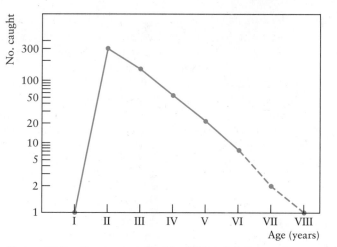

Figure 10.3   Catch curve for bluegill sunfish (Lepomis macrochirus) *from Muskellunge Lake, Indiana, 1942. (After Ricker 1958.)*

discusses these problems in detail for fish populations. All indirect measures of survival involve some assumptions and should be used only after careful evaluation.

Immigration and Emigration

Dispersal—immigration and emigration—is seldom measured in a population study. In most cases, it is assumed that the two components are equal, or else work is done in an island type of habitat, where dispersal is presumably of reduced importance. Both assumptions are highly questionable. The capacity to disperse is an essential part of the life cycle of most organisms. Dispersal helps to prevent inbreeding and is the ecological process that produces gene flow between local populations. Dispersal can set limits on geographic distributions (as we saw in Chapter 4) and affects community composition (as we shall see in Chapter 25). Some populations sustain a net emigration and export individuals; others are sustained only by a net immigration. One example is small songbirds in woodlots of eastern United States. Small woodlots are not productive for birds, and these populations can only be sustained by immigration (Wilcove and Robinson 1990). Dispersal may be a critical parameter in population changes.

Dispersal can be measured if individuals can be marked in a population. The use of radio-telemetry has revolutionized the study of animal movements, particularly for larger organisms (Kenward 1987). The major technical problem in studying dispersal is the scale of the movements involved. Many animals move distances greater than the size of study areas, and information on long distance dispersal can be lost. One of the major unsolved problems of conservation biology is how to facilitate immigration and emigration from populations in isolated parks or refuges in a fragmented landscape (Chapter 19).

LIMITATIONS OF THE POPULATION APPROACH

Two fundamental limitations restrict the methods used for studying populations. First, how can we determine what constitutes a *population* for any given species?

What are the boundaries of a population in space? In some situations, the boundaries are clear. Wildebeest populations in the Serengeti area of East Africa form five herds that rarely exchange members (Sinclair 1977). The largest herd is highly migratory and moves seasonally between plains and woodlands following the rainfall. The other four smaller herds are less migratory and breed in different areas and at slightly different times of the year.

But in many other cases, organisms are distributed in a continuum, and no boundaries are evident. White spruce trees grow in northern coniferous forests from Newfoundland to Alaska. Do all these white spruce trees belong to one population? Most population biologists would answer no to this question, but the reasons for their answer would differ. Part of the definition of a population should involve the probability of genetic exchange between members of the same population, but no one is able to specify this probability in a rigorous manner. We are left with the same general problem that troubles the systematist when someone asks how we can determine what constitutes a *species*. One pragmatic answer we can give is this: A population is a group of individuals that a population biologist chooses to study. To say this is only to say that we may have to start our study by making a completely arbitrary decision on what to call a population. To remember this decision after your population study is completed is a mark of ecological wisdom.

A second limitation is that some organisms do not come in simple units of individuals. Colonies of the social insects are one example. Many plants are also difficult to categorize because they show great variation in size and structure (Harper and White 1974). Grasses are particularly difficult to fit into anyone's definition of a single individual. Other plants such as aspen (*Populus tremuloides*) form clones, so that a whole stand of trees may really be only one genetic individual.

Organisms may be classified as *unitary* or *modular* organisms. Most higher animals are unitary organisms in which form is determinate. A kitten is recognizable as a cat, and a young giraffe is instantly recognized as a giraffe. In unitary organisms an individual is easy to recognize and each individual is a separate genetic individual. By contrast, most plants are modular organisms in which the zygote develops into a unit of construction (a module) which then produces more, similar modules (Figure 10.4). Many of the lower animals, however, such as hydrozoans, corals, and bryozoans, are also modular organisms (Figure 10.5, see color insert). Such organisms are always branched, and individuals are composed of variable numbers of modules. Modules that can exist separately are known as *ramets*. Modular organisms can have individuals extremely different in size.

To study populations, we must recognize that there are two levels of population structure of modular organisms. One level is the number of individuals that are represented by original zygotes. These individuals are called *genets*, or genetic individuals (Harper 1977). In plants, an individual genet can be a single tree seedling or a clone extending over a square kilometer. Each genet is composed of one or more modular units of construction, which vary with the type of organism. For example, grasses grow as *tillers* (equivalent to ramets), the modular units of construction, and each genetic individual of a grass may have many tillers. To describe a population of modular organisms, we must specify both the number of genets and the number of modular units, in contrast to most populations of unitary organisms, in which the individual is simultaneously the genetic unit and the modular unit. Modular organisms have added another dimension to the problem of population changes. That is, in addition to studying whole organisms, we can measure changes in modules and measure modular birth and death rates (Harper et al. 1986).

In some of these cases, we can circumvent the problem by dealing with *biomass*

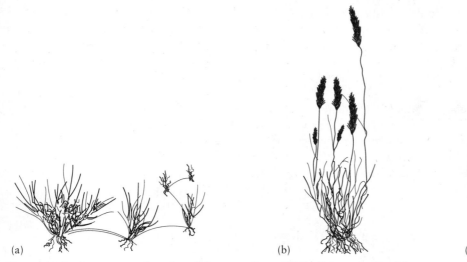

(a)   (b)   (c)

Figure 10.4   Some examples of modular organisms. (a) Buffalo grass (Buchloe
dactyloides) *in which clones spread laterally. (b) Fescue grass* (Festuca octoflora)
which forms a tussock of tightly packed modules. (c) Sea fan (Gorgonia *sp.*) *a
multiple branched coral. (Modified from Begon et al. 1990, p. 125.)*

(weight) instead of numbers. Foresters are not interested in the number of oak trees
in a woodlot but rather the sizes of the trees. In some species, we will be able to
apply population methodology only by making some arbitrary decision about what
units to count. Fortunately, in many cases, individuals and populations are easy to
recognize and study.

Populations do not exist as isolates but are imbedded in a community matrix of
associated species. When we study a population, we assume we can abstract from
the whole community a single species and the small number of species that interact
with it. Whether this abstraction can be done effectively is controversial, as we shall
see when we discuss community organization (Chapter 25), but for the moment we
will proceed with the assumption that a population can be analyzed as an abstraction
from the complex tapestry of a biological community.

COMPOSITION OF POPULATIONS

Populations are not composed of a series of identical individuals, yet we tend to
forget this heterogeneity when considering population density. Three major variables
distinguish individuals in populations: *sex*, *age*, and *size*. The composition of many
populations deviates from the expected sex ratio of 50 percent of each gender, so
we cannot assume a constant 1:1 (50 percent) sex ratio. Populations of the wood
lemming (*Myopus schisticolor*) in Finland contain only about 25 percent males
(Kalela and Oksala 1966). Adult populations of the great-tailed grackle (*Quiscalus
mexicanus*) in Texas contain about 29 percent males, although 50 percent of the
nestlings are males (Selander 1965). The sex ratio of a population clearly affects the
potential reproductive rate and may affect social interactions in many vertebrates
(Wilson 1975).

Age is a significant variable in human populations, and age effects are common to many species. Older individuals are frequently larger, and changes in size may be the main mechanism by which age effects occur. Larger fish lay many more eggs than smaller fish, and larger plants produce more seeds. Young mammals may be prone to diseases that older animals can resist. Age and size are very significant individual attributes in all animals that have social organization because they help to specify an individual's social position.

Size is a particularly significant variable in modular organisms. Size and age may be correlated in plants and animals with indeterminate growth, but there is much individual variation in size that is independent of age. For modular plants and animals size is usually the ecologically relevant variable that defines their importance in ecological processes.

Other secondary variables may distinguish individuals in some populations. Color is one obvious trait. Many phenotypic traits can affect survival, reproduction, or growth and be important to a population.

Because of these individual differences, population ecology labors under a split personality. We deal with populations as aggregates of individuals and calculate population density and other population parameters as averages over all individuals. But we recognize that to explain population processes we must understand the individuals that comprise the population and the mechanisms by which they reproduce, move, and die. In the next nine chapters we shall flip back and forth between this duality of populations as averages and populations as mixtures of heterogeneous individuals.

SUMMARY

We have looked briefly at methods of estimating population size for plants and animals. There is a crucial measurement problem here—every aspect of the problem of abundance comes down to this point: How can we estimate population size? Absolute density can be estimated by total counts or by sampling methods with quadrats or capture-recapture methods. Relative density can be estimated by many techniques, depending on the species studied. Once we obtain estimates of population size, we can investigate changes in numbers by analyzing the four primary demographic parameters of births, deaths, immigration, and emigration.

Organisms may be *unitary* or *modular*. Most animals are unitary with determinate form—cats all have four legs. In unitary organisms the individual is easily recognized and each individual is an independent genetic unit. Most plants are modular with repeated units of construction (modules)—oak trees can have any number of leaves. Individuals may be difficult to recognize in modular organisms, and a genetic individual may be a clone with many separate modules. Population changes in modular organisms involves measuring the birth rate of new modules and the death rate of old modules.

Selected References

BEGON, M. 1979. *Investigating Animal Abundance*. Edward Arnold, London.

BUSS, L. W. 1987. *The Evolution of Individuality*. Princeton University Press, Princeton.

HARPER, J. L., R. B. ROSEN, and J. WHITE (eds.). 1986. The growth and form of modular organisms. *Philosophical Transactions of the Royal Society of London*, Series B, 313:1–250.

JACKSON, J. B. C., L. W. BUSS, and R.E. COOKE (eds.). 1985. *Population Biology and Evolution of Clonal Organisms.* Yale University Press, New Haven.

KREBS, C. J. 1989. *Ecological Methodology*, chaps. 2–4. Harper and Row, New York.

MEDAWAR, P. B. 1957. Old age and natural death. In *The Uniqueness of the Individual*, chap. 1. Methuen, London.

MUELLER-DOMBOIS, D., and H. ELLENBERG. 1974. Measuring species quantities. In *Aims and Methods of Vegetation Ecology*, chap. 6. Wiley, New York.

RICKER, W. E. 1975. *Computation and Interpretation of Biological Statistics of Fish Populations*, chaps. 1–3. Fisheries Research Board of Canada, Bulletin 191.

SEBER, G. A. F. 1982. *The Estimation of Animal Abundance and Related Parameters.* Griffin, London.

Questions and Problems

1. The catch (per 100 hours of trawling) of plaice (*Pleuronectes platessa*) in the southern North Sea is given for three seasons. Plot catch curves from these data, and use the data to calculate survival rates for the various age classes.

| | CATCH | | |
Age of Plaice	1950–1951	1951–1952	1952–1953
II	39	91	142
III	929	559	999
IV	2320	2576	1424
V	1722	2055	2828
VI	389	982	1309
VII	198	261	519
VIII	93	152	123
IX	95	71	106
X	81	57	61
XI	57	60	40
XII and older	94	87	99

Source: After Gulland (1955).

2. An ecology class marking and releasing grasshoppers obtained the following data:

Morning sample: 432 marked and released
Afternoon sample: 567 caught, of which 47 were already marked

Apply the Petersen method of population estimation to the data, and discuss the necessary assumptions with particular reference to these animals. What are the implications of accidentally killing a grasshopper in the morning sample? In the afternoon sample?

3. Milne (1943) estimated the abundance of sheep ticks on farms in Scotland by dragging a wool blanket over the grass. Ticks will cling to anything that brushes against them during the spring. Does this technique measure absolute density or relative density? How might you determine this?

4. Populations of some organisms, such as ground-dwelling carabid beetles, can be estimated either by quadrat methods or by capture-recapture techniques. List some of the advantages and disadvantages of each approach.

5. Compare the definition of *population* presented here with that used in texts of statistics (Sokal and Rohlf 1981, p. 9; Zar 1984, pp. 14–15) and in texts of evolution (Futuyma 1986, pp. 25–26 and 276–277).

6. Aerial surveys for large mammals and birds are simply a type of quadrat sampling applied on a large geographic scale. Discuss the assumptions of quadrat sampling for population estimation given on page 154 for the special case of aerial surveys. Norton-Griffiths (1978) gives a detailed discussion of these problems.

7. Some modular organisms, such as trees, have a tightly controlled growth form. Others, like clover or corals, have a very opportunistic and variable pattern of growth. Discuss what environmental factors would select for these two extreme patterns of growth form. Harper (1988) discusses this question.

8. Suggest three hypotheses to explain why, in Figure 10.1, birds should in general exist at a lower density than mammals of the same size. Make predictions from each hypothesis and discuss how the predictions could be tested. Strathmann (1990) discusses this general problem.

Overview Question

Roadside counts of the numbers of a particular bird species are available for many years. Design a research program that will allow you to convert these indices of abundance to absolute densities. Identify the assumptions you must make to achieve this goal.

11 Demographic Techniques: Vital Statistics

One of the great strengths of population ecology is that it is quantitative. Population mathematics is not difficult but it is sufficiently different to merit some of your attention if you wish to achieve a more precise understanding of how and why populations change. The next two chapters provide this quantitative background for population ecology.

LIFE TABLES

Mortality is one of the four key parameters that drive population changes, as we saw in Chapter 10. We need a technique to summarize how mortality is occurring in a population. Is mortality high among juvenile organisms? Do older organisms have a higher mortality rate than younger organisms? We can answer these kinds of questions by constructing a life table. A life table is a convenient format for describing the mortality schedule of a population. Life tables were developed by human demographers, particularly those working for life insurance companies, which have a vested interest in knowing how long people can be expected to live. There is correspondingly an immense literature on human life tables, but there are few data on other animals or on plants.

Plant and animal populations may be composed of several types of individuals, and a demographer may group these together or may keep them separate in an analysis. A life insurance company gives to males a policy different from the one they give to females for good demographic reasons, and thus it may be useful for some purposes to classify individuals by sex or age.

A life table is an age-specific summary of the mortality rates operating on a cohort of individuals. A cohort may include the entire population or it may include only males, or only individuals born in, say, 1975. An example of a life table for song sparrows is given in Table 11.1. The columns of the life table are symbolized by letters, and these symbols are constantly used in ecology:

x = age interval
n_x = number of survivors at start of age interval x
l_x = proportion of organisms surviving at start of age interval x
d_x = number dying during the age interval x to $x + 1$
q_x = rate of mortality during the age interval x to $x + 1$

To set up a life table, we must decide on age intervals to group the data. For humans or trees the age interval may be 5 years, for deer, birds, or perennial plants

Table 11.1   Life Table for the Song Sparrow on Mandarte Island, British Columbia.[a]

Age (years) (x)	Observed No. Birds Alive (n_x)	Proportion Surviving at Start of Age Interval x (l_x)	No. Dying Within Age Interval x to $x + 1$ (d_x)	Rate of Mortality (q_x)
0	115	1.0	90	0.78
1	25	0.217	6	0.24
2	19	0.165	7	0.37
3	12	0.104	10	0.83
4	2	0.017	1	0.50
5	1	0.009	1	1.0
6	0	0.0	—	—

[a] Males born in 1976 were followed from hatching until all had died six years later. This is an example of a cohort life table defined on page 171.
Source: From Smith (1988).

1 year, and for annual plants or field mice 1 month. By making the age interval shorter, we increase the detail of the mortality picture shown by the life table.

The first important point to be made is that, given any one of the columns of the life table, you can calculate the rest. To put this another way, there is nothing "new" in each of the three columns l_x, d_x, and q_x. They are just different ways of summarizing one set of data. The columns are related as follows:

$$n_{x+1} = n_x - d_x$$

$$q_x = \frac{d_x}{n_x}$$

$$l_x = \frac{n_x}{n_0}$$

For example, from Table 11.1,

$$n_3 = n_2 - d_2 \qquad q_2 = \frac{d_2}{n_2} \qquad l_4 = \frac{n_4}{n_0}$$

$$= 19 - 7 = 12 \qquad = \frac{7}{19} = 0.37 \qquad = \frac{2}{115} = 0.017$$

The rate of mortality q_x is expressed as a rate for the time interval between successive census stages of the life table. For example, q_0 for the song sparrows in Table 11.1 is 0.78 for the interval between egg and one year, or per year. Thus 78 percent of the birds are lost in the nest or during their first year of life.

The most frequently used part of the life table (Table 11.1) is the n_x column, the number of survivors at the start of age x. This is often expressed from a starting cohort of 1000, but some human demographers prefer a starting cohort of 100,000. Other workers prefer to plot the l_x column to show the proportion surviving. The n_x (or l_x) data are plotted as a survivorship curve; Figure 11.1 illustrates the survivorship curves for the human population of the United States in 1989. Note that the n_x values are plotted on a logarithmic scale. Population data should be plotted this way when one is interested in per capita *rates* of change rather than absolute numerical changes. A simple numerical example shows this: If half of a population dies, we obtain

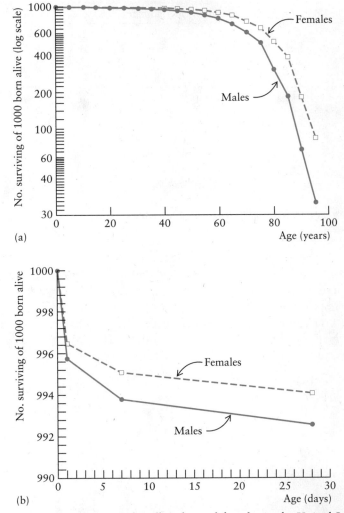

Figure 11.1 *(a) Survivorship curve for all males and females in the United States, 1989. (b) Infant survival during the first month, USA, 1989. (Data from the* Statistical Abstract of the United States *1991.)*

Starting Population Size	No. Dying	Final Population Size
1000	500	500
500	250	250
250	125	125

On a logarithmic scale, all these decreases are equal (base 10 logs):

$$\log(1000) - \log(500) = \log(500) - \log(250)$$
$$3.00 \quad - \quad 2.70 \quad = \quad 2.70 \quad - \quad 2.40$$

The *numbers* lost are greatly different, although the *rates of loss* are the same.

   The life table was introduced to ecologists by Raymond Pearl in 1921. Pearl (1928) described three general types of survivorship curves (Figure 11.2). Type I

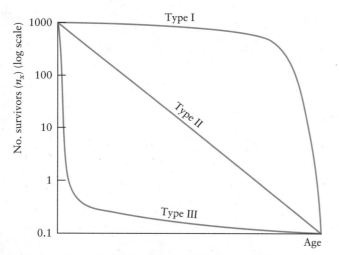

Figure 11.2   Hypothetical survivorship curves. (After Pearl 1928.)

curves are from populations with very little loss for most of the life span and then high losses of older organisms. The linear survivorship curve (type II) implies a constant rate of mortality independent of age. Type III curves indicate high loss early in life, followed by a period of much lower and relatively constant losses.

No population has a survivorship curve exactly like these idealized ones, and real curves are composites of the three types. In developed nations, for example, humans tend to have a type I survivorship curve. Many birds have a type II survivorship curve, and a large number of populations would fall in the intermediate area between types I and II. Often a period of high loss in the early juvenile stages alters these ideal type I and II curves. Type III curves occur in many fishes, marine invertebrates, and parasites.

Now that we have seen what a life table looks like, how do we get the data to construct a life table? This question brings up an important critical distinction: There are two different ways of gathering data for a life table, and they produce two different types of life tables. These are the *static* life table (stationary, time-specific, current, or vertical life table) and the *cohort* life table (generation, horizontal life table). These two life tables are different in form, except under unusual circumstances, and are always quite different in meaning (Merrell 1947).

The *static* life table, calculated on the basis of a cross section of the population at a specific time, is illustrated in Table 11.2. In this example, the census data and mortality data are for human females in Canada in 1989. A cross section of the female population in 1989 provides the number of deaths (d_x) in each age group and the number of individuals in that age group. This allows us to estimate a set of mortality rates (q_x) for each age group, and the q_x values can be used to calculate a complete life table in the way outlined earlier.

On the other hand, a *cohort* life table is calculated on the basis of a cohort of organisms followed throughout life. For example, you could, in principle, get all the birth records from New York City for 1921 and trace the history of all these people throughout their lives. This would involve following them when they move out of town and would be a very tedious task. You could then tabulate the number surviving

Table 11.2   Static Life-Table Data for the Human Female Population of Canada, 1989

Age Group (yr)	No. in Each Age Group	Deaths in Each Age Group	Mortality Rate per 1000 Persons (1000 q_x)
0–1	183,800	1189	6.5
1–4	720,600	265	0.4
5–9	897,800	162	0.2
10–14	881,600	157	0.2
15–19	913,300	331	0.4
20–24	1,006,500	460	0.5
25–29	1,194,700	555	0.5
30–34	1,178,200	705	0.6
35–39	1,066,900	904	0.8
40–44	950,700	1247	1.3
45–49	742,300	1665	2.2
50–54	625,900	2143	3.4
55–59	612,100	3350	5.5
60–64	595,100	5174	8.7
65–69	559,000	7579	13.6
70–74	429,100	9462	22.1
75–79	338,700	12,476	36.8
80–84	219,200	13,803	63.0
85 and above	178,700	25,225	141.2

Note: These data were obtained by tallying the number of females in each group by their 1989 birthdays and by tallying the number of deaths in 1989 for the same age groups.
   Source: Statistics Canada (1991).

at each age interval. Very few data like these are available for human populations.*
This procedure would give you the survivorship curve directly, and you could calculate the other life-table functions if needed. Table 11.1 is an example of a cohort life table.

These two types of life table will be identical if and only if the environment does not change from year to year and the population is at equilibrium. But normally there will be good years and bad years, variable birth and death rates, and consequently large differences between the two forms of life table. These differences can be illustrated very well for human populations. Figure 11.3 contrasts static and cohort life tables for the human male population of England and Wales in 1880. The static life table shows what the survivorship curve would have been if the population had continued surviving at the rates observed in 1880. But the continual improvement in medicine and sanitation in the last 100 years has increased survival rates, and the people born in 1880 had a generation survivorship curve unlike that of any of the years through which they lived. Static life tables assume static (stationary) populations.

---

* For human populations, unlike those of other animals and plants, it is possible to construct cohort life tables indirectly from mortality rate (q_x) data. To construct a cohort life table for the 1921 year–class of New York, we can obtain the mortality statistics for the 0- to 1-year-olds for 1921, the 1- to 5-year-olds for 1922–1925, the 6- to 10-year-olds for 1926–1930, and so on, and use these q_x rates to estimate the life-table functions. This approach was used to obtain Figure 11.3.

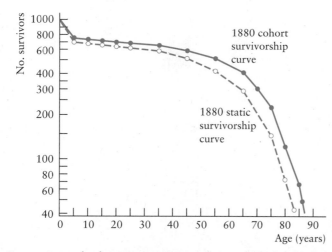

Figure 11.3   Comparison of cohort or generation survivorship of males born in 1880 in England and Wales with static or time-specific survivorship of males for 1880. (Data from Registrar General 1968.)

Insurance companies would like to have data from cohort life tables covering the future, but these are obviously impossible to get. They are definitely not interested in cohort life tables covering the past—the life table for the 1880 cohort would be of little use for predicting mortality patterns today. So they use static life tables and correct them at each census. These predictions will never be completely accurate but will be close enough for their purposes.

Life tables from nonhuman populations are harder to come by. In general, three types of data have been used to determine ecological life tables:

1. *Survivorship directly observed.* The information on survival (l_x) of a large cohort born at the same time, followed at close intervals throughout its existence, is the best to have, since it generates a cohort life table directly and does not involve the assumption that the population is stable in time. A good example of data of this type is that of Connell (1961a) on the barnacle *Chthamalus stellatus* in Scotland. This barnacle settles on rocks during the autumn. Connell did several experiments in which he removed a competing barnacle, *Balanus balanoides*, from some rocks but not from others and then counted about once a month the *Chthamalus* surviving on these defined areas (Figure 11.4). Barnacles that disappeared had certainly died; they could not emigrate.

2. *Age at death observed.* The data on age at death may be used to estimate the life-table functions for a static life table. Then we must assume that the population is stable in time and that the birth and death rates of each age group remain constant. A good example of this type of data comes from the work of Sinclair (1977) on the on the African buffalo (*Syncerus caffer*) in the Serengeti area of East Africa. Sinclair collected all the skulls he could find on his study area. He got 584 skulls of buffalo that had died from natural causes and classified them by age and sex. The age at death was determined by the annular rings on the horns. Young animals were difficult to sample properly because their fragile skulls were destroyed by weather and carnivores. Sinclair estimated the losses during the first 2 years of life by direct observations on the herd and obtained the mortality estimates shown in Figure 11.5.

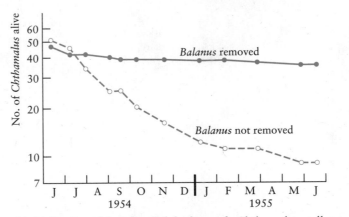

Figure 11.4 *Survivorship curves of the barnacle* Chthamalus stellatus, *which had settled naturally on the shore at Millport, Scotland, in the autumn of 1953. The survival of* Chthamalus *growing without contact with* Balanus *is compared with that in an undisturbed area.* Balanus *crowds out* Chthamalus *when the two species are together. (Data from Connell 1961a and personal communication.)*

3. *Age structure directly observed.* The ecological information on age structure, particularly of trees, birds, and fishes, is considerable and in some cases can be used to estimate a static life table. In these cases, one can often determine how many individuals of each age are living in the population. For example, if we fish a lake, we can get a sample of fish and determine the age of each from annular rings on the scales. The same type of data can be obtained from tree rings. The difficulty is that

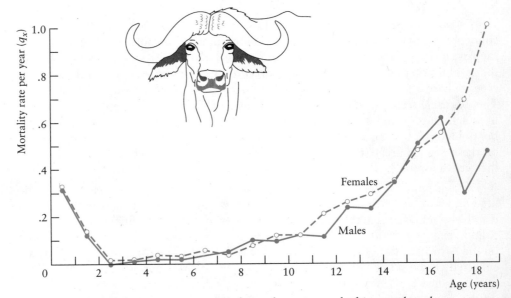

Figure 11.5 *Mortality rate per year* (q_x) *for each age interval of 1 year plotted against the midpoint of the age interval for the African buffalo. Age at death was determined from skulls of dead buffalo collected during a period of steady population increase. (Data from Sinclair 1977.)*

to produce a life table from such data, one must assume a constant age distribution, something that is rare for many populations. Consequently, data of this type are not always suitable for constructing a life table.

Attempts to gather together life-table data on organisms other than humans and to establish a general theory of mortality have all fallen short of their goal. Deevey (1947) recognized the difficulty of comparing life tables of different species because the basic data of the life table are sometimes of the static type and sometimes of the cohort type, and the point of origin of the life tables may be different—births for mammals, egg laying for insects, or settling on rocks for barnacles. Caughley (1966) restricted his remarks to mammals and suggested that most mammals might have U-shaped mortality curves similar in general shape to that of the African buffalo (Figure 11.5). He pointed out, however, that few data were available for critical comparisons.

CAPACITY FOR INCREASE IN NUMBERS

A life table summarizes the mortality schedule of a population. We must now consider the reproductive rate of a population and techniques by which we can combine reproduction and mortality estimates to determine net population changes. Students of human populations were the first to appreciate and solve these problems. One way of combining reproduction and mortality data for populations utilizes a demographic parameter called the *intrinsic capacity for increase* derived by Alfred Lotka in 1925.

Any population in a particular environment will have a mean longevity or survival rate, a mean natality or birth rate, and a mean growth rate or speed of development of individuals. The values of these means are determined in part by the environment and in part by the innate qualities of the organisms themselves. These qualities of an organism cannot be measured simply because they are not a constant, but we can measure their expression under specified conditions and thereby define for each population its intrinsic capacity for increase (also called the *Malthusian parameter*). The intrinsic capacity for increase is a statistical characteristic of a population and depends on environmental conditions.

Environments in nature vary continually. They are never consistently favorable or consistently unfavorable but fluctuate between these two extremes, for example, from winter to summer. When conditions are favorable, numbers increase; when conditions are unfavorable, numbers decrease. It is clear that no population goes on increasing forever. Darwin (1859, chap. 3) recognized the contrast between a high potential rate of increase and an observed approximate balance in nature. He illustrated this problem by asking why there were not more elephants, since he estimated that two elephants could give rise to 19 million in 750 years.

Therefore in nature we observe an actual rate of increase which is continually varying from + to − in response to changes within the population in age distribution, social structure, and genetic composition and in response to changes in environmental factors. We can however abstract from all this variation and ask what would happen to a population if it persisted in its present configuration of births and deaths. This abstraction is the ecologist's version of the perfect vacuum of introductory physics: We ask, what would happen in terms of population increase if conditions remained unchanged for a long time?

An organism's innate capacity for increase depends on its fertility, longevity, and speed of development. For any population, these are measured by the birth rate and the death rate. When the birth rate exceeds the death rate, the population will increase. If we wish to estimate quantitatively the rate at which the population increases or decreases, we need to describe how both the birth rate and the death rate vary with age.

How can we express the variations of birth and death rates with age? We have just discussed the method of expressing survival rates as a function of age. The life table includes a table of age-specific survival rates. The portion of the life table needed to compute the capacity for increase is the l_x column, the proportion of the population surviving to age x. Similarly, the birth rate of a population is best described by an age schedule of births. This is a table that gives the number of female offspring produced per unit of time per female aged x and is called a *fertility table*, or b_x function. Usually only the females are counted, and the demographer typically views populations as females giving rise to more females. Table 11.3 gives the survivorship table—the l_x schedule with which we are familiar—and the fertility table for females in the United States in 1989. In this case, the great majority of women live through the childbearing ages. The fertility table gives the expected number of *female* offspring for each female living through the five years of each age group. For example, slightly less than three women in ten between the ages of 20 and 25 will, on the average, have a female baby.

Given these data, we can obtain a useful statistic, *the net reproductive rate (R_0)*. If a cohort of females lives its entire reproductive life at the survival and fertility rates given in Table 11.3, what number will this cohort or generation leave as its female offspring? We define as the net reproductive rate:

$$\text{Net reproductive rate} = R_0 = \frac{\text{no. daughters born in generation } t + 1}{\text{no. daughters born in generation } t}$$

R_0 is thus the multiplication rate per generation* and is obtained by multiplying together the l_x and b_x schedules and summing over all age groups, as shown in Table 11.3:

$$R_0 = \sum_0^\infty l_x \, b_x$$

Thus we temper the birth rate by the fraction of expected survivors to each age. If survival were 100 percent, R_0, would just be the sum of the b_x column. In this example (Table 11.3), the human population of the United States, if it continued at these 1989 rates, would multiply 0.907 times in each generation. If the net reproductive rate is 1.0, the population is replacing itself exactly. When the net reproductive rate is below 1.0, the population is not replacing itself, and in this example if these rates continue for a long time, the population will drop about 9 percent each generation in the absence of immigration. The net reproductive rate is illustrated in Figure 11.6.

Given these two tables expressing the age-specific rates of survival and fertility, we may inquire at what rate a population subject to these rates would increase, assuming (1) that these rates remain constant and (2) that no limit is placed on

---

* A generation is defined as the mean period elapsing between the birth of parents and the birth of offspring. See page 180 and Figure 11.10.

Table 11.3 Survivorship Table (l_x) and Fertility Table (b_x) for Women in the United States, 1989.

Age Group	Midpoint or Pivotal Age x	Proportion Surviving to Pivotal Age l_x	No. Female Offspring per Female Aged x per 5-Year Time Unit (b_x)	Product of l_x and b_x
0—9	5.0	0.9895	0.0	0.0
10–14	12.5	0.9879	0.0020	0.0020
15–19	17.5	0.9861	0.1233	0.1216
20–24	22.5	0.9834	0.2638	0.2594
25–29	27.5	0.9802	0.2772	0.2717
30–34	32.5	0.9765	0.1807	0.1765
35–39	37.5	0.9712	0.0650	0.0631
40–44	42.5	0.9643	0.0125	0.0121
45–49	47.5	0.9528	0.0005	0.0005
50 and above	—	—	0.0	0.0

$$R_0 = \sum_0^\infty l_x b_x = 0.9069$$

Source: Statistical Abstract of the United States (1991).

population growth. Because these survival and fertility rates vary with age, the actual birth and death rates of the population will depend on the existing age distribution. If the whole population was over 50 years of age, it would not increase. Similarly, if all females were between the ages of 20 and 25, the rate of increase would be much higher than if they were all between 30 and 35. Before we can calculate the population's rate of increase, it would seem that we must specify (1) age-specific survival rates (l_x), (2) age-specific birth rates (b_x), and (3) age distribution.

This conclusion is not correct. Contrary to intuition, we do not need to know the age structure of the population. Lotka (1922) has shown that a population that

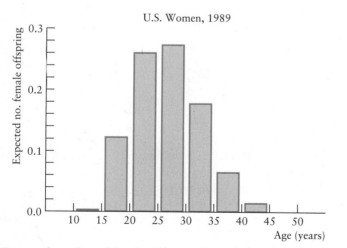

Figure 11.6 Expected number of female offspring for each female in the United States, 1989. Data from Table 11.3. The area under the histogram (shaded) is the net reproductive rate (R_0).

is subject to a constant schedule of birth and death rates will gradually approach a fixed or *stable age distribution*, whatever the initial age distribution may have been, and will then maintain this age distribution indefinitely. This theorem is one of the most important discoveries in mathematical demography. When the population has reached this stable age distribution, it will increase in numbers according to the differential equation:

$$\frac{dN}{dt} = rN$$

(Rate of change in numbers = (intrinsic capacity × (population size)
     per unit time)   for increase)

This same equation may be rewritten in integral form:

$$N_t = N_0 e^{rt}$$

where

N_0 = number of individuals at time 0
N_t = number of individuals at time t
e = 2.71828 (a constant)
r = intrinsic capacity for increase for the particular environmental conditions
t = time

This is the equation describing the curve of geometric increase in an expanding population (or geometric decrease to zero if r is negative). A simple example illustrates this equation. Let the starting population (N_0) be 100 and let $r = 0.5$ per female per year. The successive populations would be:

Year	Population Size	
0		100
1	$(100)(e^{0.5})$ =	165
2	$(100)(e^{1.0})$ =	272
3	$(100)(e^{1.5})$ =	448
4	$(100)(e^{2.0})$ =	739
5	$(100)(e^{2.5})$ =	1218

This hypothetical population growth is plotted in Figure 11.7. Note that on a logarithmic scale the increase follows a straight line, but on an arithmetic scale the curve swings upward at an accelerating rate.

To summarize to this point: (1) Any population subject to a fixed age schedule of births and deaths will increase in a geometric way, and (2) this geometric increase will dictate a fixed and unchanging age distribution called the stable age distribution.

Let us invent a simple model organism to illustrate these points. Suppose that we have a parthenogenetic animal that lives 3 years and then dies. It produces two young at exactly 1 year of age, one young at exactly 2 years of age, and no young at year 3. The life table and fertility table for this hypothetical animal are thus extremely simple:

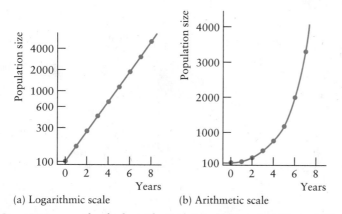

Figure 11.7   Geometric growth of a hypothetical population when $N_0 = 100$ *and* $r = 0.5$.

x	l_x	b_x	$l_x b_x$	$(x)(l_x)(b_x)$
0	1	0	0	0
1	1	2	2	2
2	1	1	1	2
3	1	0	0	0
4	0	—	—	—

$$R_0 = \sum_0^4 l_x b_x = 3 \qquad\qquad 4$$

If a population of this organism starts with one individual aged zero, the population growth will be as shown in Figure 11.8, or, in tabular form, as follows:

Year	NUMBER AT AGES				Total Population Size	% Age 0 in Total Population
	0	1	2	3		
0	1	0	0	0	1	100.0
1	2	1	0	0	3	66.67
2	5	2	1	0	8	62.50
3	12	5	2	1	20	60.00
4	29	12	5	2	48	60.42
5	70	29	12	5	116	60.34
6	169	70	29	12	280	60.36
7	408	169	70	29	676	60.36
8	985	408	169	70	1632	60.36

Note that the age distribution quickly becomes fixed or stable with about 60 percent at age 0, 25 percent at age 1, 10 percent at age 2, and 4 percent at age 3. This demonstrates Lotka's (1922) conclusion that a population growing geometrically develops a stable age distribution.

We may also use our model animal to illustrate how the capacity for increase r can be calculated from biological data. The data of the l_x and b_x tables are sufficient

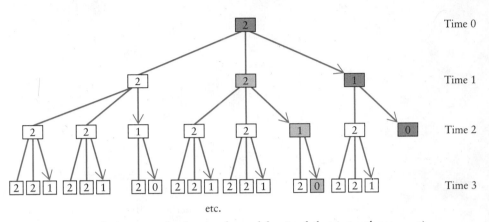

etc.

Figure 11.8   Population growth of a simple model animal that is parthenogenetic. Each box represents one live individual, and the numbers within each box indicate the number of young to be produced at the next time interval. Arrows indicate aging of individuals from one year to the next.

to allow the calculation of *r*, the capacity for increase in numbers. To do this, we first need to calculate the net reproductive rate (R_0), which we have explained earlier. For our model animal, $R_0 = 3.0$, which means that the population can triple its size each generation. But how long is a generation? The *mean length of a generation* (G) is the mean period elapsing between the birth of parents and the birth of offspring. This is only an approximate definition, since offspring are born over a period of time and not all at once. The mean length of a generation is defined approximately as follows (Dublin and Lotka 1925):

$$G = \frac{\Sigma l_x b_x x}{\Sigma l_x b_x} = \frac{\Sigma l_x b_x x}{R_0}$$

For our model organism, $G = 4.0/3.0 = 1.33$ years. Figure 11.9 illustrates the approximate meaning of generation time for a human population. Leslie (1966) has discussed some of the difficulties of applying the concept of generation time to a

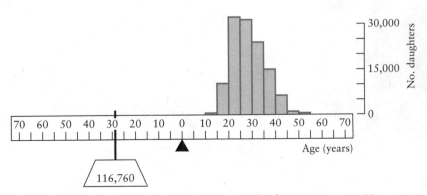

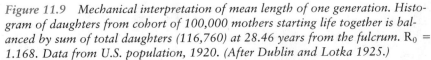

Figure 11.9   Mechanical interpretation of mean length of one generation. Histogram of daughters from cohort of 100,000 mothers starting life together is balanced by sum of total daughters (116,760) at 28.46 years from the fulcrum. R_0 = 1.168. Data from U.S. population, 1920. (After Dublin and Lotka 1925.)

continuously breeding population with overlapping generations. For organisms such as annual plants and many insects with a fixed length of life cycle, the mean length of a generation is simple to measure and to understand.

Knowing the multiplication rate per generation (R_0) and the length of a generation (G), we can now determine r directly as an instantaneous rate:

$$r = \frac{\log_e(R_0)}{G}$$

For our model organism,

$$r = \frac{\log_e(3.0)}{1.33} = 0.824 \text{ per individual per year}$$

Because the generation time G is an approximate estimate, this value of r is only an approximate estimate when generations overlap.

The innate capacity for increase can be determined more accurately by solving the characteristic equation, a formula derived by Lotka (1907, 1913):

$$\sum_{x=0}^{\infty} e^{-rx} l_x b_x = 1$$

This equation cannot be solved explicitly for r because it cannot be rearranged to have r on one side and all else on the other. By substituting trial values of r, we can solve this equation. Our model animal can again be used as an example. For our estimate of $r = 0.824$, we get

x	$l_x b_x$	$e^{-0.824x}$	$e^{-0.824x} l_x b_x$
0	0.0	1.0	0.0
1	2.0	0.44	0.880
2	1.0	0.19	0.190
3	0.0	0.08	0.0
4	0.0	0.04	0.0

$$\Sigma e^{-rx} l_x b_x = 1.070$$

Clearly, the estimate $r = 0.824$ is slightly low. We repeat with $r = 0.85$, and after several trials we find that, for this model organism, $r = 0.881$ provides

$$\Sigma e^{-rx} l_x b_x = 1.004$$

which is a close enough approximation. Birch (1948) works out another example in detail. Mertz (1970) provides a lucid derivation and discussion of these formulas.

The capacity for increase is an instantaneous rate and can be converted to the more familiar finite rate* by the formula

Finite rate of increase $= \lambda = e^r$

For example, if $r = 0.881$, $\lambda = 2.413$ per individual per year in our model organism. Thus for every individual present this year, there will be 2.413 present next year.

---

* Appendix III gives a general discussion of instantaneous and finite rates.

It should now be clear why the capacity for increase in numbers cannot be expressed quantitatively except for a particular environment. Any component of the environment, such as temperature, humidity, or rainfall, might affect the birth and death rates and hence r.

One example of the effect of the environment on the capacity for increase was shown by Birch (1953a) in his work on *Calandra oryzae*, a beetle pest that lives in stored grain. The capacity for increase in this species varied with the temperature and with the moisture content of the wheat, as shown in Figure 11.10. The practical implications of these results are that wheat should be stored where it is cool or dry to prevent losses from *C. oryzae*.

Grain beetles live in an almost ideal habitat, surrounded by food, protected from most enemies, and with relatively constant physical conditions. They are also easy to deal with in the laboratory and are thus used extensively in ecology lab experiments. Birch (1953a) studied two species, *Calandra oryzae* (a temperate species) and *Rhizopertha dominica* (a tropical species). He found that in both species, r varied with temperature and moisture (Figure 11.11). The line $r = 0$ marks the limit of the possible ecological range for each species with respect to temperature and moisture. *Calandra* is more cold-resistant; *Rhizopertha* can increase at higher temperatures and lower humidities. The distribution of the two species in Australia agrees with these results: *Rhizopertha* is a pest only in the warmer parts of the country and is absent from Tasmania, where *Calandra* occurs as a pest.

In general, the capacity for increase is not correlated with rareness and commonness of species. Species with a high r are not always common, and species with a low r are not always rare. Some species, such as the buffalo in North America, the elephant in central Africa, and the periodical cicadas, are (or were) quite common and yet have a low r value. Many parasites and other invertebrates with a high capacity for increase are nevertheless quite rare. Darwin (1859) pointed this out in *The Origin of Species*.

We can calculate how certain changes in the life history of a species would affect the capacity for increase in numbers. In general, three factors will increase r: (1)

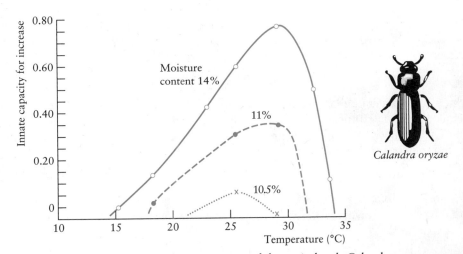

Figure 11.10 *Intrinsic capacity for increase (r) of the grain beetle* Calandra oryzae *living in wheat of different moisture contents and at different temperatures. (After Birch 1953a.)*

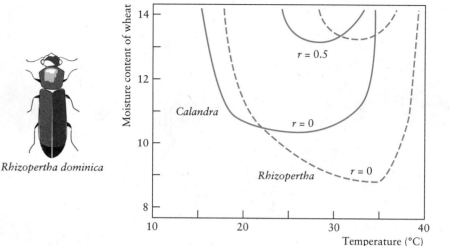

Rhizopertha dominica

Calandra oryzae

Figure 11.11 *Approximate contours of the intrinsic capacity for increase* (r) *for the two grain beetles* Calandra oryzae *and* Rhizopertha dominica *living in wheat of different moisture contents and at different temperatures. (After Birch 1953a.)*

reduction in age at first reproduction, (2) increase in number of progeny in each reproductive bout, and (3) increase in number of reproductive bouts (increased longevity). In many cases, the most profound effects are achieved by changing the age at first reproduction. For example, Birch (1948) calculated for the grain beetle *C. oryzae* the number of eggs needed to obtain $r = 0.76$ according to the age at first reproduction:

Age at Which Breeding Begins (wk)	Total No. Eggs that Must Be Laid to Produce $r = 0.76$
1	15
2	32
3	67
4	141 (actual life history)
5	297
6	564

The earlier the peak in reproductive output, the larger the *r* value, as a rule. Lewontin (1965) provides an excellent example to illustrate this in *Drosophila serrata* (Figure 11.12). The Rabaul race of this fruit fly survives poorly and lays fewer eggs than the Brisbane race, but because it begins to reproduce at an earlier age (11.7 compared with 16.0 days) and has a shorter generation length, its capacity for increase is equal to that of the longer-living, more fertile Brisbane race.

To conclude: The concept of an intrinsic capacity for increase in numbers, which we have just discussed, is an oversimplification of nature. In nature, we do not find populations with stable age distributions or with constant age-specific rates of mortality and fertility. For these reasons, the actual rate of increase we observe in natural populations varies in more complex ways than the theoretical constant *r*. The importance of *r* lies mostly in its use as a model to compare with the actual rates of increase we see in nature.

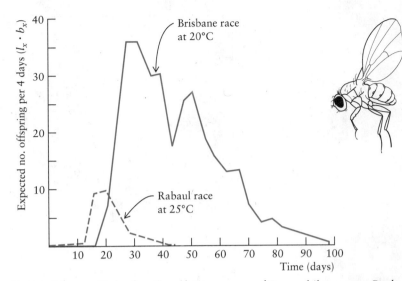

Figure 11.12   Observed $l_x \, b_x$ *functions for two races of* Drosophila serrata. *Both* $l_x \, b_x$ *functions give the same value of* r *(0.16) because of the overriding importance of earlier reproduction and shorter generation length of the Rabaul race. Brisbane females lay an average of 546 eggs at 20°C, while Rabaul flies lay only 151 eggs during their life span. (After Lewontin 1965.)*

For these reasons, the life table and fertility table can be used in the diagnosis of environmental quality because they are so sensitive to environmental conditions. Figure 11.13 illustrates the effect of varying temperature and food supply on the l_x and b_x schedules of the rotifer *Brachionus calyciflorus* in the laboratory. There is an optimal food supply of around 0.5–1.0 AD units (Figure 11.13) because the food alga *Chlorella* produces metabolites toxic to *Brachionus* at high algal densities (Halbach 1979).

REPRODUCTIVE VALUE

We can use life tables and fertility tables to determine the contribution to the future population that an individual female will make. We call this the *reproductive value* of a female aged x, and this is most easily expressed for a population that is stable in size:

$$\text{Reproductive value at age } x = V_x = \sum_{t=x}^{w} \frac{l_t}{l_x} b_t$$

where t and x are age and w is the age of last reproduction. Note that as defined here, reproductive value at age 0 is the same as net reproductive rate (R_0) defined on page 176.

Reproductive value can be partitioned into two components (Pianka and Parker 1975):

Reproductive value at age x = present progeny + expected future progeny

$$V_x = b_x + \sum_{t=x+1}^{w} \frac{l_t}{l_x} b_t$$

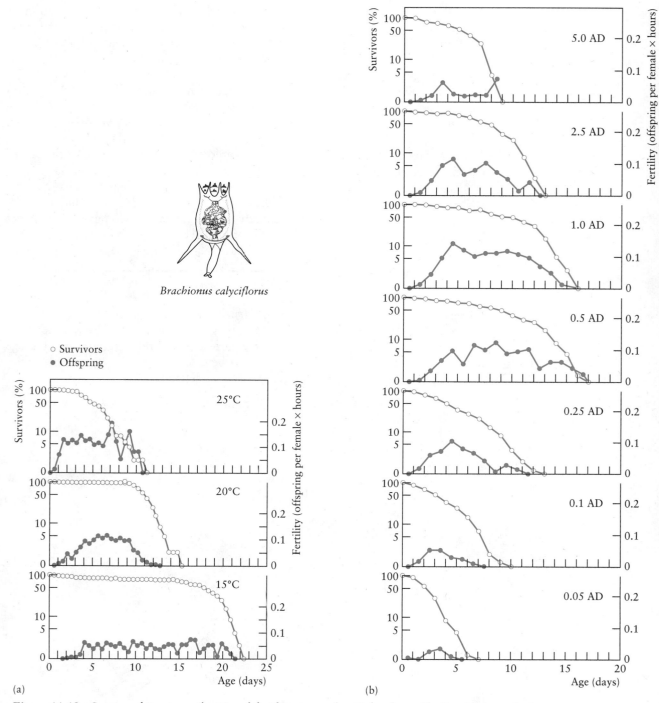

Figure 11.13 *Survivorship curves* (l_x, ○) *and fertility curves* (b_x, ●) *for the rotifer* Brachionus calyciflorus *in laboratory cultures.* (a) *Variable temperature, constant food supply.* (b) *Constant temperature, variable food supply. One AD food unit is* 10^6 Chlorella *cells per ml per 12 hours.* (After Halbach 1979.)

We call the second term *residual reproductive value,* because it measures the number of progeny on average that will be produced in the rest of an individual's life span.

Reproductive value is more difficult to define if the population is not stable (Fisher 1958, Horn 1978). In this case we must discount future reproduction if population growth is occurring because the value of one progeny is less in a larger population. Figure 11.14 illustrates the change of reproductive value with age in a red deer population in Scotland. Red deer stags defend harems and their effective breeding span is three to five years between the ages of 6 and 11 years. By contrast, red deer hinds start to produce calves at age 3 and breed until they are 15 years old or older. These differences in reproductive biology explain the shapes of the reproductive value curves in Figure 11.14.

Reproductive value is important in the evolution of life-history traits. Natural selection acts more strongly on age classes with high reproductive value and very weakly on age classes with low reproductive value. Predators will have more impact on a population if they prefer individuals of high reproductive value.

AGE DISTRIBUTIONS

We have already discussed the idea of age distribution in connection with the capacity for increase. We have noted that a population growing geometrically with constant age-specific mortality and fertility rates would assume and maintain a stable age distribution. The stable age distribution can be calculated for any set of life tables and fertility tables. The stable age distribution is defined as follows:

C_x = proportion of organisms in the age category x to $x + 1$ in a population increasing geometrically

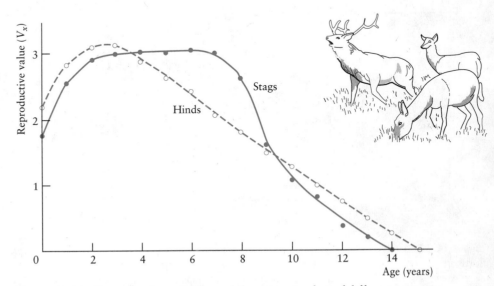

Figure 11.14   Reproductive value for red deer stags (males) of different ages compared with that of hinds (females) on the island of Rhum, Scotland. Reproductive value is calculated in terms of the number of female offspring surviving to one year of age that parents of different ages can expect to produce in the future. (From Clutton-Brock et al. 1982, p. 154.)

Mertz (1970) has shown that

$$C_x = \frac{\lambda^{-x}l_x}{\displaystyle\sum_{i=0}^{\infty} \lambda^{-i}l_i}$$

where

$\lambda = e^r$ = finite rate of increase
l_x = survivorship function from life table
x,i = subscripts indicating age

Let us go through these calculations with our model organism:

$$\lambda = e^r = e^{0.881} = 2.413$$

Age (x)	l_x	λ^{-x}	$\lambda^{-x}l_x$
0	1.0	1.0	1.0
1	1.0	0.4144	0.4144
2	1.0	0.1717	0.1717
3	1.0	0.0711	0.0711
4	0.0	0.0295	0.0

$$\sum_{x=0}^{4} \lambda^{-x}l_x = 1.6572$$

Thus to calculate C_0, the proportion of organisms in the age category 0 to 1 in the stable age distribution, we have

$$C_0 = \frac{\lambda^{-0}l_0}{\displaystyle\sum_{i=0}^{4} \lambda^{-i}l_i} = \frac{(1.0)(1.0)}{1.6572} = 0.6035$$

For C_1, we have

$$C_1 = \frac{(0.4144)(1.0)}{1.6572} = 0.250$$

In a similar way,

$$C_2 = 0.104$$
$$C_2 = 0.043$$

Compare these calculated values with those obtained empirically earlier (page 179). Birch (1953a) illustrates another method of calculating the stable age distribution for a set of l_x and b_x schedules.

Populations that have reached a constant size, in which the birth rate equals the death rate, will also assume a fixed age distribution, called a *stationary age distribution* (or *life-table age distribution*) and will maintain this distribution. The stationary age distribution is a hypothetical one and illustrates what the age composition of the population would be at a particular set of mortality rates (q_x) if the birth rate were to be exactly equal to the death rate. Figure 11.15 contrasts the stable and stationary age distributions for the short-tailed vole in a laboratory colony.

These are the only conditions in which a constant age structure is maintained in

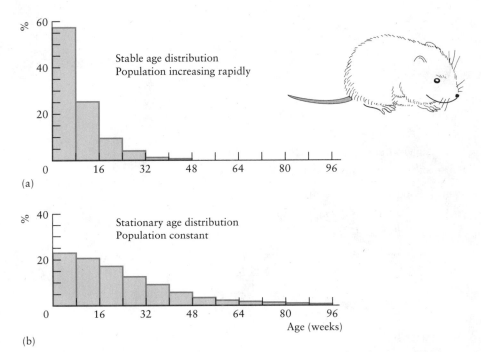

(a)

(b)

Figure 11.15   *(a) Stable age distribution and (b) stationary age distribution of the vole* (Microtus agrestis). *(After Leslie and Ranson 1940.)*

a population. Under any other circumstances, the population does not assume a constant age structure but changes over time. In natural populations, the age structure is almost constantly changing. We rarely find a natural population that has a stable age structure because populations do not increase for long in an unlimited fashion. Nor do we often find a stationary age distribution, because populations are rarely in a stationary phase for long. We can illustrate these relationships as follows:

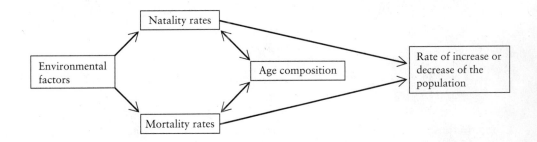

With proper care, information on age composition can be used to judge the status of a population. Increasing populations typically have a predominance of young organisms, whereas stable or declining populations do not (Figure 11.15). Figure 11.16 illustrates this contrast between the human population of Kenya, which was increasing at 3.8 percent per year in 1991 and had an average life expectancy at birth of 63 years, and that of the United States, which was increasing at 0.8 percent per year in 1991 and had an average life expectancy of about 75 years. The age structure of human populations has been analyzed in detail because of its eco-

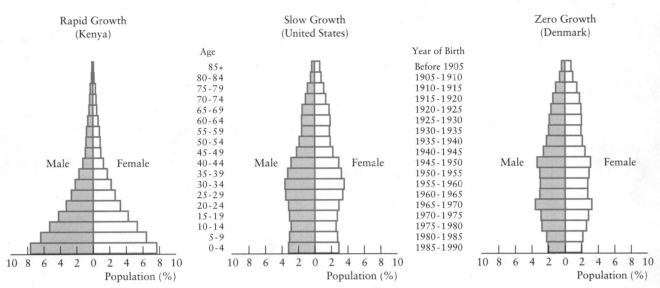

Figure 11.16   Age distribution for the rapidly increasing human population of Kenya, the slowly increasing population of the United States, and the stationary population of Denmark in 1990. (From McFalls 1991.)

nomic and sociological implications (Weeks 1989). A country with a high fertility rate and a large fraction of children—such as Kenya, with 50 percent under age 15—has a much greater demand for schools and other child services than does another country such as the United States, with 22 percent under age 15.

In natural populations, even more variation is apparent. In long-lived species such as trees and fishes, one may find *dominant year-classes*. Figure 11.17 illustrates this for Engelmann spruce and subalpine fir trees of the Rocky Mountains, in which some year-classes may be 100 times as strong as others. In these situations, the age composition can change greatly from one year to the next. Eberhardt (1988) discusses the use of age composition information in the management of wildlife population, and Ricker (1975, chap. 2) discusses this problem in exploited fish populations.

EVOLUTION OF DEMOGRAPHIC TRAITS

We can use the demographic techniques just described to investigate one of the most interesting questions of evolutionary ecology: Why do organisms evolve one type of life cycle rather than another? Only certain kinds of l_x and b_x schedules are permissible if a population is to avoid extinction. How does evolution act, within the framework of permissible demographic schedules, to determine the life cycle of a population?

Pacific salmon grow to adult size in the ocean and return to fresh water to spawn once and die. We may call this *big-bang reproduction*. Oak trees may become mature after 10 to 20 years and drop thousands of acorns for 200 years or more. We call this *repeated reproduction*. How have these life cycles evolved? What advantage might be gained by producing salmon that breed more than once or oak trees that drop one set of seeds and die?

The population consequences of life cycles were explored by Cole (1954), who

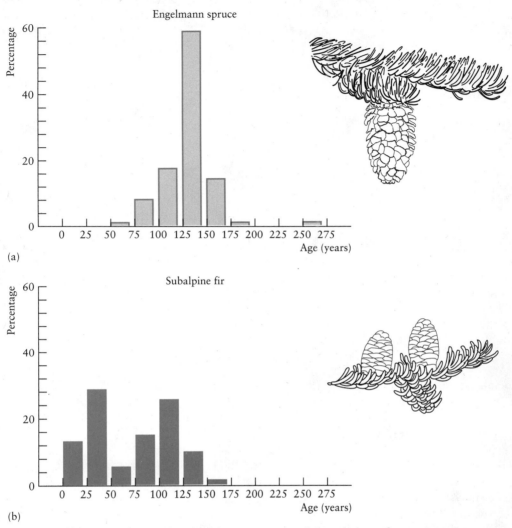

Figure 11.17   *Age structure of (a) Engelmann spruce and (b) subalpine fir in a forest stand at 3150 m elevation in northern Colorado. Neither of these tree species has an age distribution like those shown in Figure 11.15 for stable or stationary age distributions. (Data from Aplet et al. 1988, Table 2.)*

asked a simple question: What effect does repeated reproduction have on the capacity for increase (r)? Assume that we have an annual species that produces offspring at the end of the year and then dies. Assume a simple survivorship of 0.5 per year and a fertility rate of 20 offspring. The life table appears as follows:

Age (x)	Proportion Surviving (l_x)	Fertility (b_x)	Product ($l_x b_x$)
0	1.0	0	0.0
1	0.5	20	10.0
2	0.0	—	0.0
			$R_0 = 10.0$

The net reproductive rate (R_0) is 10.0, which means that the species could increase ten-fold in one year. We can determine r from the characteristic equation:

$$\sum_{0}^{\infty} e^{-rx} l_x b_x = 1$$

from which we determine

$r = 2.303$

for the annual species with big-bang reproduction. What advantage could this species gain by continuing to live and reproduce at years 2, 3, . . . , ∞? Let us assume the most favorable condition, that is, no mortality after age 1 and survival to age 100. The life table now becomes

Age (x)	Proportion Surviving (l_x)	Fertility (b_x)	Product $(l_x b_x)$
0	1.0	0.0	0.0
1	0.5	20.0	10.0
2	0.5	20.0	10.0
3	0.5	20.0	10.0
4	0.5	20.0	10.0
5	0.5	20.0	10.0
—	—	—	—
—	—	—	—
—	—	—	—
99	0.5	20.0	10.0
100	0.0	0.0	0.0

$$R_0 = \Sigma l_x b_x = 990.0$$

In the manner outlined, we determine

$r = 2.398$

for the perennial species with repeated reproduction. If we adopt repeated reproduction in our hypothetical organism, we raise the intrinsic capacity for increase only about 4 percent:

$$\frac{2.398}{2.303} = 1.04$$

Now let us work backward. What fertility rate at 1 year would equal the r of the perennial (2.398)? We can solve this problem algebraically (Cole 1954) or by trial and error. Suppose we increase the birth rate by one individual. The annual life table is now

Age (x)	Proportion Surviving (l_x)	Fertility (b_x)	Product $(l_x b_x)$
0	1.0	0.0	0.0
1	0.5	21.0	10.5
2	0.0	—	0.0

$$R_0 = 10.5$$

$$\sum_{0}^{2} e^{-rx} l_x b_x = 1.0$$

$$r = 2.351$$

This is almost the gain achieved by repeated reproduction. If we increase the fertility rate by two individuals, we get $r = 2.398$, equal to the r for the perennial. This is obviously an ideal case, since we assume no mortality after age 1 in the perennial form. Cole (1954) generalized this ideal case to a surprising conclusion: *For an annual species, the maximum gain in the intrinsic capacity for increase (r) that could be achieved by changing to the perennial reproductive habit would be equivalent to adding one individual to the effective litter size ($l_x b_x$ for age 1).* Cole assumed for his ideal case perfect survival to reproductive age (Charnov and Schaffer 1973). In our hypothetical example, we assumed that half of the organisms die before reaching reproductive age.

This simple model for the evolution of big-bang reproduction is unrealistic because it is a "cost-free" model—present reproduction is assumed to have no effect on future reproduction or future survival (Bell 1980). To be more realistic, let us assume that adult survival varies inversely with fecundity, so that an organism can produce more offspring only at the cost of being more likely to die. Figure 11.18a shows this trade-off between reproduction and survival. The critical division between big-bang reproduction and repeated reproduction is set by the survival rate of the juvenile stages. If survival of juveniles is very poor or unpredictable, selection will usually favor repeated reproduction. Let us look at one example to illustrate this theory.

The American shad (*Alosa sapidissima*) lives in the Atlantic Ocean but spawns in river systems from Florida to Quebec. Southern populations are big-bang reproducers; proceeding north, more fish are repeated spawners (Leggett and Carscadden 1978). Figure 11.18b illustrates the cost of reproduction in the shad. When there is no repeated reproduction, female shad lay about 450,000 eggs and die. Survival of these eggs to maturity is very low, estimated at only 0.0014 percent, or about 1 in 100,000. There is a very rapid drop in fecundity as repeat spawning occurs, and this

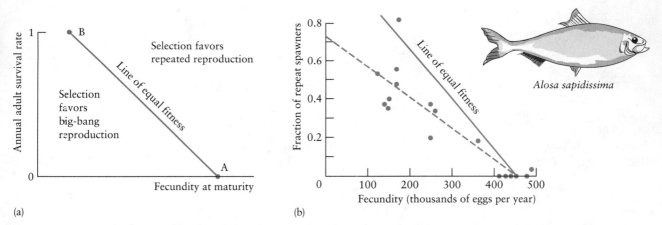

(a)   (b)

Figure 11.18   *(a) Evolution of big-bang reproduction when there is a tradeoff between future adult survival and present fecundity. Point A is the fecundity that is achieved in big-bang reproduction. Point B is the hypothetical fecundity that could be achieved with little or no cost to subsequent adult survival. The slope of this line is − (survival rate from birth to sexual maturity). (b) Data for populations of the American shad from eastern North America. Fecundity falls rapidly with repeated reproduction in this fish, and the available data fall into the zone in which big-bang reproduction would be favored. (Modified from Bell 1980.)*

high cost of survival suggests that natural selection favors big-bang reproduction in the shad, as it does in many salmon species.

The estimation of the costs associated with reproduction is very difficult. Reproductive effort at any given age can be associated with a biological *cost* and a biological *profit*. The biological cost derives from the reduction in growth or survival that occurs as a consequence of using energy to reproduce (Figure 11.19). For example, as much as 50 percent of the production of the perch (*Perca fluviatilis*) is used for reproduction (LeCren 1962). Spawning in barnacles reduces growth (Barnes 1962). Adult female carabid beetles survive better in years of little reproduction and worse in years of intensive reproduction (Murdoch 1966b). The biological profit is measured in the number of descendants left to future generations, which will be affected by the survival rate and the growth rate. The hypothetical organism must ask at each age, Should I reproduce this year or would I profit more by waiting until next year? Obviously, if the mortality rate of adults is high, it would be best to reproduce as soon as possible. But if adult mortality is low, it may pay an organism to put its energy into growth and wait until the next year to reproduce.

Many organisms do not reproduce as soon as they are physiologically capable of doing so. The key item that must be measured to predict the optimal age at maturity is the *potential fecundity cost* (Bell 1980). Individuals that reproduce in a given year will often be smaller and less fecund in the following year than an individual that has previously abstained from reproduction. This is best established in poikilotherms, such as fishes that show a reduction in growth associated with spawning. Potential fecundity costs also occur in homeotherms (Clutton-Brock et al. 1989), and the period of lactation in mammals is energetically very expensive for

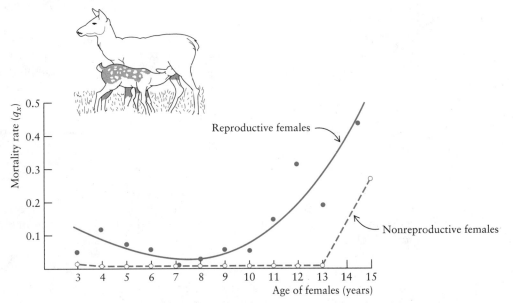

Figure 11.19   Cost of reproduction in female red deer on the island of Rhum in Scotland. Mortality in winter is always higher in females that reproduced during the previous summer, no matter what the age of the female. (After Clutton-Brock et al. 1982.)

females. Social behavior associated with reproduction can produce great differences in the costs associated with breeding in the two sexes and thus cause differences in the optimal age at maturity for males and females. Red deer stags, for example, defend harems and attain a breeding peak after 7 years of age through their fighting ability. Females mature at 3 years and live longer than males.

In birds the best estimates of the cost of reproduction can be obtained by experimental brood manipulations (Reznick 1985, Dijkstra et al. 1990). Female blue tits (*Parus caeruleus*) that raise larger broods have a reduced chance of living through the next winter. Many but not all bird studies have shown a cost for reproduction. So far, however, we do not have good measures of the overall magnitude of these costs or exactly when in the life cycle these costs are extracted (Nur 1990); nor do we understand why some birds show no cost of reproduction.

Repeated reproducers must decide in an evolutionary sense to increase, decrease, or hold constant their reproductive effort with age. In every case analyzed to date, reproductive effort increases with age (Williams 1966, Sydeman et al. 1991), and this may be a general evolutionary trend in organisms.

Why then do species bother to have repeated reproduction? The answer seems to be that repeated reproduction is an adaptation to something other than achieving maximum reproductive output. Repeated reproduction may be an evolutionary response to uncertain survival from zygote to adult stages (Murphy 1968). The greater the uncertainty, the higher the selection for a longer reproductive life. This may involve channeling more energy into growth and maintenance and less into reproduction. Thus we can recognize a simple scheme of possibilities:

	Long Life Span	Short Life Span
Steady reproductive success	?	Possible
Variable reproductive success	Possible	Not possible

Murphy (1968) analyzed data from marine fishes that support this idea, and we now believe that the advantage of repeated reproduction is that it spreads the risk of reproducing over a longer time period and thus acts as an adaptation to thwart environmental fluctuations.

SUMMARY

Population changes can be analyzed with a set of quantitative techniques first developed for human population analysis. A life table is an age-specific summary of the mortality rates operating on a population. Life tables are necessary because mortality does not fall equally on all ages, and often the very young and the old suffer high mortality.

The reproductive component of population increase can be described by a fertility table that summarizes reproduction with respect to age. The intrinsic capacity for increase of a population is obtained by combining the life table and the fertility table for specified environmental conditions. This concept leads to an important demographic principle: A population that is subject to a constant schedule of death and birth rates (1) will increase in numbers geometrically at a rate equal to the capacity for increase (r), (2) will assume a fixed or stable age distribution, and (3) will maintain this age distribution indefinitely. A hypothetical organism is invented to illustrate these conclusions empirically.

The age distribution of a population is constant and unchanging only as long as the life table and the fertility table remain constant. Populations undergoing geometric increase reach a stable age distribution and those showing a period of constant population size reach a stationary age distribution. Under other circumstances, the age distribution will shift over time, which is the usual condition in natural populations.

Demographic techniques are useful for comparing quantitatively the consequences of adopting an annual life cycle versus a perennial one. What does an organism gain by repeated reproduction? There is very little gain in potential for population increase in species that reproduce many times in each generation, and repeated reproduction seems to be an evolutionary response to conditions in which survival from zygote to adult varies from good to poor in an unpredictable way. An organism thus hedges its bets by reproducing several times.

Selected References

ANDREWARTHA, H. G., and L. C. BIRCH. 1954. The innate capacity for increase in numbers. In *The Distribution and Abundance of Animals*, chap. 3. University of Chicago Press, Chicago.

CAUGHLEY, G. 1977. *Analysis of Vertebrate Populations*. chaps. 7–9. Wiley, New York.

DEEVEY, E. S., JR. 1947. Life tables for natural populations of animals. *Quarterly Review of Biology* 22:283–314.

EBERHARDT, L. L. 1988. Using age structure data from changing populations. *Journal of Applied Ecology* 25:373–378.

REZNICK, D. 1985. Costs of reproduction: An evaluation of the empirical evidence. *Oikos* 44:257–267.

SCHAEFER, W. M., and P. F. ELSON. 1975. The adaptive significance of variations in life history among local populations of Atlantic salmon in North America. *Ecology* 56:577–590.

STEARNS, S. C. 1976. Life-history tactics: A review of the ideas. *Quarterly Review of Biology* 51:3–47.

STEARNS, S. C. 1989. Trade-offs in life history evolution. *Functional Ecology* 3:259–268.

Questions and Problems

1. Reproductive success declines in old animals. How could you determine if this decline is caused by aging (senescence) or is a reproductive cost associated with previous breeding activities? Sydeman and coauthors (1991) discuss this question for the northern elephant seal.

2. In human populations, differences in mortality rates have very little effect on age composition, while changes in birth rates have a large effect on age structure (Barclay 1958, p. 229). Why should this be? Would the same principle hold for plant and animal populations?

3. Connell (1970) gives the following data for the barnacle *Balanus glandula*:

Age (yr)	1959 SETTLEMENT		1960 SETTLEMENT	
	l_x	b_x	l_x	b_x
0	1.0	0	1.0	0
1	0.0000620	4,600	0.0000640	4,600
2	0.0000340	8,700	0.0000290	8,700
3	0.0000200	11,600	0.0000190	11,600
4	0.0000155	12,700	0.0000090	12,700
5	0.0000110	12,700	0.0000045	12,700
6	0.0000065	12,700	0.0	—
7	0.0000020	12,700	—	—
8	0.0000020	12,700	—	—

Calculate the net reproductive rate (R_0) for these two-year classes. What does this tell you about these populations? Estimate the capacity for increase from these data, and calculate the stable age distribution (C_x). Explain any difficulties and interpret the results in biological terms.

4. What additional data, if any, are required to determine the stable age distribution for the human population described in Table 11.3?

5. "A woman who gives birth to a set of twins at the age of 19, and subsequently gives birth to one other child, contributes as much to the future population of America as does a woman who produces five children, but whose age at birth of the initial child was 30" (Slobodkin 1961, p. 54). Under what conditions is this not true? Estimate the *r* value for populations with these two types of reproductive schedules. In making these estimates, assume for simplicity that the female survivorship is constant at $l_x = 0.980$ for the reproductive ages (approximately true for U.S. females, 1990).

6. The life table and the seed production of the winter annual plant *Collinsia verna* for 1983–1984 was as follows (Kalisz 1991):

Life cycle	Age Interval (mo)	Number Alive (n_x)	Average No. Seeds Produced per Plant (b_x)
Seed	0–5	23,061	0.0
Seedling	5–7	6,019	0.0
Overwintering plants	7–12	4,617	0.0
Flowering plants	12–13	2,612	0.0
Fruiting plants	13–14	692	10.754

Calculate the net reproductive rate for these plants and discuss the biological interpretation of this rate.

7. Calculate a complete life table for the data in Table 11.2.

8. Forest ecologists usually measure the *size* structure of a forest and less often

make use of the annual rings of temperate-zone trees to get the age structure of the forest. What might one learn from determining age structure in addition to size structure in a forest stand?

9. The population of the United States was growing in 1989 at 0.6 percent per year (not counting immigration), yet Table 11.3 shows a net reproductive rate less than 1.0. How can this happen?

10. Can the reproductive value of males and females at a given age differ? Discuss the data presented on red deer in Clutton-Brock et al. (1982, p. 154) as an example.

11. Read Lowe's (1969) analysis of the life table of red deer on Rhum and Caughley's (1977, p. 118) description of the *age-distribution tautology*. Discuss how to avoid this error.

Overview Question

A life table and a fertility table are available for a species of threatened plant. If you were in charge of a management plan for this plant species, what could you conclude from these two tables, and what further information on its abundance would you like to have?

12 Population Growth

Population growth is a central process of ecology. No population goes on growing forever, and this leads us to the problem of *population regulation* (Chapter 16). Species interactions like *predation*, *competition*, and *herbivory* affect population growth (Chapters 13 to 15). Population growth produces changes in *community structure* (Chapters 22 to 25). It is thus important to understand how population growth occurs.

The demographic techniques described in Chapter 11 are useful because they permit prediction of future changes in population density in a precise manner. In this chapter, we will apply these demographic parameters to the description of population growth and explore some of the difficulties of analyzing the growth of natural populations. To illustrate their utility, we will use these methods to address a practical problem in conservation biology.

MATHEMATICAL THEORY

A population that has been released into a favorable environment will begin to increase in numbers. What form will this increase take, and how can we describe it mathematically? Let us start by considering a simple case in which generations are separate, as in univoltine insects (one generation per year) or annual plants.

Discrete Generations

Consider a species with a single annual breeding season and a life span of 1 year. Let each female produce R_0 female offspring, on the average, which survive to breed in the following year. Then

$$N_{t+1} = R_0 N_t$$

where

N_t = population size of females at generation t

N_{t+1} = population size of females at generation $t + 1$

R_0 = net reproductive rate, or number of female offspring produced per female per generation

What happens to this population will very much depend on the value of R_0. Consider two cases:

1. *Multiplication rate constant.* Let R_0 be a constant. If $R_0 > 1$, the population increases geometrically without limit, and if $R_0 < 1$, the population decreases to extinction. For example, let $R_0 = 1.5$ and $N_t = 10$ when $t = 0$:

$$N_{t+1} = 1.5N_t$$

Generation	Population Size (N_t)
0	10
1	15 = (1.5)(10)
2	22.5 = (1.5)(15)
3	33.75 = (1.5)(22.5)

Figure 12.1 shows some examples of geometric population growth with different R_0 values.

2. *Multiplication rate dependent on population size.* Populations do not normally grow with a constant multiplication rate as in Figure 12.1. If we look at the trajectory of a species' population through time, we observe a variety of dynamics, including populations that fluctuate little, others that fluctuate in a chaotic manner, and still others that fluctuate in cycles. How can we explain this variety of dynamical behavior?

The simplest way is to assume that the multiplication rate changes as population density rises and falls. At high densities, birth rates will decrease or death rates will

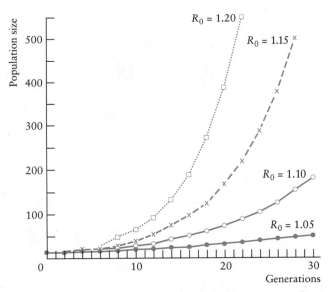

Figure 12.1 *Geometric population growth, discrete generations, reproductive rate constant. Starting population size = 10.*

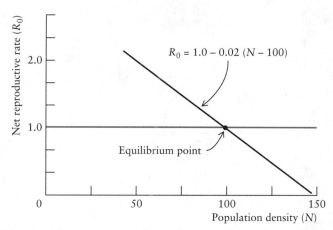

Figure 12.2 *Net reproductive rate as a linear function of population density. In this hypothetical example, equilibrium density is 100.*

increase from a variety of causes, such as food shortage or epidemic disease. At low densities birth rates will be high and losses from diseases and natural enemies low. We need to express the way in which the multiplication rate slows down as density increases. The simplest mathematical model is linear: Assume that there is a straight-line relationship between the density and multiplication rate so that the higher the density, the lower the multiplication rate (Figure 12.2). The point where the line crosses $R_0 = 1.0$ is a point of equilibrium in population density where the birth rate equals the death rate. It is convenient to measure population density in terms of deviations from this equilibrium density:

$$z = N - N_{eq}$$

where

z = deviation from equilibrium density
N = observed population size
N_{eq} = equilibrium population size (where $R_0 = 1.0$)

The equation of the straight line shown in Figure 12.2 is thus

$$R_0 = 1.0 - B(N - N_{eq})$$
$$= 1.0 - Bz$$

where

$(-)B$ = slope of line
R_0 = net reproductive rate

In Figure 12.2, $B = 0.02$ and $N_{eq} = 100$. Our basic equation can now be written

$$N_{t+1} = R_0 N_t$$
$$= (1.0 - Bz_t)N_t$$

The properties of this equation depend on the equilibrium density and the slope of the line. Let us work out a few examples to illustrate this. Consider first a simple example in which $B = 0.011$ and $N_{eq} = 100$. Start a population at $N_0 = 10$:

$$N_1 = [1.0 - 0.011(10 - 100)]10$$
$$= (1.99)(10) = 19.9$$
$$N_2 = [1.0 - 0.011(19.9 - 100)]19.9$$
$$= (1.881)(19.9) = 37.4$$

Similarly,

$$N_3 = 63.1$$
$$N_4 = 88.7$$
$$N_5 = 99.7$$

and the population density converges smoothly toward the equilibrium point of 100. A second example is worked out in Table 12.1, and three additional examples are plotted in Figure 12.3.

The behavior of this simple population model is very surprising because it generates many different patterns of population changes. If we define $L = BN_{eq}$, then

1. If L is between 0 and 1, the population approaches the equilibrium without oscillations.

2. If L is between 1 and 2, there are oscillations of decreasing amplitude to the equilibrium point (*convergent oscillations*). (Figure 12.3a)

3. If L is between 2 and 2.57, there are stable cyclic oscillations, which continue indefinitely. (Figure 12.3b)

4. If L is above 2.57, the population fluctuates in a chaotic manner, showing what appear to be random changes, depending on the starting conditions (May 1974a, Maynard Smith 1968). (Figure 12.3c)

The fact that such a simple population model can produce such a diversity of population growth trajectories is one of the most surprising results found by mathemat-

---

Table 12.1   Hypothetical Population Growth, Discrete Generations, Net Reproductive Rate a Linear Function of Density.[a]

General Formula: $N_{t+1} = [1.0 - 0.025(N_t - 100)]N_t$
$$N_1 = [1.0 - 0.025(50 - 100)]50$$
$$= (2.25)(50) = 112.5$$
$$N_2 = [1.0 - 0.025(112.5 - 100)]112.5$$
$$= (0.6875)(112.5) = 77.34$$
$$N_3 = [1.0 - 0.025(77.34 - 100)]77.34$$
$$= (1.5665)(77.34) = 121.15$$
$$N_4 = [1.0 - 0.025(121.15 - 100)]121.15$$
$$= (0.4712)(121.15) = 57.09$$

Similarly,

$$N_5 = 118.33$$
$$N_6 = 64.09$$
$$N_7 = 121.63$$
$$N_8 = 55.80$$

The population oscillates in a stable pattern.

[a] Assume that $B = 0.025$, $N_{eq} = 100$, and the starting density is 50.

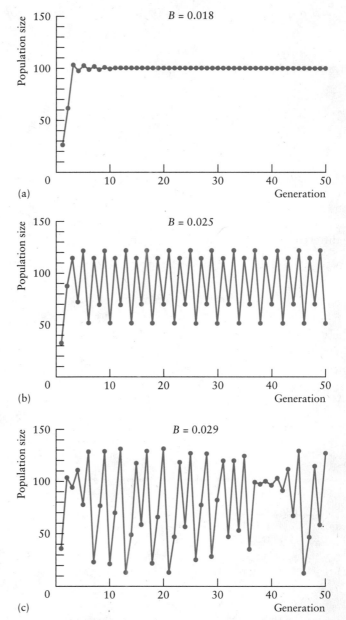

Figure 12.3   Examples of population growth with discrete generations and multi-plication rate as a linear function of population density. Starting density is 10 and equilibrium density is 100. Three examples are shown with different slopes. (a) B = 0.018; the population shows convergent oscillations to asymptotic density. (b) B = 0.025 the population oscillates in a two-generation cycle. (c) B = 0.029; the population fluctuates chaotically in an irregular pattern.

ical ecologists in this century. This model, in which the net reproductive rate decreases in a linear way with density, is the discrete-generation version of the *logistic
equation* described in the next section.

Overlapping Generations

In populations that have overlapping generations and a prolonged or continuous
breeding season, we can describe population growth more easily by the use of
differential equations. As earlier, we shall assume for the moment that the growth
of the population at time t depends only on conditions at that time and not on past
events of any kind.

1. *Multiplication rate constant.* Assume that, in any short time interval dt, an
individual has the probability $b\,dt$ of giving rise to another individual. In the same
time interval, it has the probability $d\,dt$ of dying. If b and d are instantaneous rates*
of birth and death, the instantaneous rate of population growth per capita will be

Instantaneous rate of population growth $= r = b - d$

and the form of the population increase is given by

$$\frac{dN}{dt} = rN = (b - d)N$$

where

N = population size
t = time
r = per-capita rate of population growth
b = instantaneous birth rate
d = instantaneous death rate

This is the curve of geometric increase in an unlimited environment that we discussed
with regard to the intrinsic capacity for increase (Chapter 11).

Note that we can use the geometric growth model to estimate the doubling time
for a population growing at a certain rate:

$$\frac{N_t}{N_0} = e^{rt}$$

But if the population doubles, $N_t/N_0 = 2$. Thus

$$2.0 = e^{rt}$$

or

$$\log_e(2.0) = rt$$

$$\frac{0.69315}{r} = t$$

where

t = time for population to double its size
r = rate of population growth per capita

---

* See Appendix III.

A few values for this relationship are given for illustration:

r	t
0.01	69.3
0.02	34.7
0.03	23.1
0.04	17.3
0.05	13.9
0.06	11.6

Thus if a human population is increasing at an instantaneous rate of 0.03 per year (finite rate = 1.0305), its doubling time would be about 23 years, if geometric increase prevails.

2. Multiplication rate dependent on population size. But populations do not show continuous geometric increase. When a population is growing in a limited space, the density gradually rises until eventually the presence of other organisms reduces the fertility and longevity of the population. This reduces the rate of increase of the population until eventually the population ceases to grow. The growth curve defined by such a population is *sigmoid*, or S-shaped (Figure 12.4). The S-shaped curve differs from the geometric curve in two ways: It has an upper asymptote (that is, the curve does not exceed a certain maximal level), and it approaches this asymptote smoothly, not abruptly.

The simplest way to produce an S-shaped curve is to introduce into our geometric equation a term that will reduce the rate of increase as the population builds up. We also want to reduce the rate of increase in a smooth manner. We can do this by making each individual added to the population reduce the rate of increase an equal amount. This produces the equation

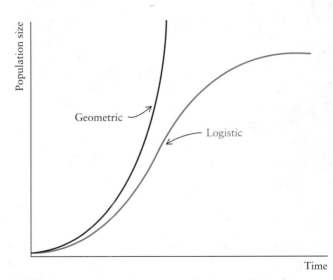

Figure 12.4   Population growth: geometric growth in an unlimited environment and logistic (sigmoid) growth in a limited environment.

$$\frac{dN}{dt} = rN \left(\frac{K - N}{K} \right)$$

where

N = population size
t = time
r = rate of population growth per capita
K = upper asymptote or maximal value of N

This equation states that

$$\begin{pmatrix} \text{Rate of} \\ \text{increase of} \\ \text{population} \\ \text{per unit time} \end{pmatrix} = \begin{pmatrix} \text{Rate of} \\ \text{population} \\ \text{growth} \\ \text{per capita} \end{pmatrix} \times \begin{pmatrix} \text{population} \\ \text{size} \end{pmatrix} \times \begin{pmatrix} \text{unutilized} \\ \text{opportunity for} \\ \text{population} \\ \text{growth} \end{pmatrix}$$

and is the differential form of the equation for the *logistic curve*. This curve was first suggested to describe the growth of human populations by Verhulst in 1838. The same equation was independently derived by Pearl and Reed (1920) as a description of the growth of the population of the United States.

Note that r is the rate of population growth per individual in the population. It is the same as the capacity for increase discussed in Chapter 11.

The integral form of the logistic equation can be written as follows:

$$N_t = \frac{K}{1 + e^{a - rt}}$$

where

N_t = population size at time t
t = time
K = maximal value of N
e = 2.71828 (base of natural logarithms)
a = a constant of integration defining the position of the curve relative to the origin
r = rate of population growth per capita

Let us look for a minute at the factor $(K - N)/K$, which has been called the "unutilized opportunity for population growth." To demonstrate that this factor does put the brakes on the basic geometric growth pattern, we consider a situation like this:

K = 100
r = 1.0
N_0 = 1.0 (starting density)

Very early in population growth, there is little difference between the curves for the logistic and the geometric equations (Figure 12.4). As we approach the middle segment of the curve, they begin to diverge more. As we approach the upper limit, the curves diverge much farther, and when we reach the upper limit, the population stops growing because $(K - N)/K$ becomes zero. The following calculations demonstrate this pattern:

r	Population Size (N)	Unutilized Opportunity For Population Growth [(K − N)/K]	Rate of Population Growth (dN/dt)
1.0	1	99/100	0.99
1.0	50	50/100	25.00
1.0	75	25/100	18.75
1.0	95	5/100	4.75
1.0	99	1/100	0.99
1.0	100	0/100	0.00

Note that the addition of one animal has the same effect on the rate of population growth at the low and at the high ends of the curve (in this example, 1/100).

The logistic equation can be written in yet another way by rearranging terms:

$$\log_e\left(\frac{K - N}{N}\right) = a - rt$$

This is the equation of a straight line in which the y coordinate is $\log_e[(K - N)/N]$, the x coordinate is time, and the slope of the line is r. This relationship can be used to fit a logistic equation to actual biological data (Pearl 1930).

Two attributes of the logistic curve makes it attractive: its mathematical simplicity and its apparent reality. The differential form of the logistic curve contains only two constants, r and K. Both these mathematical symbols can be translated into biological terms. The constant r is the per capita (or per individual) rate of population growth. It seems reasonable to attribute to K a biological meaning—the density at which the space being studied becomes "saturated" with organisms.

There are two ways of viewing the logistic curve. One is to view it as an empirical description of how populations tend to grow in numbers when conditions are initially favorable. This is the more general, more flexible viewpoint. The other way is to view the logistic as an implicit strict theory of population growth, as a "law" of population growth. The logistic curve was proposed as a strict theory of population growth, and we shall examine this theory.

Does the logistic curve fit the facts? There are two ways to test this question:

1. A colony of organisms may be reared in a constant space with a constant supply of food. From this information, a logistic curve can be calculated, and we can look to see if the data fit the curve.

2. The several assumptions on which the logistic theory rests may be examined separately and studied by experimental methods.

We shall now look into both of these approaches.

LABORATORY TESTS OF THE LOGISTIC THEORY

Many populations have been followed in the laboratory as they increase in size. Let us consider relatively simple organisms first. Gause (1934) studied the growth of populations of *Paramecium aurelia* and *P. caudatum*. He used 20 *Paramecium* to begin his experiments in a tube with 5 cubic centimeters of a salt solution buffered to pH 8. Each day Gause added a constant quantity of bacteria, which served as food. The bacteria could not multiply in the salt solution. The cultures were incu-

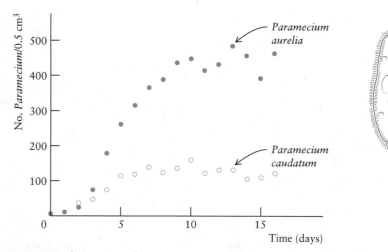

Figure 12.5 *Population growth in the protozoans* Paramecium aurelia *and* P. caudatum *at 26°C in buffered Osterhout's medium, pH 8.0, "one-loop" concentration of bacterial food. (Data form Gause 1934.)*

bated at 26°C, and every second day they were washed with fresh salt solution to remove any waste products. Therefore, Gause had a *constant environment in limited space*; the temperature, volume, and chemical composition of the medium were constant, waste products were removed frequently, and food was added in uniform amounts each day. The growth of some of Gause's *Paramecium* populations is shown in Figure 12.5. In general, the fit of these data to the logistic curve was quite good. The asymptotic density (K) was approximately 448 per 0.5 cubic centimeter for *P. aurelia* and 128 per 0.5 cubic centimeter for *P. caudatum* under these conditions.

Carlson (1913) grew yeast in laboratory cultures, and Pearl (1927) calculated logistic curves for his data. These yeast data give a very good fit to the logistic equation (Figure 12.6), and we can use them to investigate one alternative form of the logistic:

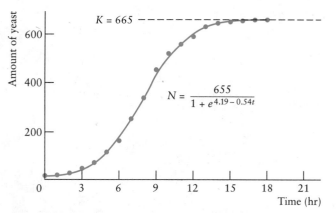

$$N = \frac{655}{1 + e^{4.19 - 0.54t}}$$

Figure 12.6 *Growth of a population of yeast cells. (After Pearl 1927; data from Carlson 1913.)*

$$\log_e\left(\frac{K - N}{N}\right) = a - rt$$

An approximate estimate of the parameters r and a of the logistic can be obtained from this equation as follows. The asymptotic density is estimated by eye or by taking the mean of some of the data points that appear to be at the equilibrium density. Once you have estimated K, you can obtain the $(K - N)/N$ term of the equation. Table 12.2 gives the raw data for Carlson's yeast, and Figure 12.7 plots these in the linear form. The slope of this line is an approximate estimate of the rate of population growth (r), and the y intercept is an estimate of a. More detailed information on the methods of fitting the logistic curve to actual data is given in Pearl (1930).

Populations of organisms with more complex life cycles may also increase in an S-shaped curve. Pearl (1927) fitted a logistic curve to the growth of *Drosophila melanogaster* in laboratory populations that he maintained in bottles with yeast as food. The fit of the data was fairly good (Figure 12.8), and Pearl ushered in the "logistic era," in which he proclaimed the logistic curve to be the universal law of population growth. But Sang (1950) criticized the application of the logistic to *Drosophila* populations and pointed out that there were complexities in the *Drosophila* cultures that Pearl did not recognize. First, the yeast that was the source of food was not constant but was itself a growing population. Hence the flies did not receive a constant amount of food. Also, the composition of the yeasts varied as the cultures aged. Second, the fruit fly has several stages in its life cycle, and it is not

Table 12.2 Growth of a Yeast Population[a]

Hours (t)	Amount of Yeast (N)	$\dfrac{K - N}{N}$	$\log_e \dfrac{K - N}{N}$
0	9.6	68.270	4.223
1	18.3	35.340	3.565
2	29.0	21.930	3.088
3	47.2	13.090	2.572
4	71.1	8.353	2.123
5	119.1	4.584	1.522
6	174.6	2.809	1.033
7	257.3	1.585	0.460
8	350.7	0.896	−0.110
9	441.0	0.508	−0.677
10	513.3	0.296	−1.219
11	559.7	0.188	−1.671
12	594.8	0.118	−2.137
13	629.4	0.056	−2.872
14	640.8	0.038	−3.276
15	651.1	0.021	−3.847
16	655.9	0.014	−4.278
17	659.6	0.008	−4.805
18	661.8	0.005	−5.332

[a] These data are plotted in Figures 12.6 and 12.7. K is 665.
Source: Data of Carlson (1913), after Pearl (1927).

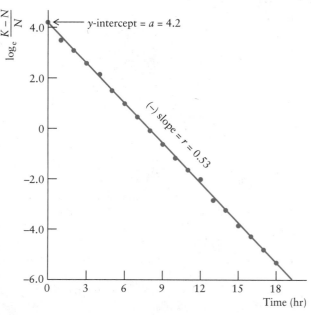

Figure 12.7   *Logistic growth of a yeast population. Data plotted in the linear form of the logistic:* $\log_e[(K - N)/N] = a - rt.$ *(See Table 12.2.)*

clear just what we should use to measure "population size." Pearl counted only the adult flies, but the adults and larvae to some extent feed on the same thing.

Beetles that live in flour (*Tribolium*) and wheat (*Calandra*) have been used very often for experimental population studies. These beetles are preferable to *Drosophila* because, even though they have as complex a life cycle (involving eggs—larvae—pupae—adults), they live in a dead food medium that can be precisely controlled. Chapman (1928), one of the first to use *Tribolium* for laboratory studies in ecology,

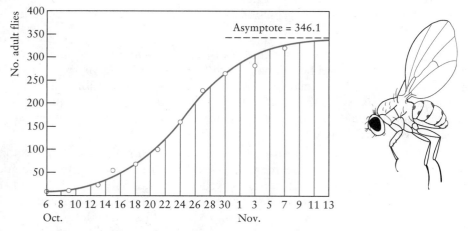

Figure 12.8   *Growth of an experimental population of the fruit fly* Drosophila melanogaster. *The circles are observed census counts, and the smooth curve is the fitted logistic. (After Pearl 1927.)*

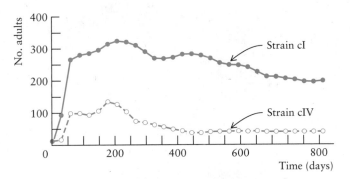

Figure 12.9   Population growth of the flour beetle Tribolium castaneum *at 29°C and 70 percent relative humidity in 8 grams of flour. There is considerable variation in population growth among different genetic strains. (Data from Park et al. 1964.)*

found that colonies of these beetles grew in a logistic fashion. Most workers stopped their cultures as soon as they reached the upper asymptote. Thomas Park, however, reared populations of *Tribolium* for several years and obtained the results shown in Figure 12.9. The upper asymptote of the logistic is imaginary—the density does not stabilize after the initial sigmoid increase but rather shows a long-term decline. Similar studies have been done by Birch (1953b) on *Calandra oryzae*, and he found logistic growth initially followed by large fluctuations in density with no indication of stabilization around an asymptote.

One important point to note here is that these populations of a single species of beetle living in a constant climate with constant food supply show wide fluctuations in numbers. These fluctuations are brought about by the influence of the animals on each other completely independent of any fluctuation in temperature, food, predators, or disease. No cases have as yet been demonstrated where the population of any organism with a complex life history comes to a steady state at the upper asymptote of the logistic curve.

What assumptions are inherent in the logistic equation? When we say that the growth of a population may be represented by a logistic curve, we imply the following four facts about that population:

1. *The population has a stable age distribution initially.* The logistic model assumes that a population beginning growth (when $[(K - N)/K]$ is very nearly 1.0) increases at a rate approximately equal to rN. But r is only realized as a rate of population increase when there is a stable age distribution. Thus all experiments on logistic growth should be started with the population in an approximate stable age structure. Few studies have taken this problem into account.

2. *The density has been measured in appropriate units.* We have already noted the difficulty of deciding whether to include only *Drosophila* adults in the population or to include eggs, larvae, and pupae as well. An additional problem arises here: Many plants and animals are smaller in size when they are raised in crowded situations. For example, with flies, we may be adding large flies at the start of population growth and small flies near the end. In these situations, it may be more accurate to measure biomass.

3. The relationship between density and the rate of increase is linear. This can be seen by rewriting the logistic equation:

$$\frac{dN}{dt}\frac{1}{N} = r - \frac{r}{K}N$$

which says that the rate of population increase per individual is a linear function of population density. Few direct experiments have been done to test this assumption, which is probably violated in many growing populations. Smith (1963) forced *Daphnia magna* populations to grow at certain predetermined rates and then measured the density achieved. The relation between rate of population growth and density was not linear in this case when he used either numbers as a measure of density or biomass (dry weight) as a measure.

4. *The depressive influence of density on the rate of increase operates instantaneously without any time lags.* It is highly unlikely that in organisms with complex life cycles the rate of population increase could respond instantaneously to changes in density because of the time lags built into every life history. For example, it may take from a week to several months for an insect larva to become an adult. With simple forms, such as *Paramecium* or bacteria, this assumption should be approximately true.

It is one of the curious facts about population ecology that interest has historically focused on the logistic model for overlapping generations and largely overlooked the simpler discrete models with their much richer dynamic behavior (May 1981). Insects comprise a large fraction of animal species and many insects have nonoverlapping generations which are described well by the simpler discrete models. This change of focus away from the logistic equation has been highlighted by data on laboratory populations and shown even more graphically by data on field populations.

FIELD DATA ON POPULATION GROWTH

Population growth does not occur continuously in field populations. Many species in seasonal environments show population growth during the favorable season each year. Long-lived organisms may show population growth only rarely, and few populations fill up a vacant habitat in nature the way they do in the laboratory. Some examples we have are from situations where animals were introduced onto islands or other new habitats and were then studied as they increased in numbers.

Reindeer have been introduced into many parts of Alaska since 1891 to replace the dwindling caribou herds in the economy of the Eskimo. In 1911 reindeer were introduced onto two of the Pribilof Islands in the Bering Sea off Alaska. Four males and 21 females were released on St. Paul Island (106 km^2) and 3 males and 12 females on St. George Island (90 km^2). The stockings were an immediate success. The subsequent history of these herds is of interest because the islands were completely undisturbed environments—there was little hunting pressure, and there were no predators. The two herds have had quite different histories on the two islands (Figure 12.10). The St. George herd reached a low ceiling of 222 reindeer in 1922 and then subsided to a small herd of 40 to 60 animals. The St. Paul herd grew

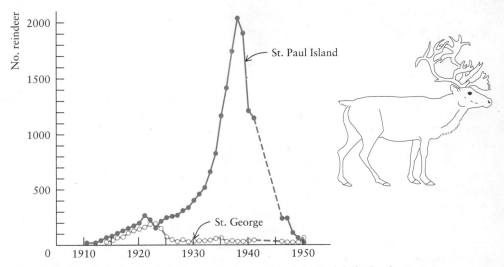

Figure 12.10   *Reindeer population growth on two of the Pribilof Islands, Bering Sea, from 1911, when they were introduced, until 1950. (After Scheffer 1951.)*

continuously to about 2000 reindeer in 1938, overgrazed the habitat, and then abruptly declined to only 8 animals in 1950. The ecological differences between these two islands appear to have been very slight (they had the same type of vegetation and the same climate), and no one understands why the two populations behaved so differently (Scheffer 1951). It is possible that illegal hunting on St. George Island was the cause of the differences shown in Figure 12.10.

Reindeer were introduced to St. Matthew Island (332 km²) in the Bering Sea in 1944. They increased from an initial 29 animals (24 females, 5 males) in 1944 to 1350 in 1957, to 6000 in 1963, and then crashed to 42 animals in 1966 (Klein 1968), thus repeating the St. Paul Island sequence in a slightly shorter time. In both these cases, winter food shortage, particularly of lichens, has been the major cause of the dramatic population crashes. Reindeer populations on islands seem to be a dramatic example of population growth according to the discrete population model when L is large (see Figure 12.3c).

The whooping crane (*Grus americana*) is a good example of an endangered species now recovering from the brink of extinction. Only 47 whooping cranes survived when this species was first protected in 1916, and only 15 birds were still alive in 1941. The whooping crane breeds in the Northwest Territories of Canada and migrates to overwinter on the Texas coast at the Aransas National Wildlife Refuge. The entire population has been counted on the wintering grounds in Texas since 1938, and the population growth curve is shown in Figure 12.11. Population growth has been irregular in the whooping crane. Binkley and Miller (1983) found that the rate of increase (r) changed around 1956 and the population began to recover more rapidly after 1956 than before. There is also a ten-year cycle superimposed on the population growth curve, possibly a spinoff of predation from the ten-year cycle of snowshoe hares in the breeding areas (Nedelman et al. 1987).

Many organisms show strong annual fluctuations in density, and thus the pattern of population growth can be observed once a year. The diatom *Asterionella formosa* shows a spring maximum in numbers in lakes of northwestern England (Lund 1950).

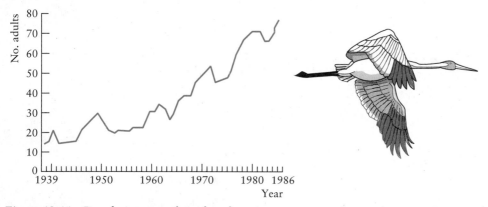

Figure 12.11   Population growth in the whooping crane, an endangered species that has recovered from near extinction in 1940. Counts of adults were made annually on the wintering grounds at Aransas, Texas. (Data from Binkley and Miller 1983, Nedelman et al. 1987 and E. Kuyt, personal communication.)

These populations increase in a general sigmoid manner (Figure 12.12) but then decline rapidly, possibly because they deplete the supply of silica.

   Very often, field data on population growth are too rough to show definitely whether or not the logistic curve is a good representation of the data. The cases we have illustrated here suggest that the logistic is only an approximate model to describe field population increases.

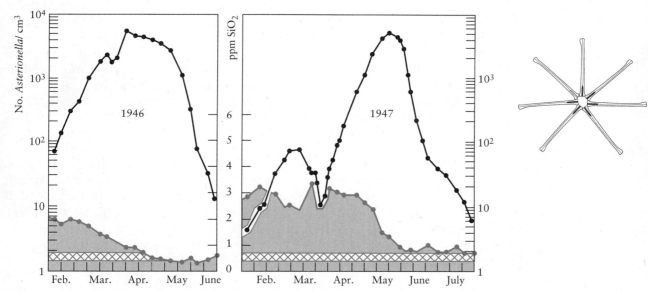

Figure 12.12   Population growth of the diatom Asterionella formosa *in the spring, Blelham Tarn, English Lake District, 1946–1947. The top line represents the number of live* Asterionella *cells per cubic centimeter in water from 0 to 5 meters deep. The solid area at the bottom represents the dissolved silica content (mg/L) of the water and the hatched zone is a lower threshold for silica content needed by diatoms. Silica may be a factor that limits population growth. (After Lund 1950.)*

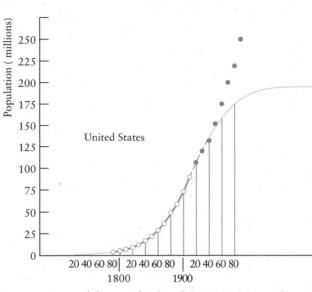

Figure 12.13 *Census counts of the population of the United States from 1790 to 1990. The smooth curve is the logistic equation fitted to the census counts from 1790 to 1910 inclusive. The broken lines show the extrapolation of the curve beyond the data to which it was fitted. In mid-1991 the U.S. population was 253 million. (After Pearl et al. 1940; census data for 1950-1990 added.)*

The logistic curve was used by Pearl and Reed (1920) to predict the future growth of the U.S. population. They fitted the logistic curve to the census data from 1790 to 1910 and projected it to asymptotic density, a value of 197 million to be reached around the year 2060 (Figure 12.13). The census data for 1920 to 1940 fit the curve very well (Pearl et al. 1940), but subsequent census data show a nearly geometric increase rather than a logistic one. The predicted asymptote of 197 million was, in fact, reached in 1968, and current estimates for the U.S. population in 2025 range from 260 to 357 million (*Statistical Abstract of the United States, 1991*).

We conclude from this analysis that population growth may sometimes be sigmoid in natural populations and thus fit the logistic model, but often it is not. The asymptotic stable density of the logistic is almost never achieved by natural populations, hence the logistic model has serious drawbacks as a general model of population growth. What can be done about this? Work on population growth models has proceeded along three lines. One has been to analyze the effect of time lags on the logistic model, since this is the assumption most clearly at odds with the biological realities of complex organisms. The second line has been to construct probabilistic (stochastic) models of population growth. The third approach has been to use more specific models based on age or size to project population changes (Leslie matrix models). Let us look briefly at these three approaches.

TIME-LAG MODELS OF POPULATION GROWTH

Consider a simple model with discrete generations, and assume that the reproductive rate at generation t depends on density in a linear manner but that, instead of depending on density at generation t (as in Figure 12.2), it depends on density at the generation past $(t - 1)$. Measure density as a deviation from the equilibrium point:

$$z = N - N_{eq}$$

where

z = deviation from equilibrium density
N = observed population size
N_{eq} = equilibrium population size (where $R_0 = 1.0$)

The reproductive rate is described by Figure 12.2 as a straight line, $R_0 = 1.0 - Bz$. The population growth model can thus be written

$$N_{t+1} = R_0 N_t$$
$$= (1 - Bz_{t-1})N_t$$

which is similar to the preceding treatment except that the reproductive rate is now defined by the density of the previous generation. The properties of this equation depend on the equilibrium density and the slope of the line. Let us work out the hypothetical case without any time lag:

$$B = 0.011$$
$$N_{eq} = 100$$

Start a population at $N_0 = 10$ (and use $N = 10$ for first-generation calculation of the time-lag term):

$$N_1 = [1.0 - 0.011(10 - 100)]10$$
$$= 19.9$$
$$N_2 = [1.0 - 0.011(10 - 100)]19.9$$
$$= 39.6$$
$$N_3 = [10 - 0.011(19.9 - 100)]39.6$$
$$= 74.4$$

Similarly,

$$N_4 = 123.9$$
$$N_5 = 158.7$$

This population oscillates more or less regularly with a period of six to seven generations between peaks in numbers, in contrast to the smooth approach to equilibrium density that occurred when there was no time lag in regulation. This contrast is shown in Figure 12.14 for a hypothetical example. A delay in feedback by one generation can change a stable population growth pattern into an unstable one. Maynard Smith (1968, p. 25) has shown that, assuming $L = BN_{eq}$,

If $0 < L < 0.25$, stable equilibrium, no oscillation
If $0.25 < L < 1.0$, convergent oscillation
If $L > 1.0$, stable cycles or divergent oscillation

Compare the results of this time-lag model with those obtained earlier without any time lags.

Laboratory populations of *Daphnia* are a good example of the effect of time lags on population growth. Pratt (1943) followed the development of *Daphnia* populations in the laboratory at two temperatures. The populations, in 50 cubic centimeters of filtered pond water, were started with two parthenogenetic females

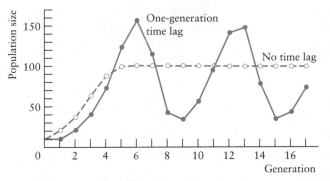

Figure 12.14 Hypothetical population growth with and without a time lag, discrete generations, reproductive rate a linear function of density. Starting density = 10, slope of reproductive curve = 0.011, equilibrium density = 100.

each. *Daphnia* were counted every two days and transferred to a fresh culture. The only food used was a green alga, *Chlorella*. Populations at 25°C showed oscillations in numbers (Figure 12.15), whereas those at 18°C were approximately stable. Oscillations occurred at 25°C because there was a delay in the depressing effect of population density on birth rates and death rates. At 25°C, the birth rate is affected first by rising density, and only later is the death rate increased. This causes the *Daphnia* population to continuously overshoot and then undershoot its equilibrium density. Note that these oscillations are intrinsic to the biological system and not caused by external environmental changes.

The biological mechanisms in *Daphnia* that account for these time lags are now well understood (Goulden and Hornig 1980). *Daphnia* store energy in the form of oil droplets, mainly as triacylglycerols, when food is superabundant. They use these energy reserves once the food supply has collapsed, so the effects of low food supply are not instantaneous but delayed. Females can thus continue to give birth even after food has become scarce. After these energy reserves are metabolized, *Daphnia* starve and die, producing the oscillations shown in Figure 12.15.

So we see that the introduction of time lags into simple models of population growth permits three possible alternatives: (1) a converging oscillation toward equilibrium, (2) a stable oscillation about the equilibrium level, or (3) a smooth approach to equilibrium density. In addition, some configurations of time lags will produce a divergent oscillation that is unstable and leads to extinction of the population. These outcomes are clearly more realistic models of what seems to occur in natural populations.

STOCHASTIC MODELS OF POPULATION GROWTH

The models we have discussed so far are *deterministic* models, which means that given certain initial conditions, the model predicts one exact outcome. But biological systems are probabilistic, not deterministic. Thus we speak of the probability that a female will have a litter in the next unit of time, or the probability that there will be a cone crop in a given year, or the probability that a predator will kill a certain number of animals within the next month. Population trends are therefore the joint outcome of many individual probabilities like this, which has led to the development of probabilistic, or *stochastic*, models.

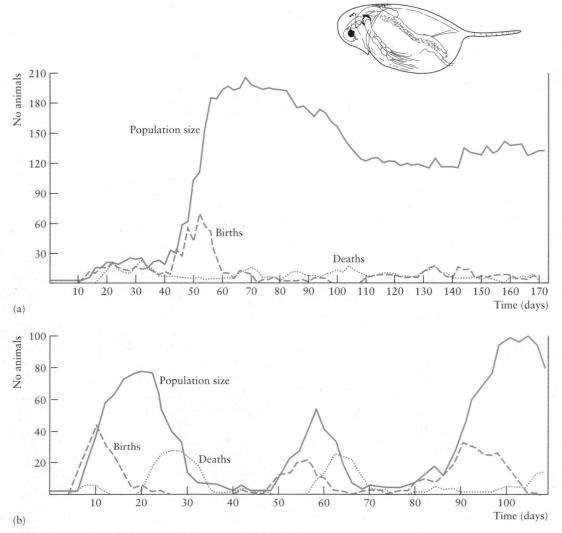

Figure 12.15   *Population growth in the water flea* Daphnia magna *in 50 cc of pond water (a) at 18°C and (b) at 25°C. The number of births and deaths have been doubled to make them visible. (After Pratt 1943.)*

We can illustrate the basic nature of stochastic models very simply. Consider the geometric growth equation, a deterministic model we developed earlier for discrete generations:

$$N_{t+1} = R_0 N_t$$

Consider an example in which the net reproductive rate (R_0) is 2.0 and the starting density is 6:

$$N_{t+1} = (2.0)(6) = 12$$

The deterministic model predicts a population size of 12 at generation 1. A stochastic model for this could be constructed as follows. Assume the probabilities of reproduction are

	Probability
One female offspring	0.5
Three female offspring	0.5

Clearly, on the average, two female parents will leave four offspring, so $R_0 = 2.0$. But now use a coin to construct some numerical examples. If the coin flips heads (h), one offspring is produced; if tails (t), three offspring:

	OUTCOME			
Parent	**Trial 1**	**Trial 2**	**Trial 3**	**Trial 4**
1	(h) 1	(t) 3	(h) 1	(t) 3
2	(t) 3	(h) 1	(t) 3	(h) 1
3	(h) 1	(t) 3	(h) 1	(h) 1
4	(t) 3	(t) 3	(t) 3	(t) 3
5	(t) 3	(t) 3	(t) 3	(h) 1
6	(t) 3	(t) 3	(h) 1	(h) 1
Total population in next generation	14	16	12	10

Some of the outcomes are above the expected value of 12, and some are below it. If we continued doing this many times, we could generate a frequency distribution of population sizes for this simple problem; an example is shown in Figure 12.16. Note that populations starting from exactly the same point with exactly the same biological parameters could, in fact, finish one generation later with either 6 or 18 members.

The population growth of species with overlapping generations can also be

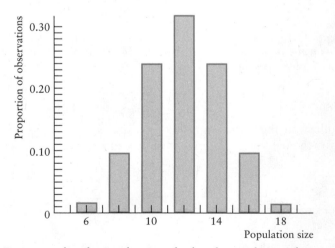

Figure 12.16 Frequency distribution for size of a female population after one generation for the example discussed in the text. $N_0 = 6$, $R_0 = 2.0$, *probability of having one female offspring = 0.5, probability of having three female offspring = 0.5.*

described by stochastic models. Geometric growth in this case follows the differential equation

$$\frac{dN}{dt} = rN$$
$$= (b - d)N$$

where

b = instantaneous birth rate
d = instantaneous death rate

In the simplest case (the pure birth process), assume that $d = 0$, so no organisms can die. If we assume a simple binary fission type of reproduction, the probability that an organism will reproduce in the next short time interval (dt) is $b\,dt$, where b is the instantaneous birth rate. Consider an example where $b = 0.5$ and $N_0 = 5$ (starting population). In one time interval, according to the deterministic model,

$$N_t = N_0 e^{rt}$$
$$N_1 = (5)e^{(0.5)(1)} = 8.244$$

For the stochastic equivalent of this simple model, we must determine from the instantaneous rate of birth:

Probability of not reproducing in one time interval $= e^{-b} = 0.6065$
Probability of reproducing at least once in one time unit $= 1.0 - e^{-b} = 0.3935$

Thus for five organisms, the chance that none of the five will reproduce in the next unit of time is

$$(0.6065)(0.6065)(0.6065)(0.6065)(0.6065) = 0.082$$

so in approximately 1 trial out of 12 there will be no population change in the unit of time ($N_1 = 5$). We could laboriously count up all the other possibilities, remembering that each individual may undergo fission more than once in each unit of time. Or we may follow a mathematician into an application of probability theory to this problem (Pielou 1969, p. 9) to reach the conclusions shown in Figure 12.17. Note again the possible variation from initial to final population size when births are considered in a probabilistic manner. Figure 12.18 illustrates these principles about stochastic models of population growth.

If we use probabilistic models and allow both births and deaths to occur in a random manner, there is a chance of a population's becoming extinct. What is the chance of extinction for a population starting with N_0 organisms and undergoing stochastic changes in size with instantaneous birth rate b and death rate d, as in Figure 12.18? Pielou (1969, p. 17) has discussed two cases:

1. *Birth rate greater than death rate.* These populations should increase geometrically but may by chance drift to extinction, particularly during the first few time periods if population size is small. The probability of extinction at some time is given by

$$\text{Probability of extinction} = \left(\frac{d}{b}\right)^{N_0}$$

as time becomes very large. For example, if $b = 0.75$ and $d = 0.25$ for $N_0 = 5$, we have

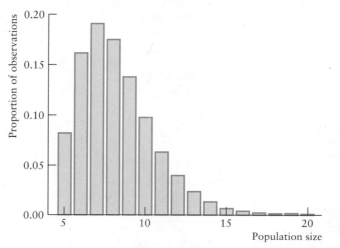

Figure 12.17   Frequency distribution of the size at time 1 of a population undergoing a pure birth process. b = 0.5, N_0 = 5. (After Pielou 1969.)

$$\text{Probability of extinction} = \left(\frac{0.25}{0.75}\right)^5 = 0.0041$$

But if $b = 0.55$, $d = 0.45$, and $N_0 = 5$,

$$\text{Probability of extinction} = 0.367$$

Thus the larger the initial population size and the greater the difference between birth and death rates, the more chance a population has of staying in existence. The effects of random fluctuations in birth and death rates on individuals is called demographic stochasticity, or demographic uncertainty (Soulé 1987). The important principle is that, even when the birth rate exceeds the death rate on average, there is a finite probability of a population going extinct.

2. *Birth rate equals death rate.* These populations are stationary in numbers, as is typical of the real world on the average, and by the formula just given,

$$\text{Probability of extinction} = \left(\frac{d}{b}\right)^{N_0} = (1.0)^{N_0} = 1.0$$

as time approaches infinity. Thus, when births equal deaths on average, extinction is a certainty for any population subject to stochastic variations in births and deaths, if we allow a long enough time span.

Stochastic models of population growth thus introduce an important idea of biological variation into the consideration of population changes. The probability approach to these ecological problems is consequently more realistic. The price we must pay for the greater realism of stochastic models is the greater difficulty of the mathematics, but part of this difficulty can be resolved by the use of computers. The variation inherent in stochastic models becomes more important as population size becomes smaller, which is pointed out clearly in human populations. Predictions about what change in size an individual family will show from one year to the next are much less certain than predictions about what change in size the world population

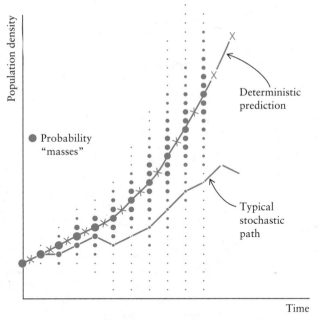

Figure 12.18   Stochastic model of geometric population growth for continuous, overlapping generations. Population predictions cannot be represented by a single value in stochastic models, and the uncertainty of the prediction grows with time. (After Skellam 1955.)

will show. If all populations were in the millions, stochastic models could be eliminated, and deterministic models would be adequate.

POPULATION PROJECTION MATRICES

One realistic way of estimating population growth was pioneered by Leslie (1945), who calculated population changes from age-specific birth and survival rates. Such an age-classified model is called a *Leslie matrix*. The essential feature of these models is that the life cycle of the plant or animal is broken down into a series of stages (Figure 12.19). Each age class is one stage in a simple Leslie matrix and this is exactly the same approach we used in Chapter 11 for life tables. Organisms pass from one stage to the next with probability P_x, and they reproduce a number of offspring F_x. In the conventional life table notation:

$$P_x = \frac{l_{x+1}}{l_x} = (1 - q_x)$$ probability that an individual of age group x will survive to enter age group $x + 1$ at the next time interval (of the life table, page 169)

$$F_x = b_x s_x =$$ number of female offspring born in one time interval per female alive aged x to $x + 1$; these offspring must survive to enter age group 0 at the next time interval

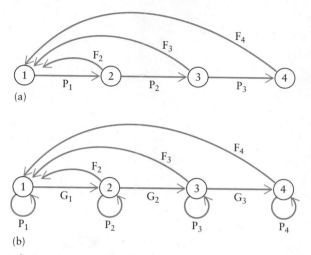

Figure 12.19 Population projection matrices. (a) The Leslie matrix, or age-classified life cycle. Four age classes are shown in this example, with different fecundities (F$_x$) for each age class and different probabilities P$_x$ of surviving from one age to the next. (b) A size- or stage-based life cycle, in which the only added complication is that an individual has a probability P$_x$ of remaining in the same life-cycle stage in one time period and a probability G$_x$ of surviving and moving on into the next stage of the life cycle. (After Caswell 1989.)

where

l_x = number of individuals alive at start of age interval x

b_x = number of births in one time interval per adult female aged x to $x + 1$

s_x = proportion of the b_x offspring that are still alive at the start of the next time interval

Begin with a population having specified age structure at time t:

N_0 = number of organisms between ages 0 and 1

N_1 = number of organisms between ages 1 and 2
(and so on to the oldest age class)

N_k = number of organisms between ages k and $k + 1$ (oldest organisms)

Time units for age are often one year but can be any fixed time unit, depending on the organism. Usually only the female population is considered.

If we assume no emigration and no immigration, the population's age structure at the next time interval is defined as follows:

new age structure
↓

Number of new organisms at
time $t + 1$ $= F_0 N_0 + F_1 N_1 + F_2 N_2 + F_3 N_3 + \ldots + F_k N_k$

$$= \sum_{x = 0}^{k} F_x N_x$$

Number of age 1 organisms at time $t + 1 = P_0 N_0$
Number of age 2 organisms at time $t + 1 = P_1 N_1$
Number of age 3 organisms at time $t + 1 = P_2 N_2$

and so on.

Leslie (1945) recognized that this problem could be cast as a simple matrix problem if one defined a transition matrix **M** as follows:

$$\mathbf{M} = \begin{bmatrix} F_0 & F_1 & F_2 & F_3 & F_4 & F_5 & . & . & F_{k-1} & F_k \\ P_0 & 0 & 0 & 0 & 0 & 0 & . & . & 0 & 0 \\ 0 & P_1 & 0 & 0 & 0 & 0 & . & . & 0 & 0 \\ 0 & 0 & P_2 & 0 & 0 & 0 & . & . & 0 & 0 \\ 0 & 0 & 0 & P_3 & 0 & 0 & . & . & 0 & 0 \\ 0 & 0 & 0 & 0 & P_4 & 0 & . & . & 0 & 0 \\ . & . & . & . & . & . & . & . & & . \\ . & . & . & . & . & . & . & . & & . \\ 0 & 0 & 0 & 0 & 0 & 0 & . & . & P_{k-1} & 0 \end{bmatrix}$$

where $F_x \geq 0$ and P_x ranges from 0 to 1. By casting the present age structure as a column vector, we get

$$\overrightarrow{N_t} = \begin{bmatrix} N_0 \\ N_1 \\ N_2 \\ N_3 \\ N_4 \\ . \\ . \\ N_k \end{bmatrix}$$

Leslie showed that the age distribution at any future time could be found by premultiplying the column vector of age structure by the transition matrix **M**:

$$\mathbf{M}N_t = N_{1+1}$$
$$\mathbf{M}N_{t+1} = N_{t+2}$$

Students who are familiar with matrix algebra will benefit from the discussion of the properties of this matrix in Leslie (1945).

Lefkovitch (1965) realized that the Leslie matrix was a special case of a more general stage-based matrix, in which life history stages replace ages. Such a stage-based or size-based model is illustrated in Figure 12.19. One new complexity is added to the age-based model: whereas all individuals of age x increase to age $x + 1$ after 1 unit of time, in a stage- or size-based model some individuals will remain in the same life cycle stage. We thus have two probabilities associated with each stage:

P_x = probability an individual will survive and remain in stage- or size-class x in the next time unit
G_x = probability an individual will survive and move up to the next stage- or size-class $x + 1$ in the next time unit

Note that we set the time unit in stage-based matrices to make it impossible for the organism to jump up two or more stages in one time step.

Table 12.3   Stage-based Life and Fecundity Tables for the Loggerhead Sea Turtle.[a]

Stage No.	Class	Size (Carapace Length) (cm)	Approximate Ages (yr)	Annual Survivorship	Fecundity (No. eggs/yr)
1	Eggs, hatchlings	<10	<1	0.6747	0
2	Small juveniles	10.1–58.0	1–7	0.7857	0
3	Large juveniles	58.1–80.0	8–15	0.6758	0
4	Subadults	80.1–87.0	16–21	0.7425	0
5	Novice breeders	>87.0	22	0.8091	127
6	First-year remigrants	>87.0	23	0.8091	4
7	Mature breeders	>87.0	24–54	0.8091	80

[a] These values assume a population declining at 3% per year.
  Source: Data from Crouse et al. (1987).

All of this seems uncomfortably abstract so let us look at one example of a size-based matrix model. Crouse and coauthors (1987) analyzed the dynamics of the loggerhead sea turtle (*Caretta caretta*), an endangered species of sea turtle from the Atlantic Ocean off the southeastern United States. Sea turtles have a long life span which can be broken down into seven stages based on size. These stages are listed in Table 12.3 along with the size and approximate age of turtles in each of the stages. Survivorship varies with size and only individuals over 87 centimeters long are sexually mature.

The population projection matrix based on this life history takes the form:

$$\begin{bmatrix} P_1 & F_2 & F_3 & F_4 & F_5 & F_6 & F_7 \\ G_1 & P_2 & 0 & 0 & 0 & 0 & 0 \\ 0 & G_2 & P_3 & 0 & 0 & 0 & 0 \\ 0 & 0 & G_3 & P_4 & 0 & 0 & 0 \\ 0 & 0 & 0 & G_4 & P_5 & 0 & 0 \\ 0 & 0 & 0 & 0 & G_5 & P_6 & 0 \\ 0 & 0 & 0 & 0 & 0 & G_6 & P_7 \end{bmatrix}$$

The best estimates of the parameters of this matrix are given in Table 12.4.

Given this model of population growth for the loggerhead sea turtle, we can ask some interesting questions about how to reverse the population decline of this endangered species. By holding all the life history parameters constant, save one, we

Table 12.4   Stage-class Population Matrix for Loggerhead Sea Turtles[a]

$$\begin{bmatrix} 0 & 0 & 0 & 0 & 4 & 127 & 80 \\ 0.6747 & 0.7370 & 0 & 0 & 0 & 0 & 0 \\ 0 & 0.0486 & 0.6610 & 0 & 0 & 0 & 0 \\ 0 & 0 & 0.0147 & 0.6907 & 0 & 0 & 0 \\ 0 & 0 & 0 & 0.0518 & 0 & 0 & 0 \\ 0 & 0 & 0 & 0 & 0.8091 & 0 & 0 \\ 0 & 0 & 0 & 0 & 0 & 0.8091 & 0.8089 \end{bmatrix}$$

[a] Estimates based on the life table presented in Table 12.3.
  Source: Data from Crouse et al. (1987).

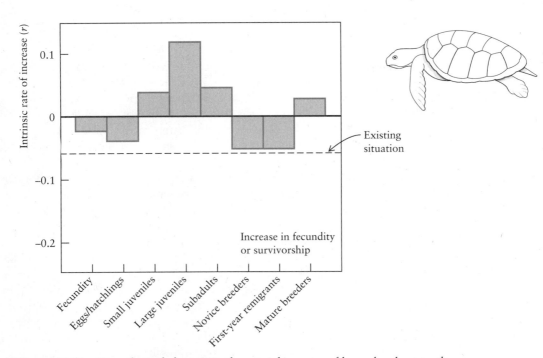

Figure 12.20   *Hypothetical changes in the rate of increase of loggerhead sea turtle populations off the southeastern United States resulting from simulated increases of 50 percent in fecundity or increases in survival to 100 percent for the stages of the life cycle. The greatest improvement for this endangered turtle would occur by improving the survival of the large juveniles. (From Crouse et al. 1987.)*

can investigate quantitatively the impact of conservation efforts. Figure 12.20 shows the results of increasing fecundity 50 percent or improving survival in each stage of the life cycle. Improving fecundity 50 percent still leaves the population declining. Maximum improvement is achieved by improving the survival of juvenile turtles. At the present time most conservation efforts on sea turtles are focused on protecting the eggs on beaches, and conservationists have found that even after 20 to 30 years of protecting nests on beaches, no increase in sea turtles has occurred (Crouse et al. 1987). In fact this is exactly what our model would predict (Figure 12.20). What is needed for conservation is an improvement of juvenile turtle survival at sea. Much juvenile loss is caused by turtles being caught in shrimp nets and drowning, and shrimp trawls are now being fitted with a device to stop the capture and drowning of sea turtles (Anonymous 1983).

Stage-based or size-based matrix models have been used extensively for plant populations in which size is a more useful measure of an individual than is age (Caswell 1989). Matrix models also permit plants to grow or to shrink in size, a useful biological assumption. The solution to matrix population growth models based on age- or size- stages is just as complex as those illustrated above for simple population growth models. Populations may increase or decrease geometrically or may show oscillations. These models assume a constant schedule of survival and reproduction and thus can be applied to natural populations only for short time periods for which this assumption is valid.

SUMMARY

The growth of a population can be described with simple mathematical models for organisms with discrete generations and for those with overlapping generations. If the multiplication rate is constant, geometric population growth occurs. Populations stabilize at a finite size only if the multiplication rate depends on population size, and large populations have lower multiplication rates than small populations. For species with overlapping generations, the logistic equation is a simple mathematical description of population growth to an asymptotic limit.

The S-shaped logistic curve is an adequate description for the laboratory population growth of *Paramecium*, yeast, and other organisms with simple life cycles. Population growth in organisms with more complex life cycles seldom follows the logistic curve very closely. In particular, the stable asymptote of the logistic curve is not achieved in natural populations; numbers fluctuate.

Three general modifications of the population growth models have been used. *Time-lag models* analyze the effects of different time lags on the curve of population growth. The introduction of time lags into the simple models of population growth can produce oscillations in population size instead of a stable asymptotic density. *Stochastic models* of population growth introduce the effects of chance events on populations. Populations starting from the same density with the same average birth and death rates may increase at different rates because of chance events. Chance events can lead to extinction as well and are particularly important in small populations. *Matrix models* of population growth can be age- or size-based and thus can be used for plants and animals. Matrix models are ideally suited to asking hypothetical questions about the contribution of specific life table parameters to population growth and exploring the consequences of alternative management plans for endangered species or for pests.

Selected References

ANDREWARTHA, H. G., and L. C. BIRCH. 1954. *The Distribution and Abundance of Animals*, chap. 9. University of Chicago Press, Chicago.

BERRYMAN, A. A. 1981. *Population Systems: A General Introduction*, chap. 2. Plenum Press, New York.

BINKLEY, C. S., and R. S. MILLER. 1983. Population characteristics of the whooping crane, *Grus americana. Canadian Journal of Zoology* 61:2768–2776.

CASWELL, H. 1989. *Matrix Population Models*. chaps 1–4. Sinauer Associates, Inc., Sunderland, Mass.

KAREIVA, P. 1989. Renewing the dialogue between theory and experiments in population ecology. In *Perspectives in Ecological Theory*, ed. J. Roughgarden, R. M. May, and S. A. Levin, pp. 68–88. Princeton University Press, Princeton, N.J.

MAY, R. M. 1981. Models for single populations. In *Theoretical Ecology*, ed. R. M. May, chap. 2. Blackwell, Oxford, England.

MAYNARD SMITH, J. 1968. *Mathematical Ideas in Biology*, chaps. 2–3. Cambridge University Press, New York.

PEARL, R. 1927. The growth of populations. *Quarterly Review of Biology.* 2:532–548.

PIELOU, E. C. 1969. *An Introduction to Mathematical Ecology*, chaps. 1–2. Wiley-Interscience, New York.

PIMM, S. L., H. L. JONES, and J. DIAMOND. 1988. On the risk of extinction. *American Naturalist* 132:757–785.

SCHAFFER, W. M., and M. KOT. 1986. Chaos in ecological systems: The coals that Newcastle forgot. *Trends in Ecology and Evolution* 1:58–63.

SILVERTOWN, J. W. 1987. *Introduction to Plant Population Ecology.* chaps. 3 and 5. John Wiley, New York.

SYMONIDES, E. 1988. Population dynamics of annual plants. In *Plant Population Ecology*, ed. A. J. Davy, M. J. Hutchings, and A. R. Watkinson, pp. 221–248. Blackwell Scientific Publications, Oxford.

Questions and Problems

1. List, for plants and animals, six reasons why the assumption (p. 203) might be incorrect that population growth at a given point in time depends only on conditions at that time and not on past events.

2. Plot the logistic data of Table 12.2 on semilog paper (logarithmic scale for population density, arithmetic scale for time). What shape of curve does this give and why?

3. Determine the population growth curve for ten generations for an annual plant with a net reproductive rate of 25 and a starting density of 2. Assume a constant reproductive rate.

4. Determine the population growth curve for this same annual plant if reproductive rate is a linear function of density of the form $R_0 = 1.0 - 0.01z$ ($z =$ deviation from equilibrium density), equilibrium density is 1000, and starting density is 2. Repeat under the assumption that there is a one-generation time lag in changing reproductive rate.

5. Determine the doubling time for the following human populations (1991 data):

Country	Instantaneous Rate of Growth (r)
Ghana	0.031
South Africa	0.024
Canada	0.006
Argentina	0.012
United Kingdom	0.002
Ireland	0.005
Hungary	−0.002

What assumptions must one make to predict these doubling times?

6. Chapman (1928) gives the following population growth data for the flour beetle *Tribolium confusum* (eggs + larvae + pupae + adults):

Days	No. in 32 g of Flour
0	2
8	47
28	192
41	256
63	768
79	896
97	1120
117	896
135	1184
154	1024

Estimate the parameters of the logistic equation for these data. Use the mean value of the last five censuses for an estimate of K. Use the logistic formula to determine the estimated density for each of the census days, and compare these estimated densities with the observed values.

7. The whooping crane may be removed from the endangered species list once it reaches a total population size of 200 birds, which should provide 50 nesting pairs. If this population is growing geometrically since 1957 at $r = 0.0416$ per year (Binkley and Miller 1983), calculate when it will be removed from the endangered list, given a population of 131 birds in 1992.

8. Discuss how the logistic pattern of population growth might be changed if K and r are not constant but vary over time. May (1981, pp. 24–27) discusses some simple examples.

9. Nelson (1978, p. 958) gives the following counts of total pairs of North Atlantic gannets on Ailsa Craig, off northwestern England:

Year	Pairs
1941	3,518
1942	4,829
1947	5,383
1948	5,190
1949	4,947
1950	6,579
1951	7,833
1952	7,987
1953	8,249
1954	8,555
1955	10,402
1956	8,063
1957	7,742
1958	9,506
1959	9,390
1960	13,532

Describe the population growth of this colony and calculate the annual rate of population growth.

10. A feral house mouse population can increase at $r = 0.0246$ per day. At this rate of increase how many days are needed for the population to double?

11. If the human population instantly adopted zero population growth ($R_0 = 1.0$), the population would continue to grow until it reached the stationary age structure. Keyfitz (1971) showed that such a population would increase by this demographic momentum as follows:

$$Q = \frac{be}{ra} \left[\frac{R_0 - 1}{R_0} \right]$$

where

> Q = finite rate of population change (1.0 = no change)
> b = crude birth rate per 1000 persons per year
> e = life expectancy at birth in years
> r = current rate of natural increase per 1000 persons per year
> a = average age at first reproduction in years
> R_0 = current net reproductive rate

A human population growing at these rates would increase Q times before it reached equilibrium, if zero population growth was instantly adopted. Calculate how much the human population of the earth would increase from current levels if zero population growth happened overnight. In 1990 the human population parameters were: $b = 28$, $e = 61$ years, $r = 17$, $a = 25$ years, and $R_0 = 1.77$. How sensitive is the estimate to changes in the birth rate? To changes in average age at reproduction?

12. Suppose a population lives in an environment generally favorable to growth but which provides the occasional very bad year. Assume that the finite rate of increase (λ) is 1.1 for nine years out of ten and 0.3 for the bad year. At what average rate will the population grow over a long time interval? Lewontin and Cohen (1969) discuss this problem.

Overview Question

Most analyses of population growth describe processes applicable to unitary organisms. Plants and other modular organisms also undergo population growth. Discuss the application of the models given in this chapter to population growth in modular organisms.

Species Interactions: Competition

Organisms do not exist alone in nature but in a matrix of other organisms of many species. Many species in an area will be unaffected by the presence or absence of one another, but in some cases two or more species will interact. The evidence for this interaction is quite direct; populations of one species change in the presence of a second species.

CLASSIFICATION OF INTERACTIONS

Interactions between populations can be classified on the basis of the *mechanism* of the interaction or on the basis of the *effects* of the interaction (Abrams 1987). Ecologists use both of these classifications and often mix them up. Here we define interactions on the basis of mechanism and recognize six interactions between individuals of different species:

1. *Competition*: Two species use the same limited resource or harm one another while seeking a resource.

2. *Predation*: One animal species eats all or part of a second animal species.

3. *Herbivory*: One species eats part or all of a plant species.

4. *Parasitism*: Two species live in an obligatory association in which the parasite depends metabolically on the host.

5. *Disease*: An association between pathogenic microorganisms and a host in which the host suffers physiologically.

6. *Mutualism*: Two species live in close association with one another to the benefit of both.

Some authors do not separate parasitism from disease, or predation from herbivory, and there is great variability in how loosely these terms are used in the ecology literature.

Interactions can also be defined on the basis of effects. The most common effects studied are on population growth and Odum (1983) categorizes effects as 0, +, and −. A zero means no effect of one species on the other, a plus means that the first population has benefited at the expense of the other, and a minus means that the first population has been adversely affected by the other. This system has fatal flaws for classifying interactions because it does not specify a time frame for recognizing effects and more importantly cannot describe many indirect interactions (Abrams 1987). Indirect interactions occur because one species affects a second which in turn affects a third species. Ecological communities are composed of many species linked in complex food webs and the plus or minus $(+, -)$ indicates a simple two-species interaction that cannot adequately summarize the many interactions possible in a web. For this reason we will define species interactions based on their mechanism, and then explore the variety of effects these mechanisms can produce in populations of plants and animals.

In this chapter, we will discuss the interactions between two species that result from competition. There are two different types of competition, defined as follows (Birch 1957):

> *Resource competition* (also called *scramble* or *exploitative competition*) occurs when a number of organisms (of the same or of different species) utilize common resources that are in short supply.
>
> *Interference competition* (also called *contest competition*) occurs when the organisms seeking a resource harm one another in the process, even if the resource is not in short supply.

Note that competition may be *interspecific* (between two or more different species) or *intraspecific* (between members of the same species). In this chapter, we will deal only with interspecific competition.

Competition occurs for resources and a variety of resources may become the center of competitive interactions. For plants, light, nutrients, and water may be important resources, but plants may also compete for pollinators or for space. Water, food, and mates are possible sources of competition for animals. Competition for space also occurs in some animals and may involve many types of specific requirements, such as nesting sites, wintering sites, or sites that are safe from predators. Thus resources are diverse and complex. Species must share resources before they can be potential competitors.

Several aspects of the process of competition must be kept clear. First, there is no need for animals to see or hear their competitors. A species that feeds by day on a plant may compete with a species that feeds at night on the same plant if the plant is in short supply. Second, many or most of the organisms that an animal does see or hear will not be its competitors. This is true even if resources are shared by the organisms. Oxygen, for example, is a resource shared by most terrestrial organisms, yet there is no competition for oxygen among these organisms because the resource is superabundant. Third, competition in plants usually occurs among individuals rooted in position and therefore differs from competition among mobile animals. The spacing of individuals is thus more important in plant competition.

Mathematical models have been used extensively to build hypotheses about what happens when two species live together, either sharing the same food, occupying the same space, or preying on or parasitizing one or the other. The classical models of these phenomena are the *Lotka-Volterra equations*, which were derived independently by Lotka (1925) in the United States and Volterra (1926) in Italy. More recent models by Tilman (1982, 1986) have provided another important perspective on competition theory.

COMPETITION FOR RESOURCES

Mathematical Model of Lotka and Volterra

Lotka and Volterra each derived two different sets of equations: One set applies to the *predator-prey* situation, the other set to *nonpredatory* situations involving competition for food or space. We are concerned here only with their second set of equations.

The Lotka-Volterra equations, which describe the competition between organisms for food or space, are based on the logistic curve. We have seen that the logistic curve is described by the following simple logistic equations: for species 1,

$$\frac{dN_1}{dt} = r_1 N_1 \left(\frac{K_1 - N_1}{K_1} \right)$$

and for species 2,

$$\frac{dN_2}{dt} = r_2 N_2 \left(\frac{K_2 - N_2}{K_2} \right)$$

where

N_1 = population size of species 1
 t = time
 r_1 = per capita rate of increase of species 1
 K_1 = asymptotic density for species 1

and similarly for species 2.

We can visualize two species interacting, that is, affecting the population growth of each other, with the following simple analogy. Consider the environment to contain a certain amount of a limiting resource, like nitrogen in the soil. Species 1 uses this resource and the environment will hold K_1 individuals of this species, when all the resource is being monopolized. But some of this resource can also be used by the competitor species 2, which in this example needs more resource to support one individual:

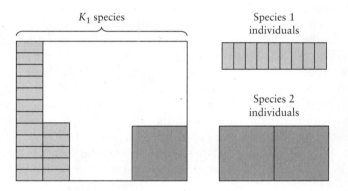

In most cases, the resources used by one individual of species 2 is not exactly the same as that used by one individual of species 1. For example, species 2 may be larger and require more of the critical resource that is contained in the environment. For this reason, we need a factor to convert species 2 individuals into an equivalent number of species 1 individuals. For this competitive situation, we define

αN_2 = equivalent number of species 1 individuals

where α is the conversion factor for expressing species 2 in units of species 1. This is a very simple assumption, which states that under all conditions of density there

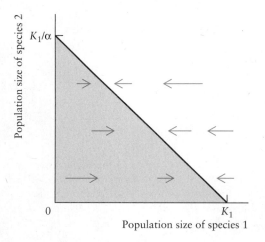

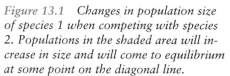

Figure 13.1   Changes in population size of species 1 when competing with species 2. Populations in the shaded area will increase in size and will come to equilibrium at some point on the diagonal line.

is a constant conversion factor between the competitors. We can now write the competition equation for species 1:

$$\frac{dN_1}{dt} = r_1 N_1 \left(\frac{K_1 - N_1 - \alpha N_2}{K_1} \right) \quad \text{population growth of species 1 in competition}$$

This is mathematically equivalent to the simple analogy we have just developed. Figure 13.1 shows this graphically for the equilibrium conditions when dN_1/dt is zero. The two extreme cases are shown at the ends of the diagonal line in Figure 13.1. All the "space" for species 1 is used (1) when there are K_1 individuals of species 1, or (2) when there are K_1/α individuals of species 2. Populations of species 1 inside this line will increase in size until they reach the diagonal line, which represents all points of equilibrium and is called the *isocline*. Note that we do not yet know where along this diagonal we will finish, but it must be somewhere at or between the points $N_1 = K_1$ and $N_1 = 0$.

Now we can retrace our steps and apply the same line of argument to species 2. We now have a volume of K_2 spaces to be filled by N_2 individuals but also by N_1 individuals. Again we must convert N_1 into equivalent numbers of N_2, and we define

βN_1 = equivalent number of species 2 individuals

where β is the conversion factor for expressing species 1 in units of species 2.* We can now write the competition equations for the second species:

$$\frac{dN_2}{dt} = r_2 N_2 \left(\frac{K_2 - N_2 - \beta N_1}{K_2} \right) \quad \text{population growth of species 2 in competition}$$

Figure 13.2 shows this equation graphically for the equilibrium conditions when dN_2/dt is zero.

Let us now try to put these two species together. What might be the outcome of this competition? Only three outcomes are possible: (1) Both species coexist, (2) species 1 becomes extinct, or (3) species 2 becomes extinct. Intuitively, we would

---

* α and β can be written more generally as α_{ij}, the effect of species j on species i. Thus $\alpha = \alpha_{12}$ and $\beta = \alpha_{21}$.

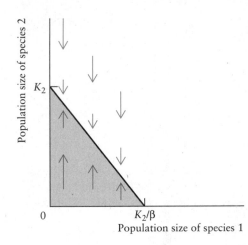

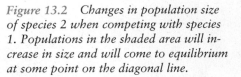

Figure 13.2   Changes in population size of species 2 when competing with species 1. Populations in the shaded area will increase in size and will come to equilibrium at some point on the diagonal line.

expect that species 1, if it had a very strong depressing effect on species 2, would win out and force species 2 to become extinct. The converse would apply for the situation where species 2 strongly affected species 1. In a situation where neither species has a very strong effect on the other, we might expect them to coexist. These intuitive ideas can be evaluated mathematically in the following way.

Solve these simultaneous equations at equilibrium:

$$\frac{dN_1}{dt} = 0 = \frac{dN_2}{dt}$$

This can be done by superimposing figures (like Figures 13.1 and 13.2) and adding the arrows by vector addition. Figure 13.3 shows the four possible geometric configurations. In each of these, I have abstracted the vector arrows, and the results can be traced by following the horizontal and vertical hatching. Species 1 will increase in areas of horizontal hatching, and species 2 will increase in areas of vertical hatching. There are a number of principles to keep in mind in viewing these kinds of curves. First, there can be no equilibrium of the two species unless the diagonal curves cross each other. Thus in cases 1 and 2 (Figure 13.3), there can be no equilibrium, because one species is able to increase in a zone in which the second species must decrease. This leads to the extinction of one competitor. Second, if the diagonal lines cross, the equilibrium point represented by their crossing may be either a *stable* point or an *unstable* point. It is stable if the vectors about the point are directed *toward* the point and unstable if the vectors are directed *away* from it. In case 4 (Figure 13.3), the point where the two lines cross is unstable because if by some small disturbance the populations move slightly downward, they reach a zone of horizontal hatching in which N_1 can increase but N_2 can only decrease, which results in species 1 coming to an equilibrium by itself at K_1. Similarly, slight movement upward will lead to an equilibrium of only species 2 at K_2.

Tilman's Model

The Lotka-Volterra equations describe competition only by its results on changes in population size of the two competing species. Competition itself is viewed in these models as a black box and no mechanisms are specified by which the effects of

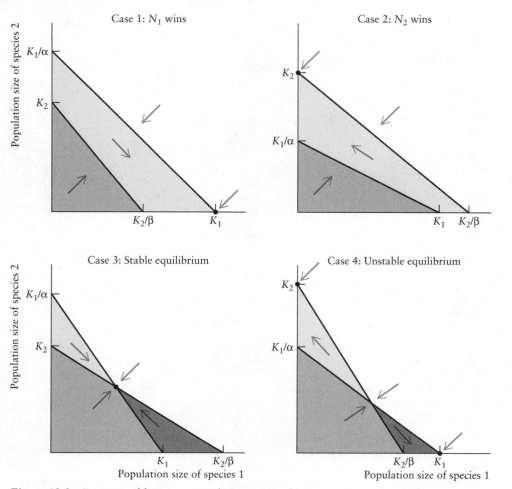

Figure 13.3   *Four possible outcomes of competition between two species. Arrows indicate direction of change in populations, and final equilibrium points are indicated by a black dot and arrow.*

competition are produced. Tilman (1987) criticized this approach to competition and emphasized that we need to study the mechanisms by which competition occurs.

Tilman (1977, 1982) presented a mathematical model of competition based on resource use. The essential features of Tilman's model can be presented graphically. Begin by considering the response of a species to two essential resources. For terrestrial plants this might be nitrogen and light, for example. Figure 13.4 illustrates the response of an organism to variations in two essential resources. If the abundance of either resource 1 or resource 2 is too low, the population declines. Conversely if both resources are abundant, the population increases. The boundary between population growth and decline is the zero growth isocline of this species.

If we repeat this analysis for a second species, we can superimpose the two zero growth isoclines. Figure 13.5 shows the four possible outcomes of this overlay for the two competing species. In the first case (a) species B needs more of both resources than species A. Species A will win out in competition and species B will go extinct.

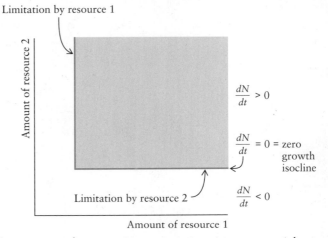

Figure 13.4   The response of an organism to variations in two essential resources (like nitrogen and water for plants). The thick line represents the zero growth isocline. Above this line in the shaded area, the population will increase in size. Below this line in the white area, the population will decline. This information is essential to using Tilman's model of competition. (After Tilman 1982.)

The second case (b) of Figure 13.5 is the mirror image of (a) and in this case species A goes extinct. In the two remaining cases (c,d) the zero growth isoclines cross so there is an equilibrium point. To determine whether this equilibrium is stable or unstable, we need to have additional information on the consumption curves for each species. At the equilibrium point in (c) and (d) species A is limited by resource 2 and species B is limited by resource 1. If species A consumes relatively more of resource 1 than species B, the equilibrium point is unstable and one or the other species will go extinct. To apply Tilman's model to a particular environment, you must know the rate of supply of the limiting resources to the populations, and the rates of consumption of these resources by each species.

Figure 13.6 illustrates five examples of competition in Tilman's model and shows that the rate of supply of the limiting nutrients is critical in determining the outcome of a competitive situation. Given a resource supply rate for a habitat, Tilman's model can predict the outcome of competition if you know for each species the concentration to which the limiting nutrient would be reduced by that species (Tilman 1990).

Tilman's model provides the same final predictions as the Lotka-Volterra model (compare Figure 13.5 with Figure 13.3), but this is deceptive because Tilman's model can be extended to make predictions at the community level about species diversity and succession (Tilman 1986). The strength of Tilman's model is in its emphasis on mechanism, and because of this it can help us understand with more precision how species interact over limited resources (Tilman 1987).

Given these mathematical formulations, we must now see if these are an adequate representation of what happens in biological systems.

Experimental Laboratory Populations

One of the first and most important investigations of competitive systems was that of a Russian microbiologist named Gause. Gause (1932) studied in detail the

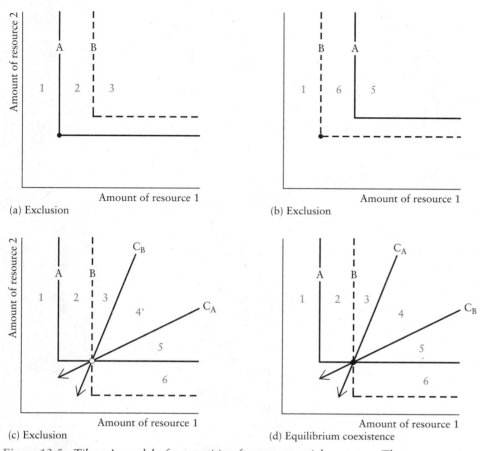

Figure 13.5 *Tilman's model of competition for two essential resources. The zero growth isoclines for species A and B are shown, along with the consumption rates for each species (C$_A$and C$_B$). For all four cases, the regions are labeled as follows: 1 = neither species can live; 2 = only species A can live; 3 = species A wins out; 4 = stable coexistence; 4' = unstable coexistence, one or the other species wins; 5 = species B wins; 6 = only species B can live. ● = stable equilibrium, ○ = unstable equilibrium. (From Tilman 1982.)*

mechanism of competition between two species of yeast, *Saccharomyces cervisiae* and *Schizosaccharomyces kephir*.* The first aspect of his investigations dealt with the growth of these two species in isolation. He found that the population growth of both species of yeast was sigmoid and could reasonably be fitted by the logistic curve.

Gause then asked, What are the factors in the environment that depress and stop the growth of the yeast population? It was known from earlier work by Richards (1928) that when the growth of yeast stops under anaerobic conditions, a considerable amount of sugar and other necessary growth substances remain. Since growth ceases before the reserves of food and energy are exhausted, something else in the

---

* These species names have changed since Gause's studies.

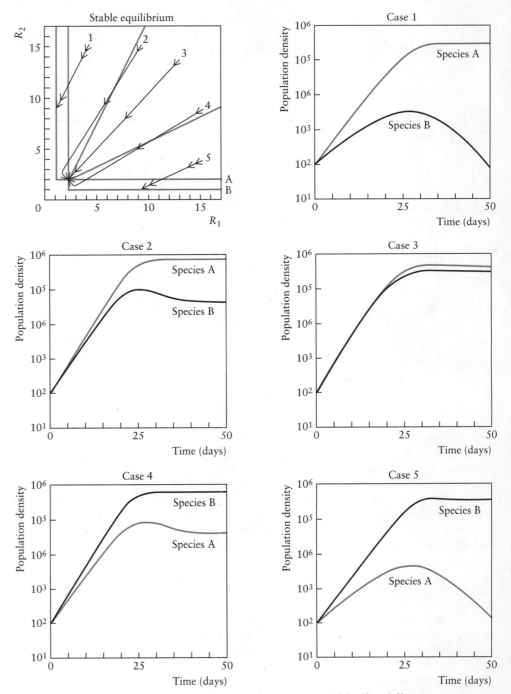

Figure 13.6 The outcome of competition for Tilman's model for five different resource supply points for Figure 13.5d, equilibrium coexistence. The resulting five population curves for the two species show the time course of competition. (From Tilman 1982.)

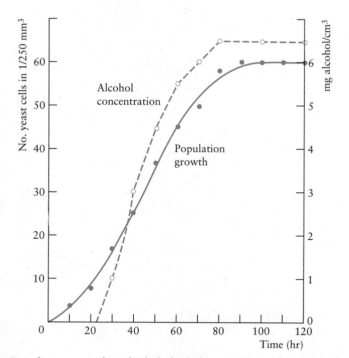

Figure 13.7   Population growth and ethyl alcohol accumulation in a yeast (Saccharomyces). *(After Richards 1928.)*

environment must be responsible. The decisive factor seems to be the accumulation of ethyl alcohol, which is produced by the breakdown of sugar for energy under anaerobic conditions (Figure 13.7). The action of the alcohol is to kill the new yeast buds just after the bud separates from the mother cell. Richards showed that the growth could be reduced by artificially adding alcohol to cultures and that changes in pH of the medium were of secondary importance. Thus with yeast we have an apparently quite simple relationship, with the population being limited principally by one factor, ethyl alcohol concentration.

When grown separately, the yeast reacted as shown in Figure 13.8. From these curves, Gause calculated logistic curves (calculated in units of volume):

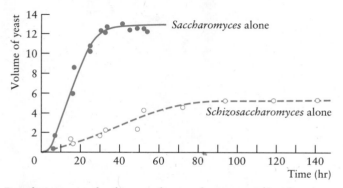

Figure 13.8   Population growth of pure cultures of two yeasts, Saccharomyces *and* Schizosaccharomyces. *(After Gause 1932.)*

	Saccharomyces	*Schizosaccharomyces*
K	13.00	5.80
r	0.22	0.06

Gause then investigated what would happen when the two yeast species were grown together. He obtained the results shown in Figures 13.9 and 13.10. Gause assumed that these data fit the Lotka-Volterra equations, and using the equations on the data from the mixed cultures, he obtained

	COMPETITION COEFFICIENTS	
Age of Culture (hr)	**α (*Saccharomyces*)**	**β (*Schizosaccharomyces*)**
20	4.79	0.501
30	2.81	0.349
40	1.85	0.467
Mean value	3.15	0.439

The influence of *Schizosaccharomyces* on *Saccharomyces* is measured by α, and this means that, in terms of competition, *Saccharomyces* finds that its K_1 spaces can be filled according to the equivalence

1 volume of *Schizosaccharomyces* = 3.15 volumes of *Saccharomyces*

Note that α and β values tend to change with the age of the culture, but as a first approximation, we can assume α and β to be constants.

If alcohol concentration is the critical limiting factor in these anaerobic yeast populations, Gause argued that we should be able to determine the competition coefficients α and β by finding the alcohol production rate of the two yeasts. He found

	Alcohol Production (% EtOH/cc yeast)
Saccharomyces	0.113
Schizosaccharomyces	0.247

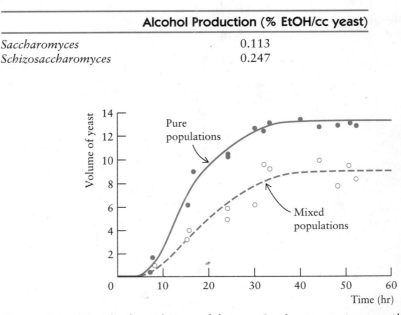

Figure 13.9   Growth of populations of the yeast Saccharomyces *in pure cultures and in mixed cultures with* Schizosaccharomyces. *(After Gause 1932.)*

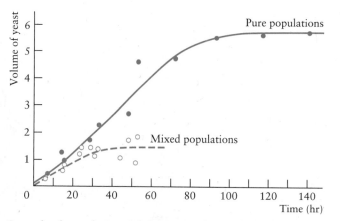

Figure 13.10   *Growth of populations of the yeast* Schizosaccharomyces *in pure cultures and in mixed cultures with* Saccharomyces. *(After Gause 1932.)*

Gause then argued that the competition coefficients, α and β, should be determined by a direct ratio of these alcohol production figures, since alcohol was the limiting factor of population growth:

$$\alpha = \frac{0.247}{0.113} = 2.18$$

$$\beta = \frac{0.113}{0.247} = 0.46$$

These independent physiological measurements agree in general with those obtained from the population data given earlier. Gause attributes the differences in the α's to the presence of other waste products affecting *Saccharomyces*. Gause assumed that the competition coefficients would be the reciprocals of each other, but this assumption need not apply to all cases of competition.

In many laboratory experiments, a species can do well when raised alone but can be driven to extinction when raised in competition with another species. Birch (1953b) raised the grain beetles *Calandra oryzae* and *Rhizopertha dominica* at several different temperatures. He found that *Calandra* would invariably eliminate *Rhizopertha* at 29°C but that *Rhizopertha* would always eliminate *Calandra* at 32°C (Figures 13.11 and 13.12). Birch found that he could predict these results from the intrinsic capacity for increase; for example,

	r	Temperature	Winner
Calandra	0.77		
Rhizopertha	0.58	29.1° C	*Calandra*
Rhizopertha	0.69		
Calandra	0.50	32.3°C	*Rhizopertha*

Thus we could change the outcome of competition by changing only one component of the environment—in this example the temperature, by only 3°C.

In all the grain beetle experiments just discussed, one species or the other has died out completely. These situations all fall under cases 1 or 2 in our treatment of

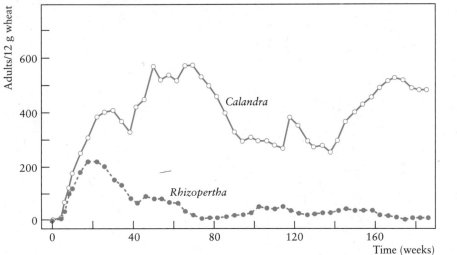

Calandra oryzae

Figure 13.11 *Population trends of adult grain beetles (*Calandra oryzae *and* Rhizopertha dominica*) living together in wheat of 14 percent moisture content at 29.1°C. (After Birch 1953b.)*

the Lotka-Volterra equations. What about case 3, where the species coexist? Gause's yeasts coexisted in his experiments. Does coexistence ever occur in grain beetles?

Under the conditions of extreme crowding in laboratory experiments, it is possible for two species to live together indefinitely if they differ slightly in their requirements. For example, Crombie (1945) reared the grain beetles *Rhizopertha* and *Oryzaephilus* in wheat and found that they would coexist indefinitely. The larvae of *Rhizopertha* live and feed inside the grain of wheat; the larvae of *Oryzaephilus* live

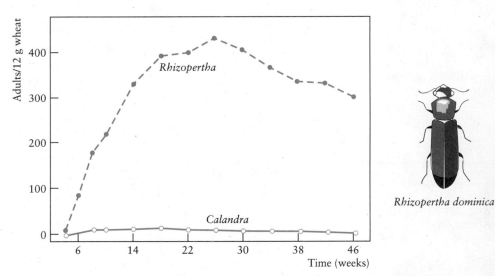

Rhizopertha dominica

Figure 13.12 *Population trends of adult grain beetles (*Calandra oryzae *and* Rhizopertha dominica*) living together in wheat of 14 percent moisture content at 32.3°C. (After Birch 1953b.)*

and feed from outside the grain. The adults of both species are the same, feeding outside the grain. Apparently these larval differences were sufficient to allow coexistence.

The green hydra (*Chlorohydra viridissima*), which has green algae in its endoderm, wins in competition with the brown *Hydra littoralis* in the light and in the absence of predation (Slobodkin 1964). Either darkness or very intense predation permits the two species to coexist in laboratory cultures.

Gause (1935) found that *Paramecium aurelia* and *P. bursaria* would coexist in a tube containing yeast. *P. aurelia* would feed on the yeast suspension in the upper layers of the fluid, whereas *P. bursaria* would feed on the bottom layers. This difference in feeding behavior between these species allowed them to coexist.

Thus by introducing only a very slight difference in the species' habits or in the environment, we can get coexistence between competing animal species under laboratory conditions.

Freshwater diatoms are often limited in abundance by silicate and by phosphate levels in lakes. Tilman (1977) cultured two diatoms *Asterionella formosa* and *Cyclotella meneghiniana* in the laboratory in continuous cultures for 30 days. He measured the consumption rates of each species in isolated cultures and obtained the isoclines plotted in Figure 13.13. He could test this model of competition by setting the rate of resource supply in his experiments. He chose four levels of supply ranging from high phosphate, low silicate to the opposite extreme of low phosphate, high silicate. Figure 13.13 shows the results of four trials. The observations were in accord with the model's predictions. *Asterionella* wins when phosphate is very limited in supply, and *Cyclotella* wins when silicate is abundant and phosphate limited (Tilman 1977).

These examples of studies of competition in diatoms illustrate the conclusions we reached from the Tilman and the Lotka-Volterra model of competition: Two species may coexist in some manner, or one may be eliminated from the mixture.

Natural Populations

We now come to the question of how these theoretical and laboratory results apply to nature. In asking this question, we come up against one controversy of modern ecology, the problem of Gause's hypothesis.

Gause (1934) wrote: "As a result of competition two similar species scarcely ever occupy similar niches, but displace each other in such a manner that each takes possession of certain peculiar kinds of food and modes of life in which it has an advantage over its competitor" (p. 19). Gause referred to Elton (1927), who had defined *niche* as follows: "The niche of an animal means its place in the biotic environment, its relations to food and enemies" (p. 64). Thus Elton used the term *niche* to describe the role of an animal in its community, so one could speak (for example) of a broad herbivore niche, which could be further subdivided.

Gause went on to say that the Lotka-Volterra equations do "not permit of any equilibrium between the competing species occupying the same 'niche,' and [lead] to the entire displacing of one of them by another" (p. 48). He continued "Both species survive indefinitely only when they occupy different niches in the microcosm in which they have an advantage over their competitors" (p. 48). Gause identifies case 3 (coexistence) with the situation of "different niches" and cases 1, 2, and 4 with the situation of "same niche."

Gause himself never formally defined what is now called *Gause's hypothesis*,

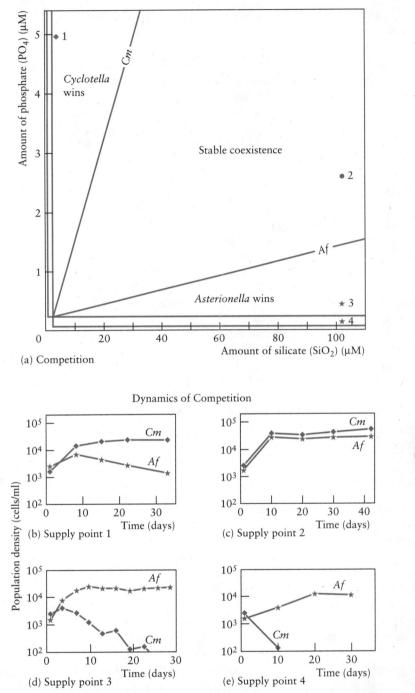

(a) Competition

Dynamics of Competition

(b) Supply point 1

(c) Supply point 2

(d) Supply point 3

(e) Supply point 4

Cyclotella

Asterionella

Figure 13.13 *Laboratory test of Tilman's competition theory using the two diatoms* Asterionella formosa *(Af)* and Cyclotella meneghiniana *(Cm). (a) The zero growth isoclines and consumption curves for each species. (b–e) Results of four competition experiments at the supply points labeled 1–4. The outcomes are as predicted by the model. (After Tilman 1977.)*

and who was the first to identify this idea with Gause is not known. In 1944 the British Ecological Society held a symposium on the ecology of closely related species. An anonymous reporter (David Lack) wrote that year in the *Journal of Animal Ecology* that "the symposium centred about Gause's contention (1934) that two species with similar ecology cannot live together in the same place . . ." (p. 176).

As is usual, several workers immediately searched out and found earlier statements of "Gause's hypothesis." Monard, a French freshwater biologist, had expressed the same idea in 1920. Grinnell, a California biologist, wrote much the same thing in 1904. The same idea was apparently in Darwin's mind but was never expressed clearly by him. It has been suggested that we drop the use of names and call this idea the *competitive exclusion principle*. Hardin (1960) states this principle succinctly: "Complete competitors cannot coexist." The competitive exclusion principle encapsulates the conclusions of the Lotka-Volterra models for competition.

The niche concept is intimately involved with the competitive exclusion principle, and we must first clarify this concept. The term *niche* was almost simultaneously defined to mean two different things. Joseph Grinnell in 1917 was one of the first to use the term *niche* and viewed it as a subdivision of the habitat (Udvardy 1959). Each niche was occupied by only one species. Elton in 1927 independently defined the niche as the "role" of the species in the community. These vague concepts were incorporated into Hutchinson's redefinition of the niche in 1958. Consider just two environmental variables, such as temperature and humidity, and determine for each species the range of values that allows the species to survive and multiply. This is illustrated in Figure 13.14a. This area in which the species can survive is part of its niche. Now introduce other environmental variables, such as pH or size of food, until all the ecological factors relative to the species have been measured. The addition of the third variable produces a volume, and ultimately we arrive at an *n*-dimensional hypervolume, which we call the *fundamental niche* of the species.

This idea of a fundamental niche has some practical difficulties. First, it has an infinite number of dimensions, and we cannot completely determine the niche of any organism. Second, we assume that all environmental variables can be linearly ordered and measured. This is particularly difficult for the biotic dimensions of the niche. Third, the model refers to a single instant in time, and yet competition is a dynamic process. MacArthur (1968) suggests one way to escape these problems: Restrict discussion to statements about differences between niches in one or two dimensions only (Figure 13.17b). Thus we can discuss the differences in the *feeding niches* of two closely related birds, and can avoid discussing the attributes of unmeasurable entries such as the entire fundamental niches of the two species.

To keep the discussion clear, Whittaker and coworkers (1973) have suggested that we use these words as follows:

Niche: The role of an organism within a community. This is Elton's and Hutchinson's idea of the niche.
Habitat: The range of environments in which a species occurs. This is Grinnell's idea, and it is essentially a distributional concept.

Given that we have now defined a niche, we can next ask whether two species in the same community can exist in a single niche. Does competitive exclusion occur in natural populations? Before answering this question, we must realize that every hypothesis has its limits, and we should be careful to set down at the start some situations in which competitive exclusion would *not* be expected to occur. These situations are (1) unstable environments that never reach equilibrium and are occu-

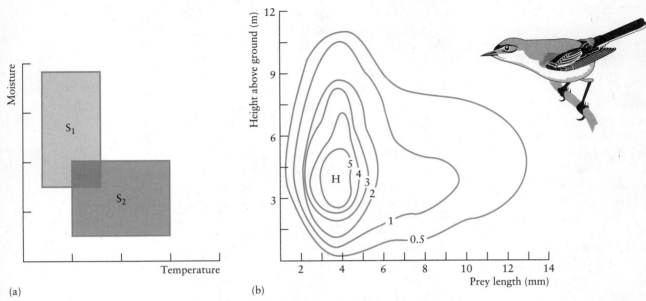

(a)

(b)

Figure 13.14 (a) Hypothetical diagram of part of the niche space of two species, S₁ and S₂. Only two environmental variables are used for illustration, but this could be extended to three or more variables to define a hypervolume, which Hutchinson (1958) called the fundamental niche of a species. Note that the niche space of organisms may overlap to a greater or lesser extent. (b) The feeding niche of the blue-gray gnatcatcher (Polioptila caerulea), represented by capture of insect prey of different sizes taken at different heights above the ground. The contour lines map the feeding frequencies (in terms of percentage of total diet) to these two niche axes for adult gnatcatchers during the incubation period in July and August, in oak woodlands in California. (Data of Root 1967 in Whittaker et al. 1973.)

pied by colonizing species, (2) environments in which species do not compete for resources, and (3) fluctuating environments that reverse the direction of competition before extinction is possible (Hutchinson 1958).

Field naturalists were the first to question Gause's hypothesis. They pointed out that one may see in the field many examples of closely related species living together and apparently in the same habitat. Anyone who has made field collections of plants or insects will attest to the great number of species living in close association. This observation is the ecological paradox of competition: How can we reconcile the frequent extinction of closely related species in laboratory cultures and the apparent coexistence of large numbers of species in field communities?

Two simple views have developed in attempting to answer this question. One holds that competition is rare in nature, and since species are not competing for limited resources, there is no need to expect evidence of competitive exclusion in natural communities. The other view holds that competition has been very common throughout the evolutionary history of communities and has resulted in adaptations that serve to minimize competitive effects.

How common is competition in nature? Much investigation has centered on closely related species on the assumption that taxonomic similarity should promote possible competition. Lack (1944, 1945), for example, studied the ecology of closely related species of birds in an attempt to test Gause's hypothesis. One example of his work was that on the cormorant (*Phalacrocorax carbo*) and the shag (*P. aristotelis*). These species are very similar in habits and appeared to overlap widely in their ecological requirements; they are both cliff nesters and feed on fish. Lack showed that the cormorant nests chiefly on flat, broad cliff ledges and feeds chiefly in shallow estuaries and harbors; the shag nests on narrow cliff ledges and feeds mainly out at sea. Thus there were significant ecological differences between these closely related species, and competition was assumed to be minimized.

Lack (1944) analyzed all the pairs of closely related species of British passerine birds. He obtained the following results:

	Cases (pairs)
Geographic separation	3
Separation by habitat	18 or more
Separation by feeding habits	4
Separation by size differences	5
Separation by different winter ranges	2
Apparent ecological overlap	5–7

Lack believed that further study would reveal differences between the five to seven pairs that apparently overlap.

The boreal forests of New England are inhabited by five warbler species of the genus *Dendroica*. All these birds are insect eaters and about the same size. Why does one species not exterminate the others by competitive exclusion? MacArthur (1958) showed that these warblers feed in different positions in the canopy (Figure 13.15), feed in different manners, move in different directions through the trees, and have slightly different nesting dates. The feeding-zone differences seem sufficiently large to explain the coexistence of the blackburnian, black-throated green, and bay-breasted warblers. The myrtle warbler is uncommon and less specialized than the other species. The Cape May warbler is different from these other species because it depends on occasional outbreaks of forest insects to provide superabundant food for its continued existence. During outbreaks of insects, the Cape May warbler increases rapidly in numbers and obtains a temporary advantage over the others. During years between outbreaks, they are reduced in numbers to low levels.

Thus closely related species of birds either live in different sorts of places or else use different sorts of foods. Lack suggested that these differences arose because of competition in the past between these closely related species. Therefore, because of Gause's hypothesis and its associated selection pressure, species either "moved" to different places and so avoided competition or changed their feeding ecology to avoid competition. There is, however, a great logical difficulty in testing Lack's hypothesis that competition has caused two species to differ (Simberloff and Boecklen 1981). Two species are always somewhat different from each other as a byproduct of speciation (Chapter 2), so observing differences between species does not necessarily mean that competition has caused the differences. This fundamental difficulty means that descriptive studies of species differences are by themselves not useful for understanding the importance of competition in natural populations.

Tropical lakes contain large numbers of closely related species of fish, and competition might be expected in these species groups. Fryer (1959) studied the ecology of *Cichlidae* fish in Lake Nyasa. He was particularly interested in the species group that lives in the rocky littoral zone. Off these rocky shores there are 12 species, very

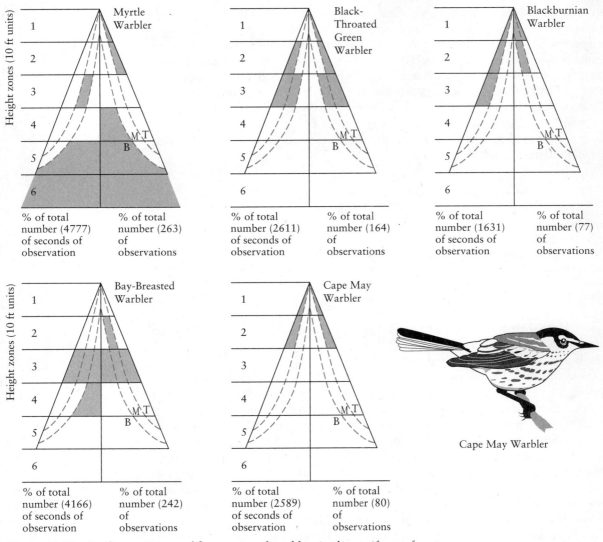

Cape May Warbler

Figure 13.15 *Feeding positions of five species of warblers in the coniferous forest of the northeastern United States. The zones of most concentrated feeding activity are shaded. B = base of branches, M = middle, T = terminal. (After MacArthur 1958.)*

closely related, that feed almost entirely on attached algae. Of these 12, seven feed only on one type of algae. Of these seven,

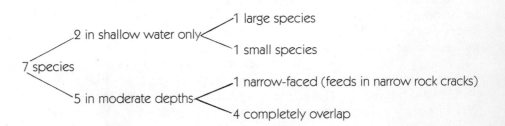

Fryer believed that the four overlapping species could live in the same place because there was no competition for food; he thought their numbers were kept down by predators and thus competitive exclusion would not occur.

The paradox of the plankton has been aptly described by Hutchinson (1961) as a possible exception to the competitive exclusion principle. The phytoplankton of marine and freshwater environments consists of a large number of plant species that utilize a common pool of nutrients and undergo photosynthesis in a relatively unstructured environment. How can all these species coexist, especially since natural waters are often deficient in nutrients and, consequently, competition should be strong? Hutchinson suggested that these species could coexist because of environmental instability; before competitive displacement could have time to occur, seasonal changes in the lake or the sea would occur. The phytoplankton may thus be viewed as a nonequilibrium community of competing species and may not be an exception to the principle of competitive exclusion.

All vascular plants require water, light and nutrients, and consequently, competition between plants over essential resources ought to be common. In arid environments root competition for water might be expected. Range managers have long been interested in improving arid zone grazing by replacing shrubs like sagebrush with grasses or legumes. Robertson (1947) tested for the effect of competition on 20 species of perennial grasses and herbs in northern Nevada. He eliminated all sage brush on one acre and on other plots he dug a trench around the sagebrush to reduce root competition. Figure 13.16 shows that the presence of sagebrush dramatically reduced the amount of growth of the other herbs and grasses. Competition for water and other nutrients in arid environments greatly affects plant growth and may exclude some plant species from the vegetation.

Interspecific competition has been analyzed in a wide variety of plants and animals during the last 30 years, and we can now ask how frequently competition occurs between species in nature. Connell (1983) and Schoener (1983) reviewed the literature to answer this question. These two reviews reached slightly different results

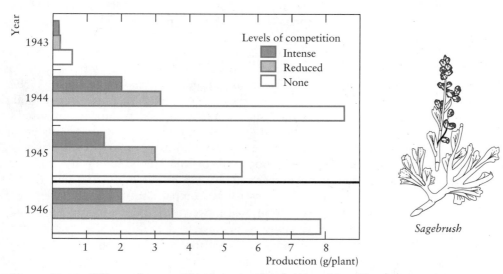

Figure 13.16 *Effects of competition from sagebrush (Artemisia tridentata) on annual production of herbage by 20 species of perennial arid-land plants in northern Nevada. (From Robertson 1947.)*

Table 13.1 Percentage of All Experiments That Showed Interspecific Competition, Broken Down in Three Taxonomic Groups and Three Habitats

	TERRESTRIAL		MARINE		FRESHWATER		TOTAL	
	No. Exp.	%	No. Exp.	%	No. Exp.	%	No. Exp.	%
Plants	205	30	31	68	2	50	238	35
Invertebrates	57	16	37	32	0	...	94	22
Vertebrates	47	23	10	90	3	67	60	37
Total	309	26	78	54	5	60	392	32

Source: Connell (1983).

because of decisions about what studies to include (Schoener 1985). Table 13.1 summarizes one survey covering papers published from 1974 to 1982 in six major ecological journals. Several general trends are shown in this table. Terrestrial organisms show a lower frequency of interspecific competition than marine organisms. One explanation may be that marine organisms in intertidal and subtidal zones often compete for space, while terrestrial organisms more often compete for nutrients or food resources, which are more heterogeneous. Additionally, invertebrates compete less than plants or vertebrates, and this seems to be explained by size differences, which make small animals more vulnerable to harsh climatic changes or higher predation losses (Schoener 1974). The overall conclusion of these surveys of interspecific competition is that it occurs frequently but not always in natural populations and that more information is needed on the mechanisms by which competition operates in nature and how large the effects of competition might be.

The fact that strong competitive interactions are found in some natural populations led MacArthur (1972) to explore the consequences of competition for the evolution of species' differences. MacArthur essentially turned Gause's hypothesis around and asked a simple question: If complete competitors cannot coexist, *how different do two species have to be in order to coexist in the same habitat?* This is a key question for natural populations. Species that come into competition will evolve differences to minimize the impact of competition. The process can be illustrated with a simple graphic model.

Let us assume that we have identified two species that may compete for food at a certain time of the year. We measure their food consumption for items of different size and obtain the *resource-utilization* curves shown in Figure 13.17. Three cases can be envisaged. (a) If the curves are completely separated, some food resources are not being utilized, and (unless some other constraints are imposed) one or both species will benefit by feeding on the unutilized food sizes. This change in feeding will lead to one of the next two cases. (b) If the curves overlap only slightly so that each species has a set of food sizes for itself, the species will each be able to survive and reproduce in the same habitat. (c) If the curves overlap a great deal so that both species eat most of the same foods, competition will be so severe that one species will be driven to extinction, will move to a different habitat to avoid competition, or, over many generations, will evolve feeding differences, as in scenario (b).

The result of this simple graphic analysis is that competing species should evolve toward case (b) as the limiting similarity possible for coexistence. A great deal of theoretical work has been devoted to attempts to quantify the limiting similarity shown in Figure 13.17b (May 1974b), but there appears to be no "ecological con-

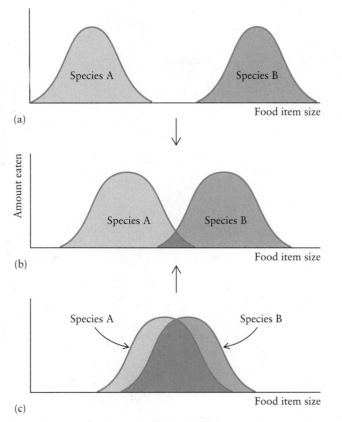

*Figure 13.17   Hypothetical resource-utilization curves for two species. Food size
is the resource for which competition may occur in this hypothetical situation.
Arrows indicate direction of evolution pressures toward case (b).*

stant" that can describe limiting similarity for all possible competitive situations
(Abrams 1975). Nevertheless resource partitioning is an important focus of study
for the analysis of species interactions.

Other difficulties confound the simple approach outlined in Figure 13.17. Niches
may have more than one dimension in which competition is possible. Most studies
have emphasized size of food as one dimension, but type of food, habitat usage, and
feeding times could also be important axes by which competitors segregate to avoid
competition. Cases of nearly complete overlap in one niche dimension (as in Figure
13.17c) may be misleading if two niche dimensions are important to the competing
organisms.

The role of competition in natural populations can be analyzed in several ways
(Wiens 1989). We can search for patterns in resource utilization, like those shown
in Figure 13.17. One good example comes from the study of the diets of five species
of terns on Christmas Island in the Pacific Ocean (Figure 13.18). Terns are ecologically segregated in their diets on Christmas Island, and these data are consistent with
the idea that competition has favored ecological divergence in diets. Schoener (1974,
1986b) has compiled data from many studies of this sort that show ecological
segregation. If such segregation occurred by evolutionary changes in the past, it is

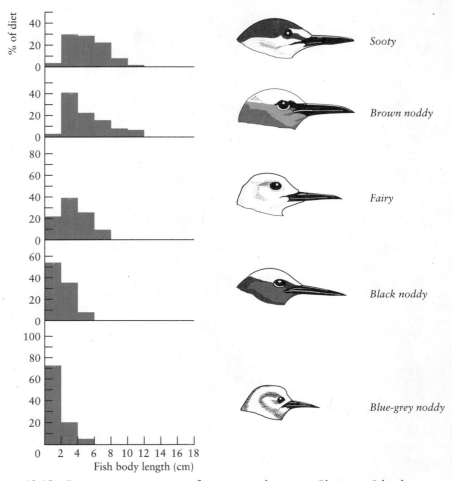

Figure 13.18 *Resource partitioning in five species of terns on Christmas Island, Pacific Ocean. Frequency distributions of fish body lengths. Terns are arranged in order of size from the larger species at the top to the smallest at the bottom. The two largest terns are nearly the same size and eat very similar sizes of fish, but the sooty tern feeds at sea several hundred kilometers from land and the brown noddy tern feeds within 100 kilometers of land. (From Ashmole 1968.)*

still an open question whether interspecific competition is operating today in these populations.

Wiens (1989) pointed out that in order to show that interspecific competition occurs, one must demonstrate that species overlap in resource use and that competition over these resources has negative effects. Table 13.2 lists criteria that ecologists use to become more certain that interspecific competition is producing negative effects in modern populations. Much of the data on resource utilization, such as Figure 13.18, satisfy criteria 1 and 2 only. A second, and better, way to analyze competition is thus through experiments of the type previously described in Chapter 6. Some of the best experiments in competition have been inadvertently carried out by humans through introductions of species into new areas.

Some ants have extended their distributions with the help of human beings and

Table 13.2   Criteria to Establish the Occurrence of Interspecific Competition with Varying Degrees of Certainty in Natural Populations

Strength of Evidence	Criteria
Weak ↓ Suggestive ↓ Convincing	1. Observed checkerboard patterns of distribution consistent with predictions 2. Species Overlap in resource use 3. Intraspecific competition occurs 4. Resource use by one species reduces availability to another species 5. One or more species is negatively affected 6. Alternative process hypotheses are not consistent with patterns

Source: Wiens (1989) p. 17.

in the process have eliminated the native ant fauna through competition. Relatively few species of ants have shown a striking ability to displace resident species, but two such highly aggressive colonizers have invaded the oceanic island of Bermuda within the last century. *Pheidole megacephala* invaded Bermuda in the latter part of the nineteenth century and apparently drove some of the native ants to extinction (Haskins and Haskins 1965). This ant is now being replaced by the Argentine ant (*Iridomyrmex humilis*), which was introduced into Bermuda around 1949 (Crowell 1968). From 1953 to 1959 *Iridomyrmex* increased its distribution in Bermuda at the rate of 394 hectares per year, replacing *Pheidole* in all habitats. Since 1959 the rate of displacement has slowed markedly, and whether or not some equilibrium will be established short of extinction of *Pheidole* is not yet known. The nature of the competition between the ants is not known; it could involve direct aggressive fighting, shortage of suitable food, or chemical repellents.

Plant ecologists have repeatedly demonstrated the effect of one plant on another for agricultural crops; one example will illustrate this. Wild oats (*Avena fatua*) is a serious weed in the northern Great Plains of North America, where it competes with the flax (*Linum usitatissimum*) crop as well as with wheat and barley. Wild oats persists in flax fields because its seeds ripen earlier than flax seeds and drop to the ground. There is a serious reduction in yields of flax at increasing oats densities (Bell and Nalewaja 1968):

Wild Oat Density (no./m^2)	FLAXSEED YIELD (bu/acre) AT FARGO IN 1966		
	Fertilized Plots	Unfertilized Plots	Average Reduction (%)
0	19.5	17.9	—
10	13.4	14.3	26
40	6.7	8.0	60
70	4.3	6.3	72
100	3.5	4.2	80
130	3.4	3.4	82
160	2.9	2.3	85

Hence agriculturalists' concern with weeds.

Two annual plants dominate the grasslands of the Great Central Valley of California, *Bromus mollis* and *Erodium botrys*. They are important grazing species, and unfertilized land is dominated by grasses like *Bromus*. *Erodium* has a low growth form but rapid root penetration in the seedling stage, so it is favored in competition for soil nutrients. *Bromus* grows taller, and if adequate nutrients are available, it becomes the superior competitor for light and shades out *Erodium*. Thus a balanced competition seems to exist in which *Erodium* is favored in poor soils and drought years and *Bromus* is favored in better soils when adequate moisture is available (McCown and Williams 1968).

Potential competitors can avoid competition by ecological segregation in time, in space, or in food resources. Animals most often segregate in space by using different habitats, segregate in diet less often, and rarely segregate by time (Schoener 1986b). Plants all require the same resources of light, water, and nutrients; consequently, they can usually partition resources only in space. Plants thus ought to be expected to compete more often than animals. Much of competition theory has been done with birds in mind (Wiens 1989), and yet other groups of organisms are not as specialized as birds. Many fishes for example have wide tolerances to habitat conditions and feeding conditions, and very plastic growth rates. This plasticity or flexibility is also found in many invertebrates and in many plants, and it contrasts with the specialized and often fixed patterns of growth and reproduction in birds and mammals. Thus poor environmental conditions, which might cause a mammal population to die out, might cause a fish population to stop growing or an invertebrate population to enter a dormant phase. Ecological segregation might be less important for these kinds of plants and animals.

We should not assume in our analysis that competition in natural populations will always be occurring. Wiens (1977) has argued that competition may be rare in some populations because of high environmental fluctuation. Populations are typically below the carrying capacity of their environment, and thus resources are plentiful, according to this argument. Occasionally a "crunch" occurs, a period of scarcity in which competition does occur. This may happen only once every five or ten years, or even less frequently, implying that competition may be hard to detect in most short-term studies.

EVOLUTION OF COMPETITIVE ABILITY

If two species are competing for a resource that is in short supply, both would benefit by evolving differences that reduce competition, as indicated in Figure 13.17. The benefit involved is a higher average population size for each species and presumably a reduced possibility of extinction. But in many cases, it will be impossible to evolve differences that reduce competition. Consider, for example, food size as a limiting resource, as in Figure 13.17. If species A evolves to use smaller food items than species B, it may run into a third species, C, which also feeds on small-sized food. Thus species may be hemmed in by a web of other possible competitors, so the option of evolving to avoid competition is not always feasible.

There is only one option available to organisms caught in a competitive net: stay and fight. To fight in the broad sense means to evolve competitive ability. The idea of competitive ability is another ecological concept that is intuitively clear but difficult to define. To understand competitive ability, we can look at the Lotka-Volterra equations for competition. These equations are based on the logistic curve for each competing species. Two parameters characterize the logistic, r (rate of increase) and

K (saturation density). We can characterize organisms by the relative importance of r and K in their life cycles.

In some environments, organisms exist near the asymptotic density (K) for much of the year, and these organisms are subject to K selection. In other habitats, the same organisms may rarely approach the asymptotic density but remain on the rising sector of the curve for most of the year; these organisms are subjected to r selection. MacArthur and Wilson (1967) defined r and K selection to be density–dependent natural selection. As a population initially colonized an empty habitat, r selection would predominate for a time and ultimately the population would come under K selection.

Species that are r-selected seldom suffer much pressure from interspecific competition, and hence they evolve no mechanisms for strong competitive ability. Species that are K-selected exist under interspecific competitive pressures that operate within as well as between species. The pressures of K selection thus push organisms to use their resources more efficiently. The individual that can convert limiting resources into reproductive adults the fastest is usually declared the superior competitor (Gill 1974).

If K selection is a complete description of competitive ability, we should be able to predict the outcome of competition in laboratory situations by knowing the K values for the two competing species. We cannot do this, however, because there is a third parameter in the Lotka-Volterra equations for competition—the competition coefficients α and β. Species can evolve competitive ability by the process of α selection (Gill 1974). Any mechanism that prevents a competitor from gaining access to limiting resources will increase α (or β) and thereby improve competitive ability. Most types of interference phenomena fall into this category. Territorial behavior in birds and allelopathic chemicals in plants are two examples of interference attributes that keep competing species from using resources.

One major evolutionary problem with α selection is that the technique of interference will often affect members of the same species as well as members of competing species, so that competitive ability is achieved only by a reduction in the species' own values of r and K. An example would be a shrub that produces chemicals to retard the germination and growth of competing plants but may suffer from auto-intoxication after several years (Rice 1984). An individual's negative effects on members of its own species present no evolutionary problem so long as those affected do not include the individual itself or its kin.

Alpha selection for interference attributes can also operate when organisms are at low density. In animals, the evolution of broad array of aggressive behaviors has been crucial in substituting ability in combat for ability to utilize resources in competition in many situations (MacArthur 1972), and we can recognize an idealized evolutionary gradient:

Low density—colonization and growth   (*r* selection)
                    ↓
High density—resource competition   (*K* selection)
                    ↓
High density—interference mechanisms   (α selection)
       prevent resource competion

Populations may exist at all points along this evolutionary gradient because competition for limiting resources is only one source of evolutionary pressure molding the life cycles of plants and animals (Boyce 1984).

CHARACTER DISPLACEMENT

One evolutionary consequence of competition between two species has been the divergence of the species in areas where they occur together. This sort of divergence is called *character displacement* (Figure 13.19) and can arise for two reasons. Since two closely related species must maintain reproductive isolation, some differences between them may evolve to reinforce reproductive barriers (see Chapter 2). In other cases, interspecific competition causes divergence, as illustrated in Figure 13.17 (c → b).

In most cases, character displacement is inferred from studies in areas where the two species occur together and where they occur alone. Figure 13.20 gives a classic example of character displacement from Darwin's finches on the Galápagos Islands. Before we conclude that this example is a good illustration of evolutionary changes in competing populations, we must satisfy four criteria (Arthur 1982):

1. The change in mean value of the character in areas of overlap should not be predictable from variation within areas of overlap or areas of isolation.

2. Sampling should be done at more than one set of locations so that local effects can be eliminated.

3. Heritability of the character must be estimated to be high so that genetic variation may be assumed to underlie the differences observed.

4. Evidence must be presented to show that the species are indeed competing and that the character measured has some relevance to the competitive process.

For Darwin's finches, criteria 1, 3 and 4 are satisfied. The change in beak size in *Geospiza fortis* in isolation on Daphne Island, for example, is much greater than one would predict from observed variation on any of the other islands. Beak characters in *G. fortis* have a very high heritability (Figure 13.21), which suggests that the variation in beak depth in *Geospiza* shown in Figure 13.20 is largely genetic in origin. There is good observational evidence of competition for food in Darwin's finches (criterion 4), as discussed by Grant (1986). Unfortunately, because of the finite size of the Galápagos Islands, we cannot replicate these findings in another set of the islands (criterion 2).

There are at present few cases in which character displacement has been conclusively demonstrated (Arthur 1982). We do not know whether this means that character displacement rarely occurs or that it is common but very difficult to demonstrate in natural populations.

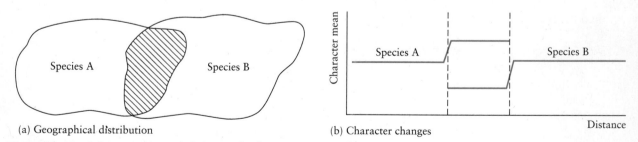

(a) Geographical distribution

(b) Character changes

Figure 13.19 *Schematic view of character displacement arising from interspecific competition in the zone of overlap of two species. This scheme is inferred as an explanation of the observations in Figure 13.20.*

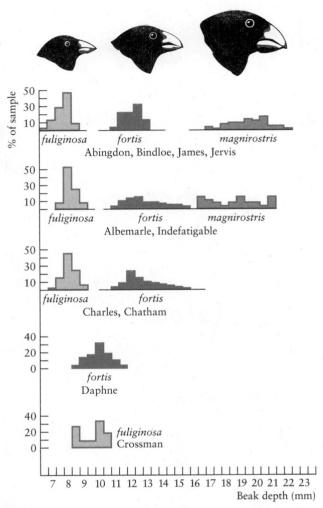

Figure 13.20   *Character displacement in beak size in Darwin's finches from the Galápagos Islands. Beak depths are given for* Geospiza fortis *and G.* fuliginosa *on islands where these two species occur together (upper three sets of islands) and alone (lower two islands).* Geospiza magnirostris *is another large finch that occurs on some islands. (After Lack 1947.)*

DIFFUSE COMPETITION AND INDIRECT EFFECTS

Competition between species is usually thought of in terms of two species interacting over limited resources. MacArthur (1972) recognized that competition could also occur between many species. He defined *diffuse competition* as the combined effects of many species upon a given species. In diffuse competition any single pair of species has very weak interactions so it may be difficult to measure any impact of one species on another. The principal ideas about competition also apply to cases of diffuse competition, and the main reason for emphasizing this idea is to remind us that species in nature rarely interact as pairs but are involved in many types of interactions with many other species (Moen 1989).

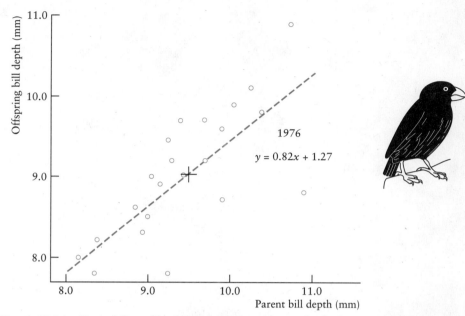

Figure 13.21   Heritability of beak depth in one of Darwin's finches, Geospiza fortis. *The slope of the regression line is an estimate of heritability, which can range from 0 (no heritability) to 1 (perfect heritability). (Modified from Boag 1983.)*

Organisms may interact directly or indirectly. Interference competition occurs by direct effects in which, for example, two species of birds fight over access to tree holes for nesting. Exploitative competition involves indirect effects because the two species have no interactions with each other but interact only through a third species

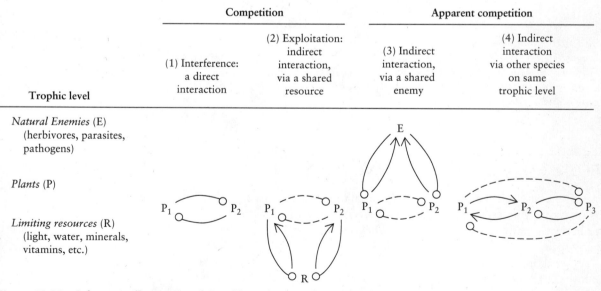

Figure 13.22   Schematic illustration of possible pathways of interspecific competition for plants. Solid lines are direct interactions, dashed lines are indirect ones. An arrowhead indicates a positive effect, a circle indicates a negative effect. A similar type of interaction scheme can be applied to animals. (From Connell 1990.)

or resource (Figure 13.22). For example, if buffalo eat grass and grasshoppers also eat the same grass, exploitative competition may occur even though buffalo have nothing directly to do with grasshoppers. Indirect effects are often surprising and can take on a variety of forms (Abrams 1987). For example, Holt (1977) recognized how indirect effects could produce apparent competition. Consider two herbivores like rabbits and pheasants that do not eat any of the same foods or compete for any essential resources. If these two species have a common predator, an increase in the abundance of rabbits could increase the abundance of the predator, which might then eat more pheasants and reduce their numbers (Figure 13.22, case 3). In systems like this one could easily be fooled into thinking that two species were competing because when one went up in numbers, the other went down and vice versa. The important idea here is that we should try to understand the mechanisms behind interactions between species and not simply describe how numbers may go up or down without knowing why (Tilman 1987).

SUMMARY

Competition between species occurs when both species strive to obtain resources that each one needs. Theoretical models of competition indicate that, in cases of competition between two similar species, one species may be displaced, or both may reach a stable equilibrium mixture. The possibility of displacement has given rise to the *competitive exclusion principle*, which states that complete competitors cannot coexist. In simple laboratory populations, one species often becomes extinct but sometimes coexists with another species. Natural communities show many examples of similar species that are coexisting, and this must be reconciled with the principle of competitive exclusion. One approach to solving this paradox is to suggest that competition is rare in nature, and hence ecological displacement is not to be expected. Another approach is to suggest that competition has occurred and that the interrelations we now see are the outcome of competition, displacement, and subsequent evolution in the past. Organisms evolve competitive ability by becoming more efficient resource users and by developing interference mechanisms that keep competing species from using scarce resources.

Experimental work with agricultural crops and range plants suggests that competitive interactions are very great in field populations. Experimental work on animal population has shown that interspecific competition is common and can exert a major influence on population size in some natural communities. Detailed studies of the mechanisms of competition between species are needed to understand multispecies systems and to predict patterns in natural and agricultural communities.

Selected References

ABRAMS, P. A. 1987. On classifying interactions between populations. *Oecologia* 73:272–281.

ARTHUR, W. 1982. The evolutionary consequences of interspecific competition. *Advances in Ecological Research* 12:127–187.

CONNELL, J. H. 1983. On the prevalence and relative importance of interspecific competition: Evidence from field experiments. *American Naturalist* 122:661–696.

GILL, D. E. 1974. Intrinsic rate of increase, saturation density, and competitive ability. II. The evolution of competitive ability. *American Naturalist* 108:103–116.

MACARTHUR, R. H. 1972. *Geographical Ecology*, chap 2. Harper & Row, New York.

PAINE, R. T. 1984. Ecological determinism in the competition for space. *Ecology* 65:1339–1348.

SCHOENER, T. W. 1986. Resource partitioning. In *Community Ecology: Pattern and Process*, ed. J. Kikkawa and D. J. Anderson, pp. 91–126. Blackwell, Melbourne, Australia.

SIMBERLOFF, D. S., and W. BOECKLEN. 1981. Santa Rosalia reconsidered: Size ratios and competition. *Evolution* 35:1206–1228.

TILMAN, D. 1986. Resources, competition and the dynamics of plant communities. In *Plant Ecology*, ed. M. J. Crawley, pp. 51–75. Blackwell, Oxford, England.

WIENS, J. A. 1989. *The Ecology of Bird Communities. Volume 2. Processes and variations.* chapter 1, pp. 3–63. Cambridge University Press, Cambridge, England.

Questions and Problems

1. The introduced house sparrow (*Passer domesticus*) competes with the native house finch (*Carpodacus mexicanus*) in western United States for nesting sites and the house finch seems to lose out more frequently in interference competition. In 1940 the house finch was introduced into the eastern United States. Discuss the potential impact of this eastern introduction on the house sparrow, and list the observations and experiments you would like to do to investigate this species interaction. Bennett (1990) summarizes the data currently available on these species.

2. In the Lotka-Volterra competition model, what is the meaning of a situation in which $\alpha = \beta$? In which $\alpha = \beta = 1$? What outcome is predicted when $\alpha = \beta = 1$ and $K_1 = K_2$? What is implied if $\alpha = 1/\beta$ and if $\alpha \neq 1/\beta$?

3. Charles Darwin in *The Origin of Species* (1859, chap. 3) states:

   As the species of the same genus usually have, though by no means invariably, much similarity in habits and constitution, and always in structure, the struggle will generally be more severe between them, if they come into competition with each other, than between the species of distinct genera.

   Discuss.

4. Review the classic case of character displacement in the rock nuthatches *Sitta neumayer* and *S. tephronota* described by Brown and Wilson (1956), the subsequent evaluation of this case by Grant (1975), and the comments on it by Schoener (1986b).

5. Schoener (1983) reviewed field experiments on interspecific competition and concluded that competition was present in 76 percent of 390 species studied. Connell (1983) reviewed the same literature and concluded that interspecific

competition was present in 55 percent of 200 species. Explain how this difference could arise. Read Schoener (1985) and evaluate his explanation for these differences.

6. Plant populations respond to crowding by reducing the biomass of individual plants. The relationship between plant density and individual biomass has been called the $-3/2$ power law of plant thinning. Review the evidence for and against this law of plant growth summarized in Weller (1987).

7. Analyze the yeast results of Gause (1932) by the use of Lotka-Volterra plots (as in Figure 13.3), and predict the outcome of this competition from the estimates of α, β, K_1, and K_2.

8. Review the work of Connell (1961a) discussed in Chapter 6, and discuss the role of competitive exclusion in affecting distributions of organisms.

9. Competition for light in trees should produce an immediate benefit for individuals that are taller than their neighbors. Discuss the factors that may affect the height to which trees grow in terms of the costs and benefits of being tall. King (1990) discusses this problem in detail.

Overview Question

Sheep, rabbits, eastern gray kangaroos and red kangaroos are possible competitors for food in the rangelands of eastern Australia. Design a research protocol for testing the presence and intensity of competition among these herbivores.

14 Species Interactions: Predation

In addition to competing for food or space, species may interact directly by predation. Predation in the broad sense occurs when members of one species eat those of another species. Often, but not always, this involves the killing of the prey. Five specific types of predation may be distinguished. *Herbivores* are animals that prey on green plants or their seeds and fruits; often the plants eaten are not killed but may be damaged. Typical predation occurs when *carnivores* prey on herbivores or on other carnivores. *Insect parasitoids* are another type of predator, which lay eggs on or near the host insect, which is subsequently killed and eaten. *Parasites* are plants or animals that depend on the host for nutrition. They differ little in their effects from herbivores as a result of their feeding. Finally, *cannibalism* is a special form of predation in which the predator and the prey are the same species. All these processes can be described with the same kind of mathematical models, and we shall begin by considering them together as "predation" in the broad sense. The effect of predation on populations has been studied theoretically and practically because it has great economic implications for humans.

Predation is an important process from three points of view. First, predation on a population may restrict distribution (Chapter 6) or reduce abundance of the prey. If the affected animal is a pest, we may think predation useful. If it is a valuable resource like a caribou, we may think the predation undesirable. Second, along with competition, predation is another major type of interaction that can influence the organization of communities, and we shall discuss the role of predation in affecting community structure in Chapters 24 and 25. Third, predation is a major selective force, and many adaptations we see in organisms, such as warning coloration, have their explanation in predator-prey coevolution.

We will begin our analysis of the predation process by constructing some simple models. All these models have the underlying assumption that we can isolate in nature a system of one predator species–one prey species.

MATHEMATICAL MODELS

Discrete Generations

Let us explore a simple model of predator-prey interactions using a discrete generation system. In seasonal environments, many insect parasitoids (predator) and their insect hosts (prey) have one generation per year and can be described by a model of the following type.

Assume that the prey population will increase in the absence of predation and this increase can be described by the logistic equation (Chapter 12)

$$N_{t+1} = (1.0 - Bz_t)N_t$$

where

> N = population size
> t = generation number
> B = slope of reproductive curve (of Figure 12.2)
> $z_t = (N_t - N_{eq})$ = deviation of present population size from equilibrium population size

In the presence of a predator, we must subtract from this equation a term accounting for the individuals eaten by predators, and this could be done in a number of ways. All the prey above a certain number (the number of *safe sites*) might be killed by predators. Or each predator might eat a constant number of prey. If, however, the abundance of the predator is determined by the abundance of prey, the whole predator population must eat proportionately more prey when prey are abundant and proportionately less prey when prey are scarce. They could do this by becoming more abundant when prey are abundant or by being very flexible in their food requirements. We subtract a term from the prey's logistic equation:

$$N_{t+1} = (1.0 - Bz_t)N_t - CN_tP_t$$

where

> P_t = population size of predators in generation t
> C = a constant measuring the efficiency of the predator

What about the predator population? We assume that the reproductive rate of the predators depends on the number of prey available. We can write this simply as

$$P_{t+1} = QN_tP_t$$

where

> P = population size of predator
> N = population size of prey
> t = generation number
> Q = a constant measuring the efficiency of utilization of prey
> for reproduction by predators

Note that if the prey population (N) were constant, this equation would describe geometric population growth (Chapter 12) for the predator.

To put these two equations together and interpret them, we must first obtain the maximum reproductive rates of the predator and the prey. When predators are absent and prey are scarce, the net reproductive rate of the prey will be, approximately,

$$N_{t+1} = (1.0 - BN_{eq})N_t$$

or

$$R = \frac{N_{t+1}}{N_t} = 1.0 - BN_{eq}$$

where R is the maximum finite rate of population increase of the prey. For the predator, when the prey population is at equilibrium and predators are scarce, predators will increase at

$$P_{t+1} = QN_{eq}P_t$$

or

$$S = \frac{P_{t+1}}{P_t} = QN_{eq}$$

where S is the maximum finite rate of population increase of the predator.

Let us now work out an example. Let $R = 1.5$ and $N_{eq} = 100$ so that the absolute value of the slope of the reproductive curve $B = 0.005$. Assume that the constant C measuring the efficiency of the predator is 0.5. Thus

$$N_{t+1} = (1.0 - 0.005z_t)N_t - 0.5N_tP_t$$

Assume that under the best conditions, the predators can double their numbers each generation ($S = 2.0$), so that the constant Q is

$$S = QN_{eq}$$
$$2.0 = Q(100)$$

or

$$Q = 0.02$$

Consequently, the second equation is

$$P_{t+1} = 0.02\, N_tP_t$$

Start a population at $N_0 = 50$ and $P_0 = 0.2$:

$$N_1 = ([1.0 - 0.005(50 - 100)]50) - [(0.5)(50)(0.2)]$$
$$= 62.5 - 5.0 = 57.5$$
$$P_1 = (0.02)(50)(0.2)$$
$$= 0.2$$

For the second generation,

$$N_2 = ([1.0 - 0.005(57.5 - 100)]57.5) - [(0.5)(57.5)(0.2)]$$
$$= 69.72 - 5.75 = 63.97$$
$$P_2 = (0.02)(57.5)(0.2)$$
$$= 0.23$$

These calculations can be carried over many generations to produce the results shown in Figure 14.1—a cycle of predator and prey numbers.

A stable oscillation in the numbers of predators and prey is only one of four possible outcomes. The other three are (1) stable equilibrium with no oscillation, (2) convergent oscillation, and (3) divergent oscillation leading to the extinction of either predator or prey. Maynard Smith (1968) has shown that the range of variables for a stable equilibrium without oscillation is very restricted. One example will illustrate this solution. Let $N_{eq} = 100$, $B = 0.005$, and $C = 0.5$ for the prey, while $Q = 0.0105$ ($S = 1.05$) for the predator. For the first generation, from a starting population of 50 prey and 0.2 predators:

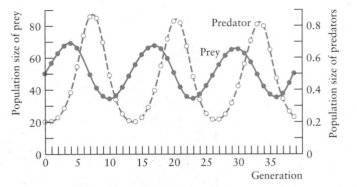

Figure 14.1 *Population changes in a hypothetical predator-prey system with discrete generations. For the prey population, N_{eq} = 100, B = 0.005, and C = 0.5. For the predator, Q = 0.02.*

$$N_1 = ([1.0 - 0.005(50 - 100)]50) - [(0.5)(50)(0.2)]$$
$$= 57.50$$
$$P_1 = (0.0105)(50)(0.2)$$
$$= 0.105$$

Similarly,

	N	**P**
Second generation	66.70	0.063
Third generation	75.70	0.044
Fourth generation	83.20	0.035
Fifth generation	88.70	0.031

The populations show decreasing small oscillations and gradually stabilize around a level of 95.2 for the prey and 0.048 for the predator.

Discrete generation predator-prey models show a variety of dynamic behaviors much like discrete population growth models (page 201).

Continuous Generations

Many predators and prey have overlapping generations, with births and deaths occurring continuously; vertebrate predators provide many examples. For the continuous-generation case, Lotka (1925) and Volterra (1926) independently derived a set of equations to describe the interaction between populations of predators and prey. The early models of Lotka and Volterra were unrealistic and they have been replaced by other models that are capable of biological realism (Taylor 1984). The best general models were developed by Rosenzweig and MacArthur (1963) as graphic models.

Consider first the population growth of the prey species in relation to predator and prey abundance (Figure 14.2). Now do an imaginary experiment and construct a series of populations at different predator and prey densities and at each point

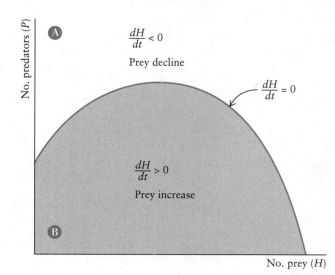

Figure 14.2   *The prey isocline of the Rosenzweig and MacArthur model for a predator-prey interaction. The curve shows all the combinations of predator and prey densities that will result in no growth of the prey population. In the shaded zone prey numbers will increase. Above the isocline, prey numbers will always decline. (After Rosenzweig and MacArthur 1963.)*

measure whether the prey increase or decline. For example, at point A in Figure 14.2 there is a large number of predators, and prey will certainly decline. At point B, there are few predators, and prey will increase. By doing a series of points you can divide the area of the graph into a zone of prey increase and a zone of prey decline.

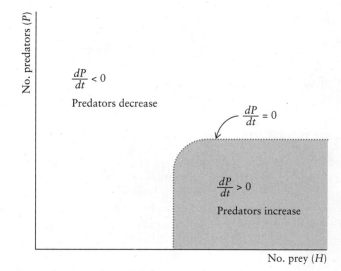

Figure 14.3   *The predator isocline of the Rosenzweig and MacArthur model for a predator-prey interaction. The curve shows all the combinations of predator and prey densities that will result in no growth of the predator population. In the shaded zone predator numbers will increase. Above the isocline, predator numbers will fall. (After Rosenzweig and MacArthur 1963.)*

This fixes the prey isocline, the boundary between these two zones at which the rate of increase of the prey population is zero. At equilibrium the prey population must exist somewhere on this line. The prey isocline always has a hump—on the right side of the hump the prey crowd themselves and reduce their own population growth, and on the left side of the hump there are too many predators eating fewer prey (Rosenzweig 1969).

Now consider the population changes of a predator that is food-limited at low prey densities and eats only a single prey species. When prey numbers are high, predator numbers should increase. But at high predator numbers, predators stop increasing because of other limitations such as territorial behavior or burrow sites. The resulting predator isocline is shown in Figure 14.3. The predator isocline will not always be this shape, and not all predators will have exactly the same shape of isocline.

By superimposing the two isoclines of Figures 14.2 and 14.3 we get a graphic model of a predator-prey interaction. In this case by examining the vectors around the equilibrium point, you can see that this is a stable equilibrium for both predator and prey (Figure 14.4). In the upper right quadrant both predators and prey are decreasing so the vector points downward and inward. In the lower right quadrant, the prey is decreasing but the predator is increasing so the vector points upward and inward. The upper left quadrant represents prey increasing and predators decreasing. The lower left quadrant represents both species increasing. This is the Rosenzweig and MacArthur model of predator-prey interactions. It is a useful model because we can explore in a graphic manner the effects of simple changes to the predator-prey system.

Consider the situation in which the predator is not restricted by any other

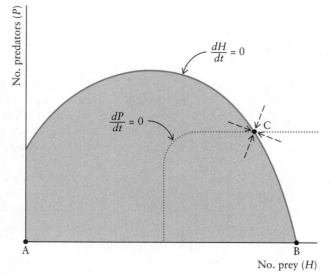

Figure 14.4   The predator and the prey isoclines superimposed for the Rosenzweig and MacArthur model for a predator-prey interaction. There are three equilibrium points indicated by the dots: at A there are no predators and no prey, at B there are no predators but only prey, and at C there is a stable equilibrium of both predators and prey. The arrows indicate the trajectories of populations near the stable equilibrium point. (After Rosenzweig and MacArthur 1963.)

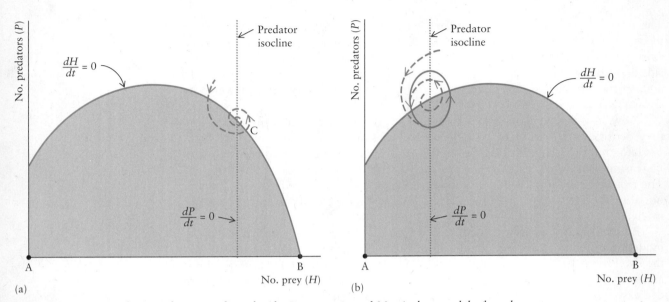

Figure 14.5 *(a) Predator and prey isoclines for the Rosenzweig and MacArthur model when the predators are limited only by food and not by other factors. When disturbed from equilibrium, the populations spiral inward, giving a convergent oscillation of both predator and prey numbers to the equilibrium point C. (b) Predator and prey isoclines when the predators are highly efficient. When disturbed from equilibrium, the populations spiral outward and reach a stable limit cycle of predator and prey numbers. Compare with Figure 14.4. (After Rosenzweig and MacArthur 1963.)*

limitations than its food supply. In this case the predator isocline is vertical and does not bend (Figure 14.5a). This system is stable and if disturbed from equilibrium will show convergent oscillations back to the equilibrium point. Now consider this same system with a more efficient predator. Predator efficiency in this graphic presentation means that the predators can subsist on lower prey numbers, so the predator isocline is moved to the left on the graph (Figure 14.5b). When the predator isocline intersects the prey isocline to the left of the hump, there is no stable equilibrium for the system and populations cycle endlessly around the hypothetical equilibrium point. The farther the equilibrium point is from the hump of the prey isocline, the larger will be the amplitude of the resulting cycles.

The Rosenzweig-MacArthur model of predator-prey interactions is thus capable of a wide variety of dynamic behavior, from stability to strong oscillations. This model provides a focus for asking simple questions about predator-prey systems, such as: What would happen if prey became less abundant? The model also serves as an entry point into understanding the more complex real world.

All these predator-prey models make a series of simplifying assumptions about the world. The models assume a homogeneous world in which there are no refuges for the prey or different habitats. The models assume that the system is one predator eating one prey. There are no alternative prey species or multiple predator species. Relaxing all of these assumptions leads to more complex and more realistic Rosenzweig-MacArthur models (Taylor 1984).

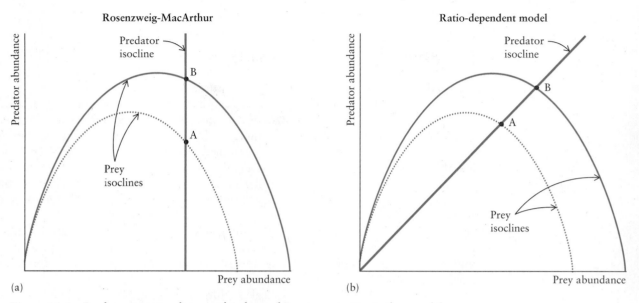

Figure 14.6 *Predator-prey isoclines in the classical Rosenzweig-MacArthur model (a) and in the ratio-dependent model (b). Two prey isoclines are shown for less productive and more productive habitats. Increasing prey productivity changes only the equilibrium predator abundance from A to B in the classical model, but changes both predator and prey abundance in the ratio-dependent model. (Modified after Hanski 1991.)*

The classical form of the Rosenzweig-MacArthur model uses a vertical predator isocline, as in Figure 14.5. In this model the equilibrium for the prey species depends only upon the predator's characteristics. In particular, if the productivity of the prey population increases, the equilibrium density of the prey does not change (Figure 14.6). All the gain in prey productivity goes to the predators, which increase in abundance. An alternative model, the ratio-dependent model, has been suggested by Arditi and coauthors (1991a, 1991b). This model postulates a predator isocline that slopes upward (Figure 14.6b). The ratio-dependent model assumes that the predation rate depends on the ratio of predators to prey, rather than just on prey numbers alone (Arditi et al. 1991a, Hanski 1991). These two models make quite different predictions about the relationship between prey abundance and predator abundance. In the ratio-dependent model, as prey productivity is increased, predator and prey equilibria both rise. In some biological systems the classical theory may be adequate, but in other systems the ratio dependent theory fits better (Hanski 1991).

The simple models of predation that we have just developed are interesting in that they indicate that oscillations may be an outcome of a simple interaction between one predator and one prey species in an idealized environment. In discrete generation systems, the outcome of a simple predation process may be stable equilibrium, oscillations, or extinction. Discrete systems are more likely to lead to extinction in a fluctuating environment (May 1976). We will now consider evidence from laboratory and field populations to see how well real predator-prey systems fit these simple models.

LABORATORY STUDIES

Laboratory systems can be set up in which the major assumptions of predator-prey models can be met, and we can investigate how these simple laboratory experiments work before we tackle the more complex natural world.

Gause (1934) was the first to make an empirical test of the models for predator-prey relations. He reared the protozoans *Paramecium caudatum* (prey) and *Didinium nasutum* (predator) together in an oat medium. In these initial experiments, *Didinium* always exterminated *Paramecium* and then died of starvation; that is, the system has a stable point, but it is at point A in Figure 14.4, and this is not very interesting biologically. This result occurred under all the circumstances Gause used for this system—making the culture vessel very large, introducing only a few *Didinium*, and so on. The conclusion was that the *Paramecium-Didinium* system did not show either a stable equilibrium or a stable limit cycle. Gause thought that stability could not be achieved because of a biological peculiarity of *Didinium*: It was able to multiply very rapidly even when prey were scarce, the individual *Didinium* becoming smaller and smaller in the process.

Gause then introduced a complication into the system: He used an oat medium with a sediment. *Paramecium* in the sediment were safe from *Didinium*, which never entered it. Gause had added a refuge for the prey to his simple system. In this type of system, the *Didinium* again eliminated the *Paramecium*, but only from the clear-fluid medium; *Didinium* then starved to death, and the *Paramecium* hiding in the sediment emerged to increase in numbers (Figure 14.7). The experiment ended with many prey and no predators. Gause had reached stable point B (Figure 14.4) predicted by the mathematical model.

Gause, quite determined, tried another system, introducing *immigrations* into the experimental setup. Every third day he added one *Paramecium* and one *Didinium*, and he got the results shown in Figure 14.7c. Gause concluded that in *Paramecium* and *Didinium*, stable oscillations in numbers of the predators and the prey are not a property of the predator-prey interaction itself, as some of the models predict, but apparently occur as a result of constant interference from outside the system.

Huffaker (1958) questioned these conclusions of Gause that the predator-prey system was inherently self-annihilating without some outside interference such as immigration. He claimed that Gause had used too simple a microcosm. Huffaker studied a laboratory system of a phytophagous mite, *Eotetranychus sexmaculatus*, as prey and a predatory mite, *Typhlodromus occidentalis*, as predator. The prey mite infests oranges, so Huffaker used these for his experiments. When the predator was introduced onto a single prey-infested orange, it completely eliminated the prey and died of starvation (like Gause's *Didinium*). Huffaker gradually introduced more and more spatial heterogeneity into his experiments. He placed 40 oranges on rectangular trays like egg cartons and partly covered some oranges with paraffin or paper to limit the available feeding area; in other cases he used rubber balls as "substitute" oranges. Huffaker could then disperse the oranges among the rubber balls or place all the oranges together. Finally, he could add whole new trays and set up artificial barriers of petroleum jelly, which the mites could not cross.

All Huffaker's simple systems eventually resulted in extermination of the populations. Figure 14.8 illustrates a population that became extinct in a moderately complex environment of 40 oranges. Finally, Huffaker produced the desired oscillation in a 252-orange universe with a complex series of petroleum-jelly barriers; in

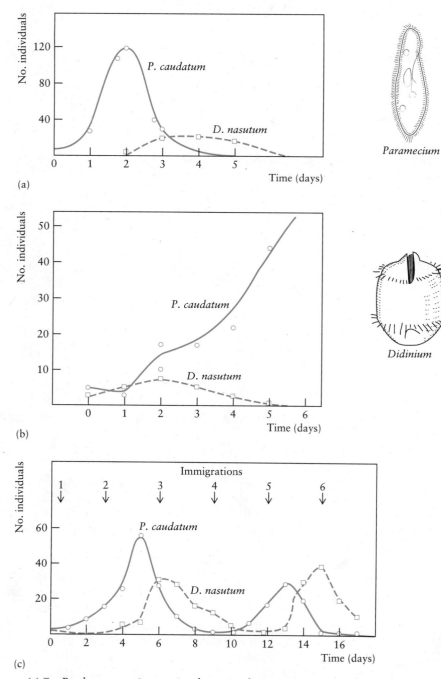

Figure 14.7   *Predator-prey interaction between the protozoans* Paramecium caudatum *and* Didinium nasutum *in three microcosms: (a) oat medium without sediment, (b) oat medium with sediment, (c) oat medium without sediment with immigrations. (After Gause 1934.)*

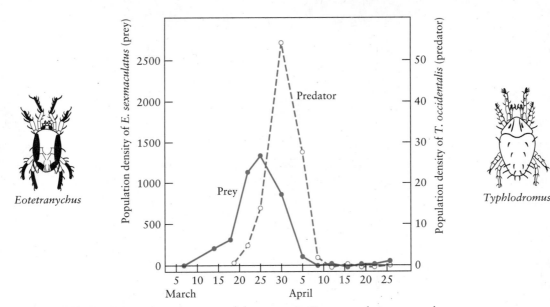

Eotetranychus

Typhlodromus

Figure 14.8 *Densities per orange area of the prey mite* Eotetranychus sexmaculatus *and the predator mite* Typhlodromus occidentalis, *with 20 small areas of food for the prey (orange surface) alternating with 20 foodless positions. (After Huffaker 1958.)*

this system, the prey were able to colonize oranges in hop, skip, and jump fashion and keep one step ahead of the predator, which exterminated each little colony of the prey it found (Figure 14.9). The predators died out after 70 weeks, and the experiment was terminated.

Huffaker concluded that he could establish an experimental system in which the predator-prey relationship would not be inherently self-destructive. He admits, however, that his system is dependent on local emigration and immigration and that a great deal of environmental heterogeneity is necessary to prevent immediate annihilation of the system.

Stable oscillations of predator-prey interactions have been obtained in several laboratory systems. Utida (1957) maintained a system of the azuki bean weevil as a host (prey) and a parasitic wasp on the larvae of the weevil as a parasite (predator) in a petri dish. Systems of this type show oscillations (Figure 14.10), which Utida followed for a maximum of 112 generations (14 complete oscillations). The oscillations were gradually damped in amplitude (convergent oscillations), and Utida noted that a long-term trend was imposed on the cycle; the host population gradually increased in density, and the parasite population gradually declined. This raised an interesting question: Can there be evolutionary changes in laboratory predator-prey systems during short experiments? Lotka and Volterra, Rosenzweig and MacArthur, and all modellers of predator-prey systems have assumed a constant and unchanging prey species and a constant and unchanging predator species, and other ecologists have often followed their lead in assuming that evolution cannot occur on an ecological time scale.

Evolutionary changes in predator-prey systems in the laboratory have been studied most thoroughly by Pimentel and his co-workers at Cornell. In one study, a

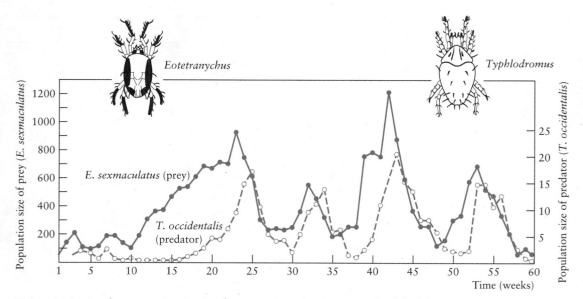

Figure 14.9 *Predator-prey interaction between two mites in a complex laboratory environment with a 252-orange system with one-twentieth of each orange exposed for possible feeding by the prey. (After Huffaker et al. 1963.)*

population system of the house fly (*Musca domestica*) and a wasp parasite (*Nasonia vitripennis*) maintained for 20 generations showed significant evolutionary changes (Pimentel et al. 1963). These changes occurred in both the host flies and the parasite wasps: The host became more resistant to the parasite, and the parasite became less virulent to the host. This is indicated in the population parameters given in Table 14.1. Selection had produced evolutionary changes in a short time to reduce the intensity of interaction between the host and the parasite. The genetic properties of both host and parasite were not constant.

Laboratory studies of predator-prey systems have carried us a long way from

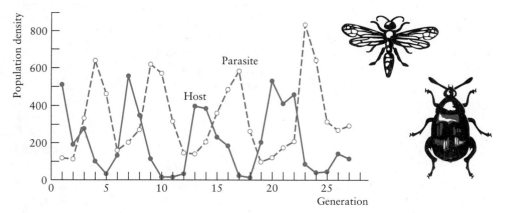

Figure 14.10 *Fluctuations in population density in a host-parasite system of the azuki bean weevil (*Callosobruchus chinensis*) and its larval parasite* Heterospilus prosopidis *(a wasp). (After Utida 1957.)*

Table 14.1 Evolutionary Changes in a Host-Parasite System of the House Fly (*Musca domestica*) and a Wasp Parasite (*Nasonia vitripennis*) After 20 Generations of Interaction in the Laboratory

	Reproductive Rate (avg. no. progeny/ female wasp)	Parasitism Rate (% fly pupae parasitized)	Longevity of Female Wasps (days)	Longevity of Male Wasps (days)
Control wasps on control flies	140	51.7	7.0	1.6
Experimental wasps on experimental flies[a]	46	39.6	4.6	1.4
Control wasps on experimental flies[b]	68	46.0	5.2	1.4
Experimental wasps on control flies[c]	123	52.6	6.6	1.7

[a] Measures evolutionary changes in both hosts and parasites.
[b] Measures evolutionary changes in host resistance.
[c] Measures evolutionary changes in parasite virulence.
   Source: After Pimentel et al. (1963).

our starting point. What might we look for in field populations of predator-prey systems? We have assumed so far that predators determine the abundance of their prey and vice versa, and we should consider first whether this generalization holds for field situations. If it does, we might expect to see evidence of stability in predator-prey associations in some natural systems. Associations of predator and prey might show evolutionary changes, and these evolutionary changes could be looked for in species that have recently come into contact in the field.

FIELD STUDIES

How can we find out whether predators determine the abundance of their prey? The suggested experiment is to remove predators from the system and to observe the prey's response. Few direct experiments like this have been done properly with adequate controls, but let us examine some case studies.

Atlantic salmon (*Salmo salar*) are important in both commercial and sport fishing along the east coast of Canada, and declining stocks have been a serious problem. One attempt to increase production came from the observation that bird predators, particularly kingfishers and mergansers, were eating a large fraction of the young salmon population. Atlantic salmon lay their eggs in fresh water, and the young salmon live two or three years in fresh water before they emigrate to sea as smolts. White (1939) removed 154 kingfishers and 56 mergansers from the Margaree River in Nova Scotia in 1937 and 1938; they obtained increased numbers of young salmon:

	No. Salmon Smolts Going to Sea
1937, before bird control	1834
1938, after bird control	4065

Two objections can be raised to this experiment: (1) Production of salmon smolts varies greatly from year to year normally, and no control data are available to counter the possibility that 1938 was just a "good" salmon year; (2) an increased number of young salmon does not necessarily mean that more adult salmon will return one or two years later.

Elson (1962) repeated this experiment on a longer time scale by removing for six years an average of 54 mergansers and 164 kingfishers a year from a 10-mile (16-km) stretch of the Pollett River in New Brunswick. He obtained these results:

Year	No. Salmon Fry Released	% Surviving to Smolts (2 Yr)
NO BIRD CONTROL		
1942	16,000	12
1943	16,000	6
1945	249,000	2
BIRD-CONTROL PROGRAM		
1947	273,000	8
1948	235,000	6
1949	243,000	8
1950	246,000	10

Some side effects of bird control were found. Most species of fish in the river approximately doubled their numbers as a result of the bird control. Elson concluded that intensive bird control could increase Atlantic salmon production from streams.

Ruffed grouse (*Bonasa umbellus*) populations fluctuate greatly in numbers, and the importance of this species in hunting has led to several predator-control experiments. A total of 557 predatory birds and mammals were removed from about 2000 acres in New York, and an adjacent area was left as a control. The results of this experiment were as follows (Edminster 1939):

	1931		1932	
	Predators Removed	No Removal	Predators Removed	No Removal
Nest loss(%)	24	51	39	72
Chick mortality (%)	57	67	54	55
Adult loss (%)	11	15	32	21
Grouse population density in fall (birds/100 acres)	13.0	9.8	18.7	18.0

Thus predator removal greatly improved nesting success, but there was no carryover to higher population densities of adults in the fall.

After a year to recover, this experiment was repeated with the control and experimental areas reversed. The same results were obtained; predator control reduced nest losses but did not alter chick mortality, and the population density on the two areas in September 1934 was 17.7 grouse per 100 acres on the predator removal area and 18.1 grouse per 100 acres on the area with no removal. Edminster (1939) concluded that predator control was not an effective means of increasing grouse densities.

There were two objections to these grouse experiments: (1) Grouse could move onto and off the experimental and control areas, and (2) predators could colonize the removal area, so the experiment might alter predator numbers over a greater area than desired. Both objections were answered by repeating the experiment on an island population of grouse (Crissey and Darrow 1949). From 1940 to 1945, predators were removed from 1050-acre Valcour Island in Lake Champlain, New York. The results were again the same: Predator removal increased nest success but had little effect on chick losses or adult losses. A substantial decline in population density of grouse occurred over the winter of 1943 to 1944 on Valcour Island, yet predator removal was almost complete by this time.

Paul Errington studied muskrats (*Ondatra zibethicus*) in the marshes of Iowa for 25 years and tried to determine the effects of predation on muskrat populations. He questioned the common assumption that if a predator kills a prey animal, the prey population must then be lower by one animal than it would have been without predation. You cannot study the effects of predation, Errington argued, by counting the numbers of prey killed; you have to determine the factors that condition predation, the factors that make certain individuals vulnerable to predation while others are protected. Mink predation on muskrats was a primary cause of death in Iowa marshes, but Errington considered that mink were removing only surplus muskrats that were doomed to die for other reasons. The numbers of muskrats were determined by the territorial hostility of muskrats toward one another, and the muskrats driven out by this hostility over space were doomed to die—if not from predators, then from disease or exposure. Predators were merely acting as the executioners for animals excluded by the social system (Errington 1963).

The role of predators in limiting the abundance of mammals is controversial. When a proper experimental design is used, the question can be clearly answered. The best example involves dingo predation on kangaroos in Australia (Caughley et al. 1980). The dingo (*Canis familiaris dingo*) is a doglike predator, the largest carnivore in Australia. Because dingoes eat sheep, some very long fences (9660 km) have been built in southern and eastern Australia to prevent dingoes from moving into sheep country. Intensive poisoning and shooting of dingoes in sheep country, coupled with the dingo fence preventing recolonization, has produced a classic experiment in predator control. Figure 14.11 gives the results for red kangaroos (*Macropus rufus*). There is a spectacular increase in the abundance of red kangaroos when dingoes are eliminated—densities of kangaroos are 166 times higher in New South Wales than in South Australia. Emus (*Dromaius novaehollandiae*), large flightless birds, are also over 20 times more abundant in the dingo-free areas. There is no alternative explanation of these sudden changes across the dingo fence (Caughley et al. 1980), and we conclude that dingo predation limits the density of red kangaroos. Dingoes are able to hold kangaroo numbers low without going extinct themselves because they have alternate prey such as rabbits and rodents to sustain them when kangaroos are in short supply.

In contrast, the Serengeti Plains of eastern Africa contain a suite of large mammals and their predators, but the predators—lions, leopards, cheetahs, wild dogs, and spotted hyenas—seem to have little impact on their large mammal prey (Bertram 1978). Most of the individuals taken by predators are doomed surplus—older, in-

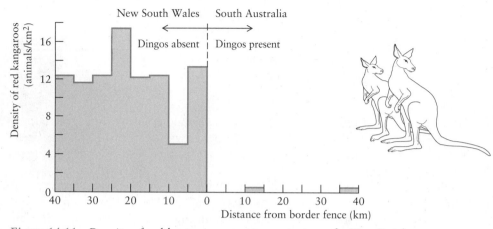

Figure 14.11   Density of red kangaroos on a transect across the New South Wales–South Australia border in 1976. The border is coincident with a dingo fence that prevents dingoes from moving from South Australia into the sheep country of New South Wales. (After Caughley et al. 1980.)

jured, or diseased animals. Also, the vast majority of the prey species are migratory, whereas most of the predators are resident. Lions, for example, seem to be limited in numbers by the resident prey species available in the dry season when the migratory ungulates are elsewhere.

Wildlife managers have instituted many predator control programs on the assumption that predation was limiting the abundance of prey populations. There has been growing hostility from the conservation movement to the idea of predator control, and there is an urgent need to resolve these issues for proper management (Gasaway et al. 1992, Boutin 1992). One of the classic controversies in North America has been about wolf predation on moose. If wolf predation affects the abundance of moose, then removing wolves should increase moose numbers. This experiment has been done five times in Canada and Alaska (Boutin 1992). There was an improvement in moose calf survival in three of the five studies but in only one study was there a significant increase in the moose population. Larsen and colleagues (1989) put radio-collars on moose calves in the Yukon and found that 58 percent of all calf deaths were caused by grizzly bears and 25 percent by wolves. Predation may affect moose numbers but it is the combined predation by grizzly bears, black bears, and wolves that is significant (Gasaway et al. 1992), and not predation by wolves alone.

Caribou herds in North America are also preyed on by wolves, and there is considerable argument about whether predation limits caribou density or whether food resources are limiting (Bergerud 1980). This controversy has important practical consequences because of pressure for wolf-control programs. Unfortunately, although many wolf-control programs have been implemented, few of them have used a proper experimental design with unmanipulated populations for comparison (Boutin 1992). Bergerud (1980) used circumstantial evidence to suggest that a combination of predation losses on young caribou and human hunting of adult caribou were jointly holding caribou populations in check. He compared the rates of increase (*r*) of 40 different caribou herds and got these average values of *r*:

Hunting Mortality	PREDATORS		
	None	Few	Normal
None	0.28	0.11	0.02
< 5% per year	0.08	0.05	−0.01
> 5% per year	−0.11	−0.19	−0.13

Predators alone, according to Bergerud (1980), would hold caribou populations around 0.4 per square kilometer (1/sq. mile). Herds without wolves or other predators increase up to 20 or more per square kilometer and then suffer starvation (e.g., Figure 12.10, page 212).

Caribou, moose, and their predators may represent an interesting example of net reproduction curves with two stable points (Figure 14.12). The lower equilibrium may be caused by predation and the upper equilibrium by food shortage (Sinclair 1981).

Spectacular examples of the influence of predators have occurred where humans have accidentally introduced a new predator. A striking example is the virtual elimination of the lake trout fishery in the Great Lakes by the sea lamprey (*Petromyzon marinus*). The marine lamprey lives on the Atlantic coast of North America and migrates into fresh water to spawn. The adult lampreys have a sucking, rasping mouth by which they attach themselves to the sides of fish, rasp a hole, and suck out body fluids. The passage of the lamprey to the upper Great Lakes was presumably blocked by Niagara Falls before the Welland Canal was built in 1829. The first sea lamprey was found in Lake Erie in 1921, in Lake Michigan in 1936, in Lake Huron in 1937, and in Lake Superior in 1945 (Applegate 1950). Lake trout catches decreased to virtually zero within about 20 years of the lamprey invasion (Figure 14.13). Control efforts have been applied to reduce the lamprey population since 1951 and lamprey are now rare, and attempts have been made to rebuild the Great Lakes fishery by releasing trout bred in hatcheries. Lake trout increased greatly in all the lakes during the 1970s and 1980s and are now approaching their former abundance (MacCallum and Selgeby 1987).

We conclude that in some but not all cases, the abundance of predators does

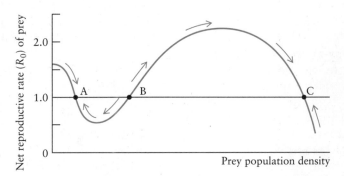

Figure 14.12 *Simple population model with two stable points. Point A is a stable equilibrium, as is point C. Point B is an unstable equilibrium that marks a boundary from which populations always diverge. This type of model might be appropriate to caribou populations in which the lower equilibrium is set by predators and the upper equilibrium by food shortage. Compare with Figure 12.2.*

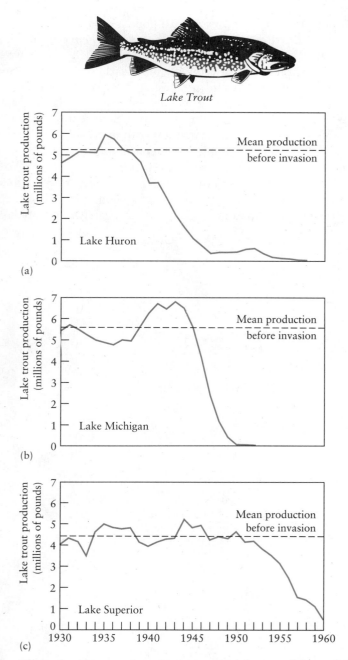

Lake Trout

(a)

(b)

(c)

Figure 14.13   *Effect of sea lamprey introduction on the lake trout fishery of the upper Great Lakes. Lampreys were first seen in (a) Lake Huron and (b) Lake Michigan in the 1930s and in (c) Lake Superior in the 1940s. (After Baldwin 1964.)*

influence the abundance of their prey in field populations. This raises an important question: *What is it about certain predators that makes them effective in controlling populations of their prey?* Can we find some type of system by which we can classify predators? This question has great economic implications both in the management of fish and wildlife populations and in agricultural pest control. It is, of course, possible to proceed in a case-by-case manner and to investigate each individual predator-prey system on its own, but this is clearly inefficient, and we would rather attempt to reach some generalizations that apply to many individual cases.

Let us begin by assuming a simple one predator–one prey system and ask how predators can respond to an increase in prey population density. Four possible responses are: (1) a *numerical* response, in which the density of predators in a given area increases by reproduction; (2) a *functional* response, in which the number of prey eaten by individual predators changes; (3) an *aggregative* response, in which

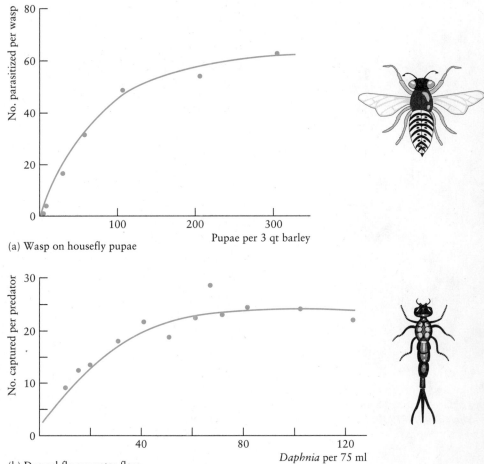

(a) Wasp on housefly pupae

(b) Damsel fly on water fleas

Figure 14.14   *Functional responses from an insect parasitoid and a predatory larva in the laboratory. (a) The parasitoid wasp* Nasonia vitripennis *parasitizing housefly pupae. (b) The larvae of the damsel fly* Ischnura elegans *feeding on the water flea* Daphnia magna. *(After Hassell 1976.)*

individual predators move into and concentrate in certain areas within the study area; and (4) a *developmental* response, in which individual predators eat more or less prey as predators grow toward maturity (Murdoch 1971). Considerable theoretical and practical work has been done on the functional, aggregative and developmental responses since the early analyses by Holling (1959).

The functional responses of many predators rise to a plateau as prey density increases. Figure 14.14 shows two examples. The upper plateau of these functional responses is fixed by *handling time*—the time it takes for a predator to catch, kill, and eat a prey organism. The curve rises rapidly when the searching capacity of the predator is high.

The developmental response occurs because predators are often growing and maturing during laboratory and field studies of predation. Figure 14.15 illustrates the effects of a functional response and the additional effects of the developmental response on the number of mosquito larvae eaten by backswimmers (*Notonecta hoffmanni*) in the laboratory. Backswimmers grow more rapidly at higher food levels and this explains the rise in the curve for total consumption in Figure 14.15 (Murdoch and Sih 1978).

A numerical response of predators can occur because of reproduction by the predator and an aggregative response by the movements or concentration of preda-

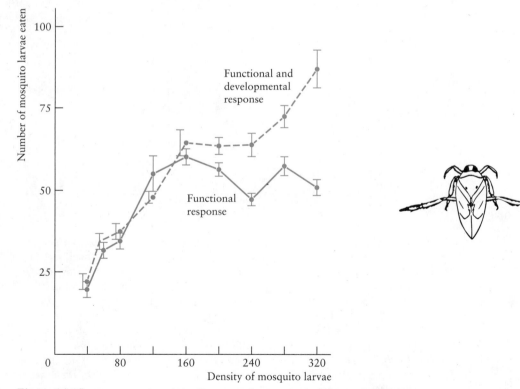

Figure 14.15 *Functional and developmental responses of the predatory insect* Notonecta *feeding on mosquito larvae in the laboratory. The prey consumption rate is measured by the number of mosquito larvae eaten per day. At high food levels* Notonecta *grow larger faster, and thus the combined functional and developmental response curve accelerates upward. (From Murdoch and Sih 1978.)*

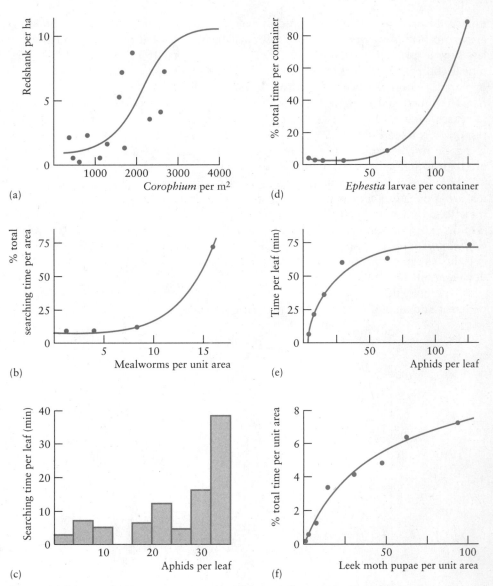

Figure 14.16 *Aggregative responses of predators and insect parasites arising from the tendency to congregate in regions of high prey density. Prey abundance is on* x *axis, predator density on the* y *axis. (a) Redshank (*Tringa totanus L.*) density in relation to the average density of its amphipod prey (*Corophium volutator *Pallas) per m². (b) Percentage of total searching time by great tits (*Parus major L.*) for different densities of mealworms (*Tenebrio mollitor L.*). (c) Searching time of a coccinellid larva (*Coccinella septempunctata L.*) for different densities of its aphid prey (*Brevicoryne brassicae L.*) per cabbage leaf. (d) Percentage of total searching time by the Ichnuemonid parasite (*Nemeritis canescens*) for different densities of its host (*Ephestia cautella Walk.*) per container. (e) Searching time of the braconid parasite (*Diaretiella rapae*) for different densities of its aphid host (*Brevicoryne brassicae) per cabbage leaf. (f) Percentage of total searching time by the parasite (*Diadromus pulchellus*) for different densities of leek moth pupae (*Acrolepia assectella Zell.*) per unit area. (After Hassell and May 1974.)*

tors in areas of high prey density. Predators are usually mobile and do not search at random but concentrate on patches of high prey density (Figure 14.16). There is an enormous literature on the behavioral ecology of predators in choosing prey patches that is part of *optimal foraging theory* (Stephens and Krebs 1986). The ability of predators to aggregate in patches of high prey density is a critical element in determining how effective the predator can be at limiting prey populations.

Within a patch, the searching efficiency of a predator becomes crucial to its success. Searching efficiency can vary with temperature and can also decrease at high predator densities because of interference (Ives 1981, Hassell 1981).

Much of the work on predator-prey models has been on laboratory populations and has proved difficult to translate to field populations (Gilbert et al. 1976). We cannot therefore give more than a vague answer to our general question about what makes some predators effective in controlling their prey. Much of the theoretical work has concentrated on single-species systems, and there is a need to consider the more complex cases of predators that feed on several prey species (Hassell 1979).

If we can measure the functional, numerical, developmental and aggregative responses for a predator-prey system, we can determine the total response of the predators (Figure 14.17). If the total response is sigmoid, the predator may limit the density of the prey at low prey densities. The key question is whether mortality imposed by the predators on the prey increases as prey density goes up. Mortality rate increases with prey density only for the first part of the sigmoid curve (up to C in Figure 14.17c). At higher prey densities, the predator will exert no controlling influence on the prey because it will be swamped by prey numbers. For example, the bay-breasted warbler (*Dendroica castanea*) increased 12-fold during an outbreak of the spruce budworm (*Choristoneura fumiferana*) in eastern Canada. Both a numerical and functional response occurred in this warbler (Figure 14.18), but an 8000-fold increase in the budworm reduced this predator to an insignificant agent of loss (Morris et al. 1958).

One important general implication of our analysis is that predators may have important effects on prey abundance when prey populations are low and become unimportant when prey densities are high. Populations of this sort can exist in two different phases, a low-density *endemic* phase and a high-density *epidemic* phase, as illustrated in Figure 14.12. Some forest insect pests, such as the spruce budworm

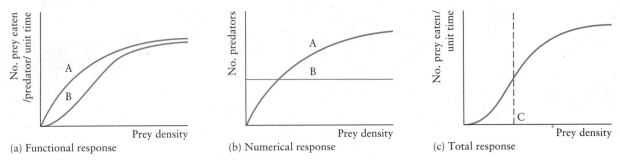

Figure 14.17 Alternative ways of obtaining a sigmoid total response between prey density and prey eaten. The total response curve may be obtained by combining functional response A or B with numerical response A or combining functional response B with any numerical response. Below prey density C the predator may control the density of the prey. This figure is equivalent to Figure 14.12. (After Hassell 1981.)

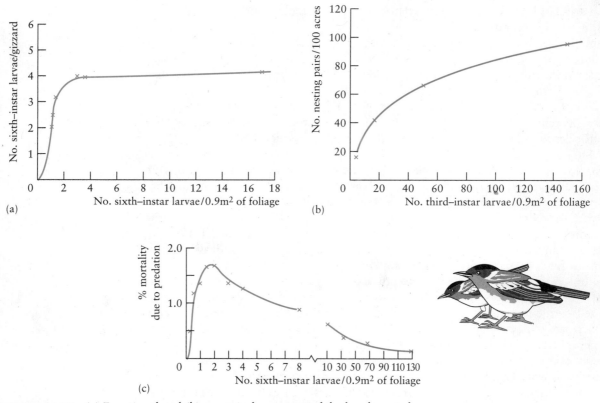

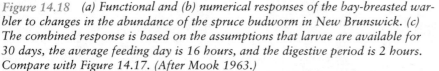

Figure 14.18 *(a) Functional and (b) numerical responses of the bay-breasted warbler to changes in the abundance of the spruce budworm in New Brunswick. (c) The combined response is based on the assumptions that larvae are available for 30 days, the average feeding day is 16 hours, and the digestive period is 2 hours. Compare with Figure 14.17. (After Mook 1963.)*

(Morris 1963), show these two density phases, and the key to the endemic phase may be the action of predators at low budworm densities.

When a predator has a choice of two different foods, the situation becomes more complex. The predator may have a fixed preference for one species of prey over the other, and this preference may not change as the composition of the available prey changes. But in other cases, the preference for one prey type depends on the proportions of prey types available, and the predator switches from preferring one prey type to preferring the other. Figure 14.19 illustrates this *switching* in the predatory snails *Nucella* (*Thais*) and *Acanthina*. These marine snails prefer to eat mussels when barnacles comprise less than 75 percent of the available prey, but change to prefer barnacles when barnacles reach 80 percent or more of the available food items (Murdoch 1969). Switching seems to occur most often between prey that are nearly equally preferred in the diet. Switching may be common in social animals that feed in flocks (Murdoch and Oaten 1975).

Switching can be important in predators that feed on several types of prey because it could act to stabilize the density fluctuations of the prey species. As one prey species increases in abundance relative to the others, the predator would concentrate its feeding on the more abundant prey species and possibly restrict the prey's

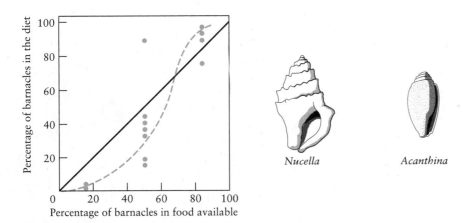

Figure 14.19   Switching in the predatory marine snails Nucella *and* Acanthina *feeding on barnacles and mussels in the laboratory. The 45 degree line is the line of no preference. (After Murdoch and Oaten 1975.)*

population growth. Conversely, switching to alternative foods may help the prey population to recover if it falls to a low level. Switching behavior could thus benefit the predator by allowing it to maintain a stable population size.

Some predators do seem to maintain stable population densities by living on alternative foods when their main prey is scarce. A tawny owl (*Strix aluco*) population near Oxford, England, was remarkably stable despite of great fluctuations in the abundance of the small rodents that served as the main prey species (Figure 14.20). The amount of successful reproduction in tawny owls was determined by the abundance of the prey rodents, but reproductive success did not influence subsequent population size of the owls.

Much of predation theory has been directed toward the question of how predators might stabilize prey populations (Murdoch and Oaten 1975, Hassell 1978, Murdoch and Bence 1987). It is clear from many field studies that predators do not necessarily stabilize prey numbers, as we have just seen with the tawny owl (Figure 14.20). The fact that some predator-prey models (Figures 14.1 and 14.6) result in oscillations appears to be strikingly applicable to some biological systems. The Canada lynx (*Lynx canadensis*) eats snowshoe hares (*Lepus americanus*) and both species show dramatic cyclic oscillations in density with peaks every 9 to 10 years. Charles Elton analyzed the records of furs traded by the Hudson's Bay Company in Canada for over 200 years and showed that the cycle is a real one that has persisted unchanged for at least 200 years (Figure 14.21) (Elton and Nicholson 1942). This lynx-hare cycle has been interpreted as an example of an intrinsic predator-prey oscillation, but this has been disputed. Keith (1983) has shown that food shortage during the winter initiates the decline in hare numbers, and predators play a secondary role in prolonging the decline to low numbers. Trostel and coauthors (1987) argue that predation may in fact be responsible for the snowshoe hare cycle. Although lynx depend on snowshoe hares, and are thus food-limited, it is not yet known if hares are predator-limited or both food- and predator-limited (Keith 1987).

Islands can provide natural experiments in predator removal to investigate predator-prey interactions. Lizards (*Anolis* spp.) and spiders are conspicuous animals on small islands in the Caribbean. Schoener and Toft (1983) surveyed 93 small

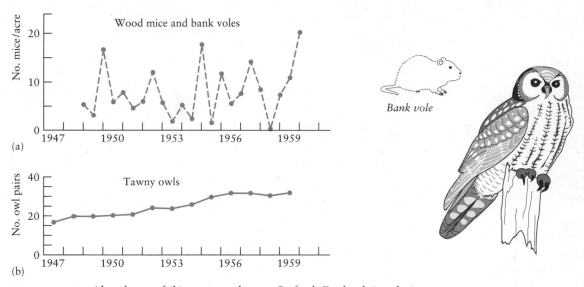

(a)

(b)

Figure 14.20   *Abundance of (b) tawny owls near Oxford, England, in relation to changes in abundance of their principal prey species, (a) the wood mouse and bank vole. Changes in prey abundance are not reflected in the avian predator's population size. (Data from Southern 1970.)*

islands in the Bahamas and counted all the orb-weaving spiders. They found ten times as many spiders on islands that had no lizards, compared with islands that had lizards present. Lizards are known to eat spiders. To analyze this system experimentally, Spiller and Schoener (1988) constructed enclosures on one island that contained both lizards and spiders, and then removed lizards from three replicate enclosures. Figure 14.22 shows the results of this experiment. Spider numbers increased dramatically in the enclosures that had no lizards. The interaction between lizards and spiders on these Caribbean islands involves both predation by lizards on

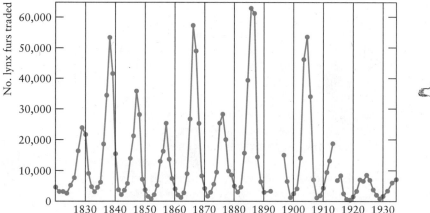

Figure 14.21   *Canada lynx fur returns of the Northern Department, Hudson's Bay Company, 1821-1913, and of the equivalent area 1915–1934. (After Elton and Nicholson 1942.)*

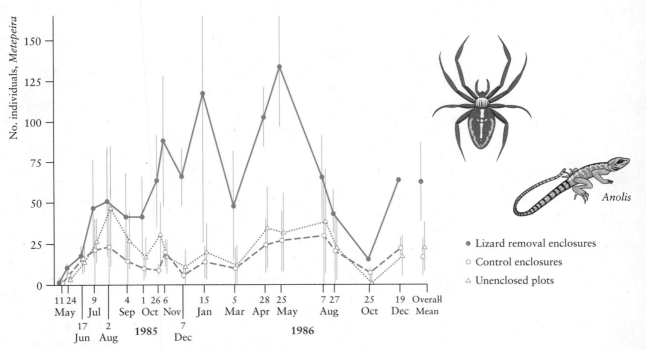

Figure 14.22 Mean density of the orb spider Metepeira datona *in lizard removal enclosures, control enclosures, and in open plots on Staniel Cay, Bahamas. Spider densities increase 3–5 times in plots from which lizards were removed. (From Spiller and Schoener 1988.)*

spiders and competition for food. The diets of lizards and spiders overlap considerably and lizards consume many of the larger insects that spiders could also eat (Spiller and Schoener 1990).

EVOLUTION OF PREDATOR-PREY SYSTEMS

One of the striking features of the simple models of predator-prey interactions is that these models are often unstable. Oscillations are common in many predator-prey models (May 1972) but are not common in the real world. One way in which we can explain the stability of real predator-prey systems is to postulate that natural selection has changed the characteristics of both predators and prey so that their interactions produce population stability. Evolutionary changes in two or more interacting species are called *coevolution*, and we are concerned in this case with the coevolution of predator-prey systems.

If one predator is better than another at catching prey, the first individual will probably leave more descendants to subsequent predator generations. Thus predators should be continually selected to become more efficient at catching prey. The problem, of course, is that by becoming too efficient, the predator will exterminate the prey and then suffer starvation. The prey at the same time are being selected to be better at escaping predation. Because of the conflicting adaptive goals of predator and prey, many evolutionists have described predator-prey evolution as an "arms race" (Dawkins and Krebs 1979). Predators may have an inherent disadvantage in the arms race because of the "life-dinner" principle. This principle states that selection will be stronger on the prey than on the predator because a prey individual that loses the race loses its life, while the unsuccessful predator loses only a meal. Dawkins (1982) suggested that the inherent disadvantage of a predator could be offset if the predator is rare and the prey is common. In this case the predator will be only a minor selective agent on the whole prey population. Abrams (1986) criticized the arms race analogy for predator-prey coevolution. He found many theoretical situations in which the arms race would not occur, but suggested that there was usually an asymmetry in the evolutionary responses of predators and prey. Prey should always increase their investment in escape mechanisms if predators invest in becoming more efficient. But the reverse is often not true—predators do not always respond to prey investment—and it depends on the details of a specific predator-prey system. The arms race analogy may not be correct in many particular cases of predator-prey coevolution.

Two obvious constraints operate in systems with several species of predators and prey. The existence of several species of predators feeding on several species of prey places limits on predator efficiency. For example, one prey species may escape by hiding under rocks, while a second species may run very fast. Clearly, a predator is constrained by conflicting pressures to get very good at turning rocks over or to get very good at running and it is difficult to be good at both these activities. Conversely, we can imagine that the prey population is always being selected for escape responses. Because of several predators with different types of hunting strategy, prey will not be able to evolve specific escape behavior suitable to all species of predators.

One persistent belief about predator-prey systems is that predators typically capture substandard individuals from prey populations, so that weak, sick, aged and injured prey are culled from prey populations (Morse 1980). Temple (1987) tested this idea by flying a trained red-tailed hawk (*Buteo jamaicensis*) and measuring the outcome of attacks on chipmunks, rabbits and gray squirrels. Table 14.2 shows that the more difficult the prey is to catch, the higher is the fraction of substandard individuals that are caught. Gray squirrels are particularly difficult for red-tailed hawks to catch, and squirrels taken were in markedly poorer condition than squirrels in the general population. The same generalization seems to hold for other vertebrate predators—substandard individuals are captured disproportionately when the type of prey is difficult to catch but not when it is easy to capture. Thus hyenas in Africa take wildebeest in poor condition because wildebeest are difficult to capture, but take gazelles at random because they are easy to catch (Kruuk 1972).

The coevolution of predator-prey systems occurs most tightly when the predators strongly affect the abundance of the prey. In some predator-prey systems, the predator does not determine the abundance of the prey, so the evolutionary pressures are considerably reduced. In some cases, the prey has refuges available where the predator does not occur, or the prey may have certain size classes that are not vulnerable to the predator. In other cases (Figure 14.20), the predators have developed territorial behavior that restricts their own density so that they cannot easily respond to excessive numbers of prey animals (Taylor 1984).

Much of the stability we see in the natural world probably results from the continued coevolution of predators and prey. Predators that do not have prudence forced on them by their prey may exist for only a short time in the evolutionary record, and we are left today with a residue of highly selected predator-prey systems.

Predators need not limit the density of their prey species to play an important role in the evolution of prey characteristics. Two antipredator defense strategies are

Table 14.2 Predation by Red-Tailed Hawk on Substandard Individuals of Three Prey Species in Wisconsin.[a]

Species of Prey	% Attacks That Failed	Difficulty of Prey Caputre	% Excess of Substandard Individuals in Kills
Eastern chipmunk	72	easy	8
Cottontail rabbit	82	moderate	21
Gray squirrel	88	difficult	33

[a] Results based on 447 attacks.
Source: Modified from Temple (1987).

common in animals and illustrate how evolutionary pressures can affect predator-prey systems.

Warning Coloration

Among animals there is a widespread correlation between conspicuous coloration and the presence of aversive qualities for potential predators. For example, poisonous snakes like coral snakes are highly colored (Figure 14.23, color insert). Many butterflies and other insects are brightly colored, and contain poisons that are distasteful to predators. The theory of warning (or aposematic) coloration is usually put forward as an explanation of this correlation (Guilford 1988).

Mechanisms of prey defense using warning coloration must evolve by increasing the chances of survival of the individuals in which they are found. But for distasteful species, the predator must first eat one individual before the predator learns to avoid other prey of similar color. If the prey are gregarious and nearby individuals are closely related, kin selection would operate to favor the warning coloration (Fisher 1958). If only a few siblings are sampled from a large brood and the predator learns to avoid other individuals of the group, an allele for distastefulness can increase in frequency by kin selection. Predators do in fact seem to learn very quickly to avoid distasteful insects (Brower 1988).

Coral snakes are highly colored with red, yellow and black rings. There are 120 species of coral snakes in tropical America, and all are extremely poisonous. Many other nonpoisonous snakes have evolved color patterns to mimic the appearance of coral snakes. These nonpoisonous snakes are called Batesian mimics* because they mimic the color patterns of unrelated poisonous species (see Figure 14.23 in color insert). Birds that live in areas occupied by coral snakes have an innate tendency to avoid snakes with these color patterns (Smith 1980), so that a predator does not need a lethal encounter to avoid the poisonous species (Pough 1988). The mimic species can profit from looking like a poisonous or unpalatable prey species, and this coevolution has been particularly well developed in tropical species groups.

Group Living

Why do birds form flocks and ungulates go about in herds? What are the advantages of group living? First, we should be clear that not all birds flock and

---

* A Batesian mimic is a sheep in wolf's clothing.

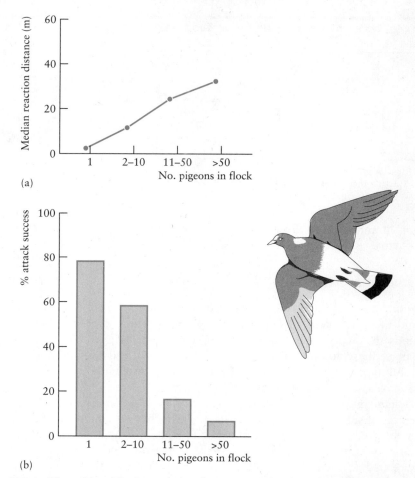

(a)

(b)

*Figure 14.24   The value of flocking in wood pigeons (*Columba palumbus*), measured by the distance at which they detected the approach of a goshawk (*Accipiter gentilis*). Goshawk attack success is much reduced against flocks. (From Kenward 1978.)*

some ungulates are solitary. Group living is not always advantageous, and when it is, there may be several reasons why. One possible reason for living in groups is to reduce the risk of predation (Alcock 1989).

There are three main advantages a prey organism can obtain by living in a group. First, early detection of predators may be facilitated if many eyes are looking. Figure 14.24 illustrates this early detection in wood pigeons. Prey in groups may thus be able to spend more time feeding and less time looking for potential predators.

Second, if the prey are not too much smaller than the predator, several of them acting together may be able to deter the predator from attacking. One possible example of this is mobbing in birds. Primate troops will also mob predators in some circumstances, and musk-oxen threatened by wolves gather into a circular defense formation.

Third, if the predator is still able to attack a group, it must select one individual in the group to capture. By fleeing in confusion, the prey may disconcert the predator,

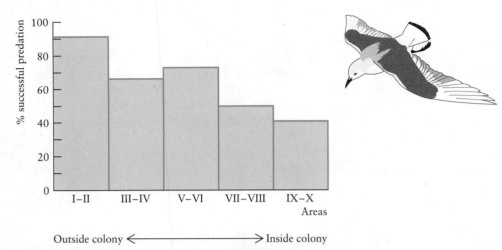

Figure 14.25   *The value of nesting near the center of a gull colony. Egg predation by crows* (Corvus corone) *in colonies of the black-headed gull* (Larus ridibundus) *in England. (Data from Kruuk 1964.)*

who may not be able to concentrate on any one individual. Alternatively, by being in the center of a group, an individual may reduce its chances of being eaten. Figure 14.25 illustrates this advantage of grouping in the case of egg predation in colonies of black-headed gulls.

Many of the most striking characteristics of animal structure and social behavior are adaptations concerned with predation.

SUMMARY

Species may interact by predation. Simple mathematical models can be used to describe this interaction. When generations are discrete, simple models can produce stable equilibria of predator and prey but usually produce oscillations in the numbers of both species. When generations are continuous, graphic models developed by Rosenzweig and MacArthur can be used to evaluate the equilibrium levels and the stability of predator-prey systems. Both stable equilibria and cyclic oscillations may occur. All these simple models make the assumption that the world is homogeneous (one habitat), there are no prey refuges, and only one predator species eats one prey species. Relaxing these assumptions leads to more complex models.

In laboratory systems of predators and prey, cyclic oscillations are produced only in complex environments, and most simple systems do not show stability but are self-annihilating. Laboratory systems may show gradual evolutionary changes toward greater stability over a short number of generations by coevolution of predators and prey.

Field populations will be models of predator-prey systems only if predators have a strong effect on the abundance of their prey. This assumption can be tested by predator-removal experiments. In some but not all cases studied, the abundance of predators does influence the abundance of prey. The properties of effective predators can be described in a general manner, but we cannot at present predict which predators will be good agents of prey control without actually doing field tests. Both the predator and the prey species are affected by many other factors in the environment, and consequently the population trends shown with simple predator-prey models are not obtained in field populations. Predator-prey systems always involve a coevolutionary race in which prey are selected for escape and predators for hunting ability. These systems stabilize most easily when several species are involved, when prey have refuges safe from predators, and when predators take old animals of little reproductive value. Many characteristic structures and behavior patterns of animals are adaptations concerned with predation.

Predation is a major process in the organization of communities of species, and we shall return to consider its impact on the structure of communities in Chapters 24 and 25.

Selected References

ABRAMS, P. A. 1986. Adaptive responses of predators to prey and prey to predators: The failure of the arms-race analogy. *Evolution* 40:1229–1247.

ARDITI, R., and L. R. GINZBURG. 1989. Coupling in predator-prey dynamics: ratio-dependence. *Journal of Theoretical Biology* 139:311–326.

CAUGHLEY, G., G. C. GRIGG, J. CAUGHLEY, and G. J. E. HILL. 1980. Does dingo predation control the densities of kangaroos and emus? *Australian Wildlife Research* 7:1–12.

HASSELL, M. P. 1981. Arthropod predator-prey systems. In *Theoretical Ecology*, ed. R. M. May, pp. 105–131. Blackwell, Oxford, England.

HOLLING, C. S. 1965. The functional response of predators to prey density and its role in mimicry and population regulation. *Memoirs of the Entomological Society of Canada* 45:1–60.

KAREIVA, P. 1987. Habitat fragmentation and the stability of predator-prey interactions. *Nature* 326:388–390.

MURDOCH, W. W., and J. BENCE. 1987. General predators and unstable prey populations. In *Predation: Direct and Indirect Impacts on Aquatic Communities*, ed. W.C. Kerfoot and A. Sih, pp. 17–30. University Press of New England, Hanover, N. H.

ROSENZWEIG, M. L., and R. H. MACARTHUR. 1963. Graphical representation and stability conditions of predator-prey interactions. *American Naturalist* 97:209–223.

SPILLER, D. A., and T. W. SCHOENER. 1988. An experimental study of the effect of lizards on web-spider communities. *Ecological Monographs* 58:57–77.

TEMPLE, S. A. 1987. Do predators always capture substandard individuals disproportionately from prey populations? *Ecology* 68:669–674.

Questions and Problems

1. Plot the data from Figure 14.10 (page 273) on a phase-plane like Figure 14.5 (page 268), and sketch on the phase plane the predator isocline and the prey

isocline. What can you conclude about the stability properties of this predator-prey system?

2. Calculate the population changes for ten generations in a hypothetical predator-prey system with discrete generations in which the parameters for the prey are $B = 0.03$, $N_{eq} = 100$, $C = 0.5$, and starting density is 50 prey and for the predators, $Q = 0.02$ (or $S = 2.0$) and starting density is 0.2. How would the prey population change in the absence of the predators?

3. Assume that in the ruffed grouse predator-control experiments (page 275), nest losses were reduced but chick and adult survival were unchanged, and population density changes were unaffected by predator removal. Discuss the demographic mechanics of how this is possible.

4. Buckner and Turnock (1965) studied bird predation on the larch sawfly in Manitoba. They obtained the following data for the chipping sparrow (*Spizella passerina*):

| | PLOT I | | PLOT II | |
	Sparrows per Acre	Sawfly Larvae per Acre	Sparrows per Acre	Sawfly Larvae per Acre
1954	—	—	3.2	235,000
1956	—	—	2.9	33,400
1957	1.4	2,138,700	2.3	40,000
1958	0.5	879,400	2.5	41,200
1959	0.4	437,800	2.2	27,300
1960	0.2	354,300	2.2	54,600
1961	0.5	199,900	2.3	15,000
1962	1.1	191,800	5.0	3,200
1963	0.2	366,800	0.3	3,900

Plot the numerical response of chipping sparrows to changes in sawfly larval abundance, and discuss the differences between plots I and II for this predator-prey system.

5. The collapse of fish populations in the Great Lakes coincided with a general increase in pollution of the lakes. Discuss the hypothesis that the collapse of fish stocks in the Great Lakes (e.g., Figure 14.13, page 279) was caused more by pollution than by the introduction of the sea lamprey. Steedman and Regier (1987) give references.

6. How does the predation by herbivores on green plants differ from the predation of insect parasites on their hosts and the predation of carnivores on herbivores? Make a list of similarities and differences, and discuss how these affect the simple models of predation that we have discussed.

7. Introduce a time lag into the simple predator-prey model for discrete generations by making the predator density change in relation to prey density at generation $t - 1$ rather than generation t. Repeat the calculations for one of the examples discussed in the text, and determine what effect this time lag has on the system's behavior.

8. One of the "textbook" examples of the effects of predator control is the Kaibab deer herd. According to several textbooks, the removal of predators from the Kaibab mule deer population allowed the deer to increase dramatically in

numbers, to overgraze their food supply, and to starve. Trace the history of this example from Rasmussen (1941) to the critique by Caughley (1970).

9. Wildebeest in the Serengeti area of East Africa have a very restricted calving season. All females give birth within a space of three weeks at the start of the rainy season (Estes 1976). How would you test the hypothesis that this restricted calving season is an adaptation to reduce predation losses of calves?

10. The graphic model of Rosenzweig and MacArthur (Figure 14.5, page 268) predicts that the predator-prey system will become unstable when nutrients are added to the prey population (the paradox of enrichment). Evaluate the evidence for the occurrence of the paradox of enrichment in laboratory and field populations. Does the same prediction follow from ratio-dependent predation theory? McCauley and Murdoch (1990) and Arditi and co-workers (1991b) provide references and a discussion of the problem.

Overview Question

When populations of moose, caribou, or deer decline in Alaska or Canada, there is typically a great public pressure to instigate wolf control programs. What data would you collect to describe and understand the dynamics of this predator-prey system? List the alternative hypotheses you would like to test and their management implications.

15

Species Interactions: Herbivory and Mutualism

Plant-animal interactions are the focus of many population interactions. *Herbivory* is a major interaction in which animals prey on plants, and herbivory is traditionally considered a profit for the animals and a loss for the plants. Many *mutualisms* involve plant-animal interactions that are by contrast a gain for both species. In this chapter we shall discuss herbivory and mutualism to assess their impact on abundance and their evolutionary origins.

Herbivory is just a special kind of predation. Over half of the macroscopic species on earth are plants and herbivores, and consequently a major part of species interactions involve herbivory. In this chapter we will cover some of the specific relationships that herbivores have with plants. The uniqueness of these relationships is often only a reflection of the simple fact that most plants cannot move, so "escape" from herbivores can be achieved only by some clever adaptations. Herbivores can be important selective agents on plants, thus the evolutionary interplay between plants and animals is a major theme in this chapter.

DEFENSE MECHANISMS IN PLANTS

The world is green, and there are three possible explanations for this. First, some herbivore populations may evolve self-regulatory mechanisms that prevent them from destroying their food supply (see Chapter 16). Or, second, other control mechanisms, such as predation, may hold herbivore abundance down. Third, all that is green may not be edible. Plants have evolved an array of defenses against herbivores, and this has set up a coevolution game with plants and herbivores trying to outwit each other in evolutionary time.

Plants may discourage herbivores by structural adaptations, as anyone who has tried to prune a rosebush will attest, but they may also use a variety of chemical weapons that we are now starting to appreciate (Feeny 1992). Plants contain a variety of chemicals that have always puzzled plant physiologists and biochemists. These chemicals are found only in some plants and not in others and have been called secondary plant substances. These substances are byproducts of the primary metabolic pathways in plants, and Figure 15.1 gives a simplified view of the origin of some of the major chemical groups of secondary plant substances. A number of these substances are familiar to us already. Juglone is an acetogenin produced by walnut trees (page 81). The characteristic spices cinnamon and cloves are phenylpropanes found in some herbs. Peppermint oil and catnip are terpenoids. Nicotine is an alkaloid found in tobacco plants. Morphine and caffeine are other alkaloid secondary plant substances.

Two views have developed about the function of secondary plant substances.

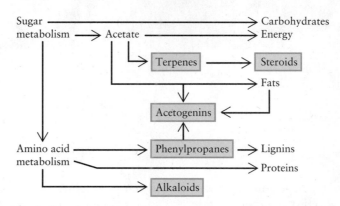

Figure 15.1   Metabolic relationships of the major groups of plant secondary substances (shown in boxes) to the primary metabolism of plants. (After Whittaker and Feeny 1971.)

One view is that these compounds are primarily waste products of plant metabolism and that the evolutionary origin of secondary plant substances can be understood as an adaptation to avoid autointoxication (Muller 1970). Excretion is a necessary part of metabolism and in plants takes much different forms from excretion in animals. According to this view, plants have evolved numerous ways of eliminating toxic organic chemicals by volatilization or leaching and other ways of rendering toxic substances harmless by chemical alterations within the plant. This done, the plant may now be in a position to use these substances for its benefit in one of two ways. First, by releasing chemicals into its immediate environment, a plant may be able to suppress competitors, which are poisoned by the excretory products. This produces allelopathic effects (Chapter 6). Second, by accumulating some chemicals in its leaves or stem, a plant may become toxic or distasteful to herbivores.

A somewhat different view of the origin of secondary plant substances is that they are chemicals specifically evolved by plants to thwart herbivores (Ehrlich and Raven 1964). This view assumes that few secondary plant substances are really needed by plants as excretory products, and most substances are actively produced at a metabolic cost to the plant. Only because plants that produce secondary substances are at a selective advantage is there such a chemical variety in different plant groups. If all animals could be removed from the community, secondary substances would not be produced by plants.

The plant-defense view argues that herbivores have a strong impact on plant fitness, and that well-defended plants are more fit. If plant defense characteristics are inherited, then all the elements needed for natural selection are present.

If plant defense has a cost in terms of plant fitness, we can make four general predictions:

1. Organisms evolve more defenses if they are exposed to much damage and fewer defenses if the cost of defense is high.

2. More defenses are allocated within an organism to valuable tissues that are at risk.

3. Defense mechanisms are reduced when enemies are absent and increased when plants are attacked.

4. Defense mechanisms are costly and cannot be maintained if plants are severely stressed by environmental factors.

The cost of defense is due to the diversion of energy and nutrients from other needs. Much evidence to support these four predictions has now accumulated. The autointoxication view is now thought incorrect, and the hypothesis that secondary compounds have an ecological role as antiherbivore compounds has become a fruitful and exciting area of research (Stamp 1992). Secondary compounds in plants are not static end products of metabolism but have rapid turnover rates in the metabolic pool.

The first general theory to attempt to explain plant-herbivore interactions was the plant apparency theory of Feeny (1976) and Rhoades and Cates (1976). The major premise of this theory is that the type of defenses employed by the plant depend upon how easily a herbivore can find the plant. The plant apparency theory suggested that plants that are easily found by herbivores evolve chemical defenses of a different type from those used by plants that are difficult for herbivores to locate. Short-lived plants may be able to escape the attention of herbivores by developing so quickly that their herbivores are unlikely to discover them—such plants are "unapparent" to their herbivores. In contrast, "apparent" plants are sure to be found by herbivores because they are long-lived.

Defense mechanisms in plants may be quantitative and secondary compounds may vary in concentration from low to high values. For example, tannins and resins in leaves may occupy up to 60 percent of the dry weight of a leaf (Feeny 1976). Quantitative plant defenses should be used by apparent plants. Other defense mechanisms may be qualitative because the compounds involved are present in very low concentrations (less than 2 percent dry weight). Examples are alkaloids and cyanogenic compounds in leaves. Qualitative defenses protect plants against generalized herbivores that are not adapted to cope with the toxic chemicals, but they do not stop specialized herbivores that have evolved detoxifying mechanisms in the digestive system. Qualitative defenses should be used by "unapparent" plants (Feeny 1976).

Both the type of defense and the amount of defense used in plants depend on the vulnerability of the plant tissues. Growing shoots and young leaves are more valuable to plants than mature leaves, so plants typically invest more heavily in the defense of growing tips and young leaves (Feeny 1976). Tannins, resins, alkaloids, and other defense chemicals are concentrated at or near the surface of the plant, thereby increasing their effectiveness.

The plant apparency theory stimulated a great deal of work on plant defense during the 1970s and 1980s, and it was soon found to be inadequate as an explanation of plant defense. Some apparent plants have qualitative, toxic defenses, and some unapparent species use quantitative defenses. For some plants it is difficult to decide whether or not they are apparent to their herbivores, not just to humans. It was clear that a more general theory was needed (Feeny 1992).

One element missing from the previous ideas was the ability of a plant to defend itself, given the resources available to it. The *resource availability hypothesis* was proposed by Coley and colleagues (1985) to take into account the plant's ability to replace tissues taken by herbivores. Fast-growing and slow-growing plants have a suite of physiological characteristics that are summarized in Table 15.1. Herbivores prefer fast-growing plants and tend to avoid slow-growing plants. Because each leaf represents a greater investment for a slow-growing plant, slow-growing plants have

Table 15.1   Characteristics of Inherently Fast-growing and Slow-growing Plant Species

Variable	Fast-growing Species	Slow-growing Species
GROWTH CHARACTERISTICS		
Resource availability in preferred habitat	High	Low
Maximum plant growth rates	High	Low
Maximum photosynthetic rates	High	Low
Respiration rates in darkness	High	Low
Leaf protein content	High	Low
Responses to pulses in resources	Flexible	Inflexible
Leaf lifetimes	Short	Long
Successional status	Often early	Often late
ANTIHERBIVORE CHARACTERISTICS		
Rates of herbivory	High	Low
Amount of defense metabolites	Low	High
Type of defense	Qualitative (alkaloids)	Quantitative (tannins)
Turnover rate of defense	High	Low
Flexibility of defense expression	More flexible	Less flexible

Source: Coley et al. (1985).

more to lose to herbivores and thus invest more in defensive chemicals. Figure 15.2 summarizes a conceptual model of the resource availability hypothesis. The higher the plant's growth rate, the lower the predicted investment in defense.

Some plants can avoid the cost of producing defensive compounds when they may not be needed. If a plant is then attacked by a herbivore, it may be induced to increase its defenses. Induction times for defensive reactions by plants have been studied for only a few species and vary from 12 hours to 1 year or more (Tallamy and Raupp 1991). Rapid defensive responses in plants have been unexpected and are now the subject of intensive interest in plant-herbivore research. If defenses are costly, we would expect a relaxation of defenses after a herbivore's attack, but few measurements have yet been made (Neuvonen and Haukioja 1991).

By concentrating on the herbivore-plant interaction, we should not forget that herbivores must also be concerned about their enemies. Herbivores can feed only on plants that allow them to avoid predators and parasites (Jeffries and Lawton 1984). The evolution of plant feeding preferences must be integrated with the evolution of enemy avoidance (Stamp 1992).

Herbivores do not, of course, sit idly by while plants evolve defense systems (Strong et al. 1984). Animals circumvent plant defenses either by evolving enzymes to detoxify plant chemicals or by timing the herbivore's life cycle to avoid the noxious chemicals of the plants. The coevolution of animals and plants can thus occur, and we shall examine three cases to illustrate this.

Tannins in Oak Trees

The common oak (*Quercus robur*) is a dominant tree in the deciduous forests of Western Europe and is attacked by the larvae of over 200 species of *Lepidoptera*,

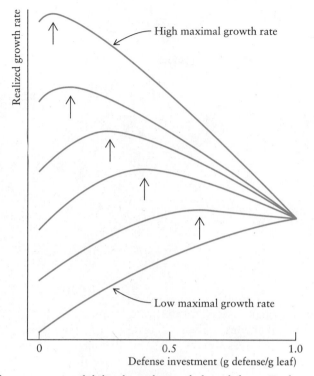

Figure 15.2   The resource availability hypothesis of plant defense. Each curve represents a plant species with a different maximum growth potential. Levels of defense that maximize growth rate are indicated by an arrow. Slow growing plants should invest much more in defense because losses to herbivores are more difficult to replace. (From Coley, Bryant, and Chapin 1985.)

more species of insect attackers than any other tree in Europe. The attack of insects is concentrated in the spring (Figure 15.3), with a smaller peak of feeding in the fall. One of the most common oak insects is the winter moth, whose larvae feed on oak leaves in May and drop to the ground to pupate in late May. Why is insect attack concentrated in the spring? One possibility is that oak leaves change with age to become less suitable insect food (Feeny 1970).

If winter moth larvae are fed "young" oak leaves, they grow well, but if larvae are fed slightly "older" leaves, they grow very poorly.

Winter Moth Larvae Diet	Mean Peak Larval Weight (mg)
May 16 oak leaves—natural diet ("young")	45
May 28—June 8 leaves ("old")	18

No adults emerged from the larvae fed older leaves. Thus some change occurs very rapidly in oak leaves in the spring to make them less suitable for winter moth larvae. The most obvious changes in oak leaves during the spring are a rapid darkening and an increase in toughness. The thin oak leaves of May become thick and more difficult to tear by early June (Figure 15.4). If leaf toughness is a sufficient explanation for

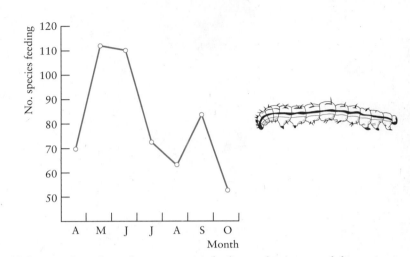

Figure 15.3 *Number of* Lepidoptera *species feeding as larvae on oak leaves in Britain from April to October. (After Feeny 1970.)*

the feeding pattern of oak insects in the spring, grinding up the older leaves should provide an adequate diet. But if chemical changes have occurred as well as toughening, ground-up older leaves should still be inadequate as a larval diet. Ground-up leaves seem to be an adequate diet, at least until early June:

Larvae Fed Ground-Up Leaves	Mean Peak Larval Weight (mg)
May 13 ("young" leaves)	37
June 1 ("old" leaves)	35

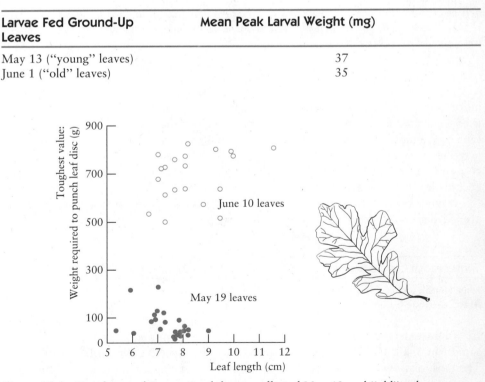

Figure 15.4 *Toughness of "young" oak leaves collected May 19 and "old" oak leaves collected June 10. (After Feeny 1970.)*

If mature oak leaves can provide an adequate diet, why was there not natural selection in favor of insect mouthparts able to cope with tough leaves? Some *Lepidoptera* do feed on summer oak leaves, so that it is possible to feed on tough leaves. If mature oak leaves later in summer are relatively poor nutritionally, compared with young spring leaves from May and early June, this would produce natural selection toward early feeding.

Two chemical changes in oak leaves seem to be significant for feeding insects. The amount of tannins in the leaves increases from spring to fall (especially after July), and the amount of protein decreases from spring to summer and remains low from June onward. Tannins are secondary plant substances that may act to reduce palatability and to discourage herbivores. Larval weights of winter moths are significantly reduced if their diet contains as little as 1 percent oak-leaf tannin. Tannins act by forming complexes with proteins and may act in oak leaves by tying up proteins in complexes that insects cannot digest and utilize.

Nevertheless, some insects have evolved ways of minimizing the effect of tannins. Insects that feed on oak leaves in the summer and fall tend to grow very slowly, which may be an adaptation to a low-nitrogen diet. Table 15.2 shows that many of the late-feeding insects on oak overwinter as larvae and complete their development on the spring leaves. Many others are leaf miners, which may avoid tannins by feeding on leaf parts that contain little tannin.

Thus the oak tree has defended itself against herbivores by the use of tannins as chemical defenses and leaf texture (toughness) as structural defense. Herbivores have compensated by concentrating feeding in the early spring on young leaves and by altering life cycles in the summer and fall.

Ants and Acacias

A mutualistic system of defense has been achieved by the swollen-thorn acacias and their ant inhabitants in the New World tropics. The ants depend on the acacia tree for food and a place to live, and the acacia depends on the ants for protection

Table 15.2 Larval Feeding Habits of Early-Feeding and Late-Feeding *Lepidoptera* Species on Leaves of the Common Oak in Britain

Feeding habit	Early-Feeding Species[a] (%)	Late-Feeding Species[b] (%)
Larvae complete growth on oak leaves in one season	92	42
Larvae complete growth on low herbs after initial feeding on oak leaves	3	11
Larvae overwinter and complete growth in following year	4	38
Larvae bore into leaf parenchyma (leaf miners)	3	26

Note: Some species exhibit more than one of the feeding habits, so the columns do not add to 100%.

[a] Early-feeding larvae are in May and June. Total of 111 species.

[b] Total of 90 species.

Source: After Feeny (1970).

(a)

(b)

(c)

Figure 15.5   *(a)* Acacia collinsii *growing in open pasture in Nicaragua. This tree had a colony of about 15,000 worker ants and was about 4 m tall. (b) Area cleared over ten years around a growing* Acacia collinsii *in Panama by ants chewing on all vegetation except the acacia. Machete in photo is 70 cm long. The area was not disturbed by other animals. (c) Swollen thorns of* Acacia cornigera *on a lateral branch. Each thorn is occupied by 20–40 immature ants and 10–15 worker ants. All the thorns on the tree are occupied by one colony. An ant entrance hole is visible in the left tip of the fourth thorn up from the bottom. (Photos courtesy of D. H. Janzen.)*

from herbivores and neighboring plants. Not all acacias (*Acacia* spp.), approximately 700 species, depend on the ants in the New World tropics, and not all the acacia ants (*Pseudomyrmex* spp.), 150 species or more, depend completely on acacia. In a few cases, a high degree of mutualism has developed, described in detail by Janzen (1966). Some of the species of ants that inhabit acacia thorns are obligate acacia ants and live nowhere else.

Swollen-thorn acacias have large, hollow thorns in which the ants live (Figure 15.5c). The ants feed on modified leaflet tips called Beltian bodies, which are the primary source of protein and oil for the ants, and also on enlarged extrafloral nectaries, which supply sugars. Swollen-thorn acacias maintain year-round leaf production, even in the dry season, providing food for the ants. If all the ants are removed from swollen-thorn acacias, the trees are quickly destroyed by herbivores and crowded out by other plants. Janzen (1966) showed that acacias without ants grew less and were often killed:

	Acacias with Ants Removed	Acacias with Ants Present
Survival rate over 10 months (%)	43	72
Growth increment		
May 25–June 16 (cm)	6.2	31.0
June 16–August 3 (cm)	10.2	72.9

Swollen-thorn acacias have apparently lost (or never had) the chemical defenses against herbivores found in other trees in the tropics.

The acacia ants continually patrol the leaves and branches of the acacia tree and immediately attack any herbivore that attempts to eat acacia leaves or bark. The ants also bite and sting any foreign vegetation that touches an acacia, and they clear all the vegetation from the ground beneath the acacia tree. Thus the swollen-thorn acacia often grows in a cylinder of space virtually free of all foreign vegetation (Figure 15.5b).

Thus the ant-acacia system is a model system of the coevolution of two species in an association of mutual benefit. The ants reduce herbivore destruction and competition from adjacent plants and thus serve as a living defense mechanism.

Spines in a Marine Bryozoan

Induced defense mechanisms are not confined to plants. Both freshwater and marine organisms produce spines or thickened shells in response to chemical cues released by the species that feed on them (Harvell 1986). These spines or thickened shells are inducible defenses and do not appear if the predator is not present in the area. In many cases these systems are predator-prey interactions, rather than plant-herbivore, but the principles of defense are the same—how can organisms respond morphologically to reduce their chances of being eaten (Myers and Bazely 1991)?

The marine bryozoan *Membranipora membranacea* produces large, chitinous spines very rapidly, within 36 hours of attack by nudibranch predators (Figure 15.6). Harvell (1986) analyzed the cues that triggered spine production in these modular organisms and how effective it was against predators. The presence of nudibranchs stimulated spine production, and this occurred even in bryozoan colonies distant

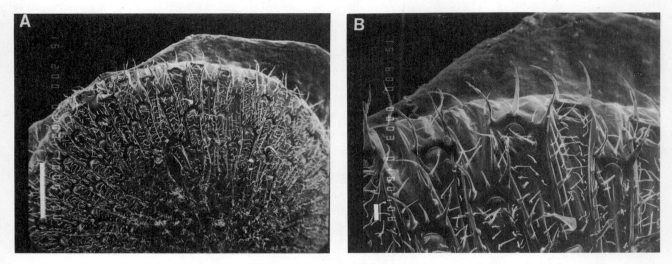

Figure 15.6   Inducible spines on colonies of the marine bryozoan Membranipora membranacea. *Spines are typically produced as a peripheral band of one or two zooids. (a) Colony view (field, 20 mm). (b) Zooid view (field, 100 μm). These spines deter feeding by nudibranch predators. (From Harvell 1986.)*

from the nudibranchs. A chemical cue released from the nudibranchs into the water is presumed to be the triggering cue for spine production. On spined colonies nudibranchs ate only 40 percent as many zooids per day as those on unspined colonies. Producing spines is thus effective for reducing damage in bryozoans. There is a cost to spine production. Harvell (1986) found that bryozoan colonies producing spines grew at only 85 percent the rate of unspined colonies. In these colonial animals, a decrease in growth is converted directly to a decrease in reproductive output. The key feature of this bryozoan-nudibranch system is that the consumer does not eat all of the bryozoan colony, and induced defenses are possible to deter further losses. But this benefit of defense has a cost in energy diverted from growth into spine production.

Thorns, spines, and prickles also occur widely on terrestrial plants, and everyone assumes that they act as physical defenses against herbivores. There is remarkably little evidence that this is true (Myers and Bazely 1991). A variety of observations are consistent with this idea, but there is little experimental evidence. For example, the cactus *Opuntia stricta* has more spines on Australian islands with cattle grazing than on islands with no grazing (Figure 15.7). If thorns and spines are herbivore defense mechanisms, they could be used to test ideas about plant defenses. The resource availability hypothesis (Figure 15.2) predicts that plants growing in nutrient-poor soils should invest more in plant defense than plants on rich soils. This is not the case for the fynbos vegetation of South Africa (Campbell 1986). Fynbos is a shrubland of sclerophyllous, evergreen plants growing on very poor soils. Only 4 percent of the total plant cover in fynbos has spines, compared with 13 percent of the plant cover in more nutrient rich nonfynbos areas. Campbell (1986) suggests that this is because the fynbos vegetation is so poor there are no large herbivores that can live on it, and consequently there is no selection for physical plant defenses like thorns.

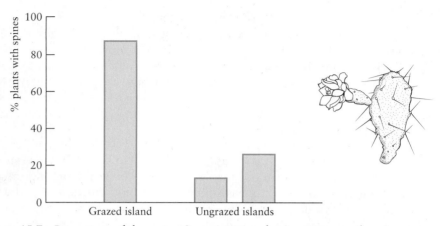

Figure 15.7 Percentage of the cactus Opuntia stricta *having spines on three islands off the coast of Queensland, Australia. Cattle were present on one island and grazing damage was observed, but were absent on the other two islands. (After Myers and Bazely 1991.)*

HERBIVORE INTERACTIONS

Because of all the defense mechanisms of plants, all that is green is not necessarily edible and herbivores may be more food-limited than they appear. The result is that herbivores may still compete for food plants, as we saw in Chapter 13. But in some cases herbivores may cooperate in the harvesting of plant matter. The grazing system of ungulates on the Serengeti Plains of East Africa is an excellent illustration of how herbivores may interact over their food supply. The Serengeti Plains contain the most spectacular concentrations of large mammals found anywhere in the world. A million wildebeest (Figure 15.8), 600,000 Thomson's gazelles, 200,000 zebras, and 65,000

Figure 15.8 Blue wildebeest and Burchell's zebras grazing on the Serengeti Plains of East Africa. A saddle-billed stork is in the foreground. (Photo courtesy of A. R. E. Sinclair.)

buffaloes occupy an area of 23,000 square kilometers (9,000 sq. miles), along with undetermined numbers of 20 other species of grazing animals (McNaughton 1976).

The dominant grazers of the Serengeti Plains are migratory and respond to the growth of the grasses in a fixed sequence (Figure 15.9). First, zebras enter the long-grass communities and remove many of the longer stems. Zebras are followed by wildebeest, which migrate in very large herds and trample and graze the grasses to short heights. Wildebeest are in turn followed by Thomson's gazelles, which feed on the short grass during the dry season (Bell 1971).

Different grazers in the Serengeti system do not select different species of grasses but instead select different parts of the grass plant (Figure 15.10). Zebras eat mostly grass stems and sheaths and almost no grass leaves. Wildebeest eat more sheaths and leaves, and Thomson's gazelles eat grass sheaths and a large fraction of herbs not touched by the other two ungulates. These feeding differences have significant consequences for the ungulates because grass stems are very low in protein and high in lignin, while grass leaves are relatively high in protein and low in lignin, so that

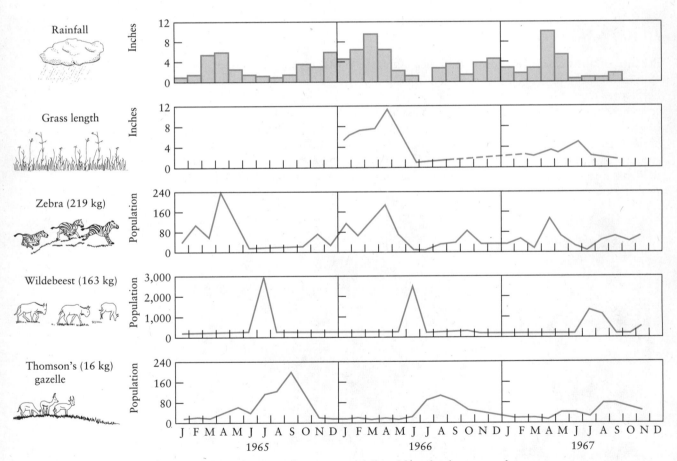

Figure 15.9   Population of migrating ungulates in relation to rainfall and length of grass on the Serengeti Plains of East Africa. The figures were obtained in western Serengeti by a series of daily transects in a strip approximately 3000 m long and 800 m wide. Successive peaks during each year mark the passage of the main migratory species in the early dry season. (From Bell 1971.)

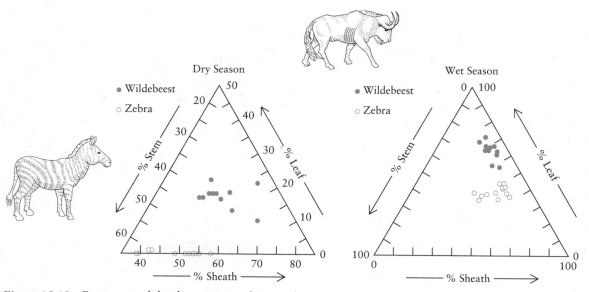

Figure 15.10 *Frequency of the three structural parts of grass in the diets of wildebeest and zebras during the dry season and the wet season, Serengeti Plains, East Africa. (After Gwynne and Bell 1968.)*

leaves provide more energy. Herb leaves typically contain even more protein and energy than grass leaves (Gwynne and Bell 1968). So zebras seem to have the worst diet and Thomson's gazelles the best.

How can zebras cope with grass stems as the major part of their diet during the dry season? Most of the ungulates in the Serengeti are ruminants, which have a specialized stomach containing bacteria and protozoa that break down the cellulose in the cell walls of plants. But the zebra is not a ruminant and is similar to the horse in having a simple stomach. Zebras survive by processing a much larger volume of plant material through their gut than ruminants do, perhaps roughly twice as much. So even though a zebra cannot extract all the protein and energy from the grass stems, it eats more and compensates by volume. Zebras also have an advantage of being larger than wildebeest and Thomson's gazelles, and larger animals need less energy and less protein per unit of weight than smaller animals. The net result of these considerations is that in times of food stress, large animals are able to tolerate low food quality better than small animals can.

Competition for food may occur between wildebeest and Thomson's gazelles because they eat the same parts of the grass. Wildebeest have what appears to be a devastating impact on the grassland as they pass through in migration. Green biomass was reduced by 85 percent and average plant height by 56 percent on sample plots. By setting out fenced areas as grazing exclosures, McNaughton (1976) was able to follow the subsequent changes in grassland areas subject to wildebeest grazing and in areas protected from all grazing (Figure 15.11). Grazed areas recovered after the wildebeest migration had passed and produced a short, dense lawn of green grass leaves. As gazelles entered the area during the dry season, they concentrated their feeding on areas where wildebeest had previously grazed and avoided areas of grassland that the wildebeest herd had missed.

Grass production was reduced by both wildebeest and gazelles but no signs of competition were found. The Serengeti ungulate populations show evidence of *graz-*

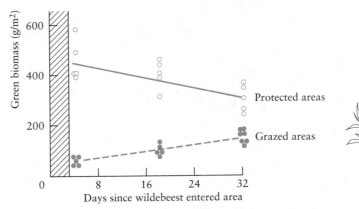

Figure 15.11   Vegetation recovery after a wildebeest migration passed through the western Serengeti Plains of Africa. The 3 to 5 days of wildebeest passage is marked by the hatched area at the start of the graph. Some areas were protected from wildebeest grazing by fencing before the migration arrived. (Modified from McNaughton 1976.)

ing facilitation, in which the feeding activity of one herbivore species improves the food supply available to a second species. Heavy grazing by wildebeest prepares the grass community for subsequent exploitation by Thomson's gazelles in the same general way that zebra feeding improves wildebeest grazing. Potential competition may be replaced by mutualism. Feeding systems of this type may be severely upset by the selective removal of one herbivore link in the sequence.

The grazing-facilitation hypothesis was tested by comparing population trends of wildebeest, zebras, and Thomson's gazelles in the Serengeti (Sinclair and Norton-Griffiths 1982). If grazing facilitation is mutualistic and obligatory, wildebeest numbers should not increase if zebra numbers do not increase, and if wildebeest numbers increase, gazelle numbers should also increase. Figure 15.12 shows that this has not happened. Wildebeest numbers more than doubled during the 1970s, while gazelle numbers fell slightly and zebra numbers remained constant. Predation may hold zebra numbers down, and these three ungulates apparently are not as closely linked as Bell (1971) suggested and may not form a mutualistic association.

Competition for grass in the Serengeti region may occur between very different types of herbivores (Sinclair 1975). In addition to the large ungulates, 38 species of grasshoppers and 36 species of rodents consume parts of the grasses and herbs. In the Serengeti Plains, most of the plant material consumed by herbivores is consumed by the large ungulates, but in some plant communities within the Serengeti, grasshoppers consumed nearly half as much grass as did the ungulates. The grazing system of the Serengeti is thus even more complex than we suggested in Figure 15.9. In any grazing system, we should realize that herbivores of greatly differing size and taxonomy may be affecting one another positively or negatively.

There is no evidence that grazing ever increases plant production or improves the fitness of plants (Belsky 1986, 1987). Plants respond to grazing by regrowth (Figure 15.11) but they never recover completely from the losses caused by grazing. The prevailing view of the plant-herbivore interaction is that it is a predator-prey type of interaction in which the herbivore gains and the plant loses. But some plant-animal interactions can be mutualistic, as we saw with the ant-acacia system. Polli-

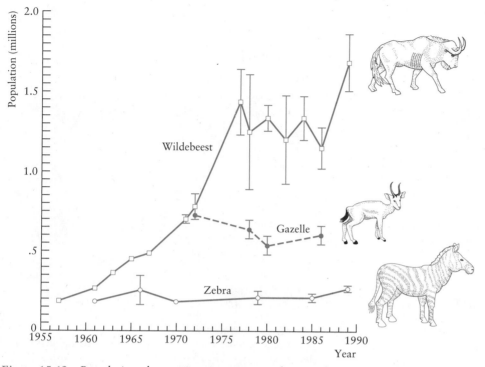

Figure 15.12   Population changes in migratory ungulates in the Serengeti Plains of East Africa. Estimates were obtained by aerial census. Vertical lines are error estimates (1 standard error). (Data from Sinclair and Norton-Griffiths 1982, Dublin et al. 1990.)

nation and fruit dispersal are two additional interactions that can be beneficial for both the plant and the herbivore.

Herbivores are commonly thought to be "lawn mowers," so it is important to recognize that they are highly selective in their feeding. This selectivity is a major reason why the world is not green for a herbivore. Figure 15.13 illustrates selective feeding in the snowshoe hare (*Lepus americanus*). Snowshoe hares feed in winter on the small twigs of woody shrubs and trees. In the southwestern Yukon, only three main plant species are available above the snow, and hares clearly prefer dwarf birch (*Betula glandulosa*) over willow (*Salix glauca*). These preferences may be caused by plant secondary compounds such as phenols (Sinclair and Smith 1984).

IRRUPTIONS OF HERBIVORE POPULATIONS

There are two basic types of herbivore-plant systems. One type we have just been discussing is called an *interactive herbivore system* because the herbivores influence the rate of growth and the subsequent fate of the vegetation. Other herbivore systems, called *noninteractive*, show no relationship between herbivore population density and the subsequent condition of the vegetation. Many herbivore systems are interactive. Serengeti ungulates provide many examples, and most grazing systems are of this type. Let us look at an example of each type to contrast the two ways in which animal populations react to their food plants.

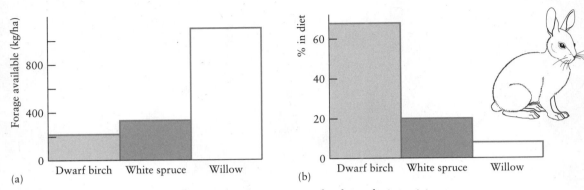

Figure 15.13 *Selective feeding of snowshoe hares on woody plants during winter. (a) Biomass of available forage. (b) Percent in diet. Hares do not eat a random sample of the plants available but are highly selective and prefer dwarf birch. Data from Kluane Lake, Yukon, winter 1979–1980. (Unpublished data, A. R. E. Sinclair and J. N. M. Smith.)*

Interactive Grazing: Ungulate Irruptions

Many ungulates introduced into new regions increase dramatically to high densities and then collapse to lower levels. The increase and subsequent collapse is called an *irruption*. Introduced reindeer populations have provided several examples. Figure 15.14 illustrates the stages of an irruption. Irruptions commonly occur when the introduced ungulate has an excellent food supply and no natural predators. As the population increases in stage 1 of an irruption (see Figure 15.14), the food resources are reduced. During stage 2 the population exceeds the carrying capacity of the habitat and food plants are overutilized and damaged. In stage 3 the population collapses because of food shortage, often aggravated by severe weather. This collapse may continue to near-extinction (see Figure 12.10, page 212) or the population may stabilize at lower numbers (Leader-Williams 1988).

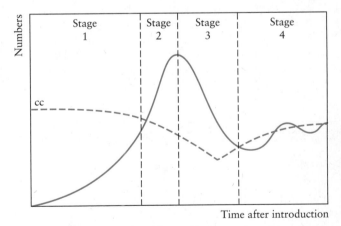

Figure 15.14 *The general pattern of ungulate irruptions often shown by populations introduced into new areas. Four general stages of the irruption can be recognized. CC = carrying capacity of the habitat based on food supplies. (After Riney 1964.)*

One of the best studied irruptions occurred in New Zealand when the Himalayan thar was introduced (Caughley 1970). The Himalayan thar is a goatlike ungulate of Asia. It was introduced into New Zealand in 1904 and has since spread over a large region of the southern Alps. As density increased, the birth rate fell only slightly and the death rate increased, primarily because of more juvenile mortality. After a period of high density, the population declined from a combination of reduced adult fecundity and further increase in juvenile losses.

What caused these population changes? Caughley (1970) suggested that grazing by the thar reduced its food supply and changed the character of the vegetation. The link between ungulates and their food plants is critical for these irruptions. The most conspicuous effect of thar grazing was found in the abundance of snow tussocks (*Chionochloa* spp.), which were the dominant vegetative cover where thar were absent but were scarce in places where thar had become common. Snow tussocks were believed to be important as food in late winter, and these evergreen perennial grasses cannot tolerate even moderate grazing pressures. When thar reach high densities, they begin to browse on shrubs in winter and may even kill some shrubs by their feeding activities.

Reindeer were introduced to South Georgia, a subantarctic island, by Norwegian whalers in 1911 primarily for sport hunting. Two separate populations were introduced. During the 1950s whales became scarce and hunting nearly stopped. Figure 15.15 shows the population history of the two reindeer herds on South Georgia (Leader-Williams 1988). The Barff herd reached a peak in 1955 and collapsed back to about two thousand animals. The Busen herd grew more slowly to a peak in the early 1970s and then also began to collapse. By this time reindeer were overgrazing tussock grasslands, which are their dominant winter food. In this simple island system, with no predators, and no other grazing animals, reindeer illustrate clearly the interplay between plants and herbivores.

The general picture of an ungulate irruption that emerges is that of a small

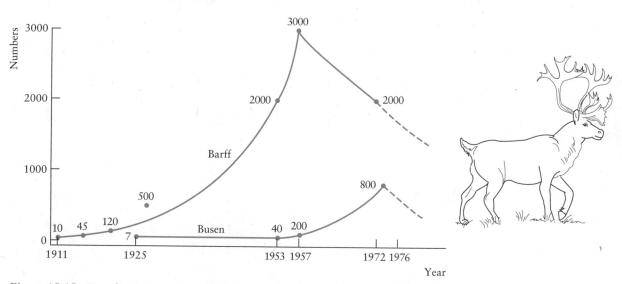

Figure 15.15   Population irruption of reindeer on South Georgia. Two separate herds were introduced on different parts of the island in the early 1900s, and estimates of their numbers have been made periodically. (After Leader-Williams 1988.)

number of animals introduced onto a range with superabundant food and a gradual increase in animal density and decrease in plant density until the animals have reduced or eliminated their best forage. Animal numbers decline until a new, lower density is reached, at which the herbivores and their plants may stabilize. Ungulate irruptions occur in both native and introduced species (Caughley 1970).

This sequence of events is similar to that predicted by some simple predator-prey models (Caughley 1976a). To make the predator-prey model more realistic as a possible description of a herbivore-plant interaction, Caughley (1976a) suggested that the plant population grows by a logistic equation instead of by a simple geometric increase. Figure 15.16 illustrates the results of this simple herbivore-vegetation model, which mimics the reindeer irruption just discussed. The behavior of simple herbivore-vegetation models is highly dependent on the rates of increase of both the plants and the herbivores and also on the feeding rates of the herbivores. Caughley (1976a) showed that such simple model systems oscillate in cycles if the grazing pressure tends to hold the amount of vegetation below about half the amount present in the ungrazed state. If the herbivore is a very efficient grazer, such simple systems can collapse completely.

In every interactive grazing system, the abundance of vegetation is affected by the abundance of herbivores. The experimental technique of setting up *exclosures* can be used to demonstrate this effect. For example, Randall (1961) set square wire exclosures, 30 centimeters (1 ft) on a side, over subtidal sections of rocky bottom off Hawaii. Herbivorous fishes, particularly surgeonfish and parrotfish, normally graze the algae growing on rocky areas. Within one month Randall found more algal growth inside the wire exclosure. After two months algae outside the exclosures averaged 1 to 2 millimeters in height, while inside the wire cages the dominant algal species reached 15 millimeters, and other algae had grown to 30 millimeters in height. Algal growth on rocks in the intertidal zone, where fishes rarely feed, is also luxuriant when compared with that in deeper waters. Randall (1961) suggested that many marine plants might be reduced in abundance by marine herbivores in tropical waters.

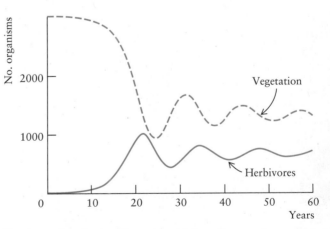

Figure 15.16 Simple model of a herbivore-plant interactive grazing system, which is similar to an ungulate irruption in showing a damped oscillation. The model used is a modified version of the Lotka-Volterra predator-prey model in which vegetation increases in a logistic manner rather than exponentially. (After Caughley 1976a.)

Many insect populations experience irruptions that damage their food plants (Myers 1987). The spruce budworm, for example, periodically irrupts to epidemic proportions in the coniferous forests of eastern Canada (Morris 1963). Budworm eat the buds, flowers, and needles of balsam fir trees. Outbreaks occur every 35 to 40 years, in association with the maturing of extensive stands of balsam fir, and during budworm outbreaks, many balsam fir trees are defoliated and killed. Populations of the large aspen tortrix moth outbreak in interior Alaska at intervals of 10 to 15 years. During these outbreaks quaking aspen trees are severely defoliated (80 to 100 percent) for 2 to 4 years (Clausen et al. 1991).

Many herbivorous insect populations may be held at low densities by a protein deficiency in their food plants. White (1984) has suggested that most plant material is not suitable food for insects because of nitrogen deficiencies. When plants are physiologically stressed—by water shortage, for example—they often respond by increasing the concentration of amino acids in their leaves and stems. Some larval insects may survive much better when more amino acids are available, and thus the stage is set for an insect outbreak. This hypothesis is called the *plant stress—insect performance* hypothesis and postulates that plants under abiotic stress become more suitable as food for herbivorous insects (Larsson 1989). This hypothesis may explain irruptions in many insect pests like the spruce budworm, which interact destructively with their food plants only occasionally (White 1984). But not all insects may respond positively to plant stress. Figure 15.17 illustrates how different types of insects may respond to plant stress (Larsson 1989). Sucking insects like aphids improve their performance as plants are stressed, while chewing insects in most cases do not improve their performance on stressed plants. Bark beetles are unique among insects in requiring a certain level of tree stress to colonize their hosts (Raffa and Berryman 1983) and thus show a unique relationship to stress levels (Figure 15.17).

Physiological stress in plants also has strong effects on plant defense compounds (Rhoades 1979). Quantitative defense substances (like tannins) usually decrease with plant stress, while qualitative compounds (like alkaloids) increase. The net result may be that, for a specialized herbivore, stressed plants are better to eat because they have more nitrogen and fewer toxic compounds.

Noninteractive Grazing: Finch Populations

European finches feed on the seeds of trees and herbs, and their feeding activities do not in any way affect the subsequent production of their food plants. They form a good example of a noninteractive system in which controls operate in only one direction:

Food plant production → herbivore density

Two groups of British finches can be distinguished by their feeding habits. One group feeds on the seeds of herbs, and their populations are quite stable (Newton 1972). A second group feeds on the seeds of trees, and their populations fluctuate greatly. Population dynamics in finches is determined by fluctuations in seed crops from year to year. Herbs in the temperate zone produce nearly the same numbers of seeds from one year to another, but trees do not. Most trees require more than one year to accumulate the reserves necessary to produce fruit. Spruce trees in Europe, for example, have moderate to large cone crops every two to three years in central Europe, every three to four years in southern Scandinavia, and every four to five years in northern Scandinavia. Good weather is also needed when the fruit buds are forming

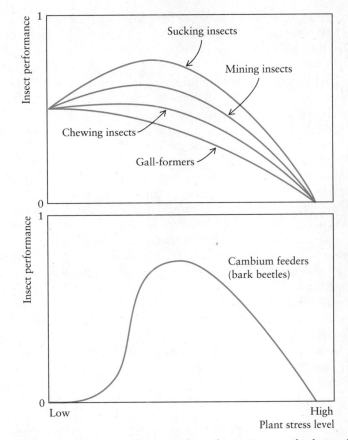

Figure 15.17 Plant stress–insect performance hypothesis. Suggested relationship between insect performance and host plant stress level for different feeding-types of insects. (From Larsson 1989.)

during the year before the seed crop is produced. The net result is that trees usually fruit in synchrony in the same geographic region. Other geographic regions may or may not be in synchrony, depending on local weather conditions.

Finches that depend on tree seed go through great irruptions in population density. They exist only by being opportunistic and moving over large areas to search for areas of high seed production. All the "irruptive" finches breed in northern areas and rely at some critical part of the year on seeds from one or two tree species (see Figure 6.11, page 89). Periodically, these finches leave their northern breeding areas and move south in large numbers. Figure 15.18 shows the dates of invasion of the common crossbill in southwestern Europe.

Mass emigration of crossbills and other finches is presumably an adaptation that avoids food shortages on the breeding range (Newton 1972). But crop failure alone is not sufficient to explain these mass movements. For example, in Sweden the spruce cone crop has been measured in all districts since 1900. Not all poor spruce crops in Sweden have resulted in crossbill movements. Very poor spruce crops occurred in 14 years between 1900 and 1963, but in only 6 of these years did crossbills move. Other evidence suggests that high population density may be necessary before large scale movements can be triggered. In some years, crossbills began to emigrate in the

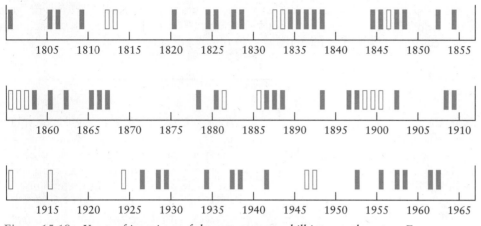

Figure 15.18   Years of invasions of the common crossbill into southwestern Europe. Solid blocks indicate large invasions; open blocks indicate small invasions. (Data from Newton 1972.)

spring, even before the new cone crop was available and discovered to be poor. Crossbills also put on additional fat before they emigrate, in the same way that migratory birds do. The suggestion is that high crossbill density is a prerequisite for large scale movements and that emigration occurs in response to the first inadequate cone crop once high densities are present.

Why emigrate? Mass emigration presumably is advantageous to the birds that stay behind, provided they find sufficient food. Emigration, by contrast, is often called suicidal, and the question arises as to how such an adaptation could exist. Emigrants might have two potential advantages. They could colonize new habitats in the south and thereby leave descendants. More likely, however, they obtain an advantage by migrating back north again after the food crisis has passed. Newton (1972) described four common crossbills that were banded in Switzerland during an irruption and were recovered a year later in northern Russia. Thus some birds return north, even though many die in the south during the irruption.

Crossbills achieve a close correlation between their breeding densities and the size of their food supply (Figure 15.19). This correspondence is obtained by having

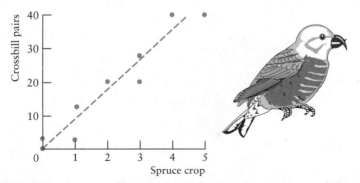

Figure 15.19   Relation between the early spring density of common crossbills in northern Finland and the relative size of the spruce cone crop. (Data of Reinikainen, cited in Newton 1972.)

great mobility so that populations can concentrate their nesting in areas with good cone crops. How this random mobility within the normal breeding zone becomes a unidirectional emigration in years of irruption is not understood.

SEED DISPERSAL

Not all plant-animal interactions are detrimental to plants, one good example of mutualism involves seed dispersal. Many plants depend on birds and mammals for the dispersal of their seeds. Coevolution between plants and their seed predators may help to explain some features of plant reproductive biology and animal feeding habits (Pijl 1969). Coevolution may be seen particularly clearly in the evolution of fruits.

Fruits are basically discrete packages of seeds with a certain amount of nutritive material. Vertebrates eat these fruits and digest part of the material, but many of the seeds pass through the digestive system unharmed. Plants advertise fruits by a ripening process in which the fruits change color, taste, and odor. Once fruits are ripe, they can be attacked by other damaging agents, like fungi and bacteria, or by destructive feeders who will not disperse the seeds.

There are four general ways for a plant to defend its fruit against damage by herbivores (Table 15.3). First, the cheapest defense is to fruit during the season of the year of minimum pest numbers. For example, in the temperate zone, insect damage to fruits will be minimal in the autumn and winter. Second, plants can reduce their exposure to fruit damage by ripening fruit more slowly. This strategy, however, will also reduce the availability of ripe fruit to dispersers and may be counterproductive to the plant. Third, by making the fruit very unbalanced nutritionally, a plant may discourage insect and fungal attack (Herrera 1982). Fruit pulp is high in carbohydrates, low in lipids, and extremely low in protein. Fruits are among the poorest protein sources in nature, and this may be a general pest-defense mechanism. Fourth, plants may adopt a chemical or mechanical defense of their fruits although this is the most expensive method. Since chemical defenses in fruits reduce herbivore attack and thus seed dispersal, it has commonly been assumed that as fruits ripen, chemical defenses should be eliminated. For example, the alkaloid tomatine in green tomatoes is degraded by a new enzyme system that is activated in ripening red tomatoes (McKey 1979). But the assumption that ripe fruits will not be chemically defended is probably a biased extrapolation from cultivated fruits that

Table 15.3   Gross Correlates of the Main Defense Methods Employed by Plants to Protect Fruits

Defense Method	Cost to the Plant in Energy and Nutrients	Effect on Disperser Feeding Response
Fruiting during the time of lowest pest pressure	None	Variable,[a] usually positive
Reducing exposure time by decreasing ripening rate	None	Variable,[a] often negative
Unbalanced and/or poor ration in the flesh	None	Negative
Chemical defense	Some	Negative

[a] Depending on disperser abundance and attractability.
Source: After Herrera (1982).

have been selected for centuries. Of all the wild European plants that produce fleshy fruits, at least one-third have fruits toxic to man (Herrera 1982). Toxic fruits are avoided by many birds and mammals, and consequently such fruits are dispersed only by a selected subset of vertebrates who can detoxify the specific chemicals. Toxic fruits are thus one way for a plant to "choose" its dispersal agents, although the detailed reasons why this is adaptive for the plants have not been discovered.

The interaction of seed predation and seed dispersal has been analyzed particularly clearly in the case of Clark's nutcracker (*Nucifraga columbiana*) and the whitebark pine (*Pinus albicaulis*). Several soft pines in North America have large, wingless seeds that are not dispersed by wind. These seeds are frequently removed from their cones by jays and nutcrackers, transported some distance, and cached in the soil as a future food source. When the birds cache more seeds than they can subsequently eat, the surplus is available for germination. This interaction can be considered a case of mutualism if both species increase their fitness because of the association.

Does the whitebark pine benefit from seed dispersal from Clark's nutcracker? Nutcrackers cache three to five seeds, on average, in each store, bury them an average of 2 centimeters, avoid damp sites, and place many of their caches in microenvironments favorable for subsequent tree growth (Tomback 1982). One nutcracker stores about 32,000 whitebark pine seeds each year, which represents three to five times the energy required by the nutcracker. Thus many caches are not utilized, and the survival of pine seedlings arising from unused caches is very high. Nutcrackers also disperse pine seeds to new areas within the subalpine forest zone, so that they increase the local distribution of whitebark pine (Tomback 1982). Many other species of birds feed on whitebark pine seeds, as do mice, chipmunks, and squirrels, but none of these species caches pine seeds in ways that favor germination (Hutchins and Lanner 1982). We conclude that the Clark's nutcracker–whitebark pine system is a coevolved mutualism in which both species profit, the nutcracker by obtaining food and the pine by achieving seed dispersal.

The great variety of seed dispersal systems used by plants has important consequences for seed-eating animals, and much more exploratory work will be needed before we know how many of these systems are mutualistic and how many are exploitative.

COMPLEX SPECIES INTERACTIONS

Species interactions are rarely one-on-one in natural communities, and the untangling of complex sets of species interactions is an important focus in ecology today. Complex interactions illustrate how difficult it can be to determine whether a single species exerts a beneficial or a harmful influence on another.

Interactions between homopterous insects and the ants that tend them or prey on them have been described for nearly a century (Buckley 1987). These plant-homopteran-ant interactions are significant because many of the world's major plant pests are homopteran insects, and many of the worst diseases of our crops are transmitted by homopterans. The manipulation of ant assemblages to control homopteran pests has been practiced in China since at least A.D. 300 (Needham 1986).

Ants that tend homopterans provide a positive benefit for the homopterans, including protection from predators or parasitoids, sanitation in honeydew removal, and transportation to new feeding sites. Ants may also remove dead individuals from homopteran populations, and provide nest sites. For the plants, plant-homopteran-

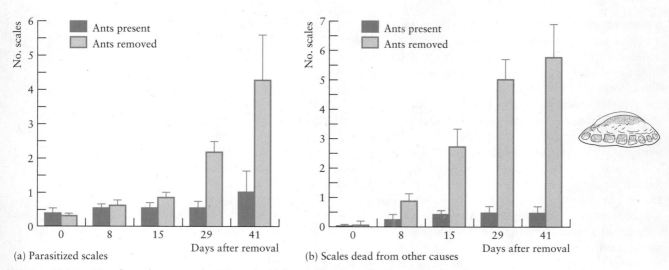

Figure 15.20   *Number of green scales (*Coccus viridis*) per leaf on plants with and without ants (*Pheidole megacephala*). Mortality of scale insects increased dramatically when ants were removed. (After Bach 1991.)*

ant interactions impose many costs and confer few if any benefits. Homopterans consume phloem sap and disadvantage the plant by removing metabolites, damaging plant tissues, and increasing water loss. Homopterans may also transmit plant pathogens. For the ants that tend homopterans, the main benefit is the food they obtain in the form of sugars from the honeydew secreted by homopterans (Buckley 1987), and ant colonies that feed on honeydew have higher populations than colonies with no honeydew source.

In Hawaii the green scale (*Coccus viridis*) is tended by the ant *Pheidole megacephala* on the host plant Indian fleabane (*Pluchea indica*). Bach (1991) analyzed this system using removal and addition experiments to measure the strength of these interactions. When she removed ants from plants, the number of parasitized green scales increased and the mortality of scales to other predators and diseases also went up (Figure 15.20). Ants removed ladybird beetle larvae introduced on to plants as possible scale predators. On plants without ants honeydew from the green scale accumulates and a sooty mold grows on the leaves. This mold reduces the photosynthetic rate of the leaf, and leaves infested with mold were often shed by the plant. Ants thus indirectly benefited the plant by removing the honeydew from the scale insects. Ants may also remove other herbivorous insects like moth larvae from plants and thus protect the plants from additional losses (Bach 1991).

Complex interactions become more difficult to unravel as the numbers of species involved grows. Interaction webs may involve herbivory, mutualism, predation, and competition and emphasize the need in ecological systems to look at both direct interactions and indirect effects in a quantitative way (Addicott 1986).

SUMMARY

Herbivory is a form of predation. Because plants are modular organisms herbivores usually eat only part of the plant, and typical herbivory thus differs from typical predation. Plants have a variety of structural and chemical defenses that discourage herbivores from eating them. Many secondary plant substances are stored in plant parts to discourage herbivores. Herbivores have responded to these evolutionary challenges by timing their life cycle to avoid the chemical threats or by evolving enzymes to detoxify plant chemicals. Two theories have attempted to specify the strategies of plant defense. The resource availability hypothesis emphasizes the differential costs and benefits of defense for slow-growing and fast-growing plants.

Some herbivores can affect the future density and productivity of their food plants. A very efficient herbivore can thus drive itself to extinction unless it has evolved some constraints that prevent overexploitation of its food plants. Most herbivore-plant systems seem to exist in a fluctuating equilibrium. Some herbivore populations track their food supply, and large fluctuations in food supply are often translated into large fluctuations in herbivore densities.

Not all herbivory is detrimental. Seed- and fruit-eating vertebrates obtain food from plants and may disperse seeds and thereby benefit the plant in a mutualistic interaction. Fruits and seeds have a variety of mechanical and chemical defenses, and plants must time their fruiting season to assist seed dispersal but retard seed destruction.

Mutualisms occur in many plant-animal interactions. Pollination and seed dispersal are two examples of processes that benefit both plant and animal species. Ants may form mutualistic relationships with plants particularly in tropical areas or with herbivorous insects like homopterans.

Selected References

BELSKY, A. J. 1986. Does herbivory benefit plants? A review of the evidence. *American Naturalist* 127:870–892.

CAUGHLEY, G., and J. H. LAWTON. 1981. Plant-herbivore systems. In *Theoretical Ecology*, ed. R. M. May, pp. 132–166. Blackwell, Oxford, England.

COLEY, P. D., J. P. BRYANT, and F. S. CHAPIN III. 1985. Resource availability and plant antiherbivore defense. *Science* 230:895–899.

FEENY, P. 1992. The evolution of chemical ecology: Contributions from the study of herbivorous insects. In *Herbivores: Their Interactions with Secondary Plant Metabolites. Vol. II. Evolutionary and Ecological Processes*, ed. G. A. Rosenthal and M. Berenbaum, pp. 1–44. Academic Press, San Diego.

HARVELL, C. D. 1986. The ecology and evolution of inducible defenses in a marine bryozoan: Cues, costs, and consequences. *American Naturalist* 128:810–823.

HOWE, H. F., and J. SMALLWOOD. 1982. Ecology of seed dispersal. *Annual Review of Ecology and Systematics* 13:201–228.

LUBCHENCO, J., and S. D. GAINES. 1981. A unified approach to marine plant-herbivore interactions. I. Populations and communities. *Annual Review of Ecology and Systematics*. 12:405–437.

MYERS, J. H. and D. BAZELY. 1991. Thorns, spines, prickles, and hairs: Are they stimulated by herbivory and do they deter herbivores? In *Phytochemical Induction by Herbivores*, ed. T. W. Tallamy and M. J. Raupp, pp. 325–344. Wiley, New York.

STRONG, D. R., J. H. LAWTON, and T. R. E. SOUTHWOOD. 1984. *Insects on Plants*. Harvard University Press, Cambridge, Mass.

TALLAMY, D. W., and M. J. RAUPP, eds. 1991. *Phytochemical Induction by Herbivores*. Wiley, New York.

WHITE, T. C. R. 1984. The abundance of invertebrate herbivory in relation to the availability of nitrogen in stressed food plants. *Oecologia* 63:90–105.

Questions and Problems

1. Three species of crossbills in northern Europe (see Figure 6.11, page 89) tend to irrupt together. But two species concentrate on larch and spruce cones, which mature in one year, while the third species feeds on pinecones, which mature in two years. Poor flowering seems to occur at the same time in pine, spruce, and larch. How can you explain this puzzle? Suggest an experiment to test your hypothesis, and compare your ideas with those of Newton (1972, p. 239).

2. Discuss the advantages and disadvantages of physical versus chemical defenses in plants.

3. How does the predation of animals on seeds and fruits differ from the predation of animals on leaves and stems of plants?

4. Wildlife managers and range ecologists both talk of the "carrying capacity" of a given habitat for a herbivore population. Write an essay on the concept of carrying capacity, how it can be measured, and how the concept can be applied to agricultural and natural situations. Dhondt (1988) discusses the definition of this term.

5. Cannabin is a secondary plant substance (a terpene) produced by the hemp plant, *Cannabis sativa* (Hollister 1971, p. 21). From information available in the literature, write an essay on the biological role of cannabin in this plant.

6. Alkaloids are plant defense chemicals, but not all plants contain alkaloids. Among annual plants, the incidence of alkaloids is nearly twice that among perennial plants. Tropical floras also contain a much higher fraction of species with alkaloids than temperate floras, and this is true for both woody and nonwoody plants. Suggest why these patterns exist, and then compare your ideas with those of Levin (1976).

7. Caughley and Lawton (1981) suggest that the growth of many plant populations will be close to logistic. Review the assumptions of the logistic equation (Chapter 12), and discuss why this suggestion is true or false.

8. *Eucalyptus* trees in Australia have high rates of insect attack on leaves, with 10 to 50 percent of the leaves eaten every year. Yet these trees also contain very high concentrations of essential oils and tannins (Morrow and Fox, 1980). Discuss how this situation could occur if *Eucalyptus*' oils and tannins are defensive chemicals.

9. Large mammal herbivores are not always present in habitats dominated by spiny plants. Why might this be? Janzen (1986) reviews the vegetation of the

Chihuahuan desert of north central Mexico and interprets the abundance of spiny cacti as reflecting the "ghost of herbivory past." Read Janzen's analysis and discuss how to test his ideas.

10. Mutualisms are usually thought to be more common in tropical areas than in temperate or polar regions. Evaluate the evidence for this generalization. Boucher and co-workers (1982) discuss this issue.

11. If a plant-homopteran-ant interaction has a net negative effect on the individual plants occupied by the homopterans, why are these plants not selectively eliminated from the population?

Overview Question

Can herbivory ever benefit a plant, so that a plant-herbivore interaction could be mutualistic? How would you test such a proposed model for a grassland grazing system like the Serengeti in Africa or the Great Plains in North America?

16 *Natural Regulation of Population Size*

We can make two fundamental observations about populations of any plant or animal. The first is that abundance varies from place to place. There are some "good" habitats where the species is, on the average, common and some "poor" habitats where it is, on the average, rare. The second observation is that no population goes on increasing without limit, and the problem is to find out what prevents unlimited increase in low- and high-density populations. This is the problem of explaining fluctuations in numbers. Figure 16.1 illustrates these two problems, which are often confused in discussions of "natural regulation."

Prolonged controversies have arisen over the problems of the natural regulation of populations. Before 1900 many authors, Malthus and Darwin included, had noted that no population goes on increasing without limit, that there are many agents of destruction that reduce the population. It was not, however, until the twentieth century that an attempt was made to analyze these facts more formally. The stimulus for this came primarily from economic entomologists, who had to deal with both introduced and native insect pests. Most of the ideas we have on natural regulation can be traced to entomologists. The basic principles of natural regulation can be derived from a simple model, taken from the models of population growth presented in Chapter 12.

A SIMPLE MODEL OF POPULATION REGULATION

If populations do not increase without limit, what stops them? We can answer this question with a simple graphic model similar to that in Figure 12.2 (page 200). A population in a closed system will increase until it reaches an equilibrium point at which

Birth rate per capita = death rate per capita

Figure 16.2 illustrates three possible ways in which this equilibrium may be defined. As population density goes up, birth rates* may fall, death rates may rise, or both changes may occur. To determine the equilibrium population size for any field population, we need to determine only the curves shown in Figure 16.2. Note that this simple model in no way depends on the shapes of the curves, provided that they are smoothly rising or falling shapes. In particular, these curves do not need to be straight lines.

We now introduce a few terms to describe the concepts shown in Figure 16.2.

---

* In all discussions of population regulation, birth rates always mean per capita birth rates, and death rates always mean per capita death rates.

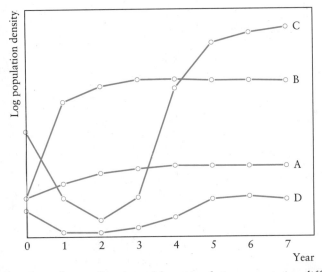

Figure 16.1   Hypothetical annual census of four populations occupying different types of habitat. Two questions may be asked about these populations: (1) Why do all fail to go on increasing indefinitely? (2) Why are there more organisms on the average in the good habitats B and C compared with the poor habitats A and D? (After Chitty 1960.)

The death rate per capita is said to be *density-dependent* if it increases as density increases (Figure 16.2a and 16.2c). Similarly, the birth rate per capita is called density-dependent if it falls as density rises (Figure 16.2a and 16.2b). Another possibility is that the birth or death rates do not change as density rises; such rates are called *density-independent rates.*

Note that Figure 16.2 does not include all logical possibilities. Birth rates might, in fact, *increase* as population density rises, or death rates might *decrease*. Such rates are called *inversely density-dependent* because they are the opposite of directly density-dependent rates. Inversely density-dependent rates can never lead to an equilibrium density; hence they are not shown in Figure 16.2. Figure 16.2 can be formalized into the first principle of natural regulation: *No closed population stops increasing unless either the per capita birth rate or death rate is density-dependent.*

We can extend this simple model to the case of two populations that differ in equilibrium density to answer the question of why abundance varies from place to place (Figure 16.3). Consider first the simple case of populations with a constant (density-independent) birth rate. Equilibrium densities vary for two reasons: (1) Either the slope of the mortality curve changes (Figure 16.3a), or (2) the general position of the mortality curve is raised or lowered (Figure 16.3b). In case 1, the density-dependent rate is changed because the slopes of the lines differ, but in case 2, only the density-independent mortality rate is changed. From this graphic model we can arrive at the second principle of natural regulation: *Differences between two populations in equilibrium density can be caused by variation in either density-dependent or density-independent per capita rates of birth and death.* This principle seems simple: It states that anything that alters birth or death rates can affect equilibrium density. Yet this principle was in fact denied by many population ecologists

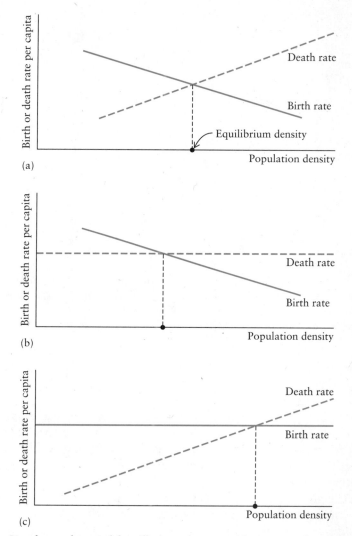

Figure 16.2   Simple graphic model to illustrate how equilibrium population density may be determined. Population density comes to an equilibrium only when the per capita birth rate equals the per capita death rate, and this is possible only if birth or death rates are density-dependent. (Modified from Enright 1976.)

for 40 years (Enright 1976, Sinclair 1989), and we now turn to review this historical controversy.

HISTORICAL PERSPECTIVE

The *balance of nature* has been a background assumption in natural history since the time of the early Greeks and underlies much of the thinking about natural regulation (Egerton 1973). The simple idea of early naturalists was that the numbers of plants and animals were fixed and in equilibrium, and observed deviations from equilibrium, such as the locust plagues described in the Bible, were the result of a

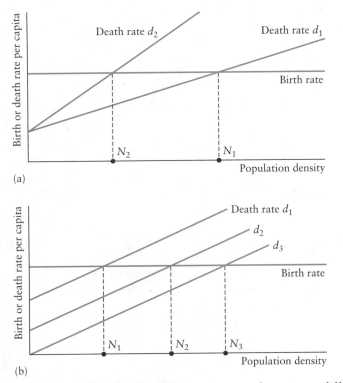

Figure 16.3   Simple graphic model to illustrate how two populations may differ in average abundance. In (a) the two populations differ in the amount of density-dependent mortality. In (b) the populations differ in the amount of density-independent mortality. Dotted lines mark the equilibrium population densities. (Modified from Enright 1976.)

punishment sent by divine powers. Only after Darwin's time did biologists try to specify how a balance of nature was achieved and how it might be restored in areas where it was upset.

A considerable amount of activity around the turn of the twentieth century centered around attempts to control insect pests by the introduction of parasites. Howard and Fiske (1911), two economic entomologists with the U.S. Department of Agriculture, studied the parasites of two introduced moths, the gypsy moth and the brown-tail moth, in an attempt to control the damage these defoliators were doing to New England trees. Howard and Fiske believed that each insect species was in a state of balance so that it maintained a constant density if averaged over many years. For this balance to exist, they argued, there must be, among all the factors that restrict the insect's multiplication, one or more *facultative agents* that exert a relatively more severe restraint when the population increases. They argued that only a very few factors, such as insect parasitism, were truly facultative.

Furthermore, Howard and Fiske said, a large proportion of the controlling factors, such as destruction by storms, high temperatures, and other climatic conditions, should be classed as *catastrophic*, since they are wholly independent in their action of whether the insect is rare or abundant. For example, a storm that kills 10 out of 50 caterpillars on a tree would undoubtedly have destroyed 20 if 100 had been there or 100 if 500 had been there. Thus the average percentage of destruction (the per capita death rate) remains the same no matter what the abundance of the insect.

Finally, Howard and Fiske noted* that other agencies, such as birds and other predators, work in a radically different manner. These agents maintain constant populations from year to year and destroy a constant number of prey. Consequently, when the prey species increases, the predators will destroy a smaller and smaller percentage of the prey (i.e., they work in a manner that is the opposite of facultative agents). Howard and Fiske did not give factors of this type a distinct name.

They concluded that a natural balance can be maintained only through the operation of facultative agencies that destroy a greater proportion of individuals as the insect in question increases in abundance. Facultative agencies thus cause the per capita death rate to increase with prey density, as in Figure 16.2 (a) and (c). Howard and Fiske believed that *insect parasitoids* were the most effective of the facultative agencies; *disease* operated only rarely, when densities got very high; and *starvation* was the ultimate facultative agency, which almost never operated.

Howard and Fiske were the prototypes of the *biotic school* of population regulation, which proposed that biotic agents, principally predators and parasitoids, were the main agents of natural regulation.

Meanwhile another school of thought, the *climate school*, was in the process of formation. Bodenheimer (1928) was one of the first to hold that the population density of insects is regulated primarily by the effects of weather on both development and survival. Bodenheimer was impressed by all the work done in the 1920s on the environmental physiology of insects, showing, for example, how low temperatures affect the rate of egg laying and speed of development. He was also impressed by the fact that weather was responsible for the largest part of the mortality of insects, often 85 to 90 percent of the insects in their early stages being killed by weather factors.

Uvarov (1931) published a large paper, "Insects and Climate," in which he reviewed the effects of climatic factors on growth, fertility, and mortality of insects. He emphasized the correlation between population fluctuations of insects and the weather, and he regarded these weather factors as the prime agents that control populations. Uvarov questioned the idea that all populations are in a stable equilibrium in nature and emphasized the instability of field populations.

Three important ideas were expressed by the early climate school: (1) Insect population parameters are strongly affected by the weather, (2) insect outbreaks could be correlated with the weather, and (3) insect population fluctuations were emphasized, not stability.

It is important to realize here that all this controversy was over *insect* populations and their regulation; work on vertebrate populations had hardly begun by 1930, and there had been no work on the populations of other invertebrates or plants.

In 1933 the *Journal of Animal Ecology* published a supplement titled "The Balance of Animal Populations" by A. J. Nicholson, an Australian economic entomologist. Nicholson was interested in the parasite-host system of insects, and he teamed up with mathematician V. Bailey to construct a model of this system. Nicholson disliked the predator-prey models of Lotka and Volterra and criticized them

---

* Incorrectly, we now know from our historical vantage point. See Figure 14.18 (page 284).

because they did not allow for time lags in the system, because they ignored age groups (assuming that all individuals are equivalent), and because Lotka and Volterra used calculus rather than finite methods of mathematical analysis. Nicholson expanded his ideas on the parasite-host system to cover all interactions between animals.

The controlling factor was always *competition*, according to Nicholson: competition for food, competition for a place to live, or the competition of predators or parasites. Nicholson's theory was predominantly a biotic one, and he is usually considered the cornerstone of the biotic school.

Nicholson's main ideas were essentially the same as those of Howard and Fiske. To these he added a mathematical model and the notion of competition as the controlling factor. Nicholson's points were given much stronger emphasis by Smith (1935), who considered the problem of population regulation in some detail. He pointed out first of all that populations are characterized both by stability and by continual change. Population densities are continually changing, but their values tend to vary about a characteristic density. This characteristic density itself may vary. Smith compared a population to the sea, the surface of which is paradoxically a universal point for altitude measurements but which is continuously being changed by tides and waves. Thus Smith reaffirmed Nicholson's ideas on balance.

Different species of animals tend to have different average densities, and the same species will have different average densities in different environments. The variations about the average density are stable because there is always a tendency to return to the average density (i.e., populations seldom become extinct or increase to infinity). This is what is loosely termed the "balance of nature."

The equilibrium position, or average density, may itself change with time. This is what causes the economic entomologists so much trouble. The equilibrium position of an introduced pest may be so high that constant damage occurs to crop plants. Smith then set out to analyze the factors that determine the equilibrium position or average density. He pointed out that the number of injurious insects is very small relative to the total number of insects and that we must study both *common* and *rare* species if we hope to understand the reasons for the abundance of species.

Smith recognized the distinction Howard and Fiske made between *facultative* and *catastrophic* agencies, and he renamed these *density-dependent* mortality factors and *density-independent* mortality factors. The average density of a population, Smith concluded, can never be determined by density-independent factors. Only if the death-rate line has a slope (i.e., a density-dependent component) can the population reach equilibrium. Thus only density-dependent mortality factors can determine the equilibrium density of a population.

He went on to point out that the density-dependent factors are mainly *biotic* in nature—disease, competition, predation—and that the density-independent factors are mainly physical, or *abiotic*, factors, mainly climate. But, Smith pointed out, we should not conclude from this that the average population densities of species are *never* determined by climate. Climate, he states, may act as a density-dependent factor under some circumstances. As an example, he suggests the case of *protective refuges*: If there are only so many of these to go around and all the unprotected individuals are killed by climate, this climatic mortality would be density-dependent.

To summarize: Smith restated the main points of Nicholson, adding the terms *density-independent* and *density-dependent*, and stated, in contrast to Nicholson, that climate might act as a density-dependent factor in some cases. By this time, then, the main tenets of the biotic school had been crystallized, that is, the idea of balance in nature, that this balance was produced by density-dependent factors, and that these factors were usually biotic agents, such as predators and diseases.

In 1954 Andrewartha and Birch, two Australian zoologists who completely disagreed with Nicholson's ideas, published an important book, *The Distribution and Abundance of Animals,* which attacked the ideas of the biotic school. They revived in their book a highly modified version of the climatic school's ideas. Andrewartha and Birch concentrated on the individual organism and based their whole approach on this question: *What are the factors that influence the animal's chance to survive and multiply?* Given this question, they proceeded to classify environmental factors.

First, they rejected the distinction Howard and Fiske and others made between the physical (abiotic) and biotic factors. For example, food and shelter may sometimes be biotic, sometimes abiotic. Hence this distinction does not help us much to classify the environment.

Second, they rejected the classification of the environment based on density-dependent factors and density-independent factors. They rejected this distinction because they believed that there is no component of the environment whose influence is likely to be independent of the density of the population (i.e., all factors are density-dependent). Here they are striking at a key principle of the biotic school. For example, they say, consider the action of frost. Between a large population and a small population there may be genetic differences in cold hardiness, and in addition, the places where the insects live may differ with respect to the degree of protection from frost. Thus large populations may be forced to occupy marginal habitats and so suffer more from frost. They conclude that density-independent factors do not exist, so there is no need to attach any special importance to density-dependent factors in classifying the effects of the environment on a population.

How then can one classify environmental factors? Andrewartha and Birch suggested that the environment may be divided into four components:

1. Weather
2. Food
3. Other animals and pathogens
4. A place in which to live

Andrewartha and Birch were only concerned with animal abundance, so plants appear principally as food in this classification. These components of the environment are nonoverlapping, and they may be subdivided if necessary, for example, into other animals of the same species and other animals of different species. Together these four components and the interactions between them completely describe the environment of any animal.

Consequently, said Andrewartha and Birch, for any given species we must ask which of the four components of environment affect the animal's chances to survive and multiply. Once we can answer this question, we will be able to determine the reasons for the animal's distribution and abundance in nature. Andrewartha and Birch presented a general theory of the numbers of animals in natural populations. First, they stated that one cannot use expressions like "balance," "steady states," "equilibrium densities," or "ultimate limits" because there is no empirical way of giving a meaning to these words. Second, they noted that one must take account of the fact that all animals are distributed patchily in nature, never uniformly. These "patches," or *local populations*, are the basic component with which they deal.

According to Andrewartha and Birch, the numbers of animals in a natural population may be limited in three ways: (1) by limited supply of material resources, such as food, places in which to make nests, and so on; (2) by inaccessibility of these material resources relative to the animal's capacities for dispersal and searching; and (3) by shortage of time when the rate of increase (*r*) is positive. Of these three ways, they believe that the last is probably the most important in nature, and the first is probably the least important. Regarding the third case, the fluctuations in the rate of increase may be caused by weather, predators, or any of the components of the environment.

Andrewartha and Birch were principally concerned with insect populations, and their field experience was with insects occupying the severe desert and semidesert areas of Australia. Their main contribution to ecology has been to reemphasize the importance of getting *empirical data* on the problems of population regulation. They continually raised the ever-bothersome question: How can a particular idea be *tested* in real populations?

The two main schools of population regulation—the biotic school and the climate school—have concentrated on the role of the *extrinsic* factors in control: food supply, natural enemies, weather, diseases, and shelter. Many of these theories tend to assume that the individuals that make up the population are all identical, like atoms or marbles. This neglect of the importance of individual differences in population regulation has been challenged by a group of workers in diverse fields who have proposed theories of self-regulation. Their rallying point has been a search for *intrinsic* changes in populations, changes that might be important in natural control.

Two basic types of changes can occur in individuals, *phenotypic* and *genotypic*, and the proponents of self-regulatory mechanisms differ in what importance they attach to each of these basic types. Of course, no matter what is the mechanism operating, it must have been evolved in the species concerned, and consequently these theories of self-regulation all become concerned with evolutionary arguments.

Chitty (1955) presented the fundamental premise underlying all ideas on self-regulatory mechanisms. Suppose, Chitty argued, that we observe a population at two times, 1 and 2, and that at time 2 the death rate (D_2) is higher than the death rate at time 1 (D_1). This death rate is the result of the interaction of the organisms (O) with their mortality factors (M). Our problem now is to determine why D_2 is greater than D_1. The first hypothesis to be explored is that on both occasions we are dealing with organisms whose biological properties are identical. In this case, we must look for a difference between the mortality factors at the two times. In other words, we might expect to find at time 2 that there are more predators or parasites or that the weather is less favorable. Some population changes can certainly be explained in this manner, but in other cases this method has failed to turn up the right clues. We must look at the matter from another angle.

Consider, Chitty continued, the possibility that the environmental conditions are much the same at all times, that there is no real difference between the mortality factors at times 1 and 2. In this case, any change in the death rate must be due to a change in the nature of the organisms, a change such that they become less resistant to their normal mortality factors. For example, the animals might die in cold weather at time 2, weather they might have survived at time 1.

These ideas can be summarized as follows:

	First Hypothesis		Second Hypothesis	
Time	1	2	1	2
Death rate	$D_1 < D_2$		$D_1 < D_2$	
Organisms	$O_1 = O_2$		$O_1 \neq O_2$	
Environment	$M_1 \neq M_2$		$M_1 = M_2$	

Changes in the individual organisms in the population may be physiological or behavioral, and they may be phenotypic changes or genotypic changes.

The first hypothesis describes the classic approach to population regulation, used for example by both the biotic school and the climate school. The second hypothesis describes an ideal self-regulatory approach to population regulation. It is unlikely in nature that this second situation would occur in such a pure form but more likely that some mixture of these two situations would be found in self-regulatory populations. Note that the concept of density dependence becomes ambiguous under the second hypothesis. The idea that the environment can be subdivided into density-dependent and density-independent factors has meaning only insofar as the properties of the individuals in the population are constant. Self-regulatory systems have added an additional degree of freedom to the system, the individual with variable properties.

Variation among the individuals in a population may be either genetically based or environmentally induced. British geneticist E. B. Ford (1931) was one of the first to point out the possible importance of genetic changes in population regulation. He suggested that natural selection is relaxed during population increases, with the result that variability increases within the population, and many inferior genotypes survive. When conditions return to normal, these inferior individuals are eliminated through increased natural selection, causing the population to decline and at the same time reducing variability within the population. Thus, Ford argued, population increase inevitably paves the way for population decline.

From a study of population fluctuations in small rodents, Chitty (1960) set up the general hypothesis that *all species are capable of regulating their own population densities without destroying the renewable resources of their environment or requiring enemies or bad weather to keep them from doing so.* Not all populations of a given species will necessarily be self-regulated, and the mechanisms evolved will be adapted only to a restricted range of environments. The species may well live in poor habitats where this mechanism seldom if ever comes into effect.

The actual mechanisms by which self-regulation can be achieved in natural populations involve some form of mutual interference between individuals or intraspecific hostility in general. Mechanisms of self-regulation do not require genetic changes and may be entirely phenotypic. This hypothesis can be applied only to species that show mutual interference or spacing behavior. The most important environmental factor for such populations is *other organisms of the same species.*

The problem of self-regulation has been approached from another angle by V. C. Wynne-Edwards, a British ecologist whose major work has been on birds. Wynne-Edwards (1962) began his analysis with the observation that most animals have highly effective mechanisms of movement. If we look in nature, we will usually find that organisms concentrate at places of abundant resources and avoid unfavorable areas. This is the first point to note—that animals are dispersed in close relation to their essential resources.

The essential resource most critical to animals is clearly *food*, Wynne-Edwards observed. Of course, many other requirements must be met before a species can survive in an area, but food is almost always the critical factor that ultimately limits population density in a given habitat. We must then study the food resource as the key to understanding population control.

Wynne-Edwards suggested that some artificial and harmless type of competition has evolved in many species as a buffer mechanism to stop population growth at a level below that imposed by food exhaustion. The best example of this kind of buffer mechanism is the territorial system of birds. The territories that birds defend so fiercely are just a parcel of ground, but the possession of a territory eliminates competition for food, since the owner and its dependents enjoy undisputed feeding rights on that area. Provided that the size of the territory varies with the productivity of the habitat, we get a perfect illustration of this model: Population density is controlled by territoriality, which ensures that the food supply will not be exhausted.

The margins of a species' range will probably not show this self-regulation, Wynne-Edwards stated. Physical factors will predominate in these harsh environments, and hence we should concentrate our attention on the more typical parts of the range, where self-regulation is the usual situation. Also, a few species will ultimately fail to be limited by food, and these will not fit into the scheme of Wynne-Edwards.

A MODERN SYNTHESIS

There has been a great deal of controversy in ecology over the concepts of population regulation (Sinclair 1989) and we need to highlight the areas of agreement and disagreement.

The definition of terms has always plagued discussions about population regulation. Let us start with a clear definition of two confusing terms:

limitation: a factor is defined to be a limiting factor if a change in the factor produces a change in average or equilibrium density. For example, a disease may be a limiting factor for a deer population if deer abundance is higher when the disease is absent.

regulation: a factor is defined to be a regulating factor if the percentage mortality caused by the factor increases with population density.* For example, a disease may be a regulating factor only if it causes a higher fraction of losses as deer density goes up.

Most experimental manipulations of populations involve studies of limitation.

The simple model of population regulation shown in Figure 16.2 is critically focused on the concept of equilibrium, and we must begin by asking whether natural populations can be equilibrial systems. Recent work on ecological stability has given us a more comprehensive view of the factors that affect stability (Figure 16.4). There

---

* Or alternatively, a factor is defined to be a regulating factor if the reproductive rate is reduced as the population rises.

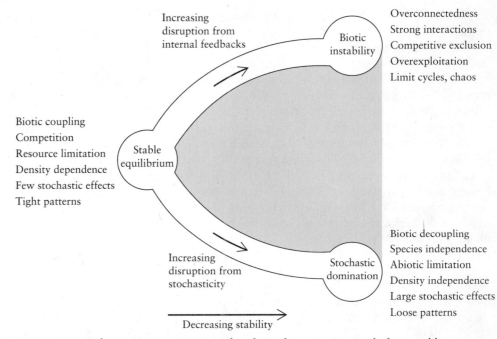

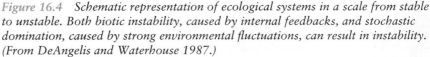

Figure 16.4   Schematic representation of ecological systems in a scale from stable to unstable. Both biotic instability, caused by internal feedbacks, and stochastic domination, caused by strong environmental fluctuations, can result in instability. (From DeAngelis and Waterhouse 1987.)

is no reason to expect all populations to show stable equilibria (Wiens 1984a). There are two sources of instability in populations. Strong environmental fluctuations in weather can produce instability, but biotic interactions may also promote instability. We have seen examples in the last chapter of predator-prey interactions that are unstable. Time lags also can affect population stability (page 214). We should expect real world populations to fall along the whole spectrum from showing stable, equilibrial dynamics to unstable, nonequilibrial dynamics. The simple model shown in Figure 16.2 will be difficult to detect in a real population that shows unstable dynamics.

The spatial scale can also be important in considerations of stability. If you study a very small population on a small area, it may fluctuate widely and even go extinct. A large population on a large study area may at the same time be stable in density. The important concept here is that of local populations linked together through dispersal into *metapopulations* (Figure 16.5). To study population regulation, you must know if a population is subdivided and how the patches are linked (Kareiva 1990). Ensembles of randomly fluctuating subpopulations, loosely linked by dispersal, will persist if irruptions at some sites occur at the same time as extinctions at other sites. The result can be that at a regional level the population appears stable while the individual subpopulations fluctuate greatly.

A third complication for the analysis of population regulation is that real world populations rarely show smooth curves like those in Figure 16.2. A more usual observation is of a cloud of points, such that density-dependence is either "vague" or absent (Strong 1984). Figure 16.6 illustrates the type of density-dependent rela-

Time 1   Time 2

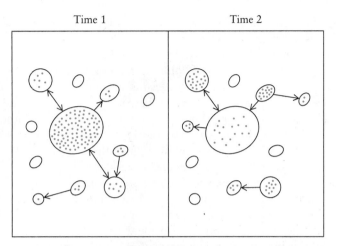

Figure 16.5 Hypothetical metapopulation dynamics. Closed circles represent habitat patches, dots represent individual plants or animals. Arrows indicate dispersal between patches. Over time the regional metapopulation changes less than each local population.

tionships that might be found in the real world. It may be very difficult to find density-dependent relationships in natural populations (Gaston and Lawton 1987).

If a population does not continue to increase, it is axiomatic that births, deaths, or movements must change at high density. The first step is to ask which of these parameters changes with population density (Sinclair 1989). Does reproductive rate decline at high density, or does mortality increase (or both)? If mortality increases, does this fall more heavily on younger or on older animals, on males or on females? These patterns of changing reproduction and mortality with population density can then be analyzed to see if they occur in a variety of populations. This should be the first step to understanding population regulation in animals.

The second step is to determine the reason for the changes in reproduction or mortality. Determining the cause of death of plants or animals in natural population is not always simple. If a fox or a bat has rabies, a fatal disease, the cause of death is clear. If a caterpillar has a tachinid parasite, it is certain to die from this parasiti-

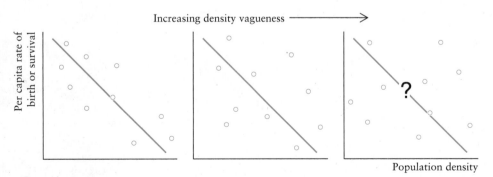

Figure 16.6 Types of density-dependent relationships likely to be found in real world populations. An increasing scatter of points makes it difficult to determine if there is an equilibrium point or where it is. (After Strong 1984.)

zation. But as you examine more complex cases, decisions about causes of death are not clear. If a moose has inadequate winter food and the snow is deep, it may be killed by wolves (Peterson 1992). Is predation the cause of death? Yes, you would answer, but it is so only in the immediate sense. Malnutrition and deep snow have increased the probability of being killed. Because many components of the environment can affect one another, mortality can be *compensatory*. The idea of compensatory mortality is one of the most important concepts needed to understand population regulation. At the two extremes, mortality may be *additive* or it may be completely *compensatory*. How can we distinguish these?

Additive mortality assumes the agricultural model of population arithmetic. If a farmer keeps sheep, and one sheep is killed by a coyote, the farmer's flock is smaller by one. Deaths are additive and to measure their total effect on a population, you simply add them up. But in natural populations when there are several causes of death, the arithmetic is not so simple. Consider, for analogy, a sheep population in which winter food is limiting so that starvation will kill many individuals by the end of winter. If a coyote kills one of these sheep, it may be doomed to die anyway from starvation. Therefore, the number of sheep left at the end of winter will be the same, regardless of whether predation occurs or not. In this hypothetical scenario predation mortality is not additive but is compensatory, and simple arithmetic does not work. Figure 16.7 illustrates how additive and compensatory effects can be recognized.

Compensatory mortality is the reason behind many ecological anomalies that puzzle the average person. If you kill pests, they will not necessarily become less abundant. Chapter 18 discusses this question of pest control. If you shoot grouse or catch fish, their numbers may not necessarily fall (Chapter 17). Compensatory mortality has practical consequences when it occurs.

In natural populations mortality agents will rarely be completely additive or completely compensatory. We can determine if a particular cause of mortality is

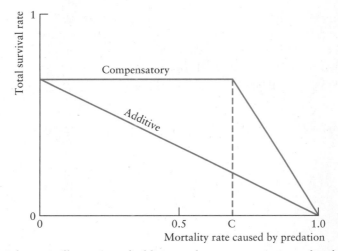

Figure 16.7 Schematic illustration of additive and compensatory mortality for losses due to predation. The additive hypothesis predicts that for an increase in predation mortality, total survival decreases an equal amount. The compensatory hypothesis predicts that, below a threshold, C, an increase in predation losses has no effect on total survival. This model can be applied to any mortality agent— predation, disease, starvation, or hunting. (After Nichols et al. 1984.)

compensatory only by doing an experiment in which total losses are measured with and without the particular cause of death. Few of these experiments have been done, and there is an unfortunate tendency to assume all losses are additive in natural populations.

If birth rates change with population density, it is often difficult to pin down the factors that cause reproduction to change. Food supply or nutrient availability may cause birth rates to change but these effects can be overridden or modified by weather or social interactions in natural populations.

Given these problems, how should one study these key questions of population regulation?

TWO APPROACHES TO STUDYING REGULATION

There are two competing paradigms about how best to study population regulation. *Key factor analysis* is a method of analyzing populations through the preparation of life tables and a retrospective analysis of year-to-year changes in mortality and reproduction. *Experimental manipulations* form a second method of analyzing population changes. Let us consider the advantages and disadvantages of these two approaches.

Key Factor Analysis

Morris (1957) developed key factor analysis as a technique for determining the cause of population outbreaks in the spruce budworm, which periodically defoliates large areas of balsam fir forests in eastern Canada. Varley and Gradwell (1960) improved Morris's method, and their approach is now used.

Key factor analysis begins with a series of life tables of the type shown in Table 16.1. The life table data are most easily obtained for organisms with one discrete generation per year. The life cycle is broken down into a series of stages (eggs, larvae, pupae, adults) on which a sequence of mortality factors operate. We define for each drop in numbers in the life table:

$$k = \log (N_s) - \log(N_e)$$

where

k = instantaneous mortality coefficient*
N_s = number of individuals starting the stage
N_e = number of individuals ending the stage

For example, 83.0 winter moth larvae entered the pupal stage in 1955 (Table 16.1), and of these, 54.6 were killed by pupal predators (shrews, mice, beetles) during late summer, which reduced the population to 28.4 per square meter. Thus

k_5 = instantaneous mortality coefficient for pupal predation
  = $\log (83.0) - \log (28.4) = 0.47$

---

* Note that these k-values are the same as the instantaneous mortality rate defined in Appendix III, without the minus sign.

Table 16.1   Life Table for the Winter Moth in Wytham Woods, Near Oxford, England, for 1955–1956

	Percentage of Previous Stage Killed	No. Killed (per m²)	No. Alive (per m²)	Log No. Alive (per m²)	k-Value
ADULT STAGE					
Females climbing trees, 1955				4.39	
EGG STAGE					
Females × 150				658.0	2.82
LARVAL STAGE					0.84 = k_1
Full-grown larvae	86.9	551.6	**96.4**	1.98	
Attacked by *Cyzenis*	6.7	**6.2**	90.2	1.95	0.03 = k_2
Attacked by other parasites	2.3	**2.6**	87.6	1.94	0.01 = k_3
Infected by microsporidian	4.5	**4.6**	83.0	1.92	0.02 = k_4
PUPAL STAGE					0.47 = k_5
Killed by predators	66.1	**54.6**	28.4	1.45	
Killed by *Cratichneumon*	46.3	**13.4**	15.0	1.18	0.27 = k_6
ADULT STAGE					
Females climbing trees, 1956				7.5	

Note. The figures in bold are those actually measured. The rest of the life table is derived from these.

Source: After Varley et al. (1973).

We do these calculations in logarithms to preserve the additivities of the mortality factors (see Appendix III). Thus we can *define generation mortality K* as

$$K = k_1 + k_2 + k_3 + k_4 + k_5 + \ldots$$

Key factor analysis assumes that all mortality factors are additive and it ignores compensatory mortality. For our sample data in Table 16.1,

$$K = \log (658) - \log (15) = 1.64$$
$$\text{(no. eggs)} \quad \text{(no. adults of both sexes)}$$

which is identical to

$$K = 0.84 + 0.03 + 0.01 + 0.02 + 0.47 + 0.27 = 1.64$$

Varley and coauthors (1973) give a detailed description of these calculations.

Given a series of life tables like Table 16.1 over several years, we can proceed to step 2 of key factor analysis. Figure 16.8 illustrates this step. We can now ask an important question: *What causes the population to change in density from year to year?* Simple visual inspection shows that in Figure 16.8, k_1 (winter disappearance) is the *key factor* causing population fluctuations. A *key factor* is defined as the component of the life table that causes the major fluctuations in population size. There is an implication in this definition that key factors could be used to predict population trends (Morris 1963).

Finally, we can use the k-values to answer a second important question: *Which mortality factors are density-dependent and thus might stop population increase?* By plotting the k-values against the population density of the life cycle stage on which they operate, we can estimate density dependence. Figure 16.9 shows these data for

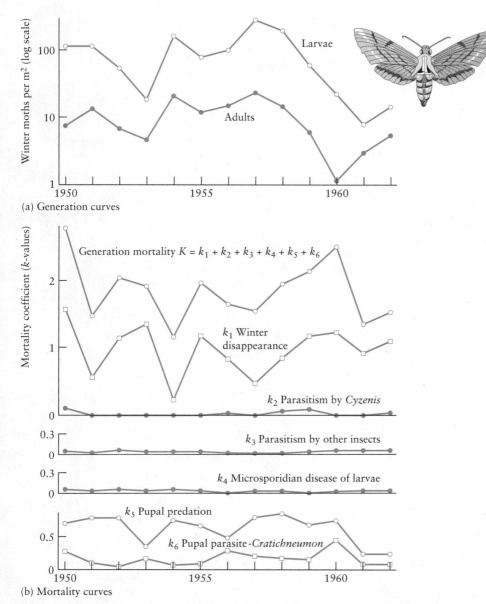

(a) Generation curves

(b) Mortality curves

Figure 16.8   Key factor analysis of the winter moth in Wytham Woods near Oxford, 1950-1962. (a) Winter moth population fluctuations for larvae and adults. (b) Changes in mortality, expressed as k-values, for the six mortality factors shown in Table 16.1. The biggest contribution to change in the generation mortality K comes from changes in k_1, winter disappearance, which is the key factor for this population. (After Varley et al. 1973.)

the winter moth, and Figure 16.10 shows the idealized types of curves that can arise from this type of key factor analysis. Note that the key factor does not need to be density-dependent and need not be involved in population regulation. In this example for the winter moth, winter disappearance is the key factor but pupal predation is the major density-dependent factor.

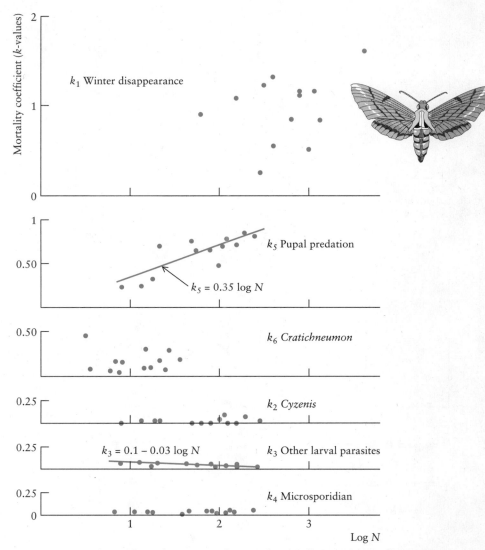

Figure 16.9   *Relationship of winter moth mortality coefficients to population density. The* k-*values for the different mortalities are plotted against the population
densities of the stage on which they acted:* k_1 *and* k_2 *are density-independent and
quite variable,* k_2 *and* k_4 *are density-independent but are constant,* k_3 *is inversely
density dependent, and* k_5 *is strongly density dependent. Compare these data with
the idealized curves in Figure 16.10. (After Varley et al. 1973.)*

Key factor analysis has been widely applied to insect populations (Varley et al. 1973) but has some important limitations. It cannot be applied to organisms with overlapping generations, like birds and mammals. Mortality factors may be difficult to separate into discrete effects that operate in a linear sequence and do not overlap and are completely additive.

Finally, density dependence may be difficult to detect if the equilibrium density (Figure 16.3) varies greatly from year to year (Moss et al. 1982). Nevertheless, key factor analysis has provided for some populations a reliable quantitative framework against which the problems of natural regulation can be discussed.

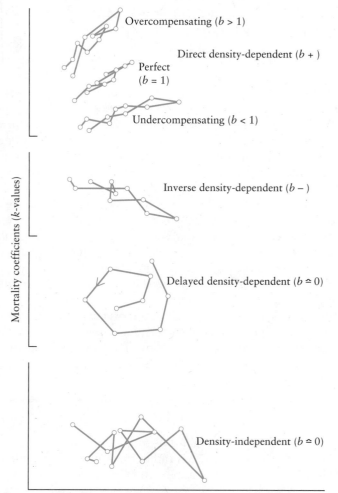

Figure 16.10   Idealized forms of the relationship possible between k-*values determined from key factor analysis and population density. The points are connected in a time sequence, and* b *is the slope of the regression line. Compare with Figure 16.9. (After Southwood 1978.)*

Experimental Analysis

An alternative approach to population regulation is to ask the empirical question, *What factors limit population density during a particular study?* This approach does not utilize the density-dependent paradigm because density dependence is often impossible to demonstrate with field data. Instead we try to identify *limiting factors* and study them experimentally. A population may be held down by one or more limiting factors, and these factors can be recognized empirically by a manipulation—adding to or reducing the relevant factor. If we suspect that food is limiting a population, we can increase the food supply and see if population size increases accordingly (Watson and Moss 1970). Alternatively, we can observe changes in population density and the supposed limiting factors over several years and see if they vary together. Observations of this type, however, always provide weaker evidence than that of experimental manipulations.

The experimental approach uses the most direct and empirical techniques for answering the two central questions of regulation—what determines average abundance and what stops population growth? If you think that parasites reduce the average abundance of pheasants, increase or reduce parasite loads and observe the changes in pheasant numbers. If you think that food shortage stops population growth in cabbage aphids, manipulate the food resources up or down and measure aphid population growth. It is important to realize that more than one factor may be involved in population limitation. Both parasite levels and food supplies may affect average abundance, and once you have shown one factor to be significant, do not assume that only one factor is involved.

Often it is not possible to manipulate a suspected factor (e.g, weather), either because it is too expensive or because it is not biologically or politically possible. Experimental analysis can be carried out without manipulations if you set up your hypothesis and make a prediction about what can be observed. For example, if you postulate that cold spring weather stops population growth in spruce budworm (Morris 1963), you can measure spring weather and population trends in several sites in several years and test this prediction.

Experimental analysis is forward-looking and oriented toward hypothesis testing about mechanisms of regulation. Key factor analysis is backward-looking and is confined to a descriptive analysis of a population. Theoretically, both methods should converge to provide an understanding about population regulation.

PLANT POPULATION REGULATION

Because most plants are modular organisms, population regulation in plants must be discussed as the regulation of biomass rather than numbers. Plant ecologists have not usually addressed the problem of population regulation in the same way as have animal ecologists (Crawley 1990), but the same principles can be applied. As a plant population increases in numbers and biomass, either reproduction or survival will be reduced by a shortage of nutrients, water or light; by herbivore damage; by parasites and diseases; or by a shortage of space. Because plants are typically fixed in one location, competition for light or nutrients is often implicated in population regulation. This competition has been described by the $-3/2$ power rule (also called Yoda's law or the self-thinning rule).

The self-thinning rule describes the relationship between individual plant size and density in even-aged populations. Mortality, or "thinning," from competition within the population is postulated to fit a theoretical line with a slope of $-3/2$:

$$\log \overline{m} = -\frac{3}{2} (\log N) + K$$

where

$\overline{m}$ = average plant weight (grams)
N = plant density (individuals per square meter)
K = a constant

This line has been suggested as an ecological law (Westoby 1984, Hutchings 1983) that applies both within one plant species and between different plant species. Figure

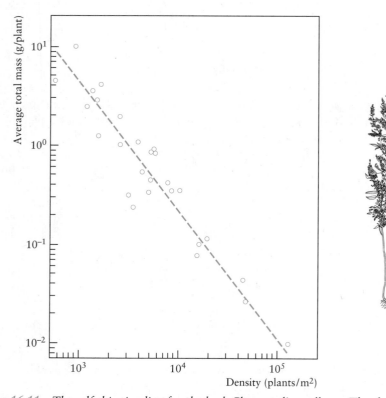

Figure 16.11 *The self-thinning line for the herb* Chenopodium album. *The slope of this line is* −1.33, *close to the theoretical* −3/2 *of the self-thinning rule. Populations started to either side of this line would be expected to move to the line and then reach equilibrium along the line. (Data from Yoda et al. 1963.)*

16.11 illustrates the −3/2 power rule. The self-thinning rule highlights the tradeoffs that can occur in organisms with plastic growth, such that the size of an individual can become smaller as density increases.

Recent evaluations of the self-thinning rule have found many exceptions to it (Weller 1987, 1991). Of 63 data sets for particular plants, only 24 fitted the −3/2 predicted slope of the self-thinning line. For gymnosperm trees, Weller (1987) found that more shade tolerant tree species had more shallow slopes than the predicted value. These results argue against a single, quantitative thinning law for all plants. The slope of the thinning line is variable, but this gives us further insight into species differences under strong competition for light and nutrients.

EVOLUTIONARY IMPLICATIONS OF NATURAL REGULATION

How are systems of natural regulation affected by evolutionary changes? We have already discussed some of the problems involved in coevolution of predator-prey systems (Chapter 14) and herbivore-plant systems (Chapter 15). In many of these interactions, evolutionary changes operate very slowly and are difficult to detect. But recent work in ecological genetics (Endler 1986) has shown that evolutionary changes may occur very rapidly, so that the evolutionary time scale approaches the ecological time scale. Natural selection may thus impinge upon natural regulation in some organisms.

Many changes in abundance can be attributed to changes in extrinsic factors, such as weather, disease, or predation. But some changes in abundance are the result of changes in the genetic properties of the organisms in a population. Such evolutionary changes are produced by natural selection (Pimentel 1961). Pimentel believes that natural population regulation has its foundation in the process of evolution.

A simple model will illustrate the type of systemic changes that could be produced by natural selection. Consider a two-species system of one plant and one herbivore, and, to make the model simple, let us focus on only one gene on one chromosome in the plant. The hypothetical gene has a major effect on (1) the ability of the plant to survive in its environment and (2) the palatability of the plant to the herbivore. Two different alleles (*A* and *a*) occur at the hypothetical gene locus, and the properties of the genotypes are as follows:

	GENOTYPE OF PLANT		
	AA	**Aa**	**aa**
Ability of plant to survive	Good	Poor	Very poor
Palatability to herbivores	High	Low	Very low

Thus plants of genotype *AA* are able to survive very well but attract many herbivores because they are desirable food. Each plant genotype can support only a limited number of herbivores before it is killed by overgrazing. Finally, we assume that the reproductive rate of the herbivore will be affected by the genotype of plant on which it lives, so that highly palatable plants are best for herbivore reproduction.

This simple model is a variant of the discrete generation predator-prey model we discussed in Chapter 14, in which the plant is the prey and the herbivore is the predator. The major change in the model is that we allow genetic variation within the plant population. Figure 16.12 illustrates one pattern of equilibrium for a hypothetical system starting with 150 herbivores and a plant population with genotype

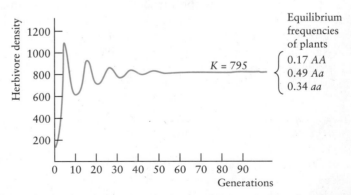

Figure 16.12   *Determination of average herbivore density resulting from the interaction of a plant and a herbivore through natural selection. Starting conditions given in the text. (After Pimentel 1961.)*

frequencies of 0.36 *AA*, 0.50 *Aa*, and 0.14 *aa*. The system stabilizes under some initial conditions (as in this example) but with other assumptions may be unstable with divergent oscillations (compare Figure 14.5, page 268).

Pimentel (1961) catalogs some spectacular examples of genetic changes of this type playing a role in population regulation. For example, the Hessian fly population was reduced drastically in Kansas after 1942 when resistant varieties of wheat were introduced. The herbivore population of Hessian flies was significantly reduced by changing the genetic makeup of the wheat plant. Another example is the myxomatosis rabbit interaction in Australia. The European rabbit was introduced into Australia in 1859 and increased to very high densities within 20 years. After World War II, an attempt was made to reduce rabbit numbers by releasing a virus disease from South America, myxomatosis. The myxoma virus was highly lethal to European rabbits, killing over 99 percent of individuals infected. Figure 16.13 shows the precipitous crash in rabbit numbers that followed the introduction of myxomatosis in 1951.

Since myxomatosis was introduced into Australia in 1951, evolution has been going on in both the virus and the rabbit. The virus has become attenuated so that it kills fewer and fewer rabbits and takes longer to cause death. Since mosquitoes are a major vector of the disease, the exposure time before death is critical to viral spread. Table 16.2 summarizes changes that have occurred in the virus. These data are obtained by testing standard laboratory rabbits against the virus, so they measure viral changes while holding rabbit susceptibility constant. Since 1951 less virulent grades of virus have replaced more virulent grades in field populations.

Rabbits have also become more resistant to the virus (Figure 16.14). By challenging wild rabbits with a constant laboratory virus source, we can detect that natural selection has produced a growing resistance of rabbits to this introduced disease.

*Figure 16.13   Population crash of the European rabbit (*Oryctolagus cuniculus*) at Lake Urana, New South Wales, after the virus disease myxomatosis was introduced in 1951. Numbers of healthy rabbits were counted on standardized transects. (After Myers et al. 1954.)*

Table 16.2 Virulence of Field Myxoma Virus Types in Rabbits in Australia

Grade of Severity	VIRULENCE TYPE GRADE					
	I	II	IIIA	IIIB	IV	V
Mean survival times of rabbits (days)	<13	14–16	17–22	23–28	29–50	—
Case mortality rate (%)	>99	95–99	90–95	70–90	50–70	<50
Australia						
1950–1951	100	—	—	—	—	—
1958–1959	0	25.0	29	27	14	5
1963–1964	0	0.3	26	33	31	9

Source: After Fenner and Myers (1978).

The evolution of resistance to the virus in rabbits is easily explained by selection operating at the individual level: rabbits that are more resistant leave more offspring. It is more difficult to explain the evolution of reduced virulence in the virus. Virulence in a virus is related to fitness because more virulent viruses make more copies of themselves. But if more virulent viruses kill rabbits more quickly, there will be less time available for transmission of the virus through mosquitoes or fleas. The result for the myxoma virus is group selection operating to reduce virulence to a moderate level (Levin and Pimentel 1981). Group selection occurred because less virulent viral colonies are favored over more virulent viral colonies because they take longer to kill the host rabbit (Table 16.2). Host-parasite systems may be ideal candidates for group selection along these lines.

We do not know if the rabbit-myxoma system has reached a stable equilibrium or whether continuing evolution will allow the rabbit population to recover to its former levels (Figure 16.13). There is some evidence that the rabbit-myxomatosis interaction in Britain is changing toward more resistant rabbits and more virulent

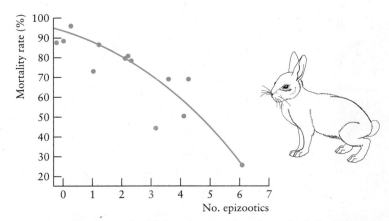

Figure 16.14 Mortality rates of wild European rabbits from the Lake Urana region of southeastern Australia after exposure to several epizootics of myxomatosis. Mortality was measured after a challenge infection with a strain of myxoma virus of grade III virulence. (After Fenner and Myers 1978.)

viruses, and the population size of rabbits in Britain seems to be increasing again (Ross and Tittensor 1981).

Pimentel's concept of genetic changes involves interspecific interactions and suggests some possible implications of these interactions to the determination of average abundance. This concept emphasizes the role of evolution in questions of average abundance, and in doing this it serves as a warning to the continual introduction of new species into ecological communities of distant areas.

Self-regulatory populations present yet another problem in evolutionary ecology. How does a population evolve the machinery to be self-regulating? Self-regulation is clearly a desirable adaptation for any population that has the potentiality of destroying its resources. The problem is that it is an adaptation that is favorable for the *population*, not necessarily for individuals. Darwinian natural selection is individual selection. How can population adaptations arise? The answer is simple, by *group selection* (see Chapter 2). Just as natural selection can operate at the level of the *individual* organism—individuals that are more fit leave more descendants in the next generation—it can also operate at the level of the group. Groups that have an adaptation may avoid extinction. Group selection was invoked by Wynne-Edwards (1962) to explain his theory of self-regulation. Most workers, however, reject group selection as a possible mechanism of evolution (Williams 1966, Wilson 1983, Endler 1986) and try to explain all adaptations on the basis of individual selection.

Group selection is typically invoked as an evolutionary force when some adaptation is good for the population but bad for an individual. In these cases, group selection pushes up the frequency of a trait, while individual selection pushes it down. For group selection to work, whole populations (groups) must become extinct while others survive. Since more individuals die than groups become extinct, individual selection is always much stronger than group selection. Obviously, if group selection is pushing the same way as individual selection, the trait will be favored at both levels, and no problem arises.

So the critical question becomes whether self-regulatory adaptations are good for populations (agreed) and good for individuals as well (questioned). If natural selection operating on individuals can favor mechanisms of self-regulation, the problem is solved, and group selection can recede into the background. How might selection favor self-regulation? The answer is very similar to that used to explain the evolution of competitive ability (see page 254). Natural selection will favor individuals that increase their fitness by means of higher reproduction or lower mortality, but it will also favor individuals that *reduce the fitness of their neighbors* by any technique of interference competition. Fitness is relative, and the mistake of many evolutionary ecologists was to assume that organisms were trapped in an upward spiral of ever-increasing fitness through ever-increasing reproductive rates. Thus self-regulatory mechanisms can be explained most easily by individual selection operating on mechanisms of interference competition within the species.

One mechanism of self-regulation is emigration of organisms from optimal to suboptimal environments. Emigration would always appear to be disadvantageous to the individuals involved (since many of them die in the process of moving), and thus, MacArthur (1972) argued, emigration cannot evolve by individual selection. This argument is not correct (Lomnicki 1978). Individual selection can favor self-regulation by emigration because not all individuals have equal access to resources in natural populations and there is both spatial and temporal variability in natural environments. There is thus no need to invoke group selection to explain self-regulation.

SUMMARY

Populations of plants and animals do not increase without limits but show more or less restricted fluctuations. Two general questions may be raised for all populations: (1) What stops population growth? (2) What determines average abundance?

Three general theories answer these two questions by focusing on the interactions between the population and the environmental factors of weather, food, shelter, and enemies (predators, parasites, diseases). The biotic school suggests that density-dependent factors are critical in preventing population increase and in determining average abundance. Natural enemies are postulated to be the main density-dependent factors in many populations. The climate school emphasizes the role of weather factors affecting population size and suggests that weather may act as a density-dependent control. By contrast, the self-regulation school focuses on events going on within a population, on individual differences in behavior and physiology. The general premise of this school is that abundance may change because the quality of individuals changes. Population increase may be stopped by a deterioration in the quality of individuals as density rises rather than by a change in environmental factors. Average abundance may be altered by genetic changes in populations. Quality and quantity are both important aspects of populations.

Population regulation theory has focused on equilibrium conditions, and many ecologists now emphasize nonequilibrium concepts and ask what factors reduce stability for populations. The spatial scale of a study affects conclusions about stability, and if a population is subdivided into local populations stability may be increased for the entire population. Mortality agents may cause additive or compensatory losses in populations. Additive losses may limit or regulate population density, but compensatory losses may be irrelevant to both limitation and regulation.

The theories of natural regulation of numbers are not mutually exclusive but overlap, and a synthesis of several approaches may be most useful in attempts to answer practical questions. The natural regulation of populations is a critical area of theoretical ecology because it is central to many questions of community ecology and because it has enormous practical consequences, which we shall explore in the next three chapters.

Selected References

CHITTY, D. 1960. Population processes in the vole and their relevance to general theory. *Canadian Journal of Zoology* 38:99–113.

DE ANGELIS, D. L., and J. C. WATERHOUSE. 1987. Equilibrium and nonequilibrium concepts in ecological models. *Ecological Monographs* 57:1–21.

HASSELL, M. P. 1985. Insect natural enemies as regulating factors. *Journal of Animal Ecology* 54:323–334.

KREBS, C. J. 1978. A review of the Chitty hypothesis of population regulation. *Canadian Journal of Zoology* 56:2463–2480.

MC NAMARA, J. M., and A. I. HOUSTON. 1987. Starvation and predation as factors limiting population size. *Ecology* 68:1515–1519.

SINCLAIR, A. R. E. 1989. Population regulation in animals. In *Ecological Concepts*, ed. J. M. Cherrett, pp. 197–241. Blackwell, Oxford, England.

STRONG, D. R. 1986. Density-vague population change. *Trends in Ecology and Evolution* 1:39–42.

WATSON, A., and R. MOSS. 1970. Dominance, spacing behaviour and aggression in relation to population limitation in vertebrates. In *Animal Populations in Relation to Their Food Resources*, ed. A. Watson, pp. 167–218. Blackwell, Oxford, England.

WELLER, D. E. 1987. A reevaluation of the $-3/2$ power rule of plant self-thinning. *Ecological Monographs* 57:23–43.

Questions and Problems

1. Morris (1957, p. 49), in discussing the interpretation of mortality data in population studies, states:

   We tend to overlook the fact that these mortality estimates do not represent an ultimate objective in population work. Long columns of percentages, which are sometimes presented only with the conclusion that high percentages indicate important mortality factors and low percentages indicate unimportant ones, contribute little to our understanding of population dynamics.

   Discuss this claim.

2. Density-dependent relationships can be looked for by studying different local populations living in different patches (spatial density dependence) or by following one local population over several years (temporal density dependence). Discuss the interpretation of these two types of data with regard to the problem of regulation.

3. Milne (1958, p. 254) states:

   Theories of natural control in mammal and bird populations are not likely to be very useful for insects. Mammals and birds, being "warm-blooded" and having more efficient water-conserving mechanisms, are far less affected by the irregular vagaries of weather, and they may also exhibit "territorial" behavior which is important in limiting density.

   Discuss whether different theories of population regulation should be developed for different taxonomic groups.

4. Darwin (1859) wrote in *The Origin of Species* (chap. 2): "Rarity is the attribute of a vast number of species of all classes, in all countries." Discuss the possible effects of rarity and commonness on population-regulation mechanisms.

5. Review the population growth models discussed in Chapter 12 for discrete generations, and relate these models to the theories discussed in this chapter. One of the principles of the biotic school is that density-independent processes cannot prevent a population from becoming extinct or from increasing to excessive numbers (starvation). Try to demonstrate this axiom with the discrete-generation growth model given on page 217.

6. Murray (1979, p. 66) states: "The fact that a particular population exhibits sigmoid growth does not constitute evidence that density-dependent factors are acting." Is this correct?

7. Read the key factor analysis of partridge populations by Blank and colleagues (1967) and then the reanalysis of these conclusions by Manly (1977). How could you decide experimentally which interpretation is correct?

8. Can a population persist without regulation? How could you determine if a population was persisting without regulation? Read Strong (1984) and Reddingius and den Boer (1970) and discuss.

9. Sinclair (1989) tabulates the causes for density dependence under six categories: space, food, predators, parasites, disease, and society. Compare and contrast these categories with those proposed by Andrewartha and Birch (1954).

10. Read Wynne-Edwards's (1986) ideas on population regulation and group selection and the review of his book by Bell (1987). Are there circumstances under which group selection could produce mechanisms of population regulation?

Overview Question

Local populations can be classified as source populations ($R_0 > 1$) or sink populations ($R_0 < 1$). How would you determine for a metapopulation which local populations were sources and which were sinks? Discuss the application of population regulation theories to a metapopulation containing sources and sinks.

17 *Applied Problems I: Harvesting Populations*

To manage a population effectively, we must have some understanding of its dynamics. Almost all of human history might be said to illustrate this idea in graphic detail, and a list of populations destroyed by inadequate management should be both a warning and a stimulus for us to achieve some understanding of harvesting principles. The central problem of economically oriented fields such as forestry, agriculture, fisheries, and wildlife management is how to produce the greatest crop without endangering the resource being harvested. The problem may be illustrated with a simple example from forestry. If you were managing a forest woodlot that was growing to maturity, you would obviously not cut the trees when they are saplings because this would give little wood production and less profit. At the other extreme, you would prevent the trees from growing too old and starting to rot because you would get little timber to sell. Somewhere between these two extremes will be some optimum point to harvest the trees, and the problem is how to locate it.

Next to forestry and agriculture, the greatest amount of work on the problem of optimum harvesting has been done in fishery biology. This is because of the tremendous economic importance of marine fisheries in particular. Many marine fisheries have dwindled in size since the 1920s because of overfishing, and this has stimulated a great deal of research on "the overfishing problem."

For any harvested population, the important unit of measure is the crop or *yield*. The yield may be expressed in *numbers* or *weight* of organisms and always involves some unit of time (often a year). We are interested in obtaining the optimum yield from any harvested population. We will begin by defining *optimum yield* very specifically, and at the end of the chapter we will reconsider other ways of defining *optimum*. The concept of *maximum sustained yield* has been the basis of scientific resource management since the 1930s (Larkin 1977). Let us consider first the simple situation in which maximum yield in biomass is defined as the optimum yield. Implicit in this concept is the idea of a sustained yield over a long time period.

Russell (1931) was one of the first to deal in detail with the overfishing problem. In any exploited population, there will usually be a portion of the population that cannot be caught by the type of gear used or that is purposely not harvested. For a fishery, interest normally centers on yield in weight. Russell pointed out that two factors decrease the weight of the catchable stock during a year: natural mortality and fishing mortality. Similarly, two factors increase the weight of the stock: growth and recruitment.* Consequently, one can write a simple equation to describe this relationship:

---

*Recruitment in fisheries is usually measured when the fish reach a certain size or age. Recruitment thus includes natality and early life history mortality.

$$S_2 = S_1 + R + G - M - F$$

where

> S_2 = weight of the catchable stock at the end of the year
> S_1 = weight of the catchable stock at the start of the year
> R = weight of new recruits
> G = growth in weight of fish remaining alive
> M = weight of fish removed by natural deaths
> F = yield to fishery

If we wish to balance the fish population, $S_1 = S_2$, and hence

$$R + G = M + F$$

This means that in an unexploited stage, in which the stock biomass remains approximately constant from one year to the next, all growth and recruitment is on the average balanced by natural mortality. When exploitation begins, the size of the exploited population is usually reduced, and the loss to the fishery is made up by compensatory changes such as (1) greater recruitment rate, (2) greater growth rate, or (3) reduced natural mortality. In some populations, none of these three occurs, and the population is exploited to extinction.

Note that stability at *any* level of population density is described by the equation

Recruitment + growth = natural losses + fishing yield

Thus the crucial question arises: What level of population stabilization safely permits the greatest weight of catch to the fishery? One of the early attempts to solve this problem was made by Graham (1935), who proposed the *sigmoid-curve theory*.

Start by considering a very small stock of fish in an empty area of the sea, said Graham. At what rate will such a stock increase in size? Graham suggested that the growth of this population would follow a sigmoid curve like the one described by the logistic equation (Figure 17.1). Initially, the population grows slowly in absolute size, reaches a maximum rate of increase near the middle of the curve, and grows slowly again as it approaches the asymptote of maximal density. We can use the terminology of the logistic equation to show that two factors interact to determine the amount of increase per year. Let $K = 200$ units and $r = 1.0$ for simplicity:

Point on Curve	Population Size (N)	$\dfrac{K - N}{K}$	rN	Amount of Increase (dN/dt) per year
S_1	20	0.90	20	18
S_2	50	0.75	50	38
S_3	100	0.50	100	50
S_4	150	0.25	150	38
S_5	180	0.10	180	18

According to the logistic, the amount of population increase is

$$\frac{dN}{dt} = rN\left(\frac{K - N}{K}\right)$$

and this is maximal at the midpoint of the curve (S_3).

If you wish to maintain the maximal yield from such a population, Graham pointed out, you should keep the stock around point S_3 of the curve. The important point here is that the highest production from such a population is not near the top

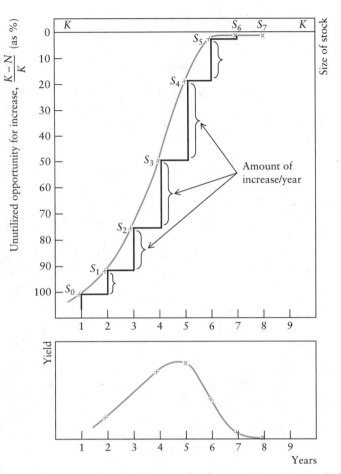

Figure 17.1 *Sigmoid curve to describe the growth of a population that could be exploited. The amount of increase per year is the yield that could be taken by the fishery. (After Graham 1939.)*

of the curve, where the fish population is relatively dense, but at a lower density. This can be expressed as a rule of exploitation: *Maximum yield is obtained from populations at less than maximum density.*

All the vital statistics of an exploited population—recruitment, growth, and natural mortality—may be a function of population density and also of age composition. Since in most fisheries we do not know how these vital statistics relate to density or age, we employ some simplifying assumptions. Two alternative approaches have been developed for determining optimum yield; these are called *logistic-type models* and *dynamic pool models* (Schaefer 1968).

LOGISTIC-TYPE MODELS

In logistic-type models,* we do not distinguish growth, recruitment, and natural mortality but combine these into a single measure, *rate of population increase*, which

---

* Also called *surplus yield models, stock production models,* or *Schaefer models.*

is a function of population size. Graham's *sigmoid-curve theory* is a classic example of this type of model. The general case can be written:

Rate of population increase = f (population size) − amount of fishing losses

If we specify that the function of population size in this equation is a simple linear function,

$$f \text{ (population size)} = r\,\frac{(K - N)}{K} = r - \frac{rN}{K}$$

we obtain the logistic equation modified for fishing losses:

$$\frac{dN}{dt} = rN\left(\frac{K - N}{K}\right) - qXN$$

where

N = population size
t = time
r = per capita rate of population growth
K = asymptotic density (in absence of fishing)
q = catchability (a constant)
X = amount of fishing effort (so qX = fishing mortality rate)

The ecological assumptions of logistic-type models are that no time lags operate in the system, that age structure has no effect on the rate of population increase and that catchability remains constant at all densities of fish. Figure 17.2 illustrates how an exploited population is postulated to respond to a series of fishing episodes in this model. This model, although crude, may be useful for populations that are in approximately steady states in the absence of fishing and that do not change greatly

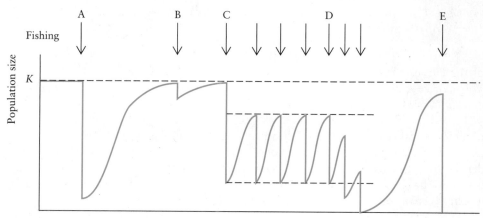

Figure 17.2 Schematic diagram of the assumed response of a fish population to exploitation according to logistic-type models. Periodic fishing of different intensity and frequency is shown at the top of the diagram. At point A fishing is intensive but the time interval is long so the fish population recovers. At C a moderate intensity fishery is operating, and at D this fishing intensity is applied more frequently, causing the stock to collapse. At E excessive fishing drives the stock to extinction. Note that in every recovery the fish population increases logistically.

from year to year. Because of their simplicity, logistic models can be used on fisheries with relatively few data available. An example will illustrate how this can be done.

The Peruvian anchovy (*Engraulis ringens*) is restricted in distribution to the area of upwelling of cool, nutrient-rich water along the coasts of Peru and northern Chile. The upwelling causes very high productivity in the coastal zone. The Peruvian anchovy is a short-lived fish, spawning first at about one year of age and rarely living beyond three years. It is a small fish, about 12 centimeters in length at one year and seldom reaching 20 centimeters in length. Young anchovies enter the fishery at only five months of age (8 to 10 cm). Anchovies occur in schools and are caught near the surface.

The Peruvian anchovy fishery was the largest fishery in the world until 1972 when it collapsed. From 1955, when the major fishery first began, the anchovy catch doubled every year until 1961. In 1970, 12.3 million metric tons were harvested, and this single-species fishery comprised 18 percent of the total world harvest of fish. Figure 17.3 shows the total catch and the total fishing effort. These two parameters were used to fit a logistic model to the fishery (Boerema and Gulland 1973). Anchovy are taken both by fishermen and by large colonies of seabirds, and the effects of the two had to be combined to measure the "catch." Figure 17.3 indicates a maximum sustained yield around 10 to 11 million metric tons, which, after subtraction of the bird share, left about 9 to 10 million tons for the fishery. From 1964 to 1971 the catch was close to the supposed maximum of Figure 17.3. Note that the estimate of maximum sustainable yield in Figure 17.3 refers to average conditions over a number of years.

In 1972 average conditions disappeared, and the Peruvian anchovy fishery col-

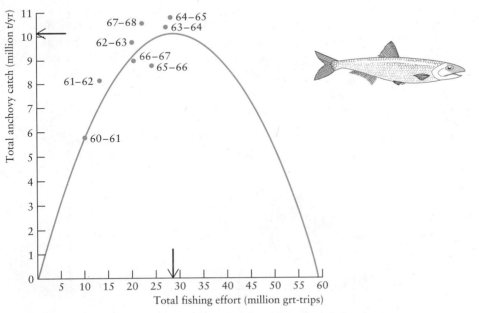

Figure 17.3   Relation between total fishing effort and total catch for the Peruvian anchovy fishery, 1960–1968. The effects of humans and seabirds are combined in these data. The parabola represents the logistic model fitted to these data, as in Figure 17.1. Arrows indicate maximum sustained yield and appropriate fishing effort. (After Boerema and Gulland 1973.)

lapsed. Early in 1972 the upwelling system off the coast of Peru weakened, and warm tropical water moved into the area. This phenomenon—known as "El Niño" (The Child) because it often happens around Christmas—occurs about every five years and greatly changes the ecology of the area (Mysak 1986). The productivity of the sea drops, seabirds starve, and anchovies move south to cooler waters and may concentrate. In early 1972 very few young fish were found; the spawning of 1971 had been very poor, only one-seventh of normal. Adult fish were highly concentrated in cooler waters in early 1972, and these concentrations produced large catches for the fishermen. By June 1972 the anchovy stocks had fallen to a low level, catches had declined drastically, and no young fish were entering the population. The fishery was suspended to allow the stocks to recover, but from 1972 to 1985 there was little sign of a return of the anchovy toward its former abundance. Catches fell to low levels and began a moderate recovery only during the late 1980s (Figure 17.4). The economic consequences of the fishery collapse were very great, and some of it might have been avoided if the fishery had been closed a few months earlier or if the fishing intensity had been slightly less than the maximum shown in Figure 17.3. The important message this collapse underlines is the fragility of the assumption that fish populations are in a state of equilibrium and that average conditions never change.

DYNAMIC POOL MODELS

Dynamic pool models of harvested populations are more biologically explicit because they include estimates of growth, recruitment, and mortality for the population being harvested. In these models, various simplifying assumptions are made. Natural mortality rate is assumed to be constant, independent of density, and the same for all ages. Growth rates are assumed to be age-specific but not related to population density. Fishing mortality (effort) is assumed to act just like natural mortality, to be independent of density, and to be constant for all ages of fish. These assumptions

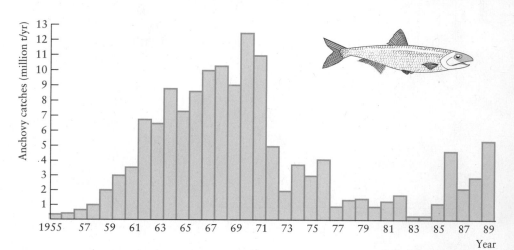

Figure 17.4   Total catch for the Peruvian anchovy fishery, 1955–1989. This fishery was the largest in the world until it collapsed in 1972. In spite of reduced fishing, there has been little recovery since then. (Data from FAO Yearbooks of Fishery Statistics.)

are unrealistic but they are useful as a starting point, and they can be relaxed later in the analysis. The object is to determine what yield a given level of fishing mortality will produce. In this simple model, the population size of R recruits after t years in the fished population is given by the formula for geometric decrease:

$$N_t = Re^{-(F + M)t}$$

where

N_t = number of recruits alive at t years after entering fishery

t = time in years since recruits entered fishery

R = number of original recruits

F = instantaneous fishing mortality rate

M = instantaneous natural mortality rate

This is the familiar curve of geometric increase (or decrease). If $R = 1$, this formula gives the fraction of recruits alive at any time since entering the fishery. The yield to the fishery in this simple model is defined as

Yield = (number in age class) × (average weight) × (fishing mortality rate)

summed over all age classes caught in the fishery. This can be written

$$Y = \sum_{t = t_c}^{\infty} FN_tW_t$$

where

Y = yield in weight for a year

F = instantaneous fishing mortality rate per year

N_t = population size of age t fish

W_t = average weight of age t fish

t_c = age at which fish enter the fishery

Let us illustrate this simple dynamic pool model with an example from the European fishery for plaice (*Pleuronectes platessa*) in the North Sea. The plaice is a shallow-water flatfish that is an important commercial species in the North Sea. Plaice spawn in midwinter when females are five to seven years old and males are four to six years old. Females can lay up to 350,000 fertile eggs, an enormous reproductive rate balanced by an equally high mortality. On the average, all but ten animals out of every million eggs must die before reaching maturity, and the actual range observed by Beverton (1962) during 26 years was between 999,970 and 999,995 dying for every million laid. Much of this loss occurs during the pelagic phase, when the eggs float in the plankton until hatching, and the larval plaice are carried about by water currents in the North Sea. After about two months the larval plaice settle out on nursery areas off the sandy coasts of the Netherlands, Denmark, and Germany. There the young plaice remain until between two and three years of age, when they begin to move off the coast and toward the middle of the North Sea. They enter the commercial fishery between three and five years of age, at a length of 20 to 30 centimeters.

The plaice population has remained fairly stable, with the exception of the periods during the world wars, when fishing was reduced and stocks increased. We can illustrate a dynamic pool model most easily in this type of near-equilibrium condition. First, we must determine growth rate with respect to age in the plaice,

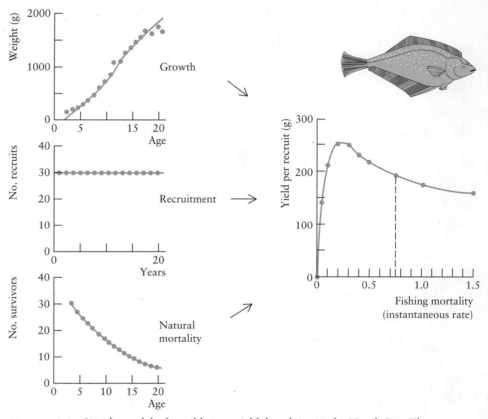

Figure 17.5   Simple model of equilibrium yield for plaice in the North Sea. The fishing intensity before World War II is indicated by the dashed line. (After Beverton and Holt 1957.)

and we can do that with samples from the fishery (Figure 17.5). We assume in this simple model that growth does not depend on the population density. Second, we need to specify recruitment, and we assume a constant number of recruits each year. For the plaice, this is not an unreasonable first approximation (Figure 17.5). Third, we must determine the natural mortality rate. We can do this by mark-and-recapture techniques (Chapter 10) or by indirect means. We assume in the simple case that natural mortality is constant at all ages and at all population densities. For the plaice, Beverton and Holt (1957) estimate $M = 0.10$, and Figure 17.5 shows how a cohort of recruits would decline according to predictions of natural mortality *only*.

Since recruitment is assumed constant, we can express the yield as yield per recruit, and by combining the three factors we obtain the yield curve shown in Figure 17.5. One example of how this yield for plaice was calculated for a fishing mortality of 0.5 is given in Table 17.1. The yield per recruit was then calculated for several values of fishing mortality to obtain the curve in Figure 17.5. This is only an approximate calculation because we should use calculus instead of finite summation to find the yield (details in Beverton and Holt 1957). Figure 17.5 also shows the pre-World War II fishing intensity ($F = 0.73$), which was clearly not at the point of optimum yield.

We have treated fishing mortality in the same way that we have treated natural

Table 17.1 Calculation of Equilibrium Yield per Recruit for North Sea Plaice for a Fishing Mortality of 0.5

Fishing Year	Age at Midpoint (yr)[a]	(1) = W_t Average Weight (g)	(2) = N_t Fraction of Recruits Surviving to This Age, $e^{-(F+M)t}$	(3) Yield to Fishery, F	Product (1) × (2) × (3)
0–1	4.2	158	0.741	0.5	58.54
1–2	5.2	237	0.407	0.5	48.23
2–3	6.2	331	0.223	0.5	36.91
3–4	7.2	435	0.122	0.5	26.54
4–5	8.2	546	0.067	0.5	18.29
5-6	9.2	664	0.037	0.5	12.28
6–7	10.2	784	0.020	0.5	7.84
7–8	11.2	904	0.011	0.5	4.97
8–9	12.2	1024	0.006	0.5	3.07
9–10	13.2	1143	0.003	0.5	1.71
				Total yield per recruit	218.38 g

Note. The average weight is obtained from the growth curve shown in Figure 17.5. The fraction of recruits is calculated by applying a constant loss per year of $(F + M)$, which in this example is $(0.5 + 0.1)$. The yield per recruit is obtained from the formula

$$Y = \sum_{t_c}^{\infty} FN_t W_t$$

[a] The age at recruitment is 3.7 years.

mortality, using humans as just another "predator" of the system. This fishing mortality rate must be converted into fishing effort before the results of a yield analysis, such as that in Figure 17.5, can be applied to an operating fishery. This application is a complex problem that revolves around the types of equipment used, the equipment's efficiency, the interactions between different units of equipment, the spatial and seasonal patterns of exploitation, and the area occupied by the stock (Winters and Wheeler 1985). Gulland (1988) discusses these problems in some detail, and the analysis depends on the type of fishery operation.

Once we have built a dynamic pool model of a fishery, we can test it by regulating the fishery accordingly. Thus in the North Sea plaice, we would predict from Figure 17.5 that an increased yield would result from lowering fishing mortality to one-third or one-half the prewar level of 0.73. This is the critical test of any model: Does it predict accurately? Alternatively, we can use the model (assuming it is accurate) to investigate the effect of various changes in the vital statistics on the yield to the fishery.

This approach can give the annual equilibrium yield of the fishery, but it has hidden in it one flaw: It assumes that a constant number of recruits enter the usable stock every year. But does any fishery in fact have a constant recruitment? A constant recruitment implies that the number of recruits does not depend on population size; to put it another way, it assumes that two adult fish could produce the same number of progeny as 10,000 adults. This is quite impossible, and thus we are led to inquire into the relationship between population size (stock) and recruitment. The relationship between stock and recruitment is just another way of discussing the problem of population regulation. Fish populations, even when exploited, are still subject to population regulation. Recruitment in exploited populations is always measured as a rate, such as the number of young fish entering a fishery per year.

Some component of the vital statistics—births, deaths, and dispersal—must be related to population density in order to prevent unlimited population growth. As we saw in Chapter 12, population growth cannot be curtailed unless the net reproduction curve is depressed below 1.0 at high population densities (Figure 12.2, page 200). This can occur by adult mortality increasing with density, but fishery ecologists think that natural mortality of adult fish is not related to density. Fecundity does decline at high population density in some fishes (Bagenal 1973), but most of the regulation in fish populations is believed to occur in the early life-cycle stages. One of the axioms of modern fisheries ecology is that the important density-dependent processes in fish occur during the first few weeks or months of life (Cushing and Harris 1973).

Figure 17.6 shows a stock-recruitment graph for a population subject to logistic population growth (compare with Figure 12.2). Two points on this curve are fixed. Where there is no stock, there is no recruitment. At some point, stock will equal recruitment and there is an equilibrium point. The shape of recruitment curves is important for fisheries management. Two general shapes may occur (Figure 17.7). Beverton and Holt (1957) suggested a curve that rises to an asymptote at very large stock densities. Maximum recruitment in the Beverton-Holt model always occurs at maximum stock size. Ricker (1975) suggested a curve that may peak below equilibrium density, so that there would be a maximum in recruitment at intermediate stock sizes.

The Beverton-Holt recruitment curve (Figure 17.7) is essentially a logistic population model and leads to a smooth asymptotic population growth curve (as Figure 12.4, page 204). The Ricker recruitment curve is more closely related to the discrete generation analogs of the logistic equation in which population growth may show large oscillations about the carrying capacity (see Figure 12.3, page 202) (Eberhardt 1977). Species that are short-lived are more likely to show Ricker-type recruitment curves, while long-lived species will show Beverton-Holt-type recruitment.

Recruitment in fish populations is highly variable from one year to the next, and

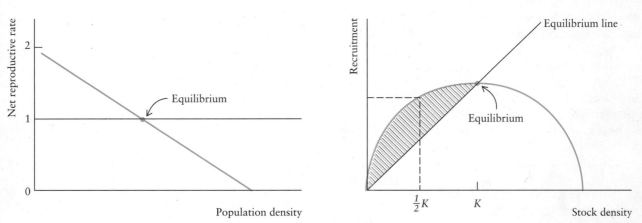

Figure 17.6 *Recruitment curve for a hypothetical population growing according to the logistic model. The left graph is the same as Figure 12.2, and the right graph is the same data plotted in a slightly different way (recruitment is population density multiplied by net reproductive rate). The hatched area shows the potential surplus that could be fished. Maximum yield is obtained at 0.5 K in the logistic model, as shown by the dashed line. An equilibrium occurs at the point where recruitment into the population just balances losses from the population.*

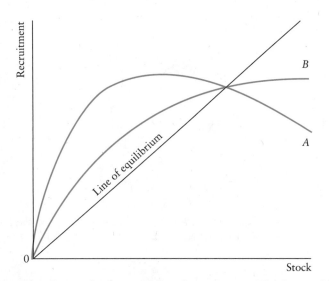

Figure 17.7   Possible relations between stock and recruitment. The diagonal line represents the equilibrium condition in which recruitment balances stock losses. Curve A with a descending upper section is the Ricker model, and curve B, which tends to plateau, is the Beverton and Holt model. (After Gulland 1962.)

most stock-recruitment relations show great scatter. Figure 17.8 illustrates this for North Sea plaice, and Figure 17.9 for sockeye salmon in Alaska. This variation is presumed to be caused by oceanographic effects on the survival of young fish. The variation seems to obscure any density effects and makes it difficult to fit the recruitment curves of Figure 17.7 to field data (Walters 1986). The variability in recruitment may be quite different in different species. For example, in the North Sea haddock, the abundance at recruitment has varied 500-fold in 30 years of study, whereas the variation in recruitment of the North Sea plaice has been only 6-fold in 26 years of study (Beverton 1962). If the amount of recruitment is highly variable, a population being exploited may be susceptible to overfishing. The Peruvian anchovy is a good example of this problem.

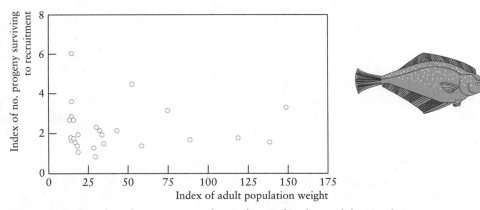

Figure 17.8   Stock and recruitment relationship in the plaice of the North Sea. The number of progeny surviving to recruitment bears no relation to the biomass of adult fish. (After Beverton 1962.)

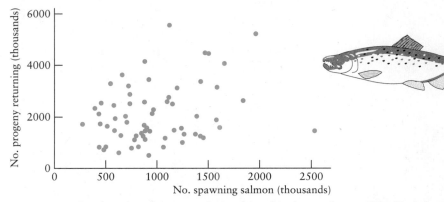

Figure 17.9 *Stock and recruitment relationship of sockeye salmon of the Karluk River, Alaska, 1887–1948. There is no clear indication that allowing more spawners past the fishery will produce a larger return. (After Rounsefell 1958.)*

Why do year-classes fail? This is one of the most important and difficult problems being addressed by fisheries ecologists. The critical period or match-mismatch hypothesis (Hjort 1914) postulates that early in the life of most fishes there is a short time period of maximum sensitivity to environmental factors. It is commonly assumed that oceanographic effects, particularly current patterns, winds, and water temperature, on food availability to newly hatched fry are critical in determining year-class size in fishes and that either temporal or spatial mismatching can produce recruitment failure.

One way to search for explanations of year-class failures in fish is to look for correlations between environmental factors, such as water temperature, and the relative success of recruitment of young fish. Fisheries ecologists have found a great number of correlations between environmental factors and recruitment success (Shepherd et al. 1984). Figure 17.10 illustrates one example of the kinds of correlations observed. Coho salmon in Oregon spend their first year of life in freshwater, then move into the ocean to grow to adult size, and return to freshwater to spawn at the age of four years. Survival during the early life history phases of salmon appear critical in determining population sizes of adult fish, and there is considerable controversy about exactly when in the larval-juvenile stage of life limitation occurs. For Oregon coho salmon Figure 17.10 would suggest that the early marine stages are the most critical. But all the correlations between fish abundance and environmental factors show a wide scatter of points that makes it difficult to predict fish production accurately. But more importantly these correlations do not specify the mechanisms by which populations are limited, and if we are to understand fish population dynamics we must uncover the mechanisms causing variations in recruitment (Rothschild 1986).

The details of how environmental factors affect recruitment are known for relatively few fishes. The English sole (*Paraphrys vetulus*) is an important commercial bottom fish off the coast of Oregon. Kruse and Tyler (1989) constructed a model of recruitment in English sole by describing the components that together determine the success or failure of young sole to survive to four years of age when they enter the fishery. Figure 17.11 illustrates their recruitment model and the relationships it incorporates. Recruitment in the English sole is limited by several factors acting on

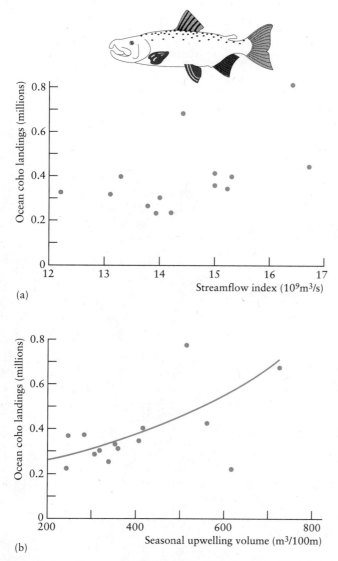

Figure 17.10   Correlations between coho salmon (Oncorhynchus kisutch*) production in Oregon and two environmental variables: (a) streamflow rates during freshwater residence of young salmon; and (b) coastal upwelling when young salmon move out into the ocean. The correlation with coastal upwelling is better than that with streamflow. (After Nickelson and Lichatowich 1983.)*

spawning adults, eggs, larval and juvenile fish, and can be understood and predicted only by studying the complex ecological relationships that determine reproduction and survival at each stage.

LABORATORY STUDIES

Animal populations vary greatly in their ability to withstand sustained losses. This ability is related to their size and to their capacity for increase. Laboratory studies

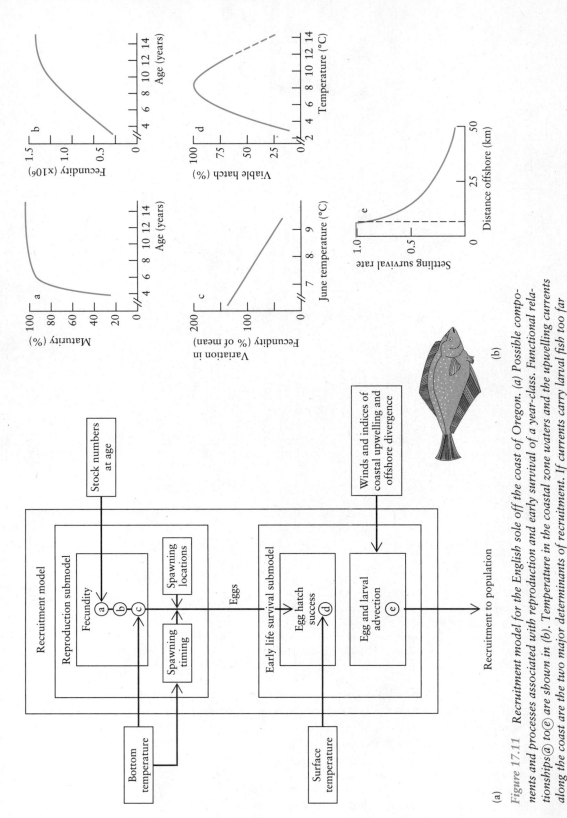

Figure 17.11 *Recruitment model for the English sole off the coast of Oregon. (a) Possible components and processes associated with reproduction and early survival of a year-class. Functional relationships (a) to (e) are shown in (b). Temperature in the coastal zone waters and the upwelling currents along the coast are the two major determinants of recruitment. If currents carry larval fish too far offshore, they cannot settle to the bottom and survive. (From Kruse and Tyler 1989.)*

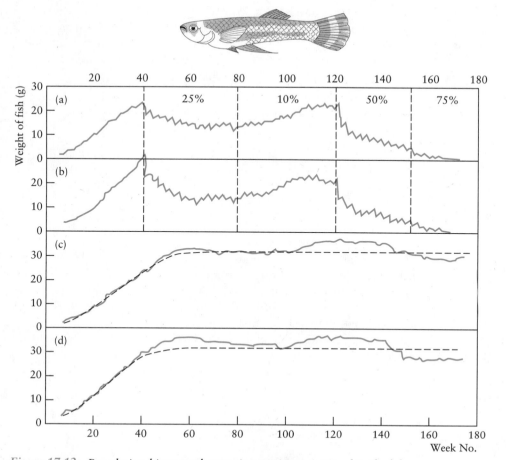

Figure 17.12   *Population biomass changes in guppies maintained in the laboratory. (a) and (b) are experimental populations subjected to harvesting after week 40 at the indicated rates. (c) and (d) are control populations that are not exploited. (After Silliman and Gutsell 1958.)*

of insects and small fishes have been particularly useful in analyzing the basic principles of harvesting theory.

A particularly clear demonstration of the relationship among rate of exploitation, population size, and yield is shown in the experimental work of Silliman and Gutsell (1958) on guppies (*Lebistes reticulatus*) in laboratory aquariums. They maintained two populations as unmanipulated controls and two populations as experimental fisheries subjected to a sequence of four rates of fishing (Figure 17.12). Populations were counted once each week and cropped every third week, so that (for example) a 25 percent cropping rate would mean that every fourth fish was removed from this population during the census at weeks 3, 6, 9, and so on.

Control guppy populations reached a stationary plateau by week 60 and remained there until the end of the experiment in week 174. Cropping at 25 percent triweekly reduced the experimental populations to about 15 grams of biomass, compared with 32 grams for the controls. Reduction of the cropping to 10 percent increased both experimental populations to about 23 grams of biomass, and the imposition of 50 percent cropping in week 121 caused a decline in population size to about 7 grams. A cropping intensity of 75 percent every third week was too great for these fish to withstand, and both experimental populations were driven extinct by "overfishing."

We can use these data to construct a yield curve directly. We weigh the fish removed at each cropping and obtain these results:

Exploitation Rate (%)	Weeks	EXPERIMENTAL POPULATION (AVG. WT., G/CROPPING)	
		a	b
25	61–76	3.35	3.58
10	100–118	2.20	2.51
50	136–148	3.88	3.58
75	163–172	0.82	0.40

Only data for the last half of each exploitation period are used, to approximate an equilibrium fishery condition. We plot exploitation rate against yield to obtain the yield curve for guppies shown in Figure 17.13. There is a maximum yield to be obtained at exploitation rates between 30 and 40 percent, with a population biomass of 8 to 12 grams, compared with 32 grams in unexploited controls.

The experiments on guppies by Silliman and Gutsell (1958) illustrate well four principles of exploitation:

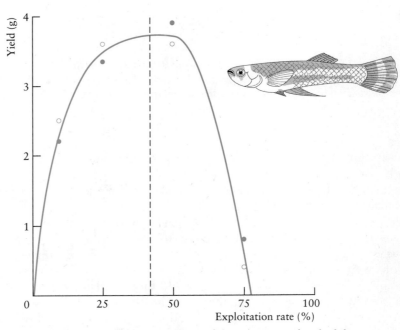

Figure 17.13   *Equilibrium yield in relation to fishing intensity for the laboratory populations of guppies studied by Silliman and Gutsell (1958). Maximum yield could be obtained at a fishing mortality of approximately* F = 0.5 *(an exploitation rate of 40 percent removal triweekly) indicated by the dashed line. Population A is represented by* •, *population B by* ○. *(After Silliman and Gutsell 1958.)*

1. Exploitation of a population reduces its abundance, and the greater the exploitation, the smaller the population becomes.

2. Below a certain level of exploitation, populations are resilient and compensate for removals by surviving or growing at increased rates.

3. Exploitation rates may be raised to a point where they cause extinction of the resource.

4. Somewhere between no exploitation and excessive exploitation is a level of maximum sustained yield.

Before we see how well these principles apply to natural populations, let us consider in more detail the concept of maximum sustained yield.

THE CONCEPT OF OPTIMUM YIELD

The concept of maximum sustained yield has dominated fisheries management since the 1930s (Larkin 1977). In many situations, maximum yield is not a desirable goal. In sport fisheries, for example, the object is to maximize recreation, and the desirable fish are often large ones. Hunters of large mammals may place more emphasis on the trophy status of the animals they harvest, and the harvesting of wildlife populations is often done without the goal of maximum sustained yield.

In any fishery that harvests several species at the same time, it is impossible to harvest at maximum sustained yield for all species. One species may be overharvested while another caught in the same nets is underharvested. Even within a single species, there are often subpopulations, or *stocks*, that have different resilience to harvesting. Harvesting of Pacific salmon operates on mixtures of stocks from different river systems and different spawning areas within one system. The result is that less productive salmon stocks are overfished, and even driven extinct, while more productive stocks are not fully utilized (Walters 1986).

Finally, any specification of optimum yield must include economic factors. The real yield from fisheries is not fish but dollars, and economists have long recognized that it is poor business to operate a fishery at maximum yield. H. Scott Gordon (1954) was one of the first to show that there is a level of harvesting associated with maximum sustained economic revenue and that this is usually at a lower fishing intensity than the maximum sustained yield. What is optimal to an economist is not necessarily optimal to a biologist.

Figure 17.14 shows a simple economic model for a fishery. Total costs are assumed to be proportional to fishing effort. The revenue or benefit from fishing is assumed to be directly proportional to the yield. Thus the yield curve of Figure 17.3 is identical to the revenue curve of Figure 17.14 in this simple model. But the important point is that *maximum sustained yield*, the peak of the curve, is not at the same point as *maximum economic rent* (total revenue − total cost). The maximum economic profit will always occur at a lower fishing intensity than maximum yield. If this simple model prevailed, the economic management of fisheries would always be a safe biological management strategy. Alas, it is not always such. Scott Gordon (1954) showed that in an unmanaged fishery the only social equilibrium that will be reached is at the point where total costs equal total revenue, beyond the point of maximum sustained yield. Clark (1990) has shown in an elegant analysis that under some situations, it will pay fishermen to deplete the fishery to extinction, as might

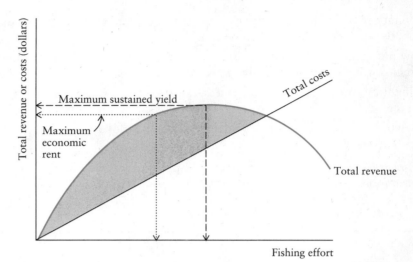

Figure 17.14   A simple economic model of a fishery in which costs are directly related to fishing effort and revenue is directly related to yield. The shaded zone is the area in which revenue exceeds costs. Maximum economic rent is achieved when the revenue-cost difference is maximal, and this is below maximum sustained yield in this simple model.

have happened to whales had the International Whaling Commission not intervened. The key economic idea in these cases is that of discounting future returns. If a fisherman can make $1000 today by overfishing, or $1500 in ten years time by delaying the harvest, most fishermen will take the money now and not wait. This type of exploitation makes perfect economic sense under our current economic theories but leads to ecological disaster and overexploited populations. Sustained yields can rarely be achieved without strong social or political controls on the allowable harvest.

Thus the concept of optimum yield must be broadened to include biological, economical, social, and political values. One of the challenges of future resource management is to weave these conflicting needs together in a satisfactory pattern, which we now see only in broad outline. The following four examples illustrate the difficulties of managing a fishery to achieve these goals.

PACIFIC SARDINE FISHERY

The Pacific sardine was not commercially exploited to any extent until World War I, when the demand for food increased. The catch rose to a peak in 1936 and has since fallen to almost nothing (Figure 17.15). The demand for sardines has always been high, and this decline in the catch is not a result of economic changes. Many sardinelike fishes are important in other world fisheries, and all seem subject to great variations in yield from year to year.

Adequate fishery statistics are available for the Pacific sardine from 1932 on, and Murphy (1966) has analyzed these. The fishery operates on two distinct "races" or stocks of sardine—a northern race, which is larger and dominated the fishery until 1949, and a southern race, which is smaller and has dominated the fishery since 1949. Coincident with this shift in dominance, the natural mortality rate increased:

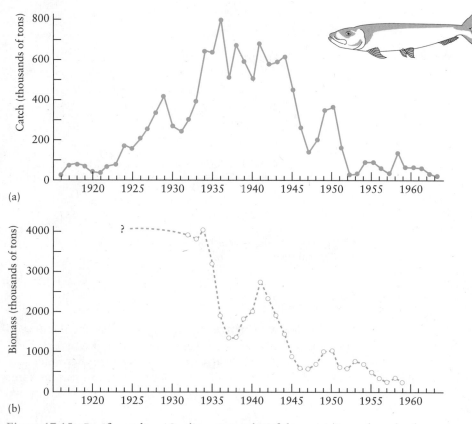

*Figure 17.15 Pacific sardine (*Sardinops caerulea*) fishery. (a) Annual catch of sardines along the Pacific Coast of North America. (b) Estimated abundance of sardines. All fish 2 years old or older are included. (After Murphy 1966.)*

	Estimated Annual Mortality Rate from Natural Causes
1932–1949 (northern race)	0.33
1950–1960 (southern race)	0.55

The reason the southern race has a higher mortality rate is not known. The northern race has a lower reproductive rate than the southern race. Only about half of two-year-old fish spawn in the northern race, but all two-year-old fish spawn in the southern race.

The size of the sardine population has declined greatly since 1932 (Figure 17.15). There is no relationship between the breeding stocks of sardines and the subsequent number of progeny recruited. The variation in the amount of recruitment from year to year is great and is presumably due to some environmental effects, but simple variables such as water temperature do not seem to be the cause of this variation (Murphy 1966).

The age structure of the sardine population has been shifted toward the younger age classes by heavy fishing pressure (Figure 17.16). This shift may explain the decline of the northern race after 1949. From 1945 to 1950 the population size was small,

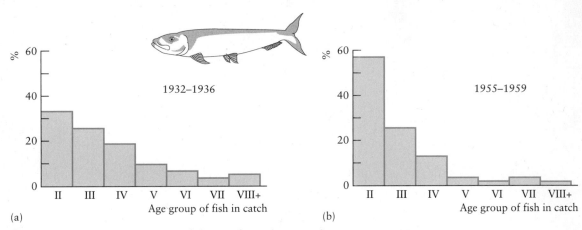

Figure 17.16   *Age composition of the Pacific sardine population (a) in 1932–1936, when the population was large and the fishery was just starting, and (b) in 1955–1959, when the population was severely reduced. Note the shift toward more fish in the younger age groups. (After Murphy 1966.)*

and over 85 percent of the sardines were of ages II and III. If we assume a failure of reproduction in the northern race for 1949 and 1950 (due to an unknown environmental catastrophe), all the 1949 and 1950 year-classes would be southern race fish. The annual loss of about one-third of the adult fish from natural causes plus some fishing mortality would thus reduce the northern race to extremely low levels in two years of no reproduction.

The sardine was a dominant species in the California Current, and the loss of this population has coincided with a rise in the density of the anchovy (*Engraulis mordax*). These two small fishes are similar and might possibly compete in some way, either through food competition or direct cannibalism of young. Alternatively, some sequence of environmental changes might have caused the sardine to decline in the presence of heavy fishing and the anchovy to increase. Murphy (1966) suggests that the rise of the anchovy *followed* the decline of the sardine, and this supports the environmental-change hypothesis. But if the anchovy population can increase only slowly, the removal of sardine competitors might not have had instantaneous effects.

The decline of the Pacific sardine population is not a unique event. Scales of both sardines and anchovies occur in the sediments of the Santa Barbara Basin off California (Soutar and Isaacs 1969). The sediment falling to the bottom in this area is largely diatoms in the summer and contains more land sediment during the winter rainy season. This alternation produces in effect a series of annual bands, or *varves*, by which one can date any position in a sediment core. Scales and bones of fish are mixed in with the other sediments, and Soutar and Isaacs (1969) were able to reconstruct, from the abundances of scales, the changes in the fish populations off California over the last 1800 years. The scales of the Pacific sardine are clumped throughout the core. There have been 12 main occurrences of sardines over the past 1800 years, each lasting 20 to 150 years, and between times sardine scales disappeared from the record for 80 years on the average. Thus the sardine population fluctuated greatly in size even before the fishery operated. The anchovy population, by contrast, maintained a more constant appearance over time but decreased steadily in abundance from 1500 years ago to the present, so that the present population is only about one-fifth that of 1500 years ago. The average density of the anchovy was considerably above that of the sardine throughout this time period. The implication of these historical data is that the Pacific sardine is subject to environmental fluctuations even without the added complications of fishing mortality. The harvesting of such species must be done cautiously without assuming that the average conditions of the past will continue to operate in the future.

Another important conclusion from this example is that we must treat an exploited species in a definite framework with other species, since the result of harvesting one species may change the environment for its competitors. The maximum sustained yield to a sardine fishery might be different if anchovies were present or absent. In this case, the relationship between sardines and anchovies could be tested by shifting the fishery to the anchovy and reducing the remaining pressure on the sardine.

KING CRAB FISHERY

The harvesting of king crabs began commercially early in this century when the Japanese began canning crabs and exporting them to the United States. The Japanese gradually moved the crab fishery east and began harvesting the eastern Bering Sea around 1930. The Japanese caught crabs in tangle nets dragged across the bottom. The king crab fishery was interrupted during World War II and then both the USSR and Japan took king crabs until the early 1970s when the United States took over the king crab fishery. King crabs are now taken only in pot traps by U.S. fishermen. The history of the king crab fishery in the eastern Bering Sea is shown in Figure 17.17. From initially low catches the crab fishery reached a peak around 9 million

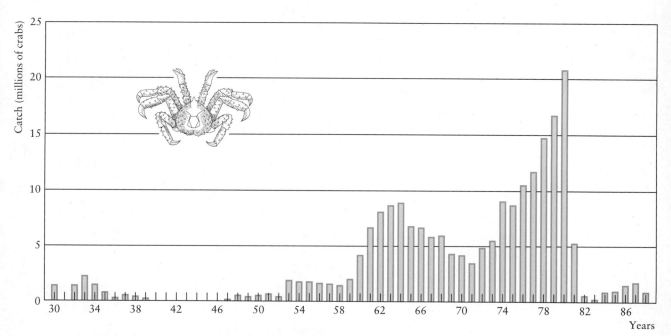

Figure 17.17   Catch of red king crab (Paralithodes camtschatica) *from the southeastern Bering Sea, 1930–1988. The catch was taken mostly by Japanese and Russian boats before 1969. (From Larkin et al. 1990.)*

crabs in 1964, and the catch was reduced over the next seven years while Japanese and USSR boats were gradually eliminated from this fishery. The catch by U.S. fishermen grew rapidly during the 1970s to a peak catch of 21 million crabs in 1980. The king crab fishery collapsed in 1981 and 1982 and has recovered only very slightly since then.

The red king crab fishery is unusual in several respects. The fishery only operates on males. The early life cycle is complex. Eggs are brooded by females for 11 months, and individual females lay from tens to hundreds of thousands of eggs (Larkin et al. 1990). The larval king crabs go through four stages while swimming in coastal waters, and then transform into the adult form at a length of 2 millimeters. Mortality is very high in the larval stages.

Growth occurs slowly (Figure 17.18) and individuals mature at about 5 years of age and 90 millimeter carapace length. Young crabs aggregate in large pods of hundreds to thousands of individuals that move as a wave across the ocean floor. King crabs are predators and scavengers on a wide variety of invertebrates and fish. Their growth rate is affected by water temperature and can change with oceanographic conditions.

Why did the king crab fishery collapse in 1981? One assumption underlying the management of this fishery has been the belief that by harvesting only males, the productivity of the population could be preserved. This assumption is true only if one mature male is able to service a large number of females. For behavioral reasons this assumption is now known to be faulty for king crabs. Fertilization occurs only after female molting and a male who has clasped a female may have to wait a few days for her to molt. After copulation the male may retain his clasp on the female. Females prefer to mate with larger males, and if a female cannot copulate within nine days of molting, the entire brood for that year is lost. Behavioral complications resulting from a male-only harvest may thus compromise the reproductive potential of the population.

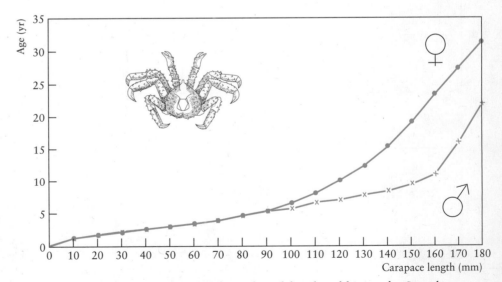

Figure 17.18   *Average growth curves for male and female red king crabs. Sexual maturity is reached at 5 years of age and 90-mm length. Only males > 136 mm in length are retained by the fishery. (From Larkin et al. 1990.)*

King crab growth and survival rates in the juvenile stages are also highly variable, perhaps due to changes in water temperature. During the 1970s the king crab fishery was relying on several very large cohorts from the 1960s, and these abundant recruits provided a false sense of optimism in the fishery managers.

Another factor affecting king crab abundance is the loss of immature crabs in pots (traps). For every legal-sized male crab (>136 mm in carapace length) taken, about seven immature males are caught and subsequently released. These immature crabs may be stressed or damaged during capture and subsequently die. Other ground-fish fisheries in the Bering Sea may capture king crabs incidentally or damage smaller crabs that are not retained in the nets. Finally, about 10 percent of the crab pots are lost each year due to storms and faulty ropes, and these pots continue to catch crabs after they are lost to add to the hidden mortality on immature and mature crabs of both sexes.

In retrospect, the collapse of the king crab fishery was unavoidable because it built up too rapidly on the basis of a series of good year-classes, so that the maximum sustained yield was overestimated. The momentum associated with large capital investments in fishing boats makes it difficult to reduce the catch quotas rapidly when environmental conditions cause a series of poor year-classes to occur. The compromise reached between socioeconomic realities and biological conservation measures almost always causes a further decline in stock abundance and consequently an even longer time before it can recover again.

ALASKA SALMON FISHERY

The decline of the Alaska salmon fishery is a good example of overfishing applied to an "inexhaustible resource." Five species of salmon are part of the Pacific salmon fishery; red (sockeye) salmon and pink salmon are the major commercial species. Coho and chinook salmon are prized catches in the sport fishery, as well as in the commercial harvest. Adult salmon are caught by the fishery along the Pacific coast as they return to spawn in freshwater streams. Many Indian tribes along the coast were dependent on salmon as a staple food before white people began fishing salmon commercially.

Commercial exploitation in Alaska began about 1800 and increased at a slow rate (Figure 17.19). Most of the salmon taken commercially have been used for canned salmon. The peak year of commercial exploitation was 1936, when 8.5 million cases of canned salmon were packed (one case is 48 lb or 22 kg net weight). Since then, the packs have declined to the point that the 1959 pack was at the same level as the 1900 pack. This increase and subsequent decline in the commercial catch also occurred in neighboring British Columbia, but on a different time scale, so the demise of the Alaska salmon fishery cannot be blamed on global ecological changes.

Throughout this time, the amount of fishing equipment used in the Alaska salmon fishery steadily increased. There were about 200 seine boats operating in 1910 and over 1000 in the 1950s. There were about 1000 gill-net boats operating in 1909 and about 9000 in 1955. This increase in equipment (and hence in fishermen) was accompanied by a striking drop in catch per unit of equipment. The average gill-net boat caught about 15,000 salmon in 1908 but only 1,500 salmon in 1954, a 90 percent decrease. Thus more and more fishermen have been catching fewer and fewer salmon. This violation of common sense is understandable only in economic terms. The real price of a can of salmon tripled from 1910 to 1955, and this permitted

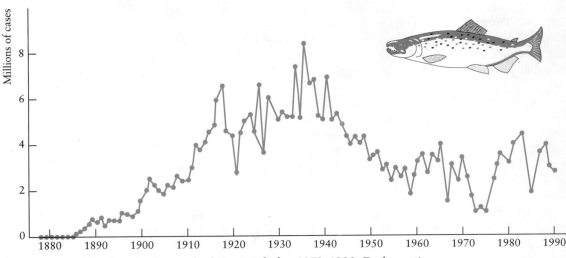

Figure 17.19 Total pack of canned salmon in Alaska, 1878–1990. Each case is 22 kg (48 lbs) net. (Data from Fishery Statistics of the United States.)

the average fisherman to increase his income, at least until 1945, when it began to decline slightly.

The core of this overfishing problem, and many others, lies in the peculiar common-property status of the fishing resource. Salmon fishing has been open to everyone who has the equipment, and the only way to limit the catch has been to limit the efficiency of fishermen and equipment. This situation reached the point in the salmon fishery where five or six days a week would be closed to fishing. Complex regulations have put a great premium on law enforcement and no premium at all on voluntary restraint of the fisherman or the cannery owners.

The "tragedy of the commons" was the term coined by Hardin (1968) to describe this type of exploitation. Whenever a resource is held in common by all the people, the best policy for every individual is to overharvest the resource. There can be no reason to stop harvesting at some optimum point because you can always make more money by overharvesting, and if you do not overharvest, your neighbor will. This tragedy of the overexploitation of common property resources can be averted only by some form of management that restricts fishing pressure, or by converting a common property resource to a private resource. Social control of harvesting is required for all large-scale fisheries, and for this reason good resource management is a creative mix of ecology, economics, and sociology.

Part of the reason for the overfishing of the Alaska salmon can be found in ecological ignorance. Almost no information on the spawning population sizes for most of the major rivers of Alaska was available until the late 1940s. Population changes in salmon are still not understood. Therefore, there is no way to know how many spawners are needed to produce an adequate return or whether in fact the population is being limited at some other part of the life cycle.

Both Alaska and Canada began programs in the 1970s to restore the former abundance of Pacific salmon. These programs are being aimed at the freshwater stages of the salmon life on the assumption that if we can improve survival and growth in fresh water, the ocean will be able to support an increased stock. These programs have been partly successful in arresting the long-term decline of the Pacific salmon. Figure 17.19 shows that since 1975 the Alaska salmon fishery has partly recovered toward the pre-1940 levels. Salmon stocks in British Columbia are also increasing as a result of stringent restrictions on commercial and sport fishing (Pearse 1982). Problems have arisen with hatcheries because of detrimental genetic interactions between wild salmon and hatchery-reared salmon (Waples 1991). Hatchery-reared salmon have been less successful than wild salmon and the reasons for hatchery failures are not yet known but are presumed to be genetic in origin.

ANTARCTIC WHALING

The exploitation of whale populations has been the subject of vigorous and heated debate during the 1970s and 1980s. At the present time almost all commercial whaling has been stopped and most whales are protected. The large whales comprise ten species divided into two unequal groups. The sperm whale was the only toothed whale hunted commercially. The other nine species were all baleen whales, which have bony plates (baleen) in the roof of the mouth. Baleen whales are filter feeders whose principal food in the Antarctic is krill, shrimplike crustaceans, and other plankton.

The history of whaling is characterized by a progression from more valuable

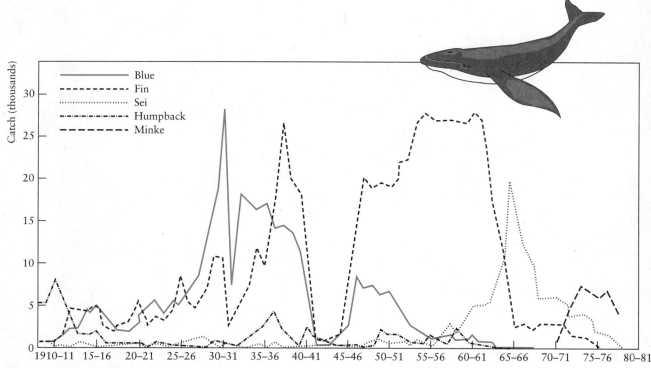

Figure 17.20   Catches of baleen whales in the Southern Hemisphere, 1910–1977. The usual lengths of whales in the commercial catches were: blue, 21–30 m; fin, 17-26 m; sei, 14-16 m; humpback, 11–15 m; and minke, 7–10 m. (After Allen 1980.)

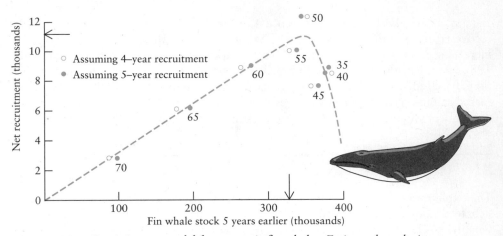

Figure 17.21   *Logistic-type model for antarctic fin whales. Estimated stock size and yield at maximum sustained yield are indicated by arrows. The logistic model would predict maximum yield at 0.5 K (200,000), but these data suggest maximum yield around 0.8 K (330,000). (After Chapman 1981.)*

species to less attractive species as stocks of the original targets were reduced. Modern whaling dates from 1868, when a Norwegian, Svend Foyn, invented the harpoon gun and the explosive harpoon. In about 1905 whalers pushed south into the Antarctic and discovered large populations of blue whales and fin whales. Blue whales dominated the catches through the 1930s, but by 1955 few were being taken (Figure 17.20). Attention was turned to the fin whale, originally the most abundant whale in the southern oceans. Fin whale numbers collapsed in the early 1960s. Sei whales were ignored as long as the bigger species were available and were not harvested until 1958. Sei whale catches were restricted after 1972 by the International Whaling Commission to prevent the collapse of these populations.

Harvesting models for whales have been developed extensively since 1961 (Allen 1980). The logistic-type models have proved inadequate (Figure 17.21). Maximum sustained yield seems to occur at a density about 80 percent of equilibrium density, rather higher than the 50 percent predicted in the simple logistic model (Chapman 1981). Complications with these simple models are not difficult to find. Figure 17.21 assumes that all fin whales in the southern ocean belong to one population; it is now known that several subpopulations occur (Chapman 1981). Whales may interact, and most whale models are single-species models that do not recognize that many different species of whales and seals feed on krill in the Antarctic.

The present management of whales is directed to measuring the recovery rate of the depleted whale populations. Paradoxically, most of the data we now have on whales has come from whaling operations, and now that commercial whaling has stopped, additional research has to be mounted to monitor how whale populations respond. Whale populations change slowly, and even ten years is a short time to estimate accurately a population's response to protection from exploitation.

The principal food of the baleen whales, krill, is now being commercially harvested in the Antarctic, and one of the emerging conservation problems of the southern oceans is to estimate the impact of krill harvesting on the recovery of whale populations in the Antarctic (Rosenberg et al. 1986).

SUMMARY

To harvest a population in an optimal way, we must understand the factors that regulate the abundance of that population. That humans so frequently mismanage exploited populations like the Pacific sardine is partly a measure of our ignorance of population dynamics. When a population is harvested by humans, its abundance must decline, and the losses caused by harvesting are compensated for by increased growth, increased reproduction, or a reduced natural mortality. Species vary greatly in the amount of harvesting they can sustain.

Maximum sustained yield is often the goal of resource managers. Simple and complex models have been developed to estimate the maximum sustained yield for fisheries and forestry. Most models contain the hidden assumption that the environment remains constant, and for this reason they often fail in practice to prevent overexploitation and collapse. Economics and politics add further difficulties to achieving maximum sustained yield for valuable populations.

Management of forestry, fishery, and wildlife resources is at present based more on rules of thumb and empirical results than on scientific knowledge and forecasting, and one of the great challenges of modern ecology is to place resource management on a scientific basis. We can all be very good at managing yesterday's populations. When will we be equally adept at managing those of tomorrow?

Selected References

CLARK, C. W. 1990. *Mathematical Bioeconomics: The Optimal Management of Renewable Resources*, chaps. 1–2. Wiley, New York.

GETZ, W. M., and R. G. HAIGHT. 1989. *Population Harvesting. Demographic Models of Fish, Forest, and Animal Resources*. Princeton University Press, Princeton, N.J.

GRAHAM, M. 1939. The sigmoid curve and the overfishing problem. *Rapport du Conseil International pour l'Exploration de la Mer* 110:15–20.

GULLAND, J. A., ed. 1988. *Fish Population Dynamics: The Implications for Management*. Wiley, New York.

LARKIN, P. A. 1977. An epitaph for the concept of maximum sustained yield. *Transactions of the American Fisheries Society* 106:1–11.

PITCHER, T. J., and P. J. B. HART. 1982. *Fisheries Ecology*. Croom Helm, London.

RICKER, W. E. 1973. Two mechanisms that make it impossible to maintain peak-period yields from stocks of Pacific salmon and other fishes. *Journal of the Fisheries Research Board of Canada* 30:1275–1286.

SILLIMAN, R. P., and J. S. GUTSELL. 1958. Experimental exploitation of fish populations. *Fishery Bulletin (U.S.)* 58(133):215–252.

WALTERS, C. 1986. *Adaptive Management of Renewable Resources*. Macmillan, New York.

Questions and Problems

1. Murphy (1966, p. 733) gives the following estimates* for the vital statistics of the Pacific sardine living under optimal conditions:

Age, x (yr)	l_x	m_x
2	1.0000	0.5147
3	0.6703	1.3618
4	0.4493	1.6819
5	0.3012	1.8816
6	0.2019	2.0257
7	0.1353	2.1358
8	0.0907	2.2357
9	0.0608	2.2686
10	0.0408	2.2686
11	0.0273	2.2686
12	0.0183	2.2686
13	0.0123	2.2686

Calculate the net reproductive rate (R_0) and the capacity for increase (r) of this sardine population. Use this value of r to calculate a logistic curve for an unexploited sardine population (assume that $K = 2,400,000$ tons and $a = 7.785$). How many years would it take for this unexploited sardine population to recover from a starting point of 10,000 tons to a level of 2,000,000 tons if it followed this logistic curve?

2. What would be the yield per recruit in the North Sea plaice fishery if the fishing mortality were infinitely large? (Use the data in Table 17.1.)

3. Suppose that there is in fact no relationship between stock and recruitment in the sockeye salmon (Figure 17.9). Discuss the implications of this with respect to the various theories of population control (Chapter 16).

4. Construct an equilibrium yield curve showing the relationship between yield (biomass) and hunting mortality rate for a hypothetical deer population with constant recruitment of 1000 fawns per year and an instantaneous natural mortality rate of 0.7 per year for age 0 to 1 and 0.4 per year for older deer. Assume for simplicity that all growth, recruitment, and losses occur at a single point each year and that hunting operates only on animals age 2 and over. The growth curve is as follows:

---

* This is a relative life table and fertility table (starting at age 2), and it can be converted to the usual format by dividing the l_x values by 71,000 and multiplying the m_x values by this amount. It is given in this way to avoid cumbersome numbers.

Age (yr)	Weight (kg)	No. Survivors (natural mortality only)
0	10	1,000
1	50	497
2	80	333
3	90	223
4	100	150
5	110	100
6	120	67
7	130	45
8	140	30
9	150	20
10	160	14
11	170	9

Calculate the equilibrium yield in numbers for this population at a harvest rate of 20 percent per year.

5. Plot the following hypothetical reproduction curves (like Figure 17.7) for a species that reproduces only once (numbers are arbitrary "egg units"):

POPULATION A		POPULATION B	
Adult Egg Production	Progeny Egg Production	Adult Egg Production	Progeny Egg Production
2	4	1	5
4	6	2	8
6	7.5	4	11
8	8	5	12
10	7.5	6	11
12	6	8	8
14	3	10	5
		12	2
		14	1

Use these plots to determine graphically population changes for 20 generations from starting adult stocks of 14, 6, and 2. Introduce random environmental variations to this simple method by flipping a coin each generation and multiplying progeny egg production by 0.5 for heads and 1.5 for tails. What effect does this random factor have on the population curves?

6. Examine the catch statistics for a fishery in your area or in an area of interest to you. Sources of data might be the *Fisheries Statistics of the United States, Fisheries Statistics of Canada,* or the *Food and Agricultural Organization's Yearbook of Fishery Statistics* published by the United Nations. If the fishery you choose has been managed, is there any evidence of overfishing?

7. Black duck populations have been declining in North America since the mid-1950s. One hypothesis for this decline is that it is due to overhunting. Review the evidence for and against this hypothesis. Nichols (1991) provides references and an overview of this management problem.

8. One of the assumptions of maximum sustained yield models is that birth, death, and growth responses to population density are repeatable, so that a given population density will always be characterized by the same vital statistics. What mechanisms may make this assumption false?

9. Ricker (1982) showed that since 1950, sockeye salmon caught in British Columbia have decreased in size by 140 to 180 grams on average, about 5 percent of their weight. Discuss mechanisms by which a fishery might select in favor of smaller salmon.

10. Ludwig and Walters (1985) showed in a computer simulation that the management of a hypothetical fishery could be done better using simple yield models like the logistic than by using more realistic, detailed models like dynamic pool models. Discuss why this might be correct for a real fishery.

Overview Question

Suppose that you are in charge of a newly established fishery. Discuss the criteria you might use to detect when the population is being overexploited and outline the relative merits of the different criteria.

18

Applied Problems II: Pest Control

Some species interfere with humans' activities, in which case they are assigned the label "pests." The most damaging pests we have are introduced species. The first response to pests is to *control* them. Control used in this context means to *control damage* and not necessarily to regulate the pest population around some equilibrium density. One of the obvious ways of controlling damage is to reduce the average abundance of the pest species, but there are other ways of reducing damage by pests without affecting abundance (such as using insect repellents).

A population is defined as being controlled when it is not causing excessive economic damage and as uncontrolled when it is. The boundary between these two states will depend on the particular pest. An insect that destroys 4 to 5 percent of an apple crop may be insignificant biologically but may destroy the grower's margin of profit. Conversely, forest insect pests may defoliate whole areas of forest without bankrupting the lumbering industry. The concept of economic thresholds must be applied to all questions of pest control. This includes the cost of the damage caused by the pest, the costs of control measures, the profit to be gained with the crop, and interactions with other pests and their associated costs.

Pest control in most agricultural systems is achieved by the use of toxic chemicals, or *pesticides*. An estimated 2.5 billion kilograms (nearly 5 billion pounds) of toxic chemicals are being used annually throughout the world to control plant and animal pests (Pimentel et al. 1992). Despite the use of these pesticides about 48 percent of the world's crops are lost to pests before and after harvesting. Despite increasing pesticide use in the last 50 years, crop losses have gone up not down (Figure 18.1). Pesticides are only a short-term solution to the problem of pest control for several reasons. First, toxic chemicals have strong effects on many species other than pests. The well-known effects of DDT on bird populations is a good example of how pesticides can degrade environmental quality. Second, many pest species are becoming genetically resistant to toxic chemicals that formerly killed them. Insects that attack cotton have evolved resistance to so many pesticides that it is no longer possible to grow cotton in parts of Central America, Mexico, and southern Texas. Third, the use of toxic chemicals in some situations can actually produce a pest problem where none previously existed. This is perhaps the most surprising effect of toxic chemicals. Figure 18.2 illustrates how lemon trees can become infested by massive outbreaks of a scale insect when sprayed with DDT. Toxic chemicals like DDT destroy many insect parasites and predators that cause mortality in the pest species, and after treatment the few pest individuals that survive can multiply without limitation.

How can we achieve pest control without these problems? There are four primary strategies for dealing with pests:

1. *Natural control.* Pest populations are exposed to naturally occurring predators, parasites, diseases and competitors.

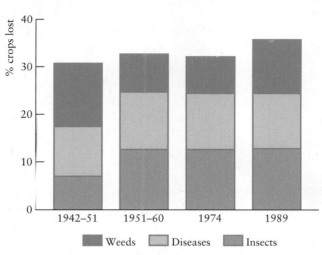

Figure 18.1   Percent of crops estimated to have been lost in the United States because of insect pests, plant diseases, and weeds, 1942–1989. During this time period pesticide use increased 33-fold. (Data from Pimentel et al. 1991.)

2. *Pesticide suppression.* Pest populations are treated with herbicides, fungicides, insecticides, or other chemical poisons to reduce their abundance.

3. *Cultural control.* Pests are reduced by agricultural manipulations involving crop rotation, strip cropping, burning of crop residues, staggered plantings, or other agricultural practices.

4. *Biological control.* Pests are reduced by biological introductions of predators, parasites, or diseases, by genetic manipulations of crops or pests, by sterilizing pests, or by mating disruption using pheromones.

Integrated control, or integrated pest management, means the use of all four of these strategies to reduce pest damage with the aim of minimizing pesticide use and maximizing natural control.

In this chapter we shall discuss the principles used in biological control and cultural control and relate these to ecological theory. Biological and cultural controls aim to reduce the average density of a pest population, and may be viewed as a practical application of the problem of what determines average abundance (Figure 18.3; see Chapter 16).

The general procedure dealt with in biological control is as follows: A pest, often an introduced species, is causing heavy damage. Efforts are then made to find predators and parasites in the pest's home country that can be introduced to its new country. If the efforts are successful, the pest population is reduced to a level at which no economic damage occurs. Let us look at several examples of biological control.

COTTONY-CUSHION SCALE (Icerya purchasi)

One of the most striking and critical successes of biological control concerned the cottony-cushion scale, a small coccid insect that sucks sap from leaves and twigs of citrus trees. This scale insect was first discovered in California in 1872, and by 1887 the whole citrus industry of southern California was threatened with destruction.

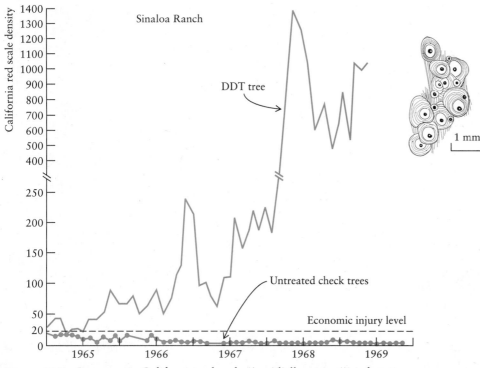

*Figure 18.2   Increases in California red scale (*Aonidiella aurantii*) infestation on lemon trees caused by monthly applications of DDT spray. Nearby untreated lemon trees suffer no damage because a variety of insect parasites and predators keep the red scale under biological control in southern California. (After DeBach 1974.)*

Because of the size of the infested area, chemical control by cyanide and other sprays was a failure. In 1888 Albert Koebele of the Division of Entomology was sent to Australia by the U.S. government to represent the State Department at an international exposition in Melbourne. Since all foreign travel had been restricted for the Division of Entomology to save money, this was the only subterfuge by which an entomologist could travel to Australia to search for parasites of the cottony-cushion scale, a native of Australia. Koebele sent two species of insects back to California, a small dipteran parasite, *Cryptochaetum iceryae*, and a predaceous ladybird beetle called the vedalia (*Rodolia cardinalis*). The dipteran parasite was thought to be an important possible agent for control, but Koebele sent the ladybird beetles along as well, apparently without thinking that they could be very useful. In late 1888 the first ladybird beetles were received in California, and by January 1889 a total of 129 individuals had been released near Los Angeles under an infested orange tree covered by a large tent. By April 1889 all the cottony-cushion scales on this tree had been destroyed; the tent was then opened. By June 1889 over 10,000 beetles had been sent to other citrus orchards from this first release point. By October 1889, scarcely one year since *Rodolia* was found in Australia by Koebele, the cottony-cushion scale was virtually eliminated from large areas of citrus orchards in southern California. Within two years it was difficult to find a single individual of the scale *Icerya*, and this control continued so that the pest was effectively eliminated—the cost: about

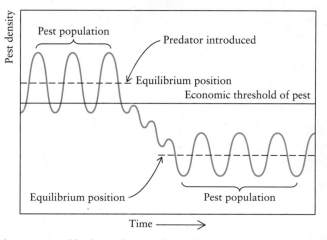

Figure 18.3 *Classic type of biological control in which the average abundance of an insect pest is reduced after the introduction of a predator. The economic threshold is determined by humans' activities, and its position is not changed by biological control programs. (After van den Bosch et al. 1982.)*

$1500; the saving: millions of dollars every year. The California legislature was impressed, and California became a center of activism in promoting the value of biological control (Doutt 1964).

The cottony-cushion scale reappeared with the advent of DDT. Infestations of the scale that had not been seen in over 50 years were found after DDT had eliminated the vedalia beetle from some local areas. Under these circumstances, the beetle has to be continuously reintroduced.

Some of the host plants of the cottony-cushion scale are not suitable for the vedalia. For example, the scale infests scotch broom (*Cytisus scoparius*) and maples in central California, but the vedalia will not become established on these plants, for unknown reasons (Clausen 1956). Such host plants serve as a refuge or reservoir for the scale, which can recolonize citrus trees.

The great success in controlling the cottony-cushion scale ushered in an era in which biological control was viewed as a panacea for all insect pest problems. Large numbers of insects were collected from all over the world and released in North America without any testing or quarantine procedures. Fortunately, only time and money were wasted with all these importations, and this dangerous policy was eventually stopped—not, however, because of its dangers but because of a sequence of repeated failures in control (Turnbull and Chant 1961).

PRICKLY PEAR (Opuntia spp.*)*

Prickly pear is a cactus native to North and South America. There are several hundred species of prickly pear, about 26 of which have been introduced to Australia for garden plants. One species, *Opuntia stricta*, has become a serious weed in Australia. In 1839 *O. stricta* was brought to Australia as a plant in a pot from the southern United States and was planted as a hedge plant in eastern Australia. It gradually got out of control and was recognized as a pest by 1880. By 1900 it occupied some

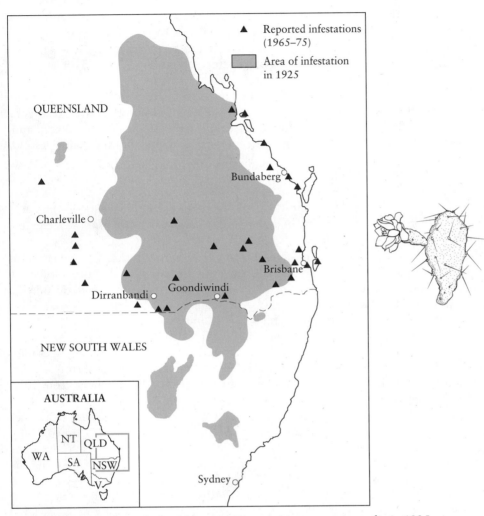

*Figure 18.4 Distribution of prickly pear (*Opuntia*) in eastern Australia in 1925 at the peak of infestation, and 1965–1975 areas of local infestation. (After White 1981.)*

40,000 square kilometers (15,600 sq. miles) and spread rapidly in Queensland and New South Wales (Figure 18.4):

| | AREA INFESTED WITH *OPUNTIA* | |
	km^2	mi^2
1900	40,000	15,600
1920	235,000	90,600
1925	243,000	93,700

About half this area was dense growth completely covering the ground, rising 1 to 2 meters in height, and too dense for anyone to walk through (Figure 18.4).

Prickly pear is propagated by seeds and by segments. The cactus pads, when detached from the parent plant by wind or people, can root and begin a new plant. Seeds are viable for at least 15 years. The problem of eradicating this weed was largely one of cost. The grazing land it occupied in eastern Australia was worth only a few dollars an acre, and poisoning the cactus cost about $25 to $100 an acre. Therefore, homesteads had to be abandoned to this invasion.

In 1912 two entomologists were sent from Australia to visit the native habitats of *Opuntia* to find possible biological control agents that could be introduced. They sent back from Sri Lanka a mealybug, *Dactylopius indicus*, which was released and, in a few years, had destroyed a minor pest, *Opuntia vulgaris*. But the major pest, *O. stricta*, continued to spread, and after World War I it was subjected to a more intensive effort of biological control. Beginning in 1920, investigations in the United States, Mexico, and Argentina resulted in 50 species' being sent back to Australia for possible control. Of these, only 12 species were released; three were of some help in controlling *O. stricta*, but only one, the moth *Cactoblastis cactorum*, was capable of eradicating it.

Cactoblastis cactorum is a moth native to northern Argentina. Two generations occur each year. Females lay about 100 eggs on the average, and the adults live about two weeks. The larvae damage the cacti by burrowing and feeding inside the pads and by introducing bacterial and fungal infections by their burrows. Two introductions of *Cactoblastis* were made. The first introduction, in 1914, failed (Osmond and Monro 1981). For the second introduction approximately 2750 eggs were shipped from Argentina in 1925, and two generations were raised in cages until March 1926, when 2 million eggs were set out at 19 localities in eastern Australia. The moth was immediately successful, and further efforts were expended from 1927 to 1930 in spreading eggs and pupae from one field area to another.

By 1928 it was obvious that *Cactoblastis* would control *O. stricta*, so further parasite introductions were curtailed. *Cactoblastis* multiplied rapidly up to 1930, and between 1930 and 1931 the *Opuntia* stands were ravaged by an enormous *Cactoblastis* population (Figure 18.5). This collapse of the prickly pear caused the moth population to fall steeply in 1932 and 1933, and the cactus then began to recover in some areas. Between 1935 and 1940 *Cactoblastis* recovered and completely controlled the cactus. Prickly pear survived after 1940 only as a scattered plant in the community (Dodd 1940, 1959). The present picture is that *Opuntia* exists in a stable metapopulation at low density maintained by *Cactoblastis* grazing. The eggs of *Cactoblastis* are not laid at random but are clumped on some plants, while other plants escape infestation entirely (Myers et al. 1981). Plants heavily loaded with larvae are subsequently completely destroyed, and many *Cactoblastis* larvae thus starve and die. Larvae cannot move from one plant to another if cacti are 2 meters or more apart. The clumping of the eggs of *Cactoblastis* thus both destroys *Opuntia* plants and ensures that not all plants are killed so that the metapopulation does not go to extinction, although local populations do disappear.

Most of the areas where prickly pear is now periodically considered a pest are outside the original area of dense cactus infestation (Figure 18.3), and plants in these areas seem to be partly resistant to *Cactoblastis* attack. Without *Cactoblastis*, prickly pear would make a rapid recovery.

Why was *Opuntia* such a successful plant in eastern Australia? Three important physiological properties of *Opuntia* determine its success (Osmond and Monro 1981). First, the tissues of this cactus are almost entirely photosynthetic. There is minimal investment in structural tissues, and the root system is shallow and small. Second, *Opuntia* is capable of crassulacean acid metabolism (CAM), a process in

Figure 18.5   (a) *Dense stand of prickly pear prior to the release of* Cactoblastis, *October 1926, Chinchilla, Queensland, Australia. (b) The same stand is shown three years later after attack by* Cactoblastis, *October 1929. (After Dodd 1940; photographs courtesy of A. P. Dodd and Commonwealth Prickly Pear Board.)*

which CO_2 fixation is largely done at night, when minimal water vapor is lost to the atmosphere. Thus photosynthesis can be done with minimal water loss. Third, CAM plants retain photosynthetically competent tissues throughout periods of stress. When the rains come, CAM plants can immediately begin to photosynthesize and grow. Because of this combination of characteristics, *Opuntia* proved a near perfect opportunist with superior competitive ability over the native plants that lacked CAM metabolism.

FLOATING FERN (Salvinia molesta)

The floating fern, *Salvinia*, is a plant native to South America. It was introduced to Sri Lanka in 1939 through the Botany Department at the University of Colombo, and then spread over the next 50 years to Africa, India, Southeast Asia, and Australia. It is a serious aquatic weed, forming mats up to 1 meter thick and covering lakes and canals, rivers and irrigation channels. Because *Salvinia* clogged the waterways (see Figure 18.6 in color insert), all water transport and fishing was disrupted, causing major problems.

Salvinia was incorrectly identified until 1972, when it was recognized to be a new species (Room 1990). Because of this taxonomic uncertainty, ecologists were not able to look for specialized herbivores of this plant in its native habitat until the plant was found in southeastern Brazil in 1978. *Salvinia molesta* is unusual in being sterile, and all ramets appear to be genetically identical no matter where they occur in its geographic range. Plants of *Salvinia* are colonies of ramets held together by horizontal, branching rhizomes. The rate of growth of *Salvinia* on the water's surface is limited by temperature and nitrogen concentration of the water.

Three insect species (a weevil, *Cyrtobagous singularis;* a moth, and a grasshopper) found attacking *S. auriculata* in Trinidad in the 1960s were introduced in Sri Lanka, India, Africa, and Fiji in the 1970s. None of these introductions had any impact on the weed. Once the correct species *S. molesta* was located in Brazil, what was thought to be the same insect species were collected and the weevil was released in north Queensland, Australia in 1981. The weevil increased dramatically and destroyed the *Salvinia* within one year (Figure 18.6). It was then discovered to be a new species of weevil, *Cyrtobagous salviniae,* and proved highly successful at control in Australia, India, Sri Lanka, Botswana and Namibia (Room 1990). The success of the weevil in controlling *Salvinia* is partly explained by its tolerance of high population densities before it shows interference competition and emigration occurs. The weevil reaches densities of 1000 adults per square meter; by feeding on the buds as adults and on roots and rhizomes as larvae, the weevil either kills the plants or reduces greatly their size.

THEORY OF BIOLOGICAL CONTROL

Most biological control has operated empirically with a few rules-of-thumb, and this approach has achieved some spectacular successes. But if we are to avoid a case-by-case approach, we need to develop some general theory of biological control that could guide empirical work (Huffaker and Messenger 1976). Most of the theory that has developed comes from the Nicholson-Bailey model of predator-prey interactions, similar to the discrete predator-prey model discussed in Chapter 14 (page 262). The premise of this approach is that successful biological control is produced by the predator imposing a low, stable host equilibrium (Figure 18.3). Theoretical evaluation of these predator-prey models by Beddington and colleagues (1976), May (1978), and May and Hassell (1981) suggest an array of properties of successful biocontrol agents: (1) host specific, (2) synchronous with the pest, (3) high intrinsic rate of increase (r), (4) able to survive with few prey available, (5) high searching ability. These properties are more typical of insect parasitoids than of predators in general. Most predators are considered by this theory to be poor candidates for biocontrol because they are generalists and not host-specific, they are rarely synchronized with the pest, they have relatively low r values, and they may feed on other, beneficial prey species. Successful control agents cause density-dependent losses in the host population, and this density-dependence may be spatial or temporal or both. Spatial density dependence occurs when predators or parasitoids cause a higher fraction of losses in dense host patches than in sparse host patches (Hassell 1978). If predators can aggregate in patches of high host density, then, according to this theory, biological control of the pest is much more likely.

The classical theory of biological control based on the Nicholson-Bailey model is an equilibrium theory. It has been challenged recently by an alternate view based on a nonequilibrium model of predator-prey interactions (Murdoch et al. 1985). The nonequilibrium model begins with the assumption that a stable equilibrium of predator and prey is not necessary for satisfactory biological control. Pest populations may fluctuate wildly without pest densities exceeding the economic threshold. Some local populations of pests may be driven extinct in this model. The nonequilibrium model is a metapopulation model and, as such, emphasizes that populations in different patches may fluctuate independently; moreover, movements between patches may play an important role in metapopulation stability.

We can test these two alternative views by comparing their predictions with observations on case histories of successful biological control. Murdoch and co-workers (1985) compiled data on several highly successful control efforts (Table 18.1) and in most cases biological control was successful but the mechanisms of success were not those predicted by the classical theory. Let us look in detail at one example, the California red scale.

The California red scale (*Aonidiella aurantii*) and its wasp parasitoid *Aphytis*

Table 18.1 Predictions of Two Alternative Models of Successful Biological Control with Observations from Four Case Studies[a]

	Stable Equilibrium of Pest Population at Low Density	Parasitoids Produce Density-dependent Mortality in Host	Parasitoids Spatially Aggregate to Host Density	Host-specific Parasitoids	Parasitoids Synchronized with Host
PREDICTIONS OF					
Classical theory	Yes	Yes	Yes	Yes	Yes
Nonequilibrium theory	No	Not necessary	Not necessary	Not necessary	No
SUCCESSFUL CASES OF BIOLOGICAL CONTROL					
Winter moth in Nova Scotia	No	?	?	No	?
Olive scale in California	No	No	No	Yes	No
Larch sawfly in Manitoba	No	Yes	?	Yes	Yes
Red scale in California	Yes	No	No	Yes	Yes
Mosquitoes in ponds and irrigation ditches and mosquitofish (*Gambusia* spp.)	No	Yes?	?	No	No

[a] The nonequilibrium theory fits these data better than the classical model.

Source: Data from Murdoch et al. (1985).

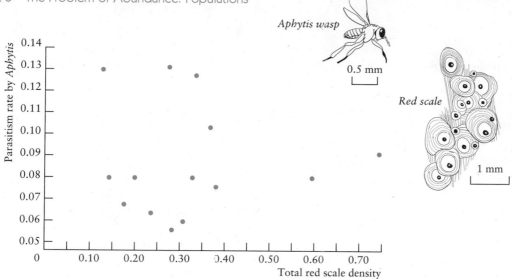

Figure 18.7 *A test for density-dependent predation by the wasp* Aphytis melinus *on California red scale over three years. There is no indication of temporal density-dependence in this predator-prey system in spite of it being a successful biological control program. (From Reeve and Murdoch 1986.)*

melinus are a classic example of successful biological control. The red scale was a serious pest of citrus crops in southern California before *Aphytis* was released in 1957. Now the red scale exists at low and constant densities, and Reeve and Murdoch (1985, 1986) analyzed three possible causes for this successful control program: (1) Mortality caused by the wasp *Aphytis* was not density-dependent in time (Figure 18.7), as classical theory would predict. (2) Stability of host numbers can be achieved by density-dependence in space as well as time (Hassell 1978), but for the California red scale there was also no sign of density-dependent mortality in space. *Aphytis* did not aggregate on high density clumps of red scale, as classical theory would predict (Figure 18.8). (3) The last possible mechanism for stability is the existence of a refuge for the prey. Reeve and Murdoch (1986) found that the red scale has a refuge in the interior branches of citrus trees, and that in this refuge the red scale is protected from parasitoids by the Argentine ant. A second refuge occurs because the red scale cannot be attacked by *Aphytis* once it has become adult (Murdoch et al. 1987). Because of these refuges, red scales do not decline to extinction on citrus trees. If ants were eliminated from trees, the red scale would have no spatial refuge and Reeve and Murdoch (1986) predict the red scale would be completely eliminated.

The results of analyzing successful biological control programs seem to favor the nonequilibrium view of control and to suggest that density-dependent interactions between predators and their prey are not necessary for successful biocontrol. Two strategies of predators could promote local pest extinction (Murdoch et al. 1985). A "lying-in-wait" strategy requires the continuous presence of the predator in local areas and a high attack rate on the pest when it reinvades the area. Predators that adopt this strategy are typically not host-specific but feed on a variety of prey. A second strategy is the "search-and-destroy" strategy in which a predator finds and attacks the pest in a hide-and-seek pattern. This strategy utilizes host-specific predators with high search rates and effective dispersal powers. It avoids extinction

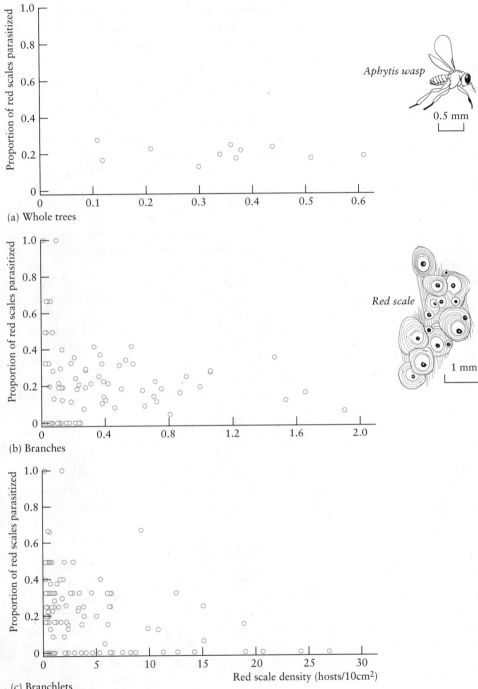

(a) Whole trees

(b) Branches

(c) Branchlets

Figure 18.8   *A test for spatial density-dependence in predation by the wasp Aphytis melinus on California red scale. Three levels of analysis were done from the whole tree (a) as a spatial unit down to the individual branchlet (c). There is no indication of density-dependence over space. (From Reeve and Murdoch 1985.)*

because of spatial patchiness or prey-refuges in the natural habitats of the prey species. The important message is that there may be several different ways to achieve success in biological control.

GENETIC CONTROL

Another alternative to chemical control by pesticides is *genetic control*. Genetic control is a type of biological control that uses two strategies to reduce pest problems. Crop plants can be manipulated to increase their resistance to pests. Alternatively, pests may be changed genetically so that they become sterile or less vigorous and thus decline in numbers.

The use of crop varieties resistant to attack by pests is one of the oldest and most useful techniques of pest control. In 1861 the grape phylloxera, an aphid that feeds on the roots of grape plants, was accidentally introduced into Europe from North America. The European grape (*Vitis vinifera*) was extremely susceptible to the phylloxera, and the wine-making industry of France was on the brink of collapse by 1880. The American grape (*Vitis labrusca*) is resistant to phylloxera attack, so European grapevines were grafted on to American rootstocks to produce an artificial hybrid grape plant that was resistant to phylloxera attack.

Resistant varieties of many crop plants have been developed by selective breeding (Maxwell and Jennings 1980). The method used is, in principle, very simple. Individual plants that are not being damaged are sought in an area where the pest species is common, and these plants are removed to the greenhouse for selective breeding. If resistance is inherited in the greenhouse lines, the new selected variety may be used for commercial production.

Selective breeding can be a two-edged sword, however, and must be used with care. For example, all species of cotton produce a plant chemical called gossypol (a sesquiterpene). This chemical occurs in the green parts and seed of the cotton plant and is toxic when fed to chickens and pigs. To increase the feed value of cottonseed, plant breeders bred strains of cotton with low gossypol content and were able to reduce the concentration of gossypol to only one-fourth that of normal cotton. But breeding gossypol out of cotton deprived the plant of much of its resistance to insect pests and also made the cotton plant susceptible to a whole set of new pests (Klun 1974).

Resistant plants do not necessarily have chemical defenses. Morphological defenses can be highly effective (Kogan 1982). Soybeans are a major crop in the midwestern United States despite the presence of a serious potential pest, the potato leafhopper (*Empoasca fabae*). The potato leafhopper will not attack soybean varieties that have leaves covered with short hairs, whereas they attack and nearly destroy soybeans that have smooth leaves with no hairs. The hairs are a mechanical defense against insect movement and are highly effective as a defense mechanism.

Breeding resistant varieties of plants has been an important factor in limiting pest damage in many crops, but the rapid adaptability of plant pathogens has compromised much effort (Lupton 1977). For example, potato blight is caused by a fungus, *Phytophthora infestans*. This disease first appeared in the 1840s in Europe, where it spread rapidly, causing famine in Ireland. An attempt has been made to introduce single genes giving a high level of resistance, derived from wild species closely related to the cultivated potato. Four genes have been used in this way, but after the commercial introduction of each new gene for resistance, new races of the fungus appeared that could attack the "resistant" potatoes. Sexual recombination or asexual mutation of fungal pathogens results in rapid evolutionary changes in field populations, so that crop resistance breaks down over time.

One promising area of intense development at the present time is the production of resistant crop plants by means of genetic engineering. Genes that produce resistance in one species can be moved into a crop plant to make the crop genetically resistant to specific pests. Alternatively, bacteria may be used as vehicles to carry biopesticide genes. *Bacillus thuringiensis* (Bt) is the main focus at present for developing insect-resistant crops (Lambert and Peferoen 1992). This bacteria normally lives in the soil and carries a gene for a toxic protein that kills the larvae of butterflies and moths. By splicing this gene into bacteria that normally live on crop plants, genetic engineers can produce insect-resistant crops. Insect pests would ingest the bacteria while feeding on the plant and thereby be poisoned. Alternatively, the Bt genes that produce the toxins can be transferred directly into the plant's genome, so the plant would protect itself. As of 1992 tobacco, potato, cotton, and tomato plants have been genetically engineered with Bt genes (Lambert and Peferoen 1992). One anticipated problem with this technology is that pest insects will become resistant to the biopesticide, just as they became resistant to chemical pesticides. Diamondback moth larvae are already showing resistance to *Bacillus thuringiensis* toxins in field crops (Tabashnik et al. 1990). The development of resistance to both natural and artificial pesticides will continue to be a major problem in pest management in agriculture (Pimentel 1991).

In addition to changing the genetic makeup of the plants, we can attempt to alter the genome of the pest species. The simplest genetic manipulation that can be carried out on a pest species is sterilization. Sterility can be produced in several ways, but usual procedure is to sterilize large numbers of pest individuals by radiation or by chemicals and then to release them into the wild, where they can mate with normal individuals. Because of sterile matings, the number of progeny produced in the next generation is greatly reduced, and control can be achieved. The sterile-insect method cannot be used on all pest populations because it requires the rearing and sterilizing of large numbers of individuals and a situation in which immigration of fertile individuals is greatly reduced.

One example of the successful use of the sterile-insect method was the suppression of the mosquito *Culex pipiens quinquefasciatus* on a small island off Florida (Patterson et al. 1970). Between 8,400 and 18,000 sterile males were released each day over a ten-week period during midsummer on a 0.3-square kilometer island, and by the end of the experiment, 95 percent of the eggs sampled on the island were sterile (Table 18.2). Thus the experiment was a success, although the island was only 3 kilometers from the mainland and recolonization by dispersing females occurred quickly after the experiment ended.

The sterile-mating technique of control may be rendered less effective if pest populations are genetically subdivided. One example of this difficulty is the screwworm control program in the southern United States (Richardson et al. 1982). Screwworms are larvae of several species of blowflies that lay their eggs on open wounds of warm-blooded animals. The larvae enter the wound and feed on the live tissue, leading to the death of the host animal (often cattle, sheep, or deer) because of physical damage or secondary infections. A program to eradicate screwworms in the southern United States was begun in the 1950s. Such programs were very effective until 1968, when a series of unexplained outbreaks began. Serious outbreaks again occurred in 1972, 1976, and 1978.

Table 18.2 Sterile-male Release Experiment with the Mosquito *Culex pipiens quinquefasciatus* on Seahorse Key, a Small Island Off Florida

Generation	Ratio of Sterile to Normal Males	Eggs Expected to Be Sterile (%)	Eggs Actually Sterile (%)	Reduction in Eggs Laid (%)
1	All normal	0	0	0
2 (begin releases)	3:1	75	62	36
3	4:1	80	85	34
4	12:1	92	82	79
5	100:1	99	84	96
6 (end)	100:1	99	95	96

Note. Each generation of mosquitoes took about 2 weeks during this summer period.
Source: Data from Patterson et al. 1970.

At least 11 chromosomal types of screwworms can be recognized (Richardson et al. 1982). Many of these types occur together geographically and could possibly be different species of screwworms. If there is a genetic mismatch between the sterile flies raised in the laboratory and the wild type causing an outbreak, clearly the sterile-mating technique will not work because the flies will not mate. All sterile screwworm flies are now produced in a single factory in Chiapas, Mexico, and recent failures of sterile-insect releases to stop screwworm outbreaks have probably been due to genetic differences between the screwworms in culture and the screwworms in the outbreak area.

INTEGRATED CONTROL

Many important pests cannot be controlled by any one technique, so biologists concerned with pest management have been forced to take a wider view of pest problems. A unified approach, called *integrated control* or integrated pest management, uses biological, chemical, and cultural methods of control in an orderly sequence. Integrated control can be achieved only if the population ecology of the pest and its associated species and the dynamics of the crop system are known. Integrated control systems are ecologically sound because they rely on natural biological control as much as possible and resort to chemical treatments only when absolutely necessary. A considerable amount of information is needed to permit the effective use of an integrated control program. Density levels of the potential pest populations, stage of plant development, and weather data are often required to enable the pest manager to predict the future development of the crop and to judge the necessity for pesticide application.

An example of an integrated control program is the alfalfa pest management project developed in Indiana (Giese et al. 1975). Alfalfa is an important crop because it produces high-quality feed for cattle and also improves the soil by fixing nitrogen. Alfalfa is a perennial crop and is relatively long-lasting. Several hundred species of insects can be found in alfalfa fields, yet only a few are serious pests. The alfalfa weevil (*Hypera postica*) is the most important single alfalfa pest in the world, and as an illustration we can design an integrated control program for this weevil (Armbrust and Gyrisco 1982).

The life cycle of the alfalfa weevil in the eastern United States is shown in Figure 18.9. Eggs are laid in the fall and winter, and they hatch in the spring. New adults

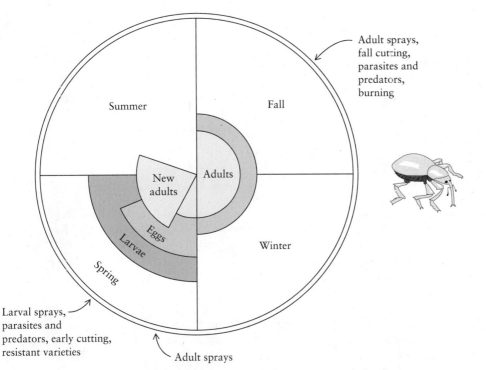

Figure 18.9   Life cycle of the alfalfa weevil living on alfalfa in the eastern United States. Severe defoliation can occur in the spring from larval feeding. In the summer, adults move into woody habitats and then return to alfalfa fields in the fall. Possible seasonal control methods are listed. (After Armbrust and Gyrisco 1982.)

that emerge in the spring feed for a short time and then move into wooded areas to estivate during the summer. In the fall, adults return to the alfalfa fields and become sexual. When many eggs are laid over the winter, larvae hatch and start to feed just as the alfalfa plant begins to grow in the spring. Damage can be severe, and spring weather is critical. Low temperatures retard larval growth more than plant growth so that little damage occurs. Higher temperatures speed larval development and increase damage.

Weather conditions are critical for determining the timing of control procedures against the alfalfa weevil. In Indiana, a computer-based weather observation system has been set up to obtain up-to-the-minute weather data from alfalfa-growing areas (Giese et al. 1975). These weather data are used in two computer models, one to predict the growth of alfalfa and another to predict the development of alfalfa weevil populations. Because temperature is the major variable for both the plant and insect populations, it is possible to predict when damage will occur and when intervention by spraying will be necessary. Farmers can be told, for example, that on a given date in spring, if fewer than 25 alfalfa stems out of 100 show larval-feeding damage, they will not yet need to spray. Timing becomes a critical element in all integrated control programs; hence a detailed understanding of the life cycle of the pest and its host is necessary. Figure 18.10 illustrates how the proper timing of an insecticide spraying can reduce the weevil population, allow the alfalfa to grow, and delay the onset of high weevil densities until later in the spring, when insect parasites can attack weevil

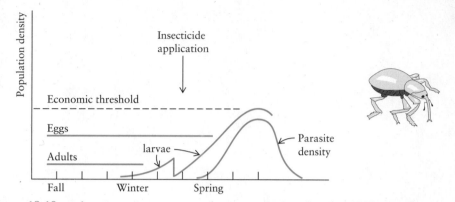

Figure 18.10   Schematic representation of integrated control in the alfalfa weevil in the eastern United States. Larval development depends on temperatures in the spring, and the timing of insecticide spraying is crucial. (After Armbrust and Gyrisco 1982.)

larvae and maintain densities below the damage threshold. Emphasis in integrated control has shifted from trying to eradicate the pest to asking how much damage can be tolerated. We cannot eliminate pests and must learn how best to live with them, taking advantage of all techniques available to keep them below the economic threshold.

Another example of an integrated control program is cotton. Cotton has always been a difficult crop because of the large number of insects that attack it. Cotton occupies 1.1 percent of the total crop acreage in the United States but in the 1970s cotton growers were using 47 percent of all the insecticides used in agriculture (Pimentel and Goodman 1978). An integrated control program in the San Joaquin Valley of California has attempted to analyze the cotton pest problem with the aim of minimizing pesticide usage (Table 18.3). For example, by planting alfalfa strips along cotton fields, lygus bugs can be attracted out of cotton, where they can do serious damage, and into alfalfa, where they do little harm. With integrated control, pesticide usage has been reduced by more than 50 percent in cotton areas of central California (van den Bosch et al. 1982). In Texas insecticide use on cotton has declined 90 percent in the last 25 years because of integrated control. Pimentel and coauthors (1991) suggest that this amount could be reduced an additional 40 percent by more effective pest monitoring, better spraying methods, and crop rotation. These uses of integrated pest management will pay for themselves through reduced pesticide costs.

Integrated control programs derive their validity from field studies and are thus empirical ecology in action. They have not been developed as theoretical strategies but as working programs, and they hold great promise for the future because they retain biological control as a core element of the integrated program.

GENERALIZATIONS ABOUT BIOLOGICAL CONTROL

Why can we not control all pests by biological control? Biological control is something akin to a gambling system—it works, sometimes. But how often? Table 18.4 summarizes data from a global appraisal of the success rate of classic biological control against insect and arachnid pests. About one-third of the parasites and predators introduced get established more or less permanently after introduction (Hall and Ehler 1979). If we define success in biological control according to economic benefits, only 16 percent of classic biological control attempts qualify as complete successes (Hall et al. 1980). Turnbull and Chant (1961) concluded that well over half of the Canadian biological control projects were failures. Why is this? What makes some biological control agents like the vedalia work so well, while others completely fail? A number of empirical generalizations have been suggested.

Most successful biological control programs have operated quickly. Clausen (1951) suggests that three generations (or a maximum of three years) is the outside limit and that if definite control is not achieved in the vicinity of the colonization point within this time, the control agent will be a failure. This rule of thumb suggests that colonization projects should be discontinued after three years if no success is achieved and that prolonged efforts at establishment are wasting money. Most of the successful biological control examples to date support this rule, which suggests that major evolutionary changes in the host-parasite system seldom occur in introduced pests (see Table 14.1, page 274). If a parasite is not already adapted to control the host, it will not evolve quickly into a successful control agent.

The unfortunate truth is that we can evaluate a biological control agent only in retrospect, and biological control programs are part gambling; we release a predator or parasite and hope for the best. A vital historical lesson is the frequency with which a critical species like the vedalia was released more on faith than on any evidence that it could control the pest. There is at the moment no evidence that biological control would not be just as successful if we were to release a random sample of the enemies of the pest species. But ecological theory can learn from the past successes of control programs and develop new insights in how to select biological control agents (Waage 1990).

Most successful biological control programs have resulted from a single species of parasite or predator, which raises the question, If one parasite species is good, are two species better? Turnbull and Chant (1961) argued that only one species should be released at a time for pest control, because two parasites might interfere with each other when the pest is reduced to low numbers. This argument follows from the observation that native insect pests have a great number of predators and parasites. The spruce budworm, for example, has over 35 species of parasites and many predators, yet it is a serious forest pest. Is the spruce budworm a pest because it has many parasites? Or does it have many parasites because it is moderately abundant?

If competitive interactions occur between introduced parasitoids and predators, biological control should be more successful when fewer enemies are released. Figure 18.11 shows that the more species of enemies released, the fewer were successful in getting established (Ehler and Hall 1982). Thus competitive exclusion of introduced natural enemies has probably occurred and has contributed to the low rate of success in biological control (see Chapter 13).

What can we conclude regarding the problem of natural regulation from these examples on biological control? This is a difficult question. Belief in the success of biological control is based on a thoroughly biased sample. Economic pressures run high in this field, since crops worth millions of dollars may be destroyed by a single pest. Hence, states like California have full-time bureaus devoted solely to searching the world for insects to control current agricultural pests. Candidates for control are carefully screened before they are released to make sure that they will not destroy the native fauna rather than the pest. However, once these control agents are intro-

Table 18.3   Comparison of Conventional and Integrated Pest Control Programs in Cotton in California's San Joaquin Valley

Pest	Conventional Control	Integrated Control
Lygus hesperus (lygus bug)	*Treatment threshold:* 10 bugs/50 net sweeps at any time of season from about June 1 until about mid-August, or prophylactic applications based on time of season, grower apprehension, sales advice, etc. *Cultural controls:* None.	*Treatment threshold:* When population levels of 10 bugs/50 net sweeps are recorded on two consecutive sampling dates (3- to 5-day interval) during flower budding period (approximately June 1 to early July). Treatments after mid-July are discouraged as being essentially useless and highly disruptive to natural enemies of lepidopterous pests. *Cultural controls:* In limited use. 1. Strip-harvesting of alfalfa hay fields to prevent migration of bugs into adjacent cotton when alfalfa is mowed. 2. Interplanting of alfalfa strips in cotton fields to attract *Lygus* out of the cotton. Both practices are based on the strong preference of *L. hesperus* for alfalfa over cotton.
Major lepidopterous pests *Heliothis zea* (bollworm)	*Treatment threshold:* 1. Four small (1/4 in. or less in length) larvae/100 inspected plants. This is a long-standing treatment level of unknown origin that has been found to be completely erroneous. Many fields in the San Joaquin Valley sustain infestations of this magnitude, and there is consequently much unnecessary treatment for bollworm control. 2. Nebulous criteria such as time of season, grower apprehension, sales pressure, etc., are also involved in insecticide treatment decisions for bollworm control.	*Treatment threshold:* 1. Fifteen small larvae/100 inspected plants where field has been previously treated with insecticides. 2. Twenty small larvae/100 inspected plants where fields have not been previously treated with insecticides. Infestations of this magnitude rarely develop in untreated cotton fields in the San Joaquin Valley. Consequently, cotton that has reached the period of major *H. zea* activity (i.e., late July and early August) free of insecticide treatment is virtually unthreatened by this pest.
Trichoplusia ni (cabbage looper)	*Treatment threshold:* There is no established economic threshold for this "pest." Instead chemical control is invoked on the basis of such nebulous criteria as "when abundant," "when damage is evident," "when defoliation is threatened." The cabbage looper is commonly abundant in conventionally managed cotton fields, where it erupts in the wake of early and midseason insecticide treatments, particularly those applied for Lygus control. Consequently, San Joaquin Valley cotton is heavily treated for its control. The insect is very difficult to kill, and as a result, heavy dosages of insecticide mixtures are used against it. Needless to say, such treatments add substantially to pest control costs and in-	*Treatment threshold:* There is no evidence that this insect causes economically significant damage to cotton. It is rarely abundant in integrated control fields, since its populations are normally controlled by predators and parasites. Therefore, in fields that have been in the integrated control program, there have been no insecticidal treatments for cabbage looper control.

Table 18.3   (*continued*)

Pest	Conventional Control	Integrated Control
	crease the health hazard and environmental pollution.	
Spodoptera exigua (beet armyworm)	*Treatment threshold:* There is no established treatment threshold for this species, and the same criteria are utilized in making decisions for its chemical control as with cabbage looper. However, ongoing research indicates that where abundant, the insect can cause economically significant injury to cotton. The beet armyworm is essentially a secondary outbreak pest; like the cabbage looper, it is very difficult to "control." Hence, in conventionally managed fields, it often receives costly chemical treatments.	*Treatment threshold:* The beet armyworm has not been abundant in the integrated control fields, and there has been no need to consider its control.
Spider mites (Tetranychidae)	*Treatment threshold:* There are no established economic thresholds for spider mites in San Joaquin Valley cotton. A number of selective and nonselective acaricides are used prophylactically and therapeutically for spider mite "control." Again, as in cabbage looper and beet armyworm control, the treatment criteria are largely nebulous.	*Treatment threshold:* No established economic thresholds. Research on this matter is currently under way, as is research on selective chemical controls and the development of spider mite-resistant cotton. In the fields that have been under the integrated control program, the growers have at times used a selective acaricide prophylactically.

Source: After van den Bosch et al. (1982).

duced, little further work is usually done. Either the agents work and the pest decreases, or they do not work and the entomologists go looking for other parasites or predators. Consequently, the literature is full of all sorts of spurious correlations that are seldom checked out.

The contrast between the restricted fluctuations of natural ecosystems and the recurrent pest outbreaks in agricultural systems suggests another way of looking at pest control problems. Why do pest species thrive in our agricultural systems? Three reasons may be suggested. First, the agricultural systems are typically monocultures, often of genetically similar plant varieties, whereas natural ecosystems have a great deal of spatial complexity. The hazards of dispersal and habitat selection are greatly reduced when the habitat becomes a monoculture. Root (1973) called this general effect the *resource concentration hypothesis* and suggested that monocultures permitted higher herbivore densities. Second, the plants, herbivores, and predators of agricultural crops do not always form a coevolved system, hence the normal processes of evolutionary integration are not achieved in agricultural systems. Third, the number of disturbances is much greater in agricultural systems than in natural systems. This leads to a reduced diversity of species in agricultural systems and makes natural communities and agricultural systems fundamentally different (Murdoch 1975). The best analog of an agricultural system may be simplified laboratory systems that include only a few species. Thus spatial complexity may be important in crop systems, just as it is in laboratory populations.

Integrated pest management, through the use of biological control with other types of control tactics, is rapidly becoming one of the most important practical

Table 18.4 Summary of the Success of Biological Control Efforts Against Insect and Arachnid Pests Throughout the World

Category	No. Attempts	% Established	OF THOSE ESTABLISHED % Partial or Complete Successes	% Complete Successes
Total	2295	34	58	16
BY ORDER OF INSECTS INTRODUCED				
Homoptera	819	43	80	30
Diptera	258	37	31	0
Hymenoptera	105	34	56	0
Lepidoptera	628	27	48	6
Coleoptera	364	23	36	4
BY DEMOGRAPHIC ORIGIN OF PEST				
Exotic pests	2163	34	60	17
Native pests	132	25	29	6
BY GEOGRAPHIC ISOLATION				
Islands	827	40	60	14
Continents	1468	30	56	17
BY HABITAT STABILITY				
Unstable habitats (vegetable and field crops)	640	28	43	3
Intermediate (orchards)	916	32	72	30
Stable habitats (forests, rangelands)	535	36	47	8

Source: Data compiled by Hall and Ehler (1979) and Hall et al. (1980).

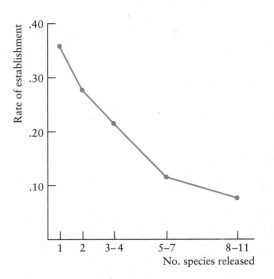

Figure 18.11 Rates of establishment of insect enemies in biological control for single-species releases and for simultaneous, multiple-species releases for 548 projects classified as failures. The same trend is seen if only the successful projects are analyzed. (After Ehler and Hall 1982.)

applications of ecological theory to modern problems of food production. We are gradually replacing an outmoded version of pest eradication through toxic chemicals by a new view of crop management with minimal environmental disturbance. To achieve this goal, we need to know the population biology of both the crops and their associated pests. The challenge is great because the payoffs are so vital.

SUMMARY

Pests are species that interfere with human activities and hence need to be controlled. Most pest control in agricultural systems is achieved in a temporary manner with pesticides, but these toxic chemicals affect other important species and become ineffective because pests develop genetic resistance to the toxins. Biological control makes use of predators and diseases to reduce the average abundance of the pest species.

There are many cases of major reductions in numbers of introduced pests by predators or by insect parasitoids that are specially introduced for purposes of control. Many other attempts have failed and have left the pest to be controlled by chemical means. We cannot adequately explain most of the successes, nor can we explain why failure is so common. The classical model of pest control through density-dependent predation does not seem to describe the field situation of successful control programs.

Genetic control of pests can be accomplished by producing resistant crop plants or by interfering with the fertility or longevity of the pest. Many techniques for the genetic control of pests have been proposed, but few have been used successfully in the field. Genetic engineering holds great promise for producing more resistant crops.

Integrated control tries to combine the best features of biological and chemical control methods and to minimize the environmental degradation that has been typical of modern chemical agriculture. To achieve integrated control, we need to understand the population dynamics of the pest species, and this is, at present, one of the greatest challenges in applied ecology.

Selected References

FLINT, M. L., and R. VAN DEN BOSCH. 1981. A history of pest control. In *Introduction to Integrated Pest Management*, pp. 51–81. Plenum, New York.

HOWARD, W. E. 1988. Rodent pest management: The principles. In *Rodent Pest Management*, ed. I. Prakash, pp. 285–293. Boca Raton, Florida.

LAMBERT, B., and M. PEFEROEN. 1992. Insecticidal promise of *Bacillus thuringiensis*: Facts and mysteries about a successful biopesticide. *Bioscience* 42:112–121.

MURDOCH, W. W., J. CHESSON, and P. L. CHESSON. 1985. Biological control in theory and practice. *American Naturalist* 125:344–366.

MYERS, J. H., C. RISLEY, and R. ENG. 1989. The ability of plants to compensate for insect attack: Why biological control of weeds with insects is so difficult. *Proceedings of the VII International Symposium for the Biological Control of Weeds*, pp. 67–73.

PIMENTEL, D., L. McLAUGHLIN, A. ZEPP, et al. 1991. Environmental and economic effects of reducing pesticide use. *Bioscience* 41:402–409.

VAN DEN BOSCH, R., P. S. MESSENGER, and A. P. GUTIERREZ. 1982. *An Introduction to Biological Control.* Plenum, New York.

WAAGE, J. 1990. Ecological theory and the selection of biological control agents. In *Critical Issues in Biological Control*, ed. M. MacKauer, L. Ehler, and J. Roland, pp. 135–157. Intercept Ltd., Andover, U.K.

Questions and Problems

1. For r-selected pests Stenseth (1981) suggests that optimal control can be achieved by reducing reproduction rather than by increasing mortality. Discuss the ecological reasons behind this recommendation.

2. Why does *Bacillus thuringiensis* produce proteins that are toxic to insects? Review the biology of this bacterium and its geographic distribution and discuss the evolution of its protein toxins. Lambert and Peferoen (1992) provide references.

3. Review the evidence for and against the idea that biological control is much more successful on islands like Hawaii than on continental areas (see DeBach 1964, p. 136, and Hall et al. 1980).

4. Elton (1958) showed that introduced species often increase enormously and then subside to a more static, lower density level. How might this occur in a species that was not the subject of introductions for biological control? How could you distinguish this case from a decline that followed the introduction of some parasitoids for biological control?

5. It is customary to obtain insect parasitoids from the home country of an introduced pest and to use only these parasitoids for possible biological control. Pimentel (1963) suggests another strategy of introducing the parasitoids of other species closely related to the pest you want to control. Review Pimentel's ideas on population regulation (Chapter 16), and discuss the rationale for this recommendation. See Room (1981) for a discussion of how these ideas apply to weed control, and Waage (1990) for a critique of this strategy.

6. Fire ants were introduced into the southern United States around 1940 and are said to be a serious pest (Lofgren et al. 1975). Efforts to control fire ants have been very controversial and of limited success. Review the biology of fire ants, and discuss the reasons for the poor success of control policies. Lewis and colleagues (1992) provide an overview of the problem in California.

7. Contact your local municipal authorities and find out how Norway rats are controlled in your local area. Discuss any ecological problems you can see with their methods and approach.

8. In selection for crops resistant to insect attack should an agricultural ecologist recommend maximal crop resistance as a goal, or could there be an advantage to having crops that are only partly resistant to pests? Discuss this issue before and after reading van Emden (1991).

Overview Question

How do pests evolve resistance to chemicals used to control them? Will this problem arise with the most recent techniques that utilize genetic engineering? How could you overcome the evolution of resistance in pest populations?

19

Applied Problems III: Conservation Biology

Conservation biology is the biology of population decline and scarcity and is a central focus of much public concern. Much of population ecology has been focused on abundant species and the factors preventing population growth. Many population biologists have pointed out that most species are rare, and rarity itself ought to be a focus for population research. Species that have become endangered or threatened are either rare or in sharp decline, and in this chapter we ask what are the causes of decline and rarity of species and what can we do to alleviate problems of threatened populations.

RARE SPECIES

When a naturalist says that a bird or a plant is rare, he or she may mean several different things (Harper 1981, Rabinowitz 1981). The concept of rarity can be dissected best at three levels: geographic range, habitat specificity, and local population size. If we classify each of these levels into two classes, we end up with Table 19.1. Of the eight classes, only one describes "common" organisms and thus the term "rare" may mean seven different things to an ecologist. If there are seven different types of rare species, we must recognize different kinds of management to protect those that are threatened with extinction.

Classic rare species are often those of small geographic range and narrow habitat specificity. Many plants of this type are restricted endemics, and are often endangered or threatened (Rabinowitz 1981). Other rare species have very large geographical ranges and occur widely in different habitats but are always at low density. These species are ecologically interesting but almost never appear on lists of endangered species. The important point is that not all rare species are problems for conservation biology.

We cannot answer the general question "Why are species rare?" because there are so many different kinds of rarity. But we can answer the more specific question of why a particular species is rare. In some cases we can observe a species becoming rare over time. Figure 19.1 illustrates the decline of the African elephant population since 1950. The decline of the elephant population is a direct result of poaching for ivory (Caughley et al. 1990). Not all species that have declined to rarity are so well understood. Some plant and animal species undergo bursts of colonization and decay so that they persist as a mosaic or metapopulation of increasing and declining populations (Harper 1981).

Rare species may have small geographic ranges or narrow habitat specificity and for these species we must be concerned with the spatial distribution of the population. The number or size of habitable sites may be small and this could be one reason a

(a)

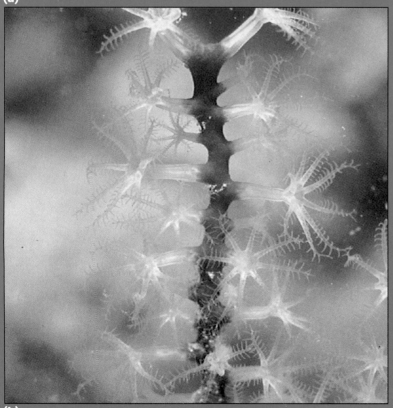

(b)

FIGURE 10.5
Gorgonid corals, an example of a modular organism.
(a) An organ-pipe gorgonian coral from the Great Barrier Reef of Australia, as viewed by a diver (photo by L. Zell).
(b) Expanded polyps on an organ-pipe gorgonian. Each polyp is a module, and in this group of corals there are 8 pinnate tentacles on each polyp (photo by G. Bull).
(Photos courtesy of the Great Barrier Reef Marine Park Authority, Townsville.)

FIGURE 14.23

Geographic color pattern variation in poisonous coral snakes and their nonvenomous mimics in Central America. The dangerous models of the genus *Micrurus* are shown on the left, and the mimics of the genus *Pliocercus* on the right are shown for five different areas (A–D, F). Simultaneous mimicry of two models is shown in E. In southern Mexico and Guatemala, the colubrid snake *Pliocercus elapoides* (center) combines elements of the patterns of two venomous coral snakes. *Micrurus diastema* (left) has broad bands of red and very narrow secondary black rings bordering the yellow rings. *Micrurus elegans* (right) has narrow red rings and broad secondary black rings. In this region, *P. elapoides* has relatively broad red rings in addition to well-developed secondary black rings.

(From Green and McDiarmid 1981, fig. 1. Illustration copyright by the artist, Frances J. Irish; used with permission.)

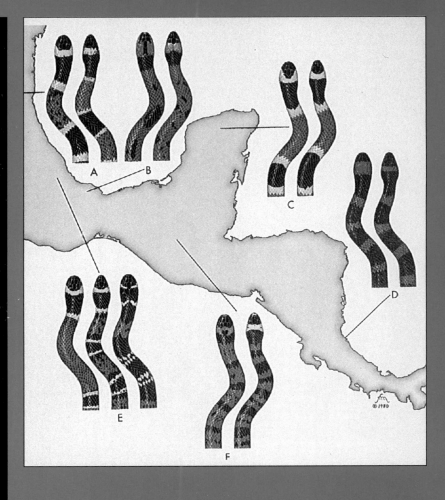

FIGURE 18.6

Biological control of the floating fern *Salvinia molesta*.
(a) Wappa Dam, Nambour, Queensland, completely covered by *Salvinia*, August 1982.
(b) The same scene in September 1983 after the population explosion of the weevil *Cyrtobagous slaviniae* and subsequent crash of both *Salvinia* and the weevil.
(c) The biological control agent for *Salvinia*: the adult weevil *Cyrtobagous slaviniae*.

(Photos courtesy of Dr. P. M. Room, CSIRO Division of Entomology.)

(a)

(b)

(c)

1981

1984

1988

1990

FIGURE 22.13
Succession on Mount St. Helens, Washington, from 1981 to 1990. There were no surviving plants on this small mudflow above Butte Camp, southwest of the crater. Intact vegetation is nearby to serve as a seed source. By 1990 eleven plants had colonized this plot including *Aster ledophyllus, Penstemon cardwellii,* and *Eriogonum pyrolifolium.*
(Photos courtesy of Roger del Moral.)

FIGURE 25.22
Loss of woodland in the Serengeti region of east Africa during the last 60 years.
(a) View of savannah woodland of Terminalia-Combretum in Burtt's valley in the northern Serengeti, photographed about 1935. The woodland is well developed.
(b) The same area photographed in 1978 showing a decrease in woodland as a result of fire.
(c) Elephants knock over large trees and also strip bark from them, as well as browse on small trees. But the major decline in the woodland in the Serengeti was caused by fire, and the woodland will recover only if both fire and elephants are removed. Even in the absence of fire the forest is not able to regenerate because of the elephants.
(Photos courtesy of A.R.E. Sinclair.)

(a)

(b)

(c)

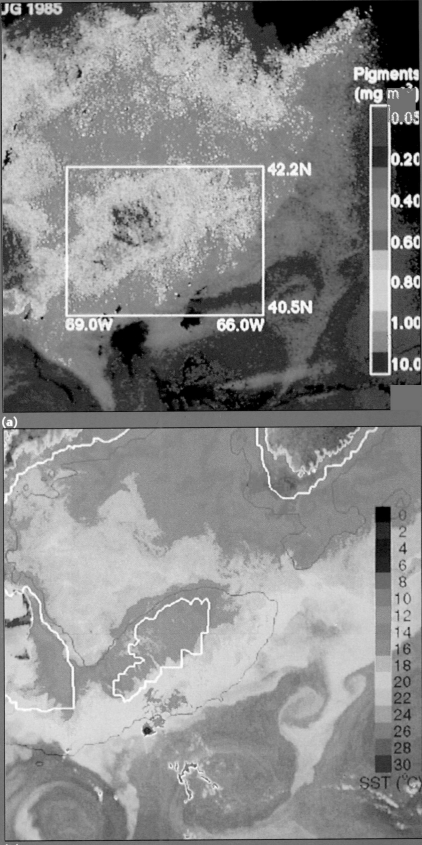

FIGURE 26. 11
Satellite images of the Georges Bank region for August 7, 1985. Southern Nova Scotia is the black area at the top right-hand side of these images and Cape Cod is at the left center. Two Gulf Stream eddies are visible in the lower portion of the images. (a) The area outlined in white occupies about 48,000 sq. km. The colors represent the distribution of chlorophyll pigments in the surface waters. Georges Bank, in the middle of the box, shows high biomass compared with the surrounding waters. (b) Sea surface temperature distribution, derived from NOAA, AVHRR data. Depth contours of 50 and 100 m are shown in white and black respectively. Georges Bank is cooler and shallower than the surrounding waters. Regions of coastal upwelling are also seen off Nova Scotia and Cape Cod. (From Sathyendranath et al. 1991.)

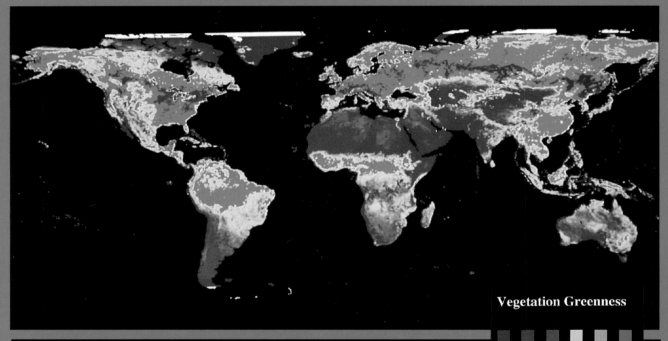

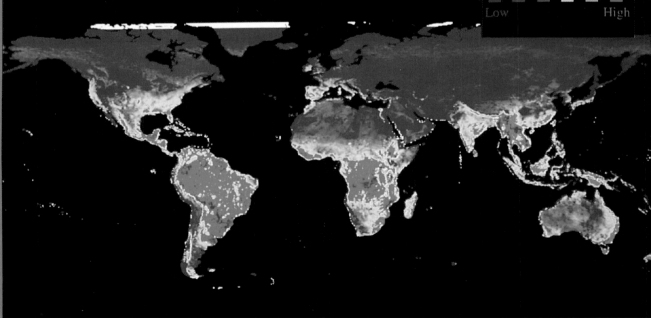

Vegetation Greenness

Low   High

FIGURE 28.11
Effects of acid rain from a copper smelter on native vegetation at Queenstown, Tasmania. The copper smelter operated from 1896 to 1922. Photos taken October 1992.
(Photos courtesy of A. J. Kenney.)

Table 19.1 A Classification of Rare Species Based on Three Characteristics, Geographic Range, Habitat Specificity, and Local Population Size

GEOGRAPHIC RANGE	Large		Small	
HABITAT SPECIFICITY	Wide	Narrow	Wide	Narrow
Large, dominant somewhere	Locally abundant over a large range in several habitats	Locally abundant over a large range in a specific habitat	Locally abundant in several habitats but restricted geographically	Locally abundant in a specific habitat but restricted geographically
Small, nondominant	Constantly sparse over a large range and in several habitats	Constantly sparse in a specific habitat but over a large range	Constantly sparse and geographically restricted in several habitats	Constantly sparse and geographically restricted in a specific habitat

Source: After Rabinowitz (1981).

species is rare. If there are many habitable sites that are not occupied, a species may be rare because of its limited dispersal powers (compare with Chapter 3). Within habitable sites interspecific competition, predation, disease or social interactions may restrict abundance. One example that illustrates these ideas is the case of the red-cockaded woodpecker.

The red-cockaded woodpecker is an endangered species endemic to southeastern United States. It was once an abundant bird from New Jersey to Texas, inland to Missouri. It is now nearly extinct in the northern and inland parts of its geographic range. The red-cockaded woodpecker is adapted to pine savannas, but most of this woodland has been destroyed for agriculture and timber production. These birds feed on insects under pine bark and nest in cavities in old pine trees. Because old pines have been mostly cut down, the availability of nesting holes has become limiting (Walters 1991).

Designing a recovery program for the red-cockaded woodpecker has been complicated by the social organization of this species. They live in groups of a breeding pair and up to four helpers, nearly all males. Helpers do not breed but assist in incubation and feeding. Young birds have a choice of dispersing or staying to help in a breeding group. If they stay, they become breeders by inheriting breeding status by the death of older birds. Helpers may wait many years before they acquire breeding status. Figure 19.2 gives the percentages of males and females that disperse and those that stay.

From a conservation viewpoint, the problem is that red-cockaded woodpeckers compete for breeding vacancies in existing groups, rather than form new groups. New groups might occupy abandoned territories or start at a new site and excavate the cavities needed for nesting. The key problem is the excavation of new breeding

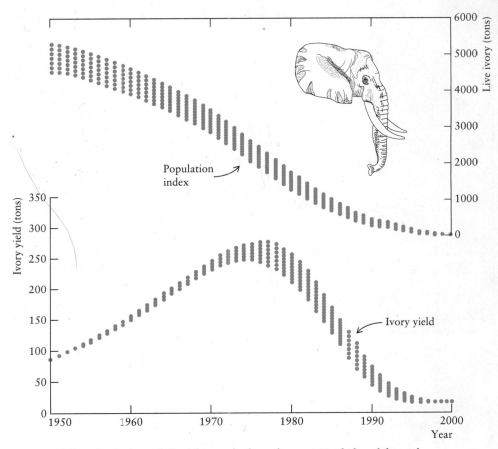

Figure 19.1 The decline of the African elephant from 1950, deduced from the ivory trade. The estimates are approximate because the ivory trade is partly illegal. The continuing decline in elephant numbers across Africa is not always reflected in changes in individual national parks which may protect local populations. (From Caughley et al. 1990.)

cavities. Because of the time and energy needed to excavate new cavities, typically several years, birds are better off to compete for existing territories than to build new ones.

To test this idea, Walters (1991) and his colleagues artificially constructed cavities in pine trees at 20 sites in North Carolina. The results were dramatic—18 of 20 sites were colonized by red-cockaded woodpeckers and new breeding groups were formed only on areas where artificial cavities were drilled. This experiment showed clearly that much suitable habitat is not occupied by this woodpecker because of a shortage of cavities. Management of this endangered species should not be directed toward reducing mortality of these birds but instead should focus on the provision of tree cavities suitable for this species.

The rescue of the red-cockaded woodpecker is a good example of how successful conservation biology must depend on a detailed understanding of population dynamics and social organization, so that limiting factors can be identified and alleviated. Rare species are a special case of the problem of what determines average abundance in a population (page 323). Since there is no general answer to this

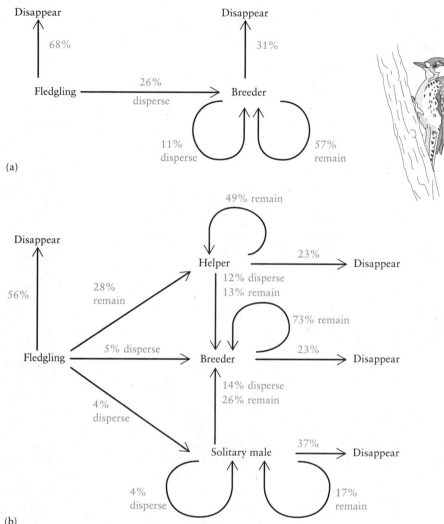

Figure 19.2   Annual transition probabilities for the red-cockaded woodpecker in the Sandhills of North Carolina: (a) females, (b) males. Individuals must decide whether to remain on the same territory or disperse to another. (From Walters 1991.)

question of average abundance, there can be no general prescription for rescuing rare species. Detailed information on resource requirements, social organization, and dispersal powers are required before recovery plans can be specified.

MINIMUM VIABLE POPULATIONS

At some population density rare species will be able to sustain their numbers. This idea has been formalized as the concept of the *minimum viable population* (MVP)— that population size that will insure at some acceptable level of risk that the population will persist for a specified time (Gilpin and Soulé 1986). The analysis of minimum viable populations involves the analysis of extinction. What factors cause a species to go extinct? Some extinctions are due to chance, and Shaffer (1981) recognized four kinds of variation that can contribute to population loss. The first is *demographic stochasticity* and reflects random variation in birth and death rates that can lead by chance to extinction (page 219). The second is *genetic stochasticity* and stresses the importance of inbreeding and loss of genetic variation as causes of extinction. The third is *environmental stochasticity*, which covers variation in weather and biotic factors as causes for species' declines. The fourth is *natural catastrophes*, such as volcanic eruptions, fires, or floods, which could eliminate species with small geographic ranges.

Not all extinctions are due to chance, and some are completely determined by some inexorable change from which there is no escape. Shaffer (1981) called these *deterministic extinctions*. Deforestation is one such change, glaciation is another. If an area is deforested, all species that require trees are eliminated. Deterministic extinctions occur when some essential resource is removed or when something lethal is introduced to the environment. Loss of habitat leads to deterministic extinctions and is a major problem in tropical rainforests where local endemics are common (Gentry 1986). Losses of habitat do not require any ecological study to be understood as causes of extinction. What is required is social and political action, not more research. The major biological problem is that the taxonomy of many groups of plants and insects is so poor for tropical areas that we do not even know what species are being lost by habitat clearing.

Detailed studies of populations going extinct are rare and the few cases known illustrate how a mix of chance events can doom a species to extinction. The heath hen (*Tympanuchus cupido*) was a common grouse from New England to Virginia when Europeans arrived in North America. The heath hen was an eastern subspecies of the prairie chicken which is still common in central and western United States. It declined gradually as European settlement occurred and by 1876 the species was confined to Martha's Vineyard, an island off Cape Cod. By 1900 there were fewer than 100 heath hens on Martha's Vineyard (Ehrlich and Ehrlich 1981). In 1907 a large refuge was established for these birds and predator control instituted on the island. By 1916 there were 800 birds and the species seemed to be rescued. But in the summer of 1916 a severe fire destroyed nests and habitat on the island, and during the winter of 1916 through 1917 a high population of goshawks drove the population even lower to between 100 and 150 birds. Disease in 1920 again cut the population in half; during the 1920s the heath hen gradually declined with increasing signs of sterility and abnormal sex ratios, and by 1932 the last bird died. In this example demographic and environmental stochasticity and natural catastrophes operated in sequence to cause the demise of this once-abundant bird.

One general principle is that the smaller the population, the greater is the risk from chance events that can lead to extinction. But how small is small, and how can we determine the minimum viable population for a particular species? The task is not easy (Soulé 1987). Demography and genetics provide the following two very different approaches for estimating minimum viable populations. At present these two approaches cannot be combined and they need to be computed separately for any species (Kinnaird and O'Brien 1991).

Demographic Extinction Model

Goodman (1987) developed an extinction model that calculates the average persistence time for a population of a given size with specified demographic parameters. This model is defined by two measurable quantities—the population growth rate and the variance of that growth rate. The equation is

$$T = \sum_{x=1}^{N_m} \sum_{y=x}^{N_m} \frac{2}{y[yV-r]} \prod_{z=y}^{y-1} \frac{Vz+r}{Vz-r}$$

where

T = persistence time for a population
r = average population growth rate (r)
V = variance in per capita growth rate (r)
N_m = maximum population size that can be reached
x, y, z = counters of population size (1, 2, 3, . . .)

Belovsky (1987) has simplified this formula and provided a computer program to calculate these estimates. One simplification is to assume that the mean per capita rate of population growth (r) is the same for all population sizes. Goodman (1987) showed that this assumption is adequate for most species and introduced little error. The variance in population growth rate is more difficult to measure since it requires information on several populations or at several times (Goodman 1987).

Let us illustrate the use of this equation on a small primate, the Tana mangabey. The Tana River crested mangabey is a rare primate that lives in small gallery forests along the lower Tana River in eastern Kenya (Kinnaird and O'Brien 1991). It has a restricted range and a small population size that is being affected by increasing agriculture. Population growth rates were estimated from three protected subpopulations of mangabeys and averaged $r = 0.11$ per year. Variance in population growth was estimated at $V = 0.20058$. Figure 19.3 illustrates the computed persistence times for this species from the equation given above. In 1975 the total mangabey population was about twelve hundred and by 1988 it had declined to 700 individuals because of habitat loss. Figure 19.3 shows a persistence time of about two hundred years for the current population. Note that this is the average persistence time for a large number of hypothetical populations, and for any one population actual persistence may well be less than or more than the average value (Belovsky 1987). In

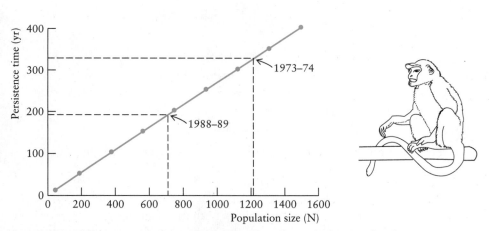

Figure 19.3   *Persistence time (years) as a function of population size for the endangered forest primate, the Tana River crested mangabey. Dashed lines indicate the expected persistence times for this population in 1973–1974 and 1988–1989. (After Kinnaird and O'Brien 1991.)*

this model the probability of persistence depends only on the demographic properties of the species and not on extrinsic agents like habitat destruction. If the gallery forests along the Tana River are cut down, the mangabey will not persist for the 200 years suggested by the model.

The demographic extinction model can be useful for special cases where demographic data are well known. Figure 19.4 shows the solutions for the extinction equation for the population size required for a mammal species to have a 95 percent chance of persisting 100 years (short term) or 1000 years (long term). Smaller populations are needed for larger mammals. Belovsky (1987) converted these population sizes into areas of habitat needed to support herbivorous mammals and carnivorous mammals, and these graphs are shown in Figure 19.4(b,c). Carnivores need larger areas to forage over than herbivores, so we must preserve larger areas to support carnivore populations of a given size. Populations with higher variation in growth rates must be larger to avoid extinction than populations with less variation.

Genetic Models

The central assumption of all genetic models for population viability is that genetic variability is necessary for long-term survival. Genetic variation is necessary for evolutionary adaptation to a changing environment. The number of individuals in a population is crucial in determining how much genetic variation can be maintained. Hence genetic models begin by trying to estimate the effective size of a population. In 1931 Sewall Wright introduced the concept of effective population size to calibrate the amount of random genetic drift in real populations. Random genetic drift is the irregular change in gene frequencies from generation to generation as a result of chance processes. Random genetic drift tends to reduce overall genetic variation, and is particularly severe in small populations. Effective population size is nearly always smaller than actual population size because of genetic drift.

Lande and Barrowclough (1987) present methods for calculating effective population size. Males and females may differ greatly in their reproductive success, and consequently we begin by estimating effective population size for each sex separately.

$$N_{em} = [(N_m K_m - 1)/(K_m + V_m/K_m) - 1]$$
$$N_{ef} = [(N_f K_f - 1)/(K_f + V_f/K_f) - 1]$$

where

N_{em}, N_{ef} = effective population size for males, females
N_m, N_f = actual number of breeding males, females
K_m, K_f = mean number of progeny produced by an individual male or
        female during its liftime
V_m, V_f = variance of K_m, K_f for each sex

Given these separate estimates for males and females, we can determine N_e:

$$N_e = 4[1/N_{em}) + (1/N_{ef})]^{-1}$$

where N_e = effective population size.

The essential feature of these equations is that, when a small proportion of individuals in a population are doing all the breeding, effective population size is small. For example, in the Tana mangabey in 1975, there were 47 (N_m) breeding males in the population and these averaged 7.41 young during their tenure as dominant males in a breeding group. There were 10 breeding females for each breeding

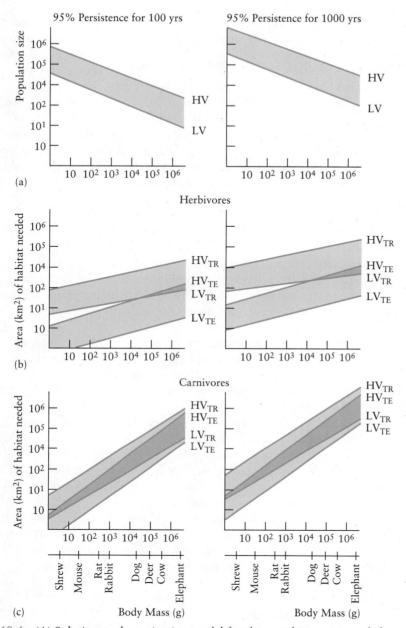

Figure 19.4   *(A) Solution to the extinction model for the population size needed to persist 100 years (left column) or 1000 years (right column) with 95 percent probability for mammals of different body sizes. HV = high variability, LV = low variability. These relationships are converted to give the area of habitat needed to sustain that population size for (b) herbivores and (c) carnivores. TR = tropical environments, TE = temperate environments. (After Belovsky 1987.)*

male and the variation in success ranged from 0 young produced to 32 offspring for individual males. This distribution of variation in progeny number provided an estimate of V_m of 44.13. From the equation given above the effective population size of males was 28. For female mangabeys variation in progeny number is less than for males ($V_f = 1.84$) and there were 458 breeding individuals (N_f). For females the estimate of effective population size was 514 (N_{ef}) and the combined estimate of effective population size was 107 animals. Note that when the sex ratio is very uneven, effective population size can be much less than actual population size.

What does all this mean for conservation? If you wish to maintain genetic variation in natural populations so that evolutionary processes can continue, you need to have an effective population size of at least 500 individuals (Lande and Barrowclough 1987). This number is a crude rule of thumb for conservation biology. More precise estimates can be obtained for species in which detailed studies of social organization, breeding system, demography, and movements have been analyzed (Crow and Kimura 1970). Most endangered species have not been studied enough to use more detailed models. Because effective population sizes are usually less than actual population sizes, this difference may translate into an actual population of a thousand or more individuals. This estimate contrasts with a much smaller number of individuals needed to avoid inbreeding depression (approximately 50).

The endangered Tana mangabey has a low effective population size, a fact that is aggravated by the isolation of breeding groups on the two sides of the Tana River in Kenya. We can predict a loss of genetic variation in this species unless management plans can be arranged to increase gene flow among the breeding groups. The amount of exchange needed to preserve genetic variation is small, only one individual per generation.

One of the most significant contributions of viability analysis has been to show that viable populations of some species are large, so large that it may be impossible to maintain the required number of animals in parks or sanctuaries (Soulé 1987). Figure 19.5 shows this for the grizzly bear in the Yellowstone–Grand Teton Park area. If we draw a biotic boundary for the grizzly bear with a minimum viable population of 500 bears, the area needed to support this population is 122,330 square kilometers, about 12 times the actual park area of 10,328 square kilometers. One of the contributions of conservation biologists has been to point out that our existing parks are far too small to maintain biotic diversity on the scale we are now used to (Newmark 1985, Soulé 1987). Areas of private land outside of parks must also contribute to the preservation of diversity.

Viability analysis has been done for relatively few plants. Furbish's lousewort (*Pedicularis furbishiae*) was once thought extinct, and when rediscovered in northern Maine it became one of the first plant species to be listed as endangered (Menges 1990). Furbish's lousewort is a herbaceous perennial that reproduces only by seed. It lives along riverbanks and only in disturbed habitats, and is found along a single river in northern Maine (Figure 19.6). Disturbance is frequent along the St. John River because of ice jams and ice scour in the spring. Ice scour is a benefit for Furbish's lousewort because it opens new habitat, but it is also a cost because it kills many plants.

To estimate population viability Menges (1990) studied 6000 individually marked and mapped plants in 15 local populations along the St. John River. He used a stage-based projection matrix (Figure 12.19, page 223) to estimate the finite rate of population change for each population. Figure 19.7 shows how the rate of population change fluctuated in different habitats and over the three years of study. Furbish's lousewort is a poor competitor so the higher the vegetative cover, the less well the population does. Woody, shrub-dominated habitats also are poor for this lousewort, as are dry soils. The prediction from the study of these local populations is that open habitats with good soil moisture will support viable populations of Furbish's lousewort.

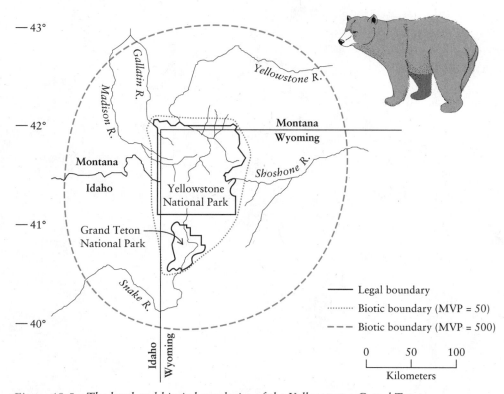

Figure 19.5   The legal and biotic boundaries of the Yellowstone–Grand Teton National Park assemblage. The biotic boundaries are defined by the entire watershed for this park and the area necessary to support a minimum viable population (MVP) (50 individuals for short-term survival, 500 individuals for long-term survival) of the grizzly bear (Ursus arctos), which has the largest home range (489 km²) of any terrestrial species found within the legal boundaries. (From Newmark 1985.)

This prediction assumes that viable populations are not destroyed by catastrophic events like ice scour. Menges (1990) surveyed 32 local populations and found that 2 to 12 percent of these disappeared each year because of ice scour or riverbank collapse. New populations were established on empty sites at a rate of 3 percent of empty sites per year. Population viability depends on the balance between extinction rates and establishment rates, and for the years of study extinction rates seemed to be higher than establishment rates.

Species like Furbish's lousewort present a challenge for conservation biologists. You cannot simply protect the best local populations and ignore others because the disturbance regime of the river causes local extinctions by chance, and new sites for colonization must be available. A reserve system that would protect only the existing populations of this plant may not provide enough recolonization sites. By contrast, too little disturbance would also doom this species because woody vegetation would take over the riverbank habitats. Table 19.2 provides estimates of the likelihood of survival of individual populations of Furbish's lousewort. There is some optimum point of disturbance. Too much or too little could be detrimental to the long-term survival of this endangered plant.

Genetic models of population viability are not relevant for Furbish's lousewort

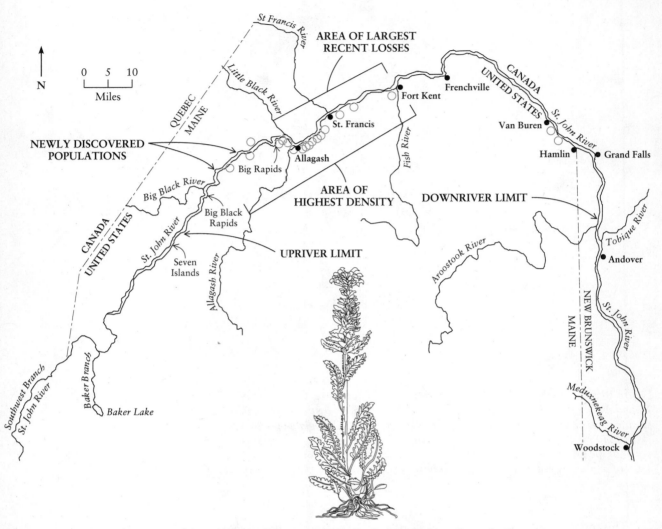

Figure 19.6 *Geographic range of the endangered plant species* Pedicularis furbishiae *along the St. John River of northern Maine. (From Menges 1990.)*

because this species does not seem to have any detectable genetic variation. Menges's (1990) analysis deals only with demographic and environmental stochasticity and natural catastrophes as potential causes of extinction for this plant species.

HABITAT FRAGMENTATION

The largest single problem that causes conservation problems is habitat loss. Humans have utilized a large fraction of the land surface for agriculture, and many plants and animals cannot survive in an agricultural landscape. Of the remaining areas, many have been fragmented, or broken up into small patches. Figure 19.8 illustrates how forest areas in southern Wisconsin were fragmented between 1831 and 1950. Forest fragmentation is occurring at a rapid rate in tropical forests, and can be

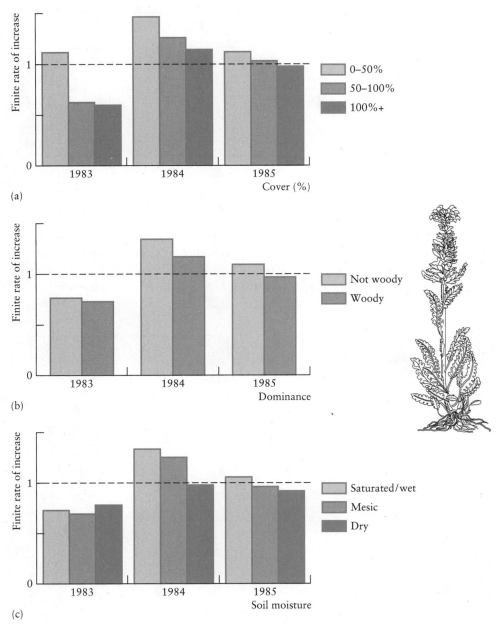

Figure 19.7   *Finite rate of population growth for Furbish's lousewort for populations growing in several habitats: (a) vegetative cover, (b) woody vegetation, and (c) soils with different moisture levels. Note the year to year changes in λ. If λ < 1, the population will decline. If λ > 1, the population will increase. (From Menges 1990.)*

Table 19.2   Probabilities of Individual Population Survival per Year for Local Populations of Furbish's Lousewort in Northern Maine

	ASSUMED ANNUAL PROBABILITY OF NATURAL CATASTROPHE				
Conditions	6%	2%	0.5%	0.1%	0%
1983–1986, all areas	0.0	0.09	0.57	0.92	1.00
1985–1986:					
Low cover	0.0	0.16	0.53	0.93	1.00
Intermediate cover	0.0	0.06	0.34	0.42	0.66
High cover	0.0	0.00	0.00	0.00	0.00

Source: Menges (1990).

documented with old air photos and more recently with satellite imagery. Figure 19.9 shows tropical rain forest losses in the Sierra de Los Tuxtlas, Veracruz, Mexico, over 20 years from 1967 to 1986. Dirzo and Garcia (1992) showed that deforestation has proceeded up from the lowlands and by 1986 about 84 percent of the original

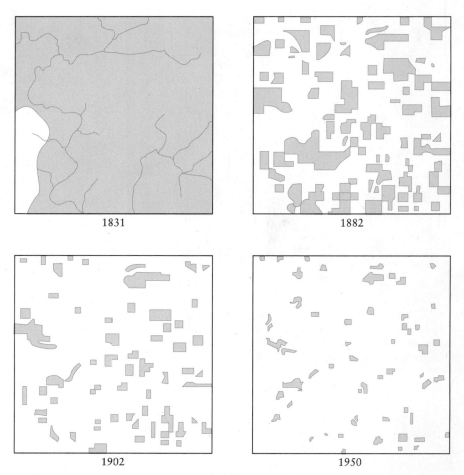

Figure 19.8   Reduction and fragmentation of the woodland in Cadiz Township, Wisconsin, 1831–1950. (After Curtis 1959.)

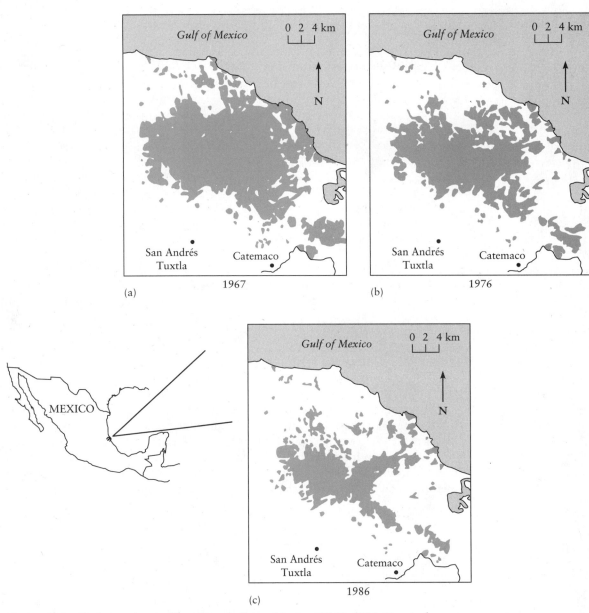

Figure 19.9   Deforestation in Veracruz, southeast Mexico, 1967–1986. Tropical rainforest (black) has been removed at a rate of 4.2 percent per year. (From Dirzo and Garcia 1992.)

forest had been lost. By the year 2000 they predict only 8.7 percent of the original forest would remain in the form of an archipelago of small forest islands. Deforestation in this area is caused mostly by clearing for cattle ranches. The human population of this region has more than doubled in the last 25 years.

Habitat fragmentation has many components with varying impacts on population dynamics (Table 19.3). The major effect of fragmentation that has concerned conservation biologists is reduced connectivity so that species losses in isolated habitat fragments cannot be reversed by immigration.

Table 19.3   Different Components of Habitat Fragmentation and How They May Influence Population Dynamics

Main Component	Habitat Change	Consequences for Population Dynamics
Population-level effects	Reduced connectivity, insularization, increased interfragment distance	Directly affects dispersal and reduces the immigration rate
	Reduced fragment size, reduced total area	Directly affects population size and increases the extinction rate
Landscape or community-level effects	Reduced interior-edge ratio	Indirectly affects mortality and production through increased pressure from predators, competitors, parasites and disease
	Reduced habitat heterogeneity within fragments	Indirectly affects population size through reduced carrying capacity within the fragment
	Increased habitat heterogeneity in surrounding matrix	Indirectly affects mortality and production through increased carrying capacity of predators, competitors, etc. in the surrounding matrix
	Loss of keystone species from the habitat	Indirect effect through disruption of mutualistic guilds or food webs

Source: From Rolstad (1991).

Fragmentation may occur in many different patterns and Figure 19.10 illustrates three different types. The impact of fragmentation is species-specific. Species like eagles that move over large areas may treat a fine-grained fragmented habitat (Figure 19.10a) as a continuous habitat. The exact same habitats may appear coarse grained to a plant with limited dispersal powers (Rolstad 1991). Scale is important in fragmentation and ecological scales are highly species-specific.

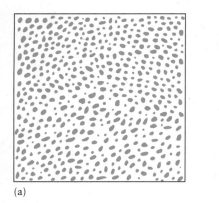
(a)

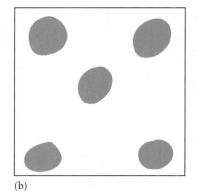
(b)

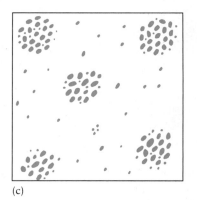
(c)

Figure 19.10   *Three different habitat fragmentation patterns: (a) fine grained, (b) coarse grained, (c) hierarchical. Fine grained fragmentation involves subdividing the habitat into patches smaller than the home ranges of an organism. In coarse grained fragmentation, each patch contains several individuals. The hierarchical pattern is probably the most common type of forest fragmentation. (After Rolstad 1991.)*

Fragmentation of habitats can be analyzed by considering the dynamics of populations subdivided into small patches. At one extreme, when patches are too small the species cannot survive. Small patches are also subject to chance extinction from weather or disease more often than large patches. In western Europe the European red squirrel *(Sciurus vulgaris)* occupies patches of forest interspersed in a mosaic of agricultural land (Verboom and van Apeldoorn 1990). Home ranges of this squirrel range from 1.5 hectares to 13.4 hectares. Figure 19.11 shows that red squirrels are almost always present in woodlots larger than 3 hectares in the eastern part of the Netherlands. Woodlots with pine trees provide more food for squirrels, and this improvement in habitat quality is also important in maintaining populations. These small woodlots may be connected by fence rows or trees along roads, and what is crucial for all fragmented populations is how readily individuals can move between patches. The study of fragmented patches thus becomes a study of metapopulations (page 332) in which populations in patches become extinct and the patch is recolonized by dispersing individuals.

Recolonization may not always occur in isolated patches. The Bogor Botanical Garden was established in 1817 on 86 hectares in west Java. Until 1936 the Botanical Garden was connected with other forest areas to the east, but for the last 60 years it has been an isolated patch of forest, with the nearest patch 5 kilometers away (Diamond et al. 1987). Of the 62 bird species recorded as breeding in the Botanical Garden from 1932 to 1952, 20 species had disappeared by 1980 through 1985 and 4 more were close to extinction. The species that were lost were the less common species, and this low abundance combined with the lack of recolonization from surrounding areas has been the main cause of extinction (Diamond et al. 1987). The

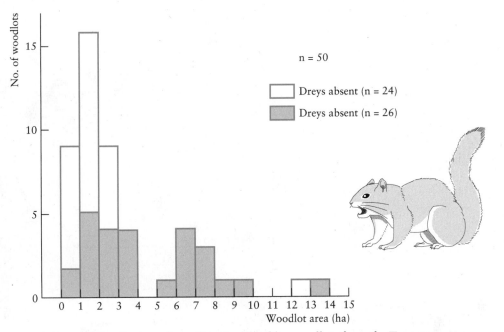

*Figure 19.11   Distribution of woodlot sizes for fifty woodlots from the Twente area of the Netherlands. Red squirrel (*Sciurus vulgaris) *presence was determined from dreys (nests). Few woodlots below 3 ha have red squirrels present. (From Verboom and Van Apeldoorn 1990.)*

result is that much of the conservation value of the Botanical Garden for birds has been lost because the Garden is too small a woodland to support by itself a secure population of many tropical forest birds.

One of the important consequences of fragmentation is to increase the amount of edge in a habitat (Table 19.3). If predators search habitat edges, there could be higher predation rates in smaller fragments because of the edge-effect. Andren and Angelstam (1988) tested this idea in central Sweden in a mosaic of conifer forest and farmland. Artificial ground nests of two brown chicken eggs were placed in farmland and in forest patches. Fifty nests were put out each year for three years. Predation rate was much higher on nests in farmland and at the edge of forest patches (Figure 19.12). This predation effect extended about 50 meters into the forest, so that nests in small forest patches were much more affected than nests in large patches.

One of the most critical variables in the dynamics of populations in fragmented habitats is migration between patches. There are at present few data on movements of animals and plants between patches. Much discussion in conservation agencies has focused on providing corridors between refuges so species may disperse from one patch to the next. Corridors, if they are used, will help to prevent inbreeding depression and will allow recolonization (Simberloff and Cox 1987). But there are potential costs to corridors, since they may transmit diseases, fires and expose individuals to increased predation risk (Table 19.4). The Florida panther (*Felis concolor*) has been reduced from approximately 1400 individuals to about 30 animals, isolated in undeveloped areas of South Florida. By providing a corridor system between wildlife refuges, managers hope to increase the effective population size of panthers (Simberloff and Cox 1987). But there are no data to determine how wide a corridor must be before large mammals like the panther will use them. Corridors may also be difficult to manage to stop poaching, and they may be expensive to purchase and

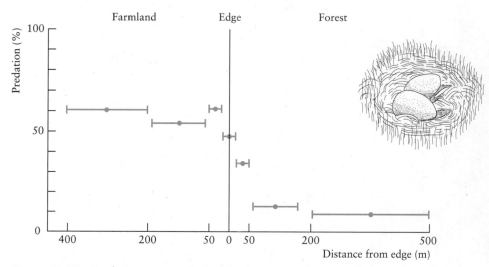

Figure 19.12   Predation rate on dummy bird nests in relation to distance from the farmland-forest edge in central Sweden, 1984–1986. Each point is based on 13 to 30 nests. Two brown chicken eggs were used as dummy nests. Generalist predators like red foxes do not seem to go more than 50 m into these coniferous forest patches. (From Andren and Angelstam 1988.)

Table 19.4 Potential Advantages and Disadvantages of Conservation Corridors

Potential Advantages of Corridors	Potential Disadvantages of Corridors

1. Increase immigration rate to a reserve, which could
   a. increase or maintain species richness and diversity (as predicted by island biogeography theory);
 	
   b. increase population sizes of particular species and decrease probability of extinction (provide a "rescue effect") or permit reestablishment of extinct local populations;
 	
   c. prevent inbreeding depression and maintain genetic variation within populations.
2. Provide increased foraging area for wide-ranging species.
 	
3. Provides predator-escape cover for movements between patches.
4. Provide a mix of habitats and successional stages accessible to species that require a variety of habitats for different activities or stages of their life cycles.
5. Provide alternative refugia from large disturbances (a "fire escape").
 	
6. Provide "greenbelts" to limit urban sprawl, abate pollution, provide recreational opportunities, and enhance scenery and land values.
	1. Increase immigration rate to a reserve, which could
	   a. facilitate the spread of epidemic diseases, insect pests, exotic species, weeds, and other undesirable species into reserves and across the landscape;
	   b. decrease the level of genetic variation among population or subpopulations, or disrupt local adaptations and coadapted gene complexes ("outbreeding depression").
	
	2. Facilitate spread of fire and other abiotic disturbances ("contagious catastrophes").
	3. Increase exposure of wildlife to hunters, poachers, and other predators.
	4. Riparian strips, often recommended as corridor sites, might not enhance dispersal or survival of upland species.
	5. Cost, and conflicts with conventional land preservation strategy to preserve endangered species habitat (when inherent quality of corridor habitat is low).

Source: From Noss (1987).

maintain. There is a need for detailed studies of the movements of individuals between patches and along corridors, and recommendations will not be the same for all species affected by fragmentation. Corridors can be an effective adjunct to planning for conservation in fragmented landscapes (Noss 1987).

Experimental manipulations of local populations could be used to test the general hypothesis that patches of remnant habitat connected to source areas by habitat corridors will be recolonized more readily than patches without corridors. Nicholls and Margules (1991) discuss this problem of how to test for a corridor effect. Figure 19.13 illustrates one idealized design that needs to be implemented. The species of interest can be experimentally driven extinct on patches that are connected by corridors and on those that have no corridor to see what effect corridors are having for this particular species. There are at present no data of this type for any animal or

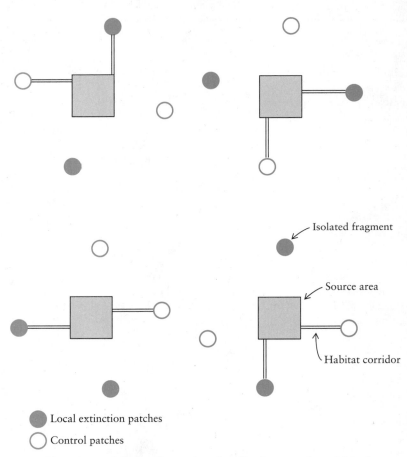

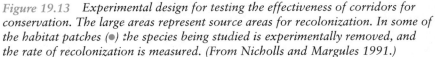

Figure 19.13   Experimental design for testing the effectiveness of corridors for conservation. The large areas represent source areas for recolonization. In some of the habitat patches (●) the species being studied is experimentally removed, and the rate of recolonization is measured. (From Nicholls and Margules 1991.)

plant, and for this reason the role of corridors in conservation is difficult to evaluate (Hobbs 1992).

When landscapes are fragmented, species on the smaller patches may begin to go extinct. If extinction occurs at random among all the species, the pattern of the remaining species will be random. But for many faunas extinction does not fall at random and the resulting patterns form nested subsets (Patterson 1987). Figure 19.14 illustrates the concept of a nested subset. A nested subset can be considered as a time series—species D which occurs only on island 4 is the first to go extinct, followed by species C, which occurs on islands 3 and 4, and the rest follow in sequence (Cutler 1991). If this is correct, extinction is more predictable than random and conservation biologists can focus on those species that need special protection.

The mammals of the coniferous forests on the tops of mountains in the Great Basin of western United States are a good example of nested subsets. Brown (1978) tabulated the occurrence of 14 mammal species on these mountain tops (Figure

Island 1 Island 2 Island 3 Island 4

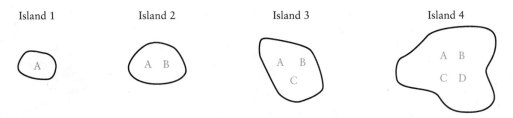

Figure 19.14 *Hypothetical island faunas forming a series of nested subsets. Islands represent any insular habitats such as true islands or habitat fragments. Letters A through D represent species occurring on the islands. All species in smaller faunas also occur in all larger faunas. Smaller faunas are therefore subsets of larger faunas. (From Cutler 1991.)*

19.15). These populations are relicts from the Ice Age when coniferous forests were more widespread. As such they now form patches or islands in a sea of unsuitable desert habitat. Table 19.5 gives the species distributions on these mountaintops. If the pattern were completely nested, there would be no "holes" or "outliers" in the table. For example, the chipmunk *Eutamias dorsalis* is "missing" from mountain

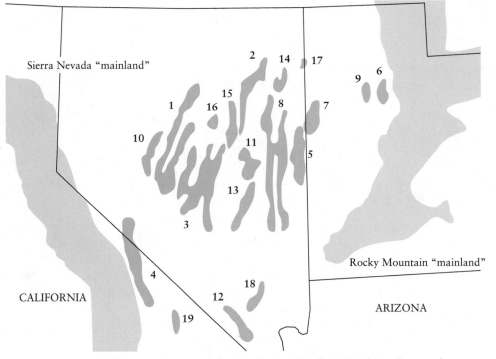

Figure 19.15 *Distribution of montane forests above 2300 m (7500 ft) in the Great Basin region studied by Brown (1971, 1978). Stippled areas indicate relict forests on isolated ranges: 1, Toiyabe-Shoshone; 2, Ruby; 3, Toquima-Monitor; 4, White-Inyo; 5, Snake; 6, Oquirrh; 7, Deep Creek; 8, Schell Creek-Egan; 9, Stansbury; 10, Desatoya; 11, White Pine; 12, Spring; 13, Grant-Quinn Canyon; 14, Spruce-South Pequop; 15, Diamond; 16, Roberts Creek; 17, Pilot; 18, Sheep; 19, Panamint. Areas with diagonal ruling indicate "mainland" forests in the Sierra Nevada and Rocky Mountains. (Modified from Brown 1978.)*

Table 19.5   Species Distribution Matrix for Boreal Mammals of Great Basin Mountain Ranges

Species	1	2	3	4	5	6	7	8	9	10	11	12	13	14	15	16	17	18	19	No. of Occurrences
Eutamias umbrinus	X	X	X	X	X	X	X	X	X	X	X	X	X	X	X	X		X		17
Neotoma cinerea	X	X	X	X	X	X	X	X	X	X	X	X		X	X		X	X	X	17
Eutamias dorsalis	X		X	X	X	X	X	X	X	X	X	X	X	X	X	X		X	X	17
Spermophilus lateralis	X	X	X	X	X		X	X		X	X	X	X	X	X		X			14
Microtus longicaudus	X	X	X	X	X	X	X	X	X	X	X	X	X							13
Sylvilagus nutalli	X	X	X	X	X		X	X		X	X	X	X						X	12
Marmota flaviventris	X	X	X	X	X	X	X	X	X	X	X									11
Sorex vagrans	X	X	X	X	X	X	X	X	X	X										10
Sorex palustris	X	X	X	X	X	X			X							X				8
Mustela erminea	X			X	X	X	X		X											6
Ochotona princeps	X	X	X	X					X											5
Zapus princeps	X	X	X			X										X				5
Spermophilus beldingi	X	X																		2
Lepus townsendi		X				X														2
No. of Species	13	12	11	11	10	10	9	8	8	8	7	6	5	4	4	4	3	3	3	

MOUNTAIN RANGES[a]

[a] Codes for mountain ranges as in Figure 19.15.

Source: Data from Brown and Gibson (1983).

range 2 (a hole) and the rabbit *Sylvilagus nutalli* is present in mountain range 19 (an outlier). The pattern shown in Table 19.5 is clearly a nonrandom pattern and a good example of a nested subset.

It is important to determine if nested subsets occur in fragmented habitats. Not all species are equally vulnerable to extinction, and it is important to direct conservation efforts toward the most vulnerable species (Cutler 1991).

A CASE STUDY—THE NORTHERN SPOTTED OWL

The northern spotted owl (*Strix occidentalis caurina*) has been the focus of intense debates and confrontations over how the remaining old growth forests of western United States should be managed. The northern spotted owl is a territorial owl that lives in old growth conifer forests. Each pair of owls utilizes about 250 to 1000 hectares (1–4 sq. miles) of valuable old growth forest, nesting in hollow trees and feeding on small mammals, birds, and insects. Heavy logging on private land in the last 40 years has destroyed most of the old growth forest upon which these owls depend. Most of the remaining old growth is on lands managed by the U.S. Forest Service and the National Park Service. The northern spotted owl is not now an endangered species, and the total population in the Pacific Northwest is roughly twenty five hundred pairs (Lande 1988).

Old growth forests are being rapidly reduced in the Pacific Northwest (Figure 19.16). A large part of the controversy over the northern spotted owl is over the question of what type of habitat this owl requires and how much its habitat can be fragmented by logging without causing a population decline. Northern spotted owls highly prefer old growth forests for feeding and for roosting (Carey et al. 1992). In fragmented forests owls move more but still feed and roost only in old growth (Figure

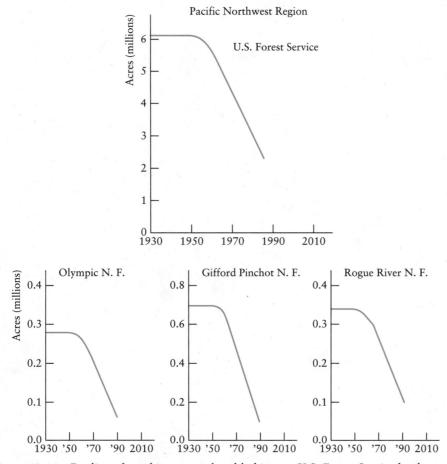

Figure 19.16   *Decline of northern spotted owl habitat on U.S. Forest Service land in Washington and Oregon outside of wilderness areas. (From Anderson et al. 1990.)*

19.17). The home range size of owls varied with the prey base. The most common prey was the northern flying squirrel (*Glaucomys sabrinus*). In Washington State owls used about 1700 hectares of old growth forests, but in Oregon they used less than half that area. These differences in home ranges were directly related to the prey base:

	Home Range (ha)	Prey Available (g/ha)
Washington	~1700	61
Oregon		
Douglas Fir	813	244
Mixed Conifer	454	338

Bart and Forsman (1992) surveyed 11,057 square kilometers throughout the range of the northern spotted owl. They found no owls in forests 50 to 80 years old, and confirmed that owls occurred only where old growth stands were present. Figure

(a)

(b)

• Owl location

◇ Old forest

Figure 19.17   Two examples of the areas used by northern spotted owls in old growth forests in southwestern Oregon. (a) Lightly fragmented area; (b) heavily fragmented old growth forest. Very little use is made of the young forest. (From Carey et al. 1992.)

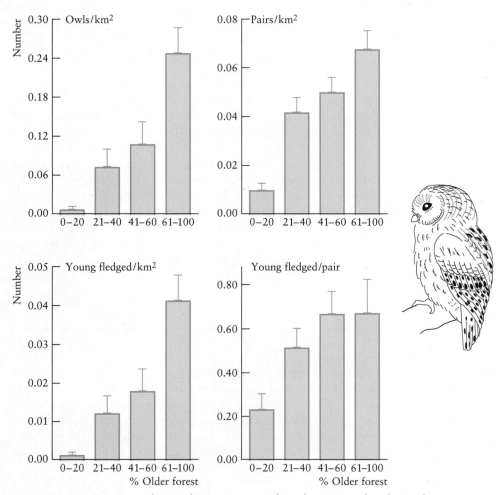

Figure 19.18   Density and reproductive success of northern spotted owls in relation to amount of older forest on 145 forest areas in Washington, Oregon, and northern California. (From Bart and Forsman 1992.)

19.18 shows that northern spotted owls were both more common and more successfully reproductively in old growth forests. Landscapes with less than 20 percent old growth forest rarely supported an owl population.

One surprising result of studies on the northern spotted owl is that wilderness areas are not very suitable as habitat for the owls (Bart and Forsman 1992). Productivity inside protected wilderness areas was only 30 to 50 percent as high as productivity in old growth forest outside these designated areas. Much of the wilderness areas as well as national parks in the Pacific Northwest are high elevation areas less suitable for these owls. The surprising result is that currently protected stands of old growth forest in parks and wilderness areas may be unable to sustain the northern spotted owl.

How much old growth forest must be kept to preserve the northern spotted owl? The key parameters needed to make this estimate are the dispersal and colonization success of young owls and the survival and reproductive rates of territorial owls living in landscapes with variable amounts of old forest. The projections of population growth rates of the northern spotted owl are most sensitive to the adult survival rate, which is estimated to be 0.942 per year (Lande 1988). This survival rate, if constant through adult life, could set a maximum age of 72 years for these owls. Lande (1988) estimated the annual finite rate of population growth (λ) for the northern spotted owl to be 0.96, very close to the equilibrium value of 1.0.

All analyses of the northern spotted owl concur in recognizing that a large part of the remaining old-growth forests in the Pacific Northwest must be preserved if we wish to have this species persist. The problem thus passes from the conservation biologist to the general public as a matter of policy. The competing land use for these forests is logging and the associated jobs in the timber industry. The conflict over the northern spotted owl is a conflict over short-term needs and long-term goals. At the current rate of harvesting, most of the old-growth forests in the Pacific Northwest will be gone within 20 years, and at that time the problems of the timber industry will still be with us but the northern spotted owl may not. The present conflict over land use in old-growth forests is one example of a much broader issue of how human populations and the earth's biota can coexist without serious disruptions. This is the central issue of the twenty-first century for conservation biology.

SUMMARY

Conservation biology is focused on the ecology of rare and declining species. There are many different kinds of rarity and thus no general answer can be given to the question of why a species is rare. Not all rare species are conservation problems. Only by detailed understanding of the population biology of an endangered plant or animal can we provide a rescue plan. In some cases, like the African elephant, the causes of population decline are clear. In many other cases we do not have the ecological understanding to recommend action.

Populations have better chances of persisting if they are large, and this maxim has been formalized in the concept of the minimum viable population (MVP). Four kinds of factors can lead to population extinction—demographic variation, genetic homogenization, environmental variation and natural catastrophes. Quantitative models of demographic and genetic processes that can lead to extinction have been valuable in giving general guidelines for practical decisions in the management of endangered species. Minimum viable populations have been estimated for relatively few organisms, and typically a thousand or more individuals are required to ensure long term survival. Existing parks and reserves are seldom large enough to contain the required numbers of individuals, particularly for larger vertebrates, and conservation efforts on private lands are essential to maintaining our flora and fauna.

Habitat fragmentation has been a side effect of agriculture and forestry and has many adverse effects on populations. Populations in isolated patches may go extinct and unless recolonization occurs, a species may be lost. Corridors between reserves may assist dispersal between patches, but some potential problems like disease spread can be aggravated with corridors.

The ecological challenge of conservation biology is to develop specific management plans for individual species, while the political challenge to the broader conservation movement is to protect large natural areas from destruction. Without parks and reserves there can be no conservation, but with them there is no guarantee of success unless conservation biology can solve the challenging ecological problems of endangered species.

Selected References

CUTLER, A. 1991. Nested faunas and extinction in fragmented habitats. *Conservation Biology* 5:496–505.

HOBBS, R. J. 1992. The role of corridors in conservation: Solution or bandwagon? *Trends in Ecology and Evolution* 7:389–392.

LEADER-WILLIAMS, N., and S. D. ALBON. 1988. Allocation of resources for conservation. *Nature* 336:533–535.

MENGES, E. S. 1990. Population viability analysis for an endangered plant. *Conservation Biology* 4:52–62.

NEWMARK, W. D. 1985. Legal and biotic boundaries of western North American national parks: A problem of congruence. *Biological Conservation* 33:197–208.

RABINOWITZ, D. 1981. Seven forms of rarity. In *The Biological Aspects of Rare Plant Conservation,* ed. H. Synge, pp. 205–217. Wiley, New York.

ROLSTAD, J. 1991. Consequences of forest fragmentation for the dynamics of bird populations: Conceptual issues and the evidence. *Biological Journal of the Linnean Society* 42:149–163.

SAUNDERS, D. A., and R. J. HOBBS. 1991. *Nature Conservation 2: The Role of Corridors.* Surrey Beatty, Chipping Norton, NSW, Australia.

SHAFFER, M. L. 1981. Minimum population sizes for species conservation. *BioScience* 31:131–134.

SIMBERLOFF, D. 1988. The contribution of population and community biology to conservation science. *Annual Review of Ecology and Systematics* 19: 473–511.

SOULÉ, M. E., ed. 1987. *Viable Populations for Conservation.* Cambridge University Press, Cambridge.

Questions and Problems

1. Mammalian carnivores in nature become extinct more often than is predicted by the model in Figure 19.4 (Belovsky 1987), whereas the model's predictions for herbivorous mammals are close to what has been observed. Why might this be?

2. Discuss the assumptions underlying the nested subset model of patch occupancy (Cutler 1991). Explain what ecological processes could produce "holes" in the data matrix (e.g., Table 19.4) and what processes could produce "outliers."

3. When organisms of the same species are brought together to breed from divergent geographic areas, outbreeding depression may occur in which the fertility or viability of the offspring is impaired (Templeton 1986). Discuss why outbreeding depression occurs and what its implications are for conservation biology.

4. The creation of an abrupt edge in a forest because of clearing land for agriculture seems to increase the number of species found in temperate zone forests

and to decrease the number of species found in tropical Amazonian forests (Lovejoy et al. 1986). What factors change when an edge is created, and why might there be different effects in tropical versus temperate forests?

5. Kirtland's warbler is an endangered species that breeds in northern Michigan jack-pine forests. Since 1951 the population of this species has been declining and it now numbers about two hundred individuals. The most important factor in the population drop seemed to be increasing parasitism of nests by brown-headed cowbirds. Cowbirds were removed from the breeding area of Kirtland's warbler starting in 1971 but no change has occurred in warbler numbers (Ryel 1981). Read Walkinshaw (1983) and Ryel (1981) and discuss why this might be and what management plan you would now recommend for this endangered species.

6. Calculate the intrinsic rate of natural increase for a northern spotted owl population in which annual fecundity is 0.24 eggs per female, the age at first breeding is 3 years, the survival probability to fledging is 0.60, the probability of successful dispersal is 0.18, the subadult annual survival rate is 0.71 and the annual adult survival rate is 0.942. How much does r change if you truncate the life table at 10 years? At 20 years? Lande (1988) discusses the life table for this species.

7. Review the history of the successful rehabilitation of the endangered Lord Howe Island woodhen (*Tricholimnas sylvestris*) on Lord Howe Island in the Pacific (Miller and Mullette 1985). Discuss the reasons for the success of this project and the general principles it illustrates for conservation problems.

8. Captive breeding programs are one technique used to rescue endangered species. Under what conditions should captive breeding be used? Discuss the limitations of captive breeding as a conservation strategy. Western and Pearl (1989) provide references.

Overview Question

Debate the following resolution: Resolved that the genetic paradigm of small populations is irrelevant to the problems of conservation biology today.

CONCLUSION

In this part we have considered a complex set of ecological questions about the abundance of populations. We developed population mathematics to illustrate how we can deal with populations in a precise, quantitative manner. Herein lies the strength and the weakness of population ecology, because to some degree we must abstract the population from the matrix of other species in the community in order to describe its dynamics.

For many populations, other species in the community are essential neighbors, and hence we need to broaden our frame of reference beyond the population level. Thus we are led to consider the whole biological community and, in particular, to ask how distribution and abundance interact to structure the biological communities that cover the globe.

20 *The Nature of the Community*

COMMUNITY AS A UNIT OF STUDY

Neither organisms nor species populations exist by themselves in nature; they are always part of an assemblage of species populations living together in the same area. We have already discussed the interactions of two or more of these species populations in predation and competition for food, but this has been with the focus on the individual species populations. Now we shall focus on the assemblage of populations in an area, the *community*. A community is *any assemblage of populations of living organisms in a prescribed area or habitat*. This is the most general definition. So we can speak of the community of animals in a rotting log or the community of plants in the beech-maple deciduous forest. A community may be of any size.

Much of plant ecology has been concerned with community studies or plant sociology; consequently, a whole series of terms has been specially devised. The fundamental unit of plant sociology is the *association*—a plant community of definite floristic composition. To plant sociologists, an association is like a species. An association is composed of a number of *stands*, which are the concrete units of vegetation observed in the field. Plant ecologists use the term *community* in a very general sense, whereas the term *association* has a very specific meaning. Zoologists, on the other hand, use the word *community* both in the general sense and in the specific sense of the botanical *association*. No end of confusion arises from this.

Various botanists and zoologists have defined the community in widely different ways, usually attempting to include in their definition a particular idea of how a community operates. Three main ideas are involved in community definitions. First, the minimum property of a community is the presence together of several species in an area. Second, some authors claim that collections of virtually the same groups of species recur in space and in time. This means that one can recognize a *community type* that has a relatively constant composition. Third, some authors claim that communities have a tendency toward dynamic stability, that this balance or steady state tends to be restored once it has been upset; that is, the community shows self-regulation, or *homeostasis*. The extreme proponents of this third idea look on the community as a type of superorganism. Both the second and the third ideas are disputed, and we shall discuss them presently.

COMMUNITY CHARACTERISTICS

Like a population, a community has a series of attributes that do not reside in its individual species components and have meaning only with reference to the community level of integration. Five traditional characteristics of communities have been measured and studied:

1. *Growth form and structure.* We can describe the type of community by major categories of growth forms: trees, shrubs, herbs, and mosses. We can further detail the growth forms into categories such as broadleaf trees and needle-leaved trees. These different growth forms determine the stratification, or vertical layering, of the community.

2. *Diversity.* We can ask what species of animals and plants live in a particular community. This species list is a simple measure of species richness or species diversity, and leads into the more critical question of what controls biodiversity.

3. *Dominance.* We can observe that not all species in the community are equally important in determining the nature of the community. Of the hundreds of species present in the community, relatively few may exert a major controlling influence by virtue of their size, numbers, or activities. Dominant species are those that are highly successful ecologically and determine to a considerable extent the conditions under which the associated species must grow.

4. *Relative abundance.* We can measure the relative proportions of different species in the community.

5. *Trophic structure.* We can ask, Who eats whom? The feeding relations of the species in the community will determine the flow of energy and materials from plants to herbivores to carnivores, and determine the organization of the community.

These attributes can all be studied in communities that appear to be in equilibrium or in communities that are changing. The changes may be temporal ones, which are called *succession* and may lead to a stable *climax community.* Or the changes may be spatial, along environmental gradients, and we may study, for example, how the characteristics of a community are altered as we move along a gradient of moisture or temperature.

Techniques of measuring the five characteristics of communities will be discussed in subsequent chapters because these characteristics are difficult to quantify, although they are intuitively clear. Let us now look at some questions that we can ask about a community.

What Is a Community?

The study of community ecology is pervaded by an important controversy over the nature of the community. Most of the discussion on this question has centered on plant communities, but the issue is equally important for animal communities. The question is this: *Is the community anything more than an abstraction made by ecologists from continuously varying vegetation?* Is vegetation more like the alphabet, so that we can recognize A, B, or C, or is it more like colors in which we see a great array of greys between *black* and *white*? Or, to put the question another way, is the community an organized system of recurrent species or a haphazard collection of populations with minimal integration? Two extreme schools have developed over this question. On the one extreme are the views of F. E. Clements and A. G. Tansley that the community is a superorganism or a quasi-organism. At the other extreme is the individualistic view of H. A. Gleason that the community is a collection of populations with the same environmental requirements. What are the consequences of these views, and how can we distinguish between them empirically?

A major assumption of many community ecologists is that some "fundamental unit" of natural communities does really exist and that this unit is natural in the sense that it is present in nature and is not a product of human classification. This assumption led some ecologists to draw the analogy between the "species" concept and the "community" concept. If fundamental units exist in nature, we should be able to discover these units and classify them, perhaps in the way we classify species. This view has been held by a majority of ecologists in Europe and North America, and the famous plant ecologists Braun-Blanquet (France), Clements (United States), and Tansley (United Kingdom) all strongly supported this assumption.

The fundamental-unit assumption of community ecology was attacked almost simultaneously by Ramensky in Russia, Gleason in the United States, and Lenoble in France (Whittaker 1962). These workers emphasized the principles of vegetational continuity and of species individuality. If you walked for a hundred kilometers, for example, through the oak-hickory forests of Missouri, you would find that the species of herbs, shrubs and trees changed gradually. There is no single oak-hickory forest community that is the same throughout its geographic extent. Each species has its own geographic range. An association can be regarded as an assemblage of populations and is an arbitrary unit, unlike a species. The individualistic school argues that communities can be recognized and classified, but any classification is for the convenience of the human observer and is not a description of the fundamental structure of nature.

Most of the argument about the nature of the plant community can be centered on two statements:

1. Plant associations (are, are not) discontinuous with one another.
2. Plant species (are, are not) organized into discrete groups corresponding to associations.

Four types of evidence may be considered as having bearing on these two questions (Whittaker 1962), and we shall discuss each in turn.

SIMILARITY AND DISSIMILARITY OF STANDS

If associations are natural units, they should consist of groups of stands that are very similar to one another but clearly different from other stands of a second association. The suggestion is to sample the vegetation without any selection of stands, but almost no one does this. Usually samples of stands are taken to represent certain associations, and this subjective selection obviously assumes beforehand the truth of the fundamental-unit assumption.

When samples are taken by unprejudiced means, many of them are "atypical," "mixed," or "transitional." For example, more than three-fourths of the weed stands measured in the Ulm district of Germany were "intermediate" to two or three previously described associations (H. Ellenberg, cited in Whittaker 1962). "Mixtures" predominated in 1029 grassland samples selected at random by W. Klapp and colleagues (Whittaker 1962), and only 25 percent of the samples represented community types. Brown and Curtis (1952) studied 55 stands of upland forest in northern Wisconsin and found that every stand differed to some degree from every other stand.

Juncus effusus is an important weed in upland pasture in Wales, and Agnew (1961) studied 99 quadrats spread through all community types that contained this weed. Species that were found less than five times in the 99 one square meter quadrats

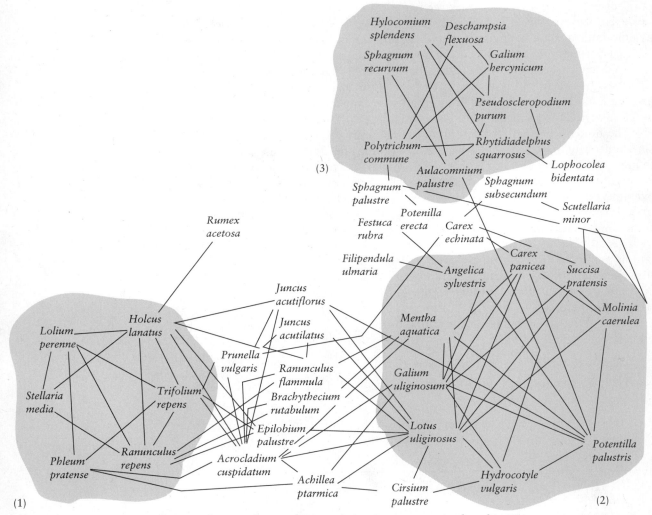

Figure 20.1   *Species "constellation" showing the positive correlation between species found in 99 samples of communities in which* Juncus effusus *occurred in northern Wales. Lines represent positive correlations between the species. Three associations (shaded) can be recognized, but they are not discontinuous. (After Agnew 1961.)*

were eliminated, and 53 plant species remained. Chi-square tests of association were run on all species pairs. These results can be shown graphically in a species constellation, as in Figure 20.1. This constellation shows only the positive associations (lines in the figure), and the position of any species in the figure is largely a matter of trial and error. Three "groups" of species can be recognized (shaded in Figure 20.1) as associations. Obviously there are intermediate species that fit none of the three associations, and the associations are not completely independent of one another. From this evidence we must conclude that species are not organized into discrete groups.

CONTINUITY AND DISCONTINUITY OF STANDS

If associations are natural units, contacts between stands of two different associations should be sharp and discontinuous. Three types of boundaries between communities might occur: sharp, diffuse, or mosaic. All three have been found in nature. The boundary between the prairie and the deciduous forest in the eastern United States was apparently very sharp. Clements would interpret this to show that associations were discrete natural units, whereas Gleason would interpret it as an artificial boundary maintained by fire disturbance. One difficulty of interpreting boundaries is that the human observer fixes on a few dominant species, often trees, for example, whereas a more analytical assessment of shrubs, herbs, or grasses might give a different view of how sharp the boundary is.

We should try to eliminate from this discussion all boundaries caused by sharp environmental discontinuities and by disturbance, but this is difficult to do. Striking breaks in soil type (Figure 8.7, page 127) will produce discontinuity, but all the competing schools of thought agree about this type of discontinuity. Other sharp boundaries are possibly due to environmental discontinuities but need more study. For example, the boreal forest in northern Canada gives way to the tundra over a broad zone of overlap. This zone of forest-tundra is a mosaic of two communities. Whenever you stand in it, you will easily recognize that you are either in boreal forest or in tundra. The sharp boundaries between forest and tundra may mark sharp breaks between areas that have soil and are well drained and areas that have little soil and are poorly drained (Marr 1948, Bliss 1991). Should we interpret forest-tundra boundaries as evidence for discrete associations *à la* Clements?

A number of plant ecologists almost simultaneously began to question the discontinuity of stands and to emphasize that vegetation was a complex *continuum* of populations rather than a mosaic of discontinuous units. This group has developed a series of techniques, called *gradient analysis*, to study the continuous variation of vegetation in relation to environmental factors (Whittaker 1967, Ter Braak and Prentice 1988).

The simplest application of gradient analysis is to take samples at intervals along an environmental gradient, such as elevation on a mountain slope. Figure 20.2 shows the relative abundance of three species of *Pinus* along an altitudinal gradient from 425 to 1430 meters (1400–4700 ft) on south-facing slopes in the Great Smoky Mountains of Tennessee. There is no discontinuity evident in Figure 20.2, and Whittaker (1956) presents data from 24 other tree species to show a continual gradation from the Virginia pine forest at low elevations, to the pitch pine heath at middle elevations, to the table-mountain pine heath at higher elevations.

Elevation is a complex environmental gradient, since it includes gradients of temperature, rainfall, wind, and snow cover. Other gradients can be used as well to show the continuity of vegetation. For example, Whittaker (1960) grouped stands along a soil-moisture gradient at a fixed elevation in the Siskiyou Mountains of southern Oregon (Figure 20.3). Some species, such as Port Orford cedar (*Chamaecyparis lawsoniana*), are found only in moist sites; others, such as Pacific madrone (*Arbutus menziesii*), are most common on dry sites.

We can use information from gradient analysis to determine which model of the community is most appropriate for natural vegetation (Austin 1985). Figure 20.4 shows four alternative models for vegetation organization. The first model is the classical community-unit model of Clements and Tansley in which communities are

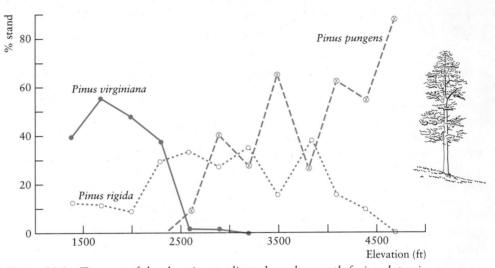

Figure 20.2   *Transect of the elevation gradient along dry, south-facing slopes in the Great Smoky Mountains of Tennessee. No boundaries separate the three community types an ecologist is likely to distinguish along this gradient:* Pinus virginiana *forest at low elevations,* Pinus rigida *heath at middle elevations, and* Pinus pungens *heath at high elevations (see Figure 20.15). (Data from Whittaker 1956.)*

discrete and have sharp boundaries (Figure 20.4a). The classical model assumes that the community is composed of dominant species (like trees) and other species that through natural selection have become coadapted to live in association with the dominant species and with each other. This concordance or coadaptation is what distinguishes the classical model from the other models. The other three models

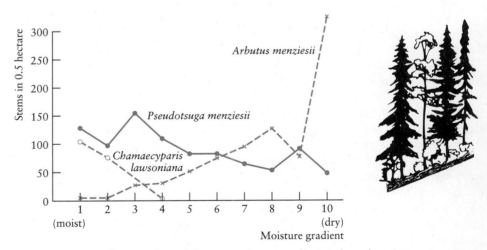

Figure 20.3   *Distribution of trees along a moisture gradient at low elevations on quartz diorite in the central Siskiyou Mountains of Oregon and California. Fifty stands were sampled between an elevation of 610 and 915 meters (2000–3000 ft). Only 3 of 20 tree species are shown to illustrate types of responses. (Data from Whittaker 1960.)*

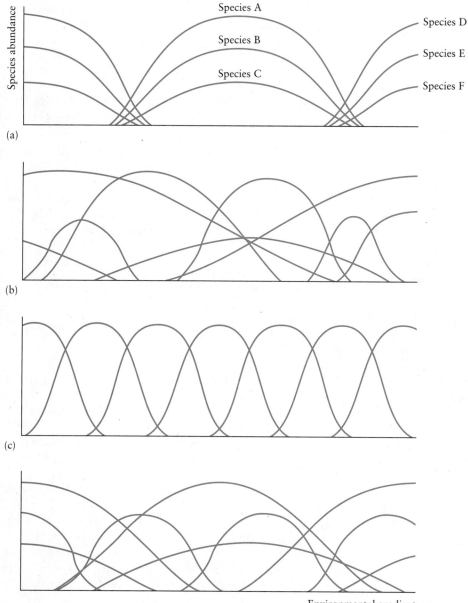

Figure 20.4   Alternative models for vegetation organization along an environmental gradient. Each curve represents one hypothetical species. (a) Fundamental unit concept of Clements and Tansley. (b) Individualistic continuum model. (c) Resource-partitioned continuum model. (d) Resource-partitioned continuum model with several strata. (After Austin 1985.)

are variants of the vegetational continuum model of Gleason and Lenoble. The individualistic-continuum model (Figure 20.4b) has neither sharp boundaries nor well-defined groups of species. Resource competition would be expected to lead to an even distribution of species along an elevational gradient (Figure 20.4c). If there are several strata in a community (like trees, shrubs, and herbs) each stratum may operate independently of the others (Figure 20.4d) and produce a pattern similar to that shown by the individualistic-continuum model. The data shown in Figures 20.2 and 20.3 are most similar to the second and the fourth models of a vegetation continuum.

Elevation gives us an easily obtained environmental gradient to test the continuum model, but in many areas we cannot find a simple gradient to measure. In this situation, we can use techniques of *ordination* to rank the samples in relation to one another (Pielou 1984). Let us consider a simple example of ordination applied to the conifer-hardwood forests of northern Wisconsin (Brown and Curtis 1952).

First, for each forest stand sampled, we determine three values for each tree species x:

$$\text{Relative density} = \frac{\text{no. individuals of species } x}{\text{total individuals of all species}} \times 100$$

$$\text{Relative frequency} = \frac{\text{frequency of species } x}{\text{sum of frequency values for all species}} \times 100$$

$$\text{Relative dominance} = \frac{\text{basal area of species } x}{\text{total basal area of all species}} \times 100$$

Frequency is defined as the probability of finding the species in any one quadrat. *Basal area* is the cross-sectional area of the tree at a point 1.4 meters (4.5 ft) above ground. The three values are summed to obtain for each species its *importance value*:

$$\begin{pmatrix} \text{Importance value} \\ \text{of species } x \end{pmatrix} = \begin{pmatrix} \text{relative} \\ \text{density} \end{pmatrix} + \begin{pmatrix} \text{relative} \\ \text{frequency} \end{pmatrix} + \begin{pmatrix} \text{relative dominance} \\ \text{of species } x \end{pmatrix}$$

Since each of the three is a percentage ranging from 0 to 100, the scale of importance values ranges from 0 to 300.

We then average together the importance values for all stands with the same leading species (with the highest importance value) to determine the results given in Table 20.1. We arrange together the stands that seem most similar. For example, stands dominated by eastern hemlock (*Tsuga canadensis*) are clearly more similar to those dominated by sugar maple (*Acer saccharum*) than to stands dominated by jack pine (*Pinus banksiana*). We recognize this by giving a rank value between 1 and 10 to each species. The end points are clear: Set *P. banksiana* at rank 1 and *A. saccharum* at 10. Intermediate ranks are more arbitrary. *T. canadensis* may be set at rank 8 because it is somewhat less similar* to *A. saccharum* than *Quercus ellipsoidalis* (rank 2) is to *P. banksiana*. These rank values were called climax adaptation numbers by Brown and Curtis (1952) and are given in Table 20.2 for all species of trees for this particular case. They measure in a quantitative manner the position of each tree along the successional sequence. The ecological interpretation is that, if you sat on a newly cleared area of upland in northern Wisconsin, you would see a sequence of trees appearing in the order of their climax adaptation numbers.

---

* Similarity can be measured by the index discussed on page 446 or by a variety of similarity indices (Krebs 1989).

Table 20.1   Average Importance Value of Trees in Stands with Given Species as Leading Dominant—104 Stands from Upland Forests of Northern Wisconsin

No. Stands	Leading Dominant	Acer saccharum	Tsuga canadensis	Betula lutea	Acer rubrum	Quercus rubra	Betula papyrifera	Pinus strobus	Pinus resinosa	Populus tremuloides	Quercus ellipsoidalis	Pinus banksiana
23	Acer saccharum	145	25	21	7	22	6	1	—	1	—	—
23	Tsuga canadensis	40	152	47	11	3	5	4	3	—	—	—
6	Quercus rubra	27	1	3	29	138	23	10	8	5	3	—
6	Betula papyrifera	48	8	7	27	16	108	19	1	29	1	—
19	Pinus strobus	12	6	2	24	12	12	150	39	9	5	—
9	Pinus resinosa	3	—	1	12	15	14	56	156	24	4	2
4	Populus tremuloides	11	—	—	10	29	34	14	19	140	—	—
4	Quercus ellipsoidalis	—	—	—	5	7	1	11	9	9	103	56
10	Pinus banksiana	—	—	—	3	3	3	13	12	14	36	213

Source: After Brown and Curtis (1952).

Finally, we determine the continuum index for each stand from the formula

Continuum index = Σ [(importance value) $\times$ (climax adaptation no.)]

where the sum is taken over all species. For example, in stand 084, Brown and Curtis obtained:

	Importance Value	Climax Adaptation No.	Product
Pinus banksiana	272	1	272
Quercus ellipsoidalis	4	2	8
Populus tremuloides	9	2	18
Pinus resinosa	12	3	36
Acer rubrum	4	6	24
		Continuum index total =	385

At the other extreme, in stand 114, they obtained:

	Importance Value	Climax Adaptation No.	Product
Acer saccharum	268	10	2680
Ostrya virginiana	7	9	63
Tilia americana	8	8	64
Betula lutea	6	8	48
Quercus rubra	3	6	18
		Continuum index total =	2873

The continuum index is assumed to measure a complex environmental gradient much like elevation up a mountainside. Thus we can plot the importance values for all the tree species against the continuum index. Typical results for the northern Wisconsin forests are shown in Figure 20.5. The continuum analysis may be viewed as a simple

Table 20.2 Climax Adaptation Numbers of Tree Species Found in Stands of Upland Forests in Northern Wisconsin

Tree Species	Climax Adaptation No.	Tree Species	Climax Adaptation No.
Pinus banksiana (jack pine)	1	*Abies balsamea*	7
		Thuja occidentalis[a]	7
Quercus ellipsoidalis	2	*Carpinus caroliniana*[a]	7
Quercus macrocarpa	2		
Populus balsamifera[a]	2	*Tsuga canadensis*	8
Populus tremuloides	2	*Betula lutea*	8
Populus grandidentata	2	*Carya cordiformis*[a]	8
		Fraxinus americana	8
Pinus resinosa	3	*Tilia americana*	8
Pinus pennsylvanica	3	*Ulmus americana*	8
Quercus alba	4	*Ostrya virginiana*	9
Prunus serotina	4		
Prunus virginiana	4	*Fagus grandifolia*	10
		Acer saccharum (sugar maple)	10
Pinus strobus	5		
Betula papyrifera	5		
Juglans cinerea[a]	5		
Acer rubrum	6		
Acer spicatum[a]	6		
Fraxinus nigra[a]	6		
Picea glauca[a]	6		
Quercus rubra	6		

[a] Climax adaptation number is tentative because of the low abundance of these species in the stands studied.
Source: After Brown and Curtis (1952).

way of quantifying the impression one gets from looking at many stands of vegetation: that discrete units with sharp boundaries are uncommon in nature.

The environmental variables responsible for the vegetation gradient can be studied in relation to the continuum index in the same way that changes in temperature and rainfall up a mountainside can be measured. Figure 20.6 shows that the moisture-holding capacity of the upper soil layer (A_1 horizon) varies with the continuum index. Jack pine stands tend to occur on soils that dry out easily, whereas sugar maple stands occur on soils that hold more moisture—but this relationship is not very tight.

At this point, you may well wonder how anyone could possibly question the continuity of vegetation and the fact that discrete stands do not occur. The advocates of the fundamental-unit view, that associations do occur as discrete units, question the whole approach of gradient analysis. They make two fundamental criticisms: (1) The stands that have been studied by gradient analysis are all disturbed stands, or stands not in equilibrium with the environment, and (2) the techniques of gradient analysis force the data into looking like a continuum. Langford and Buell (1969) and Daubenmire (1966) summarize these objections.

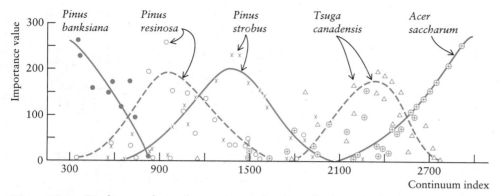

Figure 20.5   Gradient analysis of a continuum for the upland conifer-hardwood forests of northern Wisconsin. Only the dominant tree species are shown. (After Brown and Curtis 1952.)

The first objection to the gradient-analysis school is that to evaluate the hypothesis that discrete associations occur in nature, one must study stands in equilibrium, but Curtis and his co-workers have studied stands of trees that were clear-cut only 26 years before and others that have been selectively logged. Langford and Buell (1969) suggest that all the Wisconsin stands would develop to a common end point if they were undisturbed for a few hundred years. We shall discuss the ideas of vegetation change in Chapter 22; here we note that all schools agree that stages of vegetation that change toward a stable end point will always show a continuum of species.

The second objection is that the techniques of gradient analysis are insufficient to test the fundamental-unit hypothesis. Daubenmire (1966) points out that one must sample stands that differ in only one environmental variable and are otherwise homogeneous. Thus if we know that vegetation is affected by macroclimate, microclimate (slope effects), soil characteristics, and disturbances, we must hold three of these constant and study the remaining variable. Vegetation sampling that includes

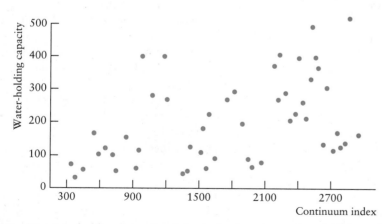

Figure 20.6   Moisture-holding capacity of the A_1 horizon (upper layer) of the soil in relation to the continuum of upland conifer-hardwood forests of northern Wisconsin illustrated in Figure 20.5. (After Brown and Curtis 1952.)

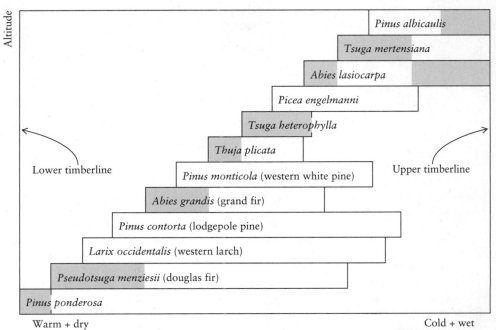

Figure 20.7 *Coniferous trees in the area centered in eastern Washington and northern Idaho, arranged vertically to show the usual order in which the species are encountered with increasing altitude. The horizontal bars designate upper and lower limits of the species relative to the climatic gradient. The shaded area indicates the portion of the altitudinal range of a species in which it can maintain a self-reproducing population in the face of intense competition. (After Daubenmire 1966.)*

all of these sources of variation mixed together must produce results in which everything overlaps with everything else, Daubenmire argued.

Gradient analysis assumes that all species are equal, and only the species names and their relative abundances are used for analysis. Daubenmire (1966) suggests that this produces erroneous conclusions because some species are dominant to others, so not all species are of equal value in determining the community. Community boundaries may be fixed by one or two species only, and there is no need to postulate a complete boundary of all species at the same geographic position, according to Daubenmire.

There is certainly a continuum in the distribution of coniferous trees with respect to altitude in eastern Washington and northern Idaho, Daubenmire (1966) states, but this is a floristic continuum and not an ecological continuum. If we examine the stands closely, we find that within certain zones, one tree species is competitively dominant over the others (Figure 20.7). This ecological fact is the basis for recognizing discrete communities in an altitudinal transect.

The advocates of the individualistic school reply that we must study vegetation as it exists now over large areas. We cannot select from a vast area only a few stands of a "homogeneous" nature to group into an association. This is too subjective, and we must employ objective quantitative techniques to eliminate human subjectivity if we are to achieve a science of community ecology. Community boundaries should

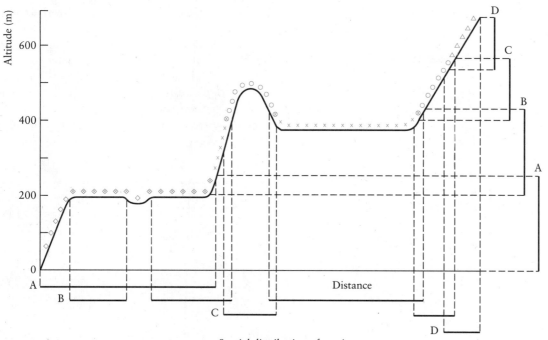

Figure 20.8   *Patterns of co-occurrence of four hypothetical species on a landscape along an environmental gradient (altitude). There is continuous variation of composition along the altitude gradient. A plant community is a landscape concept and recognition of communities depends on the frequency of environmental combinations in a particular landscape. As one moves up the mountain, the plant communities segregate cleanly, as in Figure 20.4a, and could be described as discrete associations. (From Austin and Smith 1989.)*

not be fixed by only one or two species; we must use all the species present to look for boundaries of "natural units" if the community concept is to be meaningful (Cottam and McIntosh 1966).

The controversy between the individualistic school and the community school is partly caused by two different approaches to studying vegetation (Austin and Smith 1989). Figure 20.8 illustrates a hypothetical transect up a mountainside. Supporters of the community school consider the *spatial distribution* of the species, and recognize communities A, AB, B, C, and D. Supporters of the individualistic school consider the *environmental distribution* of the four species and see a continuum of species with altitude. The important point is that communities are a function of the landscape examined, and if landscapes are similar, the communities described in them will be similar. The continuum concept applies to abstract environmental space, not to geographic space.

DISTRIBUTIONAL RELATIONS OF SPECIES

If the community is composed of a group of coadapted species (Figure 20.4a), all or many of the species must have similar geographic distributions. Plant species that comprise an association should have distributional maps that closely coincide on a local level, and the geographic limits of the species should coincide with the continental limits of the association.

Floristic provinces can be recognized on a continental scale by major vegetational changes. Figure 20.9 illustrates a subdivision of North America into ten floristic provinces. The exact position of these boundaries is often debated, and some workers recognize more or fewer provinces, but no one questions that there are large areas of similar vegetation in which the ranges of many species coincide. Figure 20.10 illustrates the coincidence of ranges for some tree species of the eastern deciduous forest province.

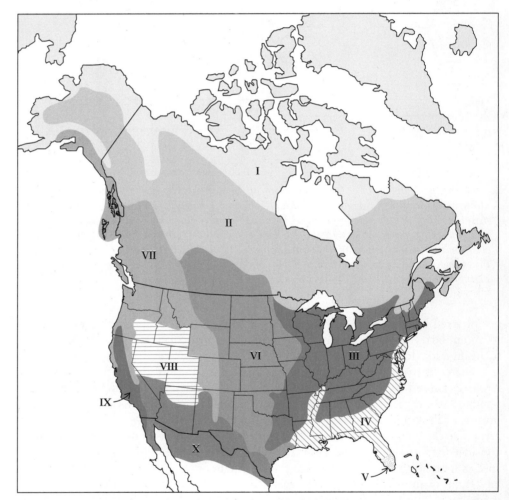

Figure 20.9 *Floristic provinces of the continental United States and Canada. I, tundra province; II, northern conifer province; III, eastern deciduous forest province; IV, coastal plain province; V, West Indian province; VI, grassland province; VII, cordilleran forest province; VIII, Great Basin province; IX, California province; X, Sonoran province. For this map, the lines between the provinces have been drawn boldly to show the general outlines rather than details. The actual boundaries are generally not sharp; they overlap and interfinger extensively, and small enclaves of one province may be wholly surrounded by another. (After Gleason and Cronquist 1964.)*

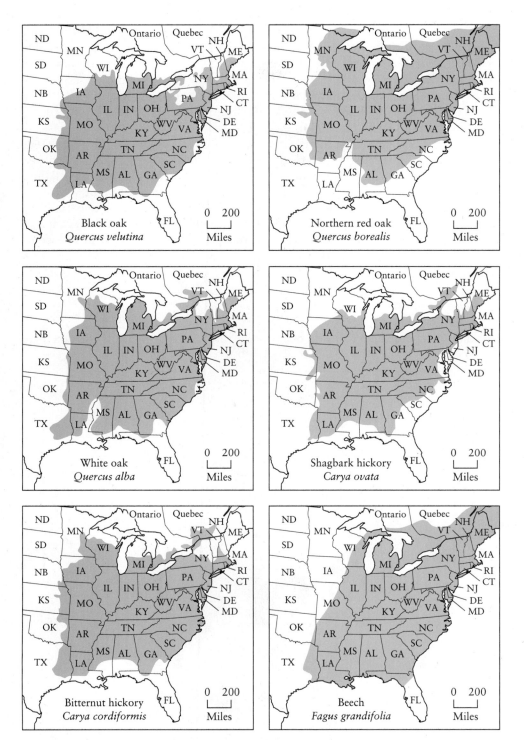

Figure 20.10   *Ranges of certain forest trees of the eastern deciduous forest of North America. (After Gleason and Cronquist 1964.)*

Boundaries between floristic provinces are called tension zones and coincide with the distributional limits of many species. Curtis (1959) has analyzed in detail a tension zone in Wisconsin between two parts of the deciduous forest, the southern prairie–hardwoods province and the northern hardwoods province. Figure 20.11 shows the range limits for 182 plant species that abut at this boundary. The width of this tension zone in Wisconsin is variable, from as little as 16 kilometers (10 mi) to as much as 48 kilometers (30 mi).

A floristic province is not a single community but is composed of many different associations, and the critical analysis of the distributions of species cannot be at the continental level of a floristic province but must be at the local level of an association. Suppose that we study a number of stands of a particular association in Wisconsin and another group of stands in Michigan. How can we compare these two samples? Several measures of community similarity are available (Greig-Smith 1983, chap. 6); we shall discuss one simple measure based on species presence only. In two communities, one with x number of species, another with y species, and with z species occurring in both communities, we define

$$\text{Index of similiarity} = \frac{2z}{x + y}$$

This index ranges from 0 to 1.0 to quantify the range from no similarity to complete

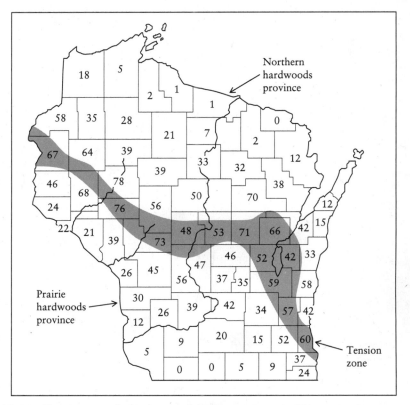

Figure 20.11   *Tension zone between two floristic provinces in Wisconsin. Summary of range limits for 182 species of plants. The figures in each county indicate the number of species attaining a range boundary there. The shaded band is the tension zone. (After Curtis 1959.)*

similarity. For example, the southern mesic forests of Wisconsin contain 26 tree species (dominated by sugar maple, basswood, beech, and red oak), and the northern mesic forests of Wisconsin contain 27 tree species (dominated by sugar maple, eastern hemlock, beech, yellow birch, and basswood). Seventeen species occur in both communities, and the index of similarity for tree species is calculated as

$$\text{Index of similiarity} = \frac{(2)(17)}{26 + 27} = 0.64$$

We can illustrate the use of the index of similarity for the animal communities of the Great Lakes.

The Great Lakes of North America form the largest inland water system in the world (Figure 20.12). Water flows from Lake Superior and Lake Michigan through Lake Huron into Lake Erie and Lake Ontario, from which the St. Lawrence River carries the flow into the North Atlantic. The Great Lakes have been studied in considerable detail by biologists in the United States and Canada because of pollution problems associated with the large cities that border the lakes.

Table 20.3 gives the records of crustacean zooplankton species for the Great Lakes. Most of the 25 species occur in all the Great Lakes, and hence the indices of similarity are all high for this species group. For example,

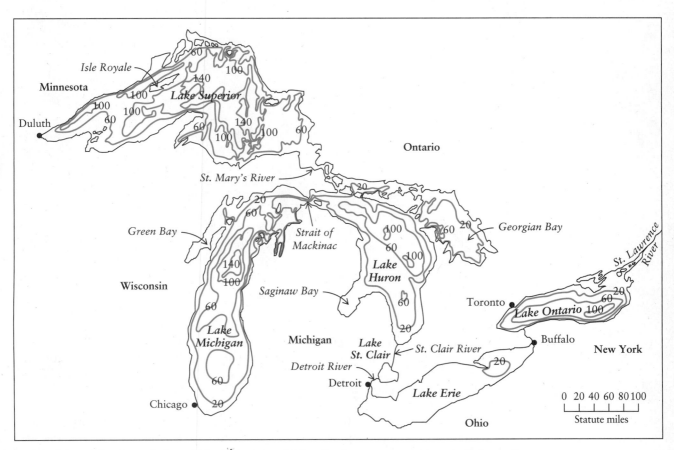

Figure 20.12   The Great Lakes of North America. Major depth contours are given in fathoms (1 fathom = 6 feet = 1.8 meters).

Table 20.3   Crustacean Zooplankton Species Recorded from the Great Lakes of North America

Species	Lake Superior	Lake Michigan	Lake Huron	Lake St. Clair	Lake Erie	Lake Ontario
Senecella calanoides Juday	*	*	*			*
Limnocalanus macrurus Sars	*	*	*	*	*	*
Eurytemora affinis (Poppe)	*	*	*	*	*	*
Epischura lacustris Forbes	*	*	*	*	*	*
Diaptomus sicilis Forbes	*	*	*	*	*	*
D. ashlandi Marsh	*	*	*	*	*	*
D. minutus Lillj.	*	*	*	*	*	*
D. oregonensis Lillj.	*	*	*	*	*	*
D. siciloides Lillj.	*	*	*	*	*	*
D. pallidus Hennrick				*	*	*
Diacyclops bicuspidatus thomasi Forbes	*	*	*	*	*	*
Acanthocyclops vernalis Fischer	*	*	*	*	*	*
Mesocyclops edax (Forbes)	*	*	*	*	*	*
Tropocyclops prasinus mexicanus Keifer		*	*	*	*	*
Osphranticum labronectum Forbes		*				
Alona spp.		*	*	*	*	*
Bosmina longirostris O.F.M.	*	*	*	*	*	*
Ceriodaphnia lacustris Birge		*	*	*	*	*
Chydorus sphaericus O.F.M.		*	*	*	*	*
Daphnia ambigua Scour.					*	
D. galeata mendotae Birge	*	*	*	*	*	*
D. longiremis Sara		*		*		
D. parvula Fordyce		*		*		
D. pulex DeGeer				*		
D. retrocurva Forbes	*	*	*	*	*	*

* Asterisk indicates the presence of a species in a particular lake.
Source: After Watson (1974).

	Index of Similarity
Lake Superior and Lake Michigan	0.81
Lake Michigan and Lake Huron	0.93
Lake Erie and Lake Ontario	0.90

High similarity does not occur in all of the species groups that comprise the freshwater community of the Great Lakes. Rotifers show more differences between the lakes (Watson 1974). For example, 15 species of rotifers are found in both Lake Erie (25 species total) and Lake Ontario (30 species total), and the index of similarity based on rotifers is 0.55 for these two lakes. Similarity among some of the lakes is even lower for some groups of fish. Of the 11 species of whitefish in the Great Lakes, 2 species are common to both Lake Erie (3 species) and Lake Ontario (7 species). The index of similarity based on whitefish is 0.4 for these two lakes.

The Great Lakes illustrate the general problem of trying to define discrete assemblages as communities. One part of the community may be very similar in two adjoining lakes, while another group of species may vary considerably from one lake to the next. The community of animals changes gradually in composition, with no sharp breakpoints.

If we repeat this kind of analysis for many different associations, we find that some are more homogeneous over large areas than others. Curtis (1959), for example, pointed out the great floristic homogeneity of the deciduous forests of eastern North America. On mesic sites, sugar maple is dominant over most of this area, and beech, basswood, and buckeye are leading dominant over large areas. By contrast, the conifer swamp community, originally thought to be very uniform, changes completely in floristic composition from west to east. Only two trees, tamarack and black spruce, remain constant; they alone produce an appearance of a similar community type to the casual observer.

DYNAMIC RELATIONS BETWEEN SPECIES POPULATIONS

If the association is a natural unit, species populations should be bound together in a network, organized by obligate interrelations. This is the basic idea of the "web of life." To evaluate how strong the network of interrelations is, we must go back to a discussion of the factors that limit the distribution and abundance of species populations. There are important practical consequences that flow from this question (Walker 1992). If species are tightly linked in natural communities, losses of species may have cascading effects on other species. Conservation biology in these cases must be concerned about community dynamics rather than single species dynamics. Within a community some species are "drivers" and others are "passengers" and the loss of some species is more critical than the loss of others.

We can recognize a gradation of possible relationships between two species of organisms on a schematic scale:

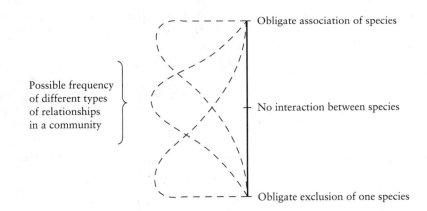

We must now ask where on this scale most of the species pairs in a community would fall. Another way of stating this question is to ask how frequently the distribution and abundance of one species is determined by interactions with other species. No one knows the answer to this at present, but we can make a few general statements.

Obligate associations may occur in certain parasites that have a single host species or certain animals that feed on only one species of plant. Few plants and only a small number of animals seem to have life cycles so tightly coupled to one other species in temperate and polar regions. Most species depend only partly on others. An insect may feed on one of several plant species, and predators may eat a variety of prey species. Partial dependence of this type seems most common in nature and grades off into a state of indifference in which species do not interact (Whittaker 1962). For example, Figure 20.1 shows that a majority of the plant species pairs in *Juncus effusus* stands show no evidence of interaction.

In tropical regions, particularly in species-rich tropical rain forests, the importance of obligate associations in less clear. The taxonomy of many species groups is unknown, and for the groups that have been properly classified there are few ecological studies to determine the sensitivity of the system to species removals.

To understand how communities operate, we need to understand the functional linkages between species. One way to do this is to define *guilds*, or groups of functionally equivalent species (see Chapter 24, page 551). The key question for conservation biology is how much redundancy there is in the biological community. If one species is lost in a community, can another take its place?

How much dynamic integration is present in a community is a critical question that arises again and again in community ecology and will be addressed in the next eight chapters. Whittaker (1962) suggests that if all species interactions were known, the distribution would be bell-shaped, with most species hovering around the middle (no interaction) and a few species at each extreme of obligate association and dissociation. If this is true, the relationships between species populations might not be strong enough to organize all species into a well-defined community and there may be considerable ecological redundancy.

Hidden in much of this discussion is an unavoidable bias. Most of the work on communities has been done in the north temperate zone; when tropical communities are studied in more detail, we may have to revise our generalizations. Tropical communities contain many more species than temperate communities (Chapter 23), and species interactions may more often be obligate in tropical areas.

The present view of the nature of the community lies closer to Gleason's individualistic view than to Clements's superorganismic interpretation. Species are distributed individualistically according to their own genetic characteristics. Populations of most species tend to change gradually along environmental gradients. Most species are not in obligatory association with other species, which suggests that associations are formed with many combinations of species and vary continuously in space and in time. To classify such associations into discrete units is a highly artificial undertaking, and attention has turned from classification of communities to analyzing community dynamics and functional organization.

A historical footnote serves to emphasize the conclusion of the individualistic nature of the community. Historical changes in vegetation can be interpreted in some detail by the use of fossil pollen grains in lake sediments. If we reconstruct the forest history of an area such as Minnesota (Wright 1968), we find a continuous series of species coming and going. Some modern forest communities have no analog in the past, and conversely, some associations found in the past do not exist anywhere at the present time. Historical evidence supports the view that we reached from analysis of modern communities.

Fossil pollen studies have proven very successful at describing the sequence of plant communities over the past 30,000 years (Pielou 1991). Because of the Pleistocene ice sheets in the Northern Hemisphere, climatic and vegetational shifts occurred around the globe. Figure 20.13 shows the pollen record of an area of the eastern United States about 500 kilometers south of the line of maximal glacial advance (Whitehead 1973). In this bay, about 5 meters of sediment has been deposited over the past 30,000 years. From 30,000 to 21,000 years before the present

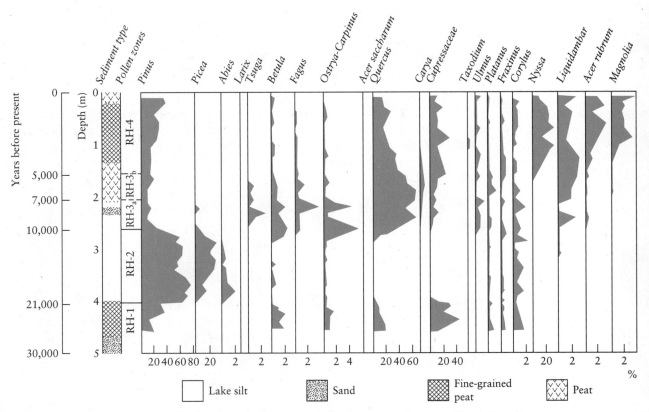

Figure 20.13   Pollen diagram for fossil tree pollen in Rockyhock Bay, northeastern North Carolina. Approximate ages are based on radiocarbon dating. (After Whitehead 1981.)

(B.P.), temperate forests of oaks, birches, and pines occupied the area. From 21,000 to 10,000 years B.P., boreal forest was present, with spruce and northern pines, and the climate was colder and drier than at present. Deciduous forests, with a predominance of oak, replaced the boreal forests about 10,000 years ago. Swamp forests began to develop about 7,000 years B.P., featuring black gum, cedar, magnolia, and red maple. The swamp forests were essentially modern by 4,000 years ago.

Pollen studies of the type shown in Figure 20.13 are important because they have established that the displacement of species by the ice sheet was individualistic—associations of species did not move north and south as a unit, but each species moved independently (Whitehead 1981). The displacement of boreal species was approximately 1300 kilometers from their present locations. Modern plant communities in the temperate zone have not existed in their present form for very many years.

How stable are communities? Fossil pollen studies clearly show changes over geological time, but do these occur over ecological time? Modern community ecologists are divided over the issue of whether communities should be considered *equilibrium* or *nonequilibrium* systems (Schoener 1987b). Closely related to this issue is whether we study a community as an *open* or a *closed* system. *Closed* systems consist of a single homogeneous patch of habitat, while *open* systems are collections of patches, connected by dispersal. *Equilibrium* models are restricted to the behavior of a system near an equilibrium point, while *nonequilibrium* models consider the transient behavior of a community as it moves from one state to another (DeAngelis and Waterhouse 1987). Many of the models we discussed in the population chapters are closed, equilibrium models; the competition equations of Lotka and Volterra (page 231) and the predation equations (page 265) are two good examples. *Stability* is a dominant preoccupation of equilibrium models.

But one characteristic of natural communities is change. Communities are open systems, and a general property of open systems is that transient, nonequilibrium conditions may persist for long periods of time. The critical question thus becomes, How long does it take a community to reach equilibrium? If there are many linkages between the species in a community, equilibrium may never be achieved. Thus it may be more useful to view communities as open, nonequilibrium systems (Paine 1983, DeAngelis and Waterhouse 1987). We will discuss this issue in detail in Chapter 25.

CLASSIFICATION OF COMMUNITIES

The problems of classifying communities are interwoven with the question of the nature of the community. If communities were discrete entities like species, we should be able to construct a taxonomy of communities. But if they vary continuously and are not discrete units, we can still classify them, provided we can use a variety of classification systems, since no single scheme is the "natural" classification. Classification is more for human convenience than for delimiting the true structure of nature.

Goodall (1963) has proposed a geometric model to conceptualize the problem we face: First, construct a geometric system in which each species in the community is one axis; then measure the abundance of the species in some manner, such as percentage cover. If we have two species in the community, we have a two-dimensional model, and this oversimplified version is shown in Figure 20.14. If we

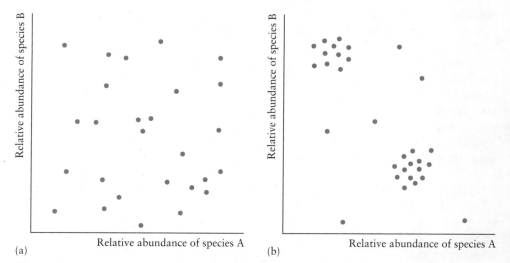

Figure 20.14 "Vegetation space" to illustrate two different conceptions of vegetation, one (a) with no clustering and the other (b) with clear clustering into compact clouds of points ("nodes") representing discrete communities.

have 200 species, we have a 200-dimensional community hyperspace, which is impossible to picture (so we think of this space in terms of two or three dimensions to get the illustration). Every stand we sample will now be a point in this community space, and we need to consider how these points are arranged. Do they form clouds of points separated by mostly empty space? Or are the points continuously distributed throughout the space? Regardless of how the points are scattered in space, we can divide the space up into little boxes. One's view of the nature of the community affects the way one interprets these little boxes.

We can use many different methods to classify communities. Three choices immediately confront a student who is attempting to classify a series of items (Pielou 1969, chap. 19).

1. Should the classification be *hierarchical* or *reticulate*? In a hierarchical classification, the familiar one of taxonomy, levels are subclasses of higher levels. In a reticulate classification, the groups are defined separately and linked in a network:

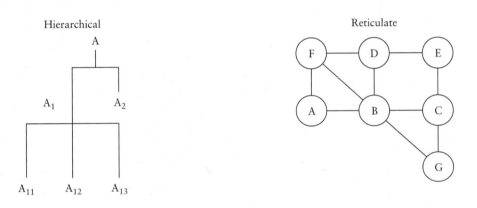

2. Should the classification be *divisive* or *agglomerative*? In a divisive classification, we begin with the whole and divide it into parts. In an agglomerative classification, we begin with the parts and add them together into classes, combining and recombining them into more inclusive groups.

3. Should the classification be *monothetic* or *polythetic*? In a monothetic classification, two closely related groups are distinguished by a single attribute. In a polythetic classification, two groups are distinguished on the basis of a number of attributes.

A variety of statistical methods have now been developed for use in classification. These are reviewed by Pielou (1984) and by Digby and Kempton (1987).

Two different traditions of classification have predominated among plant ecologists. In continental Europe, the classification system of Braun-Blanquet (1932) has been applied. As noted earlier, this system relies on diagnostic species as the way to define and distinguish community types and produces a hierarchical classification of communities. By contrast, the system used by most British and American ecologists is an informal one based on "dominance types." A *dominance type* is a group of stands with the same dominant species. The dominant species are usually the large and conspicuous ones and are almost the antithesis of the diagnostic species of Braun-Blanquet. Dominant species often have a distribution that extends well beyond the limits of their dominance types. Since there is often little association between the

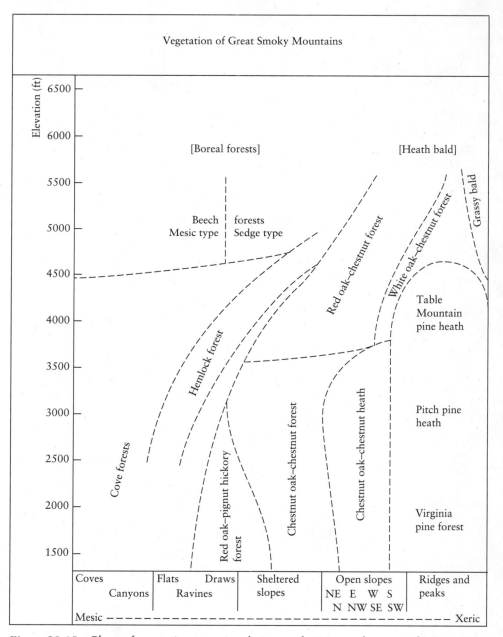

Figure 20.15 *Chart of vegetation types in relation to elevation and topography in the Great Smoky Mountains, Tennessee. The vertical axis is the complex gradient of temperature and other factors related to elevation; the horizontal axis is the complex gradient of moisture relations and other factors from moist or mesic situations on the left to dry or xeric on the right, as affected by topographic position. (After Whittaker 1956.)*

dominant species and the understory plants and animals, communities defined by dominance types have an artificiality to them. The major value of this informal system is that it is flexible and applicable to practical problems in forest and wildlife management.

Figure 20.15 illustrates the use of dominance types in describing the vegetation of the Great Smoky Mountains of Tennessee. Even though the species are distributed continuously along gradients of moisture and temperature, it is useful to describe the pattern of vegetation types. The continuum idea and the classification of communities are not incompatible concepts.

SUMMARY

A community is an assemblage of populations living in a prescribed area. A number of parameters, such as species diversity, can be estimated in a community, but their interpretation depends on the nature of the community. Two opposing schools have developed in plant ecology over the question of the nature of the community. The organismic school holds that communities are integrated units with discrete boundaries. The individualistic school holds that communities are not integrated units but collections of populations that require the same environmental conditions. The information available leans more toward the individualistic interpretation of the community. Communities are not discrete but grade continuously in space and in time, and species groups are not consistent from place to place. In spite of this continuous variation, communities can be classified, but this classification is for convenience and is not a description of the fundamental structure of nature.

Community ecologists analyze the dynamics of the community, how it is organized, the role of dominant and keystone species, and how it changes in time. By using our understanding of population ecology we can begin to build an understanding of how communities work.

Communities can be viewed as *open* or *closed* systems and as *equilibrium* or *nonequilibrium* systems. Closed, equilibrium models view community stability as the norm, while open, nonequilibrium models emphasize change as the norm. How ecologists view communities colors their approach in studying problems in community ecology.

Selected References

AUSTIN, M. P., and T. M. SMITH. 1989. A new model for the continuum concept. *Vegetatio* 83:35–47.

DAUBENMIRE, R. 1966. Vegetation: Identification of typal communities. *Science* 151:291–298.

GREIG-SMITH, P. 1979. Pattern in vegetation. *Journal of Ecology* 67:755–779.

LANGFORD, A. N., and M. F. BUELL. 1969. Integration, identity and stability in the plant association. *Advances in Ecological Research* 6:83–135.

McINTOSH, R. P. 1967. The continuum concept of vegetation. *Botanical Review* 33:130–187.

PIELOU, E. C. 1984. *The Interpretation of Ecological Data: A Primer on Classification and Ordination*. Wiley, New York.

ROUGHGARDEN, J., and J. DIAMOND. 1986. Overview: The role of species interactions in community ecology. In *Community Ecology*, ed. J. Diamond and T. J. Case, pp. 333–343. Harper and Row, New York.

WEBB, D. A. 1954. Is the classification of plant communities either possible or desirable? *Botanisk Tidsskrift* 51:362–370.

WIENS, J. A. 1984. On understanding a non-equilibrium world: Myth and reality in community patterns and processes. In *Ecological Communities*, ed. D. R. Strong Jr., D. Simberloff, L. G. Abele, and A. B. Thistle, pp. 439–457. Princeton University Press, Princeton, N.J.

Questions and Problems

1. Caswell (1978, p. 133) gives this example to illustrate how long it takes a system to reach equilibrium:

   A system consists of 100 light bulbs, each of which can be either on or off. For each bulb in each second the transition (on-off) occurs with probability 0.5. The reverse transition occurs with probability 0.5 only if at least one bulb in the on state is connected to the bulb in question. Otherwise no bulb will turn on. The equilibrium state of the system is all lights off. If the system begins with all lights on, guess how long it will take to reach equilibrium if:

   (a) there are no connections among the bulbs;
   (b) all the bulbs are connected to each other.

   Compare your guesses with the values in Caswell, and discuss the possible relevance for community ecology.

2. Whittaker (1967, p. 207) states that "gradient analysis and classification are alternative approaches to the vegetation of a landscape." Discuss your agreement or disagreement with this view.

3. Discuss the significance of ecotypes to the problem of community description and classification.

4. In discussing the classification of plant communities, Ashby (1948, p. 223) states:

   When the ecologist stops his car and decides that he has reached a "suitable place" for throwing quadrats on a community, he has already performed the major act of classification and he has performed it subjectively. Any subsequent quantitative analysis only elaborates, and possibly obscures, the original subjective decision.

   Discuss this claim.

5. Morey (1936, p. 54) studied two stands of virgin forest in northwestern Pennsylvania and found 93 species of plants common to both stands, 91 species confined to Heart's Content stand, and 57 species confined to Cook Forest stand. Calculate the index of similarity for these two virgin stands.

6. Construct an ordination of the following hypothetical data. Assign climax adaptation numbers to the eight species and plot the resulting gradient analysis
   as in Figure 20.5.

Stand	IMPORTANCE VALUE OF SPECIES							
	A	**B**	**C**	**D**	**E**	**F**	**G**	**H**
1	150	20	50	10	5	5	30	30
2	10	70	10	—	60	20	30	100
3	20	40	20	—	10	—	200	10
4	—	10	—	90	70	120	—	10
5	20	110	20	10	30	10	60	40
6	260	—	30	—	—	—	10	—
7	—	10	—	20	200	30	—	40
8	—	60	—	—	50	—	—	190
9	—	10	—	130	60	80	—	20
10	70	10	160	—	—	—	50	10
11	—	30	10	20	100	80	—	60
12	—	—	10	—	270	10	—	10

7. Discuss the relative advantages and disadvantages of field experiments and
   natural experiments for analyzing community dynamics. Natural experiments
   differ from field experiments in that the ecologist does not do a manipulation
   but instead selects sites where a manipulation is already running or has run
   (Diamond 1986). For example, natural experiments could examine the recovery of the vegetation after a volcanic explosion or a fire. Compare your analysis of field experiments and natural experiments with that of Diamond (1986).

8. An underlying assumption of many discussions of the response of plants to
   gradients (e.g., Figures 20.4 and 20.5) is that the response curves are symmet-
   ric, bell-shaped curves. What mechanisms might make this assumption questionable? Austin (1990) discusses this question.

Overview Question

What is the relationship between the concept of the ecological niche and the
concept of the community? Does the controversy over the nature of the community have any important consequences for niche theory?

21

Community Structure

Community structure can refer to the physical structure or to the biological structure of a community. We are concerned in this chapter primarily with the physical structure, which is essentially what we see when looking at a community. For example, when we visit a deciduous forest, we see a primary structure imposed by large trees, which lose their leaves seasonally, and a secondary structure in understory trees and shrubs and herbs on the forest floor. The forest soil forms the matrix for root interactions of all these plants, and the animals of this community distribute themselves within the structure defined by the plants and soil.

Another aspect of community structure is biological structure, and we will consider this aspect in Chapters 22 through 25. Biological structure involves species composition and abundance, temporal changes in communities, and relationships between species in a community. The biological structure of a community depends in part on its physical structure.

Both aspects of community structure in turn have strong influences on the functioning of a community. Function refers to how a community works as a processor of energy and nutrients, and we will consider these aspects of community ecology in Chapters 26 through 28. Communities function by means of an intricate network of species interactions, and the structure and functioning of biological communities result from a summation of the types of species interactions we described in Chapters 13 through 16.

We must remember that within this framework of community ecology, communities are integrated by the coevolution of groups of interacting species. Both the structure and the functioning of a community have been modified by natural selection acting on the individuals that make up the community.

In this chapter we will discuss three components of the physical structure of communities. Plants form the basic biological matrix of all communities, and the *growth forms* of plants are an important component of community structure. Aquatic and terrestrial systems differ greatly in their obvious structure, but many aspects of *vertical stratification* are common to both types of communities. *Seasons* change the structure of all communities, even in tropical areas, and seasonal events are critical to the functioning of natural communities.

GROWTH FORMS

Before we learned about plant taxonomy, we classified plants according to *growth forms*, which are just different classes of visible structure in plants. Trees are one growth form of plants; grasses are another. A number of characteristics of plants are used to define growth forms: There are tall and short plants, woody and nonwoody plants, evergreen and deciduous plants. Growth form can be further classified by the

Table 21.1   Major Plant Growth Forms on Land

TREES, larger woody plants, mostly well above 3 m tall

Needle-leaved (mainly conifers—pine, spruce, larch, redwood, etc.)
Broadleaf evergreen (many tropical and subtropical trees, mostly with medium-sized leaves)
Evergreen-sclerophyll (with smaller, tough, evergreen leaves)
Broadleaf deciduous (leaves shed in the temperate-zone winter or in tropical dry season)
Thorn trees (armed with spines, in many cases with compound, deciduous leaves)
Rosette trees (unbranched, with a crown of large leaves—palms and tree-ferns)

LIANAS (woody climbers or vines)

SHRUBS, smaller woody plants, mostly below 3 m in height

Needle-leaved
Broadleaf evergreen
Broadleaf deciduous
Evergreen-sclerophyll
Rosette shrubs (yucca, agave, aloe, palmetto, etc.)
Stem succulents (cacti, certain euphorbias, etc.)
Thorn shrubs
Semishrubs (upper parts of the stems and branches dying back in unfavorable seasons)
Subshrubs or dwarf shrubs (low shrubs spreading near the ground surface, less than 25 cm high)

EPIPHYTES (plants growing wholly above the ground surface on other plants)

HERBS, plants without perennial aboveground woody stems

Ferns
Graminoids (grasses, sedges, and other grasslike plants)
Forbs (herbs other than ferns and graminoids)

THALLOPHYTES

Lichens
Mosses
Liverworts

Source: After Whittaker (1975).

shapes of leaves, the form of stems, and the design of the root system. Table 21.1 lists the major plant growth forms on land. Growth forms of plants can be used as the basis of a classification system. Instead of being concerned with the species composition of a plant community, we can use its visible structure as a basis of classification (Beard 1973). The term *formation* was applied very early to vegetation of a fixed growth form, such as a meadow or a forest (Warming 1909). Formations can be defined very broadly or very narrowly and thus can be the basis of a flexible classification scheme. If we use only the major growth form, we can classify the world's vegetation into a few formations. Detailed local studies of growth forms can be useful for setting up finer classification schemes.

The great strength of the formation concept is that we can recognize *ecological equivalents* in widely separated parts of the world. Growth forms reflect environmental conditions, and similar conditions produce similar plant forms by convergent evolution. Desert plants, for example, have evolved a series of morphological features

Table 21.2 Major Formations and Their Environmental Correlates

1. TROPICAL RAIN FOREST

   Occupies regions of high and constant rainfall and temperature. The forest is many-layered, leaves are mainly evergreen, large, entire; trees tall and buttressed; epiphytes and lianas very common. The flora is very rich. Amazonia, Congo Basin, Malaysia.

2. SUBTROPICAL RAIN FOREST

   Is found in humid subtropical regions with some seasonal variation in temperature and rainfall. Luxuriance in structure and composition is reduced. Brazil, African highlands, Southeast Asia.

3. MONSOON FOREST

   Tropical and subtropical with a moderate winter dry season. Forest.is tall, many-layered, with predominance of deciduous species in the canopy. Central America, India, Southeast Asia.

4. TEMPERATE RAIN FOREST

   Expresses high and constant rainfall in cooler regions. Forest is moderately tall, dense, few-layered; leaves are evergreen, small, or coriaceous. Much moss and lichen. A variant, montane rain forest or cloud forest, is found on tropical mountains. Tasmania, New Zealand, Chile.

5. SUMMER-GREEN DECIDUOUS FOREST

   Occupies regions with a pronounced seasonal change of temperature, a cold winter with snow and a mild to warm, wet summer. Trees are tall, structure simple, leaves broad, fine, and deciduous. Eastern North America, Europe, China.

6. NEEDLE-LEAVED FOREST

   Is characteristic of cold regions with long winters and high rainfall. Trees are coniferous, needle- or scale-leaved, and may be very large in size. Western North America, northern Europe, Siberia.

7. EVERGREEN HARDWOOD FOREST

   Characterized regions of "mediterranean" climate with a dry summer and wet, mild winter. Trees are small (except in Australia) and leaves sclerophyllous. Australia, California, Mediterranean.

8. SAVANNA WOODLAND

   Appears under a summer rainfall with a long dry season, i.e., more extreme than monsoon forest. Trees are small, evergreen, in open formation, with a ground layer of tropical bunchgrasses. Brazilian and African plateau, northern Australia.

9. THORN FOREST AND SCRUB

   Tropical, dry climates. Trees are small, often thorny, and deciduous. The ground layer includes many succulents, annuals, and grasses. Brazil, Africa, India.

10. SAVANNA

    Is a moist tropical grassland, with or without trees, and may owe its origin to fire or to adverse soil conditions or both. Pantropical.

11. STEPPE AND SEMIDESERT

    Occur in dry climates with winter rainfall, i.e., more extreme that evergreen hardwood forest. Open shrublands with annual herbs and grasses, or dry grasslands. North America, Australia, Russia, Argentina.

Table 21.2   *(continued)*

12. HEATH

Like the tropical savanna, the heath in temperate regions is governed by fire, adverse soil conditions, or both. It is a formation of ericoid shrubs with scattered larger shrubs and small trees. Worldwide, locally.

13. DRY DESERT

Warm regions of very low rainfall with open vegetation and special plant forms evolved in different parts of the world, e.g., succulent Cactaceae in North America; succulent Liliaceae, Aizoaceae, Euphorbia, and Welwitschia in southern Africa; hummock grasses in Australia.

14. TUNDRA AND COLD WOODLAND

This is the semidesert of cold regions where there is a short summer growing season. Lichens are especially abundant under sedges and grasses (tundra) or under stunted trees. On rocky areas, mosses may be dominant. Northern Hemisphere in high latitudes.

15. COLD DESERT

Edge of icecaps, glaciers, and permanent snowfields. Vegetation sparse, mainly herbaceous.

Note: These formations are not confined to any one part of the world and are classified by their growth forms. Some environmental correlates of each formation are given in the table. Major formations were recognized by Schimper and von Faber (1935).

like small leaf size that reduces heat loads and water loss, and these adaptations can be found in different families of plants in North America, Africa, and Australia (Gates et al. 1968). Similar growth forms recur in different regions of the world. Schimper (1903) recognized this principle of plant geography over 80 years ago and suggested the broad classification of formations given in Table 21.2.

If formations do indeed reflect environmental conditions, we should be able to predict the structure of the vegetation once we know the important environmental factors. Figure 21.1 shows one attempt to map the formations of the world on scales of temperature and precipitation, the two master limiting factors of vegetation. Secondary influences like fire and type of soil may shift the boundaries shown in Figure 21.1. There is a zone of moderate precipitation in which grassland or woodland may predominate, and the dynamics of these communities cannot be summarized in a simple temperature-and-precipitation graph.

Formations can be seen to change dramatically as one moves along strong environmental gradients in both temperate and tropical areas. Figure 21.2 illustrates four major gradients of plant communities along temperature and moisture gradients associated with altitude and latitude. Whittaker (1975) calls these *ecoclines* because they represent complex gradients of both vegetation and environmental factors on a global scale. As we move along a gradient from a favorable to an unfavorable environment, there is a decrease in the height of the dominant plants and in the percentage of ground surface covered. Each plant growth form has a characteristic place of maximal importance along these ecoclines.

The growth form of a plant is determined in part by its leaf structure. Leaves are critical for photosynthesis and thus subject to strong evolutionary pressures. One of the most basic observations we can make about plant form is that leaves of

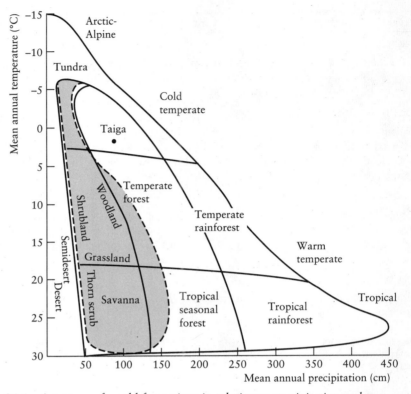

Figure 21.1   A pattern of world formations in relation to precipitation and temperature. Boundaries between formations are approximate. Soil and fire effects can shift the balance between woodland and grassland formations in the shaded region. (After Whittaker 1975.)

different plants vary in shape and size and in being deciduous or evergreen. Let us consider how natural selection operates to determine these leaf characteristics.

Two types of models have been used to predict optimal leaf size (Givnish and Vermeij 1976). The first models assumed that leaf size evolved so that it regulated leaf temperature near the optimum for photosynthesis and prevented thermal damage. These models have largely failed because the photosynthetic machinery itself can adapt to different temperature regimes (see pages 121–125). Thus some other factor determines leaf size, and the photosynthetic system can then adapt to the resulting leaf temperature.

More recent models of leaf size have been based on the observation that plants must pay for photosynthetic gas exchange by a concomitant loss of water through transpiration. Thus it might be optimal for a plant to maximize water-use efficiency, the ratio of photosynthesis to transpiration. This type of approach is similar to an economic model in which the profits-to-costs ratio is being maximized. We will now describe, in qualitative terms, a model of leaf size based on water-use efficiency.

Increasing leaf size in a sunny environment tends to increase leaf temperature and transpiration. Heat loss through convection is impeded in large leaves, and temperature consequently rises. Under shady conditions, the opposite effect occurs, and larger leaves cool below air temperature. The temperature of the leaf affects the

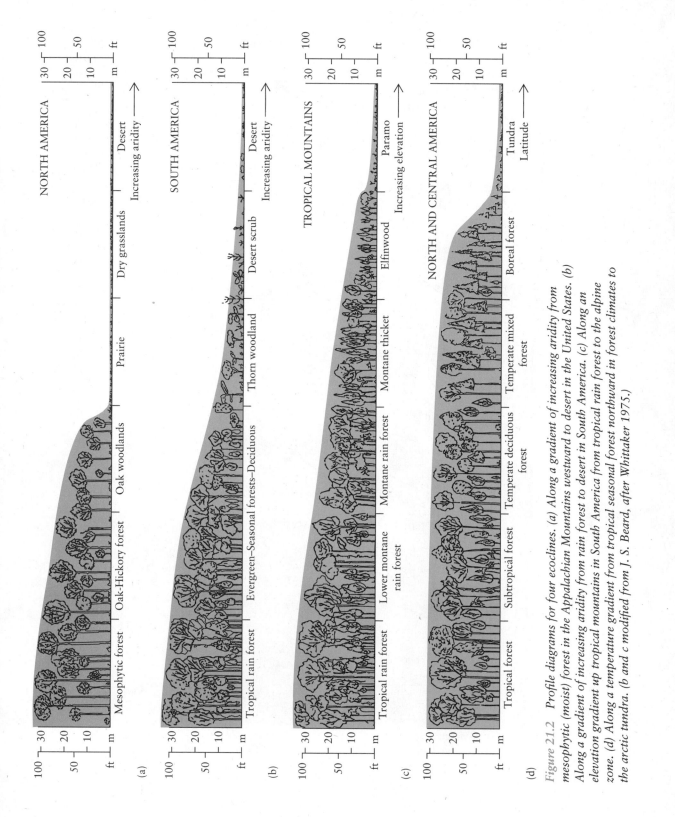

NORTH AMERICA

Mesophytic forest | Oak-Hickory forest | Oak woodlands | Prairie | Dry grasslands | Desert

Increasing aridity →

(a)

SOUTH AMERICA

Tropical rain forest | Evergreen–Seasonal forests–Deciduous | Thorn woodland | Desert scrub | Desert

Increasing aridity →

(b)

TROPICAL MOUNTAINS

Tropical rain forest | Lower montane rain forest | Montane rain forest | Montane thicket | Elfinwood | Paramo

Increasing elevation →

(c)

NORTH AND CENTRAL AMERICA

Tropical forest | Subtropical forest | Temperate deciduous forest | Temperate mixed forest | Boreal forest | Tundra

Latitude →

(d)

Figure 21.2 Profile diagrams for four ecoclines. (a) Along a gradient of increasing aridity from mesophytic (moist) forest in the Appalachian Mountains westward to desert in the United States. (b) Along a gradient of increasing aridity from rain forest to desert in South America. (c) Along an elevation gradient up tropical mountains in South America from tropical rain forest to the alpine zone. (d) Along a temperature gradient from tropical seasonal forest northward in forest climates to the arctic tundra. (b and c modified from J. S. Beard, after Whittaker 1975.)

rate of photosynthesis, but photosynthesis reaches a plateau at higher temperatures when gas exchange limits photosynthesis. Transpiration rates are not limited in a leaf as long as water is available, but they continue to rise as temperature goes up. Figure 21.3 illustrates these arguments graphically. Natural selection will favor the leaf area corresponding to the maximal (profit-cost) point, as indicated in Figure 21.3a.

This general model can be adapted to a variety of situations. Figure 21.3 illustrates two examples. In an arid environment, the photosynthetic curve is not changed, but the cost of transpiration increases greatly because of the difficulty of getting water. The result (Figure 21.3b) is that leaf size should decrease in arid environments. Conversely, in rich or moist soils, the rate of photosynthesis can be increased while

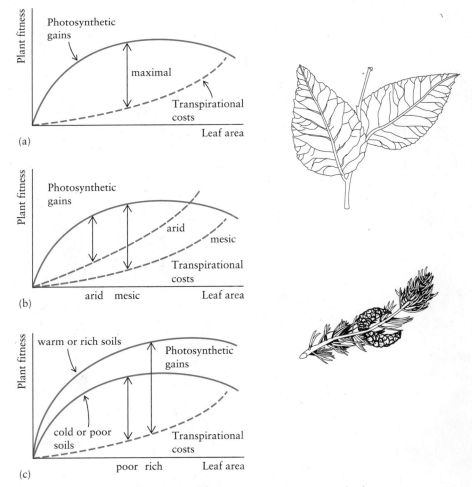

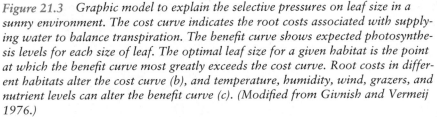

Figure 21.3 *Graphic model to explain the selective pressures on leaf size in a sunny environment. The cost curve indicates the root costs associated with supplying water to balance transpiration. The benefit curve shows expected photosynthesis levels for each size of leaf. The optimal leaf size for a given habitat is the point at which the benefit curve most greatly exceeds the cost curve. Root costs in different habitats alter the cost curve (b), and temperature, humidity, wind, grazers, and nutrient levels can alter the benefit curve (c). (Modified from Givnish and Vermeij 1976.)*

the cost of transpiration is changed very little. Hence the optimum leaf size increases for plants growing in rich or moist soils (Figure 21.3c).

Leaf data from plant communities support the broad predictions given by this model. Both desert and thorn-forest plants have small leaves because of high radiation and low water availability. Arctic plants also have small leaves.

The growth form of a plant is also strongly affected by leaf longevity. A tree or shrub is evergreen if it retains leaves throughout the year, whereas a deciduous tree or shrub sheds all its leaves and spends a portion of the year without foliage. Most ecologists have assumed that evergreen plants have leaves that live longer than leaves of deciduous plants, but this is not always true. Evergreen broad-leaved trees predominate in tropical forests, and as one moves into temperate zone forests deciduous broad-leaved trees increase in abundance. At still higher latitudes evergreen conifers predominate. How can we explain this global pattern of evergreen trees?

Figure 21.4 shows a simple model of leaf longevity. The productivity of a leaf decreases with the age of the leaf and for this reason there is an optimal time interval over which to replace the leaf (Kikuzawa 1991). Every tree, whether deciduous or evergreen, should replace leaves this way. From the graphic analysis in Figure 21.4 you can see that leaf longevity should be longer when the construction cost of the leaf is higher. Leaf longevity should also be longer when the photosynthetic rate decreases only slowly with leaf age. In a seasonal environment with a period unsuitable for plant growth, a tree will shed all its leaves if the optimal longevity from

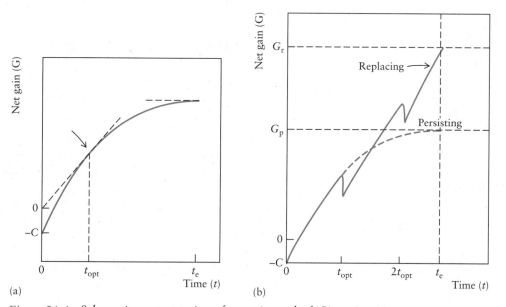

Figure 21.4   *Schematic representation of net gain per leaf (G) to time (t) curve. (a) Net gain at time zero is minus construction cost ($-C$) and increases at first rapidly and then gradually because of decrease of photosynthetic rate with time by aging. To maximize net gain by a tree the leaf must be replaced when the tangent line starting from the origin touches the curve (t_{opt}). To replace the leaf when the daily net gain is zero (t_e) does not maximize net gain of the tree. (b) A comparison of net gain of a tree by replacing and by persisting leaves. The net gain of a tree (G_r) by replacing the leaf at $t = t_{opt}$ is greater than the net gain (G_p) of a tree by retaining the leaf until $t = t_e$. (From Kikuzawa 1991.)*

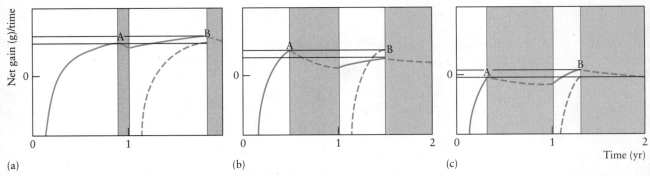

Figure 21.5 Schematic illustration of how both very favorable and very severe environments can favor evergreen leaves. These figures assume the leaf-gain model of Figure 21.4. A = net gain achieved by a persistent leaf at the end of its first growing season, B = net gain at the end of the second growing season. The unfavorable, nongrowing season is stippled. (a) With a long growing season, evergreen leaves are favored (B > A). (b) With an intermediate growing season, the loss during the non-growing season cannot be overcome in the next growing season (B < A) and deciduous leaves are favored. (c) With a short growing season, the leaf construction costs can barely be recovered during the first growing season, and plants become evergreen (B > A). These three models correspond to conditions in tropical rain forest, temperate deciduous forest, and subarctic conifer forest. Dashed lines show the hypothetical gain that would accrue by adding new leaves in year 2. (Modified from Kikuzawa 1991.)

Figure 21.4 is less than the length of the favorable period for growth. Such trees are considered deciduous trees. When the optimal longevity is greater than one year, the tree will be considered an evergreen.

In an environment with a long growing season and a short unfavorable season, evergreen leaves will be favored (Figure 21.5a) because this results in a higher net gain to the plant. This should occur in tropical rainforests and temperate rainforests. As the length of the favorable period shortens and the unfavorable period lengthens, deciduous leaves will be favored (Figure 21.5b). But as the favorable period shortens even further and the unfavorable period lengthens (Figure 21.5c), a plant cannot recover the construction costs in one growing season and thus evergreen leaves are favored in the polar regions.

These predictions from the simple model of Figure 21.4 agree with the observed pattern of leaf types—evergreen leaves predominate in tropical and subtropical regions and also in subarctic regions (Kikuzawa 1991, Chabot and Hicks 1982). This model also explains the prevalence of coniferous trees in the rainforests of the Pacific Northwest region of the United States and Canada, an area with a long growing season and a short unfavorable season.

Related to this pattern of leaf types is a much broader evolutionary problem—the decline of conifers in the fossil record. During the middle and late Cretaceous (60 to 100 million years ago), angiosperms replaced conifers over much of the earth (Bond 1989). Most paleobotanists presume that angiosperms are competitively superior to conifers, and thus replaced conifers over evolutionary time. The seedlings of conifers grow slowly in comparison with angiosperm seedlings, and competition during regeneration is postulated as the explanation of angiosperm dominance in the world today. Conifers persist only in habitats where there is restricted competition in the seedling stage (Bond 1989). Because conifers typically are evergreen, they accumulate more leaf biomass than deciduous angiosperms and this allows more rapid growth in older conifer trees, compared with deciduous angiosperms. The Achilles heel of conifers does not seem to be in the adult tree but in the seedling stage when they suffer from competition with fast-growing angiosperms.

VERTICAL STRUCTURE

Most communities show vertical structure, or stratification, but the source of vertical structuring is different in aquatic and in terrestrial systems. In both cases, vertical layering is associated with a decrease in light. Figure 21.6 illustrates the stratification common in forests throughout the world. In dense forests, less than 1 percent of the incident sunlight reaches the forest floor. Competition for light must be a critical factor in determining forest stratification.

Competition for light may occur whenever one plant casts a shadow on another, or within a single plant when one leaf shades another leaf. Competition for light has been studied most intensively by agricultural scientists, because when crop plants have plenty of water and nutrients, light becomes the main factor that limits production. Light is a peculiar resource because it is available instantaneously and must be intercepted instantaneously or be lost. The successful plant is not just the plant with the most foliage but the plant with its foliage in the best position for light interception. In many cases, height is critical for advantageous light interception (Waller 1986).

One important concept for studying vertical stratification in terrestrial vegetation is the *leaf area index*. The leaf area index is the ratio of total leaf surface to total ground surface. A leaf area index of 2.0 would mean that if you measured all the leaves hanging over 1 square meter of ground, you would have 2 square meters of

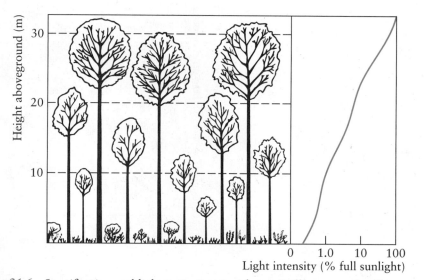

Figure 21.6   Stratification and light extinction in a forest. Different species of trees, shrubs, and herbs bear foliage at different heights above the ground (left) and are adapted to life at the different light intensities (right) that result from the absorption of sunlight by that foliage. (After Whittaker 1975.)

leaf surface. As the leaf area index increases, the stage is reached at which the lowest leaves in the vegetation cannot get enough light to produce photosynthate and therefore die.

Leaves, of course, do not lie in neat layers. Light penetrates the leaf canopy of a forest as sun flecks and is reflected from leaf to leaf. Very little light passes through leaves, and the general pattern of falloff in diffuse light is shown in Figure 21.6. Two extreme strategies of leaf arrangement can be recognized (Horn 1971). The *monolayer* arrangement places all leaves in one continuous layer. The *multilayer* arrangement places leaves in a loose scatter among several layers. An extreme monolayer has a leaf area index of 1.0 and casts a deep shade. A monolayer is more efficient at low light levels and hence should occur in the understory of a forest. A multilayer is more efficient at high light levels and should occur in the canopy of a forest. Horn (1971) measured the number of layers in an oak-hickory forest in New Jersey and found agreement with these predictions:

	Mean No. of Layers of Leaves
Canopy trees	2.7
Understory trees	1.4
Shrubs	1.1
Ground cover	1.0

Leaf height varies greatly in different herbs. What evolutionary factors determine how high an herb should grow? The most important gain from being taller is that a tall plant intercepts more light. Forest herbs typically grow in habitats where light

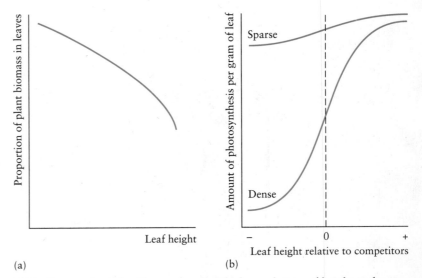

(a)   (b)

Figure 21.7 Economic tradeoffs associated with the evolution of height in forest herbs. (a) Taller plants must make more support tissue to remain mechanically stable. Consequently, they have less resources to use for leaves. (b) Taller plants have an expected photosynthetic advantage of holding leaves above a competitor. This advantage is smaller in forests with sparse cover than in forests with dense herbaceous cover. (After Givnish 1982.)

intensity limits photosynthesis (Givnish 1982). Balanced against this gain from being taller is a cost associated with the stem support a plant must make to be able mechanically to grow tall. Figure 21.7 illustrates these economic tradeoffs that affect how tall a plant should be. Is this simple model an adequate predictor of plant heights?

Figure 21.8a shows that the fraction of herb biomass in green leaves decreases as plants grow taller. Leaf height of herbs is also closely related to the amount of herbaceous cover in forests (Figure 21.8b). Leaf height roughly doubles with each 7 percent increase in herbaceous cover. Herbs growing in areas with little cover are unlikely to be shaded by a competitor and reap no benefit from being taller. Conversely, in productive communities, herbs grow taller because of competition for light. The results for forest herbs thus confirm the predictions from the model shown in Figure 21.7.

Vertical structure in aquatic systems is provided by the physical properties of water, and this contrasts dramatically with the structure imposed by plants on terrestrial communities. Water changes its density with temperature and salinity, and these physical properties of water result in considerable structuring in aquatic environments. Freshwater lakes typically stratify thermally during the summer months (Figure 21.9) and often mix during the fall or the winter. Tropical lakes maintain a year-round stratification.

Light is absorbed by water, and hence light intensity decreases sharply with depth in all water bodies, even if there are no plants to intercept the radiation. Plants that live in the open zones of lakes and in the open ocean (phytoplankton) are all small but vary greatly in size and shape. Phytoplankton face an additional problem not faced by land plants—how to float. The density of most freshwater phytoplankton organisms is 1.01 to 1.03 times that of water, so they sink slowly when placed in undisturbed water (Hutchinson 1967). Different shapes of phytoplankton sink

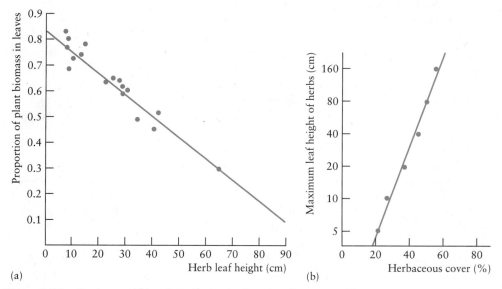

Figure 21.8 *Evolution of height in forest herbs. (a) Allocation of biomass to leaves versus support tissue in 18 species of forest herbs. Compare with figure 21.7a. (b) Relationship of average leaf height of herbs to the amount of herbaceous cover in a transect from dry oak woods to floodplain forest in Virginia. (After Givnish 1982.)*

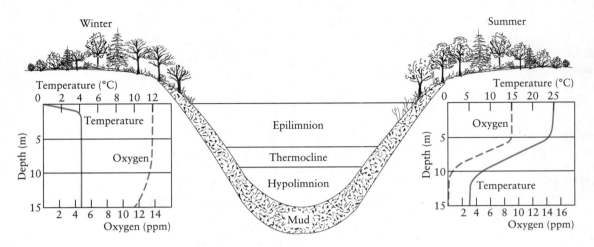

Figure 21.9   *Thermal stratification in a north temperate lake (Linsley Pond, Connecticut). Summer conditions are shown on the right, winter conditions on the left. Note that in the summer a warm, oxygen-rich circulating layer of water, the* epilimnion, *is separated from the cold, oxygen-poor* hypolimnion *waters by a broad zone, called the* thermocline, *characterized by a rapid change in temperature and oxygen with increasing depth. (After Deevey 1951.)*

more or less slowly. A cylinder, for example, falls more slowly than a sphere of similar volume.

The phytoplankton are typically concentrated in the upper part of the water column in both freshwater lakes and in the open ocean. This vertical structure can be maintained only by the turbulence of the water, which offsets the sinking tendency of the phytoplankton. Sinking slowly may be advantageous to the phytoplankton because sinking may allow more rapid nutrient uptake and easier waste disposal. The surface-to-volume ratio of a phytoplankton organism also affects the rate of nutrient and waste transfer in the aquatic medium, and selection for particular surface-to-volume ratios must restrict the shape of cells of a given biomass or restrict the biomass of cells of a given shape. Lewis (1976) showed that the observed ratios of surface to volume were restricted to a small range compared with ratios expected by a random choice of shapes (Figure 21.10). The suggestion is that the shapes of phytoplankton are a result of natural selection acting to minimize sinking and maximize surface-to-volume ratios.

Zooplankton are not confined to the lighted zone of lakes or oceans, and they exhibit a more diverse vertical structuring than the phytoplankton. Because many zooplankton can swim, the vertical distribution is not constant. Many species exhibit *vertical migrations*, in which individuals typically rise at night from deeper waters into the upper parts of the lake or ocean. A few species show reverse migration, to the surface by day, but this is unusual (Hutchinson 1967). The distance of migration is highly variable in different species.

Vertical migrations of over 100 meters are commonly undertaken each day by euphausiids (Mauchline and Fisher 1969). Euphausiids (commonly called *krill*) are marine crustaceans that are a major component of marine food chains in the polar regions. Many whales and commercially important fish, such as herring and mackerel, feed on euphausiids. Euphausiids live in large concentrations, and most species migrate vertically in response to light. Figure 21.11 shows an example of vertical migration in one species off the Scottish coast. Individuals migrate in close correlation

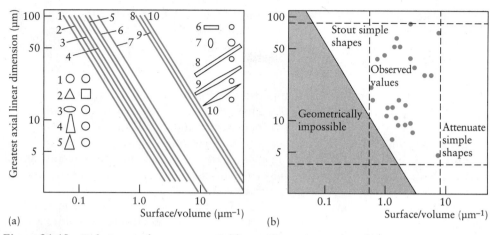

(a)   (b)

Figure 21.10   Relation to the greatest axial linear dimension and surface-to-volume ratio for (a) selected simple geometric solids and (b) the 25 most abundant phytoplankton species of Lake Lanao, Philippines. The geometric shapes in (a) are shown from perspectives perpendicular to the longest and shortest axes. Figures shown are: 1, sphere; 2, pyramid; 3, prolate ellipsoid; 4, stout cone frustrum; 5, stout cone; 6, stout cylinder; 7, oblate ellipsoid; 8, slim cylinder; 9, slim double-cone frustrum; and 10, slim double cone. (After Lewis 1976.)

with light—up to the surface in darkness and down to the bottom in the daylight. This rhythm is maintained throughout the year and results in a much shorter stay at the surface during the summer. Vertical migration is achieved by active swimming. *Meganyctiphanes norvegica* can swim upward at an average speed of 90 meters per hour and downward at 130 meters per hour.

Light is the proximate cause for vertical migrations in zooplankton. What is the selective advantage of the migrations? We should note, first of all, that migrations are a common phenomenon among the higher vertebrates, such as birds. The selective value of migration in vertebrates is that individuals move from a habitat that is becoming unfavorable for survival or reproduction toward another habitat that is more favorable for survival or reproduction. In birds, the fall migration ensures survival, and the spring migration improves reproductive potential.

Several hypotheses have been suggested to explain the adaptive value of vertical migration for aquatic organisms. The food supply of many zooplankton species is

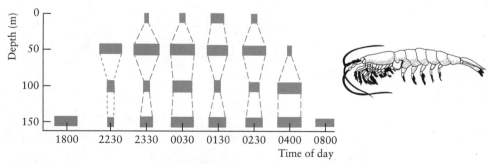

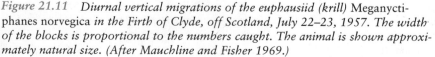

Figure 21.11   Diurnal vertical migrations of the euphausiid (krill) Meganyctiphanes norvegica *in the Firth of Clyde, off Scotland, July 22–23, 1957. The width of the blocks is proportional to the numbers caught. The animal is shown approximately natural size. (After Mauchline and Fisher 1969.)*

concentrated in the surface layers of the water, and thus vertical migration may increase food consumption. This is probably true for all species, but the critical question of why zooplankton leave the food-rich parts of the water column and migrate down to a poor feeding area remains unanswered. Why not stay near the surface all the time?

One good reason for vertical migration is to avoid visual predators that hunt by day in the surface waters. Whales and fish feed on krill in the surface waters, and migration to deeper waters probably increases an individual's chances of surviving and reproducing. The avoidance of predation is often claimed to be the major adaptive value of vertical migration (McLaren 1963) but may not operate in all cases. Moving into deeper waters may in some cases expose the animals to a whole new set of predators not present in surface waters.

McLaren (1974) suggested that the advantage of vertical migration is in the demographic advantage that animals gain from lowered metabolism in colder waters. McLaren's hypothesis relies on the observation that body size in zooplankton is larger if animals grow up in colder water. Since fecundity is related to body size in that larger individuals lay more eggs, migratory individuals will be at a selective advantage over nonmigratory individuals, but two problems may offset this gain. Respiration costs must decrease at the lower temperatures if a metabolic gain in growth is to be realized. Also, development time is increased as temperature falls, so

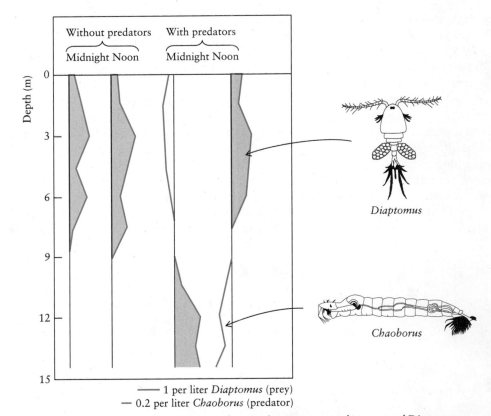

Figure 21.12 *Experimental induction of vertical migration in the copepod* Diaptomus kenai *in large enclosures in a lake. When the predatory phantom midge* Chaoborus *was added to an enclosure, the copepods immediately began vertical migration. (Modified from Neill 1990.)*

vertical migrants take longer to mature and are exposed to larval predators for a greater time. Swift (1976) did an extensive study of the energetic costs of vertical migration in the phantom midge (*Chaoborus trivittatus*), and he showed that the best strategy for maximizing net energy gain would be to stay at the surface and to avoid vertical migrations. The idea that vertical migration confers an energetic advantage on individuals is an attractive one, but it may not apply to many vertical migrants.

Vertical migration in zooplankton is not a fixed trait. When the calanoid copepod *Diaptomus kenai* is placed in a lake or an enclosure without any predators, no vertical migration occurs (Figure 21.12). Once predators are introduced (phantom midges) the copepods immediately begin to vertically migrate in a reverse pattern. They ascend during the day (when the midges descend) and descend at night (when the midges ascend). The net effect is that the copepods minimize exposure to invertebrate predators (Neill 1990).

Natural populations of zooplankton may also show plasticity in migratory behavior. Ohman (1990) found three types of vertical migration in a fjord population of the copepod *Pseudocalanus newmani* in Puget Sound, Washington (Figure 21.13).

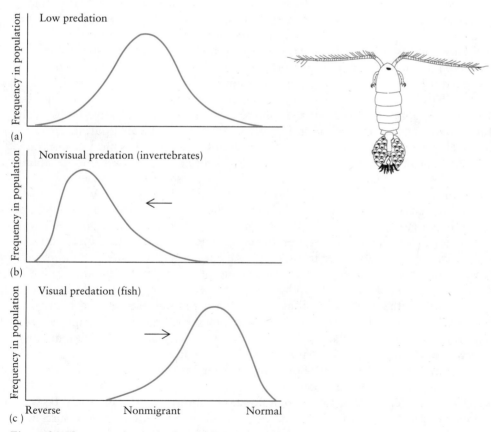

Figure 21.13   *Hypothesized relative abundance of three vertical migration phenotypes (reverse migrant, nonmigrant, and normal migrant) for* Pseudocalanus newmani, *under different predation regimes: (a) low predation pressure, (b) high predation pressure by nonvisual predators, (c) high predation pressure by visual predators. Dabob Bay, Puget Sound. (From Ohman 1990.)*

When predation pressure was low, there was no vertical migration. When fish predators that hunt visually were present, the copepods showed normal vertical migration. When invertebrate predators were abundant, the copepods showed reverse vertical migration (as illustrated in Figure 21.12). These results show clearly that vertical migration is an adaptation that reduces predation losses.

The vertical structuring of communities is an important component affecting how communities function both at the level of photosynthesis in plants and at the level of competition and predation in animals. We will investigate these effects in Chapters 26 through 28 when we discuss community metabolism.

SEASONALITY

Communities change dramatically with the seasons, and the structure of any community is not constant. Seasonality is so inscribed in the culture of temperate-zone humans that we find it difficult to think of environments that are aseasonal. The study of seasonal changes is called *phenology*. One aim of phenology has been to develop phenological calendars for different species that could be superimposed on the astronomical calendar to indicate biological events. Because seasonal events happen every day, the study of phenology has not been a high priority among ecologists (Lieth 1974). The importance of phenology can be seen once we realize that *the timing of events is critical for biological interactions*.

Flowering is one conspicuous seasonal event in terrestrial plants. Plants vary greatly in the timing and duration of the flowering period. Heinrich (1976) patiently recorded the flowering seasons of all the common plants in three habitats in central Maine (Figure 21.14). The flowering times of most native species of herbs in bogs were evenly spaced throughout the summer, whereas those of native herbs in the forest were concentrated in the spring, before the trees had leafed out. In disturbed habitats which contained many introduced species, blooming was most concentrated in the middle of the summer. The duration of flowering also differed among habitats (Figure 21.14). Woodland herbs bloomed for 18 days, on the average, bog plants for 32 days, and plants of disturbed habitats for 45 to 55 days.

What determines the length and timing of the flowering season in plants? Three broad hypotheses exist (Primack 1985). One view is that competition for pollinators is a major factor in the evolution of flowering times. If competition for pollinators occurs, species should flower in short, staggered, but slightly overlapping periods. This evolutionary argument uses the same conceptual model diagrammed in Figure 13.17 (page 251). A second hypothesis is that the flowering season is a byproduct of selection for the timing of fruit maturation, and natural selection has focused on the avoidance of seed predators and the attack by microorganisms on ripened seeds and fruits. Since in many plants a fixed period of time is needed for seed development, selection of the timing of seed maturation will automatically fix the flowering season of a plant. A third view is that flowering times are set by physiological limitations of temperature, rainfall, and light.

Flowers often depend on animals for pollination, and the interplay between plants and their pollinators is a good example of how evolutionary pressures may coordinate seasonal events in communities. Plants and pollinators are a mutualistic system in which both groups benefit from the relationship. Plants provide nectar as food for animals, which carry pollen from plant to plant and inadvertently pollinate flowers. Animal pollinators promote outcrossing in plants, the mixing of genes from

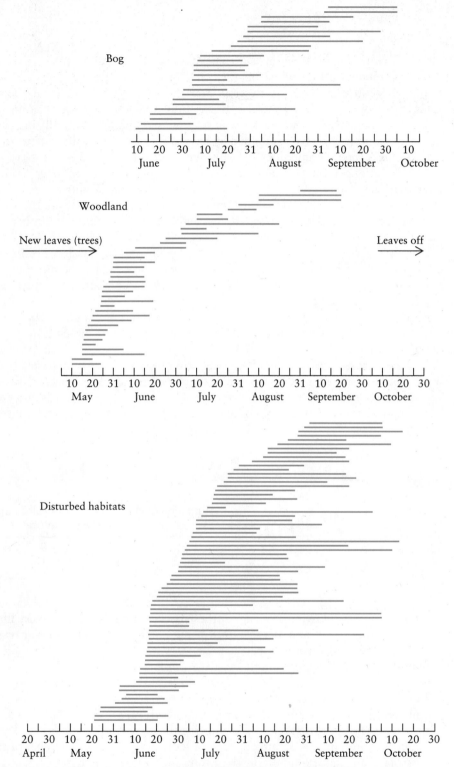

Figure 21.14 Blooming times of insect-pollinated herbs in three habitats of central Maine in 1972. Each line represents a different species. Only native species were included for bog and woodland habitats. All species were visited by bumblebees or other insects. (Modified from Heinrich 1976.)

unrelated individuals. Flowering times may have evolved under the selective pressure imposed by competition for pollinators (Primack 1985). Thus native species of bog plants flower in an unbroken progression over the growing season, most species blooming for a short time (Figure 21.14). The time available for flowering in the forest-floor habitat is very restricted, since most insect pollinators do not fly in shady habitats; woodland herbs have responded by having very short seasons of flowering. By contrast, disturbed communities contain many introduced weeds and have no history of evolutionary integration; the species therefore flower for long periods, overlap extensively, and may compete for available pollinators. Plants have evolved restricted flowering times to increase their chances of cross-pollination, and the result of this selection pressure has been to reduce competition for pollinators in natural communities.

If this view is correct, we would expect wind-pollinated plants to have longer flowering periods that are not staggered, since wind-pollinated plants do not compete for insect pollinators. Rabinowitz and colleagues (1981) compared the flowering seasons for 82 species of plants in a tallgrass prairie in Missouri. They found the same overlap between species for wind-pollinated and insect-pollinated plants, and one could not distinguish the observed flowering patterns from an artificial set of plants put together at random.

If flowering seasons are affected by competition for pollinators, better evidence might be obtained by more detailed studies on individual plant species. Waser (1978a) studied two common perennials that are visited by hummingbirds and bees in mountain meadows in Colorado. Pollination by hummingbirds and bees is essential to both these plants; seed sets are reduced by excluding pollinators:

		SEEDS PER FLOWER	
	Open Controls	**Hummingbirds Excluded**	**Hummingbirds and Bees Excluded**
Ipomopsis aggregata	10	8	0
Delphinium nelsoni	50	31	1

Figure 21.15 shows the flowering season of these two perennials. There is very little overlap in flowering times, and Waser (1978b) suggested that this could be caused by selection operating against interspecific pollen transfer. Hummingbirds gathering nectar would fly from a *Delphinium* flower to an *Ipomopsis* flower when both were available. Artificial cross-pollination between these two species reduced seed set about 25 percent in *Ipomopsis*. In natural populations, seed set in both species was reduced in the overlap period by 30 to 45 percent. Natural selection acts to reduce pollen flow between the two species by separating the flowering periods.

Pollinators visit plants to obtain energy; consequently, the energy budget of the pollinators plays an important role in the evolution of flower structure (Heinrich and Raven 1972). Flowers provide nectar at a certain rate, and specialization of pollinators by size occurs. Large animals like hummingbirds, bats, and moths require much more energy than do the smaller insects. A balance must be achieved between the energy expended by the pollinators and the caloric reward provided in the flowers if cross-pollination is to be maximal. If flowers are far apart, pollinators need addi-

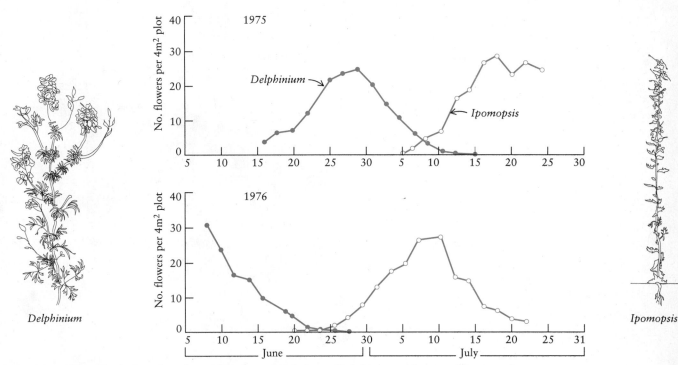

Figure 21.15   Flowering phenology of two perennial plants, Delphinium nelsoni *and* Ipomopsis aggregata, *in subalpine meadows of Colorado. Bees and hummingbirds pollinate these flowers, and natural selection has reduced overlap in flowering times. (After Waser 1978a.)*

tional energy per flower. The same is true in arctic conditions, where low temperatures increase energy demands of pollinators.

It has proved difficult to separate all the factors limiting the flowering seasons of plants, and even detailed studies of hummingbird-pollinated plants in tropical forests have failed to document displacement of flowering seasons due to competition for pollinators (Murray et al. 1987). Plants with bimodal seasons of flowering are also difficult to integrate into existing theories (Primack 1985). Detailed studies of individual plants and their experimental manipulation are needed to determine the reasons behind the seasonal phenology of flowering in plant communities. Competition for pollinators may occur between certain species pairs but in general does not seem to have played a large role in the evolution of flowering times.

Seasonal events are often believed absent from tropical regions, but this mistaken notion is based on temperature records and not on biological observations. Rainfall is critical in many tropical areas, and wet and dry seasons have strong effects on community structure. Frankie and co-workers (1974) studied the phenology of two forest sites in Costa Rica. La Selva is a wet forest site occupied by tropical rain forest. Appreciable rainfall occurs every month (Figure 21.16), and the forest remains green year round. The dry forest site is in Guanacaste Province, where there is a pronounced wet and dry season (Figure 21.16). The dry site is occupied by tropical deciduous forest.

Leaf-fall occurs seasonally in the tropical rain forest in La Selva, although it is

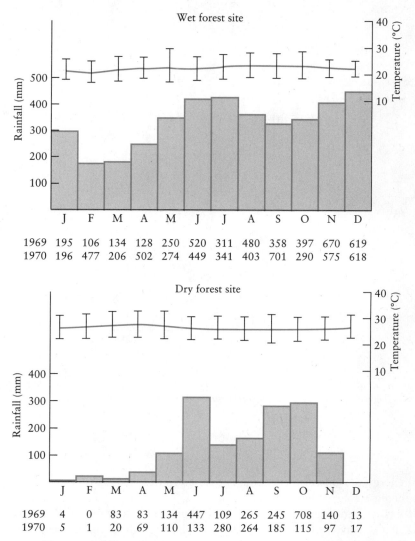

Figure 21.16 *Mean monthly temperature (upper line) and average rainfall (histogram) at the wet forest site (tropical rain forest) and at the dry site (tropical deciduous forest) in Costa Rica studied by Frankie et al. (1974). Rainfall figures (mm) for the two years of study are given along the bottom of each graph. (After Frankie et al. 1974.)*

not as pronounced as the obvious leaf-fall at the dry forest site (Figure 21.17). Individual species in the tropical rain forest vary in their leafing behavior. More overstory trees have pronounced leaf drop and new leaf growth seasonally, while more understory trees continuously shed a few leaves and add a few more.

Flowering seasons are of two types in the trees of the La Selva rain forest. *Extended flowering* occurs in approximately 40 percent of the trees, and these trees flower on the average for five to six months before they stop. *Seasonal flowering* occurs in about 60 percent of the tree species and lasts six to seven weeks. Flowering occurs throughout the year in the wet forest (Figure 21.18) but is more restricted to the dry season in the dry forest site. Only about 10 percent of the tree species in the

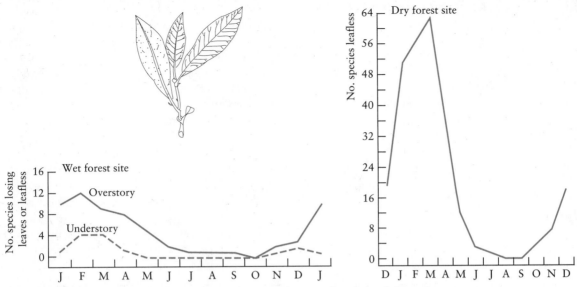

Figure 21.17 *Leaf-fall periodicity of tree species in the west forest site (left) and dry forest site (right). A seasonal periodicity is present at both sites but is much stronger in the dry forest. (After Frankie et al. 1974.)*

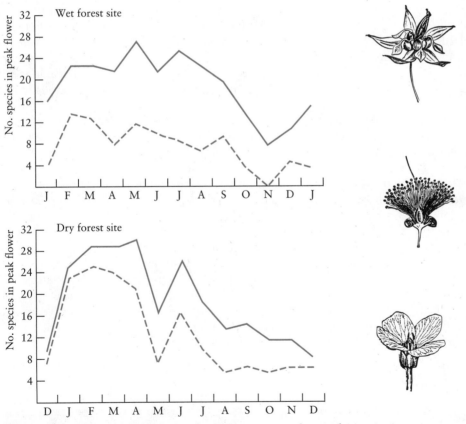

Figure 21.18 *Flowering periodicity of overstory trees at the wet forest site (upper) and dry forest site (lower), Costa Rica. The continuous line shows data for all tree species; the dashed line shows data for seasonal-flowering species only. In the tropical deciduous forest, flowering is concentrated in the dry season. (After Frankie et al. 1974.)*

dry forest have extended flowering. In the dry forest, 59 species of seasonal-flowering trees bloomed in a sequential manner through the dry season, with minimal overlap in the flowering period among the species (Frankie et al. 1974). Virtually all of these trees depend on animal pollination for fertilization and fruiting.

The dry-season peak of flowering and fruiting of trees in the tropical deciduous forests of Central America results from strong natural selection. Janzen (1967) argued that there are many advantages of dry-season flowering. Insect pollinators profit from dry sunny weather. The lack of leaves can make flowers more visible to pollinators, and for ground-nesting bees, the lack of rain prevents nest flooding. Pollinator activity can therefore be enhanced during the dry season in tropical forest communities.

Seasonality thus occurs even in the tropical rain forest, and the measurement of seasonal changes is important in the analysis of community structure.

SUMMARY

The structure of a community strongly affects its functioning. Plants form the structure of most terrestrial communities, and we can classify the vegetation of the world into six major growth forms. Similar climates select for similar growth forms, even on different continents where all the taxonomic units differ. The growth forms of plants depend on leaf size and shape and on leaf longevity. Leaf size is adjusted by natural selection to maximize the efficiency of water use. If a plant retains its leaves for more than one year, it is considered an evergreen. A simple model of leaf longevity can predict why evergreen plants predominate in tropical and polar regions but are largely absent from temperate zone areas.

Competition for light determines the vertical structure of land communities. Many forest communities have several strata of plants. The vertical structure of aquatic systems is determined by the physical properties of water. Light is absorbed in the upper layers of water, and all plants must live near the surface to get light. Many species of zooplankton in lakes and in the ocean undergo vertical migrations, surfacing at night and moving into deeper waters by daylight. Vertical migration in zooplankton is an adaptation to reducing predation. Populations may stop vertical migration if no predators are present, or undergo reverse migration if nonvisual invertebrate predators occur in the community.

Communities change with the seasons, even in tropical forests. Flowering seasons may evolve under the selective pressure imposed by competition for pollinators, with the result that flowering periods are spaced out among different plant species. Plants and their pollinators form a coadapted complex in which both benefit. In some plants the flowering season is fixed by the timing of the fruiting season, and the major selective agent is the need to optimize seed dispersal.

Selected References

BOND, W. J. 1989. The tortoise and the hare: Ecology of angiosperm dominance and gymnosperm persistence. *Biological Journal of the Linnean Society* 36:227–249.

CHABOT, B. F., and D. J. HICKS. 1982. The ecology of leaf life spans. *Annual Review of Ecology and Systematics* 13:229–259.

GIVNISH, T. J. 1982. On the adaptive significance of leaf height in forest herbs. *American Naturalist* 120:353–381.

HEINRICH, B., and P. H. RAVEN. 1972. Energetics and pollination ecology. *Science* 176:597–602.

LIETH, H., ed. 1974. *Phenology and Seasonality Modeling.* Springer-Verlag, New York.

NEILL, W. E. 1990. Induced vertical migration in copepods as a defence against invertebrate predation. *Nature* 345:524–526.

OHMAN, M. D. 1990. The demographic benefits of diel vertical migration by zooplankton. *Ecological Monographs* 60:257–281.

PRIMACK, R. B. 1985. Patterns of flowering phenology in communities, populations, individuals, and single flowers. In *The Population Structure of Vegetation*, ed. J. White, pp. 571–593. Dr. W. Junk Publishers, Dordrecht, Netherlands.

REGAL, P. J. 1982. Pollination by wind and animals: Ecology of geographic patterns. *Annual Review of Ecology and Systematics* 13:497–524.

SMITH, A. P. 1973. Stratification of temperate and tropical forests. *American Naturalist* 107:671–683.

WALLER, D. M. 1986. The dynamics of growth and form. In *Plant Ecology*, ed. M. J. Crawley, pp. 291–320. Blackwell, Oxford, England.

Questions and Problems

1. The number of vertical strata in forests is reported to decrease from tropical to temperate areas (Smith 1973). Suggest one or more possible reasons for this change, and compare them with Smith's hypotheses. How would you test your ideas?

2. Wind pollination is virtually absent from tropical trees but common in temperate and polar tree species (Whitehead 1969). Why?

3. McAllister (1969) stated:

   Vertical migration may give the additional advantage of better utilization of the growth potential of the phytoplankton, as well as permitting the unimpeded growth of plants during the daylight hours.

   Is this a possible reason for the evolution of vertical migration? Why or why not?

4. Orchids are richly represented in tropical rain forests but are typically rare so that individuals are widely spaced (Pijl and Dodson 1967). About half the orchid species produce no nectar and thus give no energy reward to their pollinators. How can this type of animal-pollination system operate?

5. What selective forces might prevent the two plants in Figure 21.15 from evolving completely distinct flowering seasons?

6. How would you test the hypothesis that the three copepod shown in Figure 21.13 represent three different genotypes? How would the ecological explanation of vertical migration differ if these behaviors are under complete genetic control or behaviorally plastic for each individual?

7. Stiles (1977) presented data to show that an orderly, staggered sequence of flowering times occurred in hummingbird-pollinated plants in a tropical rain forest. Poole and Rathcke (1979) analyzed the same data and concluded that flowering times were randomly spaced. Review this controversy, including Stiles (1979) reply to Poole and Rathcke, Cole's (1981) evaluation and Pleasants' (1990) evaluation.

8. A striking feature of dipterocarp forests in the tropical rain forests of Malaysia is mass flowering. In these aseasonal rain forests many species of trees come into flower at the same time at irregular intervals of 2 to 10 years. The ensuing large seed drop swamps all the seed predators. How do mass-flowering dipterocarps avoid competition for pollinators? How could this synchronization of flowering times evolve in an aseasonal rain forest with hundreds of tree species? Compare your ideas with the suggestions of Ashton and colleagues (1988).

9. Explain the reasons for the shape of the cost and benefit curves shown in Figure 21.3, page 464. What are the implications of a linear cost curve?

Overview Question

Conifers dominated the earth's vegetation 100 million years ago, and were then replaced by angiosperms over most of the earth. Review the reasons postulated for this vegetation change, and discuss what kinds of plants might replace angiosperms over the next 100 million years.

22 Community Change

One of the most important features of communities is change, and in this chapter we focus on the factors causing communities to change on an ecological time scale. There are two main types of temporal changes in communities: directional changes in time, called *succession*, and nondirectional changes in time, which are *cyclic* (fluctuate about a mean). We shall discuss each of these in turn and focus on two questions: (1) How predictable are community changes? (2) What factors cause community changes?

SUCCESSION

When stripped of its original vegetation by fire, flood, or glaciation, an area of bare ground does not remain devoid of plants and animals. The area is rapidly colonized by a variety of species that subsequently modify one or more environmental factors. This modification of the environment may in turn allow additional species to become established. This development of the community by the action of vegetation on the environment leading to the establishment of new species is termed *succession*. Succession is the universal process of directional change in vegetation during ecological time. It can be recognized by a progressive change in the species composition of the community.

Most observed successions are called *secondary succession*, the recovery of disturbed sites. A few successions are called *primary succession* because they occur on a new sterile area, such as that uncovered by a retreating glacier.

Understanding succession requires understanding the mechanisms that cause succession. One focus has been on the effects that early successional species have on later successional species. Early species can either help or hinder or not affect the establishment of later species. Competition between individual plants for resources such as water, light, or nitrogen may drive succession.

The concept of succession was largely developed by the botanists J. E. B. Warming (1896) and H. C. Cowles (1901), who studied the stages of sand-dune development. Successional studies have led to four major hypotheses of succession. The first is the classical theory of succession, which has been called *relay floristics* by Egler (1954) because it postulates an orderly hierarchical system of change in the community (Figure 22.1a). The classical ideas of succession were elaborated in great detail by F. E. Clements (1916, 1936), who developed a complete theory of plant succession and community development called the *monoclimax hypothesis*. The biotic community, according to Clements, is a highly integrated superorganism. It shows development through a process of succession to a single end point in any given area—the *climatic climax*. The development of the community is gradual and

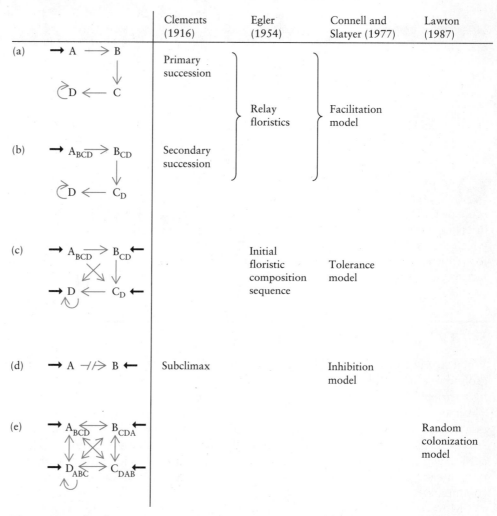

	Clements (1916)	Egler (1954)	Connell and Slatyer (1977)	Lawton (1987)
(a)	Primary succession			
		Relay floristics	Facilitation model	
(b)	Secondary succession			
(c)		Initial floristic composition sequence	Tolerance model	
(d)	Subclimax		Inhibition model	
(e)				Random colonization model

Figure 22.1   Replacement sequences in succession proposed by different authors. The letters A–D represent hypothetical vegetation types or dominant species; subscript letters indicate that species are present as minor components or as propagules. Thin arrows represent species or vegetation sequences in time; bold arrows represent alternative starting points for succession after disturbance. Circular arrows indicate that the species replaces itself. (Modified from Noble 1981.)

progressive, from simple pioneer communities to the ultimate or climax stage. This succession is due to biotic reactions only; the plants and animals of the pioneer stages alter the environment so as to favor a new set of species, and this cycle recurs until the climax is reached. Development through succession in a community is therefore analogous to development in an individual organism, according to Clements's view (Clements 1936, Phillips 1934–1935). Thus reverse succession (retrogression) is not possible unless some disturbance such as fire, grazing, or erosion intervenes. Secondary succession differs from primary succession in having a seed bank from plants that occur later in succession, so that late succession species are already present in the early stages of secondary succession (Figures 22.1a and 22.1b).

The key assumption of the classical theory of succession is that species replace one another because at each stage the species modify the environment to make it less suitable for themselves and more suitable for others. Thus species replacement is orderly and predictable and provides directionality for succession. These characteristics led Connell and Slatyer (1977) to call this the *facilitation model* of succession.

The climax community in any region is determined by climate, in Clements's view. Other communities may result from particular soil types, fire, or grazing, but these are understandable only with reference to the end point of the climatic climax. Therefore, the natural classification of communities must be based on the climatic climax, which represents the state of equilibrium for the area.

A second major hypothesis of succession was proposed by Connell and Slatyer (1977), who called it the *inhibition model*. In this view, succession is very heterogeneous because the development at any one site depends on who gets there first. Species replacement is not necessarily orderly because each species excludes or suppresses any new colonists (Figure 22.1d). Thus succession becomes more individualistic and less predictable because communities are not always converging toward the climatic climax. No species in this model is competitively superior to another. Whoever colonizes the site first holds it against all comers until it dies. Succession in this model proceeds from short-lived species to long-lived species and is not an orderly replacement.

A third major model of succession was proposed by Connell and Slatyer (1977), who called it the *tolerance model*. This model is intermediate between the facilitation model and the inhibition model. In the tolerance model, the presence of early successional species is not essential—any species can start the succession (Figure 22.1c). Some species are competitively superior, however, and these eventually come to predominate in the climax community. Species are replaced by other species that are more tolerant of limiting resources. Succession proceeds either by the invasion of later species or by a thinning out of the initial colonists, depending on the starting conditions. Egler (1954) called this model the *initial floristic composition sequence.*

A fourth model of succession was proposed by Lawton (1987) to provide a null model. The *random colonization model* suggests that succession involves only the chance survival of different species and the random colonization by new species. There is no facilitation and no interspecific competition (Figure 22.1e).

The first three hypotheses of succession agree in predicting that many of the pioneer species in a succession will appear first because these species have evolved colonizing characteristics, such as rapid growth, abundant seed production, and high dispersal powers (Table 22.1). The critical feature of the life-history traits given in Table 22.1 is that there is an inverse correlation between traits that promote success in early succession and traits that are advantageous in late succession (Huston and Smith 1987).

The critical distinction among the four hypotheses is in the mechanisms that determine subsequent establishment. In the classical facilitation model, species replacement is *facilitated* by the previous stages. In the inhibition model, species replacement is *inhibited* by the present residents until they are damaged or killed. In the tolerance and random colonization models, species replacement is *not affected* by the present residents.

The utility of these four models of succession is that they immediately suggest experimental manipulations to test them. Removing or excluding early colonizers,

Table 22.1 Physiological and Life-History Characteristics of Early- and Late-successional Plants

Characteristic	Early Succession	Late Succession
Photosynthesis		
Light-saturation intensity	high	low
Light-compensation point	high	low
Efficiency at low light	low	high
Photosynthetic rate	high	low
Respiration rate	high	low
Water-use efficiency		
Transpiration rate	high	low
Mesophyll resistance	low	high
Seeds		
Number	many	few
Size	small	large
Dispersal distance	large	small
Dispersal mechanism	wind, birds, bats	gravity, mammals
Viability	long	short
Induced dormancy	common	uncommon?
Resource-acquisition rate	high	low?
Recovery from nutrient stress	fast	slow
Root-to-shoot ratio	low	high
Mature size	small	large
Structural strength	low	high
Growth rate	rapid	slow
Maximum life span	short	long

Source: From Huston and Smith (1987).

transplanting seeds or seedlings of late succession species into earlier stages, and other experiments can shed light on the mechanisms involved in succession.

To explain a successional sequence we must add to these idealized models of succession some additional information on seed availability, insect and mammal herbivory, mycorrhizal fungi, and plant pathogens (Walker and Chapin 1987). Figure 22.2 summarizes the relative importance of different factors in succession. The primary processes underlying successional changes could be competition or mutualistic interactions between plants but these plant-plant interactions are affected by animal grazing and diseases, and by seed dispersal and storage. The resulting successional sequences are thus complex and do not proceed in a single direction to a fixed end point (Burrows 1990).

A MODEL OF SUCCESSION

We can construct a simple mathematical model of succession by assuming that succession is a replacement process (Horn 1981). For each plant, we ask a simple question: What is the probability that this plant will be replaced in a given time by another plant of the same species or by another species? A table of replacement probabilities like that in Table 22.2 can be estimated in forests by counting the

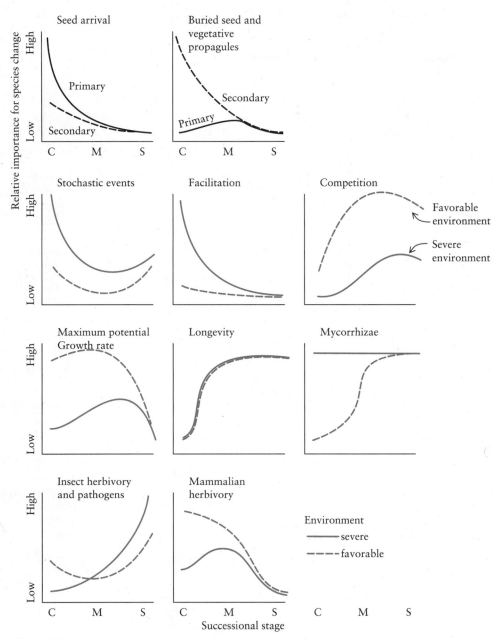

Figure 22.2   Influence of type of succession and environmental severity upon major successional processes that determine change in species composition during colonization (C), maturation (M), or senescence (S) stages of succession. (Modified after Walker and Chapin 1987.)

number of saplings of various species growing under the canopy of the mature trees. For example, in the data of Table 22.2, Horn (1975a) found a total of 837 saplings growing underneath large gray birch trees; among these, there were no gray birch saplings, 142 red maples, and 25 beech saplings (plus other species). Thus the probability that a gray birch will be replaced by another gray birch (self-replacement) is

Table 22.2   Transition Matrix for Institute Woods in Princeton, New Jersey: Saplings Under Various Species of Trees

Canopy Species	SAPLING SPECIES (%)											Total No.
	BTA	GB	SF	BG	SG	WO	RO	HI	TT	RM	BE	
Big-toothed aspen	**3**	5	9	6	6	—	2	4	2	60	3	104
Gray birch	—	—	47	12	8	2	8	—	3	17	3	837
Sassafras	3	1	10	3	6	3	10	12	—	37	15	68
Blackgum	1	1	3	**20**	9	1	7	6	10	25	17	80
Sweetgum	—	—	16	—	**31**	—	7	7	5	27	7	662
White oak	—	—	6	7	4	**10**	7	3	14	32	17	71
Red oak	—	—	2	11	7	6	**8**	8	8	33	17	266
Hickories	—	—	1	3	1	3	13	**4**	9	49	17	223
Tulip tree	—	—	2	4	4	—	11	7	**9**	29	34	81
Red maple	—	—	13	10	9	2	8	19	3	**13**	23	489
Beech	—	—	—	2	1	1	1	1	8	6	**80**	405

Note. The number of saplings of each species listed in the row at the top, where the abbreviations are self-explanatory, is expressed as a percentage of the total number of saplings (last column) found under individuals of the species listed in the first column. The entries are interpreted as the percentages of individuals of species listed on the left that will be replaced one generation hence by species listed at the top. The percentage of "self-replacements" is shown in boldface.
Source: After Horn (1975b).

0/837 = 0, by a red maple 142/837 = 0.17, and by a beech 25/837 = 0.03. These probabilities are entered as percentages in Table 22.2.

We can use these replacement probabilities to calculate what will happen to this forest community in the future. For example, working down the columns of the table, we can calculate:

Proportion of the next generation that will be white oak = 0.02 gray birch +
0.03 sassafras +
0.01 blackgum +
0.10 white oak +
0.06 red oak +
0.03 hickory +
0.02 red maple +
0.01 beech

where all the species on the right side of this equation refer to the current abundance of that species. We multiply the current abundances of each species by the replacement probabilities in Table 22.2 and add them together to get the predicted abundance of each canopy species in the next generation. We can recycle through these calculations, which are more tedious than difficult, and predict the abundances of all species any number of generations into the future (Table 22.3). After several generations, the abundances of all species settle down to a stationary distribution, which will not change over time. In this example, the stationary distribution predicted will contain 50 percent beech, 16 percent red maple, and 7 percent tulip tree.

To use this simple model of succession, we must make a number of assumptions. The most critical assumption is that the table of replacement probabilities does not change over time. Under this assumption, the community will approach a steady state that is independent of the community's initial composition and independent of the type of disturbance that starts the succession going. In this form, the model is a

Table 22.3 Theoretical Approach of the Institute Woods at Princeton to a Stationary Distribution ("Climax Forest")

	BTA	GB	SF	BG	SG	WO	RO	HI	TT	RM	BE
					SPECIES (%)						
Theoretical Generation											
0	—	49	2	7	18	—	3	—	—	20	1
1	—	—	29	10	12	2	8	6	4	19	10
2	1	1	8	6	9	2	8	10	5	27	23
3	—	—	7	6	8	2	7	9	6	22	33
4	—	—	6	6	7	2	6	8	6	20	39
5	—	—	5	5	6	2	6	7	7	19	43
·	·	·	·	·	·	·	·	·	·	·	·
·	·	·	·	·	·	·	·	·	·	·	·
·	·	·	·	·	·	·	·	·	·	·	·
Stationary distribution (%)	—	—	4	5	5	2	5	6	7	16	50
Longevity (yr)	80	50	100	150	200	300	200	250	200	150	300
Age-corrected stationary distribution (%)	—	—	2	3	4	2	4	6	6	10	63

Notes: The starting point is an observed 25-year-old forest stand dominated by gray birch. Tree species names are listed in Table 22.2. The theoretical predictions are obtained by multiplying the starting tree composition (generation 0) by the transition matrix of probabilities in Table 22.2.
Source: After Horn (1975b).

statement of any type of succession that predicts a regular, repeatable change culminating in a stable climax.

We must also assume that we can calculate replacement probabilities in a realistic way. For forests, this involves assuming that abundance in the sapling stage is a sufficient predictor of the chances of growing up to the canopy. There is also a problem of overlapping generations in different species. In this example, some trees live longer than others, and one must correct these predictions for variable life spans (Table 22.3). This correction is tedious but not difficult (Horn 1975b).

Can we alter the replacement model to describe the inhibition model of succession that does not culminate in a stable, fixed climax? If the table of replacement probabilities depends on the present composition of the forest, predictions from the model change dramatically. For example, assume that the recruitment of young plants depends on the density of their own species. The transition probabilities for any one tree are not constant under this assumption but change depending on who the neighbors are. Succession in this model does not converge to one point, but alternative communities could be produced, depending on the accidents of history, as suggested by the inhibition model.

The replacement model of succession is useful because it focuses our attention on the local regeneration of species and how species replace both themselves and other species as disturbances open up new areas for succession. It is not, however, a mechanistic model of succession, and to understand why succession is occurring we need to focus on the mechanisms controlling succession. One possible mechanism driving succession is competition for limiting resources (Tilman 1985). One way to model succession mechanistically is to use an individual–based plant model that explicitly incorporates light as the limiting resource (Huston and Smith 1987). Each individual plant is given species-specific traits of maximum size and age, maximum growth rate, and tolerance to shading. Most of these models have been used for trees

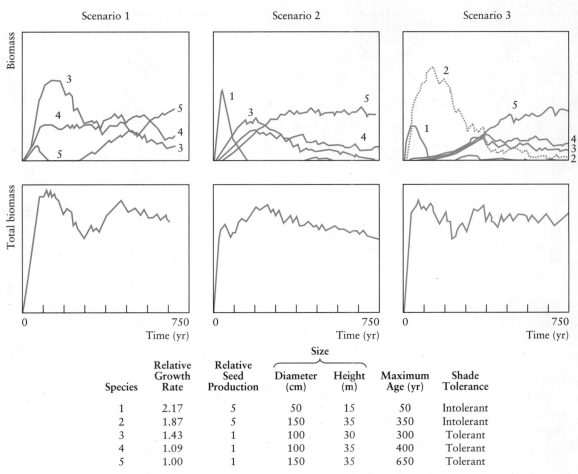

Species	Relative Growth Rate	Relative Seed Production	Size Diameter (cm)	Height (m)	Maximum Age (yr)	Shade Tolerance
1	2.17	5	50	15	50	Intolerant
2	1.87	5	150	35	350	Intolerant
3	1.43	1	100	30	300	Tolerant
4	1.09	1	100	35	400	Tolerant
5	1.00	1	150	35	650	Tolerant

Figure 22.3   *Species biomass dynamics and community biomass for hypothetical successional sequences with three, four, and five species. In scenario 1, all three species have late-successional characteristics but differ sufficiently in relative competitive abilities to produce a "typical" successional replacement. In scenario 2, an early-successional species is added to the three species in scenario 1. In scenario 3, a "super-species" (species 2) with a high growth rate and large size is added to the four species in scenario 2. (From Huston and Smith 1987.)*

to model forest succession but they could be applied to other kinds of plants as well. The key variable in these models is light availability, and each individual plant is analyzed to see how much shading is produced by its neighbors; if light is limited, growth rates and survival rates are reduced accordingly. This type of simple mechanistic model can produce successional sequences of tree species that resemble natural succession (Shugart 1984). Figure 22.3 illustrates three scenarios with species of different life-history traits. In every case the species that is most shade-tolerant in regeneration and grows to the largest size wins out in succession. The seral stages vary greatly depending on which trees are present.

The additional effects of competition for soil nitrogen can be added to these simple models, so that both light and nitrogen become the limiting resources (Tilman 1985, Shugart 1984). Mechanisms of nutrient limitation make these models more realistic but also more complex to evaluate. The most successful models of succession are for forests. Because of their economic importance a wealth of details is known about the life-history traits of individual tree species.

How well do natural communities fit the four hypotheses of succession? Does succession in a region converge to a single end point, or are there multiple stable states? Let us look at a few examples of succession. Numerous examples of succession have been described in detail, but there are few cases in which the succession can be related to a time scale. Some examples have been investigated in detail where the time scale is known, and I shall describe these briefly.

Glacial Moraine Succession in Southeastern Alaska

During the past 200 years there has been a generalized retreat of glaciers in the Northern Hemisphere. As the glaciers retreat, they leave moraines, whose age can be determined by the age of the new trees growing on them or, in the last 80 years, by direct observation. The most intensive work on moraine succession has been done at Glacier Bay in southeastern Alaska. Since about 1750 the glaciers there have retreated about 98 kilometers, an extraordinary rate of retreat (Figure 22.4)

The pattern of primary succession in this area proceeds as follows (Cooper 1939, Lawrence 1958). The exposed glacial till is colonized first by mosses, fireweed, *Dryas*, willows, and cottonwood. The willows begin as prostrate plants but later grow into erect shrubs. Very quickly the area is invaded by alder (*Alnus*), which eventually forms dense pure thickets up to 9 meters tall. This requires about 50 years. These alder stands are invaded by Sitka spruce, which, after another 120 years, forms a dense forest. Western hemlock and mountain hemlock invade the spruce stands, and after another 80 years the situation has stabilized with a climax spruce-hemlock forest. This forest, however, remains on well-drained slopes only. In areas of poor drainage, the forest floor of this spruce-hemlock forest is invaded by *Sphagnum* mosses, which hold large amounts of water and acidify the soil greatly. With the spread of conditions associated with *Sphagnum*, the trees die out because the soil is waterlogged and too oxygen-deficient for tree roots, and the area becomes a *Sphagnum bog*, or *muskeg*. The climax vegetation then seems to be muskeg on the poorly drained areas and spruce-hemlock forest on the well-drained areas.

The bare soil exposed as the glacier retreated is quite basic, with a pH of 8.0 to 8.4 because of the carbonates contained in the parent rocks. The soil pH falls rapidly with the advent of vegetation, and the rate of change depends on the vegetation type (Figure 22.5). There is almost no change in the pH due to leaching in bare soil. The most striking change is caused by alder, which reduces the pH from 8.0 to 5.0 in 30 to 50 years. The leaves of alder are slightly acid, and as they decompose they become more acid. As the spruce begins to take over from the alder, the pH stabilizes at about 5.0, and it does not change in the next 150 years.

The organic carbon and total nitrogen concentrations in the soil also show marked changes with time. Figure 22.6 shows the changes in nitrogen levels. One of the characteristic features of the bare soil is its low nitrogen content. Almost all the pioneer species begin the succession with very poor growth and yellow leaves due to inadequate nitrogen supply. The exceptions to this are *Dryas* and alder; these species have some way of fixing atmospheric nitrogen (Lawrence et al. 1967). The rapid increase in soil nitrogen in the alder stage is caused by the presence of nodules on the alder roots that contain microorganisms that actively fix nitrogen from the air. Spruce trees have no such adaptations; consequently, the soil nitrogen level falls

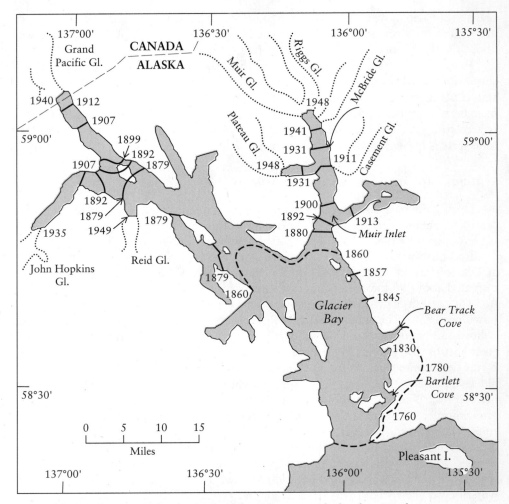

Figure 22.4   Glacier Bay fjord complex of southeastern Alaska showing the rate of ice recession since 1760. As the ice retreats it leaves moraines along the edge of the bay on which primary succession occurs. (After Crocker and Major 1955.)

when alders are eliminated. The spruce forest develops by using the capital of nitrogen accumulated by the alder.

The important point to notice here is the reciprocal interrelations of the vegetation and the soil. The pioneer plants alter the soil properties, which in turn permit new species to grow, and these species in turn alter the environment in different ways, bringing about succession. The classical facilitation theory provides a good description of glacial moraine succession.

Lake Michigan Sand-Dune Succession

Cowles (1899), from the University of Chicago, worked on the sand-dune vegetation of Lake Michigan and made a classic contribution to the ideas of plant succession.

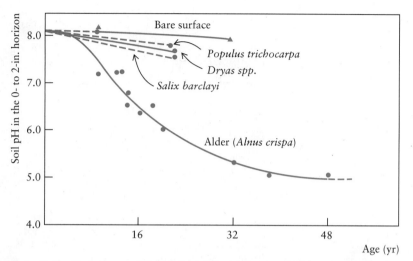

Figure 22.5   Soil pH change at Glacier Bay, Alaska, under different types of pioneer vegetation. The soil becomes acid very rapidly under alder. (After Crocker and Major 1955.)

Olson (1958) has reexamined the successional stages in this area in relation to an absolute time scale.

During and after the retreat of the glaciers from the Great Lakes area, the resulting fall in lake level left several distinct "raised beaches" and their associated dune systems. These systems, which run roughly parallel to the present shoreline of Lake Michigan, are about 7, 12, and 17 meters above the present lake level (Figure

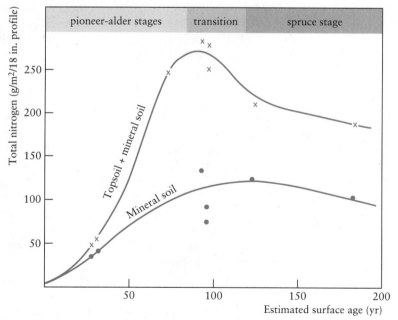

Figure 22.6   Total nitrogen content of soils recently uncovered by glacial retreat at Glacier Bay, Alaska. Plant succession is shown along the top. (After Crocker and Major 1955.)

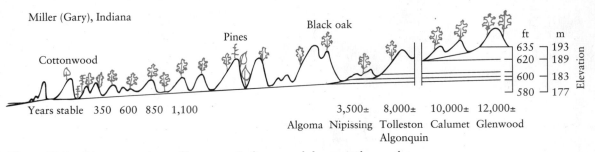

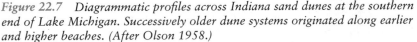

Figure 22.7   Diagrammatic profiles across Indiana sand dunes at the southern end of Lake Michigan. Successively older dune systems originated along earlier and higher beaches. (After Olson 1958.)

22.7). Olson dated the older dunes by radiocarbon techniques and the younger areas by tree-ring counts and recorded historical changes since 1893.

The dunes, like glacial moraines, offer a near-ideal system for studying plant succession because many of the complicating variables are absent. The initial substrate for all the area is dune sand, the climate for the whole area is similar, the relief is similar, and the available flora and fauna are the same. So the differences between the dunes should be due only to *time, biological processes of succession,* and *chance events* associated with dispersal and colonization.

Two processes produce bare sand surfaces ready for colonization. One is the slow process of a fall in lake level; the other is a rapid process, the *blowout* of an established dune. These blowouts result from the strong winds that come off the lake. This wind erosion sets up a moving dune that is gradually stabilized after migrating inland. The dunes are stabilized only by vegetation.

The bare sand surface is colonized first by dune-building grasses, of which the most important is marram grass *(Ammophila breviligulata)*. Marram grass usually propagates by rhizome migration, only rarely by seed. It spreads very quickly and can stabilize a bare area in six years. After the sand is stabilized, marram grass declines in vigor and dies out. The reason for this is not known, but the result is that this grass is not found in the stable dune areas after about 20 years.

Two other grasses are important in dune formation and stabilization: sand reed grass *(Calamovilfa longifolia)* and the little bluestem *(Andropogon scoparius)*. The sand cherry *(Prunus pumila)* and willows *(Salix* spp.) also play a role in dune stabilization. The first tree to appear in the young dune is usually the cottonwood *(Populus deltoides)*, which may also help to stabilize the sand.

Once the dune is stabilized, it may be invaded very quickly by jack pine and white pine if seed is available; normally pines are found after 50 to 100 years of development. Under normal conditions, black oak replaces the pines, entering the succession at about 100 to 150 years. A whole group of shrubs that require considerable light invade the early pine and oak stands; they are replaced by more shade-tolerant shrubs as the forest of black oak becomes denser.

Cowles believed that this succession to black oak might be part of the succession sequence, which would then proceed to a white oak–red oak–hickory forest and finally to the "climatic climax," beech-maple forest. But Olson questioned whether this could ever occur. The oldest dunes Olson studied (12,000 years) still had black oak associations, and he could see no tendency for any further succession. Moreover, the black oak community was very heterogeneous, and Olson recognized four dif-

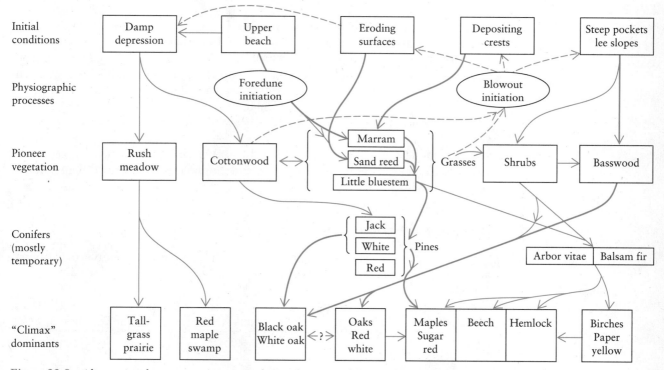

Figure 22.8   Alternative dune successions in Lake Michigan sand dunes. Beaches, foredunes, and blowout dunes provide diverse sites that undergo different successions. The center of the diagram gives an oversimplified outline of "normal" succession, from dune builders to jack or white pine to black oak–white oak with several undercover types. (After Olson 1958.)

ferent types of understory communities that could occur under black oak. Figure 22.8 summarizes the successional patterns on the dunes.

Olson also studied the changes in the soil in relation to this time sequence. The pH of the soil decreased with dune age, from high values of 7.6 at the start of succession to 4.0 after 10,000 years. The initial drop in pH is caused by carbonates being leached from the soil very quickly. Soil nitrogen increases rapidly in the first 1000 years of development, from very low values initially to approximately 0.1 percent, and then remains unchanged in older dunes. Organic carbon in the soil develops similarly.

Thus most of the soil improvements of the original barren dune sand occur within about 1000 years of stabilization. As a result of these trends in the soil, Olson pointed out, the nutritional conditions for succession toward beech and maple probably become *less* favorable with time. (These trees require more calcium, near neutral pH, and larger amounts of water.) It appears improbable that this succession will move beyond the black oak stage, contrary to what Cowles had suggested. Beech and maple associations in this area are found only in favorable situations such as moist lowlands, where the soil characteristics differ from those of the dry dunes. The low fertility of the dune soils favors vegetation, such as the black oak, that has limited nutrient and water requirements. But this sort of vegetation is ineffective in returning nutrients to the dune surface in its litter, which continues the cycle of low fertility.

As far as the dunes are concerned, probably the most striking vegetational changes occur in the first 100 years, and the system seems to become stabilized with the black oak association by about 100 years. Olson pointed out that it was a mistake to distort the different dune successions into a single linear sequence leading from pine to black oak to oak-hickory to beech-maple. Successions in the dunes go off in different directions with different destinations, depending on the various soil, water, and biotic factors involved at a particular site. Instead of convergence to a single climax, Olson suggested, we may get *divergence* of different communities on different sites.

Thus dune succession begins as the classical model suggests but changes to the inhibition model at the black oak stage and culminates in several different stable communities.

Abandoned Farmland in North Carolina

When upland farm fields are abandoned in the Piedmont area of North Carolina, a succession of plant species colonizes the area. The sequence is as follows:

Years After Last Cultivation	Dominant Plant	Other Common Species
0 (fall)	Crabgrass ↓	
1	Horseweed ↓	Ragweed
2	Aster ↓	Ragweed
3	Broomsedge ↓	
5–15	Shortleaf pine ↓	(Loblolly pine)
50–150	Hardwoods (oaks)	(Hickories)

This is initially a striking sequence of rapid replacements of herbaceous species, and Keever (1950) attempted to find out why the initial species die out and why the later colonizers are delayed in entry to the succession.

The sequence of this succession is dictated by the life history of the dominant plants. Horseweed *(Erigeron canadensis)* seeds will mature as early as August and germinate immediately. It overwinters as a rosette plant, which is drought-resistant and grows, blooms, and dies the following summer (it is an annual). The second generation of horseweed plants is stunted and does not grow well in the second year of the succession. Decaying horseweed roots inhibit the growth of horseweed seedlings. The density of horseweed individuals is much greater in second-year fields, but these individuals do not do well with increased competition from aster and other horseweed. The result is a great reduction of horseweed in second-year fields.

Aster *(Aster pilosus)* seeds mature in the fall, too late to germinate in the year that plowing ceases. Seeds germinate the following spring, and seedlings grow slowly during their first year to 5 to 8 centimeters (2–3 inches) by autumn. This slow growth is partly caused by shading by horseweed, and decaying horseweed roots stunt aster growth. So horseweeds do not pave the way for asters; if anything, they have a detrimental effect on them. Asters enter the succession in spite of horseweeds, not because of them.

Aster (a perennial) blooms in its second year, after horseweed declines, but is not drought-resistant. Seedlings of aster are present in large numbers in third-year fields but succumb to competition for moisture with the drought-resistant broom-sedge *(Andropogon virginicus)*. In fields with more available water, aster is able to last into the third year, but eventually broomsedge overwhelms it.

Broomsedge seeds will not germinate without a period of cold dormancy. A few broomsedge plants are found in one-year fields but do not drop seed until the fall of the second year. Broomsedge is a very drought-resistant perennial and competes very well for soil moisture. There are few broomsedge plants in one- and two-year fields because few seeds are present. Once some plants begin seeding, broomsedge rapidly increases in numbers (third year). Broomsedge grows better in soil with organic matter, especially in soil with aster roots. It grows very poorly in the shade.

Thus early succession in Piedmont old fields is governed more by competition than by cooperation between plants. The early pioneers do *not* make the environment more suitable for later species, and the later species achieve dominance despite changes caused by the early species rather than because of them. If seeds were available, broomsedge could colonize an abandoned field immediately rather than following horseweed and aster. Old-field succession is not described well by the facilitation or tolerance models but seems to be an excellent illustration of the inhibition model.

After this succession by herbs and grasses, abandoned farmland of the Piedmont of North Carolina is invaded in great numbers by shortleaf pine *(Pinus echinata)*. Pines seeds are blown by the wind and effectively disperse into abandoned fields. Pine seeds can germinate only on mineral soil and are able to become established in shady but bare sites among the herbs and grasses (Christensen and Peet 1981). Pine seedlings are very effective competitors for soil water and as the pine seedlings grow, they shade the herbs and grasses in the understory. The density of pines is very high but falls rapidly as the pines lose their dominance to hardwoods (Figure 22.9). After approximately 50 years several species of oaks become important trees in the understory, and the hardwoods gradually fill in the community. Reproduction of short-

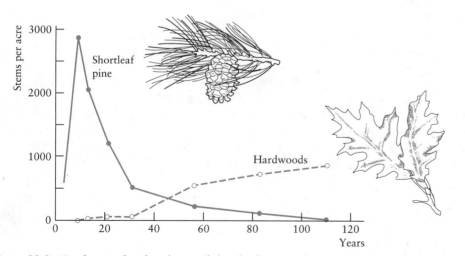

Figure 22.9   *Decline in the abundance of shortleaf pine and increase in the density of hardwood tree seedlings during succession on abandoned farmland in the Piedmont area of North Carolina. (After Billings 1938.)*

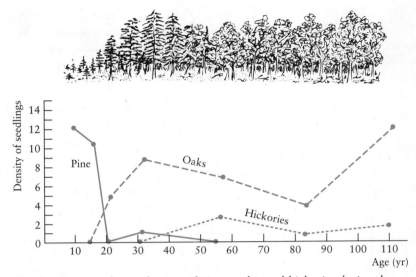

Figure 22.10   Density of reproduction of pines, oaks, and hickories during the shortleaf pine succession in the Piedmont area of North Carolina. (After Billings 1938.)

leaf pine is almost completely lacking after about 20 years (Figure 22.10) because there is no bare soil for seed germination, and pine seedlings cannot live in shade. The networks of pine roots in the soil become closed very quickly, and the accumulation of litter under the pines causes the old-field herbs to die out (Figure 22.11). Oak seedlings first appear after 20 years, when enough litter has accumulated to protect the acorns from desiccation and the soil is able to retain more moisture. Hardwood seedlings persist in the understory because they develop a root system deep enough to exploit soil water, because they are shade tolerant, and because they compete more effectively than pines for water and nutrients (see page 118).

Soil properties change dramatically along with this plant succession on Piedmont soils. Organic matter accumulates in the surface layers of the soil and increases in the deeper soil layers. Since the moisture-holding capacity of the soil increases with organic content, the soil becomes more able to hold colloidal water for use by the plants.

Thus shortleaf pine is independent of the early succession in that it requires only bare soil for germination. If all herbaceous species could be eliminated from the early succession, this would not affect colonization by pines, which fit the tolerance model of Figure 22.1 and invade old fields as soon as seeds become available. Oaks and other hardwoods, by contrast, depend on the soil changes caused by pine litter, so oak seedlings could not become established without the environmental changes produced by pines. Thus the latter part of this succession seems to fit the classical facilitation model.

Primary Succession on Mount St. Helens

Mount St. Helens in southwest Washington state erupted catastrophically on 18 May 1980. An area of about 400 meters was blown off the cone of this volcano and

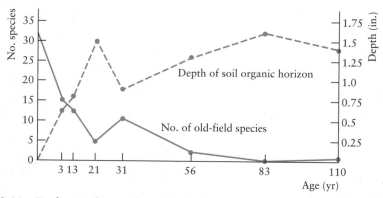

Figure 22.11   *Decline in the number of herbaceous species and increase in the soil organic matter during the succession from old-field to shortleaf pine in the Piedmont area of North Carolina. (After Billings 1938.)*

the blast from the eruption devastated a wide arc extending up to 18 kilometers north of the crater (Franklin et al. 1985). The eruption produced a variety of conditions for vegetative recolonization (del Moral and Wood 1988a). Figure 22.12 shows the three main areas affected by the eruption—the blast zone in which trees and vegetation were blown down but not eliminated, the pyroclastic or lava flow to

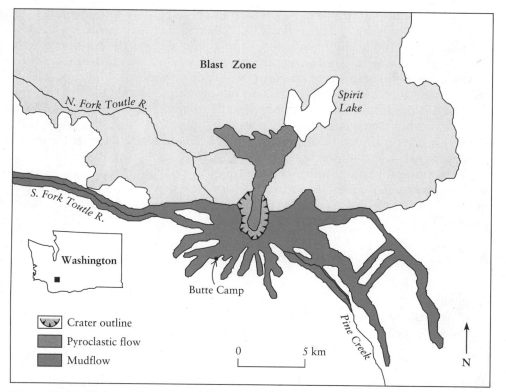

Figure 22.12   *Volcanic flows from Mount St. Helens in Washington state. Primary succession since the 1980 eruption has been very slow on both the mudflow areas and the pyroclastic (lava) flow area. The photos in Figure 22.13 (see color insert) were taken at Butte Camp. (From del Moral and Wood 1988a.)*

the north of the crater, and the extensive mudflows away from the crater. In addition, the eruption spewed tephra (ash) over thousands of square kilometers.

Primary succession following volcanic eruptions has been studied less than other forms of succession, and Mount St. Helens provided a good opportunity to study the mechanisms determining the rate of primary succession. Permanent plots have been established at several sites above the treeline around the crater, and early succession has been described by del Moral and Wood (1988a, 1993), Wood and del Moral (1988) and by del Moral (1993).

Colonization of habitats on Mount St. Helens has been slow. Figure 22.13 (see color insert) shows a time sequence from 1981 to 1990 of a small mudflow southwest of the crater. In this site there were no surviving plants in 1981, and by 1990 eleven species had colonized the area. None of the species were common.

Early primary succession on volcanic substrates rarely produces plant densities sufficient to inhibit the colonization of new species. Space or light are not limiting resources for plants in this environment. Nurse plants facilitate the establishment of other species. Lupines (*Lupinus lepidus*) have heavy seeds and are poorly dispersed but have become locally common on mudflows and pyroclastic surfaces (del Moral 1993). Before lupines get very common, other wind-dispersed plants like *Aster ledophyllus* and *Epilobium augustifolium* become established in lupine clumps and survive better in the shelter of these nurse plants. Lupines die after four to five years, and because they fix nitrogen they contribute on a local scale to increased soil nitrogen levels.

Chance events strongly affected primary succession on Mount St. Helens. Biological mechanisms are initially very weak in the severe environments produced by volcanic flows. The ability to become established in these severe environments is directly related to seed size (Wood and del Moral 1987). But dispersal ability is related inversely to seed size. Consequently, subalpine areas on Mount St. Helens receive many wind-blown seeds but almost none of these colonize the stressful conditions. When plants like lupine with large seeds colonize by chance, they become a focus of further community development. If by chance a single individual plant survives in the devastated landscape, it quickly becomes a center for seed dispersal to adjacent areas. Facilitation is important in early succession on Mount St. Helens but as the vegetative cover becomes more complete, inhibition may begin to affect further succession. Primary succession has been very slow because of erosion, low nutrient soils, and chronic drought stress, coupled with limited dispersal of larger seeds to areas distant from undisturbed vegetation.

Rocky Intertidal Algae in California

Intertidal boulder fields comprise much of the rocky shore in southern California. Algal succession on boulders occurs rapidly and can be studied experimentally by scraping rocks or putting out new boulders (Sousa 1979). The first colonists on open spaces on rocks are the "weeds" of the sea shore, the green algae *Ulva* and *Enteromorpha*. These pioneer species grow quickly and take up all the space. Some time later, large perennial brown or red algae become established and replace the green algae. Figure 22.14 shows these successional trends on three sets of concrete blocks set out in three different seasons. The perennial red alga *Gigartina canaliculata* is the late successional species in this area, and it gradually becomes dominant as succession proceeds.

What causes this succession? There are two major transition points. The first

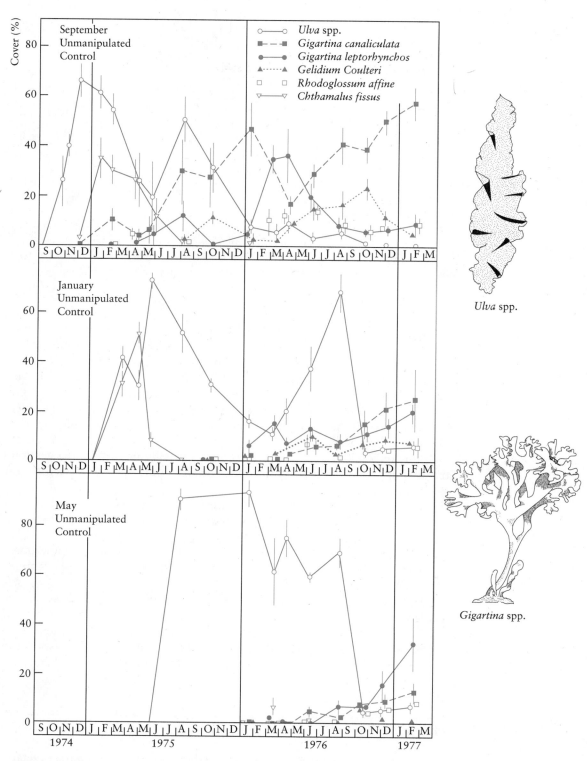

Figure 22.14 *Algal succession on concrete blocks set out in the intertidal zone near Santa Barbara, California. Sets of blocks were set out in September 1974, January 1975, and May 1975. The pioneer green alga Ulva is gradually replaced by perennial red algae. (After Sousa 1979.)*

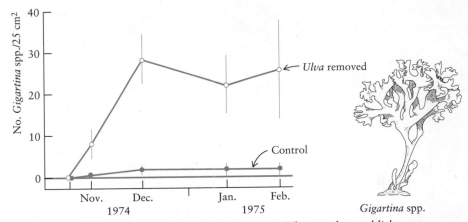

Figure 22.15 *Effects of removing the pioneer species* Ulva *on the establishment of the midsuccessional species of* Gigartina. Ulva *was manually removed from four concrete blocks and left undisturbed on four others. (After Sousa 1979.)*

occurs when *Ulva*, the early colonist, gives way to a community of red algae. Sousa (1979) removed *Ulva* to see if it was affecting the recruitment of red algae. Figure 22.15 shows that *Ulva* inhibits the establishment of *Gigartina*. With a similar removal experiment, Sousa showed that the middle successional species of red algae also inhibit the invasion of the dominant species *Gigartina canaliculata*. But if inhibition is the model for this succession, how do species replacements ever occur (Figure 22.14)? If the early successional species are more susceptible to the physical environment or to attack by herbivores than the later species, succession will proceed. *Ulva*, the pioneer species, is subject to high mortality because of desiccation stress, and it is the preferred food of the crab *Pachygrapsus*. Both these factors gradually reduce the abundance of *Ulva*. Overgrowth by epiphytes during the summer was an important cause of mortality in the middle successional species. The dominant alga, *G. canaliculata*, had few epiphytes; whether it possesses some chemical defense against epiphyte growth is not known (Sousa 1979). *Gigartina canaliculata* dominates this successional sequence because it has an effective grab-and-hold-open-space strategy. The successional events in the marine intertidal are a good example of the inhibition model.

The examples we have just discussed do not always fit the classical model of succession, as some replacements are facilitated while others are inhibited. Drury and Nisbet (1973) reviewed succession in forested regions and concluded that most forest succession does not conform to the classical model. If the classical model is not correct, we must reconsider the nature of the climax state, the "end point" of succession.

THE CLIMAX STATE

In the examples of succession described in the preceding section, the vegetation has developed to a certain stage of equilibrium. This final stage of succession is called the *climax*. Numerous definitions of the climax have been made (Phillips 1934–1935). *A climax is the final or stable community in a successional series. It is self-*

perpetuating and in equilibrium with the physical and biotic environment. There are three schools of thought about the climax state: The monoclimax school, the polyclimax school, and the climax-pattern view.

The *monoclimax* theory was an American invention of F. E. Clements (1916, 1936). According to the monoclimax theory, every region has only one climax community, toward which all communities are developing. This is the fundamental assumption of Clements—that given time and freedom from interference, a climax vegetation of the same general type will be produced and stabilized, irrespective of earlier site conditions. Climate, Clements believed, was the determining factor for vegetation, and the climax of any area was solely a function of its climate.

However, it was clear in the field that in any given area there were communities that were not climax communities but that appeared to be stable. For example, tongues of tallgrass prairie extended into Indiana from the west, and isolated stands of hemlock occurred in what is supposed to be deciduous forest. In other words, we observe communities in nature that are nonclimax according to Clements but apparently in equilibrium nevertheless. These communities are determined by topographic, edaphic (soil), or biotic factors.

The *polyclimax* theory arose as the obvious reaction to Clements's monolithic system. Tansley (1939) was one of the early proponents of the polyclimax idea—that many different climax communities may be recognized in a given area, such as climaxes controlled by soil moisture, soil nutrients, activity of animals, and other factors. Daubenmire (1966) is also a proponent of the idea that there may be several stable communities in a given area.

The real difference between these two schools of thought lies in the time factor of measuring relative stability. Given enough time, say the monoclimax students, a single climax community would develop, eventually overcoming the edaphic climaxes. The problem is, should we consider time on a geological scale or on an ecological scale? If we view the problem on a geological time scale, we would classify communities such as the coniferous forest as a seral stage to the establishment of deciduous forest. The important point here is that climate fluctuates and is never constant. We see this vividly in the Pleistocene glaciations and more recently in the advances and retreats of mountain glaciers in the last 1000 years (Figure 22.4). *So the condition of equilibrium can never be reached because the vegetation is subject not to a constant climate but to a variable one.* Climate varies on an ecological time scale as well as on a geological time scale. Succession then in a sense is continuous because we have a variable vegetation interacting with a variable climate.

Whittaker (1953) proposed a variation of the polyclimax idea, the *climax-pattern hypothesis.* He emphasized that a natural community is adapted to the whole pattern of environmental factors in which it exists—climate, soil, fire, biotic factors, and wind. Whereas the monoclimax theory allows for only one climatic climax in a region and the polyclimax theory allows for several climaxes, the climax-pattern hypothesis allows for a continuity of climax types, varying gradually along environmental gradients and not neatly separable into discrete climax types. Thus the climax-pattern hypothesis is an extension of the continuum idea and the approach of gradient analysis to vegetation (Whittaker 1953). The climax is recognized as a steady-state community with its constituent populations in dynamic balance with environmental gradients. We do not speak of a climatic climax but of prevailing climaxes that are the end result of climate, soil, topography, and biotic factors, as well as fire, wind, salt spray, and other influences, including chance. The utility of the climax as an operational concept is that similar sites in a region should produce similar climax stands. This stand-to-stand regularity should allow prediction for new sites of known environment, and we can say that a particular site should develop, in 100 years, say, a stand of sugar maple and beech of specified density.

A good example of how biotic factors may affect plant succession is found in the uplands of northwest Scotland (Miles 1987). Sheep grazing selects against trees and favors grassland (Figure 22.16). Several vegetation types occur in these Scottish uplands. Under low grazing pressure and no fire Scot's pine and birch woodlands tend to be favored. Under high grazing pressure and frequent fires only grassland is self-sustaining, and when grazing is moderate and fires are occasional both heather moor *(Calluna vulgaris)* and bracken fern *(Pteridium aquilinum)* communities are favored. There is no one climatic climax on these Scottish uplands, and the plant communities are a mosaic, changing as sheep grazing and fire frequency change. These communities are best described by Whittaker's climax-pattern hypothesis.

How can we recognize climax communities? The operational criterion is the attainment of a steady state over time. Because the time scale involved is very long, observations are lacking for most presumed successional sequences. We assume, for example, that we can determine the time course of succession for a spatial study of younger and older dune systems around Lake Michigan (Figure 22.7), but this translation of space and time may not be valid. In forests, we can use the understory of young trees to look for changes in species composition, because the large trees must reproduce themselves on a one-for-one basis if steady state has been achieved. Forest changes may be very slow. Lertzman (1992) studied a subalpine forest stand at 1100 meters elevation near Vancouver, Canada. This site was undisturbed by fire for almost 2000 years and still was not in equilibrium because the dominant hemlocks were not replacing themselves while amabilis fir seedlings were invading the understory in large numbers. Climax vegetation on this site was not reached after two millennia.

Climax forests will become monocultures if one species is able to replace itself exclusively and replace other species also. As early as 1905, French foresters suggested that in virgin forests, individual trees tended to be succeeded in time by those of another species (Fox 1977). This is called *reciprocal replacement* and is well illustrated in the beech–sugar maple forests of eastern North America. Beech seedlings are found predominantly under large sugar maples, and sugar maple seedlings occur predominantly under large beech trees. Reciprocal replacement is one mechanism for maintaining two species as dominants in climax communities and is a good example of stable coexistence between competing species (see Figure 13.3, page 235).

Some communities may appear to be stable in time and yet may not be in equilibrium with climatic and soil factors. A striking example of this occurred after the outbreak of the disease myxomatosis in the European rabbit in Britain. Before 1954, rabbits were common in many grassland areas. Myxomatosis devastated the rabbit population in 1954, and the consequent release of grazing pressure caused dramatic changes in grassland communities (Thomas 1960, 1963). The most obvious change was an increase in the abundance of flowers. Species that had not been seen for many years suddenly appeared in large numbers. There was also an increase in woody plants, including tree seedlings that were commonly grazed by rabbits. No one anticipated these effects following the removal of rabbits.

We conclude from this discussion that climax vegetation is an abstract concept that is, in fact, seldom realized, owing to the continuous fluctuations of climate. The climate of an area has clear overall control of the vegetation, but within each of the

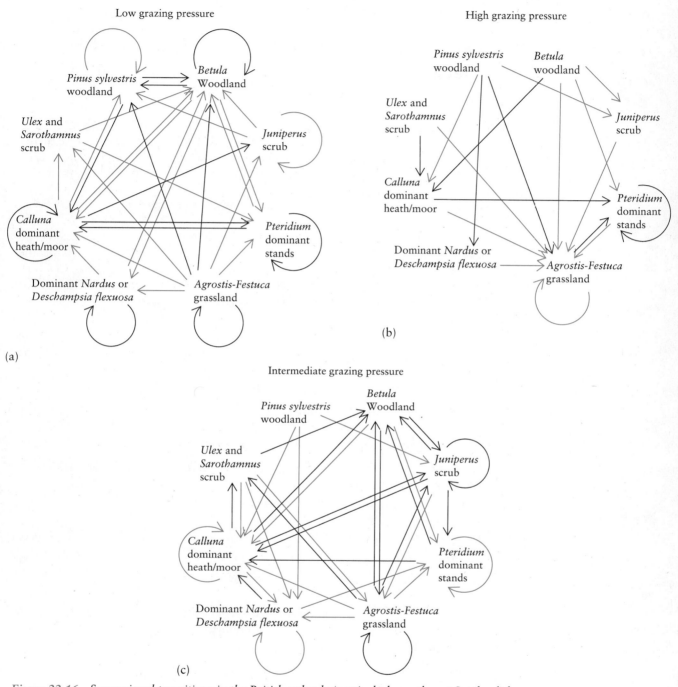

Figure 22.16 *Successional transitions in the British uplands (particularly northwest Scotland) between eight vegetation types given (a) low grazing pressures and no burning, (b) high grazing pressures and frequent burning, and (c) intermediate levels of grazing and occasional burning. Broad arrows represent common transitions, thin arrows less frequent transitions, and curved arrows self-replacement. The vegetation types are arranged so that types tending to acidify soils are on the left, and types with contrasting soil effects are on the right. (Modified after Miles 1985.)*

broad climatic zones are many modifications, caused by soil, topography, and animals, that lead to many climax situations. The rate of change in a community is rapid in early succession but becomes very slow as it approaches the climax.

CYCLIC CHANGES IN COMMUNITIES

It is usual to regard a community that is stable and in equilibrium with its environment as a static thing that will no longer change. This is misleading because there is a whole class of community changes that are nonsuccessional and cyclic and are internally caused by species interrelations. These cyclic events usually occur on a small scale and are repeated over and over in the whole of the community. They are part of the internal dynamics of the community rather than a part of succession. Three examples will be described.

Watt (1947b) has studied several examples of cyclic vegetation changes. One of these was the dwarf *Calluna* heath in Scotland. The dominant shrub is heather *(Calluna)*, which loses its vigor as it ages and is invaded by the lichens *Cladonia*. The lichen mat in time dies to leave bare ground. This bare area is invaded by bearberry *(Arctostaphylos)*, which in turn is invaded by *Calluna*.

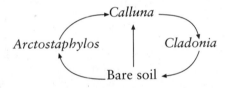

Heather *(Calluna)* is the dominant plant, and *Arctostaphylos* and *Cladonia* are allowed to occupy the area that is temporarily vacated by *Calluna*.

The cycle of change can be divided into four phases (Figure 22.17):

1. *Pioneer*: establishment and early growth in *Calluna*—open patches, with many plant species (years 6 to 10).
2. *Building*: maximum cover of *Calluna* with vigorous flowering—few associated plants (years 7 to 15).
3. *Mature*: gap begins in *Calluna* canopy and more species invade the area (years 14 to 25).
4. *Degenerate*: central branches of *Calluna* die; lichens and bryophytes become very common (years 20 to 30).

Barclay-Estrup and Gimingham (1969) describe this sequence in detail from maps of permanent quadrats in Scotland. The life history of the dominant plant *Calluna* controls the sequence.

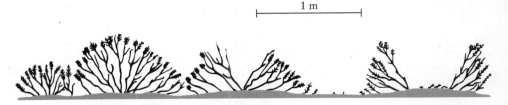

Figure 22.17   Profile of the four phases in the Calluna *cycle in Britain. Like many perennial plants, heather loses vigor with age. (After Watt 1955.)*

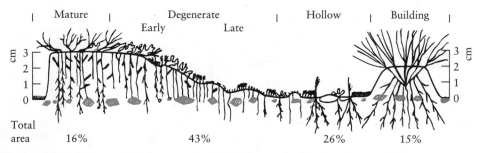

Figure 22.18   Phases of the hummock-and-hollow cycle showing change in flora and habitat and indicating the "fossil" shoot bases and detached roots of Festuca ovina *in the soil. (After Watt 1947b.)*

Watt also studied cyclic changes associated with microtopography in a grassland in England: the hummock-and-hollow cycle. The vegetation of the grassland Watt studied was very patchy, and he could recognize four stages (Figure 22.18). The whole scheme centers around the grass *Festuca ovina*. The seedlings of this grass get established in the bare soil of the hollow stage. It builds a "tussock" by trapping windborne soil particles and by its own growth. The vigor of this grass declines with age; it begins to degenerate in the mature phase and is invaded by lichens in the early degenerate phase. These lichens use the organic matter and in turn die, and the hummock is eroded down to base level, only to begin the process again.

At any given time, all four stages can be found in a *Festuca* grassland. Seedlings cannot usually get established except in the hollow and building phases. Fescue seems to be the dominant plant (Figure 22.19). Lichens dominate the degenerating phase,

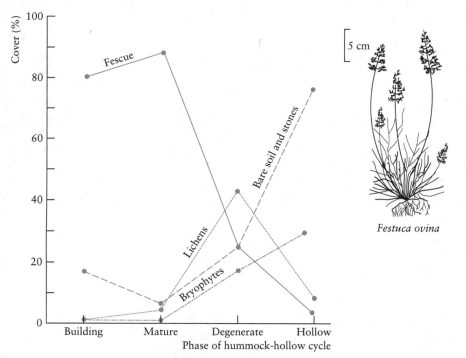

Figure 22.19   Relative abundance of the dominant grass Festuca ovina *in the different phases of the hummock-and-hollow cycle and the change in lichens, bryophytes, and bare soil. (After Watt 1947b.)*

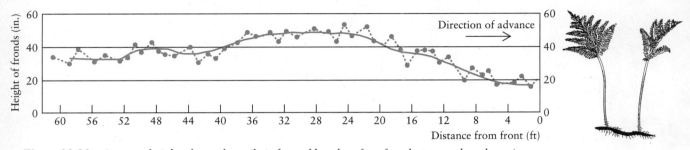

Figure 22.20   *Average height above the soil surface of bracken fern fronds across the advancing margin. (After Watt 1940.)*

when they can use the organic remains accumulated. Bryophytes seem to suffer competition from fescue and cannot get established except in the degenerate or hollow phase.

Bracken fern *(Pteridium aquilinum)* is a cosmopolitan species that lives in a great variety of soil types and climatic conditions. It forms rhizomes in the soil, and once established, it spreads vegetatively. Bracken is fire-resistant because of its rhizomes. Watt (1947a) studied bracken in the process of invading grassland and found a vigorous "front" of invading bracken but reduced vigor in older fronds (Figure 22.20). The obvious explanation for this marginal effect is that some soil nutrient is depleted by the advancing fern and is in short supply in the older stands. However, Watt (1940) could find no soil change to account for the reduced vigor, and the addition of fertilizers to sample plots in the older stands produced no effect. The significant variable seems to be rhizome age. Younger rhizomes produce more vigorous fronds. The explanation for this is not known.

Watt divided all these cycles of change into an *upgrade* series and a *downgrade*

Table 22.4   Relative Behavior of Three Tree Species that Colonize Treefall Gaps in Tropical Rain Forest, Barro Colorado Island, Panama

Species	Family	Type of Fruit	Seed Dispersal Agents
Trema micrantha	(Ulmaceae)	Drupe	Many bird species
Cecropia insignis	(Moraceae)	Seeds compressed in peduncles	Birds, bats
Miconia argentea	(Melastomataceae)	Berry	Monkeys, birds

RECRUITMENT

	First Year After Gap Formation	Second Year	Survivors to Years 8, 9
Trema	nearly all in first year	none after year 2	only those established in first year
Cecropia	most in first year	many in year 2, none after year 3	from each of these years
Miconia	begin in first year	highest in years 2–5 and continue to years 7–9	from each of these years

SIZES OF GAP OCCUPIED

	Years 1–3 (seedlings)	Years 8, 9 (saplings, poles)
Trema	50 m² and larger	Largest gaps only (> 376 m²)
Cecropia	20 m² and larger	In largest and mid-sized gaps (> 215 m²)
Miconia	20 m² and larger	All sizes of gaps down to 102 m²

Source: Data in Brokaw (1987).

series and pointed out that the total productivity of the series increases to the mature phase and then decreases. What initiates the downgrade phase? A possible explanation lies in the relationship between the general vigor of a perennial plant and its age. There seems to be a general relationship between age and performance in most perennial plants, and consequently between age and competitive ability (Kershaw 1973). Several studies on the relation of leaf diameter to age also support this idea.

For this reason, a stable community will be in a constant state of phasic fluctuation, one species becoming locally more abundant as another species reaches its degenerate phase. These dynamic interrelationships in natural communities may not be conspicuous without detailed measurement.

Tropical rain forests are an example of a community that may be a floristic mosaic of treefall-created patches undergoing cyclic changes (Richards 1952). Forest regeneration occurs by the production of a gap, colonization and growth in the gap, followed by a mature tree phase and eventual treefalls renewing the cycle. Rain and wind cause treefalls. On Barro Colorado Island in Panama gaps averaged 86 square meters and about 3 percent of the forest area was in gaps (Brokaw 1985). Gaps of this size were created at a rate of about one per hectare per year on Barro Colorado. Trees in tropical forests are loosely classified in two groups. Primary, or shade-tolerant, species live as suppressed small trees in the understory until a canopy gap opens above them. Pioneers, or shade-intolerant, species germinate only in gaps and grow rapidly. Brokaw (1987) tested the idea that each pioneer species occupies a separate regeneration niche on Barro Colorado. He studied three species that were most abundant in 30 natural gaps. Table 22.4 gives the details of his results, and Figure 22.21 shows the growth rates of the three species. *Trema micrantha* recruits

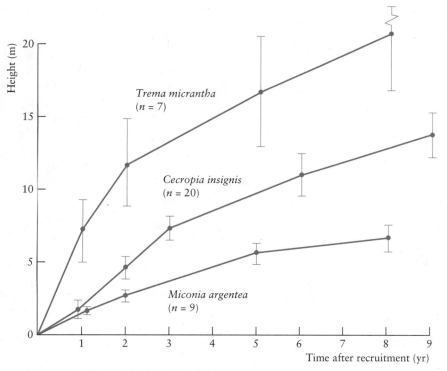

Figure 22.21   Mean height (± 1 S. D.) of three pioneer tree species at intervals after recruitment in treefall gaps on Barro Colorado Island, Panama. Sample sizes given in parentheses. (From Brokaw 1987.)

to gaps only in the first year and grows most rapidly. *Miconia argentea* grows more slowly but recruits over several years to the gaps. But there is overlap in these pioneer species, and the regeneration niches of these trees are not discrete. The floristic mosaic of lowland tropical forests is partly caused by the cyclic changes in treefall gaps.

Thus climax communities are dynamic and may change in cyclic patterns because of the life cycle of dominant species. Four phases can often be recognized: pioneer, building, mature, and degenerate. A "stable" community may be a mosaic of these four phases of cyclic change operating at a local level.

SUMMARY

Communities may change with time either in a directional way (succession) or in a nondirectional way (cyclic). Most of the work on community changes has been done on plants, but the same principles apply to animal communities.

Four models of succession have been proposed to explain directional vegetation changes. All models agree that pioneer species in a succession are usually fugitive or opportunistic species with a high dispersal rate and rapid growth. How are these pioneer species replaced? The classical model states that species replacements in later stages of succession are *facilitated* by organisms present in earlier stages. The inhibition model at the other extreme suggests that species replacements are *inhibited* by earlier colonizers and that successional sequences are controlled by who gets there first. The tolerance model suggests that species replacements are not affected by earlier colonizers and that later species in succession are those able to *tolerate* lower levels of resources than earlier species. The random colonization model is a null model that suggests that species replacements occur completely randomly with no interspecific interactions. A few parts of successional sequences fit the classical model, but more seem to fit the inhibition model. No single model explains an entire successional sequence. We now try to analyze succession as a dynamic process resulting from a balance between the colonizing ability of some species and the competitive ability of others. Succession does not always involve progressive changes from simple to complex communities.

Succession proceeds through a series of seral stages from the pioneer stage to the climax stage. The *monoclimax hypothesis* suggested that there was only a single predictable end point for whole regions and that, given time, all communities would converge to the climatic climax. This hypothesis has been superseded by the *climax pattern hypothesis,* which suggests a continuity of different climaxes varying along environmental gradients controlled by soil moisture, nutrients, herbivores, fires, or other factors.

Cyclic changes are repeated over and over again as part of the internal dynamics of the community. The life cycle of the dominant organisms dictates the cyclic changes, many of which are caused by the decline in vigor of perennial plants with age. In many forests treefall gaps create a mosaic of patches undergoing cyclic changes within a climax community.

Communities are not stable for long periods in nature because of short-term changes in climate (or other environmental factors) and the cyclic changes of growth and decay within the community. For most communities, we observe changes over time but do not know all the factors that cause the changes; therefore, we find it difficult to suggest manipulations to alleviate undersirable trends. This problem is particularly critical in communities influenced by human technology.

Selected References

BROKAW, N. V. 1987. Gap-phase regeneration of three pioneer tree species in a tropical forest. *Journal of Ecology* 75:9–19.

BURROWS, C. J. 1990. *Processes of Vegetation Change,* chaps. 12 and 13, pp. 420–489. Unwin Hyman, London.

CLEMENTS, F. E. 1949. *Dynamics of Vegetation.* Macmillan (Hafner Press), New York.

CONNELL, J. H., and R. O. SLATYER. 1977. Mechanisms of succession in natural communities and their role in community stability and organization. *American Naturalist* 111:1119–1144.

HORN, H. S. 1981. Succession. In *Theoretical Ecology,* ed. R. M. May, pp. 253–271. Blackwell, Oxford, England.

HUSTON, M., and T. SMITH. 1987. Plant succession: Life history and competition. *American Naturalist* 130:168–198.

MILES, J. 1987. Vegetation succession: Past and present perceptions. In *Colonization, Succession and Stability,* ed. A. J. Gray, M. J. Crawley, and P. J. Edwards, pp. 1–29. Blackwell, Oxford, England.

NIERING, W. A. 1987. Vegetation dynamics (succession and climax) in relation to plant community management. *Conservation Biology* 1:287–295.

SHUGART, H. H. 1984. *A Theory of Forest Dynamics: The Ecological Implications of Forest Succession Models.* Springer-Verlag, New York.

WALKER, L. R., and F. S. CHAPIN III. 1987. Interactions among processes controlling successional change. *Oikos* 50:131–135.

WATT, A. S. 1947. Pattern and process in the plant community. *Journal of Ecology* 35:1–22.

Questions and Problems

1. Discuss the inhibition, facilitation, and tolerance models of succession with respect to the following simple experiment. In this hypothetical succession, species A normally precedes species B. Two treatments are applied: (1) all of species A are removed from a series of replicate plots, (2) a portion of species B equal to the biomass of A is removed from another set of plots. Growth in biomass is then measured over several years. Interpret all the possible outcomes of this experiment. Compare your analysis with that of Botkin (1981).

2. Culver (1981) observed these transition probabilities for a forest in the Great Smoky Mountains National Park in North Carolina and Tennessee:

	SPECIES IN SAPLINGS		
Species in Canopy	**Yellow Birch**	**Red Spruce**	**Fraser Fir**
Yellow Birch	0.01	0.45	0.54
Red spruce	0.13	0.20	0.67
Fraser fir	0.12	0.28	0.61

Calculate the changes over five generations in the composition of a forest of these three species starting from equal numbers of each species.

3. Abundance within size classes in an undisturbed hemlock-beech association at Heart's Content, Pa., was measured by Lutz (1930, p. 27):

	ABUNDANCE (%) PER SIZE CLASS				
Tree Species	**0–0.9 ft[a]**	**1.0 ft[a] to 0.9 in. DBH[b]**	**1–3.5 in. DBH[b]**	**3.6–9.5 in. DBH[b]**	**≥ 10 in. DBH[b]**
Hemlock	23.4	44.0	19.2	13.5	36.1
White pine	0.2	0.1	—	—	11.1
Beech	4.9	22.7	59.9	50.3	24.0
Red maple	66.5	17.0	6.6	9.2	10.6
Chestnut	0.2	1.5	0.4	9.2	8.8
White oak	—	—	—	—	1.6
Red oak	0.1	0.2	—	0.7	1.8
Black birch	0.8	1.4	4.8	10.4	2.7
Black cherry	0.3	1.0	1.7	0.9	0.8
Yellow birch	0.5	0.1	0.6	1.7	0.1
Sugar maple	0.4	1.2	1.1	0.7	0.3
White ash	0.1	—	—	0.1	—
Miscellaneous	2.6	10.8	5.7	3.3	2.1
Total	100.0	100.0	100.0	100.0	100.0

[a] Height of tree.
[b] Diameter at breast height (1.4 m above ground)

Lutz stated that "the hemlock-beech association is believed to represent a stage in forest succession somewhat less advanced than the climatic climax of the region. Probably in the climax forest the amount of white pine entering into the stand is considerably smaller." Do these data support this conclusion? If so, how? If not, what additional data could support it?

4. Langford and Buell (1969, p. 130) state:

Whereas biotic influences play an outstanding role in determining the nature of climax vegetation in moist temperature areas, abiotic factors are outstandingly pre-eminent in controlling vegetation in arid or very cold regions. In such regions succession, which is essentially due to modification of the environment by organisms, with its direction of course somewhat variable according to the availability of various propagules, may be almost absent.

Search for data on succession and the climax for either desert or arctic plant communities, and discuss them with reference to this statement.

5. Discuss the application of the succession concept to communities in the sea.

6. Relate the adaptive strategies of species in early and late stages of succession to the *r* and *K* selection ideas discussed in Chapter 13, pages 254–255.

7. In discussing forest succession as a plant-by-plant replacement process, Horn (1975b, p. 210) states: "Copious self-replacement does not guarantee a species' abundance or even its persistence in late stages of succession." How can this be true?

8. Williamson and Black (1981) suggest that succession from pines to oaks (see Figure 22.9) can be interrupted by fire, which kills the oaks but not the pines. Natural selection has produced pine foliage and pine litter that facilitates fires and thus inhibits oaks from taking over a succession. If this is correct, suggest how this type of adaptation could evolve.

9. In the primeval forest landscape, where were the plants that are abundant today in old fields (page 496)? Discuss the evolution of colonizing ability in plants that evolved in temporary forest openings and those that evolved in persistent open, marginal habitats. Compare your conclusions with those of Marks (1983).

Overview Question

Could you construct a theory of succession in plant communities solely on the mechanism of competition between species? What would be missing from such a theory?

23

Community Organization I: Biodiversity

Ecological communities do not all contain the same number of species, and one of the currently active areas of research in community ecology is the study of species richness or biodiversity. A. W. Wallace (1878) recognized that animal life was on the whole more abundant and varied in the tropics than in other parts of the globe, and the same applies to plants. Other patterns of variation have long been known on islands; small or remote islands have fewer species than large islands or those nearer continents (MacArthur and Wilson 1967). The regularity of these patterns for many taxonomic groups suggests that they have been produced in conformity with a set of basic principles rather than as accidents of history. How can we explain these trends in species diversity?

MEASUREMENT OF BIODIVERSITY

The simplest measure of biodiversity is to count the *number of species*. In such a count we should include only resident species, not accidental or temporary immigrants. It may not always be easy to decide which species are accidentals: Is a bottomland tree species growing on a ridge top an accidental species or a resident one? The number of species is the first and oldest concept of species diversity and is called *species richness*.

A second concept of species diversity is that of *heterogeneity*. One problem with counting the number of species as a measure of diversity is that it treats rare species and common species as equals. A community with two species might be divided in two extreme ways:

	Community 1	Community 2
Species A	99	50
Species B	1	50

The first community is very nearly a monoculture, and the second community would seem intuitively to be more diverse than the first. We can combine the concepts of number of species and relative abundance into a single concept of heterogeneity. Heterogeneity is higher in a community when there are more species and when the species are equally abundant.

A difficult problem arises in trying to determine the number of species in a biological community: *Species counts depend on sample size*. Adequate sampling can usually get around this difficulty, particularly with vertebrate species, but not always with insects and other arthropods, in which species counts cannot be complete.

514

Two different strategies have been adopted to deal with these problems. First, a variety of statistical distributions can be fitted to data on the relative abundances of species. One very characteristic feature of communities is that they contain comparatively few species that are common and comparatively large numbers of species that are rare. Since it is relatively easy to sample any given area and count the *number of species* on the area and the *number of individuals* in each of these species, a great deal of information of this type has accumulated (Williams 1964). The first attempt to analyze these data was made by Fisher, Corbet, and Williams (1943).

In many faunal samples, the number of species represented by a single specimen is very large, species represented by two specimens are less numerous, and so on until only a few species are represented by many specimens. Fisher, Corbet, and Williams (1943) plotted the data and found that they fitted a "hollow curve" (Figure 23.1), and that the data could be described mathematically by the logarithmic series. The most significant ecological observation is that the largest number of species in a community fall into the "very rare" category.

The logarithmic series arises in communities with relatively few species in which a single environmental factor is of dominant importance (May 1975). It describes an extreme type of "niche preemption" in which the most successful species preempts a fraction k of the total resources, the next species a fraction k of the remaining resources, and so on. This type of niche-preemption hypothesis predicts a logarithmic series distribution or "hollow-curve" as a description of species abundances in natural communities.

The logarithmic series implies that the greatest number of species has minimal abundance, that the number of species represented by a single specimen is always maximal. This is not the case in all communities. Figure 23.2 shows the relative abundance of breeding birds in Quaker Run Valley, New York. The greatest number of bird species is represented by ten breeding pairs, and the relative abundance

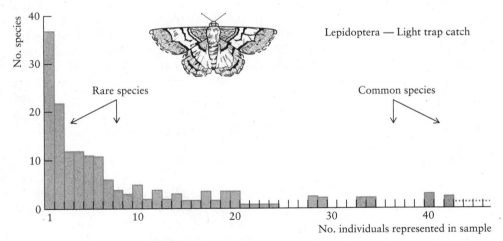

Figure 23.1   *Relative abundance of Lepidoptera (butterflies and moths) captured in a light trap in Rothamsted, England, in 1935. Not all of the abundant species are shown. There were 37 species represented in the catch by only a single specimen (rare species); one very common species was represented by 1799 individuals in the catch. A total of 6814 individuals were caught, representing 197 species. Six common species comprised 50 percent of the total catch. (Modified from Williams 1964.)*

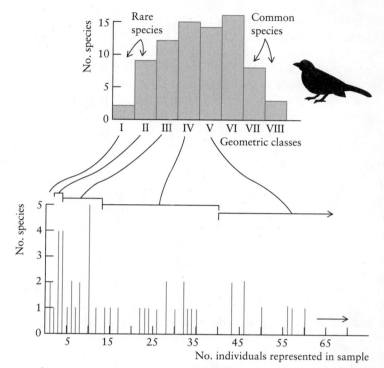

Figure 23.2 *Relative abundance of nesting bird species in Quaker Run Valley, New York. The lower figure shows the distribution on an arithmetic scale, and the upper figure shows the same data on a geometric scale with × 3 size groupings (1, 2–4, 5–13, 14–40, 41–121, etc.). (After Williams 1964.)*

pattern does not fit the hollow-curve pattern of Figure 23.1. Preston (1948) suggested expressing the x axis (number of individuals represented in sample) on a geometric (logarithmic) scale rather than an arithmetic scale. When this conversion of scale is done and the species are combined into classes whose ranges of species abundances increase geometrically, relative abundance data take the form of a bell-shaped, normal distribution, and because the x axis is expressed on a geometric or logarithmic scale, this distribution is called *log-normal*. The essential point is that populations tend to increase geometrically rather than arithmetically (Chapter 12, page 199), so that the natural way to analyze abundances is as the *logarithm* of population density.

The log-normal distribution fits a variety of data from surprisingly diverse communities. Figure 23.3 gives just a few examples of relative abundance patterns in different communities. The log-normal distribution arises in all communities in which the total number of species is large and the relative abundance of these species is determined by many factors operating independently. The log-normal distribution is thus the expected statistical distribution for many biological communities (May 1975). There is something very compelling about the log-normal distribution. The fact that moths in England, freshwater algae in Spain, snakes in Panama, and birds in New York all have a similar type of species abundance curve suggests a possible regularity in community structure. The log-normal distribution can describe all the data that fit the logarithmic series and is a more general model of species abundance patterns in natural communities.

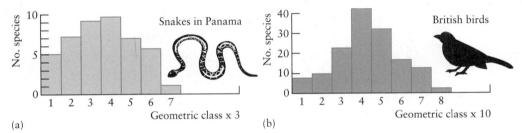

Figure 23.3   *Log-normal distribution of relative abundances in two diverse communities: (a) snakes in Panama, (b) British birds. (Data from Williams 1964.)*

The log-normal distribution was recognized as an empirical regularity before it was given a theoretical justification (Preston 1962). Sugihara (1980) has provided one explicit biological mechanism that leads to the log-normal. Assume that a community has a set of total niche requirements so that we can define a communal niche space. This niche space is likened to a unit mass that is sequentially split up by the component species so that each fragment denotes relative species abundance. Consider Sugihara's model for the simple case of a three-species community. The total niche space is broken randomly to produce two fragments (Figure 23.4). The larger of the two fragments must range in size from 0.5 to 1.0 (of the original unit mass) and will average statistically to be 0.75 units. One of these two fragments is now chosen at random and broken to yield a third fragment. If the smaller fragment is broken (breakage sequence B in Figure 23.4), we end up with three "species" with relative abundances on average of 0.75, 0.19, and 0.06. If the larger fragment is broken in the second step (breakage sequence A in Figure 23.4), we end up with three "species" with relative abundances on average of 0.57, 0.28, and 0.15 (Sugihara 1980). These relative abundance estimates are averages that would apply to many three-species communities. Any one particular community will vary because the breakage occurs at random. Multispecies communities are more difficult to do these calculations for, but the principles remain the same. The important point is that the subdividing is done sequentially and not instantaneously, which corresponds

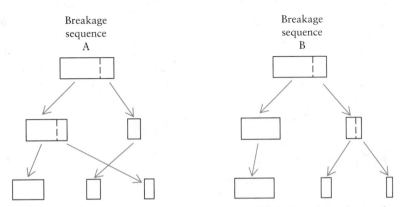

Figure 23.4   *A hypothetical illustration of the sequential breakage hypothesis of Sugihara (1980). Two possible breakage sequences are illustrated for a hypothetical three-species community. (From Sugihara 1980.)*

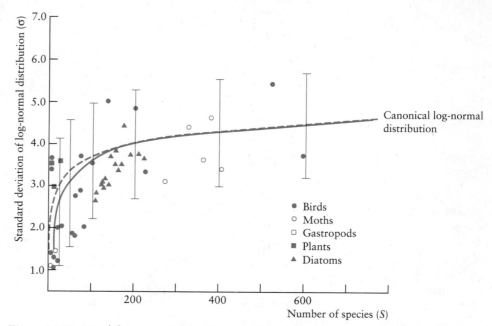

Figure 23.5 *Fit of the sequential breakage model to observed community data from a wide variety of organisms. The dashed line shows the expected relationship between the number of species and the spread of the log-normal distribution if Preston's (1962) canonical model is correct. The solid line is the average prediction of the sequential breakage hypothesis (with 2 standard deviations). The sequential breakage model is virtually identical to the canonical distribution and fits the observed data very well. (After Sugihara 1980.)*

with the biological assumption that the niche structure for communities is hierarchical. The niche space of a community has many dimensions and must not be thought of as a single resource axis.

The sequential breakage hypothesis predicts relative abundance patterns that are log-normal (Figure 23.5). Data from various communities fit this hypothesis very well, and the empirical findings of Preston (1962) can thus be biologically interpreted as a consequence of sequential niche subdivision.

A second approach to species diversity involves measures of the heterogeneity of a community. Several measures of heterogeneity are in use (Magurran 1988), and the most popular has been borrowed from information theory. The main objective of information theory is to try to measure the amount of order (or disorder) contained in a system. We ask the question, How difficult would it be to predict correctly the species of the next individual collected? This is the same problem faced by communication engineers interested in predicting correctly the name of the next letter in a message. This uncertainty can be quantified by a measure of information content, the Shannon-Wiener function, and the details of how to calculate this measure of heterogeneity are provided in Appendix IV (page 704).

To summarize, to measure biodiversity you need a combination of two types of data: (1) the number of species in the sample of the community, and (2) the relative abundance pattern of the community.

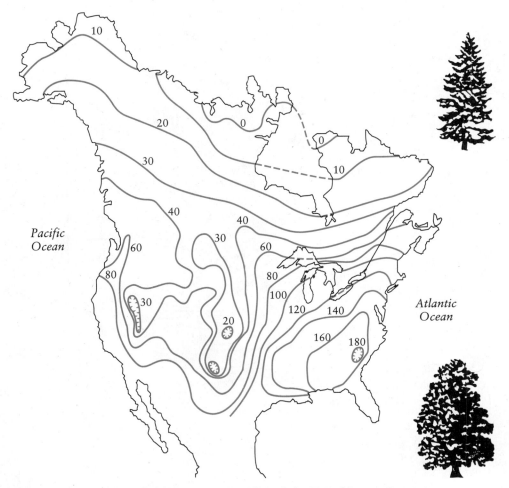

Figure 23.6   Number of tree species in Canada and the United States. Contours connect points with the same number of species. (From Currie and Paquin 1987.)

SOME EXAMPLES OF DIVERSITY GRADIENTS

Tropical habitats support large numbers of species of plants and animals, and this diversity of life in the tropics contrasts starkly with the impoverished faunas of temperate and polar areas. A few examples will illustrate this global gradient. The tropical rain forest in Malaysia may contain up to 227 species of trees on a plot of 2 hectares and 375 tree species on a plot of 23 hectares (Richards 1969). A deciduous forest in Michigan will contain 10 to 15 species on a plot of 2 hectares.

There are 620 native tree species in North America north of Mexico. Figure 23.6 shows the polar to temperate gradient in tree species. More species occur in southeastern U.S. forests than occur in western forests, and minima occur in the rain shadows just east of the Rocky Mountains and the Sierra Nevada (Currie and Paquin 1987).

Ants are much more diverse in the tropics (Fischer 1960):

Figure 23.7 Geographic pattern of biodiversity in the land birds of North and Central America. Contour lines give the numbers of species present. (From Cook 1969.)

	No. Ant Species
Brazil	222
Trinidad	134
Cuba	101
Utah	63
Iowa	73
Alaska	7
Arctic Alaska	3

There are 293 species of snakes in Mexico, 126 in the United States, and 22 in Canada. Figure 23.7 shows the number of breeding land-bird species in different parts of North America.

Freshwater fishes are much more diverse in tropical rivers and lakes. Over 1000 species of fishes have been found in the Amazon River in South America, and exploration is still incomplete in this region. By contrast, Central America has 456 fish species, and the Great Lakes of North America have 172 species (Lowe-McConnell 1969). Lake Tanganyika in East Africa has 214 fish species; all of Europe has only 192.

Marine invertebrates also have high diversity in the tropics. Figure 23.8 shows that calanoid copepods (planktonic crustaceans) are most diverse in the tropical Pacific and least diverse in the Bering Sea and Arctic Ocean.

Not all floras and faunas show a smooth trend of biodiversity from the poles to the tropics. Figure 23.9 shows the diversity of Australian carnivorous marsupials. Species diversity is highest in the arid zone of central Australia for the small carnivorous marsupials. Tropical Australia is species-poor for both small and large carnivorous marsupials (Dickman 1989).

Species diversity patterns of North American mammals were analyzed in detail by Simpson (1964) and are a good example of a complex gradient. Figure 23.10 shows that the number of land-mammal species increases from 15 in northern Canada to over 150 in Central America. Simpson recognized five notable features of this pattern:

1. *North-south gradient.* The north-south gradient is not smooth. Some mammal groups are most diverse in the temperate zone—pocket gophers, shrews, ungu-

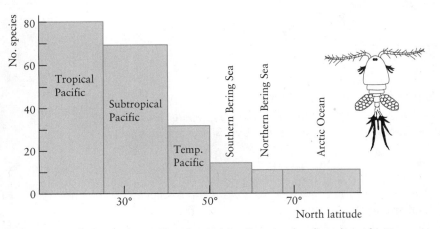

Figure 23.8 *Tropical-to-polar gradient in species diversity for the calanoid copepods of the upper 50 meters of the Pacific Ocean. (After Fischer 1960.)*

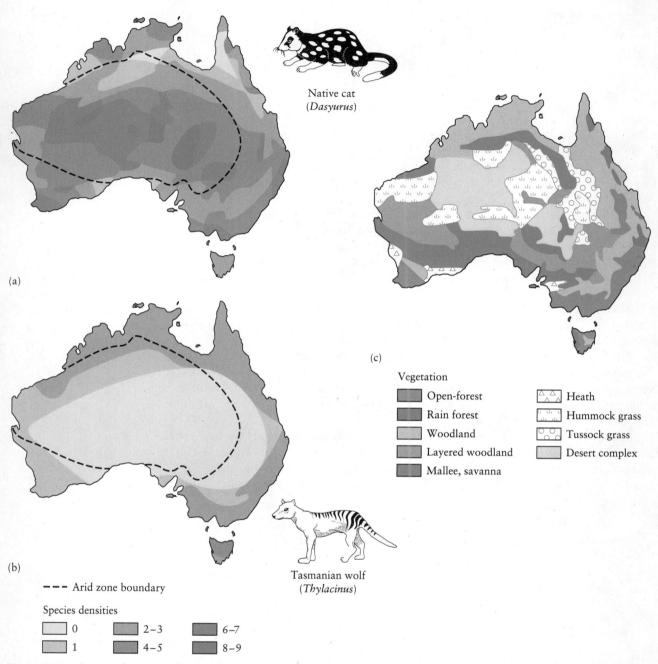

Figure 23.9 *Species diversity of Australian carnivorous marsupials. (a) Small species (<250 g). (b) Large species (>500 g). (c) Generalized vegetation map. The arid center of Australia is shown by the dashed line. (From Dickman 1989.)*

lates—and become less diverse toward the tropics. Bats contribute most of the high species richness for mammals in the tropics (Wilson 1974).

2. *Topographic relief.* Areas like the Rocky Mountains or the Appalachians support a higher than average number of mammal species.

3. *East-west trends.* Superimposed on the topographic variation is a general trend

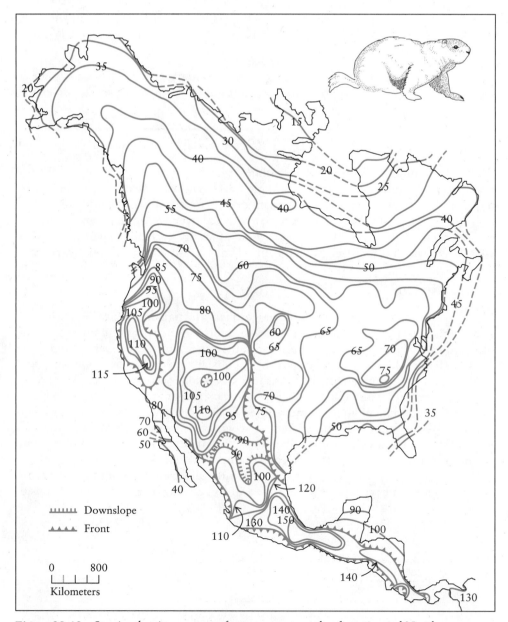

Figure 23.10 Species density contours for recent mammals of continental North America. The contour lines are isograms for numbers of continental (nonmarine and noninsular) species in quadrats 150 miles (240 km) square. The "fronts" are lines of exceptionally rapid change that are multiples of the contour interval for the given region. (After Simpson 1964.)

toward more species in the west than in the east (Figure 23.10). The topographically uniform Great Plains contain as many mammal species as the topographically diverse Appalachian Mountains.

4. *Fronts of abrupt change.* Areas of rapid change in species diversity are often but not always associated with mountain ranges.

5. *Peninsular "lows."* On peninsular areas, such as Florida, Baja California, the Alaska Peninsula, and Nova Scotia, the number of mammal species is less than on adjacent continental areas.

This brief look at some details of species diversity gradients can assist us in looking at some factors proposed to affect latitudinal gradients in species diversity.

FACTORS THAT CAUSE DIVERSITY GRADIENTS

Tropical-to-polar gradients in species richness may be produced by up to eight causal factors that are difficult to untangle (Table 23.1). These gradients represent a complex community property in which we cannot look for a single explanation involving only one causal factor. Many causes have interacted over evolutionary and ecological time to produce the assemblages we see today. For any particular diversity gradient we can ask which of these eight factors are involved and which are most important.

History Factor

This idea, proposed chiefly by zoogeographers and paleontologists, is a historical hypothesis with two main components. First, biotas in the warm, humid tropics are likely to evolve and diversify more rapidly than those in the temperate and polar regions (Figure 23.11). This is caused by a constant favorable environment and a relative freedom from climatic disasters like glaciation. Second, biotic diversity is a product of evolution and therefore is dependent on the length of time through which the biota has developed in an uninterrupted fashion (Fischer 1960). Tropical biotas are examples of mature biotic evolution, whereas temperate and polar biotas are immature communities. In short, all communities diversify in time, and older communities consequently have more species than younger ones.

The history factor may operate on an ecological or an evolutionary time scale. The ecological time scale is a shorter time scale, operating over a few generations or

Table 23.1 Factors Hypothesized to Influence Biodiversity

Factor	Rationale
1. History	More time permits more complete colonization and the evolution of new species
2. Spatial heterogeneity	Physically or biologically complex habitats furnish more niches
3. Competition	a. Competition favors reduced niche breadth b. Competitive exclusion eliminates species
4. Predation	Predation retards competitive exclusion
5. Climate	Climatically favorable conditions permit more species
6. Climatic variability	Stability permits specialization
7. Productivity	Richness is limited by the partitioning of production among species
8. Disturbance	Moderate disturbance retards competitive exclusion

Source: Modified after Pianka (1988) and Currie (1991).

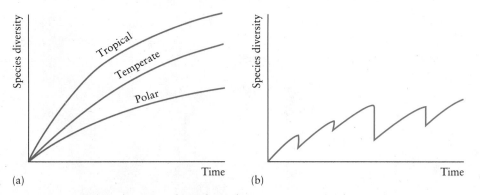

Figure 23.11 *History factor: (a) hypothetical increase in species diversity in tropical-to-polar habitats if there were no interruptions; (b) actual pattern of change in species diversity of a temperate or polar habitat subjected to glaciation and climatic variations. (After Fischer 1960.)*

over a few tens of generations. Ecological time involves situations in which a given species could occupy an environment but has not had time to disperse there. The evolutionary time scale is a longer time scale, operating over hundreds and thousands of generations. Evolutionary time applies to cases where a position in the community exists but is not occupied because of insufficient time for speciation and evolution to have occurred.

Lake Baikal in the former Soviet Union is a particularly striking illustration of the role of time in generating species diversity. Baikal is an ancient lake, one of the oldest in the world, situated in the temperate zone. Baikal contains a very diverse fauna (Kozhov 1963). For example, there are 580 species of benthic invertebrates in the deep waters of Lake Baikal. A comparable lake in glaciated northern Canada, Great Slave Lake, contains only four species in this same zone (Sanders 1968).

Some paleontological data support the assumption that species diversity increases over geological time. The number of species of terrestrial plants found as fossils appears to increase in two waves during the last 450 million years (Figure 23.12). No plateau in biodiversity has yet been reached for terrestrial plants (Knoll 1986).

Note that the species diversity of a community is a function not only of the rate of addition of species through evolution but also of the rate of loss of species through extinction or emigration. Compared with polar communities, the tropics could have a more rapid rate of evolution and also a lower rate of extinction, and these two rates act together to determine species diversity. If we accept this analysis, we are faced with a second problem: Why is the rate of evolution more rapid in the tropics, or the rate of extinction lower? The history factor can only work through one or more of the other seven ecological factors that affect species diversity. Because it often involves geological time scales and is not amenable to direct experimentation, the history factor is the most difficult of the eight causal factors to assess.

The history factor can also be used to try to explain differences in diversity on an ecological time scale. One way to do this is to study recently colonized areas and see if species richness increases over time. For example, sugarcane, which is native to New Guinea, has been introduced into at least 75 regions around the world over the last 3000 years. Strong and colleagues (1977) could find no tendency for the

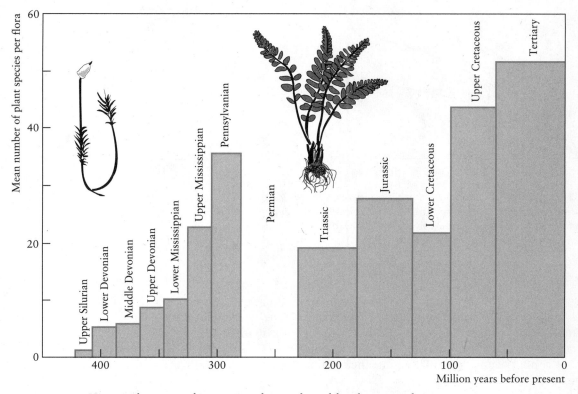

Figure 23.12   Historical pattern of increase in the number of fossil terrestrial plant species over evolutionary time. (Data from Nicklas et al. 1980.)

number of arthropod pests of sugarcane to increase with time. Pest species accumulation in this system must happen very quickly, and the history hypothesis does not explain why sugarcane has more pests in some areas and fewer pests in others. The best predictor of the number of pests on sugarcane is the area of cane under cultivation.

The history hypothesis suggests that species richness never reaches a limit but goes on rising over time. We do not know if this is a correct interpretation of the fossil record or not (Gould 1981). Because the history factor is so difficult to test, we should use it only after we have exhausted simpler hypotheses to explain diversity trends. Communities must have factors operating in ecological time to maintain biodiversity, and we need to analyze these factors to understand the best ways of protecting biodiversity in the future.

Spatial Heterogeneity Factor

There could be a general increase in environmental complexity as one proceeds toward the tropics. The more heterogeneous and complex the physical environment, the more complex are the plant and animal communities and the higher is the species diversity. This factor can be considered on both a macro and micro scale.

Topographic relief, or *macrospatial heterogeneity,* may have a strong effect on species diversity. Simpson (1964) has shown that the highest diversities of mammals in the United States occur in the mountain areas (Figure 23.10). The explanation for this seems quite simple: Areas of high topographic relief contain many different habitats and hence more species. Also, mountainous areas produce more geographic isolation of populations and so may promote speciation. But this conclusion does not fit all taxonomic groups. Neither the trees (Figure 23.6) nor the land birds (Figure 23.7) of North America show diversity patterns related to topographic relief.

MacArthur (1965) suggested that we should recognize two components in trying to analyze latitudinal gradients in species diversity: *within-habitat* diversity (also called α-diversity), and *between-habitat* diversity (called β-diversity). We can illustrate this distinction by two simple schemes to explain tropical diversity:

Hypothetical Scheme A	Temperate Country	Tropical Country
No. species per habitat	10	10
No. different habitats	10	50

All the increase in tropical diversity is caused by between-habitat or β-diversity in this hypothetical scheme.

Hypothetical Scheme B	Temperate Country	Tropical Country
No. species per habitat	10	50
No. different habitats	10	10

In this oversimplified scheme, all the increase in tropical diversity is due to within-habitat or α-diversity.

Can the factor of topographic relief provide some explanation for latitudinal variation in species diversity? There is some evidence that this is part of the reason that the tropics are so rich in species. MacArthur (1969) showed that, for land birds, 2 hectares of Panama forest supported 2 1/2 times as many bird species as 2 hectares of Vermont forest. But larger areas in the tropics support proportionately even more species. Ecuador has 7 times the number of bird species as New England, even though the areas are both approximately 260,000 square kilometers. For land birds, there are both more species per habitat in Ecuador and also more habitats per unit of area, so that both scheme A and scheme B are correct for birds.

If spatial heterogeneity means more habitats, we ought to be able to measure this by the number of plant species in a large region. One of the simplest models for animal biodiversity is that it is determined by plant biodiversity. Unfortunately this model does not seem to be correct. For North America neither mammal, nor bird, nor reptile diversity is related to plant biodiversity (Figure 23.13). Amphibian diversity is closely correlated with plant diversity, but this correlation does not appear to be a cause-effect relationship. It is probably due to a third factor related to climate that correlates well with both amphibian diversity and plant diversity (Currie 1991).

Even on a local scale, plant biodiversity may not determine animal diversity. In a classic study MacArthur and MacArthur (1961) measured bird-species diversity on a series of study areas and attempted to relate this to two aspects of the vegetation: *plant-species* diversity and *foliage-height* diversity. Foliage-height diversity is a measure of stratification and evenness in the vertical distribution of vegetation; highly stratified communities have high foliage-height diversities with dense growth of

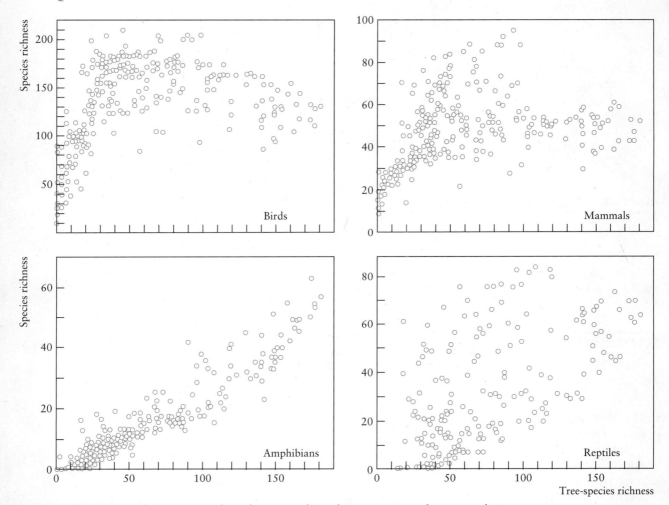

Figure 23.13 *Species richness (= number of species) of North American vertebrates in relation to tree-species richness. The number of tree species is a good index to the total number of plant species in the habitat. Only for amphibians is there any correlation with tree species richness. (From Currie 1991.)*

branches and leaves at all levels from the ground to the top of the canopy. It does not matter whether the strata are contributed by a variety of species or by a variety of age classes of one species. MacArthur found that bird-species diversity was not correlated so much with plant-species diversity as it was with foliage-height diversity (Figure 23.14). This suggests that we can predict the bird-species diversity on a forested area without any knowledge of the plant species that make up the community. Vegetation structure, or stratification, seems to be more important to birds than plant-species composition.

In shrub and grassland habitats, vertical stratification is less important in determining bird-species diversity than it is in forested habitats. In the Sonoran Desert, for example, bird diversity is related to the density of particular nest plants such as saguaro and cholla cacti (Tomoff 1974). Many different trees provide suitable nest sites in a forest, but in a desert habitat only a few growth-forms may be critical as

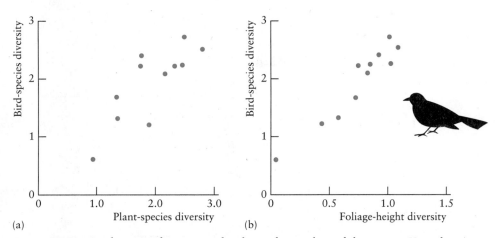

Figure 23.14   Bird-species diversity in deciduous forest plots of the eastern United States in relation to the plant-species diversity and the stratification of the plant community: (a) plant species, (b) vegetative structure. (After MacArthur and MacArthur 1961.)

nesting sites. Vertical stratification in shrub and grassland habitats is limited and may be less important than horizontal structure or *patchiness*. Bird diversity may be predicted more accurately from a knowledge of both horizontal and vertical structure in vegetation (Roth 1976).

Tropical to polar gradients in the oceans seem unlikely to be explained by spatial heterogeneity. The oceans are not uniform water masses, yet they provide rather fewer opportunities for habitat specialization. Benthic marine invertebrates become more diverse as you move from shallow waters on the continental shelf to deeper waters at the edge of the shelf (Sanders 1968). There is no obvious change in bottom sediments to explain this increase in biodiversity.

What are the ecological mechanisms that permit a large number of species to thrive in tropical habitats? Is it simply a case of more food being available? Is competition between species more intense? If it is true that spatial heterogeneity can be used to explain temperate-tropical gradients in species diversity, we must still determine the ecological machinery behind this prediction.

Competition Factor

Many naturalists have argued that natural selection in the temperate and polar zones is controlled mainly by the physical factors of the environment, whereas biological competition becomes a more important part of evolution in the tropics. For this reason, animals and plants are more restricted in their habitat requirements in the tropics, and this increases between-habitat (β) diversity. Animals may also have more restricted diets in each habitat, increasing within-habitat (α) diversity. Competition is keener in the tropics, and niches are smaller. Tropical species are more highly evolved and possess finer adaptations than do temperate species. Consequently, more species can be fitted into a given habitat in the tropics (Dobzhansky 1950).

Competition theory (Chapter 13) can be readily expanded to try to explain species diversity in an equilibrium world (Chesson and Case 1986). The key predic-

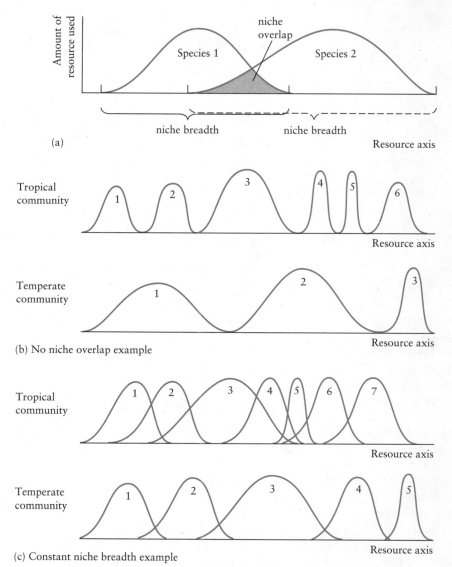

Figure 23.15   *Diagram to illustrate two hypothetical extreme cases of how niche parameters may differ in tropical and temperate communities. Both niche breadth and niche overlap are determined by competition within the communities.*

tion that emerges is that at least *n* limiting resources are needed for the coexistence of *n* species in a community. For plants there are at most four or five limiting resources (Tilman 1986); thus competition by itself cannot explain the large number of plant species in natural communities, and this hypothesis is not a possible explanation for within-habitat diversity of plants.

For animal species there are potentially many more limiting resources. The effect of competition on species richness can be visualized by looking at the niche relations of the species in a community. Consider the simple case of one resource, such as soil water for plants or food item size for animals (Figure 23.15a). Two niche measurements are critical: *niche breadth* and *niche overlap*. We can recognize two extreme cases. If there is no niche overlap between the species, the wider the average niche breadth, the fewer the number of species in the community (Figure 23.15b). At the other extreme, if niche breadth is constant, the lower the niche overlap, the fewer are the species in the community (Figure 23.15c). In this hypothetical analysis, tropical animal communities might have more species because tropical species have smaller niche breadths or higher niche overlaps. Both these arguments assume that Gause's hypothesis is true for natural communities.

To evaluate the competition factor, we must measure these niche parameters in a variety of tropical and temperate animal communities. The problems of measuring niche overlap and niche breadth are discussed in detail by Magurran (1988). The basic problem is to decide which resource axes are relevant to any particular group of species; if the resource axes can be linearly ordered and measured, these niche parameters can be measured as indicated in Figure 23.15.

In relatively few cases have the detailed measurements been made to test the schematic model of Figure 23.15. The best example is the work on *Anolis* lizards summarized by Roughgarden (1986). Lizards of the genus *Anolis* are small, insect-eating iguanid lizards that forage during the day. Most species perch on tree trunks or bushes. They are sit-and-wait predators, and food size is a critical niche dimension. *Anolis* lizards are a dominant component of the vertebrate community on islands in the Caribbean. Roughgarden (1974) tested the prediction that niche breadth would decrease as more species occurred together on an island. Figure 23.16 shows the results for two *Anolis* species. The results are consistent with the predictions from competition theory and support the suggestion that niche breadth is reduced in species-rich communities. Pacala and Roughgarden (1985) showed in enclosure ex-

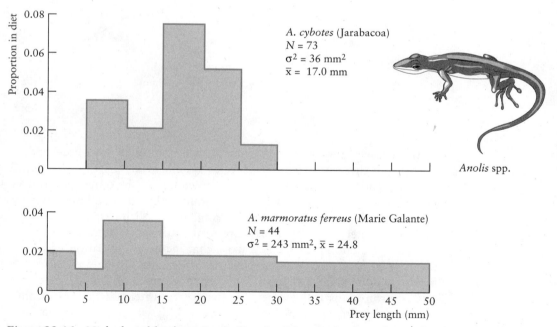

Figure 23.16   Niche breadth of two Anolis *lizard species on islands in the Caribbean.* Anolis cybotes *coexists with five other* Anolis *species on Jarabacoa and has a narrow niche.* Anolis marmoratus *is the only species on Marie Galante. (After Roughgarden 1974.)*

periments that *Anolis* lizards show strong effects of competition when their diets are similar. Competition for food is a major factor determining the species diversity of these lizards.

Competition between species is an important process in population dynamics (Chapter 13); in the next chapter we will discuss the role of competition in community organization. It is clear that competition does not play a large role in the maintenance of plant biodiversity (Austin 1990) but less clear how much it affects animal biodiversity in modern communities.

Predation Factor

Paine (1966) argues that there are more predators and parasites in the tropics than elsewhere and that these hold down their prey populations to such low levels that competition among prey organisms is reduced. This reduced competition allows the addition of more prey species, which in turn support new predators. Thus, in contrast to the competition proposal, there should be *less* competition among prey animals in the tropics. Provided we can measure "intensity of competition," we can distinguish quite clearly between these two ideas.

Paine (1966) supported his ideas with some experimental manipulations of rocky intertidal invertebrates of the Washington coast. The food web of these areas on the Pacific coast is remarkably constant:

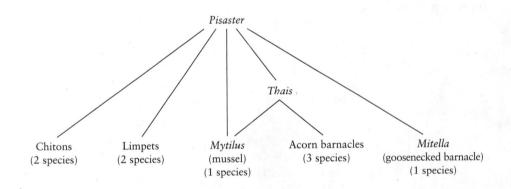

Paine removed the starfish *Pisaster* from a section of the shore and observed a *decrease* in diversity from a 15-species system to an 8-species system. A bivalve, *Mytilus*, tended to dominate the area, crowding out the other species. "Succession" in this instance is toward a simpler community. By continual predation, the starfish prevent the barnacles and bivalves from monopolizing space. Thus local species diversity in intertidal rocky zones appears to be directly related to predation intensity.

The prediction from Paine's work that increased predation will lead to higher diversity of prey species is dependent on the ability of one prey species to be competitively dominant. Addicott (1974) tested the predation model using the community of protozoans and rotifers that live in the water-filled leaves of pitcher plants (*Sarracenia purpurea*). Pitcher plant leaves fill with water and become replicate communities containing over 40 species of flagellates, ciliates and rotifers. Only one predator lives in this community, the larvae of a mosquito (*Wyeomyia smithii*). By manipulating the numbers of mosquito larvae in each pitcher plant leaf, Addicott (1974) could test the predation model. Figure 23.17 shows that as predator densities

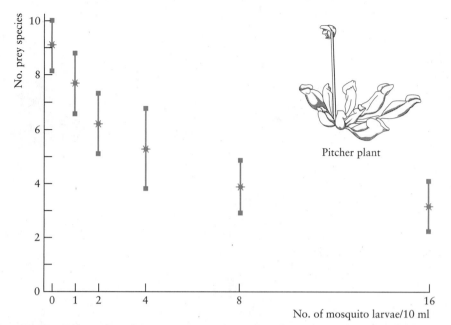

Figure 23.17   Effect of predation intensity on species richness in pitcher plant leaf communities of protozoans and rotifers. As more predators are added (mosquito larvae) more prey species are driven to extinction and diversity declines. (From Addicott 1974.)

increased, the number of species of prey decreased. This result was the opposite of that observed by Paine in the intertidal zone. The prey community in pitcher plant leaves does not seem to compete for food or space as strongly as the intertidal community, and none of the protozoans or rotifers appeared to be competitively dominant. In this kind of noninteractive community predation pressure will not increase species diversity.

For the predation hypothesis to operate on a broad scale, the predators involved must be very efficient at regulating the abundance of their prey species. In terrestrial food webs, predators are usually specialized and in some cases do not seem to regulate prey abundance (Chapter 15). Note that the predation hypothesis cannot be a sufficient explanation for tropical species diversity unless it can be applied to all trophic levels. If the species diversity of the herbivore trophic level is determined by the predators, we are left with explaining the diversity of the primary producers. Key species, such as the starfish *Pisaster*, should be more common in tropical communities, but at the moment there is no evidence that this is correct.

The predation factor can be extended to the primary-producer level. Tropical lowland forests contain many species of trees and corresponding low densities of adult trees of each species. Most adult trees of a given species are also spread out in a regular pattern in the tropical forest, and Janzen (1970) suggests that these characteristics of tropical trees can be explained by the predation hypothesis—the species that eat seeds or seedlings being analogous to the predators just discussed. Figure 23.18 shows schematically the interaction of seed production and dispersal from the parent tree and the activity of seed and seedling eaters. Many insects that eat seeds and fruits in the tropics are host-specific and tend to congregate around a source of

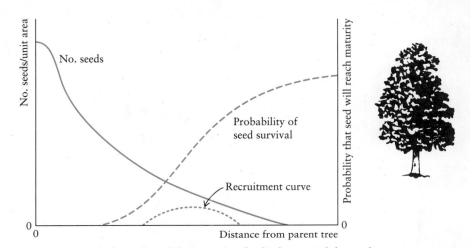

Figure 23.18   Hypothetical model to account for high tropical-forest diversity. The amount of seed dispersed falls off rapidly with distance from the parent tree, and the activity of host-specific seed and seedling herbivores is most evident near the parent tree. The product of these two factors determines a recruitment curve with a peak at the distance from the parent tree where a new adult tree is likely to appear. (After Janzen 1970.)

seeds. The probability that a seed or small seedling will be overlooked by these specific herbivores is thus increased as they move farther and farther from the parent tree and are surrounded more and more by other tree species with their own specific herbivores. Thus each tree casts a "seed shadow," in which survival of its own kind is reduced. As one moves from the lowland tropical forests to temperate forests, the seed and seedling herbivores are hypothesized to be less efficient at preventing establishment of seedlings close to the parent tree. Unfortunately, few data exist to test this suggested explanation.

Predation and competition may be complementary in their effects on species diversity (Menge and Sutherland 1976, Lubchenco 1986). Competition may be more important in maintaining high diversity among parasites and predators, while the process of predation may be more important among herbivores. Superimposed on these effects is another pattern: In complex communities with many species, predation may be the dominant interaction affecting diversity, whereas in simple communities, competition may be the dominant interaction.

Climate and Climatic Variability Factors

The environmental stability principle states that the more stable the climatic parameters and the more favorable the climate, the more species will be present. According to this idea, regions with stable climates allow the evolution of finer specializations and adaptations than do areas with erratic climates. This results in smaller niches and more species occupying a unit space of habitat. Species should be more flexible in temperate and polar areas and should be more specialized in the tropics.

The factor of climatic stability can be combined with the history factor, with which it has much in common. The *stability-time* hypothesis (Sanders 1968) emphasizes the role of all environmental parameters—temperature, moisture, salinity, oxygen, pH—in permitting diversity. Low-diversity habitats may be *severe, unpredictable,* or both. Severe environments, such as hot springs or the Great Salt Lake in Utah, can be very predictable (physical conditions are constant from day to day) but have low species diversity. A desert environment with irregular rainfall would be an example of an unpredictable and severe environment. Sanders (1968) applied the hypothesis to the marine fauna of muddy bottoms. The deep sea represents a stable environment of long time span, and the diversity of bivalve and polychaete species in bottom samples in the deep sea was almost equal to that in tropical shallow-water areas. Abele and Walters (1979) criticized Sanders's (1968) analysis and concluded that marine invertebrate biodiversity was better explained by the spatial heterogeneity of their habitats.

One of the simplest explanations of the polar-tropical gradient in terrestrial biodiversity is the *energy model* (Brown 1981). This hypothesis suggests that species diversity will correlate with available solar energy. Recent data support this hypothesis for trees (Currie and Paquin 1987), British birds (Turner et al. 1988) and vertebrates from North America (Currie 1991). Figure 23.19 illustrates the relationship between biodiversity and annual evapotranspiration (see page 96). Evapotranspiration measures the energy balance at a site and is a function of solar radiation and temperature. Vertebrate biodiversity in North America is not correlated with plant productivity or with climatic variability and only weakly correlated with spatial heterogeneity (Currie 1991). The energy hypothesis makes more specific predictions about seasonal bird migrants (Turner et al. 1988). In Britain the biodiversity of summer birds is correlated with summer temperature and the diversity of winter birds is correlated with winter temperature at 75 localities.

Favorable climates on a broad geographic scale thus support high biodiversity. This idea explains a large fraction of the global tropical-polar diversity gradient, but it will not explain local habitat-scale variations in species diversity. No single factor will explain all gradients in biodiversity from the local scale to the regional.

Productivity Factor

Connell and Orias (1964) presented the productivity idea, blended with the factor of climatic stability. The productivity hypothesis in its pure form states that greater production results in greater diversity, everything else being equal. The data available do not support this idea. For example, Tilman (1986) describes several examples in which plant biodiversity is maximal in resource-poor habitats of low productivity. Two of the world's most diverse plant communities are the fynbos of South Africa and the heath scrublands of southeastern Australia (Figure 23.20). Both of these communities occur on nutrient-poor soils, and in both cases adjacent areas with better soils and more productive vegetation have fewer species. Productivity in plant communities seems to lead to reduced biodiversity on a local scale (Tilman 1986).

Productivity could be more important for animal communities on a global scale, but again the available data do not agree with this conclusion. Currie (1991) could find no relationship between productivity and vertebrate biodiversity in North America. Productivity by itself does not seem to be the key to understanding diversity gradients on a global scale.

A common modification of the productivity factor is the idea of increased temporal partitioning in the tropics. The main argument is that the longer growing season of tropical areas allows the component species to partition the environment

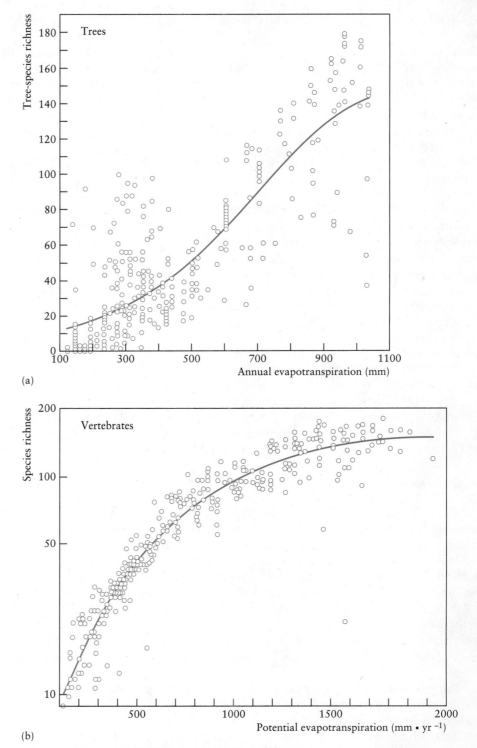

(a)

(b)

Figure 23.19   *The energy hypothesis for biodiversity. Species richness of (a) trees and (b) vertebrates from North America are related to evapotranspiration which combines solar radiation and temperature. (From Currie 1991.)*

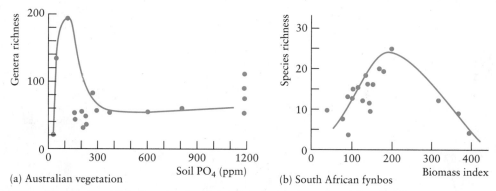

(a) Australian vegetation   (b) South African fynbos

Figure 23.20   Plant biodiversity for Australian vegetation (a) and South African fynbos (b) in relation to productivity. Higher productivity does not lead to high biodiversity. (Data from Beadle 1966 and Bond 1983.)

temporally as well as spatially, thereby permitting the coexistence of more species. This idea combines the stability hypothesis with the productivity hypothesis and suggests that the stability of primary production is a major determinant of the species diversity in a community. One means of testing this idea is by looking at primary production in different communities over an annual cycle, and another means is by looking at the way organisms thrive in different communities.

More bird species occur in tropical forests than in temperate ones because they can find completely new ways of surviving in the productive tropical forests (Orians 1969). Temperate-zone forests have no obligate fruit-eating birds like parrots, no birds of prey that eat reptiles only, no birds that follow ant swarms, and no birds that sit quietly in trees and watch for insect prey. These are "new niches" that appear only in the tropics, in part because of the stability of primary and secondary production.

Disturbance Factor

If natural communities exist at equilibrium and the world is spatially uniform, competitive exclusion ought to be the rule, and each community should come to be dominated by a few species, the best competitors (Crawley 1986). But if communities exist in a nonequilibrium state, competitive equilibrium is prevented. A whole range of factors can prevent equilibrium—predation, herbivory, fluctuations in physical factors, and catastrophes such as fires—and we lump these together as "disturbance." When disturbances occur too often, species go extinct if they have low rates of increase. When disturbances are rare, the system goes to competitive equilibrium and species of low competitive ability are lost. In between there is an intermediate level of disturbance which maximizes biodiversity (Figure 23.21). This idea is called the *intermediate disturbance hypothesis* (Connell 1978). Moreover, if population growth rates are low for all members of a community, the competitive equilibrium is approached so slowly that it is never reached. Thus we can maintain species diversity by periodic disturbance or by environmental fluctuations (Figure 23.20). Some of the factors discussed above that affect species diversity may operate through these two general mechanisms. Latitudinal gradients in species diversity can be explained by this general model if tropical populations have low growth rates and a low rate of competitive displacement (Figure 23.22).

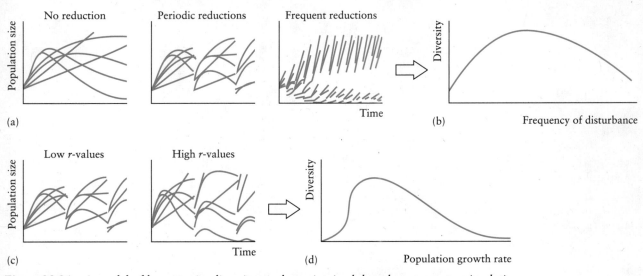

Figure 23.21   *A model of how species diversity can be maintained, based on computer simulations of the Lotka-Volterra competition equations for six species. (a) Values of rs, Ks, and αs are identical in all three simulations, and the frequency of population reductions is varied. When there is no reduction, a competitive equilibrium of low diversity is reached. When reductions are too frequent, the low-r species are unable to recover. (b) General form of the relation between species diversity and the magnitude or frequency of disturbance. (c) The effect of population growth rates in Lotka-Volterra simulations. Diversity is reduced more rapidly at higher r-values. (d) General form of the relation between species diversity and population growth rates at low-to-intermediate frequency of disturbance. (After Huston 1979.)*

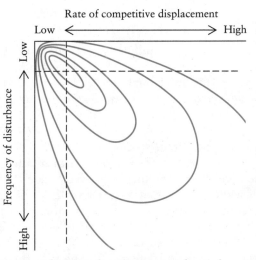

Figure 23.22   *Contour map of species diversity expected as a dynamic equilibrium between the rate of competitive displacement and the frequency or magnitude of disturbance. The highest diversity ("peak") is in the upper left corner, and this would correspond to the tropical end of the latitudinal diversity gradient. (After Huston 1979.)*

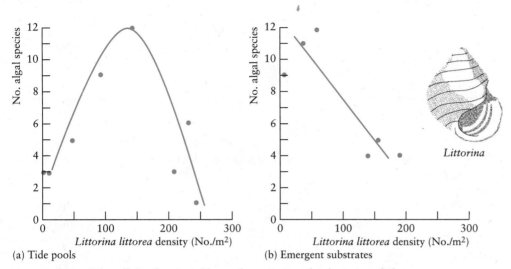

Figure 23.23   *The effect of periwinkle snail grazing on the diversity of algae in high-tide pools (a) and on emergent rocks in the low intertidal zone (b) in Massachusetts and Maine. The intermediate disturbance hypothesis applies only in the tide pools. (From Lubchenco 1978.)*

Disturbance can also operate on a local scale to produce patches that undergo succession. Within each patch on a local area the species composition may be changing (e.g., Figure 22.14, page 501), but on a larger spatial scale the species composition may be constant and include all the pioneer species as well as the climax species (Connell 1987).

Disturbance does not always produce the patterns predicted by the intermediate disturbance hypothesis (Yodzis 1986). Herbivores can increase or decrease biodiversity of plants. On rocky shores in Massachusetts the periwinkle snail *Littorina littorea* is the most common herbivore (Lubchenco 1978). In tide pools *Littorina* feeds on the algae that is competitively dominant. Moderate grazing by the snails permits many competitively inferior algae to survive (Figure 23.23a), as predicted by the intermediate disturbance hypothesis. But on emergent rocks, perennial brown and red algae are competitively superior and the snails do not eat them but feed on the competitively inferior algae. Consequently, on emergent rocks *Littorina* grazing reduced algal diversity (Figure 23.23b). The critical factors are the food preferences of the grazer and the competitive abilities of the plants. The intermediate disturbance hypothesis does not apply to all grazing systems.

CONCLUSION

Biodiversity varies greatly between habitats and regions, and systematically declines from the tropics toward the poles. Eight factors have been postulated to control biodiversity and combinations of these factors operate in many natural communities. Three factors seem most important: (1) *History,* summarizing the evolutionary history of a region, is the most difficult factor to evaluate and is potentially important on a regional scale. (2) *Climate,* particularly solar energy, operates on a regional scale and allows good predictions of diversity at least for terrestrial communities. (3) *Disturbance,* by preventing competitive exclusion, operates on a local scale to maximize diversity. There is no simple, general answer to the question *what controls biodiversity?* and we are led to consider the broader question of what controls community organization.

SUMMARY

Biodiversity can be measured by simply counting all the species in a collection or by weighting each species by its relative abundance. Most communities consist of a few common species and many rare species, and one important question about community organization is, What controls biodiversity?

Tropical environments support more species in almost all taxonomic groups than do temperate and polar areas. The mammal faunas of North America and Australia illustrate the complexity of species diversity gradients, which are not always smooth, steady trends from the equator to the poles.

Eight factors have been proposed to explain variation in species diversity. The *history* or *time* hypothesis is historical and emphasizes the time available for speciation and dispersal. *Spatial heterogeneity* is a second factor that may influence diversity through the number of habitats available per unit of area. Vegetation structure seems important for some animal groups in determining local diversity. *Competition* may be an important factor in generating high tropical diversity. Biological competition may be stronger in favorable environments where organisms can specialize and have narrow niches. *Predation* may affect diversity by holding down prey numbers and reducing competition in favorable tropical environments. *Climate* and *climatic variability* may affect diversity directly by determining the favorableness of the environment. Climatic measures correlate closely with species diversity in terrestrial communities of plants and animals. *Productivity* may also affect diversity because more productive habitats can support more individuals. But productivity does not seem to correlate well with diversity. *Disturbance* can prevent competitive dominance and an intermediate level of disturbance may increase local biodiversity.

All of these factors interact to determine biodiversity. On a regional scale, history and climate seem to be major factors, while on a local scale disturbance, spatial heterogeneity, predation and competition are more likely to be important.

To protect and preserve biodiversity we need to know what controls community organization, the central question of the next two chapters.

Selected References

CODY, M. L. 1986. Structural niches in plant communities. In *Community Ecology,* ed. J. Diamond and T. J. Case, pp. 381-405. Harper and Row, New York.

CONNELL, J. H., and E. ORIAS. 1964. The ecological regulation of species diversity. *American Naturalist* 98:399–414.

CURRIE, D. J. 1991. Energy and large-scale patterns of animal- and plant-species richness. *American Naturalist* 137:27–49.

DICKMAN, C. R. 1989. Patterns in the structure and diversity of marsupial carnivore communities. In *Patterns in the Structure of Mammalian Communities,* ed. D. W. Morris, Z. Abramsky, B. J. Fox, and M. R. Willig, pp. 241–251. Texas Tech University Press, Lubbock.

EHRLICH, P. R., and E. O. WILSON. 1991. Biodiversity studies: Science and policy. *Science* 253:758–762.

GRUBB, P. J. 1987. Global trends in species-richness in terrestrial vegetation: A view from the Northern Hemisphere. In *Organization of Communities Past and Present,* ed. J. H. R. Gee and P. S. Giller, pp. 99–118. Blackwell, Oxford, England.

HUSTON, M. 1979. A general hypothesis of species diversity. *American Naturalist* 113:81–101.

LUBCHENCO, J. 1978. Plant species diversity in a marine intertidal community: Importance of herbivore food preference and algal competitive abilities. *American Naturalist* 112:23–39.

MAGURRAN, A. E. 1988. *Ecological Diversity and Its Measurement.* Croon Helm, London.

Questions and Problems

1. The tree flora of Europe is very poor compared with that of eastern North America or eastern Asia (Grubb 1987). Why should this be? Compare your explanations with those of Grubb (1987) and of Currie and Paquin (1987).

2. The number of taxa recorded in the fossil record tends to increase irregularly but steadily with geological time (Gould 1981, Knoll 1986). Strong and coauthors (1977, p. 173) state, "Where the fossil records exist there is no evidence that species number increases inexorably over long periods in the history of communities." Discuss these different views of whether the history hypothesis fits the fossil record.

3. Marine algae along the west coast of North America do not increase in species richness toward the tropics but peak at about seventy species per 100 kilometers of coastline around 40° north latitude (Gaines and Lubchenco 1982). Along the east coast of North America, species richness is highest in tropical areas. Discuss why these patterns might hold.

4. Hurlbert (1971, p. 577) states:

   The recent literature on species diversity contains many semantic, conceptual, and technical problems. It is suggested that, as a result of these problems, species diversity has become a meaningless concept, that the term be abandoned, and that ecologists take a more critical approach to species number relations.

   Do you agree? Read Hurlbert (1971) and discuss.

5. Ferns are well known for being relatively free from insect attack in both the juvenile and the adult stages (Janzen 1970). Obtain information on species diversity gradients in ferns from floras of tropical and temperate areas, and suggest some explanations for the diversity changes you find.

6. Whittaker (1972) argues that in terrestrial plants and insects, species diversity can increase without any upper limit because the evolution of diversity is a self-augmenting process. Evaluate Whittaker's argument, and discuss its implications with respect to the analysis of diversity gradients. Cornell and Lawton (1992) also discuss this problem.

7. The longest experiment in ecology is the Park Grass Experiment begun in 1856 at Rothamsted, England. A mowed pasture was divided into 20 plots, and a series of plots were fertilized annually with a variety of nutrients including nitrogen. Discuss the predictions you would make regarding biodiversity on fertilized and unfertilized plots for this experiment, using the eight factors discussed in this chapter. Tilman (1986, pp. 62–63) shows the observed results.

8. In Antarctica species richness in soft-bottom invertebrates (sponges, bryozoans, polychaetes and amphipods) is higher than that of almost all other tropical and temperate zone soft-bottom communities (Clarke 1990). What observations or experiments would you perform to find out why this high biodiversity occurs in Antarctica?

9. Figure 23.19 shows that on a global scale species richness increases smoothly with solar energy and temperature. Why should this occur? Why is the available energy not monopolized by a few superspecies? Compare your ideas with those of Currie (1991, p. 46).

Overview Question

To preserve biodiversity how much do you need to understand about the factors that control biodiversity? Sketch the outlines of a management plan for preserving biodiversity in a large national park of tropical rainforest.

24

Community Organization II: Predation and Competition in Equilibrial Communities

Communities could be organized by physical processes and by three biological processes—competition, predation, and mutualism. Competition among plants, herbivores, and carnivores could control the diversity and abundance of the species in a community. Predation and herbivory could organize the community by who-eats-whom so that the framework of community organization is set by the animals. Mutualism is an important process that links species and could serve to increase community organization by linking species to the benefit of all. Physical processes set limits to these three biological processes, and variation in temperature, salinity, and other physical factors have potential implications for the species in a community. To study community organization, we need to look at the component species and the processes that tie them together.

When we speak of community organization, we imply that there is some regularity in the biomass or the numbers of the species that make up the community. Naturalists looking for particular birds, butterflies, or flowers have an implicit model of community organization in their heads. Conservation biologists have an implied model of community organization when they discuss the preservation of the Everglades of Florida or other natural landscapes. Natural communities could be very loosely organized, or be very tightly organized. How can we determine this for any particular community?

Communities contain so many different species that we cannot study each species separately. If we measure the species diversity of a community, we implicitly assume that each species is equal to every other species in the community. We now ask whether this is true: *Are all species of equal importance in a community?* This question is purposely vague because we must define *importance*, and we can do this in several ways. First, let us consider a species important if, when we remove it, the diversity or abundance of other species in the community changes. We cannot, of course, remove each and every species from a community, but we can take advantage of introduction experiments to investigate this question.

One way to reduce, conceptually, the complexity of a community is to group the species into broad categories. The simplest approach along these lines is to group species according to their feeding habits, so that, for example, we group all herbivores together. We can then ask how these feeding relationships affect community organization.

We will begin this analysis with the classical assumption that communities are in equilibrium. Communities are in equilibrium when species abundances remain constant over time. In most cases equilibrium means *stable equilibrium* (Figure 24.1). In different habitats the equilibrium point may differ, so that there is spatial variation in species numbers, but the key point is that at each spatial location the community is in equilibrium and remains constant (Chesson and Case 1986). In most cases the

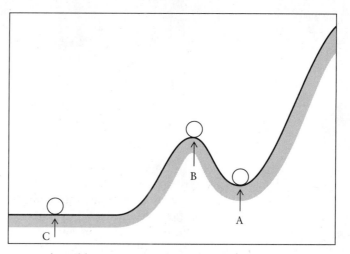

Figure 24.1   Types of equilibrium points: A is a stable equilibrium point, B an unstable equilibrium point, and C a neutrally stable equilibrium point. (Modified from DeAngelis and Waterhouse 1987.)

equilibrium is assumed to be *globally* stable, so that the system will return to the equilibrium point following any disturbance.

The classical equilibrium assumption of community ecology is an abstraction, like the frictionless pendulum of physics, and will not be found in its pure state in natural communities. Real communities will be spread along a spectrum from equilibrium to nonequilibrium (Figure 24.2). Equilibrium communities are claimed to show stability, but stability can mean at least five different things (Pimm 1991). The mathematician's idea of local stability (point A in Figure 24.1) is the simplest meaning. Stability can also mean the time it takes for a community to recover from disturbance. Stability is also commonly measured as *variability* of a community over time, or as the *persistence* of a community over time. Finally, stability has also been used to describe how much a community changes when a part of the community is altered (*resistance*). An ideal equilibrium community would score high on all these measures of stability. Such a community would have many biotic interactions involving competition and predation, and these processes would operate in a density-

NONEQUILIBRIUM	EQUILIBRIUM
Biotic decoupling	Biotic coupling
Species independence	Competition
Unsaturated	Saturated
Abiotic limitation	Resource limitation
Density independence	Density dependence
Opportunism	Optimality
Large stochastic effects	Few stochastic effects
Loose patterns	Tight patterns

Figure 24.2   Natural communities may be arrayed along a spectrum of states from equilibrium to nonequilibrium. At either extreme, several attributes of community structuring or dynamics can be anticipated. (From Wiens 1984b.)

dependent manner to regulate population size (Chapter 16). Equilibrium communities would also be saturated with species , so that species invasions would be rare. Weather catastrophes would rarely occur and the community would form a tightly coupled biotic unit.

By contrast, nonequilibrium communities would ideally score low on all these measures of stability. Species would operate individualistically, and density-dependent population regulation would be difficult to find. Climate catastrophes would occur frequently, and species would come and go regularly so the composition of the community would be highly variable. We will discuss nonequilibrium community dynamics in the next chapter and try to determine where most natural communities fall on this continuum.

There are three major equilibrium theories of community organization:

1. *Classical competition theory*. Hutchinson (1959) argued that competition is the major biological process controlling community structure. The subsequent development of this theory is reviewed by Armstrong and McGehee (1980). The essential assumptions of this theory are as follows:

   a. Population growth rates can be described with deterministic equations, and environmental fluctuations can be ignored.

   b. The environment is spatially homogeneous, and migration is unimportant.

   c. Competition between species is the only significant biological interaction.

   d. The coexistence of competing species requires a stable equilibrium point for population densities (e.g., Figure 13.3, case 3, page 235).

   This theory predicts that *n* limiting resources are required for the coexistence of *n* species. Moreover, there will be a limiting similarity of species such that species differ in their use of the available resources (Chesson and Case 1986).

2. *Competition-predation theory*. The classical theory was clearly deficient in allowing only competition to operate. By adding predation to assumption c above, a new equilibrium model can be produced that will allow *n* species to coexist on fewer than *n* resources (Levin 1970).

3. *Competition-spatial patchiness theory*. Another equilibrium model that was modified from the classical theory allowed the environment to be subdivided into patches, such that different species would be favored in different patches (Levin 1974). Each patch has its own distinct stable equilibrium, and the resulting model is similar to a metapopulation model (Figure 16.5, page 333).

In real communities competition for nutrients or food and competition for space both occur (Yodzis 1986). By adding predation and spatial patchiness we can construct more realistic models of community organization. Let us consider the detailed structure of communities and see how much we can explain by these classical equilibrium theories.

FOOD CHAINS AND TROPHIC LEVELS

One component of community organization is who-eats-whom. The transfer of food energy from its source in plants through herbivores to carnivores is referred to as the food chain. Elton (1927) was one of the first to apply this idea to ecology and to analyze its consequences. He pointed out the great importance of food to organisms, and he recognized that the length of these food chains was limited to four or five links. Thus we may have a pine tree—aphids—spiders—warblers—hawks food chain. Elton recognized that these food chains were not isolated units but were hooked together into food webs. Let us look at a few examples of food chains.

The Antarctic's pelagic food chain is a good example of a food chain found in

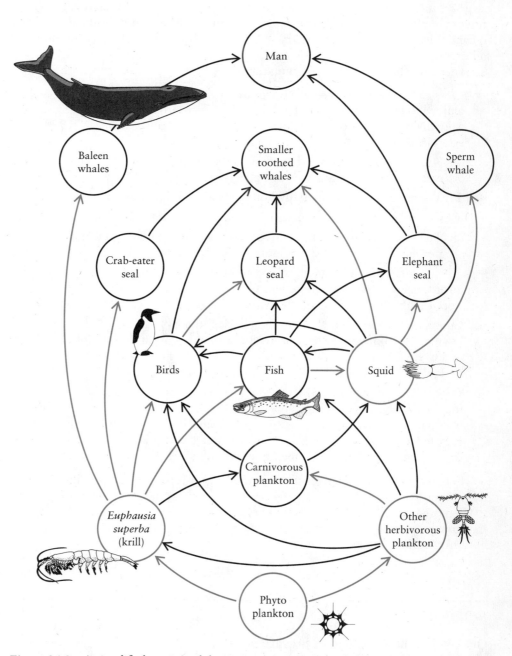

Figure 24.3   A simplified version of the Antarctic's marine food chain. Bold arrows indicate the major trophic interactions. (Modified after Knox 1970.)

seasonally productive oceans. The dominant herbivores are euphausids (krill) and copepods. These zooplankton species are fed on by an array of carnivores from penguins to seals, fish, and baleen whales (Figure 24.3). Squid, which are carnivores feeding on fish as well as zooplankton, are another important component of this food chain because they in turn are fed on by seals and the toothed whales. During the whaling years, humans became the top predator of this food chain.

In the rocky intertidal zone of the Gulf of California, Paine (1966) described a food web that contained four links of carnivores. Figure 24.4 shows that the top carnivore, the starfish *Heliaster kubiniji*, preys on two layers of predatory marine snails in addition to the herbivorous barnacles, bivalves, and gastropods. These feeding relationships are not constant. *Heliaster* can eat *Hexaplex* and *Muricanthus* only to a certain size, above which these two species also become top carnivores.

Wytham Woods, near Oxford, England, has been studied very intensively as a community by Elton (1966). A simplified food chain for this woodland is shown in Figure 24.5. Oaks *(Quercus robur)* are one dominant tree in Wytham Woods and throughout the deciduous forests of Western Europe and are fed on by more than 200 species of Lepidoptera. The winter moth *(Operophtera brumata)* is the most common oak defoliator in Wytham; it serves as food for shrews, voles, mice, titmice, and predatory ground beetles *(Philonthus, Feronia,* and *Abax)* and is parasitized by a tachinid fly, *Cyzenis*. Note that small mammals and birds occupy a variety of

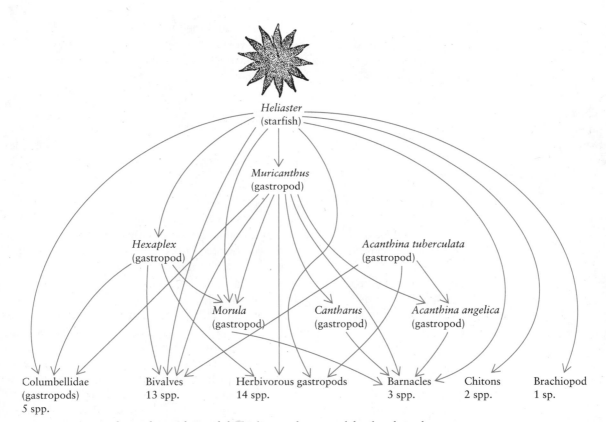

Figure 24.4   Feeding relationships of the Heliaster-*dominated food web in the northern Gulf of California. (After Paine 1966.)*

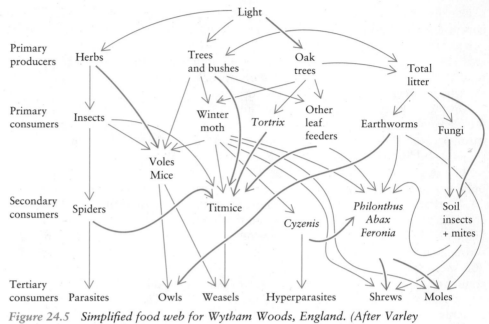

Figure 24.5   Simplified food web for Wytham Woods, England. (After Varley 1970.)

feeding levels. Birds such as the great tit and the blue tit feed on beechnuts (producer level), on winter moth females (herbivores), and on spiders and beetles (carnivores).

Within these complex food webs we can recognize several different trophic levels:

Producers	Green plants	First trophic level
Primary consumers	Herbivores	Second trophic level
Secondary consumers	Carnivores, insect parasitoids	Third trophic level
Tertiary consumers	Higher carnivores, insect hyperparasites	Fourth trophic level

The classification of organisms by trophic levels is one of *function* and not of species as such. A given species may occupy more than one trophic level. For example, male horseflies feed on nectar and plant juices, whereas the females are blood-sucking ectoparasites. Titmice in Wytham Woods feed on three trophic levels (Figure 24.5).

Size has a great effect on the organization of food chains, as Elton (1927) recognized. Animals of successive trophic levels in a food chain tend to be larger (except for parasites). There are, of course, definite upper and lower limits to the size of food a carnivorous animal can eat. The structure of an animal puts some limit on the size of food it can take into its mouth. Except in a few cases, large carnivores cannot live on very small food items because they cannot catch enough of them in a given time to provide for their metabolic needs. The one obvious exception to this is *Homo sapiens,* and part of the reason for our biological success is that we can prey upon almost any level of the food chain and can eat any size of prey.

Food webs can form a useful starting point for the theoretical analysis of community organization (Pimm et al. 1991). Table 24.1 provides definitions for terms used in food web theory. Over 200 food webs have now been published, and six

Table 24.1 Definitions of Food Web Terminology

Top predators: species eaten by nothing else in the food web.
Basal species: feed on nothing within the web (usually plants).
Intermediate species: species that have both predators and prey within the web.
Trophic species: groups of organisms that have identical sets of predators and prey.
Cycles (within a food web): species A eats species B and species B eats A.
Cannibalism: a cycle in which a species feeds upon itself.
Interactions: any feeding relationship (line on a food web diagram).
Possible interactions: between s species in a food web, there can be $[s(s − 1)/2]$ possible interactions.
Connectance: number of actual interactions in a food web divided by the number of possible interactions.
Linkage density: average number of links or interactions per species in the web.
Omnivores: species that feed on more than one trophic level.
Compartments: groups of species with strong linkages among group members but weak linkages to other groups of species.

Source: Modified from Cohen (1978) and Pimm (1982).

generalizations of features common to food webs have been postulated (Pimm et al. 1991).

One generalization that has been made about food webs is that there are limits to their complexity. As more and more species are involved in a food web, the linkage density remains constant; that is, each species tends to have approximately two trophic interactions, regardless of whether the community contains 5 or 50 species (Cohen et al. 1990). The result of this constant structure is that connectance (the ratio of actual interactions to potential interactions) falls with the number of species in the food web (Figure 24.6).

Martinez (1992) has questioned this drop in connectance with an increase in the size of food webs, and proposed an alternative hypothesis, the *constant connectance hypothesis*. In sufficiently detailed food webs, the total number of links increases in proportion to the square of the number of species in the web, and connectance remains constant. The implication is that the number of prey a predator will eat goes up in proportion to the total number of species in a community. The constant number of links per species observed in the webs analyzed by Cohen et al. (1990) may be a result of oversimplification involved in drawing detailed food webs.

A second generalization is that food chains are short (Elton 1927). If we count for each food web all the possible routes from a basal species to a top predator, we can get a set of chain lengths and a maximum chain length for the food web. For 113 food webs the most common maximum chain length is four, and the vast majority of webs have five or fewer links (Cohen et al. 1990). Why should food chains be short? There are two main hypotheses (Pimm 1982).

The *energetic hypothesis* is the most popular explanation for food chain length. It suggests that the length of food chains is limited by the inefficiency of energy transfer along the chain (Hutchinson 1959). If this were correct, food chains should be longer in habitats of high productivity. This is not correct (Pimm 1991). Marine food chains in fact are shorter in upwelling areas of high productivity.

The *dynamical stability hypothesis* explains short food chains by the fact that longer food chains are not stable so that fluctuations at lower levels are magnified

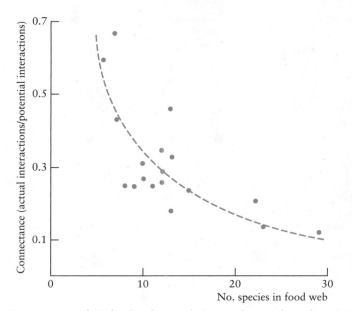

Figure 24.6   Connectance of 18 food webs in relation to the number of species in the web. There are fewer interactions in food webs as species richness increases. (From Pimm 1982.)

at higher levels and top predators go extinct. Also in a variable environment top predators must be able to recover from catastrophes. The longer the food chain, the slower the recovery rate from catastrophes for top predators. If catastrophes occur too often, the top predators will again go extinct. This hypothesis predicts shorter food chains in unpredictable environments, and this appears to be correct (Pimm 1991).

A third generalization about food webs is that there is a nearly constant proportion of species that are top predators, intermediate species, and basal species (see Table 24.1 for definitions), regardless of the size of the food web. Figure 24.7 shows this for predator/prey ratios. There is a constant ratio of 2 to 3 prey species for every predator species in food webs, regardless of the total number of species in the web (Martinez 1991).

A fourth generalization is that omnivory seems to be rare in food webs. There are some exceptions to this rule. Aquatic communities often have fishes that eat their way up the food chain as they grow in size. Also detritus feeders feed on detritus that originates in several trophic levels (Pimm et al. 1991).

There is considerable interest in developing equilibrium models of food chains (Lawton and Warren 1988). The difficulty at present is that nearly all the published food webs are only partial webs (Paine 1988). For example, Little Rock Lake in Wisconsin has a food web containing 93 species (Martinez 1991), and this complex web contains twice as many species as those analyzed by Cohen et al. (1990). Polis (1991) listed four major problems with published food webs. The number of species in real communities is far greater than the number of those portrayed in food webs like Figure 24.5 shows. There are more than 2000 species in the sandy deserts of the Coachella Valley of California; Polis (1991) concluded that even a partial compilation of this food web illustrated more complex properties than those presented by

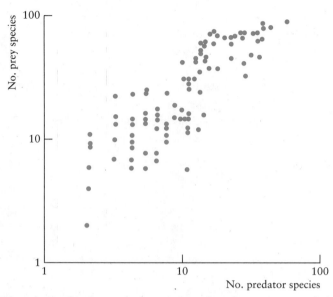

Figure 24.7   The relationship between the number of predator species and the number of prey species in 92 freshwater invertebrate food webs. The prey-predator ratio varied between 2:1 and 3.5:1 in these communities. (Data from Jeffries and Lawton 1985.)

Pimm et al. (1991) and Cohen et al. (1990). For example, food chains in the Coachella Desert had up to 18 links, and contained more prey species (10–11) per predator species. Polis (1991) suggested that current food web theory needs to be reconsidered once better descriptive data become available.

The importance of understanding food webs has been emphasized by Pimm (1991) because the structure of the food web has implications for community persistence. Some food webs can support additional species without suffering any losses, and other food webs are unstable and thus subject to species losses. If we can understand the structure of food webs better, we can design better management strategies for conservation.

FUNCTIONAL ROLES AND GUILDS

Trophic levels provide a good description of a community but by themselves are not sufficient for defining community organization. A better approach is to use the food web to subdivide each trophic level into *guilds*, which are groups of species exploiting a common resource base in a similar fashion (Root 1967). For example, hummingbirds and other nectar-feeding birds in tropical areas form a guild exploiting a set of flowering plants (Feinsinger 1976). We expect competitive interactions to be potentially strong between the members of a guild. By grouping species into guilds, we may also identify the basic functional roles played in the community.

There are four advantages to using guilds in the study of community organization (Root 1967):

1. Guilds focus attention on all sympatric competing species, regardless of their taxonomic relationship.

2. The use of *guild* clarifies the niche concept because groups of species having similar ecological roles can be members of the same guild but not occupants of the same niche.

3. Guilds allow us to compare communities by concentrating on specific functional groups. We do not need to study the entire community but can concentrate on a manageable unit.

4. Guilds might represent the basic building blocks of communities and thus help us to analyze community organization.

A community can be viewed as a complex assembly of component guilds, each containing one or more species. Guilds may interact with one another within the community and thus provide the organization we see. No one has yet been able to analyze all the guilds in a community; moreover, at present we can deal only with a few guilds that make up part of a whole community. Two examples of the organization of guilds show how this concept can be applied to communities.

Root (1973) grew collards *(Brassica oleracea* var. *acephala)* in two experimental habitats, pure stands and single rows bounded on each side by meadow vegetation. Three herbivore guilds were associated with collard stands (Figure 24.8). Pit feeders are insects that rasp small pits from the leaf surfaces, and they comprise 18 species, of which two Chrysomelid beetles were abundant. Strip feeders are insects that chew holes in the leaves and included 17 species of which only one was abundant. Sap feeders suck the juices of the collard plants and included 59 species, many of them aphids. The pit feeders usually formed the most important herbivore guild, particularly in the pure collard stands.

The species composition of the three herbivore guilds changed from year to year, and these changes were most striking in the sap feeders. The cabbage aphid *(Brevicoryne brassicae)* was the most abundant aphid in 1966 and 1968 but was absent entirely in 1967, when other aphids increased in abundance. The implication is that within some guilds, species can replace one another and carry on the same functional role.

The nectar-eating birds of successional montane forests in Costa Rica form a guild clearly organized around competition for food (Feinsinger 1976). This guild of hummingbirds is organized around the dominant species, *Amazilia saucerottei,* the blue-vented hummingbird. *Amazilia* specializes on plants that produce large quantities of nectar and sets up individual feeding territories that each bird defends against other hummingbirds. *Amazilia* is aggressively dominant over most other hummingbird species. A second common species, *Chlorostilbon canivetii,* is excluded from the rich flower resources by aggressive *Amazilia* individuals and exhibited "trap-line" feeding, following a regular route between scattered flowers. *Chlorostilbon* spends much time in flight and only rarely defends any flowers. Two other hummingbirds completed the core group of this guild. *Philodice bryantae* sets up feeding territories in rich flower areas but defends these territories only against other *Philodice.* This species seldom elicits attack behavior from the dominant *Amazilia,* partly because *Philodice* looks more like a bee than a bird. The final member of the core species in this guild is *Colibri thalassinus,* which is highly migratory and moves in to exploit seasonal flowering. Ten other hummingbird species foraged in the study area, and most of these were species that are important in adjacent communities. All these species' foraging is affected by the territorial behavior of the dominant *Amazilia,* and the high diversity of this bird guild is related to the highly migratory strategy of many of these hummingbird species.

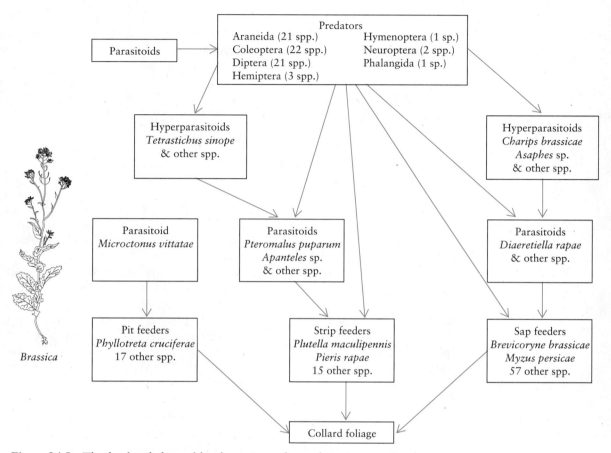

Figure 24.8   The food web formed by the major arthropod species associated with collards at Ithaca, New York. The herbivores are divided into guilds. (After Root 1973.)

Most of the hummingbird species in the guild that Feinsinger (1976) analyzed were general nectar feeders, and hence functional equivalents. If one of these species could be removed from the community, the prediction would be that the other hummingbird species would take its place, and the community would be little changed.

The guild or role concept of community organization is not yet fully developed (Simberloff and Dayan 1991). We can recognize four hypotheses that require testing in natural communities:

1. Many species form interchangeable members of a guild from the point of view of the rest of the community. These species are functional equivalents.

2. The number of functional roles within a community is small in relation to the number of species and might be constant among different communities.

3. There may be a limit on the number of species that can simultaneously fill a given functional role. A community always has a set of roles, but the guilds may be packed with different numbers of species.

4. Species within guilds fluctuate in abundance in such a way that the total biomass or density of the guild remains stable.

At present we can define roles or guilds only crudely by the analysis of food webs. There is a need to define objectively the criteria used to assign species to guilds in natural communities (Simberloff and Dayan 1991). The utility of the guild concept is that it reduces the number of components in a community and should help us to study how communities are put together. It also emphasizes that ecological units are not taxonomic units. Ants, rodents, and birds can all eat seeds in desert habitats and can thus form a single guild of great taxonomic diversity (Brown and Davidson 1979).

KEYSTONE SPECIES

A role may be occupied by a single species, and the presence of that role may be critical to the community. Such important species are called *keystone species* because their activities determine community structure (Paine 1969). Keystone species are most easily recognized by removal experiments.

The starfish *Pisaster ochraceous* is a keystone species in rocky intertidal communities of western North America (Paine 1974). When *Pisaster* was removed manually from intertidal areas, the mussel *Mytilus californianus* was able to monopolize space and exclude other invertebrates and algae from attachment sites (see page 532). *M. californianus* is an ecological dominant that is able to compete for space effectively in the intertidal zone. Predation by *Pisaster* removes this competitive edge and allows other species to use the space vacated by *Mytilus*. *Pisaster* is not able to eliminate mussels because *Mytilus* can grow too large to be eaten by starfish. Size-limited predation provides a refuge for the prey species, and these large mussels are able to produce large numbers of fertilized eggs (Paine 1974).

The lobster may be a keystone species in subtidal communities off the east coast of Canada (Mann and Breen 1972). Lobsters have been heavily exploited by fishermen, and sea urchins *(Strongylocentrotus droebachiensis)* have increased in abundance coincident with the lobster decline. Sea urchins are herbivores that can control the distribution of algae (see Figure 6.3, page 78). Population explosions of sea urchins result in the elimination of *Laminaria* and *Alaria* seaweeds and have produced large areas of nearly barren rock off the east coast. There is considerable controversy about whether the lobster-sea urchin system is a good example of a keystone predator (Elner and Vadas 1990). Figure 24.9 gives two cyclic models of subtidal communities off eastern Canada. The keystone predator model (Figure 24.9a) suggests that lobster predation holds down sea urchin numbers so that kelp abundance is high (macroalgal phase). When humans overfish lobster, the system flips to a barrens phase with few visible algae and many sea urchins. An epidemic disease swept down the Nova Scotia coast in the mid-1980s and killed most of the sea urchins, and this has led to an alternative model for this subtidal community (Figure 24.9b) in which lobsters play at best a minor role. More research is needed to determine which of these models is correct, and in particular whether lobster fishing has any significant impacts on the structure of subtidal communities in the northwestern Atlantic (Elner and Vadas 1990).

A third possible example of a keystone species is the African elephant (Laws 1970). The African elephant is a relatively unspecialized herbivore but relies on a diet of browse supplemented by grass. By their feeding activities, elephants destroy shrubs and small trees and push woodland habitats toward open grassland (Figure 24.10). Large mature trees can be destroyed by elephants feeding on the bark. As more grasses invade the woodland habitats, the frequency of fires increases and

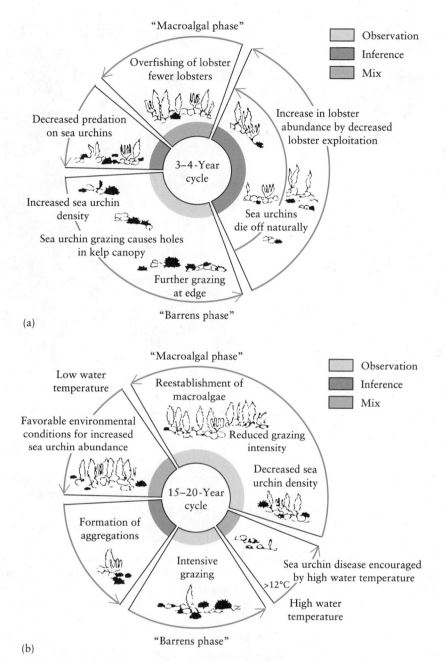

Figure 24.9   Two alternative models to explain the alternation between kelp beds and coralline barrens in the subtidal zone off the Atlantic coast of Canada. (a) Keystone predator model. (b) Water temperature–disease model. Epidemics periodically reduce sea-urchin numbers. (From Elner and Valdas 1990.)

accelerates the conversion of woods to grassland. This conversion works to the elephants' disadvantage, however, because grass is not a sufficient diet for elephants, and they begin to starve as woody species are eliminated. Other ungulates that graze the grasses are favored by the elephants' activities.

The critical impact of keystone predators is that they can reverse the outcome

Figure 24.10   African elephants browsing in open woodland, Serengeti area of East Africa. (Photo courtesy of A. R. E. Sinclair.)

of competitive interactions. The impact of keystone predators is seen very clearly in aquatic communities. Amphibians are a major component of temporary ponds. In the coastal plain of North Carolina a single pond can support 5 species of salamanders and 16 species of frogs and toads (Fauth and Resetarits 1991). Salamanders are the major predators in these temporary ponds, and the broken-striped newt *Notophthalmus viridescens* acts as a keystone predator. It selectively preys on the dominant competitors *Rana utricularia* and *Bufo americanus* and allows less competitive frogs like the cricket frog *(Hyla crucifer)* to survive.

Kangaroo rats in the Chihuahuan Desert form a keystone guild (Brown and Heske 1990). Kangaroo rats (*Dipodomys* spp.) prefer to eat large seeds. When they were excluded from areas of desert shrubland, large-seeded winter annuals increased greatly in abundance, raising the vegetative cover, which reduced the ability of ground-feeding birds to feed on seeds. After 12 years grasses increased and the area became desert grassland. Grassland species of rodents, like the cotton rat, which were previously absent, then colonized the habitat from which kangaroo rats were excluded. Fifteen species of rodents live in the Chihuahuan Desert but only the three species of kangaroo rats seem to play this keystone role.

Keystone species may be rare in natural communities, or they may be common but not recognized. At present, few terrestrial communities are believed to be organized by keystone species, but in aquatic communities keystone species may be common. Pimm (1991) has attempted to model the effects of species removals from hypothetical food chains. Trophically specialized species are at greatest risk when

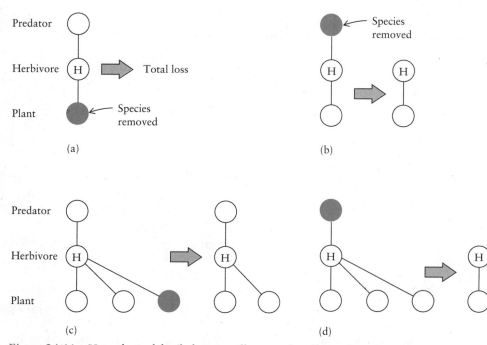

Figure 24.11 *Hypothetical food chains to illustrate the effects of species removals. (a) A plant removed from a simple food chain may cause the entire food chain to be lost. (b) A predator removed from the same food chain may have no effect. (c) A plant lost from a more complex food web may have little effect. (d) A predator lost from the same food web may lead to the loss of several plants. (After Pimm 1987.)*

plants or herbivores are lost or removed. The predictions that follow from the removal of a carnivore are much less obvious because of indirect effects. Figure 24.11 summarizes some of Pimm's (1991) analyses of potential keystone effects. The important message is that we cannot determine which species might be keystone species unless we understand the detailed structure of the food web.

DOMINANT SPECIES

Dominant species in a community may exert a powerful control over the occurrence of other species, and the concept of dominance has long been ingrained in community ecology. Dominant species are recognized by their numerical abundance or biomass and are usually defined separately for each trophic level. For example, the sugar maple is the dominant plant species in part of the climax forest in eastern North America and by its abundance determines in part the physical conditions of the forest community. Dominance usually means numerical dominance, and keystone species are not usually the dominant species in a community.

Dominance is related to the concept of species diversity, and some of the measures of species diversity discussed in Appendix IV (such as Simpson's index) could also be considered as measures of dominance. We can define a simple community dominance index as follows (McNaughton 1968):

Community dominance index = percentage of abundance contributed by the two most abundant species

$$= 100 \times \frac{y_1 + y_2}{y}$$

where

y_1 = abundance of most abundant species
y_2 = abundance of second most abundant species
y = total abundances for all species

Abundance may be measured by density, biomass, or productivity. Dominance, defined by the community dominance index, may be positively or negatively related to biodiversity. Figure 24.12 illustrates a negative relationship between dominance and diversity for annual grassland in California. In marine communities, there is a slight but significant positive relationship between species richness and dominance (Birch 1981). Common species are relatively more common in tropical, species-rich marine communities, contrary to the prediction of MacArthur (1969).

Dominant species are usually assumed to be competitive dominants and to achieve their dominance by competitive exclusion. Buss (1980) has pointed out that there are several possible configurations of competing species (Figure 24.13). The simplest case of *transitive* competition occurs when a linear hierarchy exists (species A outcompetes B and B outcompetes C). In this case, competitive exclusion can occur, as we saw in Chapter 13. A more complex case of *intransitive* competition occurs when a circular network exists in which no one species can be called dominant. In circular networks of spatial competition species A outcompetes species B, B outcompetes C, but C in turn is able to outcompete A. This type of competitive interaction is called *intransitive* because it has no end point. Competitive exclusion does not occur in circular networks, and if intransitive competition is the rule in natural communities, species diversity need not decline because of competitive exclusion (Buss and Jackson 1979).

Competitive dominance is not the only explanation for a species becoming dominant. Australian mangrove forests show a complex zonation across the intertidal zone (Smith 1987). This zonation has usually been explained by mechanisms of

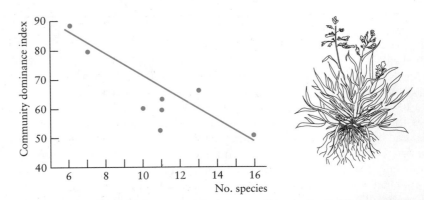

Figure 24.12   *Relationship of dominance and species diversity in California annual grasslands. Dominance is defined as the percentage of the peak standing crop contributed by the two most abundant species. (After McNaughton 1968.)*

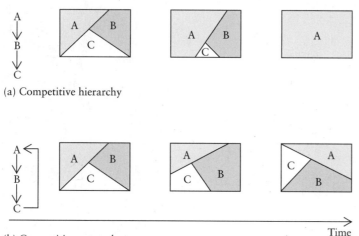

(a) Competitive hierarchy

(b) Competitive network

Time

Figure 24.13 Schematic illustration of transitive and intransitive competitive relationships between three species competing for space. Each rectangle represents a plot of ground or a rock surface in the intertidal zone. (a) The conventional competitive hierarchy showing transitive competition in which species A outcompetes species B, which in turn outcompetes species C. As time progresses, competitive exclusion will occur. (b) A competitive network showing intransitive competition in which species A outcompetes B and B outcompetes C, but C is able to outcompete A. This system changes in time but does not move toward competitive exclusion. (After Buss and Jackson 1979.)

physiological tolerance to sea water inundation or by tidal sorting of seeds by size. Smith (1987) found that seed predation by small grapsid crabs was very high, and dominance in four of five mangrove species was correlated with the amount of seed predation (Figure 24.14). Predation may override competition in some communities.

Communities that develop under similar ecological conditions in a particular geographic region are expected to be dominated by the same species. For example, a deciduous forest in Ohio is expected to be dominated by beech and sugar maple, and botanists would be surprised if a rare species such as black walnut or white ash became dominant. There has always been a question if the same pattern occurs in aquatic communities, particularly those of the open ocean. The central gyre of the North Pacific Ocean has a rich diversity of phytoplankton, zooplankton, and fish species (McGowan and Walker 1979). Because the ocean mixes and the gyre is so large, the dominant species might be expected to vary from place to place within this large area. This does not seem to occur. McGowan and Walker (1985) found that for copepods about 30 species were abundant and the dominance structure of this community remained the same in samples collected up to 16 years apart. The same constancy of community structure was evident in the phytoplankton community and the fish community of the central gyre. These oceanic communities appear to be as constant in their dominant species as are temperate zone forests.

The removal of a dominant species in a community has occurred frequently because of the impact of humans on communities, but unfortunately, few of these removals have been studied in detail. The American chestnut was a dominant tree in the eastern deciduous forests of North America before 1910, making up more than 40 percent of the overstory trees. This species has now been eliminated as a

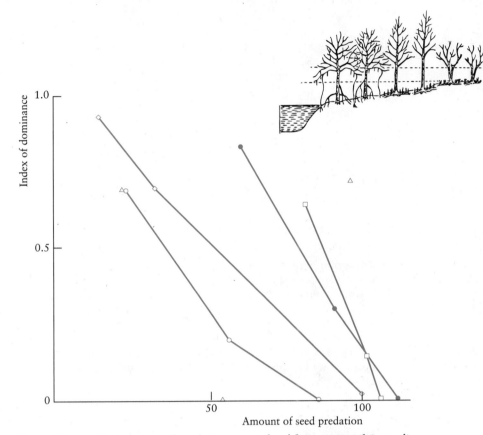

Figure 24.14   The amount of predation on seeds of five species of Australian mangrove trees in relation to community dominance. Each point represents one species on one site. Four of the five mangroves can become dominant only in areas of low seed predation by crabs. Only Ceriops tagal *is not affected.* □ Avicennia marina; ● Bruguiera exaristata; ○ B. gymnorrhiza; △ Ceriops tagal; ◇ Rhizophora stylosa. *(From Smith 1987.)*

canopy tree by chestnut blight. The impact of this removal has been negligible as far as anyone can tell, and various oaks, hickories, beech, and red maple have replaced the chestnut (Keever 1953). Fifty-six species of Lepidoptera fed on the American chestnut. Of these 7 species went extinct but the other 49 species did not rely only on the chestnut for food and they still survive (Pimm 1991).

Dominance has been studied in freshwater communities in considerable detail. The zooplankton community of many lakes in the temperate zone is dominated by large-sized species when fish are absent and by small-sized species when fish are present. Brooks and Dodson (1965) observed this change in Crystal Lake, Connecticut, after the introduction of a herringlike fish, the alewife *(Alosa pseudoharengus)* (Figure 24.15). They proposed the *size-efficiency hypothesis* as a wide-ranging explanation of the observed shift in dominance in the zooplankton community. The size-efficiency hypothesis is based on two assumptions: (1) planktonic herbivores (zooplankton) all compete for small algal cells (1–15 μm) in the open water; and (2) larger zooplankton feed more efficiently on small algae than do smaller zooplankton,

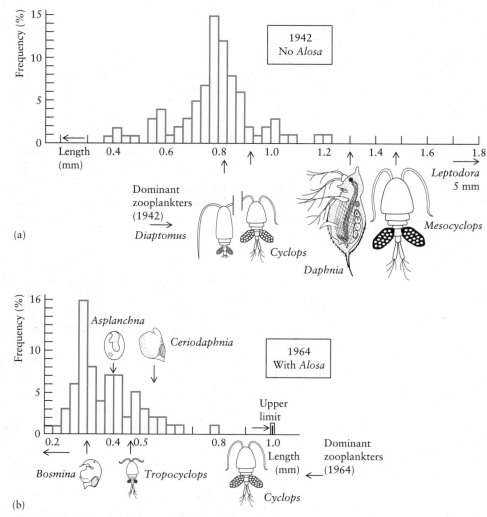

Figure 24.15   *The composition of the crustacean zooplankton of Crystal Lake, Connecticut, before (1942) and after (1964) the introduction of the alewife, a plankton-feeding fish.* Daphnia *is a large cladoceran,* Ceriodaphnia *and* Bosmina *are small cladocerans.* Mesocyclops *is a large copepod,* Tropocyclops *is a small copepod. Some of the larger zooplankton species are not represented in the 1942 histogram because they were not abundant. (After Brooks and Dodson 1965.)*

and large animals are able to eat larger algal particles that small zooplankton cannot eat.

Given these two assumptions, Brooks and Dodson (1965) made three predictions:

1. When predation is of low intensity or absent, the small zooplankton herbivores will be completely eliminated by large forms (dominance of large cladocerans and calanoid copepods).

2. When predation is of high intensity, predators will eliminate the large forms and allow the small zooplankton (rotifers, small cladocerans, small copepods) to become dominant.

3. When predation is of moderate intensity, predators will reduce the abundance of the large zooplankton so that the small zooplankton species are not eliminated by competition.

Thus competition forces communities toward larger-bodied zooplankton, while fish predation forces them toward smaller-bodied species. These three predictions of the size-efficiency hypothesis are consistent with the keystone-species idea discussed in the preceding section.

The second and third assumptions of the size-efficiency hypothesis have been tested in several lakes and the predictions seem to describe adequately the zooplankton distributions in many lakes (Kerfoot 1987). Fish predation does seem to fall more heavily on the larger zooplankton species, but invertebrate predators in the plankton seem to prey more heavily on the smaller zooplankton species. Large zooplankton may predominate in lakes with no fish either because they are superior competitors (as the size-efficiency hypothesis predicts) or because small zooplankton are selectively removed by invertebrate predators.

The first assumption of the size-efficiency hypothesis is that large-sized zooplankton are superior competitors for food resources. In a laboratory microcosm Gliwicz (1990) fed eight species of *Daphnia* on constant food levels and measured the concentration of algae needed to maintain body weight. Figure 24.16 shows that large copepods can subsist on much lower algal concentrations than small copepods, as predicted by the size-efficiency hypothesis.

The importance of fish predation in structuring zooplankton communities is now well established (Kerfoot 1987), but the competitive nature of feeding relationships in the zooplankton may not always favor large-sized species because of fluctuating food conditions in lakes and ponds. There are ponds and lakes where, in the absence of fish predators, small zooplankton predominate, and further work on the feeding strategies of zooplankton of different sizes is required (DeMott and Kerfoot 1982).

Dominance is an important component of community organization, although it

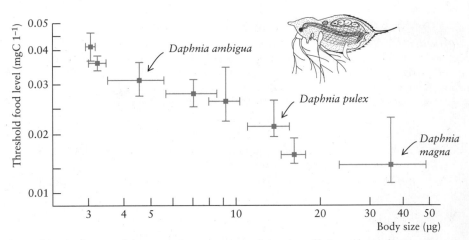

Figure 24.16   *A test of the central assumption of the size-efficiency hypothesis that large zooplankton are competitively superior to small-sized zooplankton. Eight* Daphnia *species of different sizes were raised in the laboratory on a constant food source. Larger-sized* Daphnia *can maintain weight on a lower food concentration, and are thus competitively dominant. (From Gliwicz 1990.)*

is poorly understood. Dominant species may be the focal point of interactions that structure many of the other species in a community. The characteristics of dominant species may affect the stability of the community as well as its organization.

STABILITY

Stability is a dynamic concept that refers to the ability of a system to bounce back from disturbances (page 544). If a brick is raised slightly from the floor and then released, it will fall back to its original position. This is the physicists' concept of *neighborhood stability* or local stability. The system will respond to temporary slight disturbances by returning to its original position. Thus, for example, a rabbit population may show neighborhood stability to moderate hunting pressure if it returns to its normal density after hunting is prohibited.

Physicists discuss stability in terms of small perturbations, but ecological systems are subject to large disturbances. To deal with these, we must consider a second type of stability, *global stability*. A system that has local stability shows global stability only if the system returns to the same point after large disturbances. Our brick, for example, shows both local and global stability because if we raise it 10 millimeters or 10 meters from the floor and release it, it will fall back to the floor. Global stability does not always occur; therefore, one of the problems of ecology is to map out the limits of global stability for various communities (Lewontin 1969). In Figure 24.1, if we schematically move a community at point A beyond point B, it will move to a new state somewhere near point C. In this hypothetical case community A is locally stable but not globally stable if forced beyond point B. All equilibrium theories of community organization assume that the equilibrium is stable, and the usual assumption is that the equilibrium is globally stable. Note that the shape of stability "basins" need not be circular in cross section. There may be great stability to disturbances in one direction but little stability to disturbances in other directions.

If equilibrium theories of community organization are correct, there are four important consequences for our understanding of community dynamics (Chesson and Case 1986):

1. *Community conservation.* An equilibrium community will show no tendency to lose species over time. Global stability implies that in the absence of external perturbations no losses of species will ever occur.

2. *Community recovery.* An equilibrium community can recover from events that drive any of its constituent species to low density.

3. *Community composition.* An equilibrium community can be built up by immigration of species from outside the system. Combinations of species that can coexist will increase to their equilibrium values (Figure 13.3, page 235).

4. *Independence of history.* Because of global stability, past events have no effect on community structure, and within a broad range, the order of arrival of member species is irrelevant to the final community composition.

Equilibrium communities are stable in the sense of persistence, but they may not be stable in the sense of being resilient to disturbances, particularly disturbances caused by humans.

One of the classical tenets of community ecology and a hallowed tenet of conservation biology has been that *biodiversity causes stability*. Elton (1958) suggested several lines of circumstantial evidence that support this conclusion:

1. Mathematical models of simple systems show how difficult it is to achieve numerical stability (Chapters 13 and 14).

2. Gause's laboratory experiments on protozoa confirm the difficulty of achieving numerical stability in simple systems.

3. Small islands are much more vulnerable to invading species than are continents.

4. Outbreaks of pests are most often found in simple communities on cultivated land or land disturbed by humans.

5. Tropical rain forests do not have insect outbreaks like those common to temperate forests.

6. Pesticides have caused outbreaks by the elimination of predators and parasites from the insect community of crop plants.

Many of these statements are only partly correct, and the simple, intuitive, and appealing notion that biodiversity leads to stability must be rejected as a general statement (Pimm 1984).

The intuitive argument that increasing community complexity in the food web automatically leads to increased stability was attacked by May (1973), who showed that increasing complexity *reduces* stability in general mathematical models. In hypothetical communities in which the trophic links are assembled at random, the more diverse communities are more unstable than the simple communities. Thus May cautioned community ecologists that if diversity causes stability in the real world, it is not an automatic mathematical consequence of species interactions. Natural communities are products of evolution, and evolution may have produced nonrandom assemblages of interacting species in which diversity and stability are related.

In theoretical food webs one way to preserve stability of species composition is to arrange species in compartments or blocks in which species within blocks interact much more than species between blocks. There is some evidence that compartments exist in natural food webs and this may help to preserve stability (Moore and Hunt 1988).

Is there any evidence from field or laboratory experiments that would test the diversity-stability hypothesis? A few experiments have been done on stability and diversity, and these suggest that the diversity-stability hypothesis is wrong and that there is no simple relationship between diversity and stability in ecological systems.

Small laboratory microcosms of bacteria and protozoa were analyzed by Hairston and colleagues (1968). The microcosms had two or three trophic levels:

Trophic Level	Organisms	No. Species Studied
First	Bacteria	1–3
Second	*Paramecium*	1–2
Third	*Didinium* and *Woodruffia* (predatory protozoa)	2

Stability was measured in two ways: (1) persistence in time of all species (so less stability = more extinctions) and (2) evenness of species abundance patterns.

Experiments with the first and second trophic level showed that more diversity at the first trophic level led to more stability at the second trophic level. After 20 days, fewer extinctions of *Paramecium* occurred when the bacteria were diverse:

Experiment	Cultures Showing No Extinctions of *Paramecium* After 20 Days (%)
1 species of bacteria	32 (less stability)
2 species of bacteria	61
3 species of bacteria	70 (more stability)

We can ask a second question—whether diversity within one trophic level increases stability within that trophic level. This was not true in the *Paramecium* trophic level, because the effect of adding a third *Paramecium* species to two others depended on which particular species was being added to which other two. This means that species-specific quirks can modify diversity-stability relations and that all species are not equal and interchangeable at the same trophic level.

When a third trophic level was added to the bacteria and *Paramecium*, there was a general decrease in stability of the whole system because the *Paramecium* were usually forced to extinction (see Figure 14.7, page 271), and it did not matter whether two or three species of *Paramecium* were present or whether there were one or two predator species. Thus, in simplified laboratory communities, diversity does not automatically lead to stability, and the addition of higher trophic levels may reduce community stability.

The stability of whole communities has rarely been studied in detail in spite of the great number of perturbations caused by humans. We have already discussed for a rocky intertidal community the effect of removing the top predator, *Pisaster*. This community is unstable in species diversity with respect to the removal of the starfish (page 532). In contrast, we would expect that temperate bird communities (Figure 23.14, page 529) would be unaffected by the removal of many of the plant species in their habitats.

Aquatic communities have been disturbed by pollution of human origin, and the stability of aquatic systems under pollution stress is a critical focus of applied ecology today. Several large-scale experiments have been performed by the diversion of sewage into large lakes near cities. Let us examine one such instance.

Lake Washington is a large, formerly unproductive lake in Seattle, Washington, that was used for sewage disposal until the late 1960s. In the early phases of development, Lake Washington was used for raw sewage disposal, but this practice was stopped between 1926 and 1936 (Figure 24.17). However, with additional population pressure, a number of sewage-treatment plants built between 1941 and 1959 began discharging treated sewage into the lake in increasing amounts. By 1955 it was clear that the sewage was destroying the clear-water lake, and a plan to divert sewage from the lake was voted into action. More and more sewage was diverted to the ocean from 1963 through 1968, and almost all was diverted from March 1967 onward (Figure 24.17). Thus the recent history of Lake Washington consists of two pulses of nutrient additions, followed by a complete diversion.

What happened to the organisms in Lake Washington during this time? Some information can be obtained by looking at the sediments in the bottom of the lake (Figure 24.18). After sewage had been added to the lake, the sedimentation rate rose to about 3 millimeters per year. The organic content of this sediment had progressively increased since the early 1900s, which suggests an accelerated rate of primary

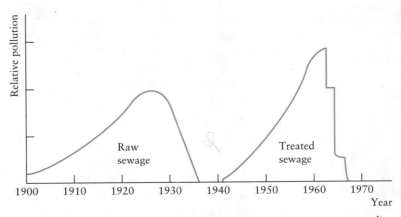

Figure 24.17   Sewage history of Seattle's Lake Washington. Raw sewage was diverted from the lake gradually over the period 1926–1936; then treated sewage was added at an increasing rate until a second diversion was made, in 1963–1967. (After Edmondson 1969.)

production. The recent lake sediments also contain a greater amount of phosphorus (Figure 24.18). Since phosphorus is one of the two main nutrients added by sewage, this is a parallel change to the organic matter. The composition of the diatom community in Lake Washington has also changed. The "shells" of diatoms are made of silica and are preserved well in sediments. Stockner and Benson (1967) showed that a group of species of diatoms (the Araphidinae) varied in abundance in association with the sewage history and consequently can be used as *indicator species* of pollution in this lake.

Since the diversion of sewage began in 1963, Edmondson (1991) has recorded the changes in Lake Washington in detail. Figure 24.19 shows the rapid drop in phosphorus in the surface waters and the closely associated drop in the standing crop of phytoplankton. Nitrogen content of the water has dropped very little, which suggests that phosphorus is a limiting nutrient to phytoplankton growth. The water of the lake has become noticeably clearer since the sewage diversion. The phosphorus tied up in the lake sediments is apparently released back into the water column rather slowly.

In 1976 Lake Washington suddenly became much clearer than had been recorded previously. This clearing of the water coincided with the colonization of Lake Washington by *Daphnia* sp. *Daphnia* are filter-feeders that sweep small green algae out of the water and in the process make lake water clearer. *Daphnia* have always been present in lakes around Lake Washington so their sudden appearance in 1976 was not a simple matter of colonization and dispersal. Two factors interacted to hold *Daphnia* numbers low before 1976. The blue-green alga *Oscillatoria* was common in Lake Washington during its polluted phase and gradually declined after 1968 when sewage was diverted. *Oscillatoria* clogs the filter-feeding apparatus of *Daphnia* because it forms long filaments. *Daphnia* could not feed properly until *Oscillatoria* became scarce. *Daphnia* are also a major prey item for the shrimp *Neomysis mercedis*. *Neomysis* decreased during the 1960s because of fish predation by increased populations of longfin smelt in the lake. Smelt had increased during the 1960s when their spawning grounds in the Cedar River were protected from dredging and con-

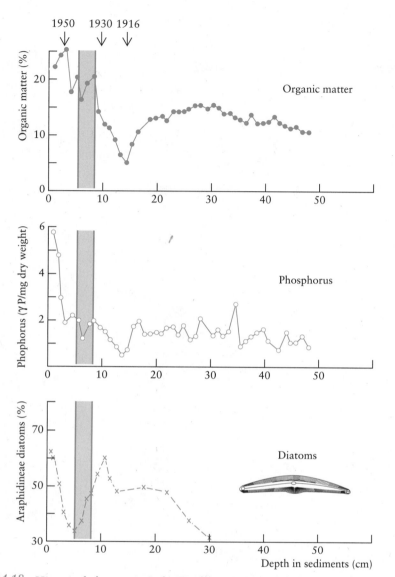

Figure 24.18 *Historical changes in Lake Washington as revealed by the sediments in the lake bottom. These core data were taken in 1958. The shaded area represents the approximate position in the core of the time period 1930–1940, when nutrient pollution from sewage was temporarily halted. (After Edmondson 1969.)*

struction in the river. The aquatic community of Lake Washington shows a complex web of predator-prey interactions that have profound impacts on community structure (Edmondson 1991).

The Lake Washington experiment is of considerable interest because it suggests that detrimental changes in lakes may be *stopped and reversed* if the input of nutrients can be stopped. That is, the Lake Washington system shows a considerable amount of global stability, and is a good example of an equilibrium community.

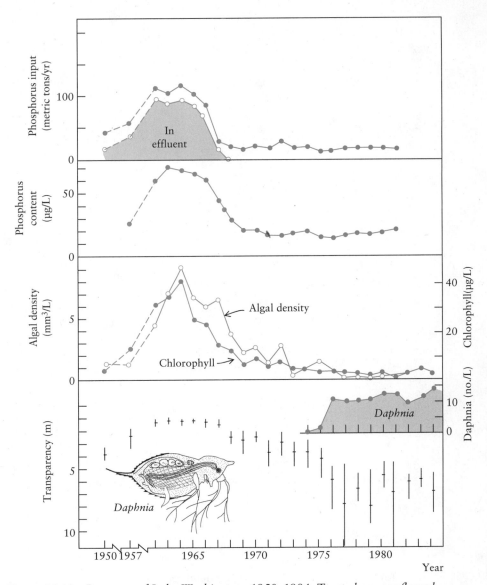

Figure 24.19 *Recovery of Lake Washington, 1950–1984. Treated sewage flowed into the lake in increasing amounts during the 1950s. Sewage was diverted from the lake gradually from 1963 to 1968. The phosphorus content of the lake decreased rapidly after sewage was diverted. Algal density dropped in parallel with phosphorus levels because phosphate is the nutrient that limits algal growth in this freshwater lake. In 1976 the small crustacean* Daphnia *increased greatly in abundance and because it eats small algae, the algal abundance fell even more, so the lake water became clearer. Transparency is the depth to which a standard white plate can be seen in the water during midsummer. (Data courtesy of W. T. Edmondson.)*

SUMMARY

Communities can be organized by competition, predation, and mutualism working within a framework set by the physical factors of the environment. Of these three processes, most emphasis has been placed on the roles of competition and predation in organizing communities.

Two broad views of communities are postulated as explanations of community organization. The classical model is that communities are in equilibrium, their species composition and relative abundances controlled by biotic interactions. Interspecific competition is the major process controlling organization, according to the equilibrium model, and predation and spatial heterogeneity can be added to this model to make it more realistic. The nonequilibrium model does not assume stable equilibria but that communities are always recovering from disturbances, and it is discussed in the next chapter.

Species in a community can be organized into food webs based on who-eats-whom. Trophic levels may be recognized in all communities from the level of the producers (green plants) to the higher carnivores and hyperparasites. Within a trophic level, we can recognize *guilds* of species exploiting a common resource base. Guilds may serve to pinpoint the functional roles species play in a community, and species within guilds may be interchangeable in some communities.

Food webs can be analyzed theoretically and empirically. Patterns of food chain length and complexity can be recognized, but all of them are of uncertain validity because most published webs are only partial webs, and the analysis of complex food webs containing more than 100 species is only beginning.

Keystone species determine community structure single-handedly and can be recognized by removal experiments. Dominant species are the species of highest abundance or biomass in a community. Dominance is often achieved by competitive superiority, and some dominant species can be removed from the community and replaced with subdominants with little effect on community organization. In aquatic communities, dominance in the zooplankton herbivores may be determined by competition when fish predators are absent and by predation when fish are present.

The characteristics of dominant species may affect community stability, in the sense of the ability of the system to return to its original configuration after disturbance. There are at least five different definitions of "stability" in the ecological literature involving concepts of variability, persistence, resistance to change, and speed of recovery from disturbance. The ecological generalization that *diversity causes stability* is not supported by field or laboratory data or by theoretical analyses, and the attributes of individual species and compartments in food webs may be more significant than biodiversity in general in promoting stability.

Selected References

BROWN, J. H., and E. J. HESKE. 1990. Control of a desert-grassland transition by a keystone rodent guild. *Science* 250:1705–1707.

CONNELL, J. H., and W. P. SOUSA. 1983. On the evidence needed to judge ecological stability or persistence. *American Naturalist* 121:789–824.

MARTINEZ, N. D. 1991. Artifacts or attributes? Effects of resolution on the Little Rock Lake food web. *Ecological Monographs* 61:367–392.

McGOWAN, J. A., and P. W. WALKER. 1985. Dominance and diversity maintenance in an oceanic ecosystem. *Ecological Monographs* 55:103–118.

PAINE, R. T. 1988. Food webs: Road maps of interactions or grist for theoretical development? *Ecology* 69:1648–1654.

PIMM, S. L. 1984. The complexity and stability of ecosystems. *Nature* 307:321–326.

PIMM, S. L., J. H. LAWTON, and J. E. COHEN. 1991. Food web patterns and their consequences. *Nature* 350:669–674.

POLIS, G. A. 1991. Complex trophic interactions in deserts: An empirical critique of food-web theory. *American Naturalist* 138:123–155.

SCHOENLY, K., R. A. BEAVER, and T. A. HEUMIER. 1991. On the trophic relations of insects: A food-web approach. *American Naturalist* 137:597–638.

ZARET, T. M. 1982. The stability/diversity controversy: A test of hypotheses. *Ecology* 63:721–731.

Questions and Problems

1. Elton (1958, p. 147) claims that natural habitats on small islands are much more vulnerable to invading species than natural habitats on continents. Find evidence that is relevant to this assertion, and evaluate its importance for the question of community stability (resistance).

2. Compare these two conclusions, and discuss why they differ:

   a. "Theoretical studies suggest that complex systems are more resilient than simple ones" (Pimm 1982, p. 199).
   b. "As a mathematical generality, increased multispecies trophic complexity makes for lowered stability" (May 1973, p. 77).

3. In discussing community organization, Hairston (1964, p. 238) states:

   It would be possible, presumably, to build a picture of community organization by separate complete studies of each species present, but such an approach would be comparable to describing an organism cell by cell.

   Is this a proper analogy?

4. Compare and contrast the statements of the evolutionist and the ecologist about species diversity and stability of biological communities:

   a. Simpson (1969, p. 175) states: "If indeed the earth's ecosystems are tending toward long-range stabilization or static equilibrium, three billion years has been too short a time to reach that condition."
   b. Recher (1969, p. 79) states: "The avifaunas of forest and scrub habitats in the temperate zone of Australia and North America have reached equilibrium and are probably saturated."

5. Conservation ideology has seized upon the diversity-stability hypothesis as a justification for various preservationist and environmentalist policies. Discuss

how alternative views of community organization might influence conservation policy if the diversity-stability hypothesis is untenable.

6. How long a time does a community need to be observed before its food web is complete? Discuss the implications of constructing a time-specific food web versus a cumulative food-web over a long period. Compare your analysis with that of Schoenly and Cohen (1991).

7. Compare the definitions of a *trophic species* (p. 549) and a *guild* (page 551). How does the aggregation of species into trophic species affect the analysis of a food web? Polis (1991) discusses this problem of aggregation.

8. In some tropical rain forests 50 to 100 percent of the canopy trees are one species (Connell and Lowman 1989). List all the possible mechanisms by which a single species can maintain its dominance in a species-rich forest system. Which of these mechanisms could not operate in species-poor temperate forests? Compare your list with that of Connell and Lowman (1989, p. 97).

9. Kenny and Loehle (1991) constructed random food webs in which the connections between "species" were drawn at random with no biological interactions assumed. They found that their random model produced data that fit the connectance graph shown in Figure 24.6. Discuss the implications of this finding.

Overview Question

You are asked to prepare a food web for a lake that is proposed to be part of a national park and to assess its stability. Discuss how you would construct this food web and measure its stability, and what operational decisions you would have to make about what to include in the web.

25 Community Organization III: Disturbance and Nonequilibrium Communities

The equilibrium model of community organization has not been accepted by many ecologists. The central issue of disagreement has not been whether equilibria are stable or unstable. Rather the issue has been whether or not it is valid to even conceive of equilibrium states for communities. The equilibrium model usually assumes a density-dependent population growth model, which has often been questioned by population biologists (Chapter 16). A more practical problem with community organization models based on equilibrium theory is that such models cannot be extrapolated to small spatial scales, the normal study areas of field ecologists (DeAngelis and Waterhouse 1987).

PATCHES AND DISTURBANCE

Two issues have dominated the discussion of nonequilibrial community organization—*patchiness* and *disturbance*. The spatial scale of a system has been recognized as a factor in how a system is described. The patchiness of different communities varies, and conclusions that apply to one spatial scale will not necessarily apply to others. We can recognize five spatial scales at which ecologists work (Wiens et al. 1986):

1. Space occupied by one plant or sessile animal, or the home range of one individual animal.
2. Local patch, occupied by many individual plants or animals.
3. Region, occupied by many local patches or local populations linked by dispersal.
4. Closed system (if it exists), or a region large enough to be closed to immigration or emigration.
5. Biogeographical scale, including zones of different climate and different communities.

At sufficiently small spatial scales all ecological systems are short-lived and can never be at equilibrium. Modeling community dynamics at small spatial scales and aggregating the resulting dynamics into a regional scale is an active focus of research (Chesson and Case 1986).

Most field studies of communities and virtually all experimental manipulations of communities are carried out at the local patch scale. There is no general definition of a "patch" that will cover all ecological communities. Note that patches do not need to be completely spatially discrete, nor do they need to be completely homogeneous (Pickett and White 1985).

A disturbance is any discrete event that disrupts community structure and changes available resources, substrate availability, or the physical environment. Note

Table 25.1   Definitions of Disturbance Regime Measures

Measure	Definition
Distribution	Spatial distribution, including relationship to geographic, topographic, environmental, and community gradients.
Frequency	Mean number of events per time period.
Return interval, or turnover time	The inverse of frequency: mean time between disturbances.
Rotation period	Mean time needed to disturb an area equivalent to the study area (the study area must be defined).
Predictability	An inverse function of variance in the return interval.
Area or size	Area disturbed. This can be expressed as area per event, area per time period, area per event per time period, or total area per disturbance type per time period.
Magnitude:	
Intensity	Physical force of the event per area per time (e.g., windspeed for hurricanes).
Severity	Impact on the community (e.g., basal area removed).
Synergism	Effects on the occurrence of other disturbances (e.g., drought increases fire intensity or insect damage increases susceptibility to windstorm).

Source: Modified from Pickett and White (1985).

that disturbances can be destructive events like fires or an environmental fluctuation like a severe frost. The notion of what is "normal" for the community is excluded from the ecologists' view of disturbance (in contrast to the everyday use of the word *disturbance*), precisely because we do not wish to assume some equilibrium "normal" structure for the community.

   Disturbances can be measured in a variety of ways to provide a time perspective and a spatial perspective (Table 25.1). Disturbances may be classified as *exogenous* (coming from outside the community, like fire) or *endogenous* (coming from biological interactions, like predation). These two classes are the end points of a continuum of types of disturbances, and many communities are affected by a combination of endogenous and exogenous disturbances.

THEORETICAL NONEQUILIBRIUM MODELS

Ecological communities can be arranged along a gradient from stable, biotically-interactive communities which are equilibrium-centered to unstable, interactive communities in which the biotic interactions do not lead to a stable equilibrium, to weakly interactive communities in which physical factors like temperature, salinity or fire prevent any stable equilibrium. Models have been developed all along this gradient to capture the ecological complexities of these systems. Chesson and Case (1986) recognize four types of nonequilibrium models of communities:

1. *Fluctuating environment models.* The simplest deviation from the classical equilibrium model of a community is a model with temporal variability. Competition is viewed in these models as the major biological interaction but the environment changes seasonally or irregularly and the competitive rankings of species also fluctuate so that no one species can win out. These models may include movements between patches and each patch may have a different environment.

Variable environment models are similar to equilibrium models in that they produce stable communities, but they differ in emphasizing the dynamics of dispersal among patches and variable life history traits.

2. *Density-independent models.* These models assume that population densities change but these fluctuations do not relate to density as the classical models assume, but are often density-independent. Density-vague dynamics (Figure 16.6, page 333) predominates in these theoretical communities, and populations are typically at levels where competition for resources is rare (Strong 1986). In some of these models spatial patchiness is added as another feature promoting fluctuations in the community.

3. *Directional changing environment models.* Variable environment models usually consider environments to fluctuate around a mean value that remains constant with time. What happens to these models when the mean itself changes? Much depends on the amount of fluctuation and the speed of the community in reacting to change. The current concern with global climate change (see Chapter 28) makes these models very significant for the future. Unlike many community models, these models cannot ignore history, and the response of a community to changing climate, for example, depends on its past history. Modern communities, these theories argue, cannot be understood only by looking at present environmental conditions. Life history characteristics and dispersal abilities strongly affect the ability of a species to track environmental changes.

4. *Slow competitive displacement models.* If competitive abilities are nearly equal, the process of competitive exclusion will be very slow and random variation in success will obscure any obvious displacements. Hubbell and Foster (1986) argue that the tropical rain forest has many species that are ecologically identical. Community composition is the net consequence of slow random-walk of tree species densities. Competition occurs in these models all the time, but because all species are identical in competitive abilities, there is no time trend or succession. Community structure, under these models, is strongly affected by chance and by history, and changes occur only on a geological time scale.

Table 25.2 draws together some of the different assumptions of these models and the equilibrium models discussed in Chapter 24, and illustrates how they differ in their predictions.

The purpose behind all of these models is to understand what enables a community to persist over time. There are five general types of hypotheses that ecologists have used to explain why communities tend to persist (Figure 25.1). The first hypotheses can be grouped as mechanisms to stabilize interactions and are usually applied as additions to classical equilibrium models of community organization. If predator functional responses are sigmoid (Figure 14.17, page 283), or if consumers were self-regulated, the stability of the community could be increased (DeAngelis and Waterhouse 1987). The second mechanism to promote community stability is (paradoxically) through disturbances. Disturbance creates nonequilibrial landscapes

Table 25.2   Emphases of Different Community Models

	EQUILIBRIUM MODELS			NONEQUILIBRIUM MODELS			
	Classical Competition Theory	Predation Extension	Spatial Variation Extension	(1) Fluctuating Environment	(2) Density-independent	(3) Directional Changing Environment	(4) Slow Competitive Displacement
ASSUMPTIONS							
No environmental fluctuations	Yes	Yes	Yes	No	No	—	Yes/no
Constant mean environment	Yes	Yes	Yes	Yes	Yes	No	Yes
Spatial homogeneity	Yes	Yes	No	Yes/no	Yes/no	—	Yes
Life history traits adequately summarized by growth rates	Yes	Yes	Yes	No	No	No	Yes
Continuous competition	Yes	Yes	Yes	Yes	No	—	Yes/no
PREDICTIONS: FACTORS INVOLVED IN COEXISTENCE OF N SPECIES							
At least n resources, limiting factors, or patches	Yes	Yes	Yes	No	No	—	No
Limits to similarity of resource use	Yes	Yes/no	No	No	No	—	No
Differential responses to fluctuating environmental variables or resources	No	No	No	Yes	Yes	—	No
Particular kinds of life history traits	No	No	No	Yes	Yes	Yes	—
Shapes of functional responses	No	Yes	No	Yes/no	—	—	No
Stability of species composition	Yes	Yes	Yes	Yes	Yes	No	No
Overall similarity of species	No	No	No	No	No	—	Yes
History	No	No	No	No	No	Yes	Yes

Notes. A dash indicates that no particular emphasis has yet emerged. Yes/no means that the theory suggests different answers in different circumstances.
  Source: Modified from Chesson and Case (1986).

and prevents competitive exclusion. The third way to stabilize a community is to have compensatory changes in reproduction, survival, or movements when populations reach low densities. Such changes could favor rare species over common ones. The fourth and fifth mechanisms that promote community persistence both involve spatial patchiness. If local populations are connected into metapopulations at the landscape level (Figure 25.1), it may not matter that each local patch is unstable. Species can recolonize by dispersal between patches. Stability in metapopulations is shown at the landscape level, not at the local population level. Species may go extinct in local patches, but as long as local patches are out of phase with one another, the species will persist in the landscape.

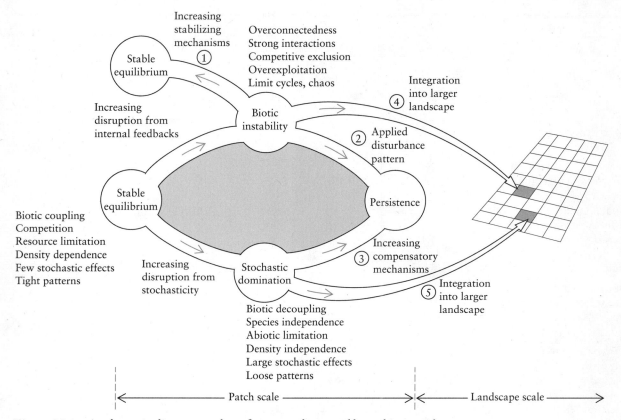

Increasing
stabilizing
mechanisms
①

Stable
equilibrium

Overconnectedness
Strong interactions
Competitive exclusion
Overexploitation
Limit cycles, chaos

Integration
into larger
landscape
④

Increasing
disruption from
internal feedbacks

Biotic
instability

Applied
disturbance
pattern
②

Stable
equilibrium

Persistence

Biotic coupling
Competition
Resource limitation
Density dependence
Few stochastic effects
Tight patterns

Increasing
disruption from
stochasticity

Stochastic
domination

Increasing
compensatory
mechanisms
③

Integration
into larger
landscape
⑤

Biotic decoupling
Species independence
Abiotic limitation
Density independence
Large stochastic effects
Loose patterns

|←———————————— Patch scale ————————————→|←——————— Landscape scale ———————→|

Figure 25.1   A schematic diagram to show five general types of hypotheses made by ecologists to explain why communities tend to persist despite the prevalence of biotic instabilities and environmental fluctuations. (From DeAngelis and Waterhouse 1987.)

How well do these models describe natural communities? Let us consider some examples.

CORAL REEF FISH COMMUNITIES

Coral reefs have been present in tropical oceans for at least 60 million years, and this long history is postulated to have produced the great diversity of organisms that exists on reefs today. For example, at One Tree Reef, a small island at the southern edge of the Great Barrier Reef, Talbot and colleagues (1978) recorded nearly 800 species of fish. At the northern edge of the Great Barrier Reef, over 1500 species of fish have been recorded. What determines the species diversity of a coral reef?

There are two extreme views about what controls the species diversity of a coral reef. The first view is the *niche-diversification hypothesis*, which suggests that coral reefs are equilibrium systems in which each species has evolved a very specific niche with respect to food and microhabitat. According to this view, competition between species is strong at the present time and maintains niche differences and species segregation (Anderson et al. 1981). The second view is the *variable recruitment* or

Table 25.3   Feeding Specializations Among 20 Species of Butterfly Fishes near Lizard Island, Great Barrier Reef, Australia

Hard Coral	Soft Coral and Some Hard Coral	Noncoralline Inver-tebrates	Generalists[a]
Chaetodon aureofasciatus	*Chaetodon kleinii*[b]	*Chaetodon auriga*	*Chaetodon citrinellus*
C. baronessa	*C. lineolatus*	*C. auriga*	*C. ephippium*
C. ornatissimus	*C. melannotus*	*Chelmon rostratus*	*C. ulietensis*
C. plebeius	*C. unimaculatus*	*Forcipiger* spp.	*C. vagabundus*
C. rainfordi			
C. speculum			
C. trifascialis			
C. trifasciatus			

[a] Plus plankton.
[b] Especially polychaetes and crustaceans.
  Source: After Anderson et al. (1981).

lottery hypothesis, which suggests that coral reefs are nonequilibrium systems in which larval recruitment is an unpredictable lottery. Competition between species is absent, and the local community present on a reef is a random sample from a common larval pool (Sale 1977). How can we distinguish these two hypotheses?

The first question we must ask is how specialized reef fishes are. Reef fish exhibit both food and habitat specialization, but there are often several species within one restricted niche; moreover, many generalist feeders are present (Sale 1977). Table 25.3 gives a sample of data on the feeding specializations of butterfly fishes. Many studies have now been done of feeding habits in coral reef fishes. Herbivorous fishes are more generalized feeders than predatory fishes. But even among the specialist feeders, it is common to find two or three species with identical specialization, as shown in Table 25.3. Feeding niches are thus not organized as tightly as the niche-diversification hypothesis would predict.

Habitat specialization could be another way that reef fishes have evolved niche differences. Adult reef fishes are very sedentary and could thus have very narrow habitat requirements, but this does not seem to be the case. Habitat partitioning does occur to the extent that few species range over all regions of the reef. Species tend to occur in broadly defined habitats, such as "reef flat" or "surge channel," but when microhabitats are assigned more carefully, there is extensive overlap of species. Thus we do not find a high degree of habitat specialization among coral reef fishes (Sale 1977).

Natural history information thus tends to go against the niche-diversification hypothesis. How can we test these hypotheses experimentally? If the variable recruitment hypothesis is correct, reef fish communities ought to be unstable in species composition and highly variable from reef to reef. Also, the species structure at a given site ought not to recover following artificial removals or additions of species.

To test the first prediction, Talbot and colleagues (1978) put out standard cement building blocks to create artificial reefs of constant size and shape. Forty-two fish species colonized these artificial reefs, averaging 17 species per reef. The similarity between replicate reefs was only 32 percent; that is to say, even though these reefs were set out in the same lagoon at the same time within a few meters of one another, only about 32 percent of the fishes colonizing them were of the same species. A survey of natural coral isolates of about the same size as the artificial reefs (0.6 m³) showed only a 37 percent similarity (Talbot et al. 1978). Moreover, as one followed the artificial reefs through time, there was very high turnover from month to month. Of the species on a reef one month, 20 to 40 percent would have disappeared by the next month and have been replaced by a new species not previously present. Clearly, reef fish communities are very unstable and highly variable from one small reef to the next.

The variable recruitment hypothesis assumes that there are no resource limitations on populations and no competitive effects. One prediction is that if recruits are experimentally added to a coral reef population, adult numbers will rise. Jones (1990) transplanted juveniles of the damselfish *Pomacentrus amboinensis* for three years to natural patch reefs, approximately 8 square meters, at the southern edge of the Great Barrier Reef in Australia. Figure 25.2 shows that adult densities increased as more recruits were added, but only to a ceiling. At high recruitment levels, density-dependent interactions between adults and potential recruits puts a ceiling on numbers, contrary to the predictions of the variable recruitment hypothesis. By manipulating both recruitment and adult fish density Forrester (1990) showed that juvenile recruitment in the humbug *Dascyllus aruanus* did not automatically translate into more breeding adults (Figure 25.3). Both recruitment and postrecruitment processes affect the abundance of coral reef fishes.

Coral reef fishes are similar to many marine organisms in having a life history that includes a larval phase in the plankton that is transported by ocean currents.

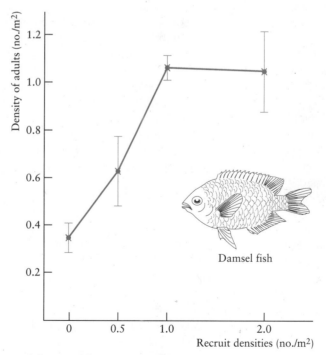

Damsel fish

Figure 25.2   Test of the variable recruitment hypothesis for coral reef fish. Four different recruitment levels were experimentally provided for the damselfish Pomacentrus ambionensis *for three years. Adult densities increased directly with recruitment up to 1 recruit/m². (From Jones 1990.)*

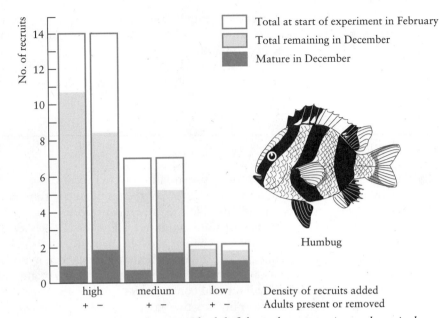

Figure 25.3   Effects of recruitment and adult fish on the maturation and survival of the humbug Dascyllus aruanus *on the Great Barrier Reef during one year. The number of recruits varies with the number added, but the number of fish reaching maturity did not depend on recruitment level but was reduced when adult fish were present on the reef. The number of recruits in natural populations is highly variable and averages about the medium level. (From Forrester 1990.)*

This type of life cycle implies that local reproduction is not linked to local recruitment, in complete contrast to the life cycle of birds and mammals. For these marine organisms the population or the community can never be a closed system, and the physical factors controlling recruitment may control the system. This has been called "supply-side ecology" by Roughgarden et al. (1987). There has been a long history of the ideas of supply-side ecology (Underwood and Denley 1984). The structure of an ecological community cannot be understood solely as a result of competition, predation, and disturbance in systems that have variable recruitment (Underwood and Fairweather 1989).

How is coexistence of so many species permitted if the variable recruitment hypothesis is correct? Sale (1982) calls the reef fish community a *lottery competition*. Individuals compete for access to units of resources (space for these fish) without which they cannot join the breeding population. A lottery competition is a type of interference competition in which an individual's chances of winning or losing are determined by who gets there first. Lottery competitive systems are very unstable but can persist if there is high environmental variability in birth rates (Chesson 1986). Because recruitment of reef fishes depends on larvae settling from the plankton, high variability is the rule, and vacated space is allocated at random to the first recruit to arrive from the larval pool (Sale 1982). The lottery competition model has two important requirements if it is to explain the coexistence of many species. First, environmental variation must be such to permit each species to have high recruitment rates at low population densities. Second, generations must overlap and adult death rates should be unaffected by competition (Chesson and Warner 1981).

Thus the high diversity of coral reef fish communities is not achieved by precise niche diversification in an equilibrium community but rather by highly variable larval recruitment, which causes a competitive lottery for vacant living spaces in which the first to arrive wins. Reef fish communities are thus not in equilibrium but continually fluctuate in species composition on a local level while retaining high diversity over a whole region.

ROCKY INTERTIDAL COMMUNITIES

The rocky intertidal zone is a tension zone between land and sea. Disturbances in the form of waves and storms are an important feature of the physical environment. Space is the key limiting resource in the intertidal, thus competition for space has been a key component of many studies of this community (Dayton 1971, Sousa 1985). Key concepts in community ecology have their roots in the rocky intertidal— for example the keystone species concept and the intermediate disturbance hypothesis—and the same is true for theories of community organization. Two examples will illustrate the ways in which rocky intertidal communities can be organized.

Seaweeds in New England can be split into two groups. Ephemeral seaweeds live for weeks or months, grow rapidly, and are eaten rapidly by herbivores like limpets. Perennial seaweeds can live for many years, grow more slowly, and except in their juvenile stages are relatively inedible (Lubchenco 1986). Seaweeds compete for space and for light, but the primary resource in the rocky intertidal is space. Using wire mesh cages to exclude herbivores, Lubchenco (1986) found that there was no simple answer to the question of what controlled seaweed abundance (Figure 25.4). On protected areas in summer limpet grazing reduced the abundance of

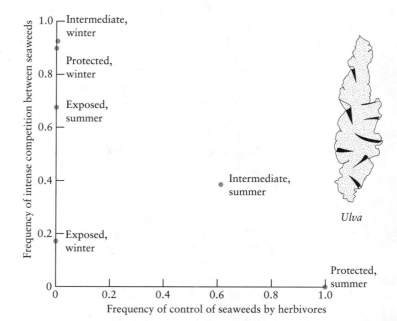

Figure 25.4 Effect of herbivores on the frequency of competition between ephemeral seaweeds (like Ulva *sp.) in the rocky intertidal zone of New England. Season and wave exposure are indicated for each point. (From Lubchenco 1986.)*

ephemeral algae so much that there was no competition for space among the algae. On wave exposed sites limpets cannot easily live, and algae themselves are washed away by wave action, so that the amount of competition among seaweeds is reduced. In this system herbivores set the stage for competitive interactions, and the exact dynamics of a small patch of rock depends on the physical environment (wave action) and the number of herbivores present.

Coralline algae are encrusting algae in the rocky intertidal that compete with each other by overgrowth. In the absence of herbivores like chitons and limpets on a smooth surface, Paine (1984) found a clear dominance of competitive interactions (Figure 25.5)—competition for space was transitive. But in natural communities with and without grazers three of the coralline algae showed intransitive competition, winning some encounters and losing others. Grazers thus act to slow down the rate of succession of coralline algae by inducing competitive uncertainty, and this acts to promote species diversity. In the absence of disturbance, a single competitive dominant would monopolize space in the rocky intertidal (Paine 1984). This algal community is not an equilibrium assemblage under natural conditions with disturbance.

CONCEPTUAL MODELS

A series of models have been proposed by field ecologists to try to capture the interrelations between physical factors and biological interactions in organizing communities. All of these models recognize that there are many different kinds of ecological communities and that the important processes will not be the same in all ecological systems (Schoener 1986a).

A comprehensive model of community organization has been proposed by Menge and Sutherland (1987). They recognize three ecological processes—physical disturbance, predation, and competition—and include variable recruitment as a part of the model (Figure 25.6). A central assumption of this model is that food web complexity will decrease with increasing environmental stress. This model makes three predictions for communities that have high recruitment. First, in stressful environments herbivores have little effect because they are rare or absent, and plants are regulated directly by environmental stress. Neither predation nor competition is significant. An example of such a community would be the arctic tundra or a desert. Second, in moderately stressful environments, consumers are ineffective at controlling plants, and plants attain high densities. Competition between plants is the dominant biological interaction in these communities. Third, in benign environments, consumers control plant numbers and plant competition is rare. Predation is the dominant biological interaction in these benign conditions. In all cases *environmental stress* is the key process in determining community organization.

The Menge-Sutherland model is a general model of community organization that incorporates some ideas from an earlier model proposed by Hairston, Smith and Slobodkin (1960). Hairston and co-workers predicted for terrestrial communities in benign environments that predators must limit herbivores, which are then unable to limit plants. Consequently, plants compete for nutrients and light but herbivores do not compete. They restricted their model to herbivores that feed on green plants and excluded seed and fruit eaters. This model predicts that herbivore removals will have little impact on plants, but predator removals will strongly affect herbivore numbers. The Hairston-Smith-Slobodkin model makes the same predic-

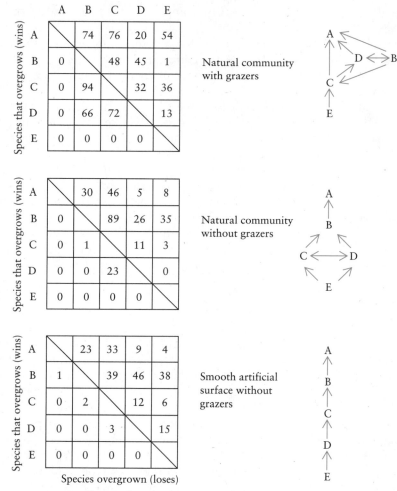

Figure 25.5 Interspecific relationships within a guild of five coralline algal species under three different conditions: (top) with grazers, on natural surface; (center) without grazers, on natural surface; (bottom) without grazers, on smooth artificial surface. Letters refer to species: A, Pseudolithophyllum lichenare; B, Lithothamnium phymatodeum; C, Pseudolithophyllum whidbeyense; D, Lithophyllum impressum; E, Bossiella sp. Numbers in the array are observed overgrowths when two guild members come into contact. The diagramed competitive interactions to the right of the ratios indicate the change of position induced under the various conditions. Arrows point from losers toward competitive winners. Two-headed arrows indicate that no significant bias exists in the interaction's direction. (From Paine 1984.)

tions as the Menge-Sutherland model for the predators but not for the herbivores or the plants. Figure 25.7 compares these two models and shows that the Menge-Sutherland model assumes omnivory to be a common feature of the food web (Pimm 1991). We can test these models by measuring the frequency of competitive effects in different communities in benign environments. Hairston and colleagues (1960) predict intense competition between plants while Menge and Sutherland predict little

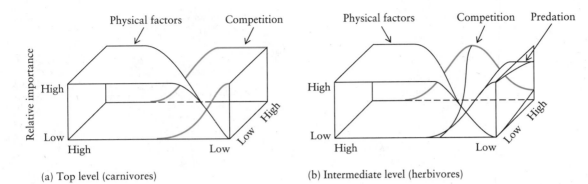

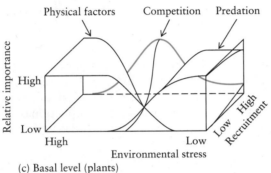

(c) Basal level (plants)

Figure 25.6 *The Menge-Sutherland model of community organization. The relative importance of physical factors, competition between species, and predation change with trophic level, as well as the harshness of the environment and the level of recruitment. (From Menge and Sutherland 1987.)*

competition between plants. Similarly, herbivores in benign environments ought to compete in the Menge-Sutherland model but rarely compete in the Hairston-Smith-Slobodkin model (Figure 25.7).

Competition experiments were reviewed by Connell (1983) and independently

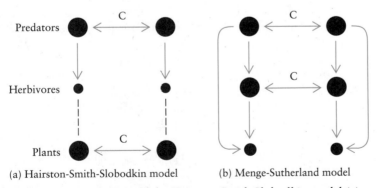

Figure 25.7 *Schematic comparison of the Hairston-Smith-Slobodkin model (a) and the Menge-Sutherland model of community organization for benign environments (b). Size of the circles indicates the relative abundance of the trophic levels. Vertical arrows indicate feeding relationships, horizontal (C) arrows indicate competition. Dashed lines indicate weak interactions. (Modified after Pimm 1991.)*

Table 25.4   Percentage of Field Experiments on Predation That Show Large Effects as a Function of Trophic Level and System

System	SPECIES REMOVED		
	Herbivore[a]	Primary Carnivore[a]	Secondary Carnivore[a]
Intertidal	84(120)	70(67)	—
Other marine	95(82)	68(57)	—
Lakes	—	73(22)	75(95)
Rivers and streams	—	61(106)	55(29)
Terrestrial	74(112)	61(36)	—

Note. Large effects are twofold or greater changes.
[a] Numbers in parentheses are sample sizes (= no. of studies).
   Source: From Sih et al. (1985).

by Schoener (1983). The evidence they summarized was inconclusive and they could not clearly discriminate between these two models. Sih et al. (1985) analyzed both predation and competition experiments and found that competition between plants was much less commonly observed that competition between herbivores. This argues against the Hairston-Smith-Slobodkin model and is consistent with the Menge-Sutherland model.

Predator removal experiments provide another way to test these two models. Herbivore removals ought to have no effects on plants, if the Hairston-Smith-Slobodkin model is correct (Figure 25.7). Sih and co-workers (1985) surveyed removal experiments with the results given in Table 25.4. The vast majority of herbivore removal experiments both in marine and terrestrial systems had large impacts in the plants. This evidence favors the Menge-Sutherland model and is contrary to the Hairston-Smith-Slobodkin model.

Freshwater ecologists have proposed several models for community organization in freshwater ecosystems. Two polar views are the *bottom-up* and the *top-down* models. The bottom-up model postulates that nutrients control community organization because nutrients control plant numbers, which in turn control herbivore numbers, and so on up the food chain. The top-down model postulates that predation controls community organization, and this model has been called the *cascading trophic interaction* model (Carpenter et al. 1985). The cascade model is similar to the Hairston-Smith-Slobodkin model (Figure 25.7). It predicts for a freshwater system that removing the top carnivores will increase the abundance of primary carnivores, decrease herbivores, and increase phytoplankton. The effects of a manipulation thus move down or up the trophic structure as a series of plus or minus (+/−) effects. McQueen and colleagues (1986) postulated a combined top-down, bottom-up theory that suggested an attenuation of cascade effects as they move through the food web. Several experiments have attempted to test these models in freshwater lakes.

An extensive winterkill of fish in Lake St. George, Ontario, allowed McQueen and colleagues (1989) to test these models. Figure 25.8 shows the changes in community structure that occurred in the five years after the winterkill. The top predators (bass, pike and yellow perch) recovered in five years to their former levels of abundance. Planktivorous fishes like bluegill (primary carnivores) increased after the winterkill, and herbivorous zooplankton declined. Phytoplankton changes (measured by chlorophyll) occurred but these were not correlated with zooplankton numbers.

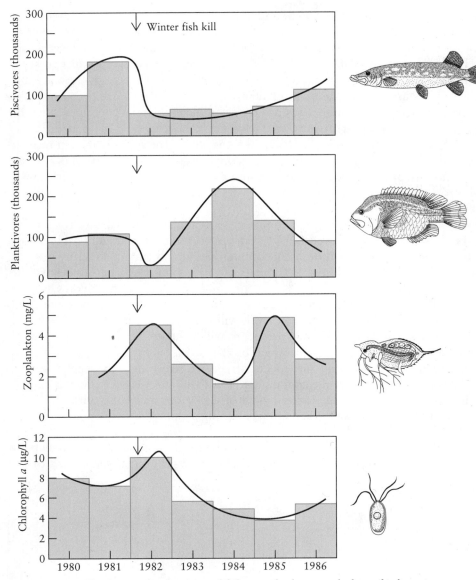

Figure 25.8 Changes in the densities of fish, zooplankton, and phytoplankton in Lake St. George, Ontario, before and after a winterkill of fish in 1982 (arrow). (From McQueen et al. 1989.)

Phosphorus levels seemed to determine phytoplankton numbers. The results shown in Figure 25.8 suggest that trophic cascades damp out as they move down the food chain, and at the level of phytoplankton and zooplankton both nutrient limitation and predation could be controlling. These data are described best by the combination top-down, bottom-up model of McQueen and colleagues (1986).

Trophic cascades can also occur in terrestrial communities. The lesser snow goose breeds in colonies in marshes along the west coast of Hudson Bay. In these marshes it is the only significant herbivore (Jeffries 1988, Kerbes et al. 1990, Hik et al. 1992). Geese numbers have increased during the last 20 years, and they have a strong impact on vegetative growth and composition. In spring the geese grub for roots and rhizomes of graminoid plants. A single goose can strip 1 square meter of sedge in one hour; there were 218,000 pairs of snow geese in 1985 on the west coast of Hudson Bay. Arctic sedges and grasses are perennials and few reproduce by seed. Disturbed sites can be recolonized by clonal growth, but when large areas are disturbed plants are not able to reclaim the area. Erosion of the bare peat can make the problem worse, and large areas are now completely bare. There appears to be no equilibrium in this community, and snow geese move to new areas as the vegetation is destroyed (Kerbes et al. 1990). Whether the plant community can recover in the absence of grazing and how long this might take are unknown. This appears to be an unusual case of a runaway trophic cascade in a simple arctic community, and Strong (1992) argues that this kind of destructive interaction is rare in diverse ecosystems.

A second integrated model of community organization based on productivity has been suggested by Fretwell (1977) and elaborated by Oksanen and coauthors (1981). They recognized that different communities would vary in structure because of the number of trophic levels. In habitats of low productivity, herbivores will be scarce and have little impact on plants. As productivity rises, herbivores will be supported so there will be two trophic levels and herbivores will suppress plants. In more productive systems three trophic levels will be present (as discussed by Hairston et al. 1960, see Figure 25.7) and predators will regulate herbivores. When four trophic levels are present, herbivores will be released to suppress the plants. In this model, predictions regarding the effects of species removals will depend on the number of trophic levels in the community. The Fretwell-Oksanen model is similar to the cascading trophic interaction model of Carpenter and colleagues (1987). It differs from the Menge-Sutherland model by using *productivity* as a major variable, instead of *environmental stress*.

River food webs provide a good model system for testing some of the predictions of community organization models. Figure 25.9 shows the food web of part of the

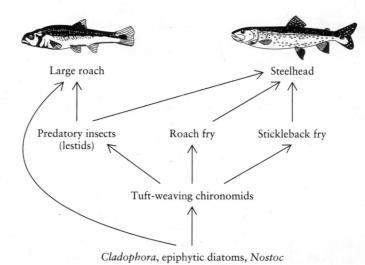

Figure 25.9 Food web of the South Fork of the Eel River in northern California during the summer period of low water flow. Four trophic levels are present. (From Power 1990.)

Table 25.5   Changes in the Abundances of the Lower Three Trophic Levels when Top Predators are Removed from the Community

| Trophic Level | PREDICTIONS OF CHANGES IN ABUNDANCE | | Observations from Eel River, California |
	Menge-Sutherland Model	Fretwell-Oksanen Model	
Plants (algae)	Increase	Increase	Increased about 3-fold for *Cladophora* and 120-fold for *Nostoc*
Herbivores (chironomids)	Increase	Decrease	Decreased about 80%
Primary carnivores (insects, fish fry)	Increase	Increase	Increased about 10-fold

Note. All changes are relative to the intact, four–trophic–level food web as illustrated in Figure 25.9.
Source: Data from Power (1990).

Eel River in California. Four trophic levels occur in this community, and by constructing cages in the river Power (1990) was able to measure the effects of removing fish on community dynamics. Table 25.5 summarizes the results of these manipulations and the predictions of the two major models of community organization. The observations fit the Fretwell-Oksanen, or cascading trophic interactions, model because the chironomids in the river were strongly reduced in numbers when predatory fishes were excluded. Fish removals in rivers can have major effects on all trophic levels.

The results of species removals are complex because the interactions between species are complex and the effects of disturbance make it doubly complex to decipher. As herbivores become more effective, for example, plants invest more in defense mechanisms or they persist in refuges where the herbivores cannot eat them (Lubchenco and Gaines 1981). No general model can describe all of these biological tricks.

THE SPECIAL CASE OF ISLAND SPECIES

Islands are special kinds of traps that catch species able to disperse there and colonize successfully. Since Darwin's visit to the Galápagos Islands, biologists have been using islands as microcosms to study evolutionary and ecological problems.

The number of species on an island is related to the area of the island. This can be seen most easily in a group of islands like the Galápagos (Figure 25.10). The relationship between species and area can be described by the simple equation

$$S = cA^z$$

or, taking logarithms,

$$\log S = (\log c) + z(\log A)$$

where

S = number of species
c = a constant measuring the number of species per unit area

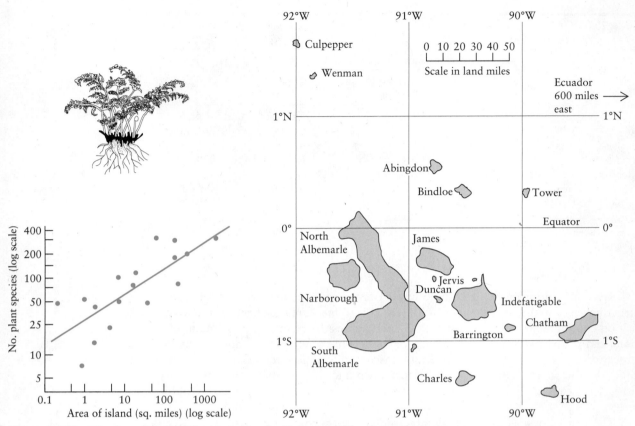

Figure 25.10   *Number of land-plant species on the Galápagos Islands in relation to the area of the island. The islands range in area from 0.2 to 2249 mi² (0.5– 5850 km²) and contain from 7 to 325 plant species. (After Preston 1962.)*

A = area of island (in square units)
z = a constant measuring the slope of the line relating S and A*

For the Galápagos land plants,

$$S = 28.6A^{0.32}$$

The species-area curve, as this relationship is called, is a fundamental one for both plants and animals. Figure 25.11 illustrates this basic principle for the amphibian and reptile fauna of the West Indies, where the relationship is

$$S = 3.3A^{0.30}$$

Preston (1962) noted that the slope of the species-area curve (z) tended to be around 0.3 for a variety of island situations, from beetles in the West Indies and ants in Melanesia to vertebrates on islands in Lake Michigan and land plants on the Galá-

---

* Many species-area curves are reported in English units rather than metric units. Because the scales are logarithmic, the z-value (slope) is independent of scale and thus does not depend on whether English or metric units are used. The c-value, however, is completely dependent on the units used to measure area.

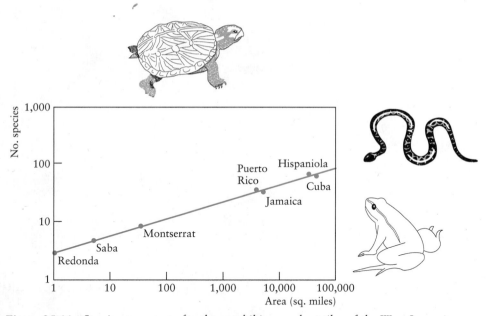

Figure 25.11   Species-area curve for the amphibians and reptiles of the West Indies. (After MacArthur and Wilson 1967.)

pagos. This raises an interesting question: What is the species-area curve for continental areas? Is it the same in slope as that for islands, and is z some sort of ecological constant?

The number of species increases with area on continental areas as well as on islands. Figure 25.12 shows the species-area curve for the breeding birds of North America. Note that the species-area curve is not a single straight line. At very small areas, the slope is greater, and the same occurs at very large areas. But there is a range from approximately 10 acres to about one million acres or more that is a straight line of the form

$$S = 40A^{0.17}$$

for North American birds. Preston (1962) noted that species-area curves for continental areas, or for *parts* of large islands, had slopes (z) that ranged from 0.15 to 0.24, a range below the z-values found in island studies. This means that as we sample larger and larger areas, we add fewer new species if we are sampling a continental area than if we are sampling a series of islands. The explanation for this is that islands are *isolates* with reduced immigration and emigration, while continental areas are under continual flux of immigrants and emigrants. Thus each sample area on the continent will probably contain some transient species from adjacent habitats, and this acts to lower the slope of the species-area curve.

The number of species living on any plot, whether an island or an area on the mainland, is a balance between immigration and extinction. If the immigration of new species exceeds the extinction of old species already present, the plot or island will gain species over time. Thus we can treat the problem of species diversity on islands by an extension of the approach used in population dynamics (Chapter 10), in which changes in population size were produced by the balance between immigration and births on the one hand and emigration and deaths on the other hand.

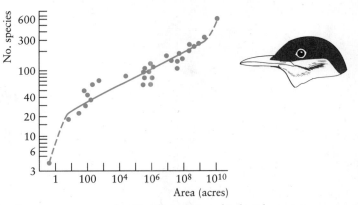

Figure 25.12   Species-area curve for North American birds. The points range from a 0.5-acre plot with three species in Pennsylvania to the whole United States and Canada (4.6 billion acres) with 625 species. (After Preston 1960.)

Figure 25.13 shows the simplest model. MacArthur and Wilson (1967) discussed this approach in detail.

The immigration rate is expressed as the number of new species per unit time. This rate falls continuously because as more species become established on the island, most of the immigrants will be from species already present. The upper limit of the immigration curve is the total fauna for the region. The extinction rate (the number of species per unit time) rises because the chances of extinction depend on the number of species already present. The point where the immigration curve crosses the extinction curve is by definition the equilibrium point for the number of species on the island.

The shape of the curves of immigration and extinction is critical for making any predictions about island situations. Assume for the moment that the only effect of distance will be on the immigration curve; near islands will receive more dispersing organisms per unit time than will distant islands. Assume also that small islands will

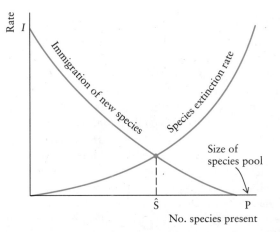

Figure 25.13   Equilibrium model of a biota of a single island. The equilibrial species number (S) is reached at the intersection point between the curve of rate of immigration of new species not already on the island and the curve of extinction of species from the island. (After MacArthur and Wilson 1967.)

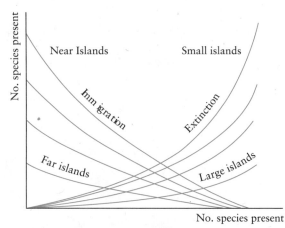

Figure 25.14   Equilibrium models of biotas of several islands of varying distances from the principal source area and of varying size. An increase in distance (near to far) lowers the immigration curve; an increase in island area (small to large) lowers the extinction curve. (After MacArthur and Wilson 1967.)

differ from large islands in their extinction rate so that the chances of becoming extinct are greater on small islands. Figure 25.14 illustrates these assumptions and shows why far islands should have fewer species than near islands (if island size is constant) and why small islands should have fewer species than large islands (if distance from the source area is constant).

The colonization of an island may go through several phases of species equilibrium (Figure 25.15). The initial colonization may occur rapidly enough so that a full complement of species exists before there is serious interaction between species. The noninteractive phase may be followed by an interactive phase in which competition and predation may reduce the species diversity. Next, there may be an assortative

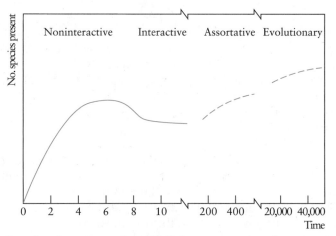

Figure 25.15   Postulated sequence of equilibria in a community of species through time. The time scale is imaginary, supplied here only to convey the notion of the vastly greater time periods required for shifts to states beyond the initial interactive equilibrium. (After Wilson 1969.)

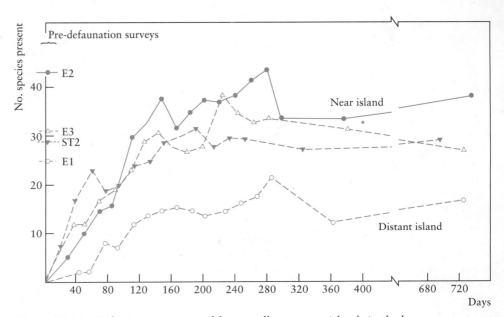

Figure 25.16   *Colonization curves of four small mangrove islands in the lower Florida Keys, whose entire faunas, consisting almost solely of arthropods, were exterminated by methyl bromide fumigation. The figures shown are the estimated numbers of species present. The number of species is an inverse function of the distance of the island from the nearest source of immigrants. This effect was evident in the predefaunation censuses and was preserved when the faunas regained equilibrium after defaunation. Thus the near island E2 has the most species, the distant island E1 the fewest, and the intermediate islands E3 and ST2 the intermediate numbers of species. (After Simberloff and Wilson 1970.)*

phase in which species are replaced through a process of succession, and this may lead to higher diversity. Finally, there may be an evolutionary trend in diversity that operates on a long time scale and results in genetic adaptation and a lowering of the extinction-rate curve.

An experimental approach to island species diversity can be used to test some of the general aspects of the equilibrium theory of MacArthur and Wilson (1967). Simberloff and Wilson (1970) fumigated six islands of mangrove off the Florida Keys in 1966 and 1967 and followed the subsequent history of colonization of these islands. The fauna of the mangroves is mostly insects and spiders, along with scorpions, isopods, and other arthropods. The species pool for the area is about 1000 species, but at any given moment only 20 to 40 species occur on each of the six mangrove islands, which are 11 to 18 meters in diameter.

The colonization curves for four of the mangrove islands are shown in Figure 25.16. In each case, the colonization curve rose rapidly in eight to nine months to a high level and then declined slightly to an equilibrium number of species that was near the original species number. The nearest island to the mainland (2 m away) reached a higher equilibrium level than the distant island (533 m away).

Unfortunately, the turnover of insect species was so fast that immigration and extinction curves could not be estimated for the mangrove islands. Species could have died out and recolonized in the time interval between sampling. Immigration and extinction rates must be high for the islands of mangrove, because the species composition changes remarkably from year to year:

Comparison	Species Present in Both Censuses (%)
Before vs. 1 year later	19
Before vs. 2 years later	30
1 year later vs. 2 years later	41

Even after three years, the species composition of the mangroves had not converged to those present before the experiment began (Simberloff 1976).

The theory of island biogeography does not fit all island data. Brown (1978) analyzed the distribution of boreal birds and mammals in the isolated mountain ranges of the Great Basin in the western United States (Figure 25.17). The mountain ranges of the Great Basin are islands of coniferous forest in a sea of desert. Species-area curves for birds and mammals from these mountain ranges differed markedly in slope. Mammals on isolated mountain ranges represent a relict, nonequilibrium community to which immigration is now zero. The mammals present now represent a set derived by extinction from the species that colonized these mountains when rainfall was higher and habitats were continuous at the end of the Pleistocene. By contrast, birds on mountaintops are not isolated because they disperse so well, and hence the species-area curve for birds rises more slowly than that for mammals (Figure 25.17). There is no effect of isolation by distance in the birds (compare Figure 25.14), as the MacArthur-Wilson model predicts, and the rate of bird faunal turnover is very low (Brown 1978).

If the equilibrium theory of MacArthur and Wilson is correct, the slopes of species-area curves for terrestrial mammals should decrease progressively from the mainland to oceanic islands (Figure 25.18a). If, however, nonequilibrial conditions apply, the slope of the species-area curve should be lower on oceanic islands than on landbridge islands (Figure 25.18b). Landbridge islands are islands close to continents that were connected to the mainland at the end of the Ice Age. Oceanic islands are distant islands never connected to the mainland. Oceanic islands are colonized less frequently than landbridge islands because of isolation by distance. Figure 25.18a illustrates the expected differences in species-area curves under the equilibrium hypothesis of Figure 25.14. On oceanic islands colonization rates are less and extinction rates are the same as landbridge islands of the same size, resulting in a lower number of species at equilibrium. If, however, species numbers on oceanic islands are not in equilibrium because they are colonization-limited, the species-area curve will be flat, as illustrated in Figure 24.18b. Oceanic islands under this hypothesis never reach the species richness predicted for them on the basis of their area (Lawlor 1986).

One critical test of these two models is to determine the slope of the species-area curve for oceanic islands. Lawlor (1986) separated mammals into those that can fly (bats) and those that cannot (all others). Figure 25.19a shows that the slope values (z) for the species-area curves of terrestrial, nonflying mammals on oceanic islands are smaller than those of mammals on landbridge islands. This relationship is illustrated in Figure 25.19b for 14 oceanic islands and 20 landbridge islands off Baja California. By contrast, bats can fly to distant islands and their species-area curves

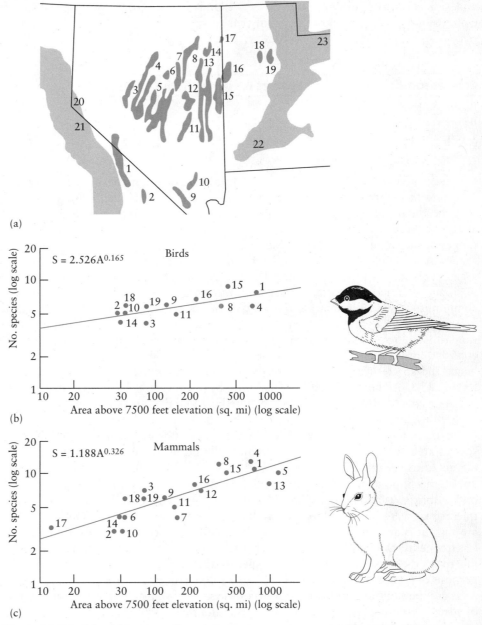

Figure 25.17 *Island biogeography applied to mountaintops. (a) Map of the Great Basin region of the western United States showing the isolated mountain ranges between the Rocky Mountains on the east and the Sierra Nevada on the west. (b) Species-area relationship for the resident boreal birds of the mountaintops in the Great Basin. (c) Species-area relationship for the boreal mammal species. Numbers refer to sample areas on the map. (After Brown 1978.)*

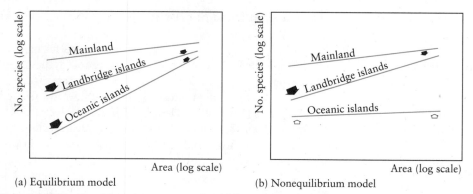

(a) Equilibrium model   (b) Nonequilibrium model

Figure 25.18   *Species-area curves predicted for terrestrial mammals on islands. Solid arrows indicate the relative impact of extinctions, open arrows indicate the relative impact of immigration. (a) Equilibrium model of MacArthur and Wilson (the same model illustrated in Figure 25.14). (b) Nonequilibrium model of Lawlor (1986). Colonization rates limit species numbers on oceanic islands. The slope of the species-area curve for oceanic islands is critical for distinguishing these two models. (From Lawlor 1986.)*

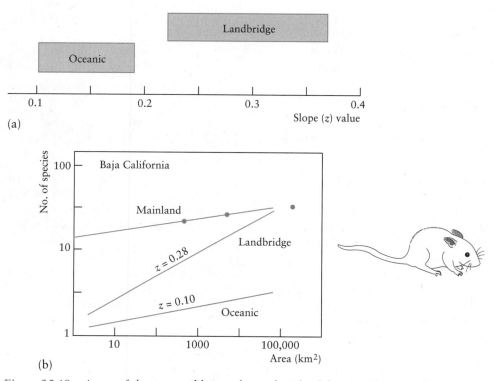

Figure 25.19   *A test of the nonequilibrium theory for island faunas of terrestrial mammals that cannot fly. (a) Range of slope values (z) for species-area curves for oceanic (6 island sets) and landbridge islands (12 island sets). Oceanic islands have lower slopes, as predicted by the nonequilibrium theory (Figure 25.18). (b) One example of species-area curves for the mammals of islands off Baja California. (From Lawlor 1986.)*

fit the equilibrium model. The mammal fauna of oceanic islands is incomplete and not an equilibrium assemblage.

The MacArthur-Wilson theory has stimulated much work on island faunas. By concentrating on predictions of the number of species, it has ignored the more difficult questions of which particular species will occur where. Habitat heterogeneity is a major cause of the species-area curve and detailed studies of habitats are needed to further our understanding of islands (Williamson 1989). Individual species differ greatly in their ability to occupy islands (Figure 25.20), and by understanding the population and community dynamics of individual species we can improve on our understanding of island faunas and floras.

MULTIPLE STABLE STATES

There is considerable interest in the idea that natural communities may exist in multiple stable states, and that community changes that appear to be nonequilibrial are instead the result of two or more alternative states for the same community (Sutherland 1990). What evidence is required to demonstrate that natural communities have multiple stable states? Connell and Sousa (1983) defined the criteria needed to determine if a community exhibits multiple stable states:

1. The community must show an equilibrium point at which it remains, or to which it returns if perturbed by a disturbing force.

2. If perturbed sufficiently, the community will move to a second equilibrium point, *at which it will remain after the disturbance has disappeared.*

3. When multiple stable states are believed to exist, the abiotic environment must be similar in the two communities.

4. The communities on both sites that are postulated to be alternate stable states must persist for more than one generation of the dominant species.

By applying these criteria Connell and Sousa (1983) questioned many of the examples that are given of multiple stable states. Many cases can be rejected because the physical environment differs on the two sites. In other cases the alternate state persists only when artificial inputs are maintained. For example, Lake Washington (page 565) is not an example of an aquatic community with multiple stable states because the enriched lake community could only be maintained by adding sewage nutrients continually to the lake. Connell and Sousa (1984) concluded that they could find no studies of natural communities that showed conclusive evidence for multiple stable states. Sutherland (1990) argued that the criteria for multiple stable states defined by Connell and Sousa were too strict, and in particular that the need for the alternate states to persist for more than one generation excluded many potential examples of multiple stable states.

Woodlands and grasslands of East Africa may represent multiple stable states of a grazing system (Dublin et al. 1990). Woodlands in parks and reserves over much of the savanna areas of East Africa have declined in the past 30 years. Three hypotheses have been proposed to explain the woodland decline (Figure 25.21).

1. Elephants eliminate woodland, and the resulting grassland is maintained by fire. There are two stable states. If fires were eliminated, woodlands would return to their former abundance.

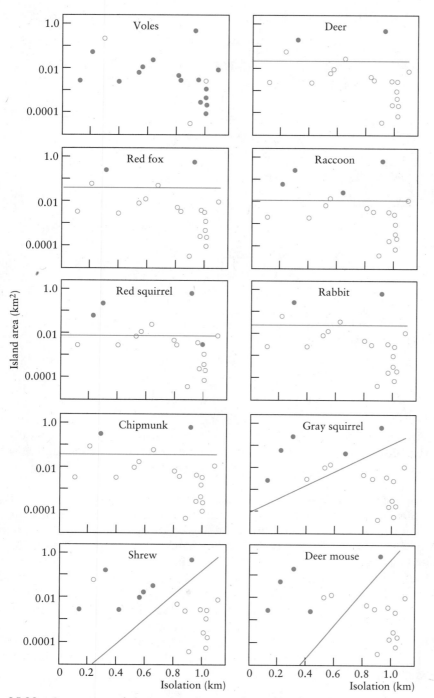

Figure 25.20   Occurrence of ten species of mammals on islands in the Thousand Islands region of the St. Lawrence River in New York. ● = present, ○ = absent. Many species occur only on islands above a certain critical size (horizontal lines). Other species like the deer mouse are affected both by island size and by distance from the mainland. (From Lomolino 1986.)

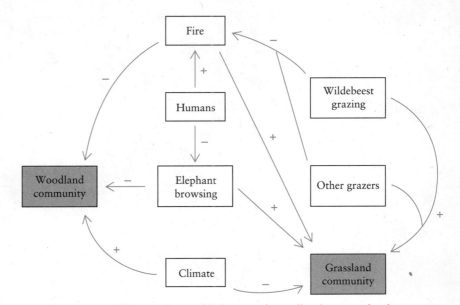

Figure 25.21 *Factors affecting the establishment of woodland or grassland communities in the Serengeti-Mara region of East Africa. See Figure 25.22 (color insert). (Based on Dublin et al. 1990.)*

2. Human-induced fires have eliminated woodland. There is one stable state, and if fires could be reduced woodlands would return.

3. Fire eliminated woodlands, and elephants hold tree regeneration in check by eating small trees so the woodlands can never return. Eliminating fire will not cause woodlands to return, and there are two stable states.

The evidence available supports the third hypothesis (Dublin et al. 1990). During the 1960s fire burned on average 62 percent of the Serengeti each year (Sinclair 1975) and even with no elephants or other browsing or grazing mammals tree recruitment would be too low to sustain woodlands. Elephant and wildebeest numbers increased in the parks and reserves by the 1980s. Wildebeest (Figure 15.8, page 305) grazed much of the grass each dry season so the fuel load in the 1980s was reduced and fires burned only 5 percent of the area each year. Elephant browsing in the 1980s was severe and by itself capable of preventing woodlands from reestablishing. If elephants and wildebeest are reduced by poaching in the future, fires will increase and woodland will not return. The Serengeti-Mara ecosystem seems to be locked into a grassland state. Woodland will return only in the absence of both fire and elephants (Figure 25.22, color insert).

Multiple stable states may occur in other communities affected by humans and may be confused with nonequilibrial communities. In some of these cases the community may revert to its original configuration once human disturbance is removed, but in others the community may become locked into a changed configuration even after the disturbance is stopped. For conservation purposes and for land management it is important to determine which model of community organization applies to natural communities.

SUMMARY

Many community ecologists question the existence of an equilibrium state for biological communities. Patchiness is an inherent property of natural communities, and the disturbances that lead to patchiness have been the main focus of nonequilibrium theories of community organization.

Several theoretical models of nonequilibrium communities have been developed. These models focus on species coexistence and ask what processes can promote coexistence without a point equilibrium. Variable and changing environments, density-independent population dynamics, and slow competitive displacement have all been proposed as mechanisms of nonequilibrial coexistence.

Conceptual models of community organization have been proposed for particular systems. The lottery model, or supply-side ecology, emphasizes variable recruitment which is a strong feature of coral reef ecosystems and some intertidal communities. Studies in the rocky intertidal zone have stimulated several models of community organization. The Menge-Sutherland model includes the roles of environmental harshness and variable recruitment in its prediction of when community interactions will be dominated by competition, predation, or physical factors. The Fretwell-Oksanen model emphasizes the role of productivity in organizing trophic interactions.

Freshwater ecologists have proposed a model like the Fretwell-Oksanen model to explain trophic interactions—the cascading trophic interactions model. When top predators are removed in freshwater systems, the effects cascade down the trophic ladder. Trophic cascades do not always occur. Some systems are affected by nutrients and the lower trophic levels may not be coupled to changes in predator abundance.

Island species are a special case of communities driven by the interaction between colonization and extinction. The species-area curve describes how biodiversity increases with island size. Not all islands are equilibrial systems, however, and historical effects are an important component of many island faunas and floras.

Some communities may exist in multiple stable states and these may be confused with nonequilibrial assemblages. If a community is perturbed sufficiently, it may change to a new configuration at which it will remain even when the disturbance is stopped. There is considerable controversy about whether multiple stable states occur in natural ecosystems, and the question is important for conservation and land management.

Selected References

CARPENTER, S. R., J. F. KITCHELL, and J. R. HODGSON. 1985. Cascading trophic interactions and lake productivity. *Bioscience* 35:634–639.

CHESSON, P. L. 1986. Environmental variation and the coexistence of species. In *Community Ecology*, ed. J. Diamond and T. J. Case, pp. 240–256. Harper and Row, New York.

DAYTON, P. K. 1971. Competition, disturbance, and community organization: The provision and subsequent utilization of space in a rocky intertidal community. *Ecological Monographs* 41:351–389.

FRETWELL, S. D. 1987. Food chain dynamics: The central theory of ecology? *Oikos* 50:291–301.

LAWLOR, T. E. 1986. Comparative biogeography of mammals on islands. *Biological Journal of the Linnean Society* 28:99–125.

MC QUEEN, D. J., M. R. S. JOHANNES, J. R. POST, et al. 1989. Bottom-up and top-down impacts on freshwater pelagic community structure. *Ecological Monographs* 59:289–309.

MENGE, B. A. 1992. Community regulation: Under what conditions are bottom-up factors important on rocky shores? *Ecology* 73:755–765.

MENGE, B. A., and J. P. SUTHERLAND. 1987. Community regulation: Variation in disturbance, competition, and predation in relation to environmental stress and recruitment. *American Naturalist* 130:730–757.

PAINE, R. T. 1984. Ecological determinism in the competition for space. *Ecology* 65:1339–1348.

UNDERWOOD, A. J., and E. J. DENLEY. 1984. Paradigms, explanations, and generalizations in models for the structure of intertidal communities on rocky shores. In *Ecological Communities*, ed. D. R. Strong, Jr., D. Simberloff, L. G. Abele, and A.B. Thistle, pp. 153–180. Princeton University Press, Princeton, N. J.

WIENS, J. A. 1986. Spatial scale and temporal variation in studies of shrubsteppe birds. In *Community Ecology*, ed. J. Diamond and T. J. Case, pp. 154–172. Harper and Row, New York.

WILLIAMSON, M. 1987. Are communities ever stable? In *Colonization, Succession and Stability*, ed. A. J. Bray, M. J. Crawley, and P. J. Edwards, pp. 353–371. Blackwell, Oxford, England.

Questions and Problems

1. The species-area curve rises continually as area is increased, and this implies that there is no limit to the number of species in any community. Is this a correct interpretation? Discuss the implications of the species-area curve for the problem of community definition (Chapter 20).

2. The number of vascular-plant species for the islands off California and the geographic parameters for each island are given by Johnson, Mason, and Raven (1968, p. 300) as follows:

Island	Area (mi²)	Maximum Elevation (ft)	Latitude (°N)	Distance from Mainland (mi)	No. Plant Species
Cedros	134.0	3950	28.2	14.0	205
Guadalupe	98.0	4600	29.0	165.0	163
Santa Cruz	96.0	2470	34.0	20.0	420
Santa Rosa	84.0	1560	34.0	27.0	340
Santa Catalina	75.0	2125	33.3	20.0	392
San Clemente	56.0	1965	32.9	49.0	235

Island	Area (mi²)	Maximum Elevation (ft)	Latitude (°N)	Distance from Mainland (mi)	No. Plant Species
San Nicolas	22.0	910	33.2	61.0	120
San Miguel	14.0	830	34.0	26.0	190
Natividad	2.8	490	27.9	5.0	42
Santa Barbara	1.0	635	33.4	38.0	40
San Martin	0.9	470	30.5	3.5	62
San Geronimo	0.2	130	29.8	6.0	4
South Farallon	0.1	360	37.7	27.0	12
Año Nuevo	0.02	60	37.1	0.25	40

Make four plots of the number of plants versus (1) area, (2) elevation, (3) latitude, and (4) distance from the mainland. Do these on arithmetic scales and repeat on logarithmic scales (log-log plot). What variable is most closely related to species numbers? Estimate graphically the slope of the species-area curve for this ensemble of islands, and compare it with those given in the text.

3. Can the lottery hypothesis explanation for coral reef fish diversity be scale-dependent? Most of the studies reported in this chapter were done on very small parts of reefs (under 1 m³). How might our perspective on this problem change with scale?

4. Can nonequilibrium models of community organization be stable? Read Chesson and Case (1986) and DeAngelis and Waterhouse (1987) and discuss the relationship between stability and equilibrium/nonequilibrium concepts.

5. Are the data in Table 25.4 (page 584) suitable for testing the Menge-Sutherland model of community organization? Why or why not?

6. In discussing the MacArthur-Wilson theory of island biogeography, Brown (1986, p. 233) states: "I suspect that the model has ultimately proven more valuable when its predictions have been refuted than when they have been supported." Is this statement valid for other models of community organization? Discuss.

7. Giller and Gee (1987, p. 540) state: "A state of equilibrium can be attained at the landscape scale even when biotic instability and stochastic domination are unrestrained at the level of the local patch." Under what conditions is this true? Discuss the impact of spatial scale on the concept of equilibrium.

8. Lawton (1984) in summarizing information on bracken fern and its insect herbivores, concluded that these insect communities were not saturated with species, and that at any particular site there were numerous vacant niches. Discuss this interpretation of community organization with respect to the nonequilibrium models of community organization discussed on pages 573–576.

9. Review the argument between Hairston (1991) and Sih (1991) over the interpretation of field data to test the predictions of the Hairston-Smith-Slobodkin (1960) model of community organization. Discuss the problem of testing hypotheses about community organization within the framework of Figure 1.3 (page 14).

10. Analyze the elephant/fire multiple stable state model of Dublin and co-workers (1990) using the criteria for the existence of multiple stable states given by

Connell and Sousa (1983). Would this example be acceptable to Connell and Sousa? To Sutherland (1990)?

Overview Question

List some of the possible manipulative experiments that could be applied to a community to test the Menge-Sutherland, Fretwell-Oksanen, and cascading trophic interaction models of community organization. List the predictions each model would make for each possible manipulation. Are some manipulations more instructive than others?

---

26 *Community Metabolism I: Primary Production*

Individual organisms require a continual input of new energy to balance losses from metabolism, growth, and reproduction. Thus individuals can be viewed as complex machines that process energy and materials. There are two major ways in which organisms pick up energy and materials. *Autotrophs* pick up energy from the sun and materials from nonliving sources. Green plants are autotrophs. *Heterotrophs* pick up energy and materials by eating living matter. Herbivores are heterotrophs that live by eating plants, and carnivores are heterotrophs that live by eating other animals. Communities are mixtures of autotrophs and heterotrophs. Energy and materials enter a biological community, are used by the individuals, and are transformed into biological structure only to be ultimately released again into the environment. The *ecosystem* level of integration includes both the organisms and their abiotic environment and is a comprehensive level at which to consider the movement of energy and materials. We could also discuss the flow of matter and energy at the individual level or at the population level. The basic unit of metabolism is always the individual organism, even when individuals are assembled into communities.

The first step in the study of community metabolism is to determine the food web of the community (Chapter 24). Once we know the food web, we must decide how we can judge the significance of the different species to community metabolism. There may be 5000 species of animals on the 5 square kilometers of Wytham Woods in Britain (Elton 1966). We feel intuitively that many of these 5000 species are not significant and that many or most of them could be removed without affecting the metabolism of the woodland.

Three measurements might be used to define relative importance in a community:

1. *Biomass.* We could use the weight or standing crop of each species as a measure of importance. This is useful in some circumstances such as the lumbering industry, but it cannot be used as a general measure. In a dynamic situation in which *yield* is important, we need to know how rapidly a community produces new biomass. When metabolic rates and reproductive rates are high, production may be very rapid, even from a low standing crop. Figure 26.1 illustrates the idea that yield need not be related to biomass.

2. *Flow of chemical materials.* We can view the community as a superorganism taking in food materials, using them, and passing them out. Note that all chemical materials can be recycled many times through the community. A molecule of phosphorus may be taken up by a plant root, used in a leaf, eaten by a grasshopper which dies, and released by bacterial decomposition to reenter the soil.

3. *Flow of energy.* We can view the community as an energy transformer that takes solar energy, fixes some of it in photosynthesis, and transfers this energy from

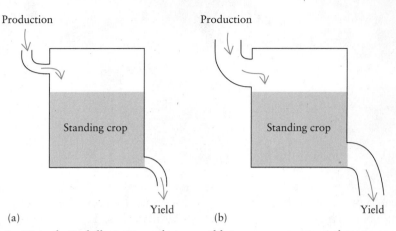

Figure 26.1   Hypothetical illustration of two equilibrium communities (where input equals output): (a) low input, low output, slow turnover; (b) high input, high output, rapid turnover. Standing crop is not related to production or yield because turnover time for all systems is not a constant.

green plants through herbivores to carnivores. Note that most energy flows through a community only once and is not recycled but is transformed to heat and ultimately lost to the system. Only the continual input of new solar energy keeps the community operating. Again we may draw the analogy between a community and an organism that processes food energy.

To study the dynamics of community metabolism, we must decide what to use as the base variable. Most ecologists have decided to use either *carbon* or *energy*. Elements other than carbon are often tied up in biological peculiarities of organisms. For example, vertebrates or mollusks contain much more calcium than most freshwater invertebrates because of the presence of bone or shell. Some marine invertebrates concentrate certain chemical elements. Even within an individual there are variations. Calcium in the teeth and bones of a mammal may be stable for long periods, while calcium in the blood serum may turn over rapidly because of ingestion and excretion. This makes the description of the calcium flow through a community very difficult. Since most energy is not recirculated, it is easier to measure than are chemical materials. Figure 26.2 illustrates the flows of energy and materials through the food chain.

Energy or carbon content is just another way of describing an individual, population, or community, and the convenience and precision of these measures should not blind us to their limitations as a way of describing organisms. The great strength (and weakness) of the energetics approach is that it can allow us to add together different species in a community. It reduces the fundamental diversity of a biological community to a single unit—either the joule (for energy) or the gram (for carbon).

PRIMARY PRODUCTION

The process of photosynthesis is the cornerstone of all life and the starting point for studies of community metabolism. The bulk of the earth's living mantle is green plants (99.9 percent by weight); only a small fraction of life consists of animals (Whittaker 1975).

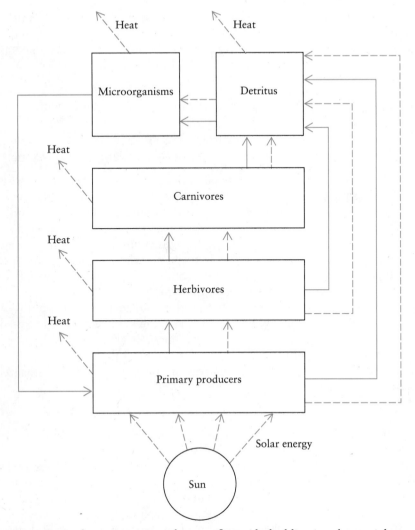

Figure 26.2   General representation of energy flows (dashed lines) and material cycles (solid lines) in the biosphere. The energy flows included are solar radiation, chemical energy transfers (in the ecological food web) and radiation of heat into space (From DeAngelis 1992.)

Photosynthesis is the process of transforming solar energy into chemical energy and can be simplified as

$$12H_2O + 6CO_2 \text{ (from air)} + \text{solar energy} \xrightarrow[\text{enzymes}]{\text{chlorophyll +}} C_6H_{12}O_6 \text{ (carbohydrate)} + 6O_2 \text{ (to air)} + 6H_2O$$

If photosynthesis were the only process occurring in plants, we could measure production by the rate of accumulation of carbohydrate, but unfortunately, at the same time, plants respire, using energy for maintenance activities. Respiration is the opposite of photosynthesis, in an overall view:

$$C_6H_{12}O_6 \text{ (carbohydrate)} + O_2 \text{ (from air)} \xrightarrow[\text{enzymes}]{\text{metabolic}} CO_2 \text{ (to air)} + H_2O + \text{energy for work and maintenance}$$

At metabolic equilibrium, photosynthesis equals respiration, and this is called the *compensation point.* Photosynthesis and respiration are rate processes, and are always measured as amount of material or energy per unit of time. If plants always existed at the compensation point, they would neither grow nor reproduce. We define two terms:

> Gross primary production = energy (or carbon) fixed in photosynthesis per unit time

> Net primary production = energy (or carbon) fixed in photosynthesis—energy (or carbon) lost by respiration per unit time

Production is always measured as a rate per unit of time. How can we measure these two aspects of primary production in natural systems?

For terrestrial plants, the direct way is to measure the change in CO_2 or O_2 concentrations in the air around plants. Most studies measure CO_2 uptake by an enclosed branch or a whole plant. During daylight conditions, CO_2 uptake rate is a measure of net production because both photosynthesis and respiration are operating simultaneously. At night, only respiration occurs, and the rate at which CO_2 is released can be used to estimate the respiration component.

Photosynthesis and respiration are both affected by temperature; photosynthesis is also affected by light intensity. Figure 26.3 illustrates this for a single tree species for one day in spring and one day in summer. The daily changes in leaf temperature and light intensity determine the net production for each day.

We can determine the energetic equivalents of photosynthesis measurements from the chemical thermodynamics of the reaction:

$$12H_2O + 6CO_2 + 2966 \text{ kJ} \xrightarrow[\text{energy}]{\text{solar}} C_6H_{12}O_6 + 6O_2 + 6H_2O$$

Thus the absorption of 6 moles (134.4 L at standard temperature and pressure) of CO_2 indicates that 2966 kJ has been absorbed.*

The measure of gas exchange around plants in the field has been used relatively little as an estimate of photosynthetic rates because it requires sophisticated and expensive instrumentation. A slightly different approach to measuring CO_2 uptake is to introduce radioactive $^{14}CO_2$ in the air surrounding a plant (covered by a transparent chamber) and after a time to harvest the whole plant and count the quantity of radioactive ^{14}C taken up by photosynthesis. This technique can be used in sites where electricity is not available.

The simplest method of measuring primary production is the harvest method. The amount of plant material produced in a unit of time can be determined from the difference between the amount present at the two times:

$$\Delta B = B_2 - B_1$$

where

> ΔB = biomass change in the community between time 1 (t_1) and time 2 (t_2)
> B_1 = biomass at t_1
> B_2 = biomass at t_2

---

* Energy units have been reported in many forms in the literature, often in calories, and can be standardized to *joules* with the following conversion factors: 1 joule (J) = 0.2390 gram calorie (cal) = 0.000239 kilocalorie (kcal); conversely, 1 gram calorie = 4.184 joules and 1 kilocalorie = 4184 joules or 4.184 kilojoules (kJ).

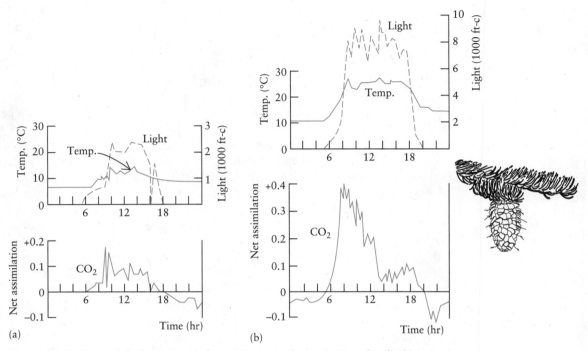

Figure 26.3   Typical daily patterns of net CO_2 assimilation in Douglas fir during (a) spring and (b) summer. In the daylight, there is an uptake of CO_2, which is the net outcome of photosynthesis and respiration. At night, CO_2 is released by respiration. (After Helms 1965.)

Two possible losses must be recognized:

L = biomass losses by death of plants or plant parts
G = biomass losses to consumer organisms

If we know these values, we can determine production:

Net primary production = $\Delta B + L + G$

This may apply to the whole plant, or it may be specified as *aerial* production or *root* production.

The net primary production in biomass may then be converted to energy by obtaining the caloric equivalent of the material in a bomb calorimeter. This should be done for each particular species studied as well as for each season of the year. Golley (1961) showed that different parts of plants have different energy content:

	MEAN OF 57 PLANT SPECIES	
	(cal/g dry wt)	**(J/g dry wt)**
Leaves	4,229	17,694
Roots	4,720	19,748
Seeds	5,065	21,192

Vegetation collected in different seasons also varied in energy content.

The harvesting technique of estimating production is used in a variety of situations. Foresters have used a modified version of it for timber estimation, and agricultural research workers use it to determine yield of crops. The application of harvesting techniques to natural vegetation involves some specialized techniques that we shall not describe here; Milner and Hughes (1968) and Newbould (1967) give details of techniques.

In aquatic systems, primary production can be measured in the same general way as in terrestrial systems. Gas-exchange techniques can be applied to water volumes, and usually oxygen release instead of carbon dioxide uptake is measured. This procedure is usually repeated with a dark bottle (respiration only) and a light bottle (photosynthesis and respiration), so that both gross and net production can be measured. Vollenweider (1974) discusses details of techniques for measuring production in aquatic habitats.

How does primary production vary over the different types of vegetation on the earth? This is the first general question we can ask about community metabolism. Table 26.1 gives some average values for net primary production in biomass for different vegetation types. In general, primary production is highest in the tropical rain forest and decreases progressively toward the poles. Productivity of the open ocean is very low, approximately the same as that of the arctic tundra, and oceans occupy about 71 percent of the total surface of the earth. Grassland and tundra areas are less productive than forests in the same general region. The standing crop of forests is very large, and green parts are a relatively small fraction of the total biomass of a forest.

How efficient is the vegetation of different communities as an energy converter? We can determine the efficiency of utilization of sunlight by the following ratio:

$$\frac{\text{Efficiency of gross}}{\text{primary production}} = \frac{\text{energy fixed by gross primary production}}{\text{energy in incident sunlight}}$$

For example, Kozlovsky (1968) calculated the efficiency of the aquatic community of Lake Mendota, Wisconsin:

$$\frac{\text{Efficiency of gross}}{\text{primary production}} = \frac{20{,}991 \text{ kJ/m}^2/\text{yr gross primary production}}{4{,}973{,}604 \text{ kJ incident solar radiation}}$$
$$= 0.42\%$$

Phytoplankton communities have very low efficiencies of primary production, usually less than 0.5 percent, although rooted aquatic plants and algae in shallow waters can have higher efficiencies. The efficiency of gross primary production is higher in forests (2.0–3.5 percent) than in herbaceous communities (1.0–2.0 percent) or in crops (less than 1.5 percent) (Kira 1975). Forest communities are relatively efficient at capturing solar energy.

How much of the energy fixed by photosynthesis is subsequently lost by respiration of the plants themselves? A great deal of energy is lost in converting solar radiation to gross primary production. Net primary production, which is what interests animals and humans, must therefore be even less efficient. In forests, 50 to 75 percent of the gross primary production is lost to respiration, so that net production may be only one-fourth that of gross production (Kira 1975). Forests have larger amounts of stems, branches, and roots to support than do herbs, and thus less energy is lost to respiration in herbaceous and crop communities (45–50 percent). The result of these losses is that for a broad range of terrestrial communities, about 1 percent

Table 26.1   Net Primary Production and Plant Biomass for the Earth

Ecosystem Type	Area (million km²)	NET PRIMARY PRODUCTIVITY PER UNIT AREA PER YEAR (g/m²) Normal Range	Mean	World Net Primary Production (billion t/yr)	BIOMASS OR STANDING CROP (kg/m²) Normal Range	Mean	World Biomass (billion t)
Tropical rain forest	17.0	1000–3500	2200	37.4	6–80	45	765
Tropical seasonal forest	7.5	1000–2500	1600	12.0	6–60	35	260
Temperate evergreen forest	5.0	600–2500	1300	6.5	6–200	35	175
Temperate deciduous forest	7.0	600–2500	1200	8.4	6–60	30	210
Boreal forest	12.0	400–2000	800	9.6	6–40	20	240
Woodland and shrubland	8.5	250–1200	700	6.0	2–20	6	50
Savanna	15.0	200–2000	900	13.5	0.2–15	4	60
Temperate grassland	9.0	200–1500	600	5.4	0.2–5	1.6	14
Tundra and alpine	8.0	10–400	140	1.1	0.1–3	0.6	5
Desert and semidesert scrub	18.0	10–250	90	1.6	0.1–4	0.7	13
Extreme desert, rock, sand, and ice	24.0	0–10	3	0.07	0–0.2	0.02	0.5
Cultivated land	14.0	100–3500	650	9.1	0.4–12	1	14
Swamp and marsh	2.0	800–3500	2000	4.0	3–50	15	30
Lake and stream	2.0	100–1500	250	0.5	0–0.1	0.02	0.05
Total continental	149		773	115		12.3	1837
Open ocean	332.0	2–400	125	41.5	0–0.005	0.003	1.0
Upwelling zones	0.4	400–1000	500	0.2	0.005–0.1	0.02	0.008
Continental shelf	26.6	200–600	360	9.6	0.001–0.04	0.01	0.27
Algal beds and reefs	0.6	500–4000	2500	1.6	0.04–4	2	1.2
Estuaries	1.4	200–3500	1500	2.1	0.01–6	1	1.4
Total marine	361		152	55.0		0.01	3.9
Full total	510		333	170		3.6	1841

Note. Units are square kilometers, dry grams or kilograms per square meter, and dry metric tons of organic matter.

Source: From Whittaker and Likens in Whittaker (1975).

of the sun's energy falling during the growing season is converted into net primary production.

FACTORS THAT LIMIT PRIMARY PRODUCTIVITY

One important question about primary production is, *What controls the rate of primary production in natural communities?* What factors could we change to increase the rate of primary production for a given community? Note that this question could be broken down into many questions of the same type for each plant-species population. The control of primary production has been studied in greater detail for aquatic systems than for terrestrial systems. Let us look first at some details of production in aquatic communities.

Marine Communities

Light is the first variable one might expect to control primary production, and the depth to which light penetrates in a lake or ocean is critical in defining the zone

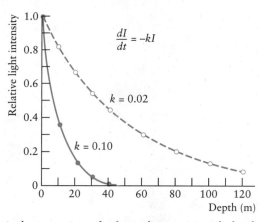

$$\frac{dI}{dt} = -kI$$

$k = 0.02$

$k = 0.10$

Figure 26.4 *Theoretical attenuation of solar radiation (z) with depth in a water column. Light intensity falls geometrically with depth, and the larger the extinction coefficient (k), the faster is the loss of light. An extinction coefficient of 0.02 would occur in pure water; one of 0.10 would occur in oceanic seawater. Coastal seawater would have higher extinction coefficients (approx. 0.30.)*

of primary production. Water absorbs solar radiation very readily. More than half of the solar radiation is absorbed in the first meter of water, including almost all the infrared energy. Even in "clear" water, only 5 to 10 percent of the radiation may be present at a depth of 20 meters. This decrease can be described reasonably well by a geometric curve of decrease in radiation:

$$\frac{dI}{dt} = kI$$

where

    I = amount of solar radiation
    t = depth
    k = extinction coefficient (a constant)

This relationship is illustrated in Figure 26.4 for two values of k, the extinction coefficient. Large k-values indicate less transparent waters. Figure 26.5 illustrates the decrease in photosynthesis with depth in three California lakes. Clear Lake is a *eutrophic* lake with high production and little light penetration. Castle Lake is a lake of intermediate productivity, in which the zone of photosynthesis extends below a depth of 20 meters. Lake Tahoe is an alpine *oligotrophic* lake of remarkably clear water in which the zone of photosynthesis extends to a depth of 100 meters, although there is little photosynthesis at any depth (Goldman 1968).

Very high light levels can inhibit photosynthesis of green plants, and this inhibition can be found in tropical and subtropical surface waters throughout the year. When surface radiation is excessive, the maximum in primary production will occur several meters beneath the surface of the sea.

Light is an important factor in limiting primary production in the ocean (Platt and Sathyendranath 1988). If you know the rate at which light decreases with depth (extinction coefficient), the amount of solar radiation, and the amount of plant chlorophyll in the water, you can calculate the net production of the phytoplankton. For any particular region of the ocean, the equations for estimating primary pro-

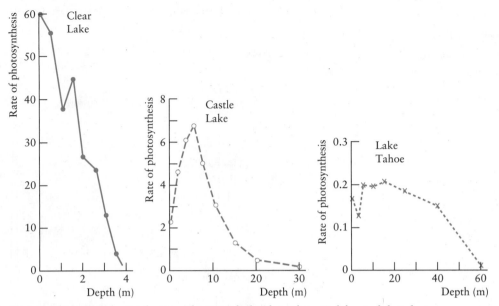

Figure 26.5   Change in photosynthesis with depth in three California lakes during the summer. Note changes in scale of depth and rate of photosynthesis. (Data from Goldman 1968.)

duction are complex because of the need to correct for absorption and backscattering of light and for cloud cover (Platt and Sathyendranath 1988).

Considerable research on ocean production has occurred in the North Pacific Central Gyre (Figure 26.6). Because the Central Gyre is large and relatively stable, biological production can be measured in the absence of external inputs. Figure 26.7 shows the vertical distribution of nutrients, temperature, and primary production in the Central Gyre. Virtually all the primary production occurs in the euphotic zone, above the 1 percent light level, at 90 meters of depth. Primary production, measured by ^{14}C, is relatively low in the surface waters (due to excessive light) and is highest in the warm waters near the surface (10–30 m depth).

If light is the primary variable limiting primary production in the ocean, there should be a gradient of productivity from the poles toward the equator. Figure 26.8 shows the global distribution of primary production in the oceans. There is no gradient of production from the poles to the equator. Some parts of the tropics and subtropics, such as the Sargasso Sea and the Central Gyre of the North Pacific, are very unproductive. In contrast, the Antarctic Ocean is one of the most productive areas on earth.

Why are tropical oceans unproductive when the light regime is good all year? *Nutrients* appear to limit primary production in tropical and subtropical seas. Primary productivity can be predicted from a knowledge of light and biomass of chlorophyll, and the action of limiting nutrients is on the biomass of chlorophyll in the phytoplankton. Two elements, nitrogen and phosphorus, often limit primary production in the oceans. One of the striking generalities of many parts of the oceans is the very low concentrations of nitrogen and phosphorus in the surface layers where the phytoplankton live (Figure 26.7), whereas the deep water contains much higher concentrations of nutrients.

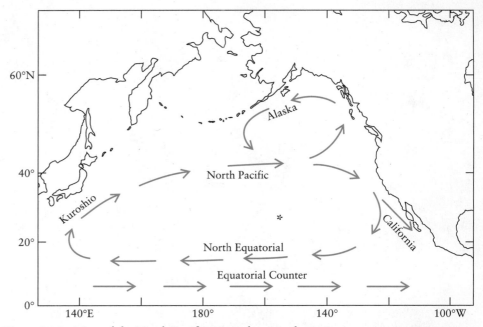

Figure 26.6   *Map of the North Pacific ocean showing the major upper ocean currents. The North Pacific Central Gyre is the large clockwise-flowing circulation system. Advection into the Central Gyre from outside is low in comparison to other regions, such as coastal boundary current systems. The biological structure of the upper layers of the central North Pacific, most intensively studied at the location indicated by the asterisk, appears to be regulated primarily by* in situ *processes. (From Hayward 1991.)*

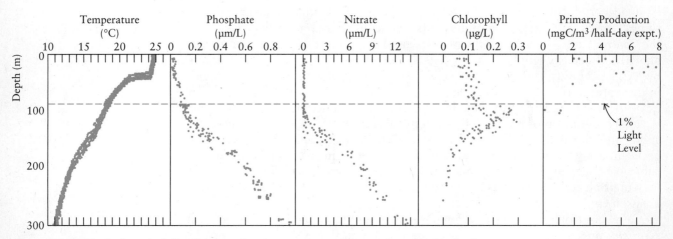

Figure 26.7   *Typical vertical distribution of temperature, nutrients, and production in the upper layer of the North Pacific Central Gyre during summer. The curves are composites of several vertical profiles made over a two-day period at a single location (28°N, 155° W) shown by the asterisk in Figure 26.6. The dashed line illustrates the depth of 1 percent of surface light (the traditional definition of the depth of the euphotic zone). Note that nitrate has been depleted to undetectable levels above the 1 percent light level, and that most of the measured primary production (from* ^{14}C *uptake) takes place above the depth where nitrate can be detected with conventional techniques. (From Hayward 1991.)*

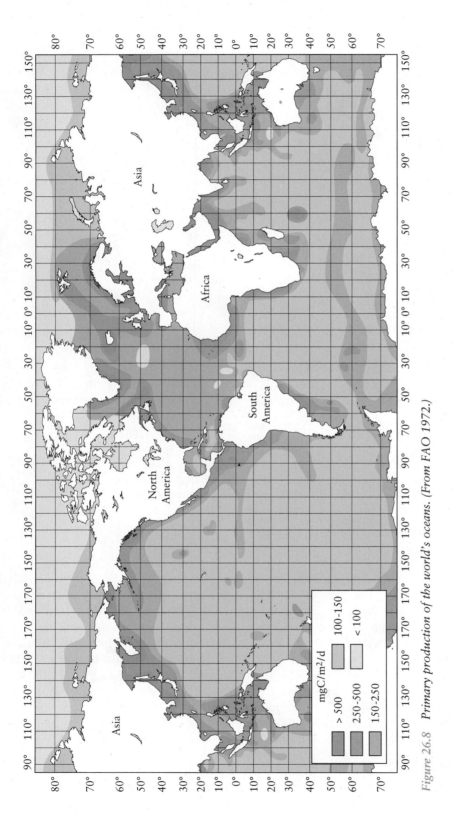

Figure 26.8   Primary production of the world's oceans. (From FAO 1972.)

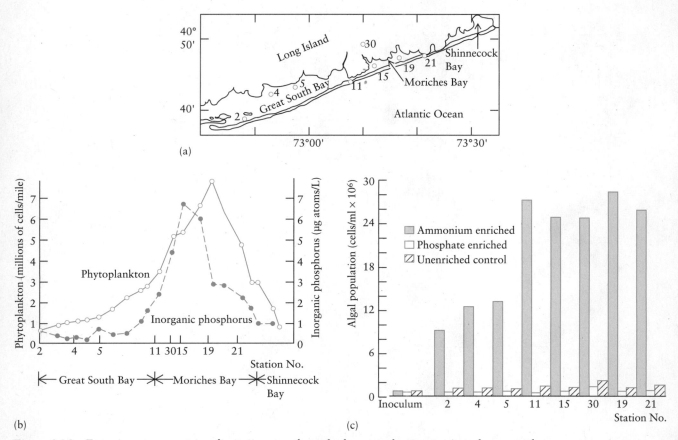

Figure 26.9   *Experiments on nutrient limitations to phytoplankton production in coastal waters of Long Island: (a) coast of Long Island, New York; (b) abundance of phytoplankton and distribution of phosphorus arising from duck farms in Moriches Bay; (c) nutrient-enrichment experiments with the alga* Nannochloris atomus *in water from the bays. Phosphorus is superabundant, and nitrogen seems to limit algal growth. (After Ryther and Dunstan 1971.)*

Nitrogen may be a limiting factor for phytoplankton in part of the ocean (Ryther and Dunstan 1971, Platt et al. 1992). Figure 26.9 illustrates this for a coastal area of New York. Pollution from duck farms along the bays of Long Island adds both nitrogen and phosphorus to the coastal water, but unlike phosphorus, the nitrogen added is immediately taken up by algae, and no trace of nitrogen can be measured in the coastal waters. That nitrogen was limiting was confirmed by nutrient-addition experiments (Figure 26.9). The addition of nitrogen (in the form of ammonium) caused a heavy algal growth in bay water, but the addition of phosphate did not induce algal growth. There are some obvious practical conclusions to this work. If nitrogen is the factor now limiting phytoplankton production, the elimination of phosphates from sewage will not help the problem of coastal pollution.

The Sargasso Sea is an area of very low productivity in the subtropical part of the Atlantic Ocean. The seawater there is among the most transparent in the world, and the surface waters are very low in nutrients. Nitrogen and phosphorus, however, do not seem to be limiting primary production, and iron seems to be critical (Menzel and Ryther 1961b). This was shown by a series of nutrient-enrichment experiments in which surface water from the Sargasso Sea was enriched with various nutrients. The following series of three-day experiments illustrates this:

Nutrients Added to Experimental Cultures	Relative Uptake of ^{14}C for Experimental Cultures (%) (Control Cultures = 100%)
N + P + metals	1290
N + P	110
N + P + metals except iron	108
N + P + iron only	1200

The addition of iron alone to Sargasso Sea water stimulated primary production, but only for a short time. This suggests that iron is the factor limiting production in the Sargasso Sea but that nitrogen and phosphorus limits are very close to that of iron. Figure 26.10 illustrates this idea of a sequence of limiting factors.

Primary production in the Sargasso Sea is highest in winter (November–April), even though solar radiation is highest in summer. High winter production is determined by mixing of surface waters by winds and storms. This mixing brings nutrients from deeper water back to the surface, where the phytoplankton is limited by the nutrients available. In this subtropical sea, light is always available for photosynthesis, but nutrients are not (Menzel and Ryther 1961a).

Compared with the land, the ocean is very unproductive; the reason seems to be that fewer nutrients are available. Rich, fertile soil contains 5 percent organic matter and up to 0.5 percent nitrogen. One square meter of soil surface can support 50 kilograms dry weight of plant matter. In the ocean, by contrast, the richest water contains 0.00005 percent nitrogen, four orders of magnitude less than that of fertile farmland soil. One square meter of rich seawater could support no more than 5 grams dry weight of phytoplankton (Ryther 1963). In terms of standing crops, the sea is a desert compared with the land. And although the maximal rate of primary production in the sea may be the same as that on land, these high rates in the sea can be maintained for a few days only, unless the water is enriched by upwelling.

Areas of upwelling in the ocean are exceptions to the general rule of nutrient limitation. The largest area of upwelling occurs in the Antarctic Ocean, where cold, nutrient-rich, deep water comes to the surface along a broad zone near the Antarctic continent (Figure 26.8). Other areas of upwelling occur off the coasts of Peru and California, as well as in many coastal areas where a combination of wind and currents moves the surface water away and allows the cold, deep water to move up to the surface. In these areas of upwelling, fishing is especially good, and in general there is a superabundance of nutrients for the phytoplankton.

The Southern Ocean occupies 20 percent of the world's ocean surface, and is one of the most productive of the ocean areas. The surface waters away from the icepack contain large amounts of nitrogen and phosphorous, in contrast to more temperate waters like the North Pacific (Figure 26.7). Martin and Fitzwater (1988) suggested that the trace element iron limits phytoplankton growth in both subarctic and antarctic waters. The *iron hypothesis* is controversial and may not apply to all parts of the Southern Ocean (Tréguer and Jacques 1992). If iron is limiting in the Southern Ocean, it might be possible to fertilize the ocean with trace amounts of

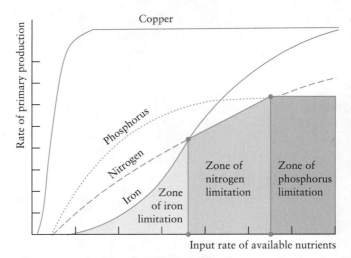

Figure 26.10   Illustration of a hypothetical sequence of nutrient factors that limit primary productivity. Such a sequence may be operating in the Sargasso Sea. The rate of primary production will follow the heavy line and be limited first by iron, then (as more iron becomes available) by nitrogen, and finally by phosphorus. Some nutrients, such as copper, may always be present in superabundant amounts.

iron and thereby increase its uptake of CO_2 to alleviate partly the greenhouse effect (Joos et al. 1991). The ocean could become a sink for CO_2 (see Chapter 28, page 677).

One of the most exciting new developments in marine ecology is the ability to estimate primary production from satellite remote sensing data (Sathyendranath et al. 1991). Chlorophyll concentration in the surface water can be estimated by spectral reflectance using blue/green ratios. Figure 26.11 (see color insert) illustrates differences in chlorophyll levels in coastal waters off New England. Remote sensing allows marine ecologists to analyze large-scale production changes in the ocean without being limited to a few measurements made off a ship. Platt and Sathyendranath (1988) provide a model to compute primary production from remotely sensed data on ocean color. These techniques promise to enlarge our understanding of primary production in the oceans and how it varies in space and time.

Total primary production in the ocean is thus limited by light in some areas and also by the shortage of nutrients, such as nitrogen, phosphorus and iron, which are critical for plant growth.

Freshwater Communities

In freshwater communities, much the same conclusion seems to hold. Solar radiation limits primary production on a day-to-day basis in lakes, and Goldman (1968) has shown that within a given lake you can predict the daily primary productivity from the solar radiation. Temperature is closely linked with light intensity in aquatic systems and is difficult to evaluate as a separate factor. Nutrient limitations also operate in freshwater lakes, and the great variety of lakes is associated with a great variety of potential limiting nutrients.

For growth, plants require nitrogen, calcium, phosphorus, potassium, sulfur, chlorine, sodium, magnesium, iron, manganese, copper, iodine, cobalt, zinc, boron, vanadium, and molybdenum. These nutrients do not all act independently, which has made the tracing of causal influences very difficult (Lund 1965). Early work suggested that nitrogen and phosphorus were the major limiting factors in freshwater lakes. This conclusion was a practical one reached by the fertilization of small farm ponds to increase fish production. Hepher (1962) showed that small ponds fertilized with phosphate and ammonium sulfate increased primary production four to five times above that of unfertilized ponds (Figure 26.12). However, doubling the amount of fertilizer did not increase primary production in the fishponds any more than a single application, indicating that, once fertilized, something other than nitrogen and phosphorus was limiting.

During the 1970s the problem of what controls primary production in freshwater lakes became acute because of increasing pollution. Nutrients added to lakes directly in sewage or indirectly as runoff had increased algal concentrations and had shifted many lakes from phytoplankton communities dominated by diatoms or green algae to those dominated by blue-green algae. This process is called *eutrophication*. Before we can control eutrophication in lakes, we have to decide which nutrients need to be controlled. Three major nutrients were suggested: nitrogen, phosphorus, and carbon. Phosphorus is now believed to be the limiting nutrient for phytoplankton production in the majority of lakes (Edmondson 1991).

The Experimental Lakes area of northwestern Ontario has been used extensively for whole-lake experiments on nutrient addition. A series of well-designed experiments in these lakes has pinpointed the role of phosphorus in eutrophication (Schindler and Fee 1974). In one experiment, lake 227 was fertilized for five years with phosphate and nitrate, and phytoplankton levels increased 50 to 100 times over those of control lakes (Figure 26.13). To separate the effects of phosphate and nitrate, lake 226 was split in half with a curtain and fertilized with carbon and nitrogen in one half and with phosphorus, carbon, and nitrogen in the other. Within two months a highly visible algal bloom had developed in the basin to which phosphorus was added (Figure 26.14). All this experimental evidence is consistent with the hypothesis that phosphorus is the master limiting nutrient for phytoplankton in lakes.

When phosphorus is added to a lake, algae may show signs of nitrogen or carbon

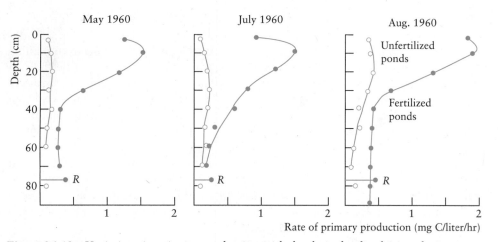

Figure 26.12   *Variations in primary production with depth, in fertilized (●) and unfertilized (○) fishponds between 1000 and 1200 hours on cloudless days.* **R** = *respiration in black bottles. (After Hepher 1962.)*

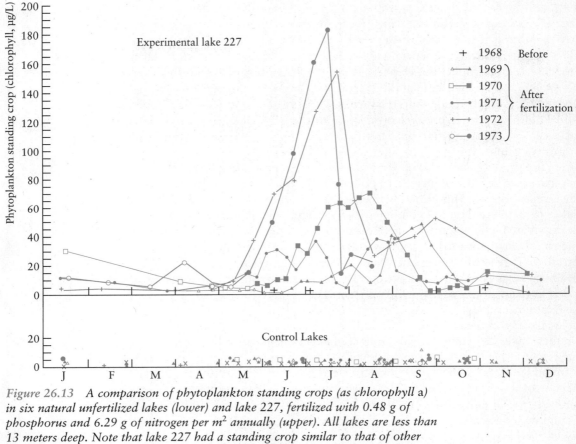

Figure 26.13 A comparison of phytoplankton standing crops (as chlorophyll a) in six natural unfertilized lakes (lower) and lake 227, fertilized with 0.48 g of phosphorus and 6.29 g of nitrogen per m² annually (upper). All lakes are less than 13 meters deep. Note that lake 227 had a standing crop similar to that of other lakes prior to fertilization. Large inputs of phosphorus and nitrogen will thus cause severe eutrophication problems regardless of how low carbon concentrations are. The necessary carbon is drawn from the atmosphere. (After Schindler and Fee 1974.)

limitation, but long-term processes cause these deficiencies to be corrected (Schindler 1977). Physical factors such as water turbulence and gas exchange seem to regulate CO_2 availability, so it rarely becomes limiting for algae. Nitrogen can be fixed by blue-green algae. These species, which are favored when nitrogen is in short supply, increase the availability of nitrogen to algae and the lake eventually returns to a state of phosphorus limitation. The net result is that the standing crop of phytoplankton is highly correlated with the total amount of phosphorus in the lake water (Figure 26.15).

The practical advice that has followed from these and other experiments is to control phosphorus input to lakes and rivers as a simple means of checking eutrophication (Likens 1972). The permissible amount of phosphorus that can be added to a lake can also be calculated so that planners can determine the desirability of human developments on a lake (Dillon and Rigler 1975).

Part of the difficulty of studying nutrient limitations of phytoplankton production is that nutrients may occur in several chemical states in aquatic systems. In some conditions, nutrients are present but not available to the organisms because they are

Figure 26.14   *Lake 226 in the Experimental Lakes area of northwestern Ontario, showing the role of phosphorus in eutrophication. The far basin, fertilized with phosphorus, nitrogen, and carbon, is covered with an algal bloom of the blue-green alga* Anabaena spiroides. *The near basin, fertilized with nitrogen and carbon, showed no changes in algal abundance. Photo taken September 4, 1973. (Photo courtesy of D. W. Schindler.)*

bound up in organic complexes in the water or mud (Wetzel and Allen 1971). This has been shown in a striking manner in acid-bog lakes, which contain large amounts of phosphorus in forms not available to the phytoplankton. Waters (1957) showed that fertilizing acid-bog lakes in Michigan with lime ($CaCO_3$) increased the pH, allowed phosphorus to be released from sediments, and greatly increased the phytoplankton abundance.

One of the changes that often accompanies eutrophication in lakes is that the blue-green algae tend to replace green algae (Figure 26.14). Blue-green algae are "nuisance algae" because they become extremely abundant when nutrients are plentiful and form floating scums on highly eutrophic lakes. Blue-green algae become dominant in the phytoplankton for several reasons. They are not grazed heavily by

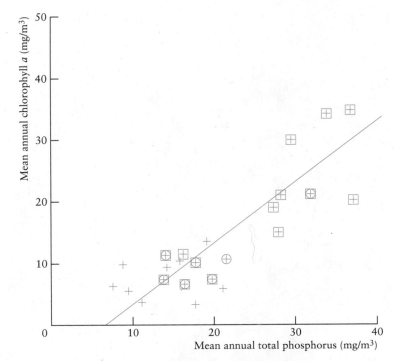

Figure 26.15   *The relationship between total phosphorus concentration and phytoplankton standing crop (measured by chlorophyll) in lakes of the Experimental Lakes area near Kenora in northwestern Ontario. Both fertilized and unfertilized lakes are included. Circles enclose fertilized lakes deficient in nitrogen; squares enclose fertilized lakes deficient in carbon. (After Schindler 1977.)*

zooplankton or fish, which prefer other algae. Zooplankton can often not manipulate the large colonies and filaments of blue-green algae. Some species of blue-greens also produce secondary chemicals that are toxic to zooplankton (DeMott and Moxter 1991). Blue-green algae are also poorly digested by many herbivores, so that they are low-quality food. Finally, many blue-green algae can fix atmospheric nitrogen, putting them at an advantage when nitrogen is relatively scarce. For example, in Lake Washington (see page 565), blue-green algae were dominant when the nitrogen-to-phosphorus ratio was less than 23 but disappeared when this ratio exceeded 25 (Tilman et al. 1982). In eutrophication, more and more phosphorus is continually loaded into a lake so that the nitrogen-to-phosphorus ratio falls and nitrogen can become a limiting factor (Figure 26.16) (Smith 1982). The phytoplankton community in many temperate freshwater lakes therefore may have two broad configurations at which it can exist, one with low nutrient levels organized by predation and dominated by green algae and one with high nutrient levels organized by competition and dominated by blue-green algae.

To summarize, in freshwater communities, primary production is limited by the following array of factors within a given lake:

1. Major controlling factors
    a. Light (and temperature)
    b. Phosphorus
    c. Silicon (for diatoms)

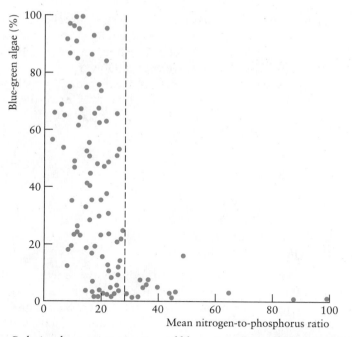

Figure 26.16 *Relation between proportion of blue-green algae in the phytoplankton and nitrogen-to-phosphorus ratios in 17 lakes around the world. Each symbol represents data from one growing season. Blue-green algae are dominant below the boundary where total nitrogen to total phosphorus equals 29 (dashed line). (After Smith 1983.)*

2. Occasional controlling factors
   a. Nitrogen
   b. Iron
   c. Manganese
   d. Molybdenum
3. Rare controlling factors
   a. Carbon
   b. Cobalt, sulfur, and the minor nutrients required for growth

Terrestrial Communities

In terrestrial habitats, temperature ranges are much greater than in aquatic habitats, and the great variation in temperature from coastal to alpine or continental areas makes it possible to uncouple the solar radiation–temperature variable, which is so closely linked in aquatic systems. The large seasonal changes in radiation and temperature are reflected in the global patterns of primary production. Using satellite imagery, we can now look at global patterns of terrestrial productivity (Figure 26.17, see color insert). Satellites, such as the NOAA* meteorological satellite operated by

---

* National Oceanic and Atmospheric Administration.

the United States, have sensors on board that record spectral reflectance in the visible and infrared region of the electromagnetic spectrum. As green plants photosynthesize, they display a unique spectral reflectance pattern in the visible (0.4–0.7 µm) and the near-infrared (0.725–1.1 µm) wavelengths (Goward et al. 1985). Vegetation indices which discriminate living vegetation from the surrounding rock, soil, or water have been developed by combining these spectral bands. One of the most common spectral vegetation indices, the normalized difference vegetation index, is a ratio of near-infrared and visible red spectral bands:

$$NDVI = \frac{NIR - RED}{NIR + RED}$$

where

$NDVI$ = normalized difference vegetation index
NIR = near-infrared reflectance (0.725–1.1 µm)
RED = red reflectance (0.6–0.7 µm)

This index is closely correlated with primary productivity (Graetz et al. 1992, p. 21). The AVHRR (advanced very high resolution radiometer) sensor on the NOAA satellite is especially useful for monitoring global vegetation because it has global coverage at a resolution of 1.1 kilometers at least once per day in daylight hours (Curran 1985).

What limits primary production in terrestrial communities and produces the patterns shown in Figure 26.17? Rosenzweig (1968) showed that actual evapotranspiration could predict the aboveground production of terrestrial communities with good accuracy (Figure 26.18). Actual evapotranspiration is a measure of solar radiation, temperature, and rainfall; it is the amount of water pumped into the atmos-

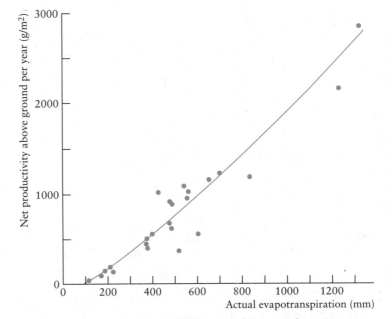

Figure 26.18   Prediction of net primary production of terrestrial communities from climatological data on solar radiation, temperature, and moisture (measured by actual evapotranspiration; see Figure 7.4). (Data from Rosenzweig 1968.)

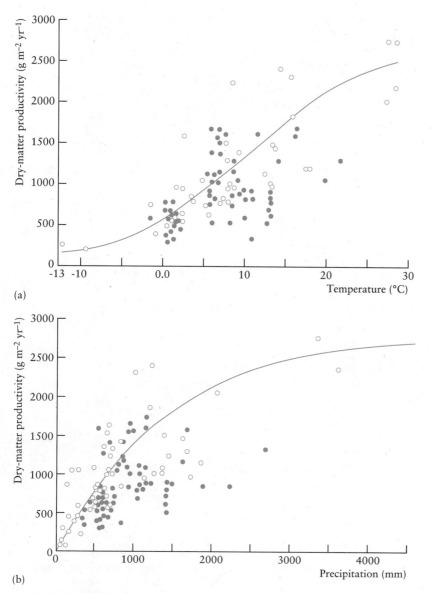

Figure 26.19   *Relationships between net primary production in forests and average temperature and precipitation data for each site. (After O'Neill and DeAngelis 1981.)*

phere by evaporation from the ground and by transpiration from the vegetation. Rosenzweig used only climax vegetation in this analysis.

Lieth (1975) assembled a variety of simple models useful for predicting net primary production in terrestrial vegetation. In addition to evapotranspiration, net production can be predicted from the length of the growing season, precipitation, or temperature (Figure 26.19). Quantities of data of this type have accumulated from studies of productivity begun by the International Biological Program (IBP) in the 1960s and 1970s. Figure 26.19 shows the relationships between net primary production of forests and temperature and precipitation.

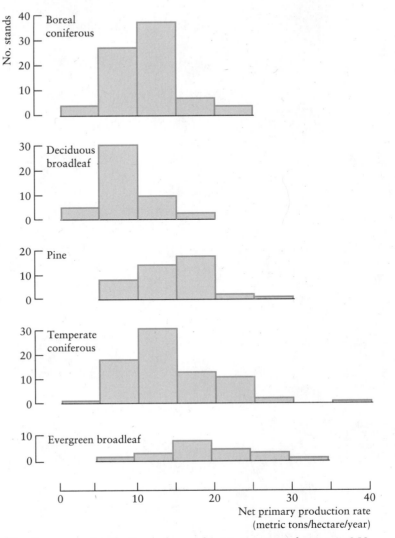

Figure 26.20   Frequency distributions of annual net primary production in 258 forest stands in Japan. Evergreen broadleaf forests of the warm-temperate zone are the most productive; in the cool-temperate zone, coniferous stands are more productive than deciduous forests. Only aboveground production is included. (After Kira 1975.)

Primary production data for different types of forests have been summarized by Kira (1975). Figure 26.20 shows the range of primary production for five growth forms of forests in Japan. Evergreen broadleaf forests of the warm temperate zone are the most productive stands. Coniferous forests are on the average more productive than deciduous forests growing under the same climatic conditions. The differences in productivity among forests are due to variation in the length of the growing season and to differences in leaf area index. Coniferous trees have a greater leaf surface area (Figure 26.21) than do deciduous trees, and by retaining their leaves, the evergreen conifers are able to achieve a longer growing season. The ratio of root biomass to shoot biomass is the same for both forest types. Figure 26.22 shows that

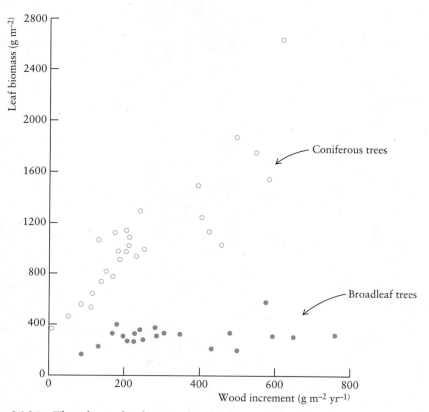

Figure 26.21   The relationship between leaf biomass and wood production for 116 forest sites studied by the International Biological Program. (From O'Neill and DeAngelis 1981.)

the gross primary production of both broadleaf and needle-leaved forests is accurately predicted by the combined effects of leaf area index and the length of the growing season. Temperature determines the length of the growing season.

Primary production in grasslands seems to be largely limited by rainfall (Sala et al. 1988), and temperature tends to be less significant as a limitation. Within the primary control exerted by moisture, secondary limitations are imposed by the type of soil and its water holding capacity, and by nutrient availability. Nutrient-addition experiments on local sites can be used to determine how much primary production can be limited by nutrients. Cargill and Jefferies (1984) added nitrate and phosphate to salt-marsh sedges and grasses in the subarctic to test for nutrient limitation. Figure 26.23 shows that in the absence of grazing the addition of nitrate doubled primary production of the sedges and grasses, and the joint addition of phosphate and nitrate quadrupled production. In this marsh, as in many terrestrial communities, nitrogen is the major nutrient limiting productivity, and when nitrogen is suitably increased, phosphorus becomes limiting (compare with Figure 26.10).

In unexploited virgin grassland or forest, all nutrients that the plants take up from the soil and hold in various plant parts are ultimately returned to the soil as litter to decompose. The net flow of nutrients must be stabilized (input = output), or the site would deteriorate over time. But in a harvested community, the situation is fundamentally different because nutrients are being continuously removed from

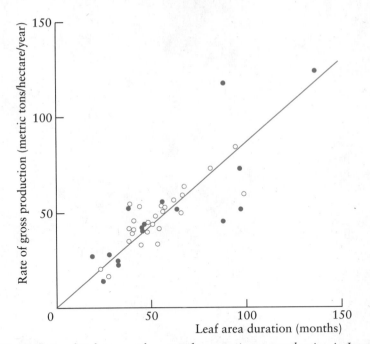

Figure 26.22 *Relationship between the rate of gross primary production in Japanese forest stands and the leaf area duration (leaf-area index times length of growing season in months). Needle-leaved forests are represented by ○, broadleaf forests by ●. (After Kira 1975.)*

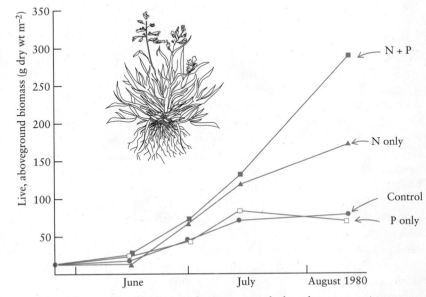

Figure 26.23 *Effects of fertilization with nitrogen and phosphorus on primary production in a salt-marsh dominated by* Carex subspathacea, *southern Hudson Bay, Canada. Four replicates were done of each treatment. (From Cargill and Jefferies 1984.)*

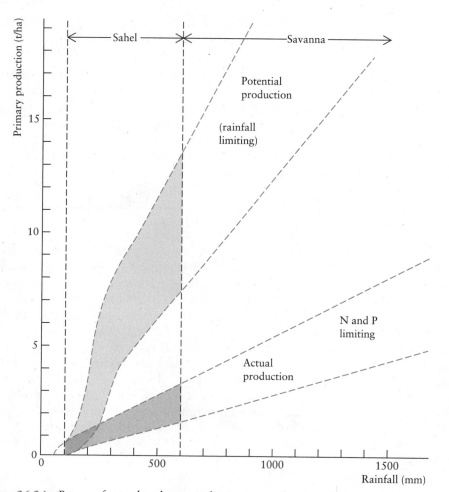

Figure 26.24 Range of actual and potential primary production of pastures in the Sahel region of North Africa in relation to rainfall. Above 250 mm annual rainfall production is limited by nitrogen and phosphorus and not water. (From Breman and deWit 1983.)

the site. This makes it necessary to study the nutrient demands of crops so that the soils will not be progressively exhausted of their nutrient capital. Coniferous forests use soil nutrients more efficiently than deciduous forests (Waring and Schlesinger 1985). Nutrients removed by harvesting trees are typically recovered from annual inputs in precipitation and in rock weathering (see page 663).

Rainfall is not always the immediate factor limiting primary production in semi-arid grasslands. The Sahel region is a semiarid transition zone between the Sahara desert and the savannas of West Africa. Drought in the Sahel during the last ten years has led to many suggestions on how to improve the traditional livestock farming in the area (Breman and deWit 1983). Assumptions about the factors limiting primary production in the Sahel have hampered aid programs. It has been assumed that rainfall limits pasture productivity in the Sahel (Figure 26.24). In fact nitrogen and phosphorus seem to be limiting primary production in the southern

Sahel, and if adequate nutrients are supplied, production can increase fivefold without any additional water. In the drier, northern part of the Sahel, water limits plant growth directly.

Terrestrial communities, especially forests, have large nutrient stores tied up in the standing crop of plants. In this way they differ from communities in the sea and in fresh water. This concentration of nutrients in the standing vegetation has important implications for nutrient cycles in forest communities. If the community is stable, the input of nutrients should equal the output, and a considerable amount of research effort is now being directed at studying nutrient cycles in terrestrial communities. We will discuss this work in Chapter 28.

SPATIAL AND TEMPORAL VARIABILITY

We have assumed in all our discussions of primary production that equilibrium conditions prevail. Recent studies, particularly in aquatic systems, have emphasized the spatial and temporal variability, or patchiness, of primary production. In a lake or ocean, horizontal and vertical diffusion operate, and these physical processes cause space and time to interact (Harris 1980). Physical processes associated with winds and currents can produce aggregation of phytoplankton. Figure 26.25 gives one example of wind-generated microzones of upwelling on lakes. Physical processes are not the only cause of patchiness in plankton. Reproduction associated with local nutrient levels can produce small blooms of phytoplankton on a local scale.

No one doubts the existence of spatial and temporal patchiness in phytoplankton communities, but views diverge on how important such patchiness is for community dynamics (Steele 1978). The equilibrium-centered view, adopted by many researchers in the past, has been to treat variability as noise that needs to be smoothed over or averaged out to reveal the underlying relationships. The nonequilibrium view, a more recent approach, has been to try to analyze the scales of variability that organ-

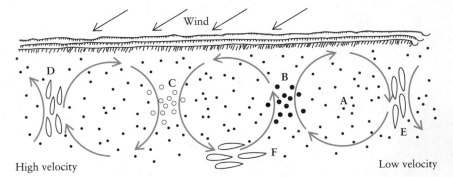

Figure 26.25   Langmuir vortices and plankton distributions in lakes. A constant wind sets up small-scale cells of water circulation that can affect plankton distribution on a local scale. A = neutrally bouyant particles, randomly distributed. B = particles tending to sink, aggregated in upwellings. C = particles tending to float, aggregated in downwellings. D = organisms aggregated in high-velocity upwelling, swimming down. E = organisms aggregated in low-velocity downwelling, swimming up. F = organisms aggregated between downwellings where there is less relative current velocity than within the vortices. (After Stavn 1971.)

isms face. We need to consider both long-term changes that may occur over years and decades as well as the rapid, short-term events on a physiological time scale.

We can illustrate the difficulties of understanding primary production in lakes by considering the effects of nutrients on production by phytoplankton. Nutrients like phosphorus can be taken up by cells in excess of their immediate needs and later used for growth when no external nutrients are available. Algae have developed a whole series of physiological adaptations that buffer them from fluctuations in the external environment. In a natural lake, algae respond within minutes to changes in nutrients, sunlight, and temperature. It is not clear whether simple equilibrium models of algal growth derived from laboratory studies can be applied to algal growth in the field (Sommer 1990).

Spatial and temporal variability in aquatic systems may be important in maintaining species diversity. Hutchinson (1961) raised this issue as the *paradox of the plankton:* How do so many species coexist in water that appears to be a homogeneous niche? One answer he gave is that aquatic systems are nonequilibrium systems that are continually changing seasonally, so competitive dominance and exclusion cannot be achieved.

We are just beginning to realize the significance of patchiness for ecological systems. The importance of *scale* in ecology is a focal issue for the 1990s as we attempt to gain a finer understanding of how ecological systems operate in the real world.

SUMMARY

A community can be viewed as a complex machine that processes energy and materials. To study community metabolism, we must determine the food web of the community and then trace the flows of chemical materials or energy through the food web. Many ecologists prefer to use energy to study community metabolism because energy is not recycled within the community.

Primary production can be measured by the amount of energy or carbon fixed by photosynthesis by green plants per unit time. Uptake of CO_2 can be measured directly with radioactive $^{14}CO_2$ or indirectly by harvesting new growth.

Only about 1 percent of solar energy is captured by the green plants and converted into primary production. Forests are relatively efficient and aquatic communities inefficient at capturing solar energy.

Primary production varies greatly over the globe; it is highest in the tropical rain forests and lowest in arctic, alpine, and desert habitats. The sea is much less productive than the land because of limitations imposed by light and nutrients, and except for coastal areas and upwelling zones, the sea is a biological desert. In freshwater communities, light, temperature, and nutrients restrict primary production, and phosphorus seems to be the master limiting nutrient in many lakes.

Terrestrial primary productivity can be predicted from the length of the growing season, temperature, or rainfall. Nutrient limitations further restrict productivity levels set by these climatic factors, and the stimulation in plant growth achieved by fertilizing forests and crops indicates the importance of studying nutrient cycling in biological communities.

Spatial and temporal variability in primary production occur particularly in aquatic systems and need to be studied on several scales if we are to understand how these systems operate. The coexistence of many species of phytoplankton may be permitted by spatial and temporal variability and the failure of aquatic systems ever to reach an equilibrium.

Remote sensing with satellites has opened up new methods for measuring the spatial and temporal variability of primary production, in the oceans and on land, on large spatial scales. Determining the factors that limit primary production is important for understanding how climatic changes will affect both natural and agricultural communities.

Selected References

BREMAN, H., and C. T. DE WIT. 1983. Rangeland productivity and exploitation in the Sahel. *Science* 221:1341–1347.

CARGILL, S. M., and R. L. JEFFERIES. 1984. Nutrient limitation of primary production in a sub-arctic salt marsh. *Journal of Applied Ecology* 21:657–668.

HARRIS, G. P. 1980. Temporal and spatial scales in phytoplankton ecology: Mechanisms, methods, models, and management. *Canadian Journal of Fisheries and Aquatic Sciences* 37:877–900.

HAYWARD, T. L. 1991. Primary production in the North Pacific Central Gyre: A controversy with important implications. *Trends in Ecology and Evolution* 6:281–284.

HUTCHINSON, G. E. 1961. The paradox of the plankton. *American Naturalist* 95:137–145.

LIETH, H. F. H., ed. 1978. *Patterns of Primary Production in the Biosphere*. Dowden, Hutchinson and Ross, Stroudsburg, Pa.

O'NEILL, R. V., and D. L. DE ANGELIS. 1981. Comparative productivity and biomass relations of forest ecosystems. In *Dynamic Properties of Forest Ecosystems*, ed. D. E. Reichle, pp. 411–449. Cambridge University Press, Cambridge, England

PLATT, T., and S. SATHYENDRANATH. 1988. Oceanic primary production: Estimation by remote sensing at local and regional scales. *Science* 241:1613–1620.

SOMMER, U. 1990. Phytoplankton nutrient competition—from laboratory to lake. In *Perspectives on Plant Competition*, ed. J. B. Grace and D. Tilman, pp. 193–213. Academic Press, New York.

TRÉGUER, P., and G. JACQUES. 1992. Dynamics of nutrients and phytoplankton, and fluxes of carbon, nitrogen and silicon in the Antarctic Ocean. *Polar Biology* 12:149–162.

Questions and Problems

1. "Red tides" are spectacular dinoflagellate blooms that occur in the sea and often lead to mass mortality of marine fishes and invertebrates. Review the evidence available about the origin of red tides, and discuss the implications

for general ideas about what controls primary production in the sea. Fiedler (1982) provides a recent discussion of this problem.

2. Calculate the attenuation of solar radiation through a water column (and the depth at which radiation falls to 1 percent of its surface value) for the following extinction coefficients:

	Extinction Coefficient (per meter)
Pure water	0.03
Coastal water, minimum	0.20
Coastal water, maximum	0.40
Eutrophic lake	1.50

3. In the central grasslands of the United States Sala and co-workers (1988) showed that net primary production could be predicted from mean annual precipitation. Read Sala and co-authors (1988) and discuss why temperature and soil nutrients are not relevant variables in this ecosystem.

4. In discussing the effect of light on primary productivity in the ocean, Nielsen and colleagues (1957, p. 108) state:

It is thus quite likely that a permanent reduction of the light intensity at the surface (to e.g., 50 percent of its normal value without the other factors being affected—a rather improbable condition in Nature) in the long run would have very little influence on the organic productivity as measured per surface area.

How could this possibly be true?

5. The concentration of inorganic phosphate in the water of the North Atlantic Ocean is only about 50 percent of that found in the other oceans. Yet the North Atlantic is more productive than most of the other oceans. How can you reconcile these observations if nutrients limit primary productivity in the oceans?

6. Review the "limiting-nutrient controversy," the argument about the relative importance of carbon and phosphorus in regulating algal growth in lakes (Likens 1972, Edmondson 1991). In particular, discuss the practical problem of what measures might be taken to reduce lake eutrophication.

7. In discussing the properties of ecosystems, Reichle and coauthors (1975, pp. 31–32) state:

Unless populations in the ecosystem contribute to specific vital system functions such as photosynthesis or cycling of nutrients, strong negative selective pressure is exerted upon the individual populations by the system. Whatever individual population responses are established, the populations establish homeostatic feedback mechanisms so that the ecosystem survives and grows to a maximum persistent biomass.

Discuss the evolutionary implications of these statements.

8. Tilman and colleagues (1982, p. 367) state:

We suggest that the spatial and temporal heterogeneity of pelagic environments will prevent us from meaningfully addressing questions on short time scales or small spatial scales.

Discuss the general issue of whether there are some questions in community ecology that we cannot answer because of scale.

9. Crop productivity has improved greatly during the last 60 years. Some of this improvement in production is due to genetic changes in the crops, and some is due to increased nutrients or water. How could you evaluate the relative contribution of these two components to increasing primary productivity in a particular crop? Boyer (1982) discusses this problem and provides some data for major U.S. crops.

10. According to the standard $^{14}CO_2$ method of measuring primary production, the central oceans are highly unproductive deserts. Recent research (Hayward 1991) has suggested that the oceans are more productive than previously thought. Read Hayward (1991) and review the technical problems of measuring primary production in ocean waters. Rivkin and Putt (1987) provide further discussion of this problem.

Overview Question

What limits primary production in agricultural systems? List the differences in the controls of primary production in natural communities and agricultural crops and discuss the implications for sustainable agriculture.

27 Community Metabolism II: Secondary Production

MEASUREMENT OF SECONDARY PRODUCTION

The biomass of plants that accumulates in a community as a result of photosynthesis can eventually go in one of two directions: to herbivores or to detritus feeders. The fate of the energy and materials captured in primary productivity can be shown most simply by looking at the metabolism of an individual herbivore.

The partitioning of food materials and energy for an individual animal can be seen as a series of dichotomies. Using energy, we have

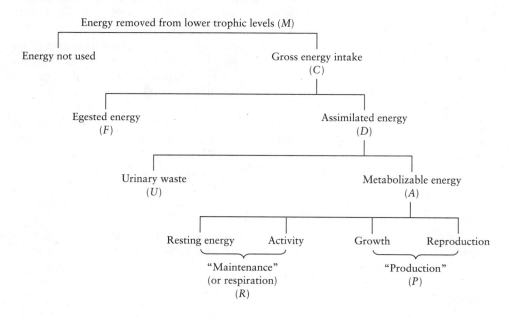

This scheme could be presented for carbon intake or any essential nutrient, and again we have the choice of using chemical materials or energy to study the system.

Let us look at this scheme in detail. Every animal will remove some energy or material from the lower trophic level for its food. Some of this energy will not be used. For example, a beaver fells a whole tree and eats only some of the bark. In this case, most of the energy removed from the plant trophic level is not used by the beaver but is left to decompose. Of the material consumed, some energy passes through the digestive tract and is lost in the feces. Of the energy assimilated, some is lost as urinary output, and the rest is available for metabolic energy. It is usually convenient to lump the energy losses of urine and feces when actually determining the metabolizable energy. We have now reached the level of assimilation, or metabolizable energy, and this energy can be subdivided into two general pathways, production or maintenance. All animals must expend energy just to subsist through the process of respiration. Production occurs by using metabolizable energy for growth and for reproduction.

How can we measure the components of secondary productivity in an animal community? Several techniques are available (Petrusewicz and Macfadyen 1970), but the general procedure is as follows. Each species of animal is considered separately. To determine the gross energy intake of the population, we must know the feeding rate. This can be measured by confining a herbivorous animal to a feeding plot and measuring herbage biomass before and after feeding. In some predators, such as birds of prey, the number of food items being consumed can be counted by direct observation. Indirect techniques such as weight of stomach contents can also be used but require knowledge of the rate of digestion and rate of feeding.

Assimilated or metabolizable energy can be measured very simply in the laboratory where the gross intake can be regulated and feces and urine can be collected, but in the field it is most difficult to estimate assimilation directly. The usual approach is to measure it indirectly by use of the relation

Assimilation rate = respiration rate + net productivity

If we can measure the rates of respiration and production, we can get assimilation rate by addition. Note that all these are rates (per unit time), and that assimilation rate is also called gross productivity.

Respiration can be measured very easily in laboratory situations by confining an animal to a small cage and measuring oxygen consumption, CO_2 output, or heat production directly. There is a minimum rate of metabolism, the basal metabolic rate, which increases with body size in warm-blooded animals. The relationship between basal metabolic rate and body size is not constant in all taxonomic groups, contrary to what is often reported (Hayssen and Lacy 1985). For example, within the mammals bats of a given body size have lower basal metabolic rates than rodents. Basal metabolism is measured under resting conditions with no food in the stomach at a temperature where the animal is not required to expend energy for extra heat production or cooling. Measured in such an abstract way, basal metabolism is not closely related to respiration losses in field situations in which activity is necessary, temperature varies, and digestion is occurring.

Energy metabolism in the field can be measured directly by means of doubly-labeled water (Nagy 1987). This method involves injecting wild animals with water in the form D_2O^{18} and then measuring the loss rates of the hydrogen (deuterium) and oxygen isotopes. The hydrogen isotope measures water loss, while the oxygen isotope is lost in both CO_2 and in water. The difference between these isotope loss rates represents CO_2 loss alone, and is a field measure of metabolic rate. Figure 27.1 shows how field metabolic rates vary with body size in mammals and birds. There is no single regression line to describe this relationship for all groups of vertebrates, although for medium-sized endotherms (250–550 grams) the field metabolic rates are similar for birds and mammals. For example, the regression for herbivorous mammals is

log (FMR) = 0.774 + 0.727 (log body mass)

where

FMR = field metabolic rate (kJ/day/individual)

and body mass is in grams.

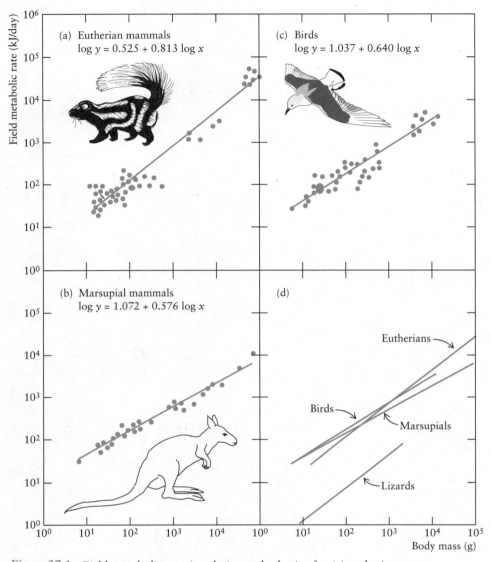

Figure 27.1   Field metabolic rate in relation to body size for (a) eutherian mammals, (b) marsupial mammals, (c) birds. In (d) the regression lines for these three groups are superimposed and contrasted with that for lizards. (From Nagy 1987.)

The energetic costs of living for birds and mammals is very high compared with that for reptiles (Figure 27.1d). A 250-gram mammal or bird will use about 320 kilojoules of energy daily whereas a 250-gram iguanid lizard will use about 19 kilojoules per day, a 17-fold difference in energy use. Respiration in cold-blooded organisms is highly dependent on temperature, as well as body size.

Net production can be measured by the growth of individuals in the population and the reproduction of new animals. We have discussed techniques for measuring changes in numbers in Chapter 10, and growth can be measured by weighing individuals at successive times. The only admonition we must make here is that sampling must be frequent enough so that individuals are not born and then die in the interval between samples. Net production is usually measured as biomass and converted to

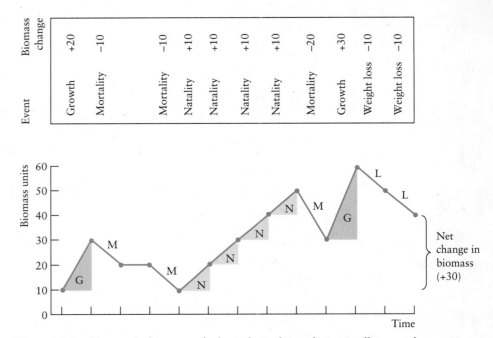

Figure 27.2 *Changes in biomass of a hypothetical population to illustrate the factors that contribute to secondary production. The net change in biomass over a given time period is the outcome of gains from growth and reproduction and losses from death and emigration.*

energy measures by determination of the caloric value of a unit of weight of the species.

Figure 27.2 gives a schematic representation of how production can be determined from information on population changes and growth. In this hypothetical population, production is the sum of growth and natality additions:

Production = growth + natality
= 20 + 10 + 10 + 10 + 10 + 30 − 10 − 10
= 70 units of biomass

Note that we can calculate production in a second way:

Production = net change in biomass + losses by mortality
= + 30 + 40 = 70 units of biomass

Losses caused by death or emigration are a part of production and should not be ignored. We can see this very clearly by looking at a population that is stable over time (net change in biomass zero): A ranch that has the same biomass of steers this year as last year does not necessarily have zero production over the year.

Let us look at an actual example of the calculations of the secondary productivity of an African elephant population. Petrides and Swank (1966) estimated the energy relations of the elephant herds in Queen Elizabeth National Park (now Ruwenzori National Park) in Uganda. To do these calculations, we must assume a stable population of elephants with a stationary age distribution. For convenience we will also assume no births or deaths occur during the study interval. The population was

Table 27.1   Elephant Life Table and Production Data, Queen Elizabeth (now Ruwenzori) National Park, Uganda, November 1956 to June 1957

Age, x	No. Alive at Beginning of Age x	Mortality Rate	Median No. Alive	Weight Average (kg)	Weight Increment (kg)	Population Weight Increment[a] (kg)
1	1,000	0.30	850.0	91	91	77,273
2	700	0.20	630.0	205	114	71,591
3	560	0.10	532.0	318	114	60,455
4	504	—	478.5	455	136	65,250
5	453	—	430.5	614	159	68,489
6	408	—	387.5	795	182	70,455
7	367	—	348.5	1,000	205	71,284
8	330	—	313.5	1,205	205	64,125
9	297	—	282.0	1,409	205	57,682
10	267	0.05	260.5	1,614	205	53,284
11	254	—	247.5	1,818	205	50,625
12	241	—	235.0	2,023	205	48,068
13	229	—	223.5	2,205	182	40,636
14	218	—	212.5	2,386	182	38,636
15	207	0.02	205.0	2,591	205	41,932
16	203	—	201.0	2,795	205	41,114
17	199	—	197.0	3,000	205	40,295
18	195	—	193.0	3,182	182	35,091
19	191	—	189.0	3,386	205	38,659
20	187	—	185.0	3,591	205	37,841
21	183	—	181.0	3,750	159	28,795
22	179	—	177.0	3,864	114	20,114
23	175	—	173.0	3,955	91	15,727
24	171	—	169.5	4,045	91	15,409
25	168	—	166.5	4,091	45	7,568
26–67	3,107[b]	0.02–0.20	3,026.5[b]	4,091	0	0
Total	10,933		10,487.5	2291[c]	4,091	1,160,398

[a] Mean no. alive times weight increment.
[b] Sum of the numbers of animals in each year-class for ages 26 to 67.
[c] Average body weight.
  Source: After Petrides and Swank (1966).

counted and the age structure estimated to construct a life table (see page 168). The maximum age was estimated at 67 years, and the survivorship schedule is given in Table 27.1. Weight growth was estimated from some records of zoo animals and a limited amount of field data on weights. The average weight at each age and the average increase in weight from one year to the next are also given in Table 27.1. If the age structure is stationary,

$$\text{Growth in biomass} \cong \sum_{\text{all ages}} \left(\begin{array}{c} \text{median no. alive} \\ \text{during age } x \text{ to } x + 1 \end{array} \right) \times \left(\begin{array}{c} \text{average weight growth} \\ \text{for ages } x \text{ to } x + 1 \end{array} \right)$$

This is calculated in the last column of Table 27.1. The caloric value of elephants is approximately 6.3 kilojoules per gram of live weight. We determine growth as follows:

From Table 27.1, 1,000 elephants lived 10,487.5 elephant-years and produced 1,160,398 kg of growth.

$$\text{Average growth (in weight)/elephant/yr} = \frac{1,160,398}{10,487.5} = 110.6 \text{ kg}$$

$$\begin{aligned} \text{Average growth (in energy)/elephant/yr} &= 110,600 \text{ g} \times 6.276 \text{ kJ/g} \\ &= 694,126 \text{ kJ} \end{aligned}$$

The population density of elephants was 2.077 elephants per square kilometer or 0.000002077 elephant per square meter. Thus

$$\text{Growth} = 694,126 \text{ kJ} \times 0.000002077 = 1.44 \text{ kJ/m}^2/\text{yr}$$

A large amount of the food consumed by elephants passes through as feces. From studies on captive elephants, an average 2,273-kilogram (5,000-lb) elephant would consume 23.59 kilograms dry weight of forage per day and produce from this 13.25 kilograms dry weight of feces. The food plants are worth approximately 16.7 kilojoules per gram dry weight, so we can calculate

$$\begin{aligned} \text{Average consumption} &= 23,590 \times 16.7 = 394,802 \text{ kJ} \\ \text{Average fecal production} &= 13,250 \times 16.7 = 221,752 \text{ kJ} \end{aligned}$$

Counting these for a whole year and multiplying by the number of elephants per square meter, we obtain

$$\begin{aligned} \text{Food consumed} &= 394,802 \text{ kJ/day/elephant} \times 365 \text{ days} \\ &\quad \times 0.000002077 \text{ elephant/m}^2 \\ &= 299 \text{ kJ/m}^2/\text{yr} \\ \text{Feces produced} &= 221,752 \text{ kJ/day/elephant} \times 365 \text{ days} \\ &\quad \times 0.000002077 \text{ elephant/m}^2 \\ &= 168.1 \text{ kJ/m}^2/\text{yr} \end{aligned}$$

We know that

$$\begin{aligned} \text{Food energy consumed} &= \text{feces} + \text{growth} + \text{maintenance} \\ 299 &= 168.1 + 1.44 + \text{maintenance} \end{aligned}$$

Maintenance must be about 130 kilojoules per square meter per year if we ignore the losses due to urine production and the production due to newborn animals.

We can also estimate maintenance from Figure 27.1 for the field metabolic rate of a standard 2,273-kilogram (5,000-lb) elephant from the equation given in Nagy (1987) for herbivores:

$$\begin{aligned} \text{log (field metabolic rate)} &= 0.774 + 0.727 \text{ log (weight)} \\ &= 0.774 + 0.727 \text{ (log 2,273,000 g)} \\ &= 5.395248 \end{aligned}$$

or   field metabolic rate = 248,455 kJ/day/elephant

$$\begin{aligned} \text{Estimated maintenance} &= 248,455 \text{ kJ/day} \times 365 \text{ days} \\ &\quad \times 0.000002077 \text{ elephant/m}^2 \\ &= 188 \text{ kJ/m}^2/\text{yr} \end{aligned}$$

This is slightly greater than the maintenance estimate of 130 kilojoules per square meter per year obtained above.

Finally, we can determine the standing crop of elephants in energetic terms:

Standing crop = 0.000002077 elephant/m^2 × 2,291,000 g × 6.3 kJ

= 30 kJ/m^2

A rough estimate of primary productivity by the harvest method produced an estimate of net primary productivity of 3,125 kilojoules per square meter per year for the foraging area of the elephants.

We can summarize these estimates for the African elephant population of Queen Elizabeth Park:

	Energy (kJ/m²/yr)
Net primary production	3125[a]
Secondary production	
Food consumed	299
Fecal energy lost	168
Maintenance metabolism	130
Growth	1.44
Standing crop of elephants	30

[a] Probably a low estimate; compare Table 26.1, page 609.

Clearly, the greatest part of the energy intake of these elephants is used in maintenance or lost in fecal production.

The details of estimating secondary production will obviously vary from species to species, and the number of assumptions that must be made will depend on how well the species is studied. The procedure is to repeat these calculations for all dominant species in the community and by addition to obtain the secondary production of the community. This procedure is more tedious than conceptually difficult, and we can now consider the results of this kind of analysis.

PROBLEMS IN ESTIMATING SECONDARY PRODUCTION

In principle ecologists can apply the kind of techniques just described for elephants to all the major species of consumers in a community. In practice these calculations have led ecologists into three conceptual problems (Cousins 1987). The first difficulty is that individuals of particular species do not fit clearly into discrete trophic levels. At the producer level, plants are usually easily assigned to trophic level 1. But herbivores are often animals that eat plants as well as other animals (Figure 27.3). House mice (*Mus musculus*) are thought of as herbivores yet they consume substantial amounts of insects at some seasons of the year. Carnivores like the red fox (*Vulpes fulva*) do eat herbivores like rabbits but they can also eat plant material, detritus, and other carnivores. In general the higher up the food chain, the less clear it is how to categorize a species.

A second problem in estimating secondary production is what to do with detritus. The normal procedure has been to put detritus and dung in with the first trophic level, to treat detritus as plant material, but this is not correct. Detritus from herbivores should be kept separate from plant detritus. There is typically a complex food web within detritus itself that is difficult to assign to the producers-herbivores-carnivores type of trophic organization.

The third major difficulty in estimating secondary production is the practical

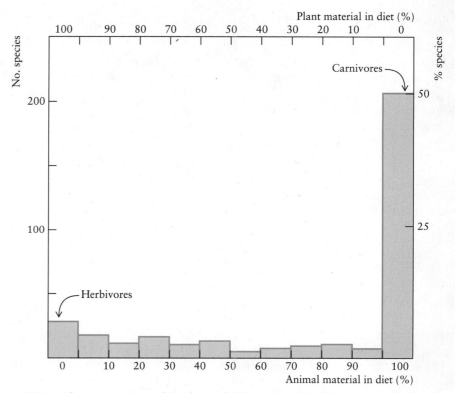

Figure 27.3   *The composition of the diets of 430 species of North American birds and mammals. Over 40 percent of these species are not strict herbivores or strict carnivores. (Modified from Peters 1977.)*

one of sampling a complex community adequately, and in allowing for nonequilibrium conditions in natural ecosystems. Aquatic ecologists have been particularly aware of these problems (Kokkinn and Davis 1986), and the rapid changes that occur in plankton populations and benthic invertebrates may render accurate estimates of production impossible on a finite budget.

Given these problems, ecologists have moved away from trying to estimate secondary production for whole trophic levels, and have begun to analyze parts of food webs taxonomically (Cousins 1987). We can analyze single species like the elephant in detail and measure their impact on the community. We will now consider how efficient different species are in their use of energy.

ECOLOGICAL EFFICIENCIES

If we view animals as energy transformers, we can ask questions about their relative efficiencies. A large number of ecological efficiencies can be defined (Kozlovsky 1968), and we shall be concerned here with efficiency within a species on a particular trophic level. A useful measure of efficiency is defined as follows:

$$\text{Production efficiency} = \frac{\text{net productivity of species } n}{\text{assimilation of species } n}$$

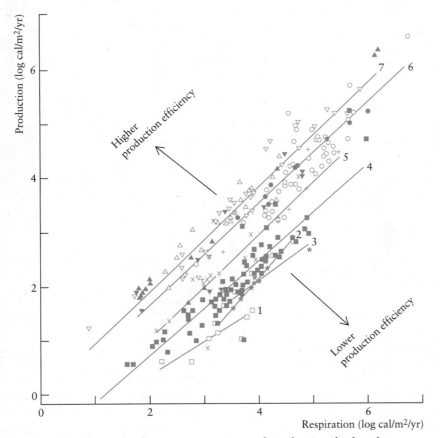

Figure 27.4   *The relationship between respiration and production (both as log$_{10}$ cal/m^2/yr) in natural populations of animals. The regression lines of the seven derived groups are shown: 1 = insectivores, 2 = small mammal communities, 3 = birds, 4 = other mammals, 5 = fish and social insects, 6 = noninsect invertebrates, and 7 = nonsocial insects. Symbols are: □ insectivores, ▣ small mammal communities, ▪ other mammals, * birds, + fishes, x social insects, ○ molluscs, ● Crustacea, ▼ other noninsect invertebrates, △ Orthoptera, ▲ Hemiptera, and ▽ other nonsocial insects. (After Humphreys 1979.)*

Data on production efficiency for individual species are readily obtained. Figure 27.4 shows the relationship between respiration and production in 235 energy budgets measured in natural populations. Production efficiency is the ratio of production and respiration, so that on Figure 27.4 the points to the upper left represent higher efficiencies and those to the lower right represent lower efficiencies. The first question we can ask about these data is whether different taxonomic groups have different efficiencies. Homeotherms separate into four groups—insectivores, birds, small mammal communities, and other mammals. Poikilotherms separate into three groups—fish and social insects, nonsocial insects, and other invertebrates. The slopes of all seven lines in Figure 27.4 are nearly equal to 1.0, so as respiration goes up 1 kilojule, production likewise goes up by 1 kilojule in all groups. This means that production efficiencies P/ (R + P) are the same for all sizes of animals within the seven groups. Table 27.2 lists these estimates for production efficiencies. In general,

Table 27.2   Average Production Efficiencies in Order of Increasing Efficiency

Group	Production Efficiency (%)	No. Studies
Insectivores	0.86	6
Birds	1.29	9
Small mammal communities	1.51	8
Other mammals	3.14	56
Fish and social insects	9.77	22
Other invertebrates (excluding insects)	25.0	73
Herbivores	20.8	15
Carnivores	27.6	11
Detritivores	36.2	23
Nonsocial insects	40.7	61
Hebivores	38.8	49
Detritivores	47.0	6
Carnivores	55.6	5

Notes. Data from 235 natural populations. A breakdown into trophic groups is presented for two of the groups with adequate data.
Source: After Humphreys (1979).

respiration seems to utilize 97 to 99 percent of the energy assimilated in mammals and birds, and consequently only 1 to 3 percent of the energy goes to net production in these groups. For insects the loss is less, approximately 59 to 90 percent of the energy assimilated being used for respiration. This difference between insects and mammals may be a reflection of the cost of homeothermy. There appears to be no variation in production efficiency between animals in different habitats. Aquatic and terrestrial poikilotherms seem to have equal production efficiencies (Humphreys 1979).

Studies on the energetics of populations do not suggest constant ecological efficiencies. The amount of energy available for growth and reproduction, the converse of the amount used for respiration, can vary with the type of diet and with the amount of food consumed. Figure 27.5 gives one example from a laboratory study of fish metabolism, in which the proportion of the food consumed that is used for respiration declines at high feeding rates. This type of complication makes it difficult to estimate energy flow through a community.

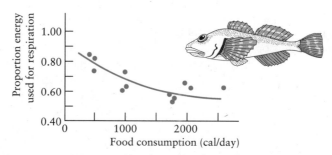

*Figure 27.5   Proportion of energy used in respiration in yearling sculpins (*Cottus perplexus*) held separately in aquariums and fed measured amounts of midge larvae in the fall. (Data from Warren and Davis 1967.)*

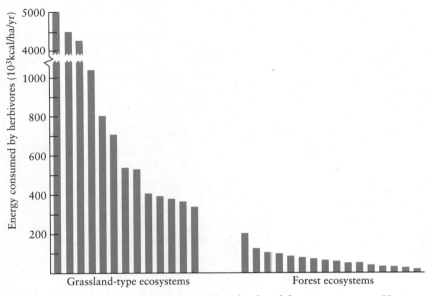

Figure 27.6   Consumption by herbivores in grassland and forest ecosystems. Herbivores have significantly more impact on vegetation in grasslands than they do in forests. (Data summarized from 29 studies by Petrusewicz and Grodzinski 1975.)

For animals in different vegetation types we can aggregate assimilation and ask whether total consumption varies with habitat type Herbivores in grassland ecosystems consume a higher fraction of the primary production than they do in forest ecosystems (Figure 27.6). Zooplankton herbivores in aquatic food webs consume an even higher fraction of the net primary production, and thus we can distinguish ecosystems dominated by grazing from those dominated by decomposers. Whittaker (1975) gives these average values:

	Net Primary Production Going to Animal Consumption (%)
Tropical rain forest	7
Temperate deciduous forest	5
Grassland	10
Open ocean	40
Oceanic upwelling zones	35

Thus in forest ecosystems, almost all of the primary production goes into the decomposer food chain.

Although much of the work on secondary production has centered on energy flow, an increasing amount of research on nutrient cycles is being done, because work on individual populations has suggested that nutrients and not energy may be limiting animal populations (Cousins 1987). We shall discuss nutrient cycling in Chapter 28.

One consequence of low ecological efficiencies is that organisms at the base of the food web are much more abundant than those at higher trophic levels. Elton recognized this in 1927, and the resulting pyramids of numbers or biomass have

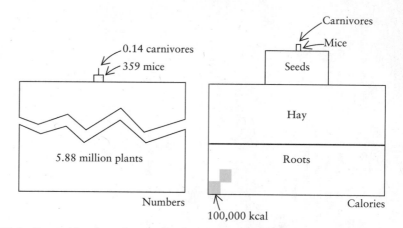

Figure 27.7   *Pyramids of numbers and caloric content on 1 acre of annual grassland in California at the time of peak plant and animal standing crops. (After Pearson 1964.)*

been called *Eltonian pyramids* in his honor. Figure 27.7 illustrates a pyramid of numbers for an annual grassland in California. Note that pyramids can be constructed on the basis of numbers, biomass, or energy of standing crop. They illustrate graphically the rapid loss of energy as one moves from plants to herbivores to carnivores, a biological illustration of the second law of thermodynamics.

WHAT LIMITS SECONDARY PRODUCTION?

One of the critical questions we need to answer is, what limits secondary production? As a first approximation, we could state that secondary production is limited by primary production and the second law of thermodynamics. The second law of thermodynamics states that no process of energy conversion is 100 percent efficient. Figure 27.8 illustrates these broad patterns for a large array of 69 terrestrial communities from tundra to tropical forests. Herbivore biomass and consumption rise rapidly with increasing primary productivity, and secondary production increases with primary production in a 1:1 ratio (McNaughton et al. 1989). There is considerable scatter in these relationships shown in Figure 27.8, and, as you would predict from Figure 27.6, forest communities tend to fall below the regression line while grassland communities fall above the line. To understand what limits secondary production in individual communities, we need to study the details of energy and nutrient flows. From 1965 to 1980 the International Biological Program carried out a series of studies of production in specific types of plant communities. Let us look at an example.

Grassland Ecosystems

Grassland is the potential natural vegetation on 25 percent of the earth's land surface. Grasslands occur in a great diversity of climates and are defined by having a period of the year when soil water availability falls below the requirement for forest. In 1968 the International Biological Program (IBP) began a series of studies on grassland sites throughout the world (Coupland 1979). The grassland biome

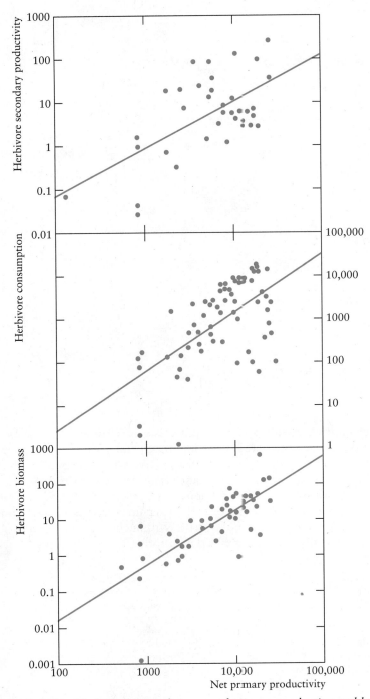

Figure 27.8   *Relationship between net aboveground primary production and herbivore biomass, consumption, and net secondary productivity. Biomass measured as kJ/m², all others as kJ/m²/yr. Data from 69 studies from arctic tundra to tropical forests. (From McNaughton et al. 1989.)*

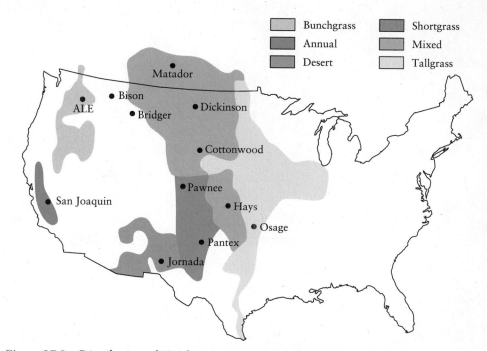

Figure 27.9 *Distribution of North American grasslands and the intensive study sites of the International Biological Program. (From French 1979.)*

study of the U.S. IBP was a major attempt to analyze how natural grasslands work (French 1979). Twelve sites were studied intensively in North America (Figure 27.9).

The primary productivity of grassland increases with precipitation, but the relevant variable is periodic drought. Grasslands range from arid desert grasslands with nearly continuous drought to tropical humid grasslands with nearly no drought. The average yearly primary production of these six climatic types ranges from 100 to 600 grams per square meter (Figure 27.10). Within these ranges, North American grasslands tend to fall at the lower end (Lauenroth 1979):

	Aboveground Net Primary Production (g/m²/yr)
Tallgrass prairie	500
Annual grassland	400
Mixed grassland	300
Shortgrass prairie	200
Bunchgrass and desert grassland	100

Eltonian pyramids were calculated to determine the distribution of biomass among producers, consumers, and decomposers. Figure 27.11 illustrates the trophic pyramid for a tallgrass prairie site. There was little variation between years in these pyramids, and the most striking finding is that biomass belowground in these grasslands greatly exceeds that aboveground.

Herbivores and carnivores were grouped into taxonomic groups in order to

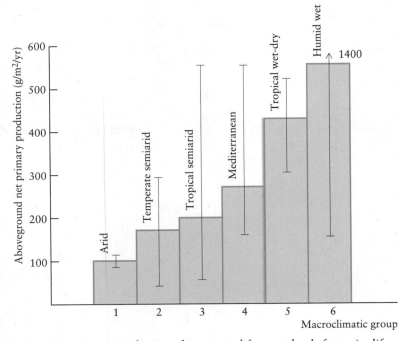

Figure 27.10   Net primary production aboveground for grasslands from six different climatic types. Vertical lines indicate the range of values within each group. (After Lauenroth 1979.)

summarize consumption and production figures for these ecosystems. Figure 27.12 shows the energy flow through tallgrass and shortgrass prairie sites. Several results stand out in these data. The most conspicuous animals are the least important energetically. Birds and mammals contribute almost nothing to production and very little to consumption (Scott et al. 1979). Birds consume 0.05 percent of the aboveground primary production in tallgrass prairie sites, and mammals consume 2.5 percent. The aboveground insects consume a greater amount, but the most important consumers are the soil animals, especially nematodes. More than half of the plant tissue consumed is taken by nematodes. Nematodes are a major factor controlling total grassland primary production.

Only a small fraction of the primary production in grasslands is consumed by animals. Table 27.3 shows that, aboveground, only 2 to 7 percent of primary production is eaten by herbivores, but belowground, 7 to 26 percent is eaten. In contrast, virtually all of the secondary production by herbivores seems to be eaten by carnivores in grasslands. Predators may control consumer populations, at least aboveground. Both production and consumption percentages increase in moving from shortgrass to tallgrass prairie (Table 27.3).

The hypothesis that emerges from this analysis of grassland ecosystems is that grassland plants may be limited by nematode consumption of roots, by soil water, and by competition for nutrients and light, while consumer populations are controlled by predators (Scott et al. 1979).

This hypothesis was tested on a shortgrass prairie by experimentally providing water and nitrogen to 1-hectare plots for six years (Dodd and Lauenroth 1979). Primary production increased dramatically in both water treatments, especially in

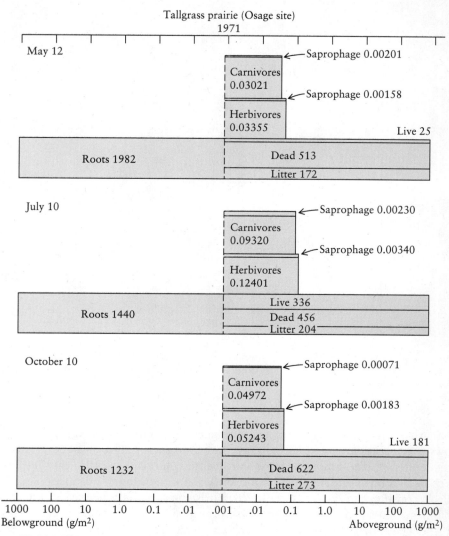

Figure 27.11   Eltonian pyramids as biomass for a tallgrass prairie site for three dates during the growing season. The base of each pyramid represents biomass (g dry wt/m²) of producers; the middle level, herbivores; and the top level, carnivores. Aboveground (right) and belowground (left) biomass are separated. (After French et al. 1979.)

the water-plus-nitrogen plots (Figure 27.13). Nematode numbers increased about fourfold on the water and water-plus-nitrogen treated areas. Small mammals (ground squirrels, voles, and mice) responded dramatically to the water-plus-nitrogen plots because of the increased herbage cover. These experimental results tend to confirm the hypothesis that both water and nitrogen limit production in grassland ecosystems.

Game Ranching in Africa

The savanna and plains areas of Africa support a great diversity of ungulates, and the conservation of these ecosystems has been a primary concern of conserva-

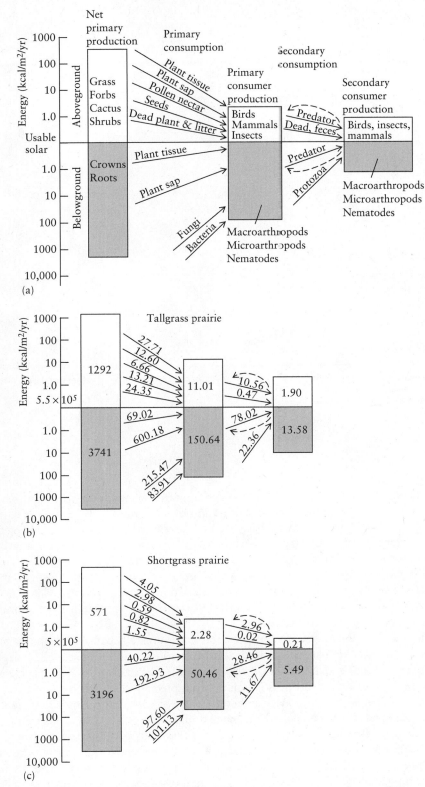

Figure 27.12 *Primary and secondary production and consumption for a tallgrass and a shortgrass prairie site in the United States. All values are expressed as energy. Much of the energy flow in these systems occurs underground. (After Scott et al. 1979.)*

Table 27.3   Plant Production Eaten and Wasted by Herbivores and Herbivore Production Eaten and Wasted by Carnivores for Four Grassland Sites in the Western United States

	Desert Grassland (%)	Shortgrass Prairie (%)	Mixed (%)	Tallgrass Prairie (%)
PLANT PRODUCTION				
Eaten by herbivores				
Aboveground	4.3	1.7	3.6	6.5
Belowground	—	7.3	26.4	17.9
Eaten and wasted				
Aboveground	6.3	3.4	8.4	10.2
Belowground	—	13.0	41.1	28.9
HERBIVORE PRODUCTION				
Eaten by carnivores				
Aboveground	110.9[a]	119.7[a]	51.1	85.4
Belowground	—	50.9	76.1	47.5
Eaten and wasted				
Aboveground	120.5[a]	124.5[a]	60.5	97.2
Belowground	—	61.0	91.3	57.0

[a] Since consumption cannot exceed 100%, these values are too high, possibly because consumption was slightly overestimated or production underestimated.

Source: Scott et al. (1979).

tionists during the last 25 years. The history of human settlement on most continents has been a repeating sequence of the elimination of wild animals by domestic cattle and sheep, and during the 1950s Africa seemed next on the list. Against this background, Fraser Darling in 1960 proposed that in many areas of Africa, game animals were more productive than domestic cattle and sheep, and hence a sustained yield of game would be more profitable than a sustained yield of cattle or sheep. Game-cropping or game-ranching schemes have been tried in many parts of Africa during the last 20 years, and these provide a practical example of the problem of what limits secondary production in ecosystems.

Game ranching has been attractive to conservationists because it seems to rest firmly on an interlocking set of theoretical postulates (Caughley 1976b). The argument can be summarized as follows. African wildlife has evolved within its ecosystems for millions of years, and thus is uniquely adapted to the African environment. The diversity of herbivores in Africa therefore use the vegetation more efficiently and are more productive than a cattle or sheep monoculture would be (Figure 27.14). Thus African wildlife should attain a higher biomass than cattle or sheep on native African ranges. Furthermore, natural selection must have ensured that the different game species partition their food resources such that competition is reduced. So the diverse complex of wild species should cause less overgrazing than cattle or sheep and also be more resistant to disease. In brief, African wildlife should provide a higher sustained yield and higher net revenue to the African people.

Caughley (1976b) makes two comments on this theory. First, it is theoretically

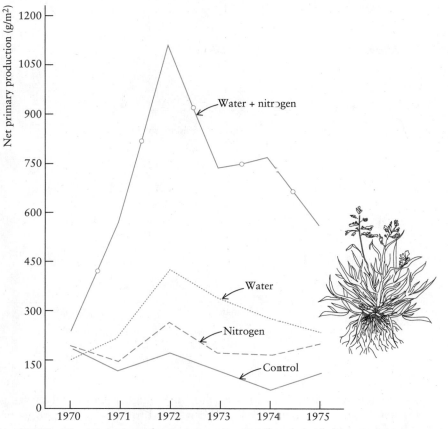

Figure 27.13 Net primary production over six years for plots of shortgrass prairie in north-central Colorado subjected to nitrogen fertilization, irrigation, and a combination of irrigation and fertilization. (After Dodd and Lauenroth 1979.)

sound and eminently reasonable. Second, attempts to demonstrate its validity since the 1960s have been unsuccessful. There does not appear to be a single case in which a sustained yield of game animals has been shown to be more valuable economically than a sustained yield of livestock in a comparable area. How can we resolve the paradox of Caughley's comments?

The initial enthusiasm for game ranching in Africa was derived from estimates of the carrying capacity of wild herbivores in African habitats (Petrides and Swank 1966). *Carrying capacity* is defined as the number or weight of animals of a single or mixed population that can be supported permanently on a given area (Sharkey 1970). Unfortunately, to an ecologist and an economist, carrying capacity means different things. *Ecological* carrying capacity is the maximum density of animals that can be sustained in the absence of harvesting without inducing trends in vegetation. *Economic* carrying capacity is the density of animals that will allow maximal sustained harvesting and is always lower than the ecological carrying capacity (Caughley 1976b). The confusion over these two types of carrying capacity has nullified some of the comparisons made between natural and domestic ecosystems in Africa.

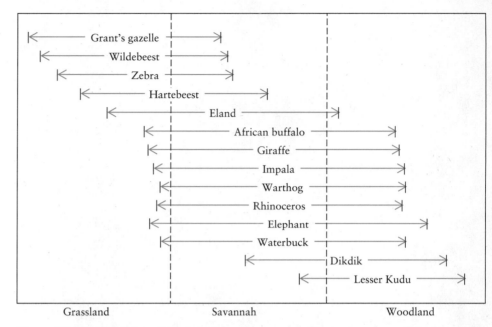

Figure 27.14 *Habitat partitioning and overlap among 14 species of ungulates in the Serengeti region, Tanzania. (From MacNab 1991.)*

Unfortunately, the early comparisons showing a higher ecological carrying capacity for wildlife than for livestock have been found to be in error, and, contrary to what Petrides and Swank (1966) report, the carrying capacity of domestic ecosystems is not necessarily different from that of wildlife ecosystems (Walker 1976). Table 27.4 lists the standing crop of some wildlife and pastoral areas in the African savanna. In all these ecosystems, the standing crop of large herbivores, secondary production, and primary production are related to rainfall. Coe and colleagues (1976) found that the standing crop of herbivores on pastoral areas is always above that predicted for wildlife areas. Figure 27.15 shows that the same occurs in South America—for a given plant production, about ten times more secondary production occurs in areas stocked with cattle (Oesterheld et al. 1992).

Attempts have been made to domesticate wild ungulates as a superior alternative to cattle. One example is the eland (*Taurotragus oryx*), a large antelope that is easy to tame and requires less water than cattle. The eland has a higher reproductive rate than domestic cattle and has excellent carcass characteristics for marketing. Domesticated herds of eland have been established in Kenya, South Africa, Zimbabwe, Zambia, and the former USSR. A series of studies on meat production and growth efficiency of the eland (Skinner 1972) show that the eland is less efficient than cattle at converting plants into meat. If the eland is to be used effectively in game ranching, it should be used in arid regions (King and Heath 1975). Wild herbivores in general are not as efficient as cattle for meat production (Walker 1976).

Game cropping is thus not an efficient way to provide cheap meat to low-income native peoples. From a conservation viewpoint, game cropping may have adverse impacts. Geist (1988) points out that the creation of a market for game-meat provides an outlet for illegal trade in these species. From an economic viewpoint game farming

Table 27.4   Standing Crop, Energy Expenditure, and Production of Large Herbivores from 30 Savanna Areas of East and Southern Africa

Locality	Large Herbivore Biomass (kg/km²)	Energy Expenditure (kJ/km²/hr)	Production (Live Weight) (kJ/km²/acre)	Annual Precipitation (mm)
WILDLIFE AREAS				
1. Rwindi plain, Albert National Park, Zaire	17,448	62,479	1,936	863
2. Ruwenzori National Park, Uganda	19,928	76,253	2,554	1,010
3. Bunyoro North, Uganda	13,261	43,979	1,145	1,150
4. Manyara National Park, Tanzania	19,189	65,905	2,405	915
5. Ngorongoro Crater, Tanzania	7,561	56,105	1,503	893
6. Lake Nakuru National Park, Kenya	6,688	35,498	1,409	878
7. Amboseli Game Reserve, Kenya	4,848	22,970	934	350
8. Lochinvar ranch, Zambia	7,568	40,591	1,983	813
9. Lake Rudolf (East) Kenya	405	2,436	87	165
10. Samburu-Isiolo, Kenya	2,018	9,745	402	375
11. Nairobi National Park, Kenya	4,824	24,728	1,008	844
12. Tsavo National Park (East, north of Voi R., Kenya)	4,033	12,968	351	553
13. Tsavo National Park (East, south of Voi R., Kenya)	4,388	15,239	438	553
14. Mkomasi Game Reserve, Tanzania	1,731	6,154	147	425
15. Loliondo Controlled Area, Tanzania	5,423	26,962	1,134	784
16. Serengeti National Park, Tanzania	8,352	43,063	1,743	803
17. Ruaha National Park, Tanzania	3,909	12,474	364	625
18. Akagera National Park, Rwanda	3,980	19,251	871	785
19. Sengwa Wildlife Research Area, Zimbabwe	4,315	18,299	722	597
20. Hendersons' ranch, Zimbabwe	2,869	15,148	684	406
21. Kruger National Park, northern section, South Africa	984	4,414	162	312
22. Kruger National Park, southern section, South Africa	3,783	20,552	884	650
23. William Pretorius Nature Reserve, South Africa	3,344	18,427	765	520
24. Mfolosi Game Reserve, South Africa	4,385	17,017	605	650
PASTORAL AREAS				
25. Mandera district, Kenya	1,901	—	—	228
26. Wajir district, Kenya	1,151	—	—	218
27. Garissa district, Kenya	3,818	—	—	398
28. Turkana district, Kenya	2,406	—	—	330
29. Samburu district, Kenya	6,514	—	—	500
30. Kaputei district, Kenya	7,884	—	—	710

Source: After Coe et al. (1976).

in fenced areas can provide luxury products (meat) and services (tourism) to foreigners at a substantial economic advantage. But private game farms typically do not promote conservation of the entire community. Large predators are not welcome on game farms and grassland species are often preferred over forest species. The net result is that private game farms conserve only a few species in the ecosystem (MacNab 1991).

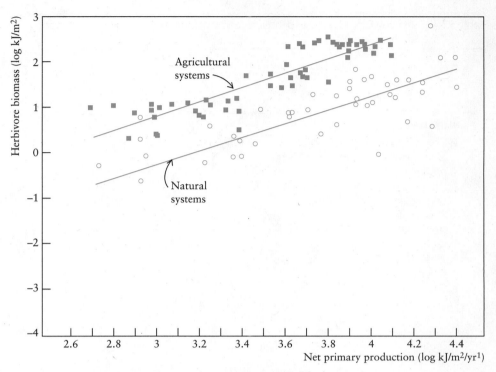

Figure 27.15   Relationship between net aboveground primary productivity and herbivore biomass for 51 natural ecosystems and 67 agricultural ecosystems for southern South America. (From Oesterheld et al. 1992.)

SUSTAINABLE ENERGY BUDGETS

The energy budget of herbivores and carnivores combine to determine the energy budget of the community. Taxonomic approaches to energy transfers have focused on the constraints that determine the maximum energy budgets of animals (Weiner 1992). Food limitations could operate in two quite different ways for animals. The absolute amount of high-quality food may be limited, and the limitations of food could operate on the time animals have to search and their searching strategy. Alternatively, food may be limiting because of physiological limits on the rate of food conversion into usable energy (Karasov 1986).

The upper limit to sustainable energy budgets is set by the digestive tract's capacity to assimilate nutrients and energy from food. In vertebrates the gut consists of several compartments where food can be stored, digested, and absorbed. In ruminants the forestomach is a fermentation chamber where a microbial flora breaks down the cellulose in plant cell walls.

The estimation of maximal rates of energy expenditure in animals is difficult. An initial estimate by Drent and Daan (1980) was that for birds a maximal energy expenditure was approximately four times the basal metabolic rate. Kirkwood (1983) estimated maximal energy expenditures of about five times basal metabolic rate. Weiner (1992) estimates maximal sustained energy budgets of seven times basal metabolism. Figure 27.16 illustrates some field energy budgets for birds and mammals measured by the doubly-labeled water technique, in relation to estimated re-

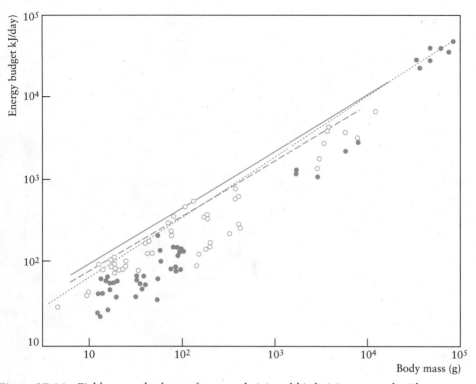

Figure 27.16 Field energy budgets of mammals (●) and birds (○) compared with three regression estimates of the maximum possible energy budgets for homeotherms. Some animals appear to be operating at maximum intensity with no safety margin. (From Weiner 1992.)

gressions for the maximum sustainable energy budget. Lactating mammals and cold-exposed birds have provided the type of data used to estimate these lines (Weiner 1992, Kirkwood 1983). It seems likely that maximum energy budgets will be found to vary seasonally and perhaps with age, so that Figure 27.16 should be viewed as a first approximation only.

The integration of studies on energy use by animals and the physiological constraints that operate on energy acquisition will help community ecologists to understand the mechanisms behind the efficiency of secondary productivity in different ecosystems (Karasov 1986).

SUMMARY

The capital or organic matter produced by green plants is used by a food web of herbivores and carnivores. A great deal of the matter and energy that animals eat is lost in feces and urine, or used for maintenance metabolism. In warm-blooded vertebrates, often 98 percent of the energy taken in is used for maintenance. Invertebrates and fish use less energy for maintenance, typically 60 to 90 percent of the energy taken in.

Data on the efficiency of secondary production of individual species is relatively easy to obtain, but the aggregation of this information into trophic levels is problematic. Many species of animals do not feed on only one trophic level. Typical "herbivores" may obtain part of their energy from other animals, and typical "carnivores" may feed on plant, herbivores, and other carnivores. The trophic level concept cannot be mapped directly on to species.

Considerable energy is lost at each step of the food chain, and thus for a given biomass of green plants, only a much smaller biomass of animals can be supported. Many of the animal species in the community that we as humans think are important are a small component of the energy flow. Herbivores consume a higher fraction of the primary production in aquatic ecosystems than they do in forest or grassland ecosystems. Much of the energy flow in terrestrial systems goes directly from plants to the decomposer food chain.

Secondary production may be limited by a variety of interacting factors. Secondary production in grasslands is limited by water and nitrogen, and a large fraction of the secondary production occurs underground. Levels of secondary production in grazing agricultural ecosystems are up to ten times higher than those achieved in natural grazing ecosystems. Until we understand the factors that limit primary and secondary production, we cannot predict the impact of a change in the environment on the community. At present there are few communities for which this understanding has been worked out.

Selected References

BURNS, T. P. 1989. Lindeman's contradiction and the trophic structure of ecosystems. *Ecology* 70:1355–1362.

COUSINS, S. 1987. The decline of the tropic level concept. *Trends in Ecology and Evolution* 2:312–316.

FRENCH, N. R. 1979. *Perspectives in Grassland Ecology*. Springer-Verlag, New York.

HUMPHREYS, W. F. 1979. Production and respiration in animal populations. *Journal of Animal Ecology* 48:427–453.

KOKKINN, M. J., and A. R. DAVIS. 1986. Secondary production: Shooting a halcyon for its feathers. In *Limnology in Australia*, ed. P. De Deckker and W. D. Williams, pp. 251–261. Dr. W. Junk Publishers, Dordrecht, Netherlands.

MAC NAB, J. 1991. Does game cropping serve conservation? A reexamination of the African data. *Canadian Journal of Zoology* 69:2283–2290.

MC NAUGHTON, S. J., M. OESTERHELD, D. A. FRANK, and K. J. WILLIAMS. 1989. Ecosystem-level patterns of primary productivity and herbivory in terrestrial habitats. *Nature* 341:142–144

NAGY, K. A. 1987. Field metabolic rate and food requirement scaling in mammals and birds. *Ecological Monographs* 57:111–128.

PETERS, R. H. 1977. Unpredictable problems with tropho-dynamics. *Environmental Biology of Fishes* 2:97–101.

POWER, M. E. 1992. Top-down and bottom-up forces in food webs: Do plants have primacy? *Ecology* 73:733–746.

STRAYER, D. 1988. On the limits to secondary production. *Limnology and Oceanography* 33:1217–1220.

WEINER, J. 1992. Physiological limits to sustainable energy budgets in birds and mammals: Ecological implications. *Trends in Ecology and Evolution* 7:384–388.

Questions and Problems

1. In assessing the ways in which ecologists have studied ecosystems, O'Neill and colleagues (1986, p. 67) state:

    It is now well recognized that the trophic-level concept is most useful as a heuristic device and tends to obscure, rather than illuminate, organizational principles of ecosystems.

    Read O'Neill and colleagues (1986) and discuss this claim.

2. Does the concept of a maximum sustainable energy budget as illustrated in Figure 27.16 imply an evolutionary assumption that all animals will be selected to operate at maximum energy budgets? Discuss the evolution of assimilation efficiency for a particular species of your choice.

3. In discussing the reality of trophic levels, Murdoch (1966a, p. 219) states:

    Unlike populations, trophic levels are ill-defined and have no distinguishable lateral limits; in addition, tens of thousands of insect species, for example, live in more than one trophic level either simultaneously or at different stages of their life histories. Thus trophic levels exist only as abstractions, and unlike populations they have no empirically measurable properties or parameters.

    Discuss.

4. Compile a list of the efficiency of some of our common physical machines, such as automobiles, electric lights, electric heaters, and bicycles.

5. How would it be possible to have an inverted Eltonian pyramid of numbers in which, for example, the standing crop of herbivores is larger than the standing crop of plants? In what types of communities could this occur? Survey the ecological literature on those communities, and find out if inverted pyramids do, in fact, occur.

6. List the ecological factors that might account for the scatter around the regression lines in Figure 27.8.

7. Why should agricultural grazing systems be more efficient than natural grazing systems at utilizing primary production? Read MacNab (1991) and Oesterheld and coauthors (1992) and discuss the problems of measuring efficiency in both kinds of ecosystems.

8. The basis for estimating secondary production is the estimation of population size, biomass, and growth, and the accuracy of any estimate of production depends on the accuracy of these three measurements. Read Morgan (1980), and discuss the relative difficulty of measuring these three variables in freshwater ecosystems for zooplankton, benthic invertebrates, and fish.

9. How would you expect trophic level biomass to change as the primary productivity of the community increases? Review the hypotheses of community organization discussed in Chapter 25, and discuss the assumptions underlying your predictions. Compare your expectations with those of Power (1992).

Overview Question

Debate the following resolution: Resolved that ecological efficiencies between trophic levels cannot in principle be measured and that ecological efficiencies can be measured only for populations.

28

Community Metabolism III: Nutrient Cycles

Living organisms are constructed from chemical elements, and one way to describe an ecosystem is to follow the transfer of elements between the living and the nonliving worlds. Interest in the nutrient content of plants and animals has been an important focus in agriculture for over 100 years. Nutrients often set some limitation on the primary or secondary productivity of a population or a community, and nutrient additions as fertilizer have become increasingly common in agriculture and forestry. In this chapter, we will describe how nutrients cycle and recycle in natural systems, linking together the living and the dead material in the ecosystem.

NUTRIENT POOLS AND EXCHANGES

Nutrients can be used as an organizing focus in ecosystem studies. We can view the biological community as a complex processor in which individuals move nutrients from one site to another within the ecosystem. These biological exchanges of nutrients interact with physical exchanges, and for this reason, nutrient cycles are also called *biogeochemical cycles*. Chemical elements that cycle through living organisms are called *bioelements*. Figure 28.1 illustrates the general pattern of bioelement cycles on a global scale. Nutrient cycles are closed on a global scale. The individual atoms that make up the cycle are indestructible and can be recycled in plants and animals. Ecologists are interested in understanding and measuring global nutrient cycles because human activities are altering these cycles with possible impacts on global climate. An analysis of nutrient cycling thus ends with an assessment of climate change and its consequences for animals and plants.

Global nutrient cycles represent the summation of local events occurring in different biotic communities, and to make progress in understanding global nutrient cycles we must begin at the level of the local community. A simple example of a nutrient cycle in a lake is shown in Figure 28.2. All nutrients reside in *compartments*, which represent a defined space in nature. Compartments can be defined very broadly or very specifically. Figure 28.2 includes all of the plants in the ecosystem as one compartment, but we could recognize each species of plant as a separate compartment or even the leaves and the stem of a single plant as separate compartments. A compartment contains a certain quantity, or *pool*, of nutrients. In the simple lake ecosystem shown in Figure 28.2, the phosphorus dissolved in the water is one pool, and the phosphorus contained in the bodies of herbivores is another pool.

Compartments exchange nutrients, and thus we must measure the uptake and outflow of nutrients for each compartment. The rate of movement of nutrients between two compartments is called the *flux rate* and is measured as the quantity of nutrient passing from one pool to another per unit of time. The flux rates and pool

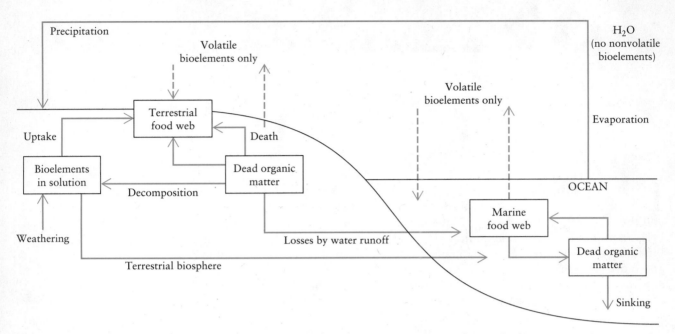

Figure 28.1   General schematic of bioelement cycling on a global scale. Movement of nonvolatile elements, such as phosphorus, is largely one way, toward ocean sediments. (From DeAngelis 1992.)

sizes together define the nutrient cycle within any particular ecosystem. Ecosystems are not isolated from one another, and nutrients come into an ecosystem through meteorological, geological, or biological transport mechanisms and leave an ecosystem via the same routes. Meteorological inputs include dissolved matter in rain and snow, atmospheric gases, and dust blown by the wind; geological inputs include elements transported by surface and subsurface drainage; and biological inputs include movements of animals between ecosystems.

Nutrient cycles can be studied by the introduction of radioactive tracers into laboratory or natural ecosystems. Studies on the movement of radioactive phosphorus in small aquariums illustrate this approach (Whittaker 1975). Figure 28.3 shows the changes that followed the introduction of 100 microcuries (μc) of ^{32}P-labeled phosphoric acid in a 200-liter aquarium. There is a very rapid initial uptake of ^{32}P by the phytoplankton. One-half of the ^{32}P had been taken up by the phytoplankton within 2 hours, and within 12 hours there was an equilibrium of uptake and excretion of ^{32}P between the phytoplankton and the water. Filamentous algae attached to the sides and bottom of the aquarium slowly picked up ^{32}P, and crustaceans grazing on phytoplankton began to accumulate ^{32}P even more slowly. As the experiment progressed, an increasing fraction of the radioactive tracer began to accumulate in the bottom mud and was tied up in less-active or bound form in the sediments. Some ^{32}P does move from the sediment back into the water column, but more moves down into the sediment and thus accumulates.

The nutrient cycle of phosphorus in aquariums is broadly similar to that in natural lakes. Phosphorus and other nutrients tend to accumulate in the sediment of lakes so that continual nutrient inputs are required to maintain high productivity. The pattern of movement of phosphorus shown in Figure 28.3 helps to explain the

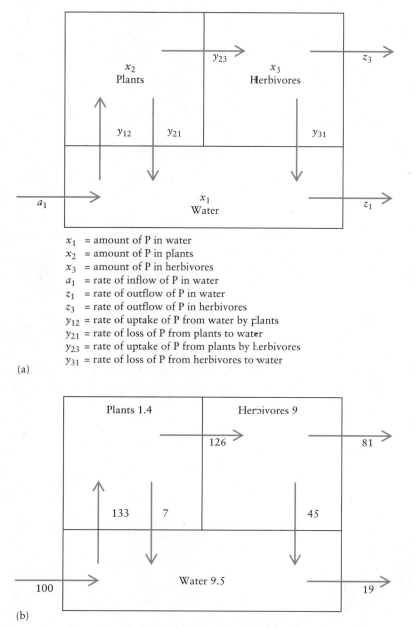

x_1 = amount of P in water
x_2 = amount of P in plants
x_3 = amount of P in herbivores
a_1 = rate of inflow of P in water
z_1 = rate of outflow of P in water
z_3 = rate of outflow of P in herbivores
y_{12} = rate of uptake of P from water by plants
y_{21} = rate of loss of P from plants to water
y_{23} = rate of uptake of P from plants by herbivores
y_{31} = rate of loss of P from herbivores to water

(a)

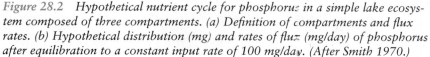

(b)

Figure 28.2   Hypothetical nutrient cycle for phosphorus in a simple lake ecosystem composed of three compartments. (a) Definition of compartments and flux rates. (b) Hypothetical distribution (mg) and rates of flux (mg/day) of phosphorus after equilibration to a constant input rate of 100 mg/day. (After Smith 1970.)

results of the lake-fertilization experiments described on page 617. A continued input of phosphate is needed to sustain high availability for phytoplankton. These results are also critical for understanding how lakes can recover from the effects of nutrient additions from pollution (Figure 26.14, page 619). Thus the sediments of lakes become nutrient-rich deposits.

Nutrient cycles may be subdivided into two broad types. The phosphorus cycle

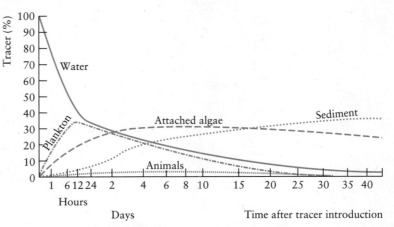

Figure 28.3   *Movement of radiophosphorus in an aquarium microcosm. Percentage of the tracer present at a given time (after correction for radioactive decay) is on the vertical axis, time after tracer introduction (on a square-root scale) is on the horizontal. (After Whittaker 1961.)*

we have just described is an example of a sedimentary or *local* cycle, which operates within an ecosystem. Local cycles involve the less mobile elements that have no mechanism for long-distance transfer. By contrast, the gaseous cycles of nitrogen, carbon, oxygen, and water are called *global* cycles because they involve exchanges between the atmosphere and the ecosystem. Global nutrient cycles link together all

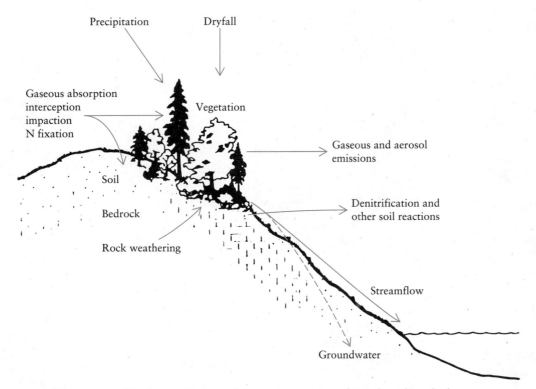

Figure 28.4   *A model of the pathways of nutrient movement through undisturbed forest ecosystems. (From Waring and Schlesinger 1985.)*

of the world's living organisms in one giant ecosystem called the *biosphere*, the whole earth ecosystem.

NUTRIENT CYCLES IN FORESTS

The harvesting of forest trees removes nutrients from a forest site, and this continued nutrient removal could result in a long-term decline in forest productivity unless nutrients are somehow returned to the system. Because of the economic question of forest productivity, an increasing amount of work is being directed toward the analysis of nutrient cycles in forests. Figure 28.4 shows the factors that must be quantified in order to describe the nutrient cycle in a forest. Some examples of nutrient cycles in forest stands will illustrate these concepts.

During forest development, nutrients accumulate in leaves and wood. Figure 28.5 illustrates the rapid accumulation of five different nutrients in a stand of jack pine in eastern Canada. As trees increase in size during succession, the soil accumulates nutrients in the surface litter and in the soil organic matter (humus) dispersed through the lower soil horizons. As forest stands age, there is a systematic change in the uptake of nutrients. Figure 28.6 illustrates changes in nitrogen cycling in a spruce forest in Russia. In this forest the spruce canopy becomes more open after 70 years and understory vegetation increases in volume and importance in nutrient cycling. Not all forest successions will produce the same pattern of changes in nutrient cycling, but the general principle will be valid that nutrient cycling varies with forest

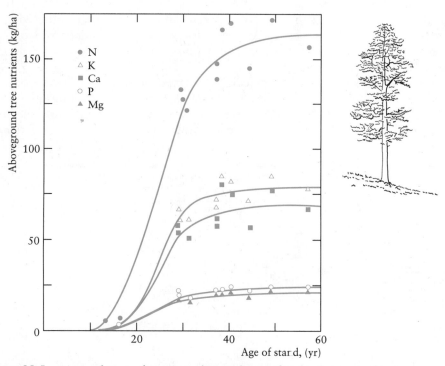

Figure 28.5 Accumulation of nutrients during the postfire development of jack pine (Pinus banksiana) *stands in New Brunswick, Canada. (From MacLean and Wein 1977.)*

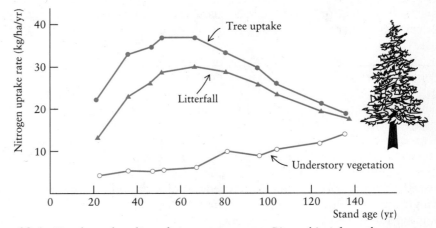

*Figure 28.6   Uptake and cycling of nitrogen in spruce (*Picea abies*) from the age of 22 to 138 years. (After Kazimirov and Morozova 1973.)*

age. In old growth forests there is little accumulation of nutrients and ecosystem inputs and outputs should be balanced.

The cycling of different nutrients in forest ecosystems is highly variable (Cole and Rapp 1981). Nutrients that are in short supply are recycled far more efficiently than those present in excess of requirements. In most forest sites nitrogen is a major limiting factor and is present at deficiency levels. Table 28.1 gives the organic matter and nitrogen amounts present in the aboveground component of 32 forests from different climatic zones. In the boreal forests of Alaska only about 20 percent of the organic matter is present in the trees aboveground. Low decomposition rates in these cold Alaskan forests causes most of the nitrogen and organic matter to be tied up in the soil. Coniferous forests have the largest forest floor accumulation of organic matter of all forests, on average about four times the biomass of tropical forests (Waring and Schlesinger 1985).

Nutrient cycles operate more quickly in warmer forests than in colder ones. If we assume as an approximation that forest soil nutrients are in equilibrium for the short time they can be studied (three to ten years), we can calculate average turnover time for each nutrient. Turnover time represents the time an average atom will remain

Table 28.1   Aboveground Tree Accumulation of Organic Matter and Nitrogen for the Various Forest Regions

Forest Region	No. of Sites	ORGANIC MATTER (kg/ha)			NITROGEN (kg/ha)		
		In Trees	Total	% Aboveground	In Trees	Total	% Aboveground
Boreal coniferous	3	51,000	226,000	19	116	3,250	4
Boreal deciduous	1	97,000	491,000	20	221	3,780	6
Temperate coniferous	13	307,000	618,000	54	479	7,300	7
Temperate deciduous	14	152,000	389,000	40	442	5,619	8
Mediterranean	1	269,000	326,000	83	745	1,025	73
Average		208,000	468,000	45	429	5,893	7

Source: From Cole and Rapp (1981).

in the soil before it is recycled into the trees or shrubs. Table 28.2 gives the mean turnover times for five elements. All northern forests have very slow turnover of nutrients. On average, a boreal conifer forest retains nitrogen 100 times longer than a Mediterranean evergreen oak forest (Cole and Rapp 1981). Deciduous forests turn over nutrients more rapidly than coniferous forests. Coniferous forests use nutrients more efficiently because they retain their needles and do not need to replace all their foliage each year.

Nutrients are lost from forest ecosystems in several ways. Streams transport both dissolved and particulate matter, and measurements of streamwater chemistry can provide a good way to monitor overall forest function. Anaerobic soil bacteria produce methane and hydrogen sulfide gases. Plants release hydrocarbons such as terpenes from their leaves, and these compounds may add to atmospheric haze in summer. Both ammonia and hydrogen sulfide can be released from plant leaves (Waring and Schlesinger 1985). During forest fires nutrients are released in both gases and in particles. Finally, forest harvesting removes nutrients in wood from the ecosystem.

One of the most extensive studies of nutrient cycling in forests has been carried out at the Hubbard Brook Experimental Forest in New Hampshire. The Hubbard Brook forest is a nearly mature, second-growth hardwood ecosystem. The area is underlain by rocks that are relatively impermeable to water, so all runoff occurs in small streams. The area is subdivided into several small watersheds that are distinct yet support similar forest communities, and these watersheds are good experimental units for study and manipulation.

Nutrients enter the Hubbard Brook forest ecosystem in precipitation, and the precipitation input was measured in rain gauges scattered over the study area. Nutrients leave the ecosystem primarily in stream runoff, and this loss was estimated by measuring stream flows. Table 28.3 gives the concentrations of dissolved substances in precipitation and streamwater for the Hubbard Brook watersheds. For most dissolved nutrients, the streamwater leaving the system contains more nutrients than the rainwater entering the system. About 60 percent of the water that enters as precipitation turns up as stream flow; most of the remaining 40 percent is transpired by plants or evaporated. The chemical composition of the precipitation and the stream discharges changed very little from year to year.

Table 28.4 gives the annual nutrient budgets for watersheds in the Hubbard Brook system, based on the difference between precipitation input and stream out-

Table 28.2   Mean Turnover Time in Years for the Forest Floor and Its Mineral Elements by Forest Regions. A Steady State Condition Is Assumed

| Forest Region | No. of Sites | MEAN TURNOVER TIME (YR) | | | | | |
		Organic Matter	N	K	Ca	Mg	P
Boreal coniferous	3	353	230.0	94.0	149.0	455.0	324.0
Boreal deciduous	1	26	27.1	10.0	13.8	14.2	15.2
Temperate coniferous	13	17	17.9	2.2	5.9	12.9	15.3
Temperate deciduous	14	4	5.5	1.3	3.0	3.4	5.8
Mediterranean	1	3	3.6	0.2	3.8	2.2	0.9
All stands	32	12	34.1	13.0	21.8	61.4	46.0

Source: From Cole and Rapp (1981).

Table 28.3   Average Concentrations of Various Dissolved Substances in Bulk Precipitation and Streamwater for Undisturbed Watersheds 1–6, Hubbard Brook Experimental Forest, 1963–1969

	Precipitation (mg/L)	Streamwater (mg/L)
Calcium	0.21	1.58
Magnesium	0.06	0.39
Potassium	0.09	0.23
Sodium	0.12	0.92
Aluminum	([a])	0.24
Ammonium	0.22	0.05
Sulfate	3.10	6.40
Nitrate	1.31	1.14
Chloride	0.42	0.64
Bicarbonate	([a])	1.90
Dissolved silica	([a])	4.61[b]

[a] Not determined, but very low.
[b] Watershed 4 only.
   Source: After Likens and Bormann (1972).

flow. Eight ions show net losses from the ecosystem: calcium, magnesium, potassium, sodium, aluminum, sulfate, silica, and bicarbonate. Three ions showed an average net gain: nitrate, ammonium, and chloride. If we assume that these nutrient budgets should be in equilibrium in this ecosystem, the net losses must be made up by chemical decomposition of the bedrock and soil.

With this background, Bormann and co-workers (1974) studied the effect of logging on the nutrient budget of a small watershed at Hubbard Brook. One 15.6-hectare watershed was logged in 1966, and the logs and branches were left on the ground so that nothing was removed from the area. Great care was taken to prevent disturbance of the soil surface to minimize erosion. For the first three years after logging the area was treated with a herbicide to prevent any regrowth of vegetation. This deforested watershed was then compared with an adjacent intact watershed.

Runoff in the small streams increased immediately after the logging, and annual runoff in the deforested watershed was, respectively, 41, 28, and 26 percent above the control in the three years after treatment. Detritus and debris in the stream outflow increased greatly after deforestation, particularly two to three years after logging. Correlated with this increase was a large increase in streamwater concentrations of all major ions in the deforested watershed. Nitrate concentrations in particular increased 40- to 60-fold over the control values (Figure 28.7). For two years the nitrate concentration in the streamwater of the deforested site exceeded the health levels recommended for drinking water. Average streamwater concentrations increased 417 percent for calcium, 408 percent for magnesium, 1558 percent for potassium, and 177 percent for sodium in the two years after deforestation (Likens et al. 1970).

The net result of deforestation in the Hubbard Brook forest is that the ecosystem is simultaneously irrigated and fertilized, so that for a short time after normal logging occurs, primary production could be stimulated. An array of species has evolved to exploit these transient nutrient-rich situations following a disturbance by fire or logging. These transients help to prevent further nutrient losses and to restore some of the nutrient capital lost by logging or fire.

Table 28.4   Annual Nutrient Budgets (kg/ha) for Watersheds of the Hubbard Brook Experimental Forest, New Hampshire

	1963–1964	1964–1965	1965–1966	1966–1967	1967–1968	1968–1969	Mean 1963–1969
Calcium							
Input	3.0	2.8	2.7 ± 0.07	2.7 ± 0.03	2.8 ± 0.05	1.6 ± 0.02	2.6
Output	12.8 ± 0.8	6.3 ± 0.4	11.5 ± 0.6	12.3 ± 0.7	14.2 ± 0.7	13.8 ± 1	11.8
Net	−9.8	−3.5	−8.8	−9.6	−11.4	−12.2	−9.2
Magnesium							
Input	0.7	1.1	0.7 ± 0.02	0.5 − 0.005	0.7 ± 0.013	0.3 ± 0.004	0.7
Output	2.5 ± 0.06	1.8 ± 0.09	2.9 ± 0.07	3.1 ± 0.08	3.7 ± 0.08	3.3 − 0.12	2.9
Net	−1.8	−0.7	−2.2	−2.6	−3.0	−3.0	−2.2
Potassium							
Input	2.5	1.8	0.6 ± 0.02	0.6 ± 0.008	0.7 ± 0.01	0.6 ± 0.009	1.1
Output	1.8 ± 0.1	1.1 ± 0.08	1.4 ± 0.09	1.7 ± 0.1	2.2 ± 0.15	2.2 ± 0.16	1.7
Net	+0.7	+0.7	−0.8	−1.1	−1.5	−1.6	−0.6
Sodium							
Input	1.0	2.1	2.0 ± 0.04	1.3 ± 0.01	1.7 ± 0.03	1.1 ± 0.02	1.5
Output	5.9 ± 0.3	4.5 ± 0.3	6.9 ± 0.5	7.3 ± 0.6	9.1 ± 0.6	7.6 ± 0.6	6.9
Net	−4.9	−2.4	−4.9	−6.0	−7.4	−6.5	−5.4
Aluminum							
Input	—	[a]	[a]	[a]	[a]	[a]	[a]
Output	—	1.2 ± 0.18	1.7 ± 0.75	1.9 ± 0.87	2.1 ± 1.00	2.2 ± 1.04	1.8
Net	—	—	−1.7	−1.9	−2.1	−2.2	−1.8
Ammonium							
Input	—	2.1	2.6 ± 0.06	2.4 ± 0.04	3.2 ± 0.06	3.1 ± 0.07	2.7
Output	—	0.27 ± 0.03	0.92 ± 0.03	0.45 ± 0.07	0.24 ± 0.02	0.16 ± 0.06	0.4
Net	—	+1.83	+1.7	+2.0	+3.0	+2.9	+2.3
Nitrate							
Input	—	6.7	17.4 ± 0.3	19.9 ±0.2	22.3 ± 0.3	15.3 ± 0.3	16.3
Output	—	5.6 ± 0.4	6.5 + 0.02	6.6 ± 0.4	12.7 ± 0.3	12.2 ± 0.4	8.7
Net	—	+1.1	+10.9	+13.3	+9.6	+3.1	+7.6
Sulfate							
Input	—	30.0	41.6 ± 0.3	42.0 ± 0.3	46.7 ± 0.3	31.2 ± 0.3	38.3
Output	—	30.8 ± 0.4	47.8 ± 0.4	52.5 ± 0.4	58.5 ± 0.4	53.3 ± 0.3	48.6
Net	—	−0.8	−6.2	−10.5	−11.8	−22.1	−10.3
Dissolved silica							
Input	—	[a]	[a]	[a]	[a]	[a]	[a]
Output	—	20.8 ± 0.1	36.1 ± 4.8	41.6 ± 4.8	42.1 ± 5.7	35.0 ± 6.0	35.1
Net	—	—	−36	−42	−42	−35	−35
Bicarbonate							
Input	—	—	[a]	[a]	[a]	[a]	[a]
Output	—	—	12.0	16.8	16.5[a]	13.1[a]	14.6
Net	—	—	−12	−17	−17	−13	−14.6
Chloride							
Input	—	—	2.6 ± 0.04	6.7 ± 0.12	5.0 ± 0.09	6.4 ± 0.05	5.2
Output	—	—	4.3 ± 0.05	4.8 ± 0.1	5.3 ± 0.01	5.2 ± 0.08	4.9
Net	—	—	−1.7	+1.9	−0.3	±1.2	+0.3

Note. Error limits are 1 standard deviation of the mean.

[a] Not measured, but very small.

Source: After Likens et al. (1971).

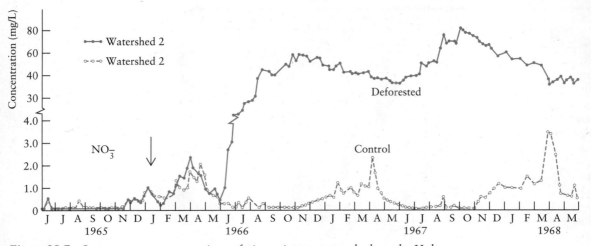

Figure 28.7 *Streamwater concentrations of nitrate in two watersheds at the Hub-bard Brook Experimental Forest, New Hampshire. The arrow marks the comple-tion of the cutting of the trees on watershed 2. Note the change in scale for the nitrate concentration. (After Likens et al. 1970.)*

The Hubbard Brook experiment has been repeated in four other forest types from Costa Rica to British Columbia (Table 28.5). Losses of nutrients in streamwater were high for a tropical rain forest in Costa Rica but were not large at three temperate forest sites. The total nutrient losses at these three sites were mostly due to the harvest of wood. Precipitation input of nutrients must be the main source of nutrient input to these forested systems, along with rock weathering which operates very slowly. Table 28.5 gives the replacement times required for each nutrient. If harvesting occurs more frequently than nutrient replacement time, the site must run down unless fertilizers are used.

The work on nutrient cycling in forests has shown the need for guidelines to specify sound management procedure in forestry. For example, bark is relatively rich in nutrients, and hence lumbering operations ought to be designed to strip the bark from the trees at the field site and not at some distant processing plant. The conser-vation of nutrients in forest ecosystems can be done intelligently only when we understand how nutrient cycles operate in these systems.

EFFICIENCY OF NUTRIENT USE

Large areas of the Northern Hemisphere have been glaciated, and the soils, derived from till in which the bedrock has been pulverized, are very fertile, with a high availability of nutrients. Areas of volcanic activity can also have rich soils. But in much of the world, soils are very old, highly weathered, and basically infertile. For example, the continents derived from Gondwanaland—Australia, South America, and India—have large areas covered with very old, poor soils. The vegetation sup-ported on these soils has adapted remarkably well to efficient nutrient use by recy-cling within the plant and by leaf fall and reabsorption.

Australian soils are typical of very old, highly weathered soils and contain almost no phosphorus. Eucalyptus trees are adapted to grow on soils of low phosphorus content (Attiwill 1981). Table 28.6 gives the biomass and nutrient content of forest

Table 28.5 Comparative Losses of Nutrients As a Result of Timber Harvest In Different Ecosystems

Local Process	N	P	K	Ca	References
BRITISH COLUMBIA (western hemlock-Douglas fir)					
Losses (kg/ha)					
Harvested products	234	34	168	260	
Streamflow (2 yr)[a]	10	0	8	0	
Total	244	34	172	260	Feller and Kimmins (1984)
Precipitation input (kg/ha/yr)	4	0	1	7	
Replacement time (yr)[b]	61		176	37	
MINNESOTA (aspen)					
Losses (kg/ha)					
Harvested products	454	43	355	1034	
Streamflow (5 yr)[a]	0	0	0	62	
Total	454	43	355	1096	Silkworth and Grigal (1982)
Precipitation input (kg/ha/yr)	7	3	10	5	
Replacement time (yr)[b]	65	14	36	219	
FLORIDA (slash pine)					
Losses (kg/ha)					
Harvested products	179	15	73	128	
Streamflow[a]	4	0	1	0	Pritchett and Comerford (1983), Riekirk
Total	183	15	74	128	(1983)
Precipitation input (kg/ha/yr)	5	0	3	5	
Replacement time (yr)[b]	37	75	25	26	
COSTA RICA (tropical rain forest)					
Losses (kg/ha)					
Harvested products[c]	111	4	0	96	
Streamflow (11 mo.)[a]	329	11	0	392	
Total	440	15	0	488	Ewel et al. (1981), Lewis (1981)
Precipitation input (kg/ha/yr)	7	2	4	9	
Replacement time (yr)	63	8		54	

[a] Streamflow losses in excess of losses on control watersheds during the interval indicated.
[b] Replacement of losses assuming atmospheric inputs only.
[c] Firewood harvest only.

ecosystems from temperate areas growing on good soils and from Australian sites on poor soils. There is no suggestion from Table 28.6 that eucalyptus growing on poor soils have lower amounts of nutrients than trees from other sites, with the single exception of phosphorus. Eucalyptus have only one-half to one-fifth the amount of phosphorus in their tissues as do trees from Northern Hemisphere forests.

Plants growing in nutrient-poor soils, one would think, ought to contain fewer nutrients than plants in fertile soils. In fact, the opposite is true (Chapin 1980). Plants from infertile habitats consistently have higher nutrient concentrations than plants from fertile habitats when grown under the same controlled conditions. Plants from nutrient-poor habitats may achieve this nutrient-rich status by being more efficient than plants from nutrient-rich habitats (Vitousek 1982). An abundance of data is available from forests to investigate nutrient use efficiency. For a forest we can define:

$$\text{Nutrient use efficiency} = \frac{\text{grams of organic matter lost from plants}}{\text{grams of nutrient lost}}$$

Table 28.6 Aboveground Living Biomass and Its Nutrient Content of Temperate Forest and Australian Eucalyptus Forest Ecosystems

Forest	Location	Age (yr)	Biomass (t/ha)	N	P	K	Na	Mg	Ca
CONIFEROUS									
1. *Abies amabilis—* *Tsuga mertensiana*	Washington	175	469	372	67	980	—	160	1046
2. *Abies balsamea—* *Picea rubens*	Quebec	various	132	387	52	159	—	36	413
3. *Abies mayriana*	Japan	?	129	527	63	278	—	415	515
4. *Cryptomeria japonica*	Japan	?	114	386	36	244	—	118	314
5. *Larix leptolepis*	Japan	?	100	230	28	212	—	40	107
6. *Picea abies*	England	47	140	331	37	161	5	39	212
7. *Picea abies*	Sweden	55	308	770	87	437	38	69	459
8. *Picea glauca*	Minnesota	40	151	383	58	230	—	41	720
9. *Picea mariana*	Quebec	65	107	167	42	84	—	27	276
10. *Pinus banksiana*	Ontario	30	81	171	15	85	—	19	114
11. *Pinus banksiana*	Minnesota	40	150	276	27	106	—	40	226
12. *Pinus muricata*	California	88 (mean)	409	510	418	321	83	140	589
13. *Pinus radiata*	New Zealand	35	305	319	40	324	—	—	187
14. *Pinus resinosa*	Minnesota	40	204	373	46	205	—	62	335
15. *Pinus sylvestris*	Southern Finland	28	21	82	9	40	—	—	50
		45	79	194	19	98	—	—	115
		47	45	136	14	58	—	—	63
16. *Pinus sylvestris*	England	47	164	364	34	314	8	43	204
17. *Pseudotsuga menziesii*	Washington	36	173	326	67	227	—	—	342
18. *Pseudotsuga menziesii*	Oregon	450	540	371	49	265	—	—	664
19. *Sequoia sempervirens*	California (slopes)	?	1155	1474	616	941	102	232	1068
20. *Sequoia sempervirens*	California (flats)	?	3190	3846	1695	2412	271	569	2574
HARDWOOD									
1. *Alnus rubra*	Washington	34	185	516	40	224	—	114	334
2. *Betula verrucosa*	England	22	63	264	33	76	3	36	327
3. *Betula verrucosa—* *Betula pubescens*	Southern Finland	40	91	232	24	136	—	—	185
4. *Betula platyphylla*	Japan	?	114	265	20	84	—	95	401
5. Evergreen broadleaf forest	Japan	?	114	329	52	249	—	94	259
6. *Fagus sylvatica*	England	39	134	285	38	187	4	42	151
7. *Fagus sylvatica*	Sweden	a. 90	314	800	53	458	24	118	980
		b. 90	324	1050	84	452	32	105	602
		c.100	226	640	65	318	17	85	478
8. *Populus tremuloides*	Minnesota	40	170	383	48	297	—	62	881
9. *Quercus alba*	Missouri	35–92	100	204	20	115	—	35	601
10. *Quercus alba—* *Quercus rubra—* *Acer saccharum—* *Carya glabra*	Illinois	150	190	478	29	310	—	105	1603
11. *Quercus robur*	England	47	130	393	35	246	5	45	257
12. *Quercus robur—* *Carpinus betulus—* *Fagus sylvatica*	Belgium	30–75	121	406	32	245	—	81	868

Table 28.6   (*continued*)

Forest	Location	Age (yr)	Biomass (t/ha)	NUTRIENT CONTENT (kg/ha)					
				N	P	K	Na	Mg	Ca
13. *Quercus robur—Fraxinus excelsior*	Belgium	115–160	328	947	63	493	—	126	1338
14. *Quercus stellata—Quercus marilandica*	Oklahoma	various	195	902	75	1093	—	230	3895
EUCALYPTUS									
1. *E. regnans*	Australia	38	654	399	38	1389	138	192	849
2. *E. regnans*	Australia	27	831	—	17	—	—	—	—
3. *E. obliqua—E. dives*	Australia	38	373	426	17	111	103	71	264
4. *E. obliqua*	Australia	51	316	—	31	256	—	204	336
5. *E. sieberi*	Australia	27	929	—	14	—	—	—	—
6. Mixed dry sclerophyll	Australia	?	176	395	—	—	—	—	—
7. *E. signata—E. umbra*	Australia	?	104	456	18	192	169	77	344
8. *E. diversicolor*	Australia	37	263	473	27	296	82	211	1133
9. *E. diversicolor—E. calophylla*	Australia	?	305	449	31	424	125	344	1266

Source: After Feller (1980).

Figure 28.8 plots these two variables for a variety of forests from tropical and temperate areas. Line A in Figure 28.8 shows the expected relationship under the hypothesis of constant nutrient-use efficiency. The data fit this hypothesis illustrated by line A with some scatter. Nutrient-use efficiency is clearly above expectation in habitats with low nitrogen levels because most of the observed data fall above line A, and below expectation in habitats rich in nitrogen (most of the data fall below the line).

One consequence of this is that forest productivity may be high on soils with low nutrient levels. A classic example is the tropical rain forest of the Amazon Basin, which represents one type of nutrient cycling pattern (Jordan and Herrera 1981). The *oligotrophic* pattern occurs on nutrient-poor soils, like the Amazon Basin, and the *eutrophic* pattern occurs on nutrient-rich soils. In the temperate zone, where most forest research has been done, forests are usually of the eutrophic type on rich soils. Oligotrophic ecosystems on poor soils constitute a much greater proportion of the tropics. But exceptions occur, and not all tropical forests are oligotrophic nor are all temperate forests eutrophic. Table 28.7 compares two eutrophic and two oligotrophic forest ecosystems. There are some striking differences between these forest types. Oligotrophic systems have a large biomass in the humus layer of the soil, and this layer of fine roots and humus is critical for nutrient cycling and nutrient conservation in these systems.

Productivity and nutrient cycling do not differ greatly in oligotrophic and eutrophic forests, as long as these ecosystems are not disturbed (Jordan and Herrera 1981). But when the forest is cleared for agriculture, the nutrient-poor systems quickly lose their productive potential, while the nutrient-rich ones do not. Once the humus and root layer on top of the mineral soil is disturbed in oligotrophic systems, the mechanism of efficient nutrient recycling is lost, and nutrients are leached out of the system. Oligotrophic ecosystems cannot be used for crop production unless critical nutrients are supplied in fertilizers (Sanchez et al. 1982).

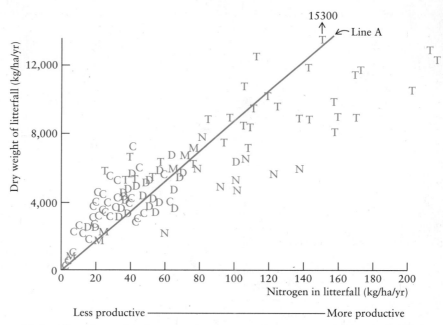

Figure 28.8   The relationship between the amount of nitrogen in the fine litterfall and the dry mass of that litterfall in forest ecosystems. Coniferous forests are represented by C, temperate deciduous forests by D, evergreen tropical forests by T, Mediterranean-type ecosystems by M, and temperate-zone sites dominated by symbiotic nitrogen fixers by N. Line A represents the expected value of the relationship assuming constant nutrient-use efficiency. (After Vitousek 1982.)

ACID RAIN: THE SULFUR CYCLE

Human activity through the combustion of fossil fuels has altered the sulfur cycle more than any of the other nutrient cycles. While human-produced emissions of carbon dioxide and nitrogen are only about 5 to 10 percent of the level of natural emissions, for sulfur we produce about 160 percent of the level of natural emissions (Likens et al. 1981). One clear manifestation of this alteration of the sulfur cycle is the widespread problem of acid rain in Europe and North America. Acid precipitation is defined as rain or snow that has a pH of less than 5.6. Low pH values are caused by strong acids (sulfuric acid, nitric acid) that originate as combustion products from fossil fuels. Over large areas of Western Europe and eastern North America, annual pH values of precipitation average between 4.0 and 4.5, and individual storms may produce acid rain of pH 2 to 3 (Figure 28.9).

Sulfur released into the atmosphere is oxidized to sulfate (SO_4) quickly, and redeposited rapidly on land or in the oceans (Schlesinger 1991). Figure 28.10 illustrates the sources and sinks of the global sulfur cycle. Short-term events like volcanic eruptions contribute to the global sulfur cycle and make it difficult to estimate the equilibrium state of the atmosphere. Human emissions are the largest component of additional sulfur to the atmosphere. Ore smelters and electrical generating plants have increased emissions during the past 100 years. To offset local pollution problems, smelters and generating plants have built taller stacks, which reduce pollution at ground level. Tall stacks (over 300 m) now are the standard, and they have

Table 28.7 Comparison of Biomass, Production, and Calcium Distribution in Eutrophic and Oligotrophic Forest Ecosystems from Temperate and Tropical Areas

	EUTROPHIC		OLIGOTROPHIC	
	Temperate Mixed Mesophytic, Oak Ridge, Tenn.	Tropical Montane Tropical Rain Forest, Puerto Rico	Temperate Oak-Pine, Brookhaven, Long Island	Tropical Amazonian Rain Forest, Venezuela
SOIL				
1. Exchangeable calcium (meq/100 g)	3.3	1–5	<0.22	0.38
2. Exchangeable calcium (kg/ha: soil depth, cm)	3784:75	1900:40	<990:150	306:40
BIOMASS				
3. Leaves (t/ha)	4.0	8.0	4.0	8.0
4. Stems and branches (t/ha)	158.0	190.0	61.0	158.0
5. Roots (t/ha)	17.0	65.0	36.0	132.0
6. Living total (t/ha)	179.0	263.0	101.0	298.0
7. Litter (t/ha)	5.0	1–6	16.0	6.0
8. Humus (t/ha)	—	—	47.0	137.0
9. Total organic (t/ha)	134.0	266.0	164.0	441.0
PRODUCTION				
10. Wood (g/m^2/yr)	432.0	486.0	451.0	440.0
11. Litter (g/m^2/yr)	358.0	547.0	406.0	572.0
CALCIUM CONCENTRATION				
12. Leaves (mg/g dry wt)	14.5	6.5	5.9	8.7
13. Stems and branches (mg/g dry wt)	8.0	2.1	0.9	1.9
14. Roots (mg/g dry wt)	7.5	5.0	2.1	1.2
CALCIUM STANDING CROP				
15. Leaves (kg/ha)	72.0	55.0	26.0	70.0
16. Stems and branches (kg/ha)	539.0	380.0	55.0	306.0
17. Roots (kg/ha)	128.0	325.0	76.0	153.0
18. Living total (kg/ha)	796.0	760.0	157.0	529.0
19. Litter (kg/ha)	72.0	11.0	148.0	3.0
20. Humus (kg/ha)	—	—	36.0	33.0
21. Total organic (kg/ha)	861.0	771.0	341.0	565.0
CALCIUM FLUX				
22. Precipitation (P) (kg/ha/yr)	10.5	21.8	3.3	16.0
23. Subsurface runoff (R) (kg/ha/yr)	27.4	43.1	9.7	2.8
24. $P - R$ (kg/ha/yr)	−16.9	−21.3	−6.4	13.2

Source: After Jordan and Herrera (1981).

exported the pollution problem downwind. Ice cores from Greenland show large increases in SO_4 deposition from the atmosphere in the last 50 years (Mayewski et al. 1986).

The destructive impact of sulfate pollution on vegetation has been known for over a century. Queenstown, Tasmania offers a striking example (Figure 28.11 in color insert). From 1896 to 1922 sulfur-rich smoke poured from the copper smelter

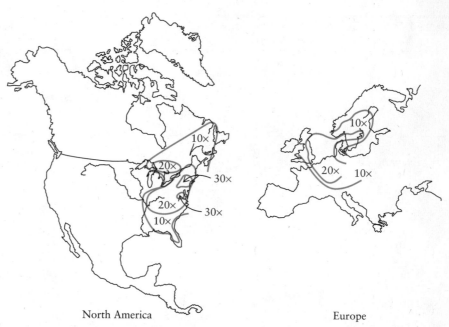

North America   Europe

Figure 28.9   Distribution of acid precipitation in North America and Europe. Areas designated 10, 20, and 30× receive 10, 20, and 30 times more acid in precipitation than expected if the pH were 5.6. (After Likens et al. 1981.)

of the Mount Lyell Mining and Railway company that created Queenstown. A typical ton of ore from the mine consisted of 48 percent sulfur, 40 percent iron, and less than 3 percent copper. More than half of the sulfur was lost in the emissions from the stacks, and the fumes from the smelter were toxic enought to cause plant death in one bad day. The prevailing winds carried the pollution to the east and southeast. Once the vegetation was killed, soil erosion from heavy rains along with fire removed all the topsoil, and the landscape now resembles the moon, 70 years after the smelter closed (Figure 28.11). Acid rain is not a new phenomenon.

Current estimates show that there is a net transport of SO_4 from the land to the oceans. The ocean is also a large source of aerosols that contain SO_4. Dimethylsulfide $[(CH_3)_2S]$ is the major gas emitted by phytoplankton in the sea, and it is oxidized quickly to SO_4 and then redeposited in the ocean. Sulfate is abundant in ocean waters (12×10^{20} g of elemental S), and the mean residence time for a sulfur molecule in the sea is over 3 million years (Schlesinger 1991).

The effects of acid rain on the environment are the subject of much current research. Some effects are very clear already. Freshwater ecosystems seem to be particularly sensitive. In areas underlain by granite and granitoid rocks, which are highly resistant to weathering, the acid rain is not neutralized in the soil, so lakes and streams become acidified. Lakes in these bedrock areas typically contain soft water of low buffering capacity. Thus bedrock can be an initial guide to sensitive areas. The Precambrian Fennoscandian Shield in Scandinavia, the Canadian Shield, all of New England, the Rocky Mountains, and other areas are thus potential trouble spots.

The clearest effects of acid precipitation have been on fish populations in Scandinavia and eastern Canada. Fish populations have been reduced or eliminated in

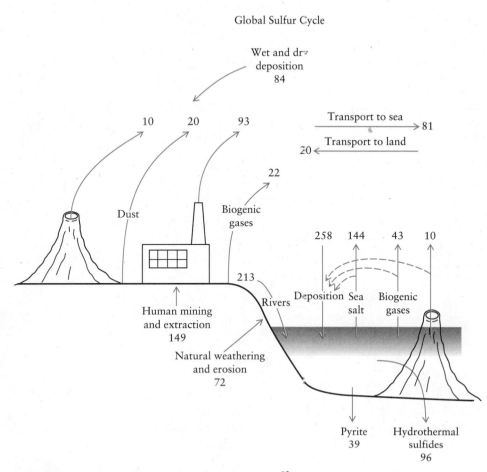

Global Sulfur Cycle

Figure 28.10   The global S cycle. All values are 10^{12} g S/yr. (From Brimblecombe et al. 1989.)

many thousands of lakes in southern Norway and Sweden once the pH in these waters fell below pH 5 (Likens et al. 1979).

The effect of acid precipitation on terrestrial ecosystems is more complex and difficult to unravel. A good example of this complexity is found in forest declines in Europe (Schulze 1989). Forest declines have been particularly severe in Central Europe where needle yellowing and needle loss in Norway spruce (*Picea abies*) have brought public attention to the problem. Large scale damage to spruce trees first became apparent in the late 1970s in Bavaria, Germany, and by the early 1980s about 25 percent of all European forests were classified as moderately or severely damaged from unknown causes. A variety of interacting causes associated with air pollution have now been shown to cause forest declines (Figure 28.12).

The forest region of Bavaria in southern Germany is exposed to high concentrations of SO_2, and both ozone levels and nitrous oxide levels are often high enough to cause direct damage to leaves. But these gaseous pollutants do not affect needles directly. Photosynthetic rates of Norway spruce needles are equal in Germany and in unpolluted New Zealand (Schulze 1989). The main effects of air pollution are on the forest soils. Soil acidification occurs because of nitrate and sulfate deposition,

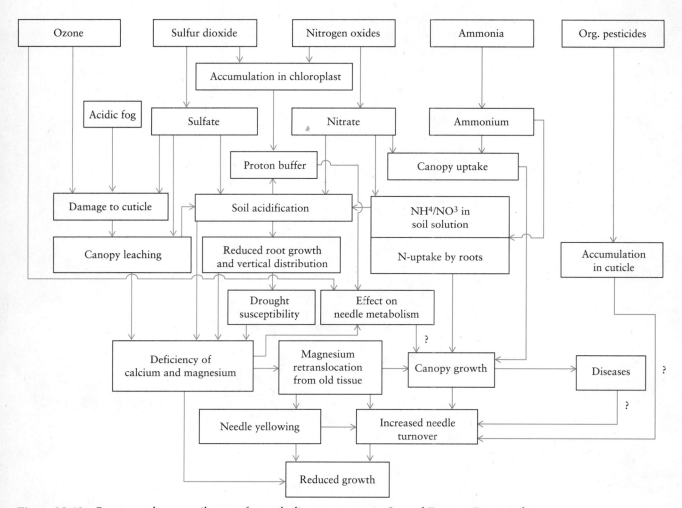

Figure 28.12   Processes that contribute to forest decline symptoms in Central Europe. Boxes indicate the air pollutants and their effects on soils and plants. (After Schulze 1989.)

and acid soil water reduces the amount of calcium, magnesium, and potassium that roots can absorb. Interactions among ions in the soil are particularly complex. For example, root uptake of magnesium is suppressed in the presence of aluminum or ammonium ions. Spruce seedlings growing in acid soils develop magnesium deficiencies, especially when ammonium is present. Spruce trees are thus stimulated to increased growth by nitrogen fertilization in the polluted air, but the acidification of the soil that accompanies the added nitrates and sulfates reduces the availability of magnesium and calcium and these nutrient deficiencies result in needle yellowing and loss. Plant diseases seem to have played only a secondary role in forest declines, and, once a tree is weakened by nutrient imbalances, fungal diseases or insect attacks may increase.

The changes which humans have made to the sulfur cycle thus have the potential to change nutrient cycling in natural ecosystems in a great variety of ways we cannot yet understand, much less predict. We cannot continue this aerial bombardment of ecosystems in the naive belief that nutrient cycles have infinite resilience to human inputs. Recent efforts to curtail sulfate emissions from fossil fuels have slowed the rise in SO_4 levels, and, once acidic precipitation is reduced, both forest and lake ecosystems can begin to recover from the damage inflicted.

THE CARBON CYCLE

Plants and animals are primarily composed of carbon and the global carbon cycle is a reflection of primary and secondary production. The fixation of carbon by plants in photosynthesis through geological time accounts for the oxygen in the earth's atmosphere. Humans have affected the global carbon cycle nearly as much as they have the sulfur cycle, and intense public interest now focuses on the resulting greenhouse effect and climate change.

Figure 28.13 shows the modern carbon cycle. The carbon cycle is mostly the carbon dioxide cycle. The largest fluxes of the global carbon cycle are between the atmosphere and land vegetation and the atmosphere and the oceans. These two fluxes are nearly equal, and the mean residence time of a molecule of carbon in the atmosphere is about three years.

The atmospheric oscillations of CO_2 are the result of the seasonal uptake of CO_2

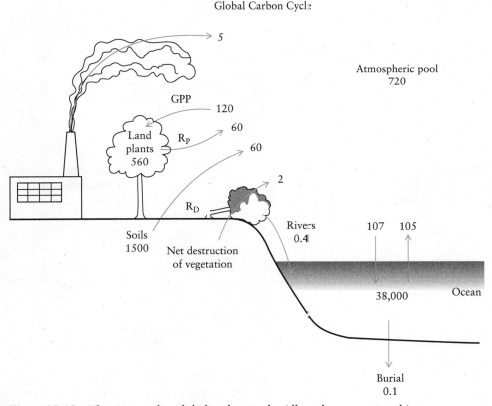

Figure 28.13   The present-day global carbon cycle. All pools are expressed in units of 10^{15} g C and all annual fluxes in units of 10^{15} g C/yr. (From Schlesinger 1991.)

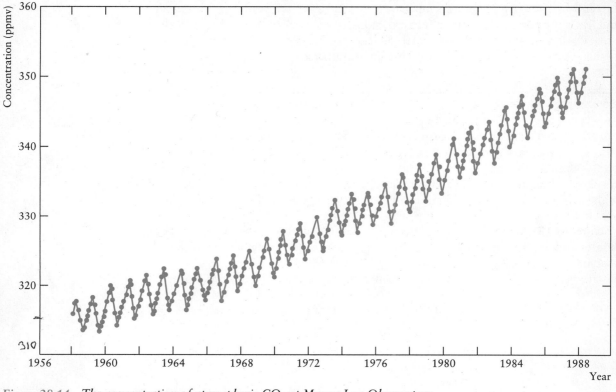

Figure 28.14   The concentration of atmospheric CO_2 at Mauna Loa Observatory in Hawaii, expressed as a mole fraction in parts per million of dry air. The annual oscillation reflects the seasonal cycles of photosynthesis and respiration by land biota in the Northern Hemisphere, while the overall increase is largely due to the burning of fossil fuels. (From Keeling 1986 and Scripps Institute of Oceanography.)

by plants in photosynthesis and seasonal differences in fossil fuel use and CO_2 exchange with the oceans. The majority of terrestrial vegetation occurs in environments with seasonal growth cycles, and particularly in the Northern Hemisphere CO_2 levels go down in summer (Figure 28.14). The atmospheric oscillations of CO_2 are superimposed on a long-term increase of 0.4 percent per year (1.5 ppm). Some of this increase comes from the burning of fossil fuels. If all the fossil fuel CO_2 accumulated in the atmosphere, CO_2 should be increasing about 0.7 percent per year. Only about 58 percent of the CO_2 released from fossil fuel is accumulating in the atmosphere (Dewar 1992). What happens to the remainder?

We cannot at the present time balance the global carbon budget to answer this question. The sizes of major sources and sinks for carbon are not well understood, and we do not understand the processes that control carbon fluxes (Dewar 1992). Oceanographers believe that about 40 percent of the CO_2 from fossil fuels enters the ocean each year (Quay et al. 1992). CO_2 is exchanged only at the ocean's surface, and much of the carbon in the oceans is in the deeper waters. Turnover of carbon for the entire ocean occurs about every 350 years (Schlesinger 1991).

Another source of atmospheric carbon dioxide is the destruction of terrestrial vegetation, often by clearing and burning for agriculture and especially in the tropics

(Detwiler and Hall 1988). There is considerable disagreement among scientists as to whether terrestrial vegetation is a source of CO_2 or a sink for CO_2. Land use changes in tropical areas can form a basis for estimating forest loss. Shifting cultivation is a dominant form of land use in tropical countries, and about three-quarters of all land use changes fall under this heading. Shifting cultivation is less destructive of forest because after one to three years the farmers move on and abandon the fields to secondary succession. Such temporary clearing of land contributes less CO_2 to the atmosphere than does permanent conversion of forest land to pastures. The best estimates indicate that tropical areas were a net source of between 0.4 and 1.6 $\times$ 10^{15} grams of carbon in 1980 (Detwiler and Hall 1988).

The data now available provide the following global budget for carbon (Sarmiento and Sundquist 1992): (units $= 10^{15}$ g carbon per year)

Emissions		$=$	Changes		
fossil fuel $+$	destruction of terrestrial plants	$=$	atmospheric increases	$+$ oceanic uptake	$+$ unknown sinks
5.3 $+$	1.0	$=$	2.9	$+$ 2.2	$+$ 1.2

We cannot account for almost 20 percent of the global carbon flux, and unknown sinks must be absorbing approximately one billion tons of carbon each year! In trying to balance the global carbon budget, it is important to remember that we are focusing on the annual movements of carbon, not on the amount stored in the various reservoirs. The ocean contains the largest pool of carbon but most of this carbon turns over very slowly. Desert soil carbonates contain more carbon than all the terrestrial plants, but there is virtually no exchange of carbon between the atmosphere and desert soils.

Recent work has focused on terrestrial vegetation as a sink for CO_2 (Kauppi et al. 1992, Dewar 1992). If biomass is increasing in terrestrial vegetation, and if this increase is rapid enough, we might have located the "unknown sink" of the global carbon cycle.

Much scientific effort is now underway to define the limits of the fluxes of the global carbon cycle and to define more clearly and more locally the sources and sinks for CO_2. All of this is important because of the implications of the carbon cycle for climate change, to which we now turn.

CLIMATE CHANGE

Global warming has become one of the major environmental issues of the 1990s. Global warming is a function of the greenhouse effect, one of the most well established theories of atmospheric chemistry (Schneider 1989). Figure 28.15 illustrates the greenhouse effect, which arises because the earth's atmosphere traps heat near the surface. Water vapor, CO_2, and other trace gases absorb the longer, infrared wavelengths emitted by the earth. An increase in the concentration of greenhouse gases thus tends to warm the earth by reradiation. None of this is controversial and the greenhouse principles apply equally well on Earth as they do on Venus (dense CO_2 atmosphere, very hot) and Mars (thin CO_2 atmosphere, very cold). What is controversial is how to predict from the greenhouse effect exactly how much the earth's temperature will rise for a specified change in greenhouse gases like CO_2.

One way to project climate changes is to look backward. Ice cores have provided one way of doing this. Air is trapped by snow as it is transformed into glacial ice,

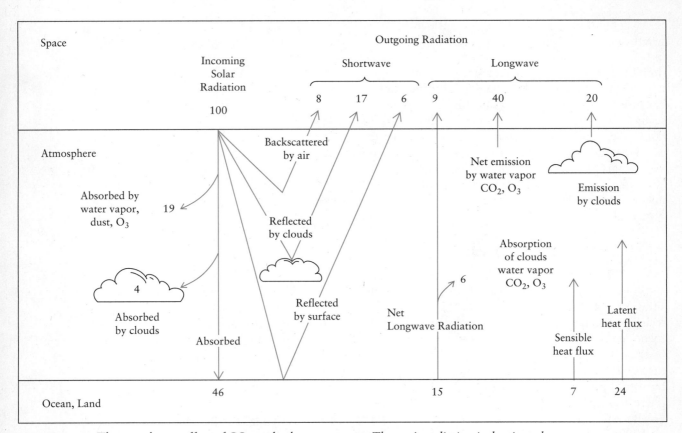

Figure 28.15   The greenhouse effect of CO_2 and other trace gases. The sun's radiation is dominated by short wavelengths, which are reflected or absorbed by the Earth's surface. The radiation absorbed is reradiated at longer wavelengths, which can be absorbed by atmospheric gases including CO_2. Higher concentrations of these gases in the atmosphere reduce the net emission of longwave radiation to space, warming the earth. (From MacCracken 1985).

and by taking ice cores one can sample the atmosphere back in time. The most spectacular example of this method is a 2083-meter long ice core collected by the Soviet Antarctic Expedition at Vostok, Antarctica (Barnola et al. 1987). This ice core spans 160,000 years. Temperature at the time of ice formation can be determined by the ratio of oxygen-18 to oxygen-16 in the ice (Lorius et al. 1985). The resulting time series of changes in the Vostok ice core are shown in Figure 28.16. There is a close correlation between carbon dioxide in the air and global temperatures over the past 160,000 years. But there are three difficulties in extrapolating these correlations into the future. First, our present CO_2 levels exceed the levels found in nature during the last 160,000 years. We do not know if we can extrapolate with the same relationships found in the past. Second, we are changing CO_2 levels very rapidly from year to year, whereas the historical changes in CO_2 were very slow. We do not know the rates at which an equilibrium can be established between CO_2 sources and sinks. Third, Figure 28.16 shows only a correlation between CO_2 and temperature and we do not yet know which is cause and which is effect (Schneider 1989).

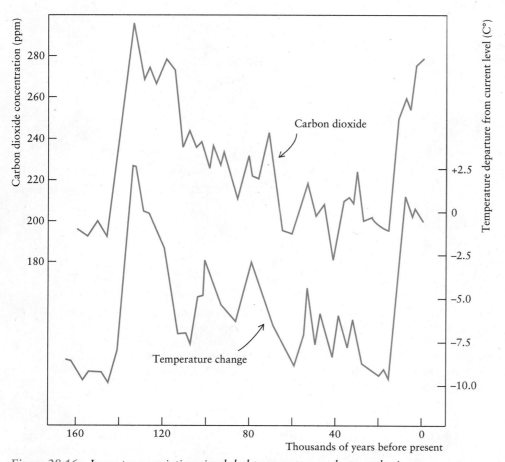

Figure 28.16   *Long-term variations in global temperature and atmospheric carbon dioxide concentration, determined from the Vostok ice core, Antarctica. Carbon dioxide can be measured from air trapped in the ice as it forms. The Vostok core is 2083 meters long. (From Barnola et al. 1987.)*

Other greenhouse gases can contribute to global warming as well as CO_2. The most important of these are methane, nitrous oxides, ozone, and chlorofluorocarbons, and these trace greenhouse gases may together be as important as CO_2 in the greenhouse effect of the next century.

How will changes in CO_2 levels and the suggested climatic warming affect the earth's biota? There is an urgent need to answer this question, and yet the difficulty of providing specific answers for agricultural and natural ecosystems is very great. Let us consider this question at the level of the individual and at the community and ecosystem level.

Individual Plant Responses

The effects of rising CO_2 and temperature have been analyzed in great detail for individual species of plants because this work can be done relatively easily in greenhouses. When other resources are available in adequate amounts, additional CO_2 can increase growth of C_3 plants over a wide range of CO_2 levels. By contrast, C_4

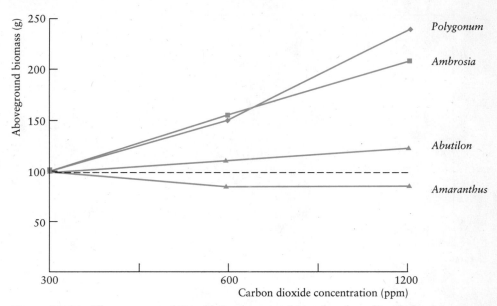

Figure 28.17   The response of C_3 and C_4 plants to elevated carbon dioxide levels in a greenhouse. Aboveground biomass for all species is scaled to 100 at ambient CO_2 levels. Growth enhancement differs among the three C_3 species (Polygonum pensylvanicum, Ambrosia artemisiifolia, *and* Abutilon theophrasti) *but is always positive. The C_4 species* Amaranthus retroflexus *does not change in biomass with CO_2 concentration. (Data from Bazzaz and Carlson 1984.)*

plants do not increase photosynthetic rates at higher CO_2 levels (Bazzaz 1990). Figure 28.17 illustrates these differences for four species of annual plants. Growth enhancement in C_3 plants is not always simple. Some plants acclimate to high CO_2 levels and show a decline in photosynthetic rate with time. Other resources for growth—water, light, and nutrients—must be present in excess to observe the impact of elevated CO_2 levels. Increased growth with increased CO_2 is also found in coniferous and broadleaved trees, and doubling CO_2 levels increased tree growth on average 40 percent (Eamus and Jarvis 1989).

In most plants stomatal conductance declines as CO_2 levels rise. Because stomata are closed more often at high CO_2 levels, water loss through transpiration is reduced, so the overall result is improved water use efficiency (Bazzaz 1990). Type C_4 crop plants may improve growth under elevated CO_2 levels because of increased water use efficiency, even when photosynthetic rates are unaffected by CO_2 levels (Rogers et al. 1983).

The effect of CO_2 on plant reproduction has been studied much less than effects on growth. There is no clear pattern of effects, and species differ in response to CO_2 levels. The onset of flowering may be accelerated or retarded under elevated CO_2, more or fewer flowers may be formed, and seed numbers and size may go up or down. At present we can predict responses only for particular species that have been studied in detail (Garbutt et al. 1990).

Plant Community Responses

To predict the global effects of CO_2 increases, ecologists must take these laboratory findings out into natural communities. This has been done for only a few ecosystems, and in reviewing these findings we can begin to appreciate the task at hand for the coming decade.

Arctic ecosystems may be the ecosystems most sensitive to climate change. High latitudes may be more severely affected by rising temperatures than lower latitudes (Schneider 1989). Arctic plant communities have several characteristics that make them susceptible. The active soil layer is shallow because of underlying permafrost, so that most root systems occur in the top 10 to 15 cm of the soil. Permafrost layers in the soil also contain large amounts of frozen organic matter that is not available to the decomposers. Rising CO_2 levels will lead to increased temperatures and increased evaporation from arctic tundra. Billings and colleagues (1983) postulated that rising temperatures could change arctic tundra communities from a sink for CO_2 to a source of CO_2. To test this idea they took 8-centimeter diameter cores of arctic tundra into a greenhouse and measured net exchange of CO_2 from these cores at two levels of CO_2 and two levels of water table. Figure 28.18 shows the results of these experiments. Under present water-table conditions, carbon dioxide enrichment has little impact on the carbon cycle in tundra vegetation, indicating that in this ecosystem CO_2 is not a limiting factor. By contrast, water-table levels strongly affect the carbon cycle of tundra. As the water table falls in the tundra, more decomposition of organic matter occurs and this is further exacerbated if temperature rises as well (Billings et al. 1983, 1984). The net result of Billings's work is that a doubling of CO_2 and the associated climatic warming will convert the wet tundra ecosystem of northern Alaska from a CO_2 sink to a CO_2 source. Plant growth in tundra communities is limited more by nitrogen availability and this limitation prevents rising CO_2 levels from enhancing the growth of individual plants.

The ecosystem consequences of CO_2 enrichment are difficult to predict for forest communities because of the time scale. Few studies with trees have extended more than two years because of greenhouse constraints. Studies of single tree species in greenhouse pots are of limited utility for extrapolation to natural communities in which competition for resources occurs. Williams and co-workers (1986) grew six species of hardwood saplings in competition under three levels of CO_2. There was no effect of elevated CO_2 levels on plant biomass during these one-year experiments (Figure 28.19). Bottomland trees (sycamore, green ash and silver maple) responded the same way as upland trees (red oak, hickory, and tulip tree). There is no support in the available forest data for the belief that CO_2 enrichment will increase plant growth in natural communities. Bazzaz (1990) suggests that late successional tree species may profit most from CO_2 enrichment, but this generalization may not fit all forest communities.

Animal Community Responses

There is no evidence that animals will be affected directly by changing CO_2 levels in the next century. But there is concern that herbivorous animal populations and communities may be affected indirectly by changes in their food plants. Nitrogen content of plant leaves tends to fall with increased CO_2 levels, and this may result in a reduction of larval insect growth in nitrogen-limited insect species (Figure 28.20). Insect herbivores are often limited by plant-nitrogen in natural communities (White 1974) and reduced nitrogen may reduce insect numbers on CO_2 enriched vegetation. To compensate for low nitrogen levels, insect herbivores feeding on plants grown with high CO_2 levels may increase their feeding rate by 20 to 80 percent (Fajer 1989). Plant damage could increase even if insect numbers fall as CO_2 levels rise in the future.

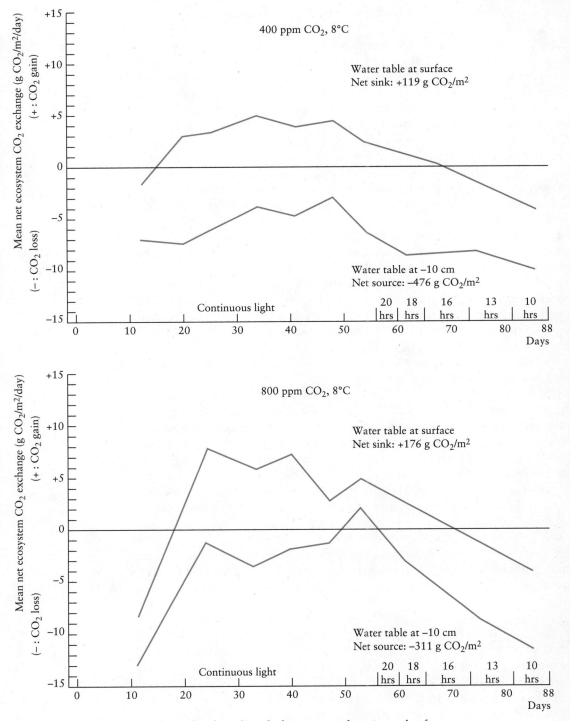

Figure 28.18   Net gain or loss of carbon dioxide from cores of arctic tundra from Barrow, Alaska, grown in a greenhouse through a simulated arctic summer. The water table is presently at the surface in this tundra community, but is predicted to drop with global warming. (From Billings et al. 1983.)

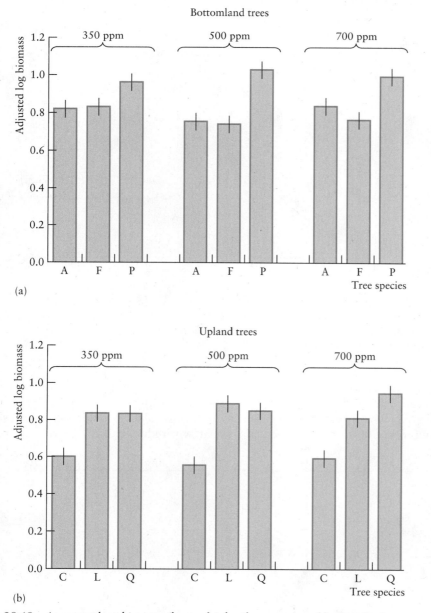

Figure 28.19   Average plant biomass (log scale) for three species of bottomland trees (a) and three species of upland trees (b) from Illinois, grown at three levels of CO_2 enrichment. All trees were grown under high light levels for one year. There is no tendency for these trees to grow more under high CO_2 levels, when they are in competition. (P = sycamore, F = green ash, A = silver maple, Q = red oak, C = hickory, L = tulip tree.) (From Williams et al. 1986.)

Climatic warming could also disrupt the timing of hatching in insects (Dewar and Watt 1992). Insects that feed on newly emerged foliage are very sensitive to the age of the foliage. Maximum overlap between bud burst and larval emergence results in good survival and growth of the insects. In Scotland the emergence of the winter moth is strongly affected by spring temperatures, and consequently these larvae will

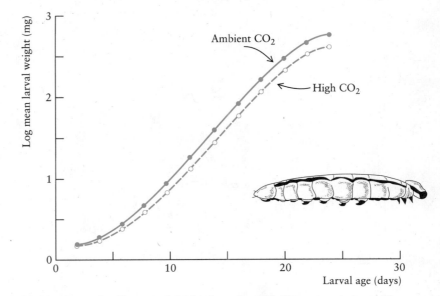

Figure 28.20 *Growth of larvae of the buckeye butterfly* Junonia coenia *on Plantago* lanceolata *grown at ambient and elevated* CO_2 *levels. High* CO_2 *plants contain less nitrogen in their leaves, which reduces larval growth. (From Fajer et al. 1989.)*

emerge earlier if climatic warming occurs. By contrast, the vegetative buds of sitka spruce, their food plant, are not greatly affected by spring temperatures and will open only slightly earlier when the climate warms. The net result will be a mismatch between the moth and its food plant that will reduce moth survival and growth

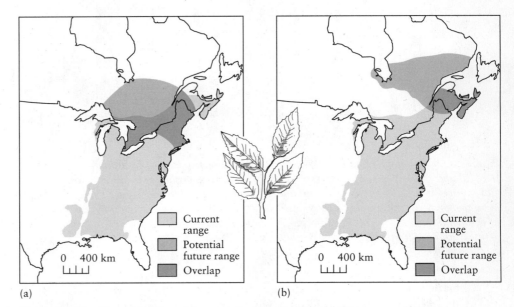

Figure 28.21 *Current geographic range for American beech (Fagus americana) in eastern North America, and the potential future range in 100 years according to two climatic warming models: (a) a milder scenario, and (b) a more severe scenario. (From Roberts 1989.)*

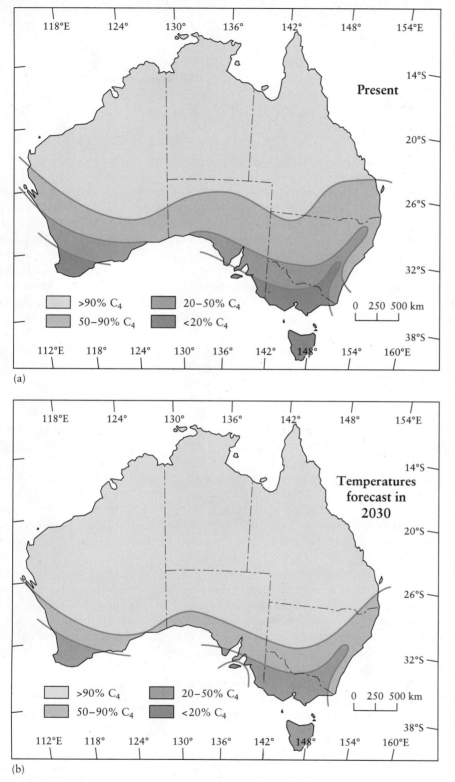

(a)

(b)

Figure 28.22 Map of Australia showing 90%, 50%, and 20% isoclines for C$_4$ native grass species (a) at the present time and (b) at temperatures forecast for the year 2030. The change in position of the isoclines is based on continental temperatures forecast for 2030 by CSIRO's Climate Impact Group. The geographic distribution of C$_4$ grasses is limited by summer temperature (see Chapter 8, page 122). (From Henderson et al. 1993.)

(Dewar and Watt 1992). These short-term effects could be alleviated by natural selection in the long term, but the disruption of life cycles in insect herbivores could be a major impact of global warming.

There is no work yet available on the impact of CO_2 enriched plants on herbivorous mammals or birds.

Changes in Species Distributions

One of the most difficult impacts of global warming to alleviate will be the resulting changes in species distributions. If species ranges are currently controlled by climate, a shift in climate implies a shift in geographic range. Shifts in range have occurred many times in the history of the earth, and consequently at first glance one may think that this problem will take care of itself. Two factors argue against complacency. First, the speed of change in climate is now many times greater than it has ever been in the past. The critical question then becomes how fast can species move? Second, human changes in land use have interrupted many possible corridors of movement for both plants and animals, so that dispersal may no longer be possible because of barriers.

The geographic distribution of many trees is probably limited by climate, and trees are thus a good group to use for studying this question. Figure 28.21 illustrates the possible magnitude of range shifts that may occur in the next 100 years (Davis 1989). Two different climate change models were used to develop these predictions, which are coarse approximations. The range changes are very large and there is no precedent in the past for such a rapid shift. At the end of the last Ice Age North American trees dispersed north at rates varying from 10 to 45 kilometers per 100 years. Beech trees moved at an average rate of 20 kilometers per 100 years (Woods and Davis 1989). But if the predictions indicated in Figure 28.21 are correct, beech trees will have to move north about 40 times faster than they did at the end of the Ice Age. This could only occur with human help. One of the major conservation problems of the next century for North America and Europe could be the reestablishment of plant and animal communities at more northerly sites as climatic warming proceeds.

The geographic distribution of C_3 and C_4 grasses in Australia is determined by summer temperatures (Hattersley 1983). Figure 28.22 shows the present isoclines for C_4 grasses in Australia and the predicted changes in these isoclines in 2030 under the current model of climatic warming for Australia. There will be major shifts in grasslands species composition in a very short time period and the ecological consequences of this shift are not easily predicted (Henderson et al. 1993).

If plant communities are disrupted by climatic warming during the next century, the animal communities on which they depend will be equally disrupted. There are at present little data available on which to judge these potential effects, and the estimation and monitoring of climatic effects on both plants and animals are part of the research agenda for all ecologists in the coming decade.

SUMMARY

Nutrients cycle and recycle in ecosystems, and tracing these nutrient cycles is another way of studying fundamental ecosystem processes. Human activities are changing cycles on a global basis, with consequences for climatic change. Nutrients reside in *compartments* and are transferred between compartments by physical or biological processes. Compartments can be defined in any operational way to include one or more species or physical spaces in the ecosystem. Nutrient cycles may be *local* or *global*. Global cycles, such as the nitrogen cycle, include a gaseous phase that is transported in the atmosphere. Local cycles include less mobile elements like phosphorus.

Nutrient cycles in forests have been studied because of nutrient losses associated with logging. The input of nutrients must equal the outflow for any ecosystem, or it will deteriorate over the long run. Logging can result in high nutrient losses even if soil erosion is absent. An undisturbed forest site recycles nutrients efficiently.

The sulfur cycle is an example of a global nutrient cycle that is strongly affected by human activities. Burning of fossil fuels adds a large amount of SO_2 to the atmosphere, and results in acid rain. Acid rain in combination with other airborne pollutants has caused forest declines in Europe and has eliminated fish populations from many lakes in eastern Canada and in Scandinavia.

The carbon cycle is also strongly affected by human activities. Rising levels of CO_2 in the atmosphere have occurred for the last century due to fossil fuel burning and the destruction of native vegetation. The global carbon budget does not balance and there is a large sink missing from the overall equation.

Climate change is one practical side effect of the carbon cycle. Greenhouse gases like CO_2 trap heat at the earth's surface and increase global temperatures. How increasing CO_2 will affect the earth's biota is a focus of intensive current research. Individual plants typically grow more when CO_2 levels are elevated in a greenhouse, but in natural communities other factors like nitrogen or water may limit productivity. Arctic ecosystems are particularly susceptible to the impacts of climatic warming. Animal communities will be affected indirectly through their food plants.

If geographic distributions are limited by climate, climatic warming in the next century will have dramatic effects on the distribution of native animals and plants. Ecosystem restoration may be the major global conservation problem of the next century.

Selected References

BAZZAZ, F. A. 1990. The response of natural ecosystems to the rising global CO_2 levels. *Annual Review of Ecology and Systematics* 21:167–196.

BILLINGS, W. D., J. O. LUKEN, D. A. MORTENSEN, and K. M. PETERSON. 1983. Increasing atmospheric carbon dioxide: Possible effects on arctic tundra. *Oecologia* 58:286–289.

DEANGELIS, D. L. 1992. *Dynamics of Nutrient Cycling and Food Webs*. Chapman and Hall, New York.

DETWILER, R. P., and C. A. S. HALL. 1988. Tropical forests and the global carbon cycle. *Science* 239:42–47.

EAMUS, D., and P. G. JARVIS. 1989. The direct effects of increase in the global atmospheric CO_2 concentration on natural and commercial temperate trees and forests. *Advances in Ecological Research* 19:1–55.

JORDAN, C. F., and R. HERRERA. 1981. Tropical rain forests: Are nutrients really critical? *American Naturalist* 117:167–180.

KAREIVA, P. M., J. G. KINGSOLVER, and R. B. HUEY. (eds.) 1993. *Biotic Interactions and Global Change*. Sinauer Associates, Sunderland, Massachusetts.

OJIMA, D. S., T. G. F. KITTEL, T. ROSSWALL, and B. H. WALKER. 1991. Critical issues for understanding global change effects on terrestrial ecosystems. *Ecological Applications* 1:316–325.

SCHLESINGER, W. H. 1991. *Biogeochemistry: An Analysis of Global Change*. Academic Press, New York.

SCHNEIDER, S. H. 1989. The greenhouse effect: Science and policy. *Science* 243:771–781.

SCHULZE, E. D. 1989. Air pollution and forest decline in a spruce (*Picea abies*) forest. *Science* 244:776–783.

Questions and Problems

1. Slash-and-burn agriculture is common in many tropical countries. Forests are cut and burned, and crops are planted in the cutover areas. Yields are usually good in the first year but decrease quickly thereafter. Why should this be? Compare your ideas with those of Jordan and Kline (1972, pp. 45–46), and evaluate the impact of slash-and-burn agriculture on the global carbon cycle.

2. Peatlands in boreal and subarctic regions store a large pool of organic matter not yet decomposed. Review the distribution of peatlands and their potential role in climatic change scenarios. Gorham (1991) provides a recent overview.

3. Discuss the relative merits of making a compartment model of a nutrient cycle very coarse with only a few compartments versus making it very fine with many compartments.

4. Pastor and Post (1988, p. 55) state, "The carbon and nitrogen cycles are strongly and reciprocally linked." Review the nitrogen cycle in Schlesinger (1991, chap. 12) and discuss these linkages.

5. Schultz (1969, p. 92) states:

   The idea of one cause–one effect is left over from the nineteenth century when physics dominated science. The whole notion of causality is under question in the ecosystem framework. Does it make sense to say that high primary production causes a rich organic soil and a rich organic soil causes high production? This kind of reasoning leads up a blind alley.

   Discuss.

6. Vitousek (1982) concluded that nitrogen-use efficiency was higher in low-nitrogen forests. Chapin (1980, p. 246) concluded that plants from infertile habitats had a lower efficiency of use of limiting nutrients than that of rapidly growing plants. Discuss why these two authors have reached apparently opposite conclusions.

7. Norby and colleagues (1992) grew yellow poplar (*Liriodendron tulipifera*) trees for three years at three levels of CO_2 enrichment. Photosynthetic rate nearly doubled at high CO_2 levels, but there was no difference found in aboveground biomass of trees grown at normal or elevated CO_2 levels after three years. How is this possible?

8. In discussing models that will allow us to predict the impact of climatic change on primary production, Ågren and coauthors (1991, p. 134) state:

   > We believe that in spite of gaps in the knowledge used to build the models used to predict the effects of climate change on forests and grasslands, the models are sufficiently well substantiated and a number of dominant processes well enough understood for the predictions made with the models to be taken seriously.

   In contrast, Long and Hutchin (1991, p. 139) conclude:

   > We conclude that there is insufficient information to predict accurately the response of primary production to climate change.

   Read these papers and discuss why there are such divergent views on this important question.

9. In agricultural landscapes farmers have the choice of managing roadside verges by mowing, burning, grazing, or doing nothing to them. Discuss the implications of these four treatments for the global carbon cycle.

Overview Question

One suggested ecological response to help arrest increasing carbon dioxide levels in the atmosphere is to plant trees. Explain the mechanisms behind this recommendation and discuss how you could calculate how many trees you would need to plant to achieve this policy goal.

Epilogue

We have progressed from the simple problem of distribution to the complex problem of how organisms are integrated in biological communities. There is a great deal we have omitted in our survey. The applied problems of pesticide contamination and pollution have been only briefly discussed; they are dealt with in a wealth of books specifically oriented toward human activities and the environment.

In this book I have attempted to sketch the framework of ecology as a pure science, and my purpose is to develop a particular view of the world, an "ecological consciousness." Two possible reasons for studying ecology are (1) to increase our understanding of the world in which we live and (2) to provide a basis for practical action on ecological problems.

Ecologists at the present time are being called upon for judgments of environmental impacts from a variety of proposals for pipelines, harbors, refineries, ski areas, and housing developments. One major task of applied ecology in the years ahead is to make environmental assessment more a science than an art, and to achieve this goal, we need sound ecological theory and solid local data. Human effects on ecosystems can be viewed as large-scale perturbation experiments, and ecologists who have adopted the experimental approach can use these activities to gain critical insights into the behavior of perturbed ecosystems.

We already understand how to ameliorate many of our environmental problems; these problems persist for political and economic reasons rather than for lack of clear ecological advice. Ecologists rarely have much to say in policy decisions, and there is a danger that as scientists we may be used as technicians to monitor the demise of the world's ecosystems.

During the last 40 years, ecology has matured very rapidly as a science. Formerly I had to search to find a single good illustration of some ecological principle; now I can find a full complement of examples to choose from. We have progressed rapidly from a descriptive science with vague generalizations to an experimental science with precise, quantitative, and testable hypotheses about ecological processes.

Theoretical ecology is particularly active at present and is making additional inroads into dealing with the complex systems ecologists must study. The pages of the *American Naturalist* and *Oikos* testify to the arguments that arise between different theoreticians and between theoreticians and field ecologists. This is a good sign of vitality: Where there is argument, there is life! Most of the ideas of ecology are still hypotheses with an unknown range of applicability. These hypotheses do have consequences, and the job of the ecologists must be to translate these hypotheses into practical suggestions. Some ideas will turn out to be completely wrong; others will stand the practical test. Through all this, controversies will flare, contradictions will be published, and tempers will explode. It is all a very human activity and in some respects a microcosm of the world.

Appendixes

Appendix I
A Précis on Population Genetics

Population genetics is the algebraic description of evolution, of how allelic frequencies change over time. I present here a very abbreviated basic statement of this approach. See Tamarin (1993, chap. 17–18) for more details.

We begin with individuals distinguished by a single locus with two possible alleles (A, B). Every individual is thus one of three genotypes (AA, AB, BB). The most important theorem in population genetics is called the Hardy-Weinberg law. In 1908 a British mathematician, G. H. Hardy, and a German physician, G. Weinberg, independently discovered that an equilibrium will arise and be maintained in both allelic and genotypic frequencies in any diploid population that is large in size, undergoes random mating, has no mutation or migration of individuals, and is subject to no selection. We can illustrate this law simply with an example. Consider a hypothetical population with this composition:

Genotype	Genotypic Frequency (%)	No. Individuals
AA	25	1000
AB	0	0
BB	75	3000

We define:

$$\text{Allelic frequency of } A = \frac{\text{number of } A \text{ alleles in population}}{\text{total number of alleles in population}}$$

$$= \frac{2(1000)}{2(1000) + 2(3000)} = \frac{2000}{8000} = 0.25$$

$$\text{Allelic frequency of } B = \frac{\text{number of } B \text{ alleles in population}}{\text{total number of alleles in population}}$$

$$= \frac{2(3000)}{2(1000) + 2(3000)} = \frac{6000}{8000} = 0.75$$

Under the given assumptions, the Hardy-Weinberg law states that (1) allelic frequencies will not change over time and (2) genotypic frequencies will come into equilibrium within one generation and will be

Genotypic frequency of AA = (allelic frequency of A)^2
Genotypic frequency of BB = (allelic frequency of B)^2
Genotypic frequency of AB = 2(allelic frequency of A)
(allelic frequency of B)

For our example,

Genotypic frequency of $AA = (0.25)^2$ $\qquad = 0.0625$
Genotypic frequency of $BB = (0.75)^2$ $\qquad = 0.5625$
Genotypic frequency of $AB = 2(0.25)(0.75) = \underline{0.3750}$
$\qquad\qquad\qquad\qquad\qquad\qquad\qquad$ Sum $= 1.0000$

If you apply the definition again for a second generation, a third generation, and so on, you will find that none of these frequencies change. The Hardy-Weinberg law produces an equilibrium because no matter what the starting genotypic frequencies are, you will always come to the same result (for one set of allelic frequencies). You can demonstrate that to yourself by doing these same calculations with genotype frequencies of $AA = 0$ percent, $AB = 50$ percent, and $BB = 50$ percent.

The important conclusion is that gene frequencies, or *genetic variability,* in a population will be maintained over time without change *unless outside forces are applied.* Evolution results from these outside forces. A brief comment on these outside forces follows.

Mutations Genetic mutations are relatively rare events and do not typically cause shifts from Hardy-Weinberg equilibrium. Mutations are important as the source of genetic variation on which natural selection acts.

Migration Migration, or movements of individuals between populations, can be a source of adding or subtracting alleles in a population. This can be critical in preventing or aiding adaptation in local populations.

Population Size In small populations, chance may be a critical element. For example, if two individuals in the earlier example colonize an island, it may occur by chance that both are BB individuals. This is called *random genetic drift* and must be considered when populations are small in size.

Random Mating Individuals may mate on the basis of similarity (or dissimilarity) so that *assortative mating* would result. For example, AA individuals may prefer to mate with AA individuals. If the choice of mates involves relatives, then either *inbreeding* or *outbreeding* may also be involved. In either case, mating is not random, and the Hardy-Weinberg equilibrium is disturbed.

Natural Selection Selection produces adaptation by altering allelic frequencies by eliminating individuals that are less fit. We can introduce the idea of natural selection into our simple model population by defining the notion of *fitness.* Fitness is defined as the relative reproductive success of a given genotype. A simple example shows how fitness can be calculated:

Genotype	AA	AB	BB
No. in generation 1	3000	6000	3000
No. in generation 2	2500	6000	3500

$$\text{Relative reproductive success of } AA = \frac{\text{no. in generation 2}}{\text{no. in generation 1}} = \frac{2500}{3000} = 0.83$$

$$\text{Relative reproductive success of } AB = \frac{6000}{6000} = 1.00$$

Relative reproductive success of $BB = \dfrac{3500}{3000} = 1.17$

By convention, the genotype with the highest relative reproductive success has a fitness equal to 1.0, and we thus define *relative fitness* of the three genotypes as follows:

Relative fitness of $BB = \dfrac{\text{relative reproductive success of } BB}{\text{highest relative reproductive success observed}}$

$= \dfrac{1.17}{1.17} = 1.00$

Relative fitness of $AB = \dfrac{1.00}{1.17} = 0.86$

Relative fitness of $AA = \dfrac{0.83}{1.17} = 0.71$

We can now define the *selection coefficient* for each genotype:

Selection coefficient for genotype $x = 1.0 -$ relative fitness of genotype x

The selection coefficient against the best genotype is always zero. Note that fitness (equivalent to relative fitness, relative Darwinian fitness, and adaptive value) is always *relative,* and evolution always deals with how fit one genotype is relative to other genotypes. If there is only one genotype in a population, you cannot define fitness.

Data obtained from a multiple census should be cast in the form of a *method B table*. This table is constructed by asking for each individual caught: (1) Was it marked or unmarked when caught? (2) If marked, when was it last captured? The method B table summarizes the answers to these questions:

		Time of Capture					
		1	**2**	**3**	**4**	**5**	**6**
Time of last capture	1		10	3	5	2	2	←Z_3
	2			34	18	8	4	
	3				33	13	8	←R_3
	4					30	20	
	5						43	
Total marked		0	10	37	56	53	77	
Total unmarked		54	136	132	153	167	132	
Total caught		54	146	169	209	220	209	
Total released		54	143	164	202	214	207	

These data are reported in Jolly (1965), and we will use his formulas to estimate the size of the marked population. We use these symbols:

M_i = marked population size at time i

m_i = marked animals actually caught at time i

S_i = total animals released at time i

Z_i = number of individuals marked before time i, not caught in the ith sample but caught in a sample after time i

R_i = number of the S_i individuals released at time i that are caught in a later sample

Some examples from the table will illustrate these definitions with respect to the method B table:

$m_2 = 10$

$m_5 = 53$

$S_3 = 164$

$S_4 = 202$

$Z_2 = 3 + 5 + 2 + 2 = 12$

$Z_4 = 2 + 2 + 8 + 4 + 13 + 8 = 37$

$R_1 = 10 + 3 + 5 + 2 + 2 = 22$

$R_3 = 33 + 13 + 8 = 54$

The formula to estimate the size of the marked population is

$$M_i = \frac{S_i Z_i}{R_i} + m_i$$

For example,

$$M_3 = \frac{(164)(39)}{54} + 37 = 155.4$$

$$M_4 = \frac{(202)(37)}{50} + 56 = 205.5$$

From the discussion in Chapter 10, we know that

$$\text{Total population size} = \frac{\text{marked population size}}{\text{proportion of animals marked}}$$

Consequently, we can calculate estimates for

$$\text{Total population at time 3} = \frac{155.4}{37/169} = 709.8 \text{ animals}$$

$$\text{Total population at time 4} = \frac{205.5}{56/209} = 767.0 \text{ animals}$$

Note that this is only one possible way of estimating the size of the marked population, and it may not be the best technique for all circumstances (Manly 1970, Seber 1982).

Appendix III
Instantaneous and Finite Rates

The concept of *rates* is critical for quantitative work in ecology, and students may find a brief review useful, especially for the discussions in Chapter 11.

A *rate* is a numerical proportion between two sets of things. For example, the number of rotten apples per bushel might be measured to be six rotten apples per bushel. The number of students failing an examination might be 27 of 350, a failure rate of 7.7 percent. In ecological usage, a rate is usually expressed with a standard time base. Thus if 8 seedlings of 12 die within one year, the mortality rate will be 66.7 percent *per year*. If a population grows from 100 to 150 within one month, the rate of population increase will be 50 percent *per month*.

We usually think in terms of *finite rates,* which are simple expressions of observed values. Some ecological examples are

$$\text{Annual survival rate} = \frac{\text{no. alive at end of year}}{\text{no. alive at start of year}}$$

$$\text{Annual rate of population change} = \frac{\text{population size at end of year}}{\text{population size at start of year}}$$

Rates can also be expressed as *instantaneous rates,* in which the time base becomes very short rather than a year or a month. The general relationship between finite rates and instantaneous rates is

$$\text{Finite rate} = e^{\text{instantaneous rate}}$$
$$\text{Instantaneous rate} = \log_e \text{finite rate}$$

where $e = 2.71828. \ldots$

The idea of an instantaneous rate can be explained most simply by the use of compound interest. Suppose we have a population of 100 organisms increasing at a *finite* rate of 10 percent per year. The population size at the end of year 1 will be

$$100(1 + \tfrac{1}{10}) = 110$$

At the end of year 2, it will be

$$110(1 + \tfrac{1}{10}) = 121$$

And at the end of year 3,

$$121(1 + \tfrac{1}{10}) = 133.1$$

In general, for an interest rate of $1/m$ carried out n times,

$$y_n = y_0\left(1 + \frac{1}{m}\right)^n$$

where

y_n = amount at end of the nth operation

y_0 = amount at start

We can repeat these calculations for a finite interest rate of 5 percent per half year. Everyone who has invested money in a savings account knows that 5 percent interest per half year is a *better* interest rate than 10 percent per year. We can see this quite simply. At six months our population size will be

$$100(1 + \tfrac{1}{20}) = 105$$

At one year, it will be

$$105(1 + \tfrac{1}{20}) = 110.25$$

and similarly at two years, 121.55, and at three years, 134.01. Compare these values with those obtained earlier for 10 percent annual interest.

Biological systems often operate on a time schedule of hours and days, so we may be more realistic in using rates that are instantaneous, rates that break up a year into very many short time periods. Let us repeat the first calculation with an *instantaneous* rate of increase of 10 percent per year. If we divide the year into 1000 short time periods, each time period having a rate of increase of 0.10/1000, or 0.0001, we have for the first 1000th of the year,

$$100(1 + 0.0001) = 100.01$$

For the second 1000th of the year,

$$100.01(1 + 0.0001) = 100.020001$$

If we repeat this for all 1000 time intervals, we end with 110.5 organisms at the end of one year.

Instantaneous rates and finite rates are nearly complementary when rates are very small. The following table shows how they diverge as the rates become large and illustrates the change in size of a hypothetical population that starts at 100 organisms and increases or decreases at the specified rate for one time period:

Change (%)	Finite Rate	Instantaneous Rate	Hypothetical Population at End of One Time Period
−75	0.25	−1.386	25
−50	0.50	−0.693	50
−25	0.75	−0.287	75
−10	0.90	−0.105	90
−5	0.95	−0.051	95
0	1.00	0.000	100
5	1.05	0.049	105
10	1.10	0.095	110
25	1.25	0.223	125
50	1.50	0.405	150
75	1.75	0.560	175
100	2.00	0.693	200
200	3.00	1.099	300
400	5.00	1.609	500
900	10.00	2.303	1000

	decreases	increases		
0	1.00		$+\infty$	Finite rates
$-\infty$	0.0		$+\infty$	Instantaneous rates

This illustrates one difference between finite rates (always positive or zero) and instantaneous rates (range from $-\infty$ to $+\infty$).

Mortality rates can be expressed as finite rates or as instantaneous rates. If the number of deaths in a short interval of time is proportional to the total population size at that time, the rate of drop in numbers can be described by the geometric equation

$$\frac{dN}{dt} = iN$$

where

N = population size
i = instantaneous mortality rate
t = time

In integral form, we have

$$N_t = N_0 e^{it}$$

or

$$\frac{N_t}{N_0} = e^{it}$$

where

N_0 = starting population size
N_t = population size at time t*

Taking logs, if $t = 1$ time unit, we obtain

$$\log_e\left(\frac{N_t}{N_0}\right) = i$$

Since N_t/N_0 is the finite survival rate by definition, we have obtained

$$\log_e(\text{finite survival rate}) = \text{instantaneous mortality rate}$$

We thus obtain the following relationships for expressing mortality rates:

$$\text{Finite survival rate} = 1.0 - \text{finite mortality rate}$$
$$\log_e(\text{finite survival rate}) = \text{instantaneous mortality rate}$$
$$\text{Finite survival rate} = e^{\text{instantaneous mortality rate}}$$
$$\text{Finite mortality rate} = 1.0 - e^{\text{instantaneous mortality rate}}$$

Why do we need to use instantaneous rates? The principal reason is that instantaneous rates are easier to deal with mathematically. A simple example will illustrate

---

* Note that instantaneous rates are determined for a specific time base (per year, per month, etc.), even though the rate applies to a very short time interval. The examples on page 703 illustrate this.

this property. Suppose you have data on an insect population and know that the mortality rate is 50 percent in the egg stage and 90 percent in the larval stages. How can you combine these mortalities? If they are expressed as finite mortality rates, we cannot add them because a 50 percent loss followed by a 90 percent loss is obviously not 140 percent mortality but only 95 percent mortality. If, however, the mortality is expressed as instantaneous rates, we can add them directly:

	Instantaneous Mortality Rate
Egg stage (50%)	-0.693
Larval stages (90%)	-2.303
Combined loss	-2.996

We can convert back to a finite mortality rate by the formula given earlier:

$$\text{Finite mortality rate} = 1.0 - e^{\text{instantaneous mortality rate}}$$
$$= 1.0 - e^{-2.996}$$
$$= 0.950$$

and the combined mortality is seen to be 95 percent.

Four examples will illustrate some of these ideas, and students are referred to Ricker (1975, chap. 1) for further discussion.

Example 1. A population increases from 73 to 97 within one year. This can be expressed as
   a. Finite rate of population growth $= 97/73 = 1.329$ per head per year (or, the population grew 32.9 percent in one year).
   b. Instantaneous rate of population growth $= \log_e(97/73) = 0.284$ per head per year.

Example 2. A population decreases from 67 to 48 within one month. This can be expressed as
   a. Finite rate of population growth $= 48/67 = 0.716$ per head per month (or, the population decreased 28.4 percent over the month).
   b. Instantaneous rate of population growth $= \log_e(48/67) = -0.333$ per head per month.

Example 3. A cohort of trees decreases in number from 24 to 19 within one year. This can be expressed as
   a. Annual survival rate (finite) $= 19/24 = 0.792$.
   b. Annual mortality rate (finite) $= 1.0 -$ annual survival rate $= 0.208$.
   c. Instantaneous mortality rate $= \log_e(19/24) = -0.234$ per year.

Example 4. A cohort of fish decreases in number from 350,000 to 79,000 within one year. This can be expressed as
   a. Annual survival rate (finite) $= 79,000/350,000 = 0.2257$.
   b. Annual mortality rate (finite) $= 1.0 - 0.2257 = 0.7743$.
   c. Instantaneous mortality rate $= \log_e 0.2257 = -1.488$ per year.

Appendix IV
Species Diversity Measures of Heterogeneity

There are several different measures of species diversity that are sensitive to both the number of species in the sample and the relative abundances of the species (Krebs 1989). I discuss here only two of the most commonly used measures of heterogeneity.

The Shannon-Wiener function approaches the measure of species diversity through information theory. We ask the question, How difficult would it be to predict correctly the species of the next individual collected? This is the same problem faced by communication engineers interested in predicting correctly the name of the next letter in a message. This uncertainty can be measured by the Shannon-Wiener function.*

$$H = -\sum_{i=1}^{S} (p_i)(\log_2 p_i)$$

where

H = information content of sample (bits/individual) = index of species diversity
S = number of species
p_i = proportion of total sample belonging to the ith species

Information content is a measure of the amount of uncertainty, so the larger value of H, the greater the uncertainty. A message such as *bbbbbbb* has no uncertainty in it, and $H = 0$. For our example on page 514 of two species of 99 and 1 individuals,

$$H = -[(p_1)(\log_2 p_1) + (p_2)(\log p_2)]$$
$$= -[0.99 (\log_2 0.99) + 0.01 (\log_2 0.01)] = 0.081 \text{ bit/individual}$$

For a sample of two species with 50 individuals in each,

$$H = -[0.50 (\log_2 0.50) + 0.50 (\log_2 0.50)]$$
$$= 1.00 \text{ bit/individual}$$

This agrees with our intuitive feeling that the second sample is more diverse than the first sample.

Strictly speaking, the Shannon-Wiener measure of information content should be used only on random samples drawn from a large community in which the total number of species is known. Pielou (1966) discusses information measures appropriate to other circumstances.

---

* This function was derived independently by Shannon and Wiener. It is sometimes mislabeled the Shannon-Weaver function.

Two components of diversity are combined in the Shannon-Wiener function: (1) number of species and (2) equitability or evenness of allotment of individuals among the species (Lloyd and Ghelardi 1964). A greater number of species increases species diversity, and a more even or equitable distribution among species will also increase species diversity measured by the Shannon-Wiener function. Equitability can be measured in several ways. The simplest approach is to ask, What would be the species diversity of this sample if all S species were equal in abundance? In this case,

$$H_{max} = -S\left(\frac{1}{S} \log_2 \frac{1}{S}\right) = \log_2 S$$

where

H_{max} = species diversity under conditions of maximal equitability
   S = number of species in the community

Thus, for example, in a community with two species only,

$$H_{max} = \log_2 2 = 1.00 \text{ bit/individual}$$

as we observed earlier. Equitability can now be defined as the ratio

$$E = \frac{H}{H_{max}}$$

where

   E = equitability (range 0–1)
   H = observed species diversity
H_{max} = maximum species diversity = $\log_2 S$

Table IV.1 presents a sample calculation illustrating the use of these formulas.

Other measures of species diversity can be derived from probability theory. Simpson (1949) suggested this question: What is the probability that two specimens picked at random in a community of infinite size will be the same species? If a person went into the boreal forest in northern Canada and picked two trees at random, there is a fairly high probability that they would be the same species. If a person went into the tropical rain forest, by contrast, two trees picked at random would have a low probability of being the same species. We can use this approach to determine an index of diversity:

Simpson's index of diversity = probability of picking two organisms at
                   random that are different species
              = 1 − (probability of picking two organisms
                 that are the same species)

If a particular species i is represented in the community by p_i (proportion of individuals), the probability of picking two of these at random is the joint probability $[(p_i)(p_i)]$ or p_i^2. If we sum these probabilities for all the i species in the community, we get Simpson's diversity (D):

$$D = 1 - \sum_{i=1}^{S} (p_i)^2$$

where

D = Simpson's index of diversity

p_i = proportion of individuals of species i in the community

For example, for our two-species community with 99 and 1 individuals,

$$D = 1 - [(0.99)^2 + (0.01)^2] = 0.02$$

Simpson's index gives relatively little weight to the rare species and more weight to the common species. It ranges in value from 0 (low diversity) to a maximum of $(1 - 1/S)$, where S is the number of species.

A more detailed discussion of these and other measures of species diversity is given in Magurran (1988) and Krebs (1989).

Table IV.1   Sample Calculations of Species Diversity and Equitability Through the Use of the Shannon-Wiener Function

Tree Species	PROPORTIONAL ABUNDANCE	
	(p_i)	$(p_i)(\log_2 p_i)^a$
Hemlock	0.521	0.490
Beech	0.324	0.527
Yellow birch	0.046	0.204
Sugar maple	0.036	0.173
Black birch	0.026	0.137
Red maple	0.025	0.133
Black cherry	0.009	0.061
White ash	0.006	0.044
Basswood	0.004	0.032
Yellow poplar	0.002	0.018
Magnolia	0.001	0.010
Total	1.000	$H = 1.829$

$$H_{max} = \log_2 S = \log_2 11 = 3.459$$

$$\text{Equitability} = E = \frac{H}{H_{max}} = \frac{1.829}{3.459} = 0.53$$

Note: Based on the composition of large trees (over 21.5 m tall) in a virgin forest in northwestern Pennsylvania.

$^a \log_2 x = \dfrac{\log_e x}{\log_e 2} = \dfrac{\log_{10} x}{\log_{10} 2}$

Note that there is no special theoretical reason to use $\log_2$ instead of $\log_e$ or $\log_{10}$. The $\log_2$ usage gives us information units in "bits" (binary digits) and is preferred by information theorists. See Pielou (1969, p. 229).

Source: Hough (1936).

Glossary

Ecological jargon is often carried to extremes in an attempt to confuse the amateur. This book tries to avoid most of this jargon, and the glossary contains words used in the text that might be unfamiliar to students.

abiotic factors characterized by the absence of life; include temperature, humidity, pH, and other physical and chemical influences.

accidental species species that occur with a lower degree of fidelity in a community type; not good species for use in community definition; see *characteristic species.*

aggregation coming together of organisms into a group, as in locusts.

allele one of a pair of characters that are alternative to each other in inheritance, being governed by genes situated at the same locus in homologous chromosomes.

allelopathy influence of plants, exclusive of microorganisms, upon each other, caused by products of metabolism; "antibiotic" interaction between plants.

association major unit in community ecology, characterized by essential uniformity of species composition.

autecology study of the individual in relation to environmental conditions.

autotroph organism that obtains energy from the sun and materials from inorganic sources; contrast with *heterotroph.*

biogeography branch of biology that deals with the geographic distribution of plants and animals.

biological control use of organisms or viruses to control parasites, weeds, or other pests.

biosphere the whole earth ecosystem.

biota species of all the plants and animals occurring within a certain area or region.

biotic factors environmental influences caused by plants or animals; opposite of *abiotic factors.*

bryophytes plants in the phylum Bryophyta comprising mosses, liverworts, and hornworts.

canonical distribution particular configuration of the log-normal distribution of species abundances.

carnivore flesh eater; organism that eats other animals; contrast with *herbivore.*

catastrophic agents term used by Howard and Fiske (1911) to describe agents of destruction in which the percentage of destruction is not related to population density; synonymous with *density-independent factors.*

characteristic species species that are rigidly limited to certain communities and thus can be used to identify a particular type of community.

climax kind of community capable of perpetuation under the prevailing climatic and edaphic conditions.

community group of populations of plants and animals in a given place; ecological unit used in a broad sense to include groups of various sizes and degrees of integration.

compensation point depth in a body of water at which the light intensity is such that the amount of oxygen produced by a plant's photosynthesis equals the oxygen it absorbs in respiration; the point at which respiration equals photosynthesis so that net production is zero.

competition occurs when a number of organisms of the same or of different species utilize common resources that are in short supply (*exploitation*); if the resources are not in short supply, competition occurs when the organisms seeking that resource harm one another in the process (*interference*).

connectance used to describe food web complexity; the fraction of potential interactions in a food web that actually exist.

contingency table frequency distribution of an *n*-way statistical classification.

continuum index measure of the position of a community on a gradient defined by the species composition.

deme interbreeding group in a population; also known as *local population.*

density number of individuals in relation to the space in which they occur.

deterministic model mathematical model in which all the relationships are fixed and the concept of probability does not enter; a given input produces one exact prediction as output; opposite of *stochastic model.*

diapause period of suspended growth or development and reduced metabolism in the life cycle of many insects, in which the organism is more resistant to unfavorable environmental conditions than in other periods.

dilution rate general term to describe the rate of additions to a population from birth and immigration.

dominance condition in communities or in vegetational strata in which one or more species, by means of their number, coverage, or size, have considerable influence upon or control of the conditions of existence of associated species.

dynamic pool model type of optimum-yield model in which the yield is predicted from the components of growth, mortality, recruitment, and fishing intensity; contrast with *logistic-type model*.

dynamics in population ecology, the study of the reasons for *changes* in population size; contrast with *statics*.

ecological longevity average length of life of individuals of population under stated conditions.

ecosystem biotic community and its abiotic environment; the whole earth can be considered as one large ecosystem.

ecotone transition zone between two diverse communities (e.g., the tundra-boreal forest ecotone).

ecotype subspecies or race that is especially adapted to a particular set of environmental conditions.

edaphic pertaining to the soil.

elfinwood (Krummholz) scrubby, stunted growth form of trees, often forming a characteristic zone at the limit of tree growth in mountains.

environment all the biotic and abiotic factors that actually affect an individual organism at any point in its life cycle.

epidemiology branch of medicine dealing with epidemic diseases.

epipelic algae algae living in or on the sediments of a body of water.

equitability evenness of distribution of species abundance patterns; maximum equitability occurs when all species are represented by the same number of individuals.

estivation condition in which an organism may pass an unfavorable season and in which its normal activities are greatly curtailed or temporarily suspended.

evapotranspiration sum total of water lost from the land by evaporation and plant transpiration.

facultative agents term used by Howard and Fiske (1911) to describe agents of destruction that increase their percentage of destruction as population density rises; synonymous with *density-dependent factors*.

fecundity an ecological concept based on the number of offspring produced during a unit of time.

fertility a physiological notion indicating that an organism is capable of breeding.

fidelity degree of regularity or "faithfulness" with which a species occurs in certain plant communities, expressed on a five-part scale: (5) exclusive, (4) selective, (3) preferetial,(2) companion, indifferent, (1) accidental, strangers.

food chain figure of speech for the dependence for food of organisms upon others in a series, beginning with plants and ending with the largest carnivores.

genecology study of population genetics in relation to the habitat conditions, the study of species and other taxa by the combined methods and concepts of ecology and genetics.

genotype entire genetic constitution of an organism; contrast with *phenotype*.

global stability ability to withstand perturbations of a large magnitude and not be affected; compare with *neighborhood stability*.

gradocoen totality of all factors that impinge on a population, including biotic agents and abiotic factors.

gross production production before respiration losses are subtracted; photosynthetic production for plants and metabolizable production for animals.

growth form morphological categories of plants, such as trees, shrubs, and vines.

herbivore organism that eats plants; contrast with *carnivore*.

heterotroph organism that obtains energy and materials by eating other organisms; contrast with *autotroph*.

homeostasis maintenance of constancy or a high degree of uniformity in functions of an organism or interactions of individuals in a population or community under changing conditions, because of the capabilities of organisms to make adjustments.

homeothermic pertaining to warm-blooded animals that regulate their body temperature; contrast with *poikilothermic*.

host organism that furnishes food, shelter, or other benefits to another organism of a different species.

hydrophyte plant that grows wholly or partly immersed in water; compare with *xerophyte* and *mesophyte*.

importance value sum of relative density, relative dominance, and relative frequency for a species in the community; scale from 0 to 300; the larger the importance value, the more dominant a species is in the particular community.

index of similarity ratio of the number of species found in common in two communities to the total number of species that are present in both.

indifferent species species occurring in many different communities; not good species for community classification.

innate capacity for increase (r_m) measure of the rate of increase of a population under controlled conditions.

interspecific competition competition between members of different species.

isotherm line drawn on a map or chart connecting places with the same temperature at a particular time or for a certain period.

life table tabulation presenting complete data on the mortality schedule of a population.

littoral shallow-water zone of lakes or the sea, with light penetration to the bottom; often occupied by rooted aquatic plants.

local population see *deme.*

logistic equation model of population growth described by a symmetrical S-shaped curve with an upper asymptote.

logistic-type model type of optimum-yield model in which the yield is predicted from an overall descriptive function of population growth without a separate analysis of the components of mortality, recruitment, and growth; contrast with *dynamic pool model.*

log-normal distribution frequency distribution of species abundances in which the *x* axis is expressed on a logarithmic scale; *x* axis is (log) number of individuals represented in sample, *y* axis is number of species.

loss rate general term to describe the rate of removal of organisms from a population by death and emigration.

mesic moderately moist.

mesophyte plant that grows in environmental conditions that are medium in moisture conditions.

metapopulation a set of local populations linked together through dispersal.

monogamy mating of an animal with only one member of the opposite sex.

monothetic positing but one essential element; in classifications, defining

groups on the basis of a single key character.

morphology study of the form, structure, and development of organisms.

multivoltine refers to an organism that has several generations during a single season; contrast with *univoltine.*

mutualism interaction between two species in which both benefit from the association and cannot live separately.

neighborhood stability ability to withstand perturbations of small magnitude and not be affected; compare with *global stability.*

net production production after respiration losses are subtracted.

niche role or "profession" of an organism in the environment; its activities and relationships in the community.

obligate predator or *parasite* predator or parasite that is restricted to eating a single species of prey.

oligochaetes any of a class or order (Oligochaeta) of hermaphroditic terrestrial or aquatic annelids lacking a specialized head; includes earthworms.

optimum yield amount of material that can be removed from a population that will maximize biomass (or numbers, or profit, or any other type of "optimum") on a sustained basis.

ordination process by which plant or animal communities are ordered along a gradient.

parasite organism that benefits while feeding upon, securing shelter from, or otherwise injuring another organism (the host); insect parasitoids are usually fatal to their host and behave more like vertebrate predators.

parthenogenesis development of the egg of an organism into an embryo without fertilization.

phenology study of the periodic (seasonal) phenomena of animal and plant life and their relations to the weather

and climate (e.g., the time of flowering in plants).

phenotype expression of the characteristics of an organism as determined by the interaction of its genic constitution and the environment; contrast with *genotype.*

photoperiodism response of plants and animals to the relative duration of light and darkness (e.g., a chrysanthemum blooming under short days and long nights).

photosynthesis synthesis of carbohydrates from carbon dioxide and water by chlorophyll using light as energy with oxygen as a byproduct.

physiological longevity maximum life span of individuals in a population under specified conditions; the organisms die of senescence.

phytoplankton plant portion of the plankton; the plant community in marine and freshwater situations that floats free in the water and contains many species of algae and diatoms.

poikilothermic of or pertaining to cold-blooded animals, organisms that have no rapidly operating heat-regulatory mechanism; contrast with *homeothermic.*

polyandry mating of a single female animal with several males.

polygyny mating of one male animal with several females.

polythetic positing many essential elements; in classifications, defining groups on the basis of many characteristics, not just one.

population group of individuals of a single species.

primary production production by green plants.

production amount of energy (or material) formed by an individual, population, or community in a specific time period; includes growth and reproduction only; see *primary production, secondary production, gross production, net production.*

productivity a general concept denoting all processes involved in production studies—consumption, rejection, respiration, etc.; some authors use as a synonym for *production,* but this should be avoided since *production* is a narrower term.

promiscuous not restricted to one sexual partner.

proximate factors in evolutionary terms, the mechanisms responsible for an adaptation with reference to its physiological and behavioral operation; the mechanics of how an adaptation operates; opposite of *ultimate factors.*

recruitment increment to a natural population, usually from young animals or plants entering the adult population.

respiration complex series of chemical reactions in all organisms by which energy is made available for use; carbon dioxide, water, and energy are the end products.

saprophyte plant that obtains food from dead or decaying organic matter.

secondary production production by herbivores, carnivores, or detritus feeders; contrast with *primary production.*

self-regulation process of population regulation in which population increase is prevented by a deterioration in the quality of individuals that make up the population; population regulation by internal adjustments in behavior and physiology within the population rather than by external forces such as predators.

senescence process of aging.

seral referring to a series of stages that follow one another in an ecological succession.

serotinous cones cones of some pine trees that remain on the trees for several years without opening and require a fire to open and release the seeds.

sessile attached to an object or fixed in place (e.g., barnacles).

sigmoid curve S-shaped curve (e.g., the logistic curve).

stability absence of fluctuations in populations; ability to withstand perturbations without large changes in composition.

statics in population ecology, the study of the reasons of equilibrium conditions or average values; contrast with *dynamics.*

stenoplastic having little or no modificational plasticity (steno- means narrow); opposite of *euryplastic.*

steppe extensive area of natural, dry grassland; usually used in reference to grasslands in southwestern Asia and southeastern Europe; equivalent to *prairie* in North American usage.

sterol any of a group of solid, mostly unsaturated polycyclic alcohols, such as cholesterol or ergosterol, derived from plants and animals.

stochastic model mathematical model based on probabilities; the prediction of the model is not a single fixed number but a range of possible numbers; opposite of *deterministic model.*

sublittoral lower division in the sea from a depth of 40 to 60 meters to about 200 meters; below the littoral zone.

succession replacement of one kind of community by another kind; the progressive changes in vegetation and animal life that may culminate in the climax.

symbiosis in a broad sense, the living together of two or more organisms of different species; in a narrow sense, synonymous with *mutualism.*

synecology study of groups of organisms in relation to their environment; includes population, community, and ecosystem ecology.

taiga the northern boreal forest zone, a broad band of coniferous forest south of the arctic tundra.

thermoregulation maintenance or regulation of temperature, specifically the maintenance of a particular temperature of the living body.

trace element chemical element used by organisms in minute quantities and essential to their physiology.

triclad any of an order Tricladida of turbellarian platyhelminths, distinguished by having one anterior and two posterior branches of the intestine.

trophic level functional classification of organisms in a community according to feeding relationships; the first trophic level includes green plants; the second trophic level includes herbivores; and so on.

tundra treeless area in arctic and alpine regions, varying from a bare area to various types of vegetation consisting of grasses, sedges, forbs, dwarf shrubs, lichens, and mosses.

ultimate factors in evolutionary terms, the survival value of the adaptation in question; the evolutionary reason for the adaptation; opposite of *proximate factors.*

univoltine refers to an organism that has only one generation per year.

vector organism (often an insect) that transmits a pathogenic virus, bacterium, protozoan, or fungus from one organism to another.

wilting point measure of soil water; the water remaining in the soil (expressed as percentage of dry weight of the soil) when the plants are in a state of permanent wilting from water shortage.

xeric deficient in available moisture for the support of life (e.g., desert environments).

xerophyte plant that can grow in dry places (e.g., cactus).

zooplankton animal portion of the plankton; the animal community in marine and freshwater situations that floats free in the water, independent of the shore and the bottom, moving passively with the currents.

Mathematical Symbols

All the symbols used in equations in this book are defined here. I have tried to minimize duplication of letters. There is no universal agreement on the use of symbols, and other books and papers will not agree with my usage in some cases.

α 1. competition coefficient measuring the effect of species 2 on species 1; converts species 2 into equivalent units of species 1 (page 232)

                  2. α-diversity = within habitat species diversity (page 527)

a in the logistic equation, a constant of integration that defines the position of the curve relative to the origin (page 205)

A area of island for species-area curve (page 587)

$AVHRR$ advanced very high resolution radiometer; a sensor aboard the NOAA satellite (page 622)

β 1. competition coefficient measuring the effect of species 1 on species 2; converts species 1 into equivalent units of species 2 (page 233)

                  2. β-diversity = between habitat species diversity (page 527)

b instantaneous birth rate (page 203)

b_x in a fertility table, number of offspring produced per unit of time per female aged x; called m_x by some authors (page 176)

B slope of line in a plot of net reproductive rate (y) on population density (x) (page 200)

B_1 and B_2 biomass in the community at time t_1 and time t_2 (page 606)

ΔB biomass change in the community between time t_1 and time t_2 (page 606)

c slope of line for a plot of instantaneous death rate (y axis) on population density (x axis) (page 325)

C constant measuring the efficiency of the predator (page 263)

C_x proportion of organisms in the age category x to $x + 1$ in a population increasing geometrically; the stable age distribution (page 187)

d instantaneous death rate (page 203)

d_x number of organisms dying in the interval x to $x + 1$ in a life table (page 168)

D Simpson's measure of species diversity (page 705)

e_x	mean expectation of life for organisms alive at start of age x in the life table (page 168)
E	equitability of species abundance patterns (page 705)
F	1. yield to fishery (page 350) 2. instantaneous fishing mortality rate (page 355)
G	1. generation time; called T by many authors (page 180) 2. growth in weight of fish in optimum-yield equations (page 350) 3. in production studies, biomass losses to consumer organisms (page 607)
H	species diversity measured by the information content of sample (Shannon-Wiener function) (page 704)
I	amount of solar radiation (page 610)
I_z	solar radiation at depth z in a water column (page 610)
k	extinction coefficient for light passing into water (page 610)
K	upper asymptote or maximal density reached by a population growing in a logistic manner (page 205)
K_1 and K_2	in the Lotka-Volterra competition equations, upper asymptote for population density of species 1 and species 2 when growing separately (page 232)
λ	finite rate of increase, equal to e^r (page 181)
l_x	proportion of individuals surviving to start of age interval x in a life table (page 168)
L	1. in population growth model, $L = BN_{eq}$ (page 201) 2. in production studies, biomass losses by death of plants or plant parts (page 607)
L_x	stationary age distribution of the life table; number of individuals alive on the average during the age interval x to $x + 1$ (page 187)
M	1. weight of fish removed by natural deaths (page 350) 2. instantaneous natural mortality rate in yield equations (page 355)
n_x	number of survivors at start of age interval x in a life table (page 168)
N	number of organisms in a population (page 156); may be used with a subscript: N_t = number of organisms at time t (page 198); N_1 = number of organisms of species 1 (page 232)
NDVI	normalized difference vegetation index used to measure relative primary production or vegetation greenness from satellite images (page 622)
N_{eq}	equilibrium population size, when $R_0 = 1.0$ (page 200)
$NOAA$	meteorological satellite operated by National Oceanic and Atmospheric Administration of the U.S. (page 622)
p_i	proportion of the total sample that belongs to the ith species in species-abundance studies (page 704)
P	population size of predator in predator-prey models (page 263)

q	constant in logistic-type optimum-yield equations, conversion factor for converting fishing effort to fishing mortality rate (page 352)
q_x	rate of mortality during the age interval x to $x + 1$ in a life table (page 168)
Q	constant measuring the efficiency of utilization of prey for reproduction by predators (page 263)
r	intrinsic capacity for increase (= per capita rate of population increase) (pages 175, 178)
r_1 and r_2	per-capita rate of increase of species 1 and species 2 in competition models (page 232)
R	1. maximum reproductive rate of prey in predator-prey models (page 263) 2. weight of new recruits in optimum-yield equations (page 350)
R_0	net reproductive rate (page 176)
S	1. maximum reproductive rate of the predator in predator-prey models (page 264) 2. number of species in a sample (page 704)
S_1 and S_2	weight of catchable stock at start of year (S_1) and at end of year (S_2) (page 350)
t	time (page 203)
t_c	age at which fish enter the fishery (page 355)
V_x	reproductive value at age x (page 184)
W_t	average weight of age t fish (page 355)
x	subscript to denote age in life tables (page 168)
X	amount of fishing effort in logistic-type equation for optimum yield (page 352)
Y	yield in weight to fishery in optimum-yield equations (page 355)
z	deviation from equilibrium density in population growth ($N - N_{eq}$) (page 200)

ÅNGREN, G. I., R. E. McMURTRIE, W. J. PARTON, J. PASTOR, and H. H. SHUGART. 1991. State-of-the-art of models of production-decomposition linkages in conifer and grassland ecosystems. *Ecological Applications* 1:118–138.

ABELE, L. G., and K. WALTERS. 1979. Marine benthic diversity: A critique and alternative explanation. *Journal of Biogeography* 6:115–126.

ABRAMS, P. 1975. Limiting similarity and the form of the competition coefficient. *Theoretical Population Biology* 8:356–375.

ABRAMS, P. A. 1986. Adaptive responses of predators to prey and prey to predators: The failure of the arms-race analogy. *Evolution* 40:1229–1247.

ABRAMS, P. A. 1987. On classifying interactions between populations. *Oecologia* 73:272–281.

ADDICOTT, J. F. 1974. Predation and prey community structure: An experimental study of the effect of mosquito larvae on the protozoan communities of pitcher plants. *Ecology* 55:475–492.

ADDICOTT, J. F. 1986. On the population consequences of mutualism. In *Community Ecology*, ed. J. Diamond and T. J. Case, pp. 425–436. Harper & Row, New York.

AGNEW, A. D. Q. 1961. The ecology of *Juncus effusus* L. in North Wales. *Journal of Ecology* 49:83–102.

ALCOCK, J. 1989. *Animal Behavior: An evolutionary approach*. Sinauer Associates, Sunderland, Mass.

ALLEN, K. R. 1980. *Conservation and Management of Whales*. University of Washington Press, Seattle.

ANDERBRANT, O. and F. SCHLYTER. 1987. Ecology of the Dutch elm disease vectors *Scolytus laevis* and *S. scolytus* (Coleoptera: Scolytidae) in southern Sweden. *Journal of Applied Ecology* 24:539–550.

ANDERSON, D. R., J. BART, T. C. EDWARDS, Jr., C. B. KEPLER, and E. C. MESLOW. 1990. *1990 Status Review: Northern Spotted Owl*. U.S. Fish and Wildlife Service, Washington, D.C.

ANDERSON, G. R. V., A. H. EHRLICH, P. R. EHRLICH, J. D. ROUGHGARDEN, B. C. RUSSELL, and F. H. TALBOT. 1981. The community structure of coral reef fishes. *American Naturalist* 117:476–495.

ANDERSON, R. C. 1982. An evolutionary model summarizing the roles of fire, climate, and grazing animals in the origin and maintenance of grasslands. In *Grasses and Grasslands: Systematics and Ecology*, ed. J. R. Estes, R. J. Tyrl and J. N. Brunken, pp. 297–308. Univiversity of Oklahoma Press, Norman.

ANDERSON, R. C. 1990. The historic role of fire in the North American grassland. In *Fire in North American Tallgrass Prairies*, ed. S. L. Collins and L. L. Wallace, pp. 8–18. University of Oklahoma Press, Norman.

ANDERSON, S. 1977. Geographic ranges of North American terrestrial mammals. *American Museum Novitates* 2629:1–15.

ANDERSON, S. 1984. Geographic ranges of North American birds. *American Museum Novitates* 2785:1–17.

ANDERSON, S. 1985. The theory of range-size (RS) distributions. *American Museum Novitates* 2833:1–20.

ANDREN, H., and P. ANGELSTAM. 1988. Elevated predation rates as an edge effect in habitat islands: Experimental evidence. *Ecology* 69:544–547.

ANDREWARTHA, H. G. 1961. *Introduction to the Study of Animal Populations.* University of Chicago Press, Chicago.

ANDREWARTHA, H. G., and L. C. BIRCH. 1954. *The Distribution and Abundance of Animals.* University of Chicago Press, Chicago.

ANTONOVICS, J., A. D. BRADSHAW, and R. G. TURNER. 1971. Heavy metal tolerance in plants. *Advances in Ecological Research* 7:1–85.

APLET, G. H., R. D. LAVEN, and F. W. SMITH. 1988. Patterns of community dynamics in Colorado Engelmann spruce—subalpine fir forests. *Ecology* 69:312–319.

APPLEGATE, V. C. 1950. Natural history of the sea lamprey, *Petromyzon marinus*, in Michigan. U. S. Fish and Wildlife Service, Special Scientific Report, Fisheries No. 55, Washington, D.C.

ARDITI, R., and L. R. GINZBURG. 1989. Coupling in predator-prey dynamics: Ratio-dependence. *Journal of Theoretical Biology* 139:311–326.

ARDITI, R., L. R. GINZBURG, and H. R. AKCAKAYA. 1991b. Variation in plankton densities among lakes: A case for ratio-dependent predation models. *American Naturalist* 138:1287–1296.

ARDITI, R., N. PERRIN, and H. SAÏAH. 1991a. Functional responses and heterogeneities: An experimental test with cladocerans. *Oikos* 60:69–75.

ARITA, H. T., J. G. ROBINSON, and K. H. REDFORD. 1990. Rarity in neotropical forest mammals and its ecological correlates. *Conservation Biology* 4:181–192.

ARMBRUST, E. J., and G. G. GYRISCO. 1982. Forage crops insect pest management. In *Introduction to Insect Pest Management*, ed. R. L. Metcalf and W. Luckman, pp. 443–463. Wiley, New York.

ARMSTRONG, R. A., and R. McGEHEE. 1980. Competitive exclusion. *American Naturalist* 115:151–170.

ARTHUR, W. 1982. The evolutionary consequences of interspecific competition. *Advances in Ecological Research* 12:127–187.

ASHBY, E. 1948. Statistical ecology. II. A reassessment. *Botanical Review* 14:222–234.

ASHMOLE, N. P. 1968. Body size, prey size, and ecological segregation in five sympatric tropical terns (Aves: Laridae). *Systematic Zoology* 17:292–304.

ASHTON, P. S., T. J. GIVNISH, and S. APPANAH. 1988. Staggered flowering in the Dipterocarpaceae: new insights into floral induction and the evolution of mast fruiting in the aseasonal tropics. *American Naturalist* 132:44–66.

ATTIWILL, P. 1981. Energy, nutrient flow, and biomass. *Proceedings of the Australian Forest Nutrition Workshop* 1:131–144. CSIRO, Melbourne.

AUSTIN, M. P. 1985. Continuum concept, ordination methods, and niche theory. *Annual Review of Ecology and Systematics* 16:39–61.

AUSTIN, M. P. 1990. Community theory and competition in vegetation. In *Perspectives on Plant Competition*, ed. J. B. Grace and D. Tilman, pp. 215–238. Academic Press, San Diego.

AUSTIN, M. P., and T. M. SMITH. 1989. A new model for the continuum concept. *Vegetatio* 83:35–47.

AVERS, C. J. 1989. *Process and Pattern in Evolution*. Oxford University Press, New York.

BACH, C. E. 1991. Direct and indirect interactions between ants (*Pheidole megacephala*), scales (*Coccus viridis*) and plants (*Pluchea indica*). *Oecologia* 87:233–239.

BAGENAL, T. B. 1973. Fish fecundity and its relations with stock and recruitment. *Rapport du Conseil International pour l'Exploration de la Mer* 164:186–198.

BAKER, M. C., D. B. THOMPSON, G. L. SHERMAN, M. A. CUNNINGHAM, and D. F. TOMBACK. 1982. Allozyme frequencies in a linear series of song dialect populations. *Evolution* 36:1020–1029.

BAKUS, G. J., N. M. TARGETT, and B. SCHULTE. 1986. Chemical ecology of marine organisms: An overview. *Journal of Chemical Ecology* 12:951–985.

BALDWIN, N. S. 1964. Sea lamprey in the Great Lakes. *Canadian Audubon Magazine* Nov.-Dec. 1964, pp. 2–7.

BARBOUR, M. G., J. H. BURK, and W. D. PITTS. 1980. *Terrestrial Plant Ecology*. Benjamin/Cummings, Menlo Park, Calif.

BARCLAY, G. W. 1958. *Techniques of Population Analysis*. Wiley, New York.

BARCLAY-ESTRUP, P., and C. H. GIMINGHAM. 1969. The description and interpretation of cyclical processes in a heath community. I. Vegetational change in relation to the *Calluna* cycle. *Journal of Ecology* 57:737–758.

BARNES, H. 1962. So-called anecdysis in *Balanus balanoides* and the effect of breeding upon the growth of calcareous shell of some common barnacles. *Limnology and Oceanography* 7:462–473.

BARNOLA, J. M., D. RAYNAUD, Y. S. KOROTKEVICH, and C. LORIUS. 1987. Vostok ice core provides 160,000–year record of atmospheric CO_2. *Nature* 329:408–414.

BART, J., and E. D. FORSMAN. 1992. Dependence of northern spotted owls *Strix occidentalis caurina* on old-growth forests in the western USA. *Biological Conservation* 62:95–100.

BARTHOLOMEW, B. 1970. Bare zone between California shrub and grassland communities: The role of animals. *Science* 170:1210–1212.

BARTHOLOMEW, G. A. 1958. The role of physiology in the distribution of terrestrial vertebrates. In *Zoogeography*, ed. C. L. Hubbs, pp. 81–95. American Association for the Advancement of Science Publication No. 51, Washington, D.C.

BAZZAZ, F. A. 1990. The response of natural ecosystems to the rising global CO_2 levels. *Annual Review of Ecology and Systematics* 21:167–196.

BAZZAZ, F. A. 1991. Habitat selection in plants. *American Naturalist* 137(suppl.):S116–S130.

BAZZAZ, F. A., and R. W. CARLSON. 1984. The response of plants to elevated CO_2. I. Competition among an assemblage of annuals at two levels of soil moisture. *Oecologia* 62:196–198.

BEADLE, N. C. W. 1966. Soil phosphate and its role in molding segments of the Australian flora and vegetation, with special reference to xeromorphy and sclerophylly. *Ecology* 47:992–1007.

BEARD, J. S. 1973. The physiognomic approach. In *Ordination and Classification of Communities*, ed. R. H. Whittaker, pp. 355–386. Dr. W. Junk Publishers, The Hague.

BEDDINGTON, J. R., C. A. FREE, and J. H. LAWTON. 1976. Concepts of stability and resilience in predator-prey models. *Journal of Animal Ecology* 45:791–816.

BEGON, M. 1979. *Investigating Animal Abundance*. Edward Arnold, London.

BEGON, M., J. L. HARPER, and C. R. TOWNSEND. 1990. *Ecology: Individuals, Populations, and Communities*. 2d ed. Blackwell, Oxford, England.

BEILMANN, A. P., and L. G. BRENNER. 1951. The recent intrusion of forests in the Ozarks. *Annals of the Missouri Botanical Garden* 38:261–282.

BELL, A. R., and J. D. NALEWAJA. 1968. Competitive effects of wild oat in flax. *Weed Science* 16:501–504.

BELL, G. 1980. The costs of reproduction and their consequences. *American Naturalist* 116:45–76.

BELL, G. 1987. Review of "Evolution through Group Selection" by V. C. Wynne-Edwards. *Heredity* 59:145–147.

BELL, R. H. V. 1971. A grazing ecosystem in the Serengeti. *Scientific American* 225(1):86–93.

BELOVSKY, G. E. 1987. Extinction models and mammalian persistence. In *Viable Populations for Conservation*, ed. by M. E. Soulé, pp. 35–57. Cambridge University Press, Cambridge, England.

BELSKY, A. J. 1986. Does herbivory benefit plants? A review of the evidence. *American Naturalist* 127:870–892.

BELSKY, A. J. 1987. The effects of grazing: Confounding of ecosystem, community, and organism scales. *American Naturalist* 129:777–783.

BENNETT, W. A. 1990. Scale of investigation and the detection of competition: An example from the house sparrow and house finch introductions in North America. *American Naturalist* 135:725–747.

BERGERON, Y., and J. BRISSON. 1990. Fire regime in red pine stands at the northern limit of the species' range. *Ecology* 71:1352–1364.

BERGERUD, A. T. 1980. A review of the population dynamics of caribou and wild reindeer in North America. In *Proceedings of the 2nd International Reindeer/Caribou Symposium*, pp. 556–581. Roros, Norway.

BERRYMAN, A. A. 1981. *Population Systems: A General Introduction*. Plenum Press, New York.

BERTRAM, B. C. R. 1978. Living in groups: Predators and prey. In *Behavioural Ecology*, ed. J. R. Krebs and N. B. Davies, pp. 64–96. Sinauer Associates, Sunderland, Mass.

BEURDEN, E. VAN. 1981. Bioclimatic limits to the spread of *Bufo marinus* in Australia: A baseline. *Proceedings of the Ecological Society of Australia* 1:143–149.

BEVERTON, R. J. H. 1962. Long-term dynamics of certain North Sea fish populations. In *The Exploitation of Natural Animal Populations*, ed. E. D. LeCren and M. W. Holdgate, pp. 242–259. Blackwell, Oxford, England.

BEVERTON, R. J. H., and S. J. HOLT. 1957. *On the Dynamics of Exploited Fish Populations*. H. M. Stationery Office, London.

BILLINGS, W. D. 1938. The structure and development of old-field shortleaf pine stands and certain associated physical properties of the soil. *Ecological Monographs* 8:437–499.

BILLINGS, W. D. 1950. Vegetation and plant growth as affected by chemically altered rocks in the western Great Basin. *Ecology* 31:62–74.

BILLINGS, W. D., J. O. LUKEN, D. A. MORTENSEN, and K. M. PETERSON. 1983. Increasing atmospheric carbon dioxide: Possible effects on arctic tundra. *Oecologia* 58:286–289.

BILLINGS, W. D., K. M. PETERSON, J. O. LUKEN, and D. A. MORTENSEN. 1984. Interaction of increasing atmospheric carbon dioxide and soil nitrogen on the carbon balance of tundra microcosms. *Oecologia* 65:26–29.

BINKLEY, C. S. and R. S. MILLER. 1983. Population characteristics of the whooping crane, *Grus americana*. *Canadian Journal of Zoology* 61: 2768–2776.

BIRCH, D. W. 1981. Dominance in marine ecosystems. *American Naturalist* 118:262–274.

BIRCH, L. C. 1948. The intrinsic rate of natural increase of an insect population. *Journal of Animal Ecology* 17:15–26.

BIRCH, L. C. 1953a. Experimental background to the study of the distribution and abundance of insects. I. The influence of temperature, moisture, and food on the innate capacity for increase of three grain beetles. *Ecology* 34:698–711.

BIRCH, L. C. 1953b. Experimental background to the study of the distribution and abundance of insects. III. The relations between innate capacity for increase and survival of different species of beetles living together on the same food. *Evolution* 7:136–144.

BIRCH, L. C. 1957. The meanings of competition. *American Naturalist* 91:5–18.

BIRCH, L. C., and P. R. EHRLICH. 1967. Evolutionary history and population biology. *Nature* 214: 349–352.

BJORKMAN, O. 1973. Comparative studies on photosynthesis in higher plants. In *Photophysiology*, vol. 8, ed. A. C. Giese, pp. 1–63. Academic Press, New York.

BJORKMAN, O. 1975. Inaugural address. In *Environmental and Biological Control of Photosynthesis*, ed. R. Marcell, pp. 1–16, Dr. W. Junk Publishers, The Hague.

BJORKMAN, O., and J. BERRY. 1973. High efficiency photosynthesis. *Scientific American* 229(4):80–93.

BLACK, C. C. 1971. Ecological implications of dividing plants into groups with distinct photosynthetic production capacities. *Advances in Ecological Research* 7:87–114.

BLANK, T. H., T. R. E. SOUTHWOOD, and D. J. CROSS. 1967. The ecology of the partridge. I. Outline of the population process with particular reference to chick mortality and nest density. *Journal of Animal Ecology* 36:549–556.

BLISS, L. C. 1991. Arctic ecosystems: Patterns of change in response to disturbance. In *The Earth in Transition: Patterns and Processes of Biotic Impoverishment*, ed. G. M. Woodwell, pp. 347–366. Cambridge Press, New York.

BOAG, P. T. 1983. The heritability of external morphology in Darwin's ground finches (*Geospiza*) on Isla Daphne Major, Galápagos. *Evolution* 37:877–894.

BOAG, P. T., and P. R. GRANT. 1978. Heritability of external morphology in Darwin's finches. *Nature* 274:793–794.

BOAG, P. T., and P. R. GRANT. 1981. Intense natural selection in a population of Darwin's finches (Geospizinae) in the Galápagos. *Science* 214:82–84.

BOCK, C. E. 1987. Distribution-abundance relationships of some Arizona landbirds: A matter of scale? *Ecology* 68:124–129.

BODENHEIMER, F. S. 1928. Welche Faktoren regulieren die Individuenzahl einer Insektenart in der Natur? *Biologisches Zentralblatt* 48:714–739.

BOEREMA, L. K., and J. A. GULLAND. 1973. Stock assessment of the Peruvian anchovy (*Engraulis ringens*) and management of the fishery. *Journal of the Fisheries Research Board of Canada* 30:2226–2235.

BOND, W. 1983. On alpha diversity and the richness of the Cape Flora: A study in Southern Cape fynbos. In *Mediterrean-type Ecosystems: The Role of Nutrients,* ed. F. J. Kruger, D. T. Mitchell, and J. U. M. Jarvis, pp. 337–356. Springer-Verlag, Berlin.

BOND, W. J. 1989. The tortoise and the hare: Ecology of angiosperm dominance and gymnosperm persistence. *Biological Journal of the Linnean Society* 36:227–249.

BORCHERT, J. R. 1950. The climate of the central North American grassland. *Annals of the Association of American Geographers* 40:1–39.

BORMANN, F. H., G. E. LIKENS, T. G. SICCAMA, R. S. PIERCE, and J. S. EATON. 1974. The export of nutrients and recovery of stable conditions following deforestation at Hubbard Brook. *Ecological Monographs* 44:255–277.

BOTKIN, D. B. 1981. Causality and succession. In *Forest Succession*, ed. D. C. West, H. H. Shugart, and D. B. Botkin, pp. 36–55. Springer-Verlag, New York.

BOUCHER, D. H., S. JAMES, and K. H. KEELER. 1982. The ecology of mutualism. *Annual Review of Ecology and Systematics* 13:315–347.

BOUTIN, S. 1992. Predation and moose population dynamics: A critique. *Journal of Wildlife Management* 56:116–127.

BOUTTON, T. W., G. N, CAMERON, and B. N. SMITH. 1978. Insect herbivory on C_3 and C_4 grasses. *Oecologia* 36:21–32.

BOYCE, M. S. 1984. Restitution of r- and K-selection as a model of density-dependent natural selection. *Annual Review of Ecology and Systematics* 15:427–447.

BOYER, J. S. 1982. Plant productivity and environment. *Science* 218: 443–448.

BRADSHAW, A. D. and K. HARDWICK. 1989. Evolution and stress—genotypic and phenotypic components. *Biological Journal of the Linnean Society* 37:137–155.

BRASIER, C. M. 1988. Rapid changes in genetic structure of epidemic populations of *Ophiostoma ulmi. Nature* 332:538–541.

BRAUN-BLANQUET, J. 1932. *Plant Sociology*, transl. G. D. Fuller and H. S. Conard. McGraw-Hill, New York.

BREMAN, H., and C. T. DE WIT. 1983. Rangeland productivity and exploitation in the Sahel. *Science* 221:1341–1347.

BRETT, J. R. 1956. Some principles in the thermal requirements of fishes. *Quarterly Review of Biology* 31:75–87.

BRIMBLECOMBE, P., C. HAMMER, H. RODHE, A. RYABOSHAPKO, and C. F. BOUTRON. 1989. Human influence on the sulphur cycle. In *Evolution of the Global Biogeochemical Sulphur Cycle*, ed. P. Brimblecombe and A. Y. Lein, pp. 77–121. Wiley, New York.

BROCK, T. D. 1966. *Principles of Microbial Ecology.* Prentice-Hall, Englewood Cliffs, N.J.

BROKAW, N. V. 1985. Treefalls, regrowth, and community structure in tropical forests. In *The Ecology of Natural Disturbance and Patch Dynamics*, ed. S. T. A. Pickett and P. S. White, pp. 53–69. Academic Press, New York.

BROKAW, N. V. 1987. Gap-phase regeneration of three pioneer tree species in a tropical forest. *Journal of Ecology* 75:9–19.

BROOKS, J. L., and S. I. DODSON. 1965. Predation, body size, and composition of plankton. *Science* 150:28–35.

BROWER, L. P. 1988. Avian predation on the monarch butterfly and its implications for mimicry theory. *American Naturalist* 131(suppl.):S4–S6.

BROWN, E. S. 1951. The relation between migration rate and types of habitat in aquatic insects, with special reference to certain species of Corixidae. *Proceedings of the Zoological Society of London* 121:539–545.

BROWN, J. H. 1971. Mechanisms of competitive exclusion between two species of chipmunks. *Ecology* 52:305–311.

BROWN, J. H. 1978. The theory of insular biogeography and the distribution of boreal birds and mammals. *Great Basin Naturalist Memoirs* 2:209–227.

BROWN, J. H. 1981. Two decades of homage to Santa Rosalia: Toward a general theory of diversity. *American Zoologist* 21:877–888.

BROWN, J. H. 1984. On the relationship between abundance and distribution of species. *American Naturalist* 124: 255–279.

BROWN, J. H. 1986. Two decades of interaction between the MacArthur-Wilson model and the complexities of mammalian distributions. *Biological Journal of the Linnean Society* 28:231–251.

BROWN, J. H., and D. W. DAVIDSON. 1979. An experimental study of competition between seed-eating desert rodents and ants. *American Zoologist* 19:1129–1145.

BROWN, J. H., and A. C. GIBSON. 1983. *Biogeography*. C. V. Mosby, St. Louis.

BROWN, J. H., and E. J. HESKE. 1990. Control of a desert-grassland transition by a keystone rodent guild. *Science* 250:1705–1707.

BROWN, R. T., and J. T. CURTIS. 1952. The upland conifer-hardwood forests of northern Wisconsin. *Ecological Monographs* 22:217–234.

BROWN, W. L., and E. O. WILSON. 1956. Character displacement. *Systematic Zoology* 5:49–64.

BUCKLEY, R. C. 1987. Interactions involving plants, homoptera, and ants. *Annual Review of Ecology and Systematics* 18:111–135.

BUCKNER, C. H., and W. J. TURNOCK. 1965. Avian predation on the larch sawfly, *Pristiphora erichsonii* (Htg.), (Hymenoptera: Tenthredinidae). *Ecology* 46:223–236.

BULL, C. M. 1991. Ecology of parapatric distributions. *Annual Review of Ecology and Systematics* 22:19–36.

BUMP, G. 1963. History and analysis of Tetraonid introductions into North America. *Journal of Wildlife Management* 27:855–867.

BURNHAM, C. R. 1988. The restoration of the American chestnut. *American Scientist* 76:478–487.

BURNS, T. P. 1989. Lindeman's contradiction and the trophic structure of ecosystems. *Ecology* 70:1355–1362.

BURROWS, C. J. 1990. *Processes of Vegetation Change*. Unwin Hyman, London.

BUSS, L. W. 1980. Competitive intransitivity and size-frequency distributions of interacting populations. *Proceedings of the National Academy of Science (USA)* 77:5355–5359.

BUSS, L. W. 1987. *The Evolution of Individuality*. Princeton University Press, Princeton, N.J.

BUSS, L. W., and J. B. C. JACKSON. 1979. Competitive networks: Nontransitive competitive relationships in cryptic coral reef environments. *American Naturalist* 113:223–234.

CAIN, S. A. 1944. *Foundations of Plant Geography*. Harper & Row, New York.

CAMPBELL, B. M. 1986. Plant spinescence and herbivory in a nutrient poor ecosystem. *Oikos* 47:168–172.

CAREY, A. B., S. P. HORTON, and B. L. BISWELL. 1992. Northern spotted owls: Influence of prey-base and landscape character. *Ecological Monographs* 62:223–250.

CARGILL, S. M., and R. L. JEFFERIES. 1984. Nutrient limitation of primary production in a sub-arctic salt marsh. *Journal of Applied Ecology* 21:657–668.

CARLQUIST, S. 1965. *Island Life*. Natural History Press, New York.

CARLQUIST, S. 1974. *Island Biology*. Columbia University Press, New York.

CARLSON, T. 1913. Über Geschwindigkeit und Grösse der Hefevermehrung in Würze. *Biochemische Zeitschrift* 57:313–334.

CARPENTER, S. R., J. F. KITCHELL, and J. R. HODGSON. 1985. Cascading trophic interactions and lake productivity. *Bioscience* 35:634–639.

CARPENTER, S. R., J. F. KITCHELL, J. R. HODGSON, P. A. COCHRAN, J. J. ELSER, M. M. ELSER, D. M. LODGE, D. KRETCHMER, X. HE, and C. N. VON ENDE. 1987. Regulation of lake primary productivity by food web structure. *Ecology* 68:1863–1876.

CARTER, R. N., and S. D. PRINCE. 1988. Distribution limits from a demographic viewpoint. In *Plant Population Ecology*, ed. A. J. Davy, M. J. Hutchings, and A. R. Watkinson, pp. 165–184. Blackwell, Oxford, England.

CASWELL, H. 1978. Predator-mediated coexistence: A nonequilibrium model. *American Naturalist* 112:127–154.

CASWELL, H. 1989. *Matrix Population Models*. Sinauer Associates, Sunderland, Mass.

CAUGHLEY, G. 1966. Mortality patterns in mammals. *Ecology* 47:906–918.

CAUGHLEY, G. 1970. Eruption of ungulate populations, with emphasis on Himalayan thar in New Zealand. *Ecology* 51:53–72.

CAUGHLEY, G. 1976a. Plant-herbivore systems. In *Theoretical Ecology*, ed. R. M. May, pp. 94–113. Saunders, Philadelphia.

CAUGHLEY, G. 1976b. Wildlife management and the dynamics of ungulate populations. *Applied Biology* 1:183–246.

CAUGHLEY, G. 1977. *Analysis of Vertebrate Populations*. Wiley, New York.

CAUGHLEY, G., H. DUBLIN, and I. PARKER. 1990. Projected decline of the African elephant. *Biological Conservation* 54:157–164.

CAUGHLEY, G., D. GRICE, R. BARKER, and B. BROWN. 1988. The edge of the range. *Journal of Animal Ecology* 57:771–785.

CAUGHLEY, G., G. C. GRIGG, J. CAUGHLEY, and G. J. E. HILL. 1980. Does dingo predation control the densities of kangaroos and emus? *Australian Wildlife Research* 7:1–12.

CAUGHLEY, G., and J. H. LAWTON. 1981. Plant-herbivore systems. In *Theoretical Ecology*, ed. R. M. May, pp. 132–166. Blackwell, Oxford, England.

CAUGHLEY, G, N. SHEPHERD, and J. SHORT. 1987. *Kangaroos: Their Ecology and Management in the Sheep Rangelands of Australia*. Cambridge University Press, Cambridge, England.

CAUGHLEY, G., J. SHORT, G. C. GRIGG, and H. NIX. 1987. Kangaroos and climate: An analysis of distribution. *Journal of Animal Ecology* 56:751–761.

CAVERS, P. B., and J. L. HARPER. 1967. Studies in the dynamics of plant populations. I. The fate of seed and transplants introduced into various habitats. *Journal of Ecology* 55:59–71.

CHABOT, B. F., and D. J. HICKS. 1982. The ecology of leaf life spans. *Annual Review of Ecology and Systematics* 13:229–259.

CHAPIN, F. S., III. 1980. The mineral nutrition of wild plants. *Annual Review of Ecology And Systematics* 11:233–260.

CHAPMAN, D. G. 1981. Evaluation of marine mammal population models. In *Dynamics of Large Mammal Populations*, ed. C. W. Fowler and T. D. Smith, pp. 277–296. Wiley, New York.

CHAPMAN, R. N. 1928. The quantitative analysis of environmental factors. *Ecology* 9:111–122.

CHARNOV, E. L., and W. M. SCHAFFER. 1973. Life-history consequences of natural selection: Cole's result revisited. *American Naturalist* 107:791–793.

CHESSON, P. L. 1986. Environmental variation and the coexistence of species. In *Community Ecology*, ed. J. Diamond and T. J. Case, pp. 240–256. Harper & Row, New York.

CHESSON, P. L., and T. J. CASE. 1986. Overview: Nonequilibrium community theories: Chance, variability, history, and coexistence. In *Community Ecology*, ed. J. Diamond and T. J. Case, pp. 229–239. Harper & Row, New York.

CHESSON, P. L., and R. R. WARNER. 1981. Environmental variability promotes coexistence in lottery competitive systems. *American Naturalist* 117:923–943.

CHITTY, D. 1955. Allgemeine Gedankengänge über die Dicteschwankungen bei der Erdmaus (*Microtus agrestis*). *Zeitschrift für Säugetierkunde* 20:55–60.

CHITTY, D. 1960. Population processes in the vole and their relevance to general theory. *Canadian Journal of Zoology* 38:99–113.

CHRISTENSEN, N. L., and R. K. PEET. 1981. Secondary forest succession on the North Carolina Piedmont. In *Forest Succession: Concepts and Application*, ed. D. C. West, H. H. Shugart, and D. B. Botkin, pp. 230–245. Springer-Verlag, New York.

CLARK, C. W. 1990. *Mathematical Bioeconomics: The Optimal Management of Renewable Resources*. Wiley, New York.

CLARKE, A. 1990. Temperature and evolution: Southern Ocean cooling and the antarctic marine fauna. In *Antarctic Ecosystems*, ed. K. R. Kerry and G. Hempel, pp. 9–22. Springer-Verlag, Berlin.

CLAUSEN, C. P. 1951. The time factor in biological control. *Journal of Economic Entomology* 44:1–9.

CLAUSEN, C. P. 1956. *Biological Control of Insect Pests in the Continental United States*. U.S. Department of Agriculture Technical Bulletin No. 1139, Washington, D.C.

CLAUSEN, J. 1965. Population studies of alpine and subalpine races of conifers and willows in the California High Sierra Nevada. *Evolution* 19:56–68.

CLAUSEN, J., D. D. KECK, and W. M. HIESEY. 1948. *Experimental Studies on the nature of species. III. Environmental Responses of Climatic Races of Achillea*. Carnegie Institute Publication No. 581, Washington, D. C.

CLAUSEN, T. P., P. B. REICHARDT, J. P. BRYANT, and R. A. WERNER. 1991. Long-term and short-term induction in quaking aspen: Related phenomena? In *Phytochemical

Induction by Herbivores, ed. D. W. Tallamy and M. J. Raupp, pp. 71–83. Wiley, New York.

CLEMENTS, F. E. 1916. *Plant Succession: An Analysis of the Development of Vegetation.* Carnegie Institute Publication No. 242, Washington, D. C.

CLEMENTS, F. E. 1936. Nature and structure of the climax. *Journal of Ecology* 24:252–284.

CLEMENTS, F. E. 1949. *Dynamics of Vegetation.* Macmillan (Hafner Press), New York.

CLUTTON-BROCK, T. H., ed. 1988. *Reproductive Success.* University of Chicago Press, Chicago.

CLUTTON-BROCK, T. H., S. D. ALBON, and F. E. GUINNESS. 1989. Fitness costs of gestation and lactation in wild mammals. *Nature* 337:260–262.

CLUTTON-BROCK, T. H., F. E. GUINNESS, and S. D. ALBON. 1982. *Red Deer: Behavior and Ecology of Two Sexes.* University of Chicago Press, Chicago.

COCHRAN, W. G. 1977. *Sampling Techniques.* 3d ed. Wiley, New York.

CODY, M. L. 1985. An introduction to habitat selection in birds. In *Habitat Selection in Birds*, ed. M. L. Cody, pp. 3–56. Academic Press, New York.

CODY, M. L. 1986. Structural niches in plant communities. In *Community Ecology,* ed. J. Diamond and T. J. Case, pp. 381–405. Harper & Row, New York.

COE, M. J., D. H. CUMMING, and J. PHILLIPSON. 1976. Biomass and production of large African herbivores in relation to rainfall and primary production. *Oecologia* 22:341–354.

COHEN, D. 1967. Optimization of seasonal migratory behavior. *American Naturalist* 101:5–17.

COHEN, J. E. 1978. *Food Webs and Niche Space.* Princeton University Press, Princeton, N.J.

COHEN, J. E., F. BRIAND, and C. M. NEWMAN. 1990. *Community Food Webs: Data and Theory.* Springer-Verlag, New York.

COLE, B. J. 1981. Overlap, regularity, and flowering phenologies. *American Naturalist* 117:993–997.

COLE, D. W., and M. RAPP. 1981. Elemental cycling in forest ecosystems. In *Dynamic Properties of Forest Ecosystems*, ed. D. E. Reichle, pp. 341–409. Cambridge University Press, Cambridge, England.

COLE, L. C. 1954. The population consequences of life history phenomena. *Quarterly Review of Biology* 29:103–137.

COLE, L. C. 1958. Sketches of general and comparative demography. *Cold Spring Harbor Symposia on Quantitative Biology* 22:1–15.

COLEY, P. D., J. P. BRYANT, and F. S. CHAPIN, III. 1985. Resource availability and plant antiherbivore defense. *Science* 230:895–899.

COLINVAUX, P. 1973. *Introduction to Ecology.* Wiley, New York.

COLLINS, S. L., and L. L. WALLACE. 1990. *Fire in North American Tallgrass Prairies.* University of Oklahoma Press, Norman.

CONNELL, J. H. 1961a. The effects of competition, predation by *Thais lapillus,* and other factors on natural populations of the barnacle, *Balanus balanoides. Ecological Monographs* 31:61–104.

CONNELL, J. H. 1961b. The influence of interspecific competition and other factors on the distribution of the barnacle *Chthamalus stellatus. Ecology* 42:710–723.

CONNELL, J. H. 1970. A predator-prey system in the marine intertidal region. I. *Balanus glandula* and several predatory species of *Thais*. *Ecological Monographs* 40:49–78.

CONNELL, J. H. 1978. Diversity in tropical rain forests and coral reefs. *Science* 199:1302–1310.

CONNELL, J. H. 1983. On the prevalence and relative importance of interspecific competition: Evidence from field experiments. *American Naturalist* 122:661–696.

CONNELL, J. H. 1987. Change and persistence in some marine communities. In *Colonization, Succession and Stability,* ed. A. J. Gray, M. J. Crawley, and P. J. Edwards, pp. 339–352. Blackwell, Oxford, England.

CONNELL, J. H. 1990. Apparent versus "real" competition in plants. In *Perspectives on Plant Competition,* ed. by J. B. Grace and D. Tilman, pp. 9–26. Academic Press, San Diego.

CONNELL, J. H., and M. D. LOWMAN. 1989. Low-diversity tropical rain forests: Some possible mechanisms for their existence. *American Naturalist* 134:88–119.

CONNELL, J. H., and E. ORIAS. 1964. The ecological regulation of species diversity. *American Naturalist* 98:399–414.

CONNELL, J. H., and R. O. SLATYER. 1977. Mechanisms of succession in natural communities and their role in community stability and organization. *American Naturalist* 111:1119–1144.

CONNELL, J. H., and W. P. SOUSA. 1983. On the evidence needed to judge ecological stability or persistence. *American Naturalist* 121:789–824.

COOK, F., and C. S. FINDLAY. 1982. Polygenic variation and stabilizing selection in a wild population of lesser snow geese (*Anser caerulescens caerulescens*). *American Naturalist* 120:543–547.

COOKE, F., P. TAYLOR, C. FRANCIS, and R. ROCKWELL. 1990. Directional selection and clutch size in birds. *American Naturalist* 136: 261–267.

COOK, R. E. 1969. Variation in species density of North American birds. *Systematic Zoology* 18:63–84.

COOPER, W. S. 1939. A fourth expedition to Glacier Bay, Alaska. *Ecology* 20:130–155.

CORMACK, R. M. 1968. The statistics of capture-recapture methods. *Oceanography and Marine Biology Annual Review* 6:455–506.

CORNELL, H. V. and J. H. LAWTON. 1992. Species interactions, local and regional processes, and limits to the richness of ecological communities: A theoretical perspective. *Journal of Animal Ecology* 61:1–12.

COTTAM, G., and J. T. CURTIS. 1956. The use of distance measures in phytosociological sampling. *Ecology* 37:451–460.

COTTAM, G., and R. P. McINTOSH. 1966. Vegetational continuum. *Science* 152:546–547.

COUPLAND, R. T. 1979. *Grassland Ecosystems of the World: Analysis of Grasslands and Their Uses.* Cambridge University Press, Cambridge, England.

COUSINS, S. 1987. The decline of the trophic level concept. *Trends in Ecology and Evolution* 2:312–316.

COWLES, H. C. 1899. The ecological relations of the vegetation on the sand dunes of Lake Michigan. *Botanical Gazette* 27:95–117, 167–202, 281–308, 361–391.

COWLES, H. C. 1901. The physiographic ecology of Chicago and vicinity. *Botanical Gazette* 31:73–108, 145–181.

CRAWLEY, M. J. 1986. The structure of plant communities. In *Plant Ecology*, ed. M. J. Crawley, pp. 1–50. Blackwell, Oxford, England.

CRAWLEY, M. J. 1990. The population dynamics of plants. *Philosophical Transactions of the Royal Society of London*, Series B, 330:125–140.

CRISSEY, W. F., and R. W. DARROW. 1949. *A Study of Predator Control on Valcour Island*. New York State Conservation Dept., Division of Fish and Game, Res. Ser. No. 1.

CRITCHFIELD, H. J. 1966. *General Climatology*, 2d ed. Prentice-Hall, Englewood Cliffs, N.J.

CROCKER, R. L., and J. MAJOR. 1955. Soil development in relation to vegetation and surface age at Glacier Bay, Alaska. *Journal of Ecology* 43:427–448.

CROMBIE, A. C. 1945. On competition between different species of graminivorous insects. *Proceedings of the Royal Society of London* 132:362–395.

CROUSE, D. T., L. B. CROWDER, and H. CASWELL. 1987. A stage-based population model for loggerhead sea turtles and implications for conservation. *Ecology* 68:1412–1423.

CROW, J. F., and M. KIMURA. 1970. *An Introduction to Population Genetics Theory*. Harper & Row, New York.

CROWCROFT, P. 1991. *Elton's Ecologists: a History of the Bureau of Animal Population*. University of Chicago Press, Chicago.

CROWELL, K. L. 1968. Rates of competitive exclusion by the Argentine ant in Bermuda. *Ecology* 49:551–555.

CULVER, D. C. 1970. Analysis of simple cave communities. I. Caves as islands. *Evolution* 24:463–474.

CULVER, D. C. 1981. On using Horn's Markov succession model. *American Naturalist* 117:572–574.

CURRAN, P. J. 1985. *Principles of Remote Sensing*. Longman, London.

CURRIE, D. J. 1991. Energy and large-scale patterns of animal- and plant-species richness. *American Naturalist* 137:27–49.

CURRIE, D. J., and V. PAQUIN. 1987. Large-scale biogeographical patterns of species richness of trees. *Nature* 329:326–327.

CURTIS, J. T. 1959. *The Vegetation of Wisconsin*. University of Wisconsin Press, Madison.

CUSHING, D. H., and J. G. K. HARRIS. 1973. Stock and recruitment and the problem of density dependence. *Rapport du Conseil International pour l'Exploration de la Mer* 164:142–155.

CUTLER, A. 1991. Nested faunas and extinction in fragmented habitats. *Conservation Biology* 5:496–505.

DAHL, K. 1919. Studies of trout and troutwaters in Norway. *Salmon and Trout Magazine* 18:16–33.

DARLINGTON, P. J., JR. 1965. *Biogeography of the Southern End of the World*. Harvard University Press, Cambridge, Mass.

DAUBENMIRE, R. 1966. Vegetation: Identification of typal communities. *Science* 151:291–298.

DAUBENMIRE, R. F. 1954. Alpine timberlines in the Americas and their interpretation. *Butler University Botanical Studies* 2:119–136.

DAVIS, E. F. 1928. The toxic principle of *Juglans nigra* as identified with synthetic juglone, and its toxic effects on tomato and alfalfa plants. *American Journal of Botany* 15:620.

DAVIS, M. B. 1981. Quaternary history and the stability of forest communities. In *Forest Succession: Concepts and Application*, ed. D. C. West, H. H. Shugart, and D. B. Botkin. Springer-Verlag, New York.

DAVIS, M. B. 1986. Climatic instability, time lags, and community disequilibrium. In *Community Ecology*, ed. J. Diamond and T. J. Case, pp. 269–284. Harper & Row, New York.

DAVIS, M. B. 1989. Lags in vegetation response to greenhouse warming. *Climatic Change* 15:79–82.

DAWKINS, R. 1982. *The extended phenotype.* Oxford University Press, Oxford, England.

DAWKINS, R., and J. R. KREBS. 1979. Arms races between and within species. *Proceedings of the Royal Society of London,* Series B, 205:489–511.

DAYTON, P. K. 1971. Competition, disturbance, and community organization: The provision and subsequent utilization of space in a rocky intertidal community. *Ecological Monographs* 41:351–389.

DAYTON, P. K., W. A. NEWMAN, and J. OLIVER. 1982. The vertical zonation of the deep-sea antarctic acorn barnacle, *Bathylasma corolliforme* (Hoek.): Experimental transplants from the shelf into shallow water. *Journal of Biogeography* 9:95–109.

DE ANGELIS, D. L. 1992. *Dynamics of Nutrient Cycling and Food Webs.* Chapman & Hall, New York.

DE ANGELIS, D. L. and J. C. WATERHOUSE. 1987. Equilibrium and nonequilibrium concepts in ecological models. *Ecological Monographs* 57:1–21.

DE BACH, P., ed. 1964. *Biological Control of Insect Pests and Weeds.* Chapman and Hall, London.

DE BACH, P. 1974. *Biological Control by Natural Enemies.* Cambridge University Press, London.

DEEVEY, E. S., JR. 1947. Life tables for natural populations of animals. *Quarterly Review of Biology* 22:283–314.

DEEVEY, E. S., JR. 1951. Life in the depths of a pond. *Scientific American* 185:68–72.

DEL MORAL, R. 1993. Mechanisms of primary succession on volcanoes: The view from Mount St. Helens. In J. Miles (ed.), *Primary Succession on Land*, pp. 79–100, Blackwell, Oxford, England.

DEL MORAL, R., and D. M. WOOD. 1988a. The high elevation flora of Mount St. Helens, Washington. *Madroño* 35:309–319.

DEL MORAL, R., and D. M. WOOD. 1988b. Dynamics of herbaceous vegetation recovery on Mount St. Helens, Washington, USA, after a volcanic eruption. *Vegetatio* 74:11–27.

DEL MORAL, R., and D. M. WOOD. 1993. Early primary succession on the volcano Mount St. Helens. *Journal of Vegetation Science* 4: (in press)

DELCOURT, P. A., and H. R. DELCOURT. 1987. *Long-Term Forest Dynamics of the Temperate Zone: A Case Study of Late-Quaternary Forests in Eastern North America.* Springer-Verlag, New York.

DEMOTT, W. R., and W. C. KERFOOT. 1982. Competition among cladocerans: Nature of the interaction between *Bosmina* and *Daphnia*. *Ecology* 63:1949–1966.

DEMOTT, W. R., and F. MOXTER. 1991. Foraging on cyanobacteria by copepods: Responses to chemical defenses and resource abundance. *Ecology* 72:1820–1834.

DESOWITZ, R. S. 1991. *The Malaria Capers: More Tales of Parasites and People, Research and Reality*. Norton, New York.

DETWILER, R. P., and C. A. S. HALL. 1988. Tropical forests and the global carbon cycle. *Science* 239:42–47.

DEVLIN, R. M. 1969. *Plant Physiology*. Van Nostrand Reinhold, New York.

DEWAR, R. C. 1992. Inverse modeling and the global carbon cycle. *Trends in Ecology and Evolution* 7:105–107.

DEWAR, R. C., and A. D. WATT. 1992. Predicted changes in the synchrony of larval emergence and budburst under climatic warming. *Oecologia* 89:557–559.

DHONDT, A. A. 1988. Carrying capacity: A confusing concept. *Acta Oecologica* 9:337–346.

DHONDT, A. A., F. ADRIAENSEN, E. MATTHYSEN, and B. KEMPENAERS. 1990. Nonadaptive clutch sizes in tits. *Nature* 348: 723–725.

DIAMOND, J. 1986. Overview: Laboratory experiments, field experiments, and natural experiments. In *Community Ecology*, ed. J. Diamond and T. J. Case, pp. 3–22. Harper & Row, New York.

DIAMOND, J. M. 1975. Assembly of species communities. In *Ecology and Evolution of Communities*, ed. M. L. Cody and J. M. Diamond, pp. 342–444. Belknap Press, Cambridge, Mass.

DIAMOND, J. M., K. D. BISHOP, and S. VAN BALEN. 1987. Bird survival in an isolated Javan woodland: Island or mirror. *Conservation Biology* 1:132–142.

DICKMAN, C. R. 1989. Patterns in the structure and diversity of marsupial carnivore communities. In *Patterns in the Structure of Mammalian Communities,* ed. D. W. Morris, Z. Abramsky, B. J. Fox, and M. R. Willig, pp. 241–251. Texas Tech University Press, Lubbock.

DIGBY, P. G. N., and R. A. KEMPTON. 1987. *Multivariate Analysis of Ecological Communities*. Chapman and Hall, London.

DIJKSTRA, C., A. BULT, S. BIJLSMA, S. DAAN, T. MEIJER, and M. ZIJLSTRA. 1990. Brood size manipulations in the kestrel (*Falco tinnunculus*): Effects on offspring and parent survival. *Journal of Animal Ecology* 59:269–285.

DILLER, J. D., and R. B. CLAPPER. 1969. Asiatic and hybrid chestnut trees in the eastern United States. *Journal of Forestry* 67:328–331.

DILLON, P. J., and F. H. RIGLER. 1975. A simple method for predicting the capacity of a lake for development based on lake trophic status. *Journal of the Fisheries Research Board of Canada* 32:1519–1531.

DINGLE, H., ed. 1978. *Evolution of Insect Migration and Diapause*. Springer-Verlag, New York.

DIRZO, R., and M. C. GARCIA. 1992. Rates of deforestation in Los Tuxtlas, a Neotropical area in southeast Mexico. *Conservation Biology* 6:84–90.

DIX, R. L. 1957. Sugar maple in forest succession at Washington, D. C. *Ecology* 38:663–665.

DOBZHANSKY, T. 1950. Evolution in the tropics. *American Scientist* 38:209–221.

DODD, A. P. 1940. *The Biological Campaign Against Prickly-Pear*. Commonwealth Prickly Pear Board, Brisbane.

DODD, J. L., and W. K. LAUENROTH. 1979. Analysis of the response of a grassland ecosystem to stress. In *Perspectives in Grassland Ecology*, ed. N. R. French, pp. 43–58. Springer-Verlag, New York.

DOUTT, R. L. 1964. The historical development of biological control. In *Biological Control of Insect Pests and Weeds*, ed. P. DeBach, pp. 21–42, Chapman and Hall, London.

DRENT, R. H. and S. DAAN. 1980. The prudent parent: Energetic adjustments in avian breeding. *Ardea* 68:225–252.

DRURY, W. H., and I. C. T. NISBET. 1973. Succession. *Journal of the Arnold Arboretum of Harvard University* 54:331–368.

DUBLIN, H. T., A. R. E. SINCLAIR, S. BOUTIN, E. ANDERSON, M. JAGO, and P. ARCESE. 1990. Does competition regulation ungulate populations? Further evidence from Serengeti, Tanzania. *Oecologia* 82:283–288.

DUBLIN, H. T., A. R. E. SINCLAIR, and J. McGLADE. 1990. Elephants and fire as causes of multiple stable states in the Serengeti-Mara woodlands. *Journal of Animal Ecology* 59:1147–1164.

DUBLIN, L. I., and A. J. LOTKA. 1925. On the true rate of natural increase as exemplified by the population of the United States, 1920. *Journal of the American Statistical Association* 20:305–339.

EAMUS, D., and P. G. JARVIS. 1989. The direct effects of increase in the global atmospheric CO_2 concentration on natural and commercial temperate trees and forests. *Advances in Ecological Research* 19:1–55.

EBENHARD, T. 1988. Introduced birds and mammals and their ecological effects. *Swedish Wildlife Research* 13(4):1–107.

EBERHARDT, L. L. 1977. Relationship between two stock-recruitment curves. *Journal of the Fisheries Research Board of Canada* 34:425–428.

EBERHARDT, L. L. 1988. Using age structure data from changing populations. *Journal of Applied Ecology* 25:373–378.

EDMINSTER, F. C. 1939. The effect of predator control on ruffed grouse populations in New York. *Journal of Wildlife Management* 3:345–352.

EDMONDSON, W. T. 1944. Ecological studies of sessile Rotatoria. I. Factors limiting distribution. *Ecological Monographs* 14:31–66.

EDMONDSON, W. T. 1969. Cultural eutrophication with special reference to Lake Washington. *Mitteilungen Internationale Vereinigung Limnologie* 17:19–32.

EDMONDSON, W. T. 1991. *The Uses of Ecology*. University of Washington Press, Seattle.

EGERTON, F. N., III. 1968a. Ancient sources for animal demography. *Isis* 59:175–189.

EGERTON, F. N., III. 1968b. Leeuwenhoek as a founder of animal demography. *Journal of the History of Biology* 1:1–22.

EGERTON, F. N., III. 1968c. Studies of animal populations from Lamarck to Darwin. *Journal of the History of Biology* 1:225–259.

EGERTON, F. N., III. 1969. Richard Bradley's understanding of biological productivity: A study of eighteenth-century ecological ideas. *Journal of the History of Biology* 2:391–410.

EGERTON, F. N., III. 1973. Changing concepts of the balance of nature. *Quarterly Review of Biology* 48:322–350.

EGLER, F. E. 1954. Vegetation science concepts. I. Initial floristic composition, a factor in old-field vegetation development. *Vegetatio* 14:412–417.

EHLER, L. E., and R. W. HALL. 1982. Evidence for competitive exclusion of introduced natural enemies in biological control. *Environmental Entomology* 11:1–4.

EHRLICH, P. R. and A. EHRLICH. 1981. *Extinction.* Random House, New York.

EHRLICH, P. R., and P. H. RAVEN. 1964. Butterflies and plants: A study in coevolution. *Evolution* 18:586–608.

EHRLICH, P. R., and P. H. RAVEN. 1969. Differentiation of populations. *Science* 165:1228–1232.

EHRLICH, P. R., and E. O. WILSON. 1991. Biodiversity studies: Science and policy. *Science* 253:758–762.

ELNER, R. W., and R. L. VADAS. 1990. Inference in ecology: The sea urchin phenomenon in the northwestern Atlantic. *American Naturalist* 136:108–125.

ELSON, P. F. 1962. Predator-prey relationships between fish-eating birds and Atlantic salmon. *Fisheries Research Board of Canada Bulletin* 133:1–87.

ELTON, C. 1927. *Animal Ecology.* Sidgwick and Jackson, London.

ELTON, C. 1966. *The Pattern of Animal Communities.* Methuen, London.

ELTON, C. S. 1958. *The Ecology of Invasions by Animals and Plants.* Methuen, London.

ELTON, C. S., and M. NICHOLSON. 1942. The ten-year cycle in numbers of the lynx in Canada. *Journal of Animal Ecology* 11:215–244.

EMDEN, H. F. VAN, V. F. EASTOP, R. D. HUGHES, and M. J. WAY. 1969. The ecology of *Myzus persicae. Annual Review of Entomology* 14:197–270.

ENDLER, J. A. 1986. *Natural Selection in the Wild.* Princeton University Press, Princeton, N. J.

ENRIGHT, J. T. 1976. Climate and population regulation: The biogeographer's dilemma. *Oecologia* 24:295–310.

Environmental Assessment of a Program to Reduce Incidental Take of Sea Turtles by the Commercial Shrimp Fishery in the Southeast United States. 1983. National Marine Fisheries Service, United States Department of Commerce. St. Petersburg, Fl.

ERRINGTON, P. L. 1963. *Muskrat Populations.* Iowa State University Press, Ames.

ESTES, R. D. 1976. The significance of breeding synchrony in the wildebeest. *East African Wildlife Journal* 14:135–152.

EVANS, F. C. 1956. Ecosystem as the basic unit in ecology. *Science* 1233:1127–1128.

EVANS, G. C. 1976. A sack of uncut diamonds: The study of ecosystems and the future resources of mankind. *Journal of Applied Ecology* 13:1–39.

EWEL, J., C. BERISH, B. BROWN, N. PRICE, and J. RAICH. 1981. Slash and burn impacts on a Costa Rican wet forest site. *Ecology* 62:816–829.

EYLES, D. E. 1944. *A Critical Review of the Literature Relating to the Flight and Dispersion Habits of Anopheline Mosquitoes.* U. S. Public Health Service, Public Health Bulletin No. 287, Washington, D.C.

FAJER, E. D. 1989. The effects of enriched CO_2 atmospheres on plant-insect herbivore interactions: Growth responses of larvae of the specialist butterfly, *Junonia coenia* (Lepidoptera: Nymphalidae). *Oecologia* 81:514–520.

FAJER, E. D., M. D. BOWERS, F. A. BAZZAZ. 1989. The effects of enriched carbon dioxide atmospheres on plant-insect herbivore interactions. *Science* 243:1198–1200.

FAO (FOOD AND AGRICULTURAL ORGANIZATION OF THE UNITED NATIONS). 1972. *Atlas of the Living Resources of the Seas.* FAO Department of Fisheries, Rome.

FARR, W. 1843. Causes of mortality in town districts. *Fifth Annual Report of the Registrar General Bulletin of the Births, Deaths, and Marriages in England*, 2d ed., pp. 406–435.

FAUTH, J. E., and W. J. RESETARITS, Jr. 1991. Interactions between the salamander *Siren intermedia* and the keystone predator *Notophthalmus viridescens*. *Ecology* 72:827–838.

FEENY, P. P. 1970. Seasonal changes in oak leaf tannins and nutrients as a cause of spring feeding by winter moth caterpillars. *Ecology* 51:565–581.

FEENY, P. P. 1976. Plant apparency and chemical defence. *Recent Advances in Phytochemistry* 10:1–40.

FEENY, P. P. 1992. The evolution of chemical ecology: contributions from the study of herbivorous insects. In *Herbivores: Their Interactions with Secondary Plant Metabolites. Vol. II. Evolutionary and Ecological Processes*, ed. G. A. Rosenthal and M. Berenbaum, pp. 1–44. Academic Press, San Diego.

FEINSINGER, P. 1976. Organization of a tropical guild of nectarivorous birds. *Ecological Monographs* 46:257–291.

FELLER, M. C. 1980. Biomass and nutrient distribution in two eucalypt forest ecosystems. *Australian Journal of Ecology* 5:309–333.

FELLER, M. C., and J. P. KIMMINS. 1984. Effects of clearcutting and slash burning on streamwater chemistry and watershed nutrient budgets in southwestern British Columbia. *Water Resources Research* 20:29–40.

FENNER, F., and K. MYERS. 1978. Myxoma virus and myxomatosis in retrospect: The first quarter century of a new disease. In *Viruses and Environment*, ed. E. Kurstak and K. Maramorosch, pp. 539–570. Academic Press, New York.

FIEDLER, P. C. 1982. Zooplankton avoidance and reduced grazing responses to *Gymnodinium splendens* (Dinophyceae). *Limnology and Oceanography* 27:961–965.

FISCHER, A. G. 1960. Latitudinal variations in organic diversity. *Evolution* 14:64–81.

FISHER, R. A. 1958. *The Genetical Theory of Natural Selection.* Dover, New York.

FISHER, R. A., A. S. CORBET, and C. B. WILLIAMS. 1943. The relation between the number of species and the number of individuals in a random sample of an animal population. *Journal of Animal Ecology* 12:42–58.

FITTER, A. H. and R. K. M. HAY. 1987. *Environmental Physiology of Plants,* 2d ed. Academic Press, London.

FLEW, A. 1957. The structure of Malthus' population theory. *Australasian Journal of Philosophy* 35:1–20.

FLINT, M. L., and R. VAN DEN BOSCH. 1981. A history of pest control. In *Introduction to Integrated Pest Management*, pp. 51–81. Plenum, New York.

FOERSTER, R. E. 1968. The sockeye salmon *Oncorhynchus nerka*. *Fisheries Research Board of Canada Bulletin* 162.

FOGLEMAN, J. C., W. B. HEED, and H. W. KIRCHER. 1982. *Drosophila mettleri* and senita cactus alkaloids: Fitness measurements and their ecological significance. *Comparative Biochemistry and Physiology* 71A:413–417.

FORBES, S. A. 1887. The lake as a microcosm. *Bulletin of the Peoria Science Association* 1887:77–87. (Reprinted in *Bulletin Illinois Natural History Survey* 15:537–550, 1925.)

FORD, E. B. 1931. *Mendelism and Evolution.* Methuen, London.

FORMAN, R. T. T. 1964. Growth under controlled conditions to explain the hierarchical distributions of a moss, *Tetraphis pellucida. Ecological Monographs* 34:1–25.

FORRESTER, G. E. 1990. Factors influencing the juvenile demography of a coral reef fish. *Ecology* 71:1666–1681.

FOSTER, B. A. 1971. On the determinants of the upper limit of intertidal distribution of barnacles (Crustacea: Cirripedia). *Journal of Animal Ecology* 40:33–48.

FOX, J. F. 1977. Alternation and coexistence of tree species. *American Naturalist* 111:69–89.

FRAENKEL, G., and D. L. GUNN. 1940. *The Orientation of Animals: Kineses, Taxes, and Compass Reactions.* Oxford University Press, New York.

FRANKIE, G. W., H. G. BAKER, and P. A. OPLER. 1974. Comparative phenological studies of trees in tropical wet and dry forests in the lowlands of Costa Rica. *Journal of Ecology* 62:881–919.

FRANKLIN, J. F., J. A. MacMAHON, F. J. SWANSON, and J. R. SEDELL. 1985. Ecosystem responses to the eruption of Mount St. Helens. *National Geographic Society Research* 1:198–216.

FRENCH, N. R. 1979. *Perspectives in Grassland Ecology.* Springer-Verlag, New York.

FRENCH, N. R., R. K. STEINHORST, and D. M. SWIFT. 1979. Grassland biomass trophic pyramids. In *Perspectives in Grassland Ecology*, ed. N. R. French, pp. 59–87. Springer-Verlag, New York.

FRETWELL, S. D. 1972. *Populations in a Seasonal Environment.* Princeton University Press, Princeton, N. J.

FRETWELL, S. D. 1977. The regulation of plant communities by the food chains exploiting them. *Perspectives in Biology and Medicine* 20:169–185.

FRETWELL, S. D. 1987. Food chain dynamics: The central theory of ecology? *Oikos* 50: 291–301.

FRYER, G. 1959. The trophic interrelationships and ecology of some littoral communities of Lake Nyasa and a discussion of the evolution of a group of rock-frequenting Cichlidae. *Proceedings of the Zoological Society of London* 132:153–281.

FUTUYMA, D. J. 1986. *Evolutionary Biology.* Sinauer Associates, Sunderland, Mass.

FUTUYMA, D. J., and M. SLATKIN, eds. 1983. *Coevolution.* Sinauer Associates, Sunderland, Mass.

GAINES, S. D., and J. LUBCHENCO. 1982. A unified approach to marine plant-herbivore interactions. II. Biogeography. *Annual Review of Ecology and Systematics* 13:111–138.

GARB, S. 1961. Differential growth inhibitors produced by plants. *Botanical Review* 27:422–443.

GARBUTT, K., W. E. WILLIAMS, and F. A. BAZZAZ. 1990. Analysis of the differential response of five annuals to elevated CO_2 during growth. *Ecology* 71:1185–1194.

GASAWAY, W. C., R. D. BOERTJE, D. V. GRANGAARD, D. G. KELLEYHOUSE, R. O. STEPHENSON, and D. G. LARSEN. 1992. The role of predation in limiting moose at low densities in Alaska and Yukon and implications for conservation. *Wildlife Monographs* 120:1–59.

GASTON, K. J. 1988. Patterns in the local and regional dynamics of moth populations. *Oikos* 53:49–59.

GASTON, K. J. 1990. Patterns in the geographical ranges of species. *Biological Reviews* 65:105–129.

GASTON, K. J. 1991. How large is a species' geographic range? *Oikos* 61:434–438.

GASTON, K. J., and J. H. LAWTON. 1987. A test of statistical techniques for detecting density-dependence in sequential censuses of animal populations. *Oecologia* 74:404–410.

GATES, D. M., R. ALDERFER, and E. TAYLOR. 1968. Leaf temperatures of desert plants. *Science* 159:994–995.

GAUSE, G. F. 1932. Experimental studies on the struggle for existence. I. Mixed population of two species of yeast. *Journal of Experimental Biology* 9:389–402.

GAUSE, G. F. 1934. *The Struggle for Existence.* Macmillan (Hafner Press), New York. (Reprinted 1964.)

GAUSE, G. F. 1935. Experimental demonstration of Volterra's periodic oscillation in the numbers of animals. *Journal of Experimental Biology* 12:44–48.

GEAR, A. J., and B. HUNTLEY. 1991. Rapid changes in the range limits of Scots pine 4000 years ago. *Science* 251: 544–547.

GEIST, V. 1978. On weapons, combat, and ecology. In *Aggression, Dominance, and Individual Spacing*, ed. L. Krames, P. Pliner, and T. Alloway, pp. 1–30. Plenum, New York.

GEIST, V. 1988. How markets in wildlife meat and parts, and the sale of hunting privileges, jeopardize wildlife conservation. *Conservation Biology* 2:15–26.

GENTRY, A. H. 1986. Endemism in tropical versus temperate plant communities. In *Conservation Biology*, ed. M. E. Soulé, pp. 153–181. Sinauer Associates, Sunderland, Mass.

GETZ, W. M., and R. G. HAIGHT. 1989. *Population Harvesting. Demographic Models of Fish, Forest, and Animal Resources.* Princeton University Press, Princeton, N.J.

GIESE, R. L., R. M. PEART, and R. T. HUBER. 1975. Pest management. *Science* 187:1045–1052.

GILBERT, N., A. P. GUTIERREZ, B. D. FRAZER, and R. E. JONES. 1976. *Ecological Relationships.* Freeman, San Francisco.

GILL, D. E. 1974:. Intrinsic rate of increase, saturation density, and competitive ability. II. The evolution of competitive ability. *American Naturalist* 108:103–116.

GILLER, P. S., and J. H. R. GEE. 1987. The analysis of community organization: The influence of equilibrium, scale and terminology. In *Organization of Communities Past and Present*, ed. J. H. R. Gee and P. S. Giller, pp. 519–542. Blackwell, Oxford, England.

GILPIN, M. E., and M. E. SOULÉ. 1986. Minimum viable populations: processes of species extinction. In *Conservation Biology*, ed. M. E. Soulé, pp. 19–34. Sinauer Associates, Sunderland, Mass.

GIVNISH, T. J. 1982. On the adaptive significance of leaf height in forest herbs. *American Naturalist* 120:353–381.

GIVNISH, T. J. 1988. Adaptation to sun and shade: A whole-plant perspective. *Australian Journal of Plant Physiology* 15:63–92.

GIVNISH, T. J., and G. J. VERMEIJ. 1976. Sizes and shapes of liana leaves. *American Naturalist* 110:743–778.

GLEASON, H. A., and A. CRONQUIST. 1964. *The Natural Geography of Plants.* Columbia University Press, New York.

GLIWICZ, Z. M. 1990. Food thresholds and body size in cladocerans. *Nature* 343:638–640.

GODFRAY, H. C. J., L. PARTRIDGE, and P. H. HARVEY. 1991. Clutch size. *Annual Review of Ecology and Systematics* 22:409–429.

GOLDMAN, C. R. 1968. Aquatic primary production. *American Zoologist* 8:31–42.

GOLLEY, F. B. 1961. Energy values of ecological materials. *Ecology* 42:581–584.

GOOD, N. F. 1968. A study of natural replacement of chestnut in six stands in the Highlands of New Jersey. *Bulletin of the Torrey Botanical Club* 95:240–253.

GOOD, R. 1964. *The Geography of the Flowering Plants.* Longmans, London.

GOODALL, D. W. 1963. The continuum and the individualistic association. *Vegetatio* 11:297–316.

GOODLAND, R. J. 1975. The tropical origin of ecology: Eugen Warming's jubilee. *Oikos* 26: 240–245.

GOODMAN, D. 1987. The demography of chance extinction. In *Viable Populations for Conservation,* ed. M. E. Soulé, pp. 11–34. Cambridge University Press, Cambridge, England.

GORDON, H. S. 1954. The economic theory of a common property resource: The fishery. *Journal of Political Economics* 62:124–142.

GORHAM, E. 1991. Northern peatlands: Role in the carbon cycle and probable responses to climatic warming. *Ecological Applications* 1:182–195.

GOTELLI, N. J. and D. SIMBERLOFF. 1987. The distribution and abundance of tallgrass prairie plants: A test of the core-satellite hypothesis. *American Naturalist* 130:18–35.

GOULD, S. J. 1981. Palaeontology plus ecology as palaeobiology. In *Theoretical Ecology,* ed. R. M. May, pp. 295–317. Blackwell, Oxford, England.

GOULD, S. J., and R. C. LEWONTIN. 1979. The spandrels of San Marco and the Panglossian paradigm: A critique of the adaptationist programme. *Proceedings of the Royal Society of London Series B,* 205:581–598.

GOULDEN, C. E., and L. L. HORNIG. 1980. Population oscillations and energy reserves in planktonic cladocera and their consequences to competition. *Proceedings of the National Academy of Science (USA)* 77:1716–1720.

GOWARD, S. N., C. J. TUCKER, and D. G. DYE. 1985. North American vegetation patterns observed with the NOAA-7 advanced very high resolution radiometer. *Vegetatio* 64:3–14.

GRAETZ, R. D., R. FISHER, and M. WILSON. 1992. *Looking Back: The Changing Face of the Australian Continent, 1972–1992.* CSIRO, Canberra, Australia.

GRAHAM, M. 1935. Modern theory of exploiting a fishery, and application to North Sea trawling. *Journal du Conseil Permanente International pour l'Exploration de la Mer* 10:264–274.

GRAHAM, M. 1939. The sigmoid curve and the overfishing problem. *Rapport du Conseil International pour l'Exploration de la Mer*: 110:15–20.

GRANT, P. R. 1975. The classical case of character displacement. *Evolutionary Biology* 8:237–337.

GRANT, P. R. 1986. *Ecology and Evolution of Darwin's Finches.* Princeton University Press, Princeton, N.J.

GRAUNT, J. 1662. *Natural and Political Observations Mentioned in a Following Index, and Made upon the Bills of Mortality.* Roycroft, London.

GREENE, H. W., and R. W. Mc DIARMID. 1981. Coral snake mimicry: Does it occur? *Science* 213:1207–1212.

GREIG-SMITH, P. 1979. Pattern in vegetation. *Journal of Ecology* 67:755–779.

GREIG-SMITH, P. 1983. *Quantitative Plant Ecology.* 3d ed. Blackwell, Oxford, England.

GRIGGS, R. F. 1938. Timberlines in the northern Rocky Mountains. *Ecology* 19:548–564.

GRIME, J. P. 1965. Comparative experiments as a key to the ecology of flowering plants. *Ecology* 46:513–515.

GRIME, J. P. 1966. Shade avoidance and shade tolerance in flowering plants. In *Light as an Ecological Factor,* ed. R. Bainbridge, G. C. Evans, and O. Rackham, pp. 187–207. Blackwell, Oxford, England.

GRIME, J. P. 1979. *Plant Strategies and Vegetation Processes.* Wiley, New York.

GRIME, J. P., and J. G. HODGSON. 1969. An investigation of the ecological significance of lime-chlorosis by mean of large-scale comparative experiments. In *Ecological Aspects of the Mineral Nutrition of Plants,* ed. I. H. Rorison, pp. 67–99. Blackwell, Oxford, England.

GRIMM, W. C. 1967. *Familiar Trees of America.* Harper & Row, New York.

GRUBB, P. J. 1987. Global trends in species-richness in terrestrial vegetation: A view from the Northern Hemisphere. In *Organization of Communities Past and Present,* ed. J. H. R. Gee and P. S. Giller, pp. 99–118. Blackwell, Oxford, England.

GUILFORD, T. 1988. The evolution of conspicuous coloration. *American Naturalist* 131(Suppl.):S7–S21.

GULLAND, J. A. 1955. Estimation of growth and mortality in commercial fish populations. *U.K. Ministry of Agriculture and Fisheries, Fisheries Investigations, Series 2,* 18(9):1–46.

GULLAND, J. A. 1962. The application of mathematical models to fish populations. In *The Exploitation of Natural Animal Populations,* ed. E. D. LeCren and M. W. Holdgate, pp. 204–217. Blackwell, Oxford, England.

GULLAND, J. A., ed. 1988. *Fish Population Dynamics: The Implications for Management.* Wiley, New York.

GWYNNE, M. D., and R. H. V. BELL. 1968. Selection of vegetation components by grazing ungulates in the Serengeti National Park. *Nature* 220:390–393.

HAIRSTON, N. G. 1964. Studies on the organization of animal communities. *Journal of Animal Ecology* 33(Suppl.):227–239.

HAIRSTON, N. G. 1980. The experimental test of an analysis of field distributions: Competition in terrestrial salamanders. *Ecology* 61:817–826.

HAIRSTON, N. G. 1991. The literature glut: Causes and consequences (reflections of a dinosaur). *Bulletin of the Ecological Society of America* 72:171–174.

HAIRSTON, N. G., J. D. ALLEN, R. K. COLWELL, D. J. FUTUYMA, J. HOWELL, M. D. LUBIN, J. MATHIAS, and J. H. VANDERMEER. 1968. The relationship between species diversity and stability: An experimental approach with protozoa and bacteria. *Ecology* 49:1091–1101.

HAIRSTON, N. G., F. E. SMITH, and L. B. SLOBODKIN. 1960. Community structure, population control, and competition. *American Naturalist* 94:421–425.

HALBACH, U. 1979. Introductory remarks: Strategies in population research exemplified by rotifer population dynamics. *Fortschritte der Zoologie* 25:1–27.

HALL, E. R. 1946. *Mammals of Nevada.* University of California Press, Berkeley.

HALL, R. W., and L. E. EHLER. 1979. Rate of establishment of natural enemies in classical biological control. *Bulletin of the Entomological Society of America* 25:280–282.

HALL, R. W., L. E. EHLER, and B. BISABRI-ERSHADI. 1980. Rate of success in classical biological control of arthropods. *Bulletin of the Entomological Society of America* 26:111–114.

HALLIGAN, J. P. 1976. Toxicity of *Artemesia californica* to four associated herb species. *American Midland Naturalist* 95:406–421.

HAMILTON, W. D., and R. M. MAY. 1977. Dispersal in stable habitats. *Nature* 269:578–581.

HANSKI, I. 1982. Dynamics of regional distribution: The core and satellite species hypothesis. *Oikos* 38:210–221.

HANSKI, I. 1991. The functional response of predators: Worries about scale. *Trends in Ecology and Evolution* 6:141–142.

HANSKI, I., J. KOUKI, and A. HALKKA. 1993. Three explanations of the positive relationship between distribution and abundance of species. In *Historical and Geographical Determinants of Community Diversity,* ed. R. Ricklefs and D. Schluter. University of Chicago Press, Chicago.

HARDIN, G. 1960. The competitive exclusion principle. *Science* 131:1292–1297.

HARDIN, G. 1968. The tragedy of the commons. *Science* 162:1243–1248.

HARPER, J. L. 1977. *Population Biology of Plants.* Academic Press, New York.

HARPER, J. L. 1981. The meanings of rarity. In *The Biological Aspects of Rare Plant Conservation,* ed. H. Synge, pp. 189–203. Wiley, London.

HARPER, J. L. 1988. An apophasis of plant population biology. In *Plant Population Ecology,* ed. A. J. Davy, M. J. Hutchings, and A. R. Watkinson, pp. 435–452. Blackwell, Oxford, England.

HARPER, J. L., R. B. ROSEN, and J. WHITE, eds. 1986. The growth and form of modular organisms. *Philosophical Transactions of the Royal Society of London, Series B,* 313:1–250.

HARPER, J. L., and J. WHITE. 1974. The demography of plants. *Annual Review of Ecology and Systematics* 5:419–463.

HARRIS, G. P. 1980. Temporal and spatial scales in phytoplankton ecology: Mechanisms, methods, models, and management. *Canadian Journal of Fisheries and Aquatic Sciences* 37:877–900.

HARRIS, V. T. 1952. An experimental study of habitat selection by prairie and forest races of the deer mouse, *Peromyscus maniculatus. Contributions of the Laboratory of Vertebrate Biology, University of Michigan* 56:1–53.

HARRISON, S. 1991. Local extinction in a metapopulation context: An empirical evaluation. *Biological Journal of the Linnean Society* 42:73–88.

HARVELL, C. D. 1986. The ecology and evolution of inducible defenses in a marine bryozoan: Cues, costs, and consequences. *American Naturalist* 128:810–823.

HASKINS, C. P., and E. F. HASKINS. 1965. *Pheidole megacephala* and *Iridomyrmex humilis* in Bermuda: Equilibrium or slow replacement? *Ecology* 46:736–740.

HASLER, A. D. 1966. *Underwater Guideposts: Homing of Salmon.* University of Wisconsin Press, Madison.

HASSELL, M. P. 1976. *The Dynamics of Competition and Predation*. Edward Arnold, London.

HASSELL, M. P. 1978. *The Dynamics of Arthropod Predator-Prey Systems*. Princeton University Press, Princeton, N. J.

HASSELL, M. P. 1979. The dynamics of predator-prey interactions: Polyphagous predators, competing predators and hyperparasitoids. In *Population Dynamics*, ed. R. M. Anderson, B. D. Turner, and L. R. Taylor, pp. 283–306. Blackwell, Oxford, England.

HASSELL, M. P. 1981. Arthropod predator-prey systems. In *Theoretical Ecology*, ed. R. M. May, pp. 105–131. Blackwell, Oxford, England.

HASSELL, M. P. 1985. Insect natural enemies as regulating factors. *Journal of Animal Ecology* 54:323–334.

HASSELL, M. P., and R. M. MAY. 1974. Aggregation in predators and insect parasites and its effect on stability. *Journal of Animal Ecology* 43:567–594.

HATTERSLEY, P. W. 1983. The distribution of C_3 and C_4 grasses in Australia in relation to climate. *Oecologia* 57:113–128.

HAY, M. E. 1986. Functional geometry of seaweeds: Ecological consequences of thallus layering and shape in contrasting light environments. In *On the Economy of Plant Form and Function*, ed. T. J. Givnish, pp. 635–666. Cambridge University Press, Cambridge, England.

HAYSSEN, V., and R. C. LACY. 1985. Basal metabolic rates in mammals: Taxonomic differences in the allometry of BMR and body mass. *Comparative Biochemistry and Physiology* 81A:741–754.

HAYWARD, T. L. 1991. Primary production in the North Pacific Central Gyre: A controversy with important implications. *Trends in Ecology and Evolution* 6:281–284.

HEED, W. B. 1978. Ecology and genetics of Sonoran desert *Drosophila*. In *Ecological Genetics: The Interface*, ed. P. F. Brussard, pp. 109–126. Springer-Verlag, New York.

HEED, W. B., and H. W. KIRCHER. 1965. Unique sterol in the ecology and nutrition of *Drosophila pachea*. *Science* 149:758–761.

HEINRICH, B. 1976. Flowering phenologies: Bog, woodland, and disturbed habitats. *Ecology* 57:890–899.

HEINRICH, B., and P. H. RAVEN. 1972. Energetics and pollination ecology. *Science* 176:597–602.

HELMS, J. A. 1965. Diurnal and seasonal patterns of net assimilation in Douglas fir, *Pseudotsuga menziesii* (Mirb.) Franco, as influenced by environment. *Ecology* 46:698–708.

HENDERSON, S., P. HATTERSLEY, S. von CAEMMER, and B. OSMOND. 1993. Are C_4 pathway plants threatened by global climatic change? In *Ecophysiology of Photosynthesis*, ed. by E.-D. Schulze, and M. Caldwell, Springer-Verlag, Berlin.

HENGEVELD, R. 1989. *Dynamics of Biological Invasions*. Chapman and Hall, London.

HEPHER, B. 1962. Primary production in fishponds and its application to fertilization experiments. *Limnology and Oceanography* 7:131–136.

HERRERA, C. M. 1982. Defense of ripe fruit from pests: Its significance in relation to plant-disperser interactions. *American Naturalist* 120:218–241.

HESLOP-HARRISON, J. 1964. Forty years of genecology. *Advances in Ecological Research* 2:159–247.

HESSE, R., W. C. ALLEE, and K. P. SCHMIDT. 1951. *Ecological Animal Geography*. 2d ed. Wiley, New York.

HIK, D. S., R. L. JEFFERIES, and A. R. E. SINCLAIR. 1992. Foraging by geese, isostatic uplift and asymmetry in the development of salt-marsh plant communities. *Journal of Ecology* 80:395–406.

HJORT, J. 1914. Fluctuations in the great fisheries of northern Europe, viewed in the light of biological research. *Rapport du Conseil International pour l'Exploration de la Mer* 20:1–228.

HOAR, W. S. 1975. *General and Comparative Physiology,* 2d ed. Prentice-Hall, Englewood Cliffs, N. J.

HOBBS, R. J. 1992. The role of corridors in conservation: Solution or bandwagon? *Trends in Ecology and Evolution* 7:389–392.

HOCKER, H. W., JR. 1956. Certain aspects of climate as related to the distribution of loblolly pine. *Ecology* 37:824–834.

HOCKING, B. 1953. The intrinsic range and speed of flight of insects. *Transactions of the Royal Entomological Society of London* 104:223–346.

HOLLING, C. S. 1959. The components of predation as revealed by a study of small mammal predation of the European pine sawfly. *Canadian Entomologist* 91:293–320.

HOLLING, C. S. 1965. The functional response of predators to prey density and its role in mimicry and population regulation. *Memoirs of the Entomological Society of Canada*: 45:1–60.

HOLLISTER, L. E. 1971. Marihuana in man: Three years later. *Science* 172:21–29.

HOLLOWAY, J. K. 1964. Host specificity of a phytophagous insect. *Weeds* 12:25–27.

HOLT, R. D. 1977. Predation, apparent competition and the structure of prey communities. *Theoretical Population Biology* 11:197–229.

HOLT, R. D. 1985. Population dynamics in two-patch environments: Some anomalous consequences of an optimal habitat distribution. *Theoretical Population Biology* 28:181–208.

HORN, H. S. 1971. *Adaptive Geometry of Trees.* Princeton University Press, Princeton, N.J.

HORN, H. S. 1975a. Forest succession. *Scientific American* 232(5):90–98.

HORN, H. S. 1975b. Markovian properties of forest succession. In *Ecology and Evolution of Communities*, ed. M. L. Cody and J. M. Diamond, pp. 196–211. Harvard University Press, Cambridge, Mass.

HORN, H. S. 1978. Optimal tactics of reproduction and life history. In *Behavioural Ecology*, ed. J. R. Krebs and N. B. Davies, pp. 411–429. Sinauer Associates, Sunderland, Mass.

HORN, H. S. 1981. Succession. In *Theoretical Ecology,* ed. R. M. May, pp. 253–271. Blackwell, Oxford, England.

HOSNER, J. F., and S. G. BOYCE. 1962. Tolerance to water-saturated soil of various bottomland hardwoods. *Forest Science* 8:180–186.

HOUGH, A. F. 1936. A climax forest community on East Tionesta Creek in northwestern Pennsylvania. *Ecology* 17:9–28.

HOWARD, L. O., and W. F. FISKE. 1911. The importation into the United States of the parasites of the gypsy-moth and the brown-tail moth. *U. S. Department of Agriculture, Bureau of Entomology Bulletin* 91.

HOWARD, W. E. 1988. Rodent pest management: The principles. In *Rodent Pest Management*, ed. I. Prakash, pp. 285–293. Boca Raton, Fla.

HOWE, H. F., and J. SMALLWOOD. 1982. Ecology of seed dispersal. *Annual Review of Ecology and Systematics* 13:201–228.

HUBBELL, S. P., and R. B. FOSTER. 1986. Biology, chance, and history and the structure of tropical rain forest tree communities. In *Community Ecology*, ed. J. Diamond and T. J. Case, pp. 314–329. Harper & Row, New York.

HUFFAKER, C. B. 1958. Experimental studies on predation: Dispersion factors and predator-prey oscillations. *Hilgardia* 27:343–383.

HUFFAKER, C. B., and C. E. KENNETT. 1969. Some aspects of assessing efficiency of natural enemies. *Canadian Entomologist* 101:425–447.

HUFFAKER, C. B., and P. S. MESSENGER, eds. 1976. *Theory and Practice of Biological Control.* Academic Press, New York.

HUFFAKER, C. B., K. P. SHEA, and S. G. HERMAN. 1963. Experimental studies on predation: Complex dispersion and levels of food in an acarine predator-prey interaction. *Hilgardia* 34:305–330.

HUMPHREYS, W. F. 1979. Production and respiration in animal populations. *Journal of Animal Ecology* 48: 427–453.

HURLBERT, S. H. 1971. The nonconcept of species diversity: A critique and alternative parameters. *Ecology* 52:577–586.

HURLEY, A. C. 1973. Larval settling behaviour of the acorn barnacle (*Balanus pacificus* Pilsbry) and its relation to distribution. *Journal of Animal Ecology* 42:599–609.

HUSTON, M. 1979. A general hypothesis of species diversity. *American Naturalist* 113:81–101.

HUSTON, M., and T. SMITH. 1987. Plant succession: Life history and competition. *American Naturalist* 130:168–198.

HUTCHINGS, M. J. 1983. Ecology's law in search of a theory. *New Scientist* 98:765–767.

HUTCHINS, H. E., and R. M. LANNER. 1982. The central role of Clark's nutcracker in the dispersal and establishment of whitebark pine. *Oecologia* 55:192–201.

HUTCHINSON, G. E. 1958. Concluding remarks. *Cold Spring Harbor Symposia on Quantitative Biology* 22:415–427.

HUTCHINSON, G. E. 1959. Homage to Santa Rosalia, or why are there so many kinds of animals? *American Naturalist* 93:145–159.

HUTCHINSON, G. E. 1961. The paradox of the plankton. *American Naturalist* 95:137–145.

HUTCHINSON, G. E. 1967. *A Treatise on Limnology*, vol. 2: *Introduction to Lake Biology and the Limnoplankton.* Wiley, New York.

HUTCHINSON, G. E. 1970. The chemical ecology of three species of *Myriophyllum* (Angiospermae, Haloragaceae). *Limnology and Oceanography* 15:1–5.

HUTCHINSON, J. B. 1965. Crop-plant evolution: A general discussion. In *Essays on Crop Plant Evolution*, ed. J. B. Hutchinson, pp. 166–181. Cambridge University Press, New York.

HUTTO, R. L. 1985. Habitat selection by nonbreeding, migratory land birds. In *Habitat Selection in Birds*, chap. 16, ed. M. L. Cody. Academic Press, Orlando.

IVES, P. M. 1981. Estimation of coccinellid numbers and movement in the field. *Canadian Entomologist* 113:981–997.

JACKSON, J. B. 1981. Interspecific competition and species distributions: The ghosts of theories and data past. *American Zoologist* 21: 889–902.

JACKSON, J. B. C., L. W. BUSS, and R. E. COOKE, eds. 1985. *Population Biology and Evolution of Clonal Organisms*. Yale University Press, New Haven.

JAENIKE, J. and R. D. HOLT. 1991. Genetic variation for habitat preference: Evidence and explanations. *American Naturalist* 137(Suppl.):S67–S90.

JANES, S. W. 1985. Habitat selection in raptorial birds. In *Habitat Selection in Birds*, ed. M. L. Cody, pp. 159–188. Academic Press, Orlando.

JANZEN, D. H. 1966. Coevolution of mutualism between ants and acacias in Central America. *Evolution* 20:249–275.

JANZEN, D. H. 1967. Synchronization of sexual reproduction of trees within the dry season in Central America. *Evolution* 21:620–637.

JANZEN, D. H. 1970. Herbivores and the number of tree species in tropical forests. *American Naturalist* 104:501–528.

JANZEN, D. H. 1986. Chihuahuan desert nopaleras: Defaunated big mammal vegetation. *Annual Review of Ecology and Systematics* 17:595–636.

JARVIS, M. S. 1963. A comparison between the water relations of species with contrasting types of geographical distribution in the British Isles. In *The Water Relations of Plants*, ed. A. J. Rutter and F. H. Whitehead, pp. 289–312. Blackwell, London.

JAYNES, R. A. 1968. Progress with chestnuts. *Horticulture* 46(12):16–17, 48.

JEFFERIES, R. L. 1988. Vegetational mosaics, plant-animal interactions and resources for plant growth. In *Plant Evolutionary Biology*, ed. L. D. Gottlieb and S. K. Jain, pp. 341–369. Chapman and Hall, London.

JEFFRIES, M. J., and J. H. LAWTON. 1984. Enemy free space and the structure of ecological communities. *Biological Journal of the Linnean Society* 23:269–286.

JEFFRIES, M. J., and J. H. LAWTON. 1985. Predator-prey ratios in communities of freshwater invertebrates: The role of enemy free space. *Freshwater Biology* 15:105–112.

JENKINS, D. 1961. Population control in protected partridges (*Perdix perdix*). *Journal of Animal Ecology* 30:235–258.

JOHNSON, C. G. 1969. *Migration and Dispersal of Insects by Flight*. Methuen, London.

JOHNSON, M. P., L. G. MASON, and P. H. RAVEN. 1968. Ecological parameters and plant species diversity. *American Naturalist* 102:297–306.

JOHNSTON, M. C. 1963. Past and present grasslands of southern Texas and northeastern Mexico. *Ecology* 44:456–466.

JONES, G. P. 1990. The importance of recruitment to the dynamics of a coral reef fish population. *Ecology* 71:1691–1698.

JOOS, F., J. L. SARMIENTO, and U. SIEGENTHALER. 1991. Estimates of the effect of Southern Ocean iron fertilization on atmospheric CO_2 concentrations. *Nature* 349:772–775.

JORDAN, C. F., and R. HERRERA. 1981. Tropical rain forests: Are nutrients really critical? *American Naturalist* 117:167–180.

JORDAN, C. F., and J. R. KLINE. 1972. Mineral cycling: Some basic concepts and their application in a tropical rain forest. *Annual Review of Ecology and Systematics* 3:33–50.

KALELA, O., and T. OKSALA. 1966. Sex ratio in the wood lemming, *Myopus schisticolor*

(Lilljeb.), in nature and in captivity. *Annales Universitatis Turkuensis, Series A, II, Biologica-Geographica* 37:5–24.

KALISZ, S. 1991. Experimental determination of seed bank age structure in the winter annual *Collinsia verna*. *Ecology* 72:575–585.

KARASOV, W. H. 1986. Energetics, physiology and vertebrate ecology. *Trends in Ecology and Evolution* 1:101–104.

KAREIVA, P. 1987. Habitat fragmentation and the stability of predator-prey interactions. *Nature* 326:388–390.

KAREIVA, P. 1989. Renewing the dialogue between theory and experiments in population ecology. In *Perspectives in Ecological Theory*, ed. J. Roughgarden, R. M. May, and S. A. Levin, pp. 68–88. Princeton University Press, Princeton, N. J.

KAREIVA, P. 1990. Population dynamics in spatially complex environments: Theory and data. *Philosophical Transactions of the Royal Society of London*, Series B, 330:175–190.

KAREIVA, P. M., J. G. KINGSOLVER, and R. B. HUEY, eds. 1993. *Biotic Interactions and Global Change*. Sinauer Associates, Sunderland, Mass.

KARN, M. N., and L. S. PENROSE. 1951. Birth weight and gestation time in relation to maternal age, parity, and infant survival. *Annals of Eugenics* 16:147–164.

KAUPPI, P. E., K. MIELIKÄINEN, and K. KUUSELA. 1992. Biomass and carbon budget of European forests, 1971 to 1990. *Science* 256:70–74.

KAZIMIROV, N. I., and R. N. MOROZOVA. 1973. *Biological Cycling of Matter in Spruce Forests of Karelia*. Nauka Publishing House, Leningrad.

KEELING, C. D. 1986. Atmospheric CO_2 concentrations—Mauna Loa Observatory, Hawaii 1958–1986. NDP-001/RI Carbon Dioxide Information Center, Oak Ridge National Laboratory, Oak Ridge, Tenn.

KEEVER, C. 1950. Causes of succession on old fields of the Piedmont, North Carolina. *Ecological Monographs* 20:229–250.

KEEVER, C. 1953. Present composition of some stands of the former oak-chestnut forest in the southern Blue Ridge Mountains. *Ecology* 34:44–54.

KEITH, L. B. 1983. Role of food in hare population cycles. *Oikos* 40:385–395.

KEITH, L. B. 1987. Dynamics of snowshoe hare populations. *Current Mammalogy* 2:119–195. (ed. H. H. Genoways, Plenum, New York.)

KENNY, D., and C. LOEHLE. 1991. Are food webs randomly connected? *Ecology* 72:1794–1799.

KENWARD, R. E. 1978. Hawks and doves: Factors affecting success and selection in goshawk attacks on woodpigeons. *Journal of Animal Ecology* 47:449–460.

KENWARD, R. E. 1987. *Wildlife Radio Tagging*. Academic Press, London.

KERBES, R. H., P. M. KOTANEN, and R. L. JEFFERIES. 1990. Destruction of wetland habitats by lesser snow geese: A keystone species on the west coast of Hudson Bay. *Journal of Applied Ecology* 27:242–258.

KERFOOT, W. C., ed. 1987. *Predation: Direct and Indirect Impacts on Aquatic Communities*. New England University Press, Hanover, N.H.

KERSHAW, K. A. 1973. *Quantitative and Dynamic Ecology*. Edward Arnold, London.

KETTERSON, E. D., and V. NOLAN, JR. 1982. The role of migration and winter mortality in the life history of a temperate-zone migrant, the dark-eyed junco, as determined from demographic analyses of winter populations. *Auk* 99:243–259.

KEYFITZ, N. 1971. On the momentum of population growth. *Demography* 8:71–80.

KIKUZAWA, K. 1991. A cost-benefit analysis of leaf habit and leaf longevity of trees and their geographical pattern. *American Naturalist* 138:1250–1263.

KING, D. A. 1990. The adaptive significance of tree height. *American Naturalist* 135:809–828.

KING, J. M., and B. R. HEATH. 1975. Game domestication for animal production in Africa. *World Animal Review* 16:23–30.

KINGSLAND, S. E. 1985. *Modeling Nature: Episodes in the History of Population Ecology*. University of Chicago Press, Chicago.

KINNAIRD, M. F., and T. G. O'BRIEN. 1991. Viable populations for an endangered forest primate, the Tana River Crested Mangabey (*Cercocebus galeritus galeritus*). *Conservation Biology* 5:203–213.

KIRA, T. 1975. Primary production of forests. In *Photosynthesis and Productivity in Different Environments*, ed. J. P. Cooper, pp. 5–40. Cambridge University Press, London.

KIRKWOOD, J. K. 1983. A limit to metabolisable energy intake in mammals and birds. *Comparative Biochemistry and Physiology* 75A:1–3.

KITCHING, J. A., and F. J. EBLING. 1961. The ecology of Lough Ine. XI. The control of algae by *Paracentrotus lividus* (Echinoidea). *Journal of Animal Ecology* 30:373–383.

KITCHING, J. A., and F. J. EBLING. 1967. Ecological studies at Lough Ine. *Advances in Ecological Research* 4:197–291.

KLEIN, D. R. 1968. The introduction, increase, and crash of reindeer on St. Matthew Island. *Journal of Wildlife Management* 32:350–367.

KLOMP, H. 1970. The determination of clutch size in birds: A review. *Ardea* 58:1–124.

KLOPFER, P. 1963. Behavioural aspects of habitat selection: The role of early experience. *Wilson Bulletin* 75:15–22.

KLOPFER, P. H., and J. P. HAILMAN. 1965. Habitat selection in birds. *Advances in the Study of Behavior* 1:279–303.

KLUN, J. A. 1974. Biochemical basis of resistance of plants to pathogens and insects: Insect hormone mimics and selected examples of other biologically active chemicals derived from plants. In *Proceedings of the Summer Institute on Biological Control of Plant Insects and Diseases*, ed. F. G. Maxwell and F. A. Harris, pp. 463–484. University Press of Mississippi, Jackson.

KNOLL, A. H. 1986. Patterns of change in plant communities through geological time. In *Community Ecology*, ed. J. Diamond and T. J. Case, pp. 126–141. Harper & Row, New York.

KNOX, E. A. 1970. Antarctic marine ecosystems. In *Antarctic Ecology*, ed. M. W. Holdgate, pp. 69–96. Academic Press, London.

KOGAN, M. 1982. Plant resistance in pest management. In *Introduction to Insect Pest Management*, ed. R. L. Metcalf and W. Luckmann, pp. 103–146. Wiley, New York.

KOKKINN, M. J., and A. R. DAVIS. 1986. Secondary production: Shooting a halcyon for its feathers. In *Limnology in Australia*, ed. P. De Deckker and W. D. Williams, pp. 251–261. Dr. W. Junk Publishers, Dordrecht, Netherlands.

KOTLER, B. P. and J. S. BROWN. 1988. Environmental heterogeneity and the coexistence of desert rodents. *Annual Review of Ecology and Systematics* 19:281–307.

KOZHOV, M. 1963. Lake Baikal and its life. *Monographiae Biologicae* 11:1–344.

KOZLOVSKY, D. G. 1968. A critical evaluation of the trophic level concept. I. Ecological efficiencies. *Ecology* 49:48–60.

KOZLOWSKI, T. T., P. J. KRAMER, and S. G. PALLARDY. 1991. *The Physiological Ecology of Woody Plants.* Academic Press, San Diego.

KREBS, C. J. 1978. A review of the Chitty hypothesis of population regulation. *Canadian Journal of Zoology* 56:2463–2480.

KREBS, C. J. 1988. *The Message of Ecology.* Harper & Row, New York.

KREBS, C. J. 1989. *Ecological Methodology.* Harper & Row, New York.

KRIEBEL, H. B. 1957. Patterns of genetic variation in sugar maple. *Ohio Agricultural Experiment Station, Research Bulletin* 791:1–56.

KRUCKEBERG, A. R. 1951. Intraspecific variability in the response of certain native plant species to serpentine soil. *American Journal of Botany* 38:408–419.

KRUCKEBERG, A. R. 1967. Ecotypic response to ultramafic soils by some plant species of northwestern U. S. *Brittonia* 19:133–151.

KRUSE, G. H., and A. V. TYLER. 1989. Exploratory simulation of English sole recruitment mechanisms. *Transactions of the American Fisheries Society* 118:101–118.

KRUUK, H. 1964. Predators and anti-predator behaviour of the black-headed gull (*Larus ridibundus* L.). *Behaviour* (Suppl.) 11:1–129.

KRUUK, H. 1972. *The Spotted Hyena: A Study of Predation and Social Behavior.* University of Chicago Press, Chicago.

LACK, D. 1933. Habitat selection in birds with special references to the effects of afforestation on the Breckland avifauna. *Journal of Animal Ecology* 2:239–262.

LACK, D. 1937. The psychological factor in bird distribution. *British Birds* 31: 130–136.

LACK, D. 1944. Symposium on "The Ecology of Closely Allied Species. " *Journal of Animal Ecology* 13:176–177.

LACK, D. 1945. The ecology of closely related species with special reference to cormorant (*Phalacrocorax carbo*) and shag (*P. aristotelis*). *Journal of Animal Ecology* 14:12–16.

LACK, D. 1947. *Darwin's Finches.* Cambridge University Press, Cambridge, England.

LACK, D. 1954. *The Natural Regulation of Animal Numbers.* Oxford University Press, New York.

LACK, D. 1965. Evolutionary ecology. *Journal of Animal Ecology* 34:223–231.

LAMBERT, B., and M. PEFEROEN. 1992. Insecticidal promise of *Bacillus thuringiensis*: Facts and mysteries about a successful biopesticide. *Bioscience* 42:112–121.

LANDE, R. 1988. Demographic models of the northern spotted owl (*Strix occidentalis caurina*). *Oecologia* 75:601–607.

LANDE, R., and G. F. BARROWCLOUGH. 1987. Effective population size, genetic variation, and their use in population management. In *Viable Populations for Conservation*, ed. M. E. Soulé, pp. 87–123. Cambridge University Press, Cambridge, England.

LANGFORD, A. N., and M. F. BUELL. 1969. Integration, identity and stability in the plant association. *Advances in Ecological Research* 6:83–135.

LARKIN, P. A. 1977. An epitaph for the concept of maximum sustained yield. *Transactions of the American Fisheries Society* 106:1–11.

LARKIN, P. A., B. SCOTT, and A. W. TRITES. 1990. *The Red King Crab Fishery of the Southeastern Bering Sea.* Fisheries Management Foundation, Seattle, Wash.

LARSEN, D. G., D. A. GAUTHIER, and R. L. MARKEL. 1989. Causes and rate of moose mortality in the southwest Yukon. *Journal of Wildlife Management* 53:548–557.

LARSSON, S. 1989. Stressful times for the plant stress–insect performance hypothesis. *Oikos* 56:277–283.

LAUENROTH, W. K. 1979. Grassland primary production: North American grasslands in perspective. In *Perspectives in Grassland Ecology*, ed. N. R. French, pp. 3–24. Springer-Verlag, New York.

LAWLOR, T. E. 1986. Comparative biogeography of mammals on islands. *Biological Journal of the Linnean Society* 28:99–125.

LAWRENCE, D. B. 1958. Glaciers and vegetation in southeastern Alaska. *American Scientist* 46:89–122.

LAWRENCE, D. B., R. E. SCHOENIKE, A. QUISPEL, and G. BOND. 1967. The role of *Dryas drummondii* in vegetation development following ice recession at Glacier Bay, Alaska, with special reference to its nitrogen fixation by root nodules. *Journal of Ecology* 55:793–813.

LAWS, R. M. 1970. Elephants as agents of habitat and landscape change in East Africa. *Oikos* 21:1–15.

LAWTON, J. H. 1984. Non-competitive populations, non-convergent communities, and vacant niches: The herbivores of bracken. In *Ecological Communities*, ed. D. R. Strong, Jr., D. Simberloff, L. G. Abele, and A. B. Thistle, pp. 67–100. Princeton University Press, Princeton, N. J.

LAWTON, J. H. 1987. Are there assembly rules for successional communities? In *Colonization, Succession and Stability*, ed. A. J. Gray, M. J. Crawley, and P. J. Edwards, pp. 225–244. Blackwell, Oxford, England.

LAWTON, J. H., and P. H. WARREN. 1988. Static and dynamic explanations for patterns in food webs. *Trends in Ecology and Evolution* 3:242–245.

LE CREN, E. D. 1962. The efficiency of reproduction and recruitment in freshwater fishes. In *The Exploitation of Natural Animal Populations*, ed. E. D. LeCren and M. W. Holdgate, pp. 283–296. Blackwell, Oxford, England.

LEADER-WILLIAMS, N. 1988. *Reindeer on South Georgia. The Ecology of an Introduced Population*. Cambridge University Press, Cambridge, England.

LEADER-WILLIAMS, N., and S. D. ALBON. 1988. Allocation of resources for conservation. *Nature* 336:533–535.

LEFKOVITCH, L. P. 1965. The study of population growth in organisms grouped by stages. *Biometrics* 21:1–18.

LEGGETT, W. C., and J. E. CARSCADDEN. 1978. Latitudinal variation in reproductive characteristics of American shad (*Alosa sapidissima*): Evidence for population-specific life history strategies in fish. *Journal of the Fisheries Research Board of Canada* 35:1469–1478.

LERTZMAN, K. P. 1992. Patterns of gap-phase replacement in a subalpine, old-growth forest. *Ecology* 73:657–669.

LESLIE, P. H. 1945. On the use of matrices in certain population mathematics. *Biometrika* 33:183–212.

LESLIE, P. H. 1966. The intrinsic rate of increase and the overlap of successive generations in a population of guillemots (*Uria aalge* Pont.) *Journal of Animal Ecology* 35:291–301.

LESLIE, P. H., and R. M. RANSON. 1940. The mortality, fertility, and rate of natural increase of the vole (*Microtus agrestis*) as observed in the laboratory. *Journal of Animal Ecology* 9:27–52.

LESSELLS, C. M. 1986. Brood size in Canada geese: A manipulation experiment. *Journal of Animal Ecology* 55: 669–689.

LEVER, C. 1987. *Naturalized Birds of the World*. Wiley, New York.

LEVIN, D. A. 1976. Alkaloid-bearing plants: An ecogeographic perspective. *American Naturalist* 110:261–284.

LEVIN, S. A. 1970. Community equilibria and stability, and an extension of the competitive exclusion principle. *American Naturalist* 104:413–423.

LEVIN, S. A. 1974. Dispersion and population interactions. *American Naturalist* 108:207–228.

LEVIN, S., and D. PIMENTEL. 1981. Selection of intermediate rates of increase in parasite-host systems. *American Naturalist* 117:308–315.

LEWIS, J. R. 1972. *The Ecology of Rocky Shores*, 2d ed. English Universities Press, London.

LEWIS, V. R., L. D. MERRILL, T. H. ATKINSON, and J. S. WASBAUER. 1992. Imported fire ants: Potential risk to California. *California Agriculture* 46:29–31.

LEWIS, W. M., JR. 1976. Surface/volume ratio: Implications for phytoplankton morphology. *Science* 192:885–887.

LEWIS, W. M. 1981. Precipitation chemistry and nutrient loading by precipitation in a tropical watershed. *Water Resources Research* 17:169–181.

LEWONTIN, R. C. 1965. Selection of colonizing ability. In *The Genetics of Colonizing Species*, ed. H. G. Baker and G. L. Stebbins, pp. 77–94. Academic Press, New York.

LEWONTIN, R. C. 1969. The meaning of stability. *Brookhaven Symposia in Biology* 22:13–24.

LEWONTIN, R. C., and D. COHEN 1969. On population growth in a randomly varying environment. *Proceedings of National Academy of Science (USA)* 62:1056–1060.

LIETH, H., ed. 1974. *Phenology and Seasonality Modeling*. Springer-Verlag, New York.

LIETH, H. 1975. Primary productivity in ecosystems: Comparative analysis of global patterns. In *Unifying Concepts in Ecology*, ed. W. H. van Dobben and R. H. Lowe-McConnell, pp. 67–88. Dr. W. Junk Publishers, The Hague.

LIETH, H. F. H., ed. 1978. *Patterns of Primary Production in the Biosphere*. Dowden, Hutchinson and Ross, Stroudsburg, Pa.

LIKENS, G. E., ed. 1972. *Nutrients and Eutrophication: The Limiting Nutrient Controversy*. American Society of Limnology and Oceanography, Lawrence, Kansas. Special Symposium, vol. 1.

LIKENS, G. E., and F. H. BORMANN. 1972. Nutrient cycling in ecosystems. In *Ecosystem Structure and Function*, ed. J. A. Wiens, pp. 25–67. Oregon State University Press, Corvallis.

LIKENS, G. E., F. H. BORMANN, and N. M. JOHNSON. 1981. Interactions between major biogeochemical cycles in terrestrial ecosystems. In *Some Perspectives of the Major Biogeochemical Cycles*, ed. G. E. Likens, pp. 93–123. Wiley, New York.

LIKENS, G. E., F. H. BORMANN, N. M. JOHNSON, D. W. FISHER, and R. S. PIERCE. 1970. Effects of forest cutting and herbicide treatment on nutrient budgets in the Hubbard Brook watershed-ecosystem. *Ecological Monographs* 40:23–47.

LIKENS, G. E., F. H. BORMANN, R. S. PIERCE, and D. W. FISHER. 1971. Nutrient-hydrologic cycle interaction in small forested watershed-ecosystems. In *Productivity of Forest Ecosystem*. Proceedings of the Brussels Symposium, 1969, pp. 553–563. UNESCO.

LIKENS, G. E., R. F. WRIGHT, J. N. GALLOWAY, and T. J. BUTLER. 1979. Acid rain. *Scientific American* 241(4):43–51.

LLOYD, M. 1967. Mean crowding. *Journal of Animal Ecology* 36:1–30.

LLOYD, M., and R. J. GHELARDI. 1964. A table for calculating the "equitability" component of species diversity. *Journal of Animal Ecology* 33:217–225.

LOFGREN, C. S., W. A. BANKS, and B. M. GLANCEY. 1975. Biology and control of imported fire ants. *Annual Review of Entomology* 20:1–30.

LOMNICKI, A. 1978. Individual differences between animals and the natural regulation of their numbers. *Journal of Animal Ecology* 47:461–475.

LOMOLINO, M. V. 1986. Mammalian community structure on islands: The importance of immigration, extinction and interactive effects. *Biological Journal of the Linnean Society* 28:1–21.

LONG, J. L. 1981. *Introduced Birds of the World*. Davis and Charles, London.

LONG, S. P., and P. R. HUTCHIN. 1991. Primary production in grasslands and coniferous forests with climate change: An overview. *Ecological Applications* 1:139–156.

LORIUS, C., J. JOUZEL, C. RITZ, L. MERLIVAT, N. I. BARKOV, Y. S. KOROTKEVICH, and V. M. KOTLYAKOV. 1985. A 150,000–year climatic record from antarctic ice. *Nature* 316:591–596.

LOTKA, A. J. 1907. Studies on the mode of growth of material aggregates. *American Journal of Science* 24:199–216.

LOTKA, A. J. 1913. A natural population norm. *Journal of the Washington Academy of Science* 3:241–248, 289–293.

LOTKA, A. J. 1922. The stability of the normal age distribution. *Proceedings of the National Academy of Science (USA)* 8:339–345.

LOTKA, A. J. 1923. Contribution to the analysis of malaria epidemiology: Summary. *American Journal of Hygiene* 3 (Jan. Suppl.): 113–121.

LOTKA, A. J. 1925. *Elements of Physical Biology*. (Reprinted in 1956 by Dover Publications, New York.)

LOVEJOY, T. E., R. O. BIERREGAARD, JR., A. B. RYLANDS, J. R. MALCOLM, C. E. QUINTELA, L. H. HARPER, K. S. BROWN, JR., A. H. POWELL, G. V. N. POWELL, H. O. R. SCHUBART, and M. B. HAYS. 1986. Edge and other effects of isolation on Amazon forest fragments. In *Conservation Biology*, ed. M. E. Soulé, pp. 257–285. Sinauer Associates, Sunderland, Mass.

LOWE, V. P. W. 1969. Population dynamics of the red deer (*Cervus elaphus* L.) on Rhum. *Journal of Animal Ecology* 38:425–457.

LOWE-McCONNELL, R. H. 1969. Speciation in tropical freshwater fishes. *Biological Journal of the Linnean Society* 1:51–75.

LUBCHENCO, J. 1978. Plant species diversity in a marine intertidal community: Importance of herbivore food preference and algal competitive abilities. *American Naturalist* 112:23–39.

LUBCHENCO, J. 1986. Relative importance of competition and predation: Early colonization by seaweeds in New England. In *Community Ecology*, ed. J. Diamond and T. J. Case, pp. 537–555. Harper & Row, New York.

LUBCHENCO, J., and S. D. GAINES. 1981. A unified approach to marine plant-herbivore interactions. I. Populations and communities. *Annual Review of Ecology and Systematics* 12:405–437.

LUBINA, J. A., and S. A. LEVIN. 1988. The spread of a reinvading species: Range expansion in the California sea otter. *American Naturalist* 131:526–543.

LUDWIG, D., and C. J. WALTERS. 1985. Are age-structured models appropriate for catch-effort data? *Canadian Journal of Fisheries and Aquatic Sciences* 42:1066–1072.

LUND, J. W. G. 1950. Studies on *Asterionella formosa* Hass. II. Nutrient depletion and the spring maximum. *Journal of Ecology* 38:1–35.

LUND, J. W. G. 1965. The ecology of the freshwater phytoplankton. *Biological Reviews* 40:231–293.

LUPTON, F. G. H. 1977. The plant breeders' contribution to the origin and solution of pest and disease problems. In *Origins of Pest, Parasite, Disease and Weed Problems*, ed. J. M. Cherrett and G. R. Sagar, pp. 71–81. Blackwell, Oxford, England.

LUTZ, H. H. 1930. The vegetation of Heart's Content, a virgin forest in northwestern Pennsylvania. *Ecology* 11:1–29.

MACAN, T. T. 1974. *Freshwater Ecology*, 2d ed. Longmans, London.

MAC ARTHUR, R. H. 1958. Population ecology of some warblers of northeastern coniferous forests. *Ecology* 39:599–619.

MAC ARTHUR, R. H. 1965. Patterns of species diversity. *Biological Reviews* 40:510–533.

MAC ARTHUR, R. H. 1968. The theory of the niche. In *Population Biology and Evolution*, ed. R. C. Lewontin, pp. 159–176. Syracuse University Press, Syracuse, N. Y.

MAC ARTHUR, R. H. 1969. Patterns of communities in the tropics. *Biological Journal of the Linnean Society* 1:19–30.

MAC ARTHUR, R. H. 1972. *Geographical Ecology*. Harper & Row, New York.

MAC ARTHUR, R. H., and J. W. MAC ARTHUR. 1961. On bird species diversity. *Ecology* 42:594–598.

MAC ARTHUR, R. H., and E. O. WILSON. 1967. *The Theory of Island Biogeography*. Princeton University Press, Princeton, N. J.

MAC CALLUM, W. R., and J. H. SELGEBY. 1987. Lake Superior revisited 1984. *Canadian Journal of Fisheries and Aquatic Sciences* 44(Suppl. 2):23–36.

MAC CRACKEN, M. C. 1985. Carbon dioxide and climate change: Background and overview. In *Projecting the Climatic Effects of Increasing Carbon Dioxide*, ed. M. C. MacCracken and F. M. Luther, pp. 1–23. U. S. Department of Energy, Er-0237, Washington, D.C.

MAC FADDEN, B. J. 1985. Patterns of phylogeny and rates of evolution in fossil horses: Hipparions from the Miocene and Pliocene of North America. *Paleobiology* 11:245–257.

MAC LEAN, D. A., and R. W. WEIN. 1977. Nutrient accumulation for post fire jack pine and hardwood successional patterns in New Brunswick. *Canadian Journal of Forest Research* 7:562–578.

MAC NAB, J. 1991. Does game cropping serve conservation? A reexamination of the African data. *Canadian Journal of Zoology* 69:2283–2290.

MAC NAIR, M. R. 1981. Tolerance of higher plants to toxic materials. In *Genetic Consequences of Man-Made Change*, ed. J. A. Bishop and L. M. Cook, pp. 177–207. Academic Press, New York.

MAGURRAN, A. E. 1988. *Ecological Diversity and Its Measurement*. Croon Helm, London.

MAJOR, J. 1963. A climatic index to vascular plant activity. *Ecology* 44:485–498.

MALTHUS, T. R. 1798. *An Essay on the Principle of Population.* (Reprinted by Macmillan, New York.)

MANLY, B. F. J. 1977. The determination of key factors from life table data. *Oecologia* 31:111–117.

MANN, K. H., and P. A. BREEN. 1972. The relation between lobster abundance, sea urchins, and kelp beds. *Journal of the Fisheries Research Board of Canada* 29:603–605.

MARKS, P. L. 1983. On the origin of the field plants of the northeastern United States. *American Naturalist* 122:210–228.

MARR, J. W. 1948. Ecology of the forest-tundra ecotone on the east coast of Hudson Bay. *Ecological Monographs* 18:117–144.

MARTIN, J. H., and S. E. FITZWATER. 1988. Iron deficiency limits phytoplankton growth in the north-east Pacific subarctic. *Nature* 331:341–343.

MARTINEZ, N. D. 1991. Artifacts or attributes? Effects of resolution on the Little Rock Lake food web. *Ecological Monographs* 61:367–392.

MARTINEZ, N. D. 1992. Constant connectance in community food webs. *American Naturalist* 139:1208–1218.

MASSEY, A. B. 1925. Antagonism of the walnuts (*Juglans nigra* L. and *J. cinerea* L.) in certain plant associations. *Phytopathology* 15:773–784.

MAUCHLINE, J., and L. R. FISHER. 1969. The biology of euphausiids. *Advances in Marine Biology* 7:1–454.

MAXWELL, F. G., and P. R. JENNINGS. 1980. *Breeding Plants Resistant to Insects.* Wiley, New York.

MAY, R. M. 1972. Limit cycles in predator-prey communities. *Science* 177:900–902.

MAY, R. M. 1973. *Stability and Complexity in Model Ecosystems.* Princeton University Press, Princeton, N.J.

MAY, R. M. 1974a. Biological populations with nonoverlapping generations: Stable points, stable cycles, and chaos. *Science* 186:645–647.

MAY, R. M. 1974b. On the theory of niche overlap. *Theoretical Population Biology* 5:297–332.

MAY, R. M. 1975. Patterns of species abundance and diversity. In *Ecology and Evolution of Communities,* ed. M. L. Cody, and J. M. Diamond, pp. 81–120. Harvard University Press (Belknap Press), Cambridge, Mass.

MAY, R. M. 1976. Models for two interacting populations. In *Theoretical Ecology*, ed. R. M. May, pp. 49–70. Saunders, Philadelphia.

MAY, R. M. 1978. Host-parasitoid systems in patchy environments: A phenomenological model. *Journal of Animal Ecology* 47:833–844.

MAY, R. M. 1981. Models for single populations. In *Theoretical Ecology*, ed. R. M. May, chap. 2. Blackwell, Oxford, England.

MAY, R. M., and M. P. HASSELL. 1981. The dynamics of multiparasitoid-host interactions. *American Naturalist* 117:234–261.

MAYEWSKI, P. A., W. B. LYONS, M. J. SPENCER, M. TWICKLER, W. DANSGAARD, B. KOCI, C. I. DAVIDSON, and R. E. HONRATH. 1986. Sulfate and nitrate concentrations from a south Greenland ice core. *Science* 232:975–977.

MAYNARD SMITH, J. 1968. *Mathematical Ideas in Biology.* Cambridge University Press, New York.

MAYR, E. 1964. The nature of colonization in birds. In *The Genetics of Colonizing Species*, ed. H. G. Baker and G. L. Stebbins, pp. 29–43. Academic Press, New York.

MAYR, E. 1982. *The Growth of Biological Thought.* Harvard University Press, Belknap Press, Cambridge, Mass.

MC ALLISTER, C. D. 1969. Aspects of estimating zooplankton production. *Journal of the Fisheries Research Board of Canada* 26:199–220.

MC ALLISTER, D. E., S. P. PLATANIA, F. W. SCHUELER, M. E. BALDWIN, and D. S. LEE. 1986. Ichthyofaunal patterns on a geographic grid. In *Zoogeography of Freshwater Fishes of North America*, ed. C. H. Hocutt and E. D. Wiley, pp. 17–51. Wiley, New York.

MC CAULEY, E., and W. W. MURDOCH. 1990. Predator-prey dynamics in environments rich and poor in nutrients. *Nature* 343:455–457.

MC COWN, R. L., and W. A. WILLIAMS. 1968. Competition for nutrients and light between the annual grassland species *Bromus molllis* and *Erodium botrys. Ecology* 49:981–990.

MC FALLS, J. A., JR. 1991. Population: A lively introduction. *Population Bulletin* 46(2):1–43.

MC GOWAN, J. A., and P. W. WALKER. 1979. Structure in the copepod community of the North Pacific Central Gyre. *Ecological Monographs* 49:195–226.

MC GOWAN, J. A., and P. W. WALKER. 1985. Dominance and diversity maintenance in an oceanic ecosystem. *Ecological Monographs* 55:103–118.

MC INTOSH, R. P. 1967. The continuum concept of vegetation. *Botanical Review* 33:130–187.

MC INTOSH, R. P. 1985. *The Background of Ecology: Concept and Theory.* Cambridge University Press, Cambridge, England.

MC KEY, D. 1979. The distribution of secondary compounds within plants. In *Herbivores: Their Interaction with Secondary Plant Metabolites*, ed. G. A. Rosenthal and D. H. Janzen, pp. 55–133. Academic Press, New York.

MC LAREN, I. A. 1963. Effects of temperature on growth of zooplankton and the adaptive value of vertical migration. *Journal of the Fisheries Research Board of Canada* 20:685–727.

MC LAREN, I. A. 1974. Demographic strategy of vertical migration by a marine copepod. *American Naturalist* 108:91–102.

MC NAMARA, J. M. and A. I. HOUSTON. 1987. Starvation and predation as factors limiting population size. *Ecology* 68:1515–1519.

MC NAUGHTON, S. J. 1968. Structure and function in California grasslands. *Ecology* 49:962–972.

MC NAUGHTON, S. J. 1976. Serengeti migratory wildebeest: Facilitation of energy flow by grazing. *Science* 191:92–94.

MC NAUGHTON, S. J., M. OESTERHELD, D. A. FRANK, and K. J. WILLIAMS. 1989. Ecosystem-level patterns of primary productivity and herbivory in terrestrial habitats. *Nature* 341:142–144

MC NEILL, W. H. 1976. *Plagues and Peoples.* Basil Blackwell, Oxford, England.

MC QUEEN, D. J., M. R. S. JOHANNES, J. R. POST, T. J. STEWART, and D. R. S. LEAN. 1989. Bottom-up and top-down impacts on freshwater pelagic community structure. *Ecological Monographs* 59:289–309.

MC QUEEN, D. J., J. R. POST, and E. L. MILLS. 1986. Trophic relationships in freshwater pelagic ecosystems. *Canadian Journal of Fisheries and Aquatic Sciences* 43:1571–1581.

MEDAWAR, P. B. 1957. Old age and natural death. In *The Uniqueness of the Individual*, chap. 1. Methuen, London.

MENGE, B. A. 1992. Community regulation: Under what conditions are bottom-up factors important on rocky shores? *Ecology* 73: 755–765.

MENGE, B. A., and J. P. SUTHERLAND. 1976. Species diversity gradients: Synthesis of the roles of predation, competition, and temporal heterogeneity. *American Naturalist* 110:351–369.

MENGE, B. A., and J. P. SUTHERLAND. 1987. Community regulation: Variation in disturbance, competition, and predation in relation to environmental stress and recruitment. *American Naturalist* 130:730–757.

MENGES, E. S. 1990. Population viability analysis for an endangered plant. *Conservation Biology* 4: 52–62.

MENZEL, D. W., and J. H. RYTHER. 1961a. Annual variations in primary production of the Sargasso Sea off Bermuda. *Deep Sea Research* 7:282–288.

MENZEL, D. W., and J. H. RYTHER. 1961b. Nutrients limiting the production of phytoplankton in the Sargasso Sea, with special reference to iron. *Deep Sea Research* 7:276–281.

MERRELL, M. 1947. Time-specific life tables contrasted with observed survivorship. *Biometrics* 3:129–136.

MERTZ, D. B. 1970. Notes on methods used in life-history studies. In *Readings in Ecology and Ecological Genetics*, ed. J. H. Connell, D. B. Mertz, and W. W. Murdoch, pp. 4–17. Harper & Row, New York.

MERTZ, D. B. and D. E. McCAULEY. 1980. The domain of laboratory ecology. *Synthese* 43:195–255.

METCALF, R. 1986. Coevolutionary adaptations of rootworm beetles (Coleoptera: Chrysomelidae) to cucurbitacins. *Journal of Chemical Ecology* 12:1109–1124.

METCALF, R. L., and W. H. LUCKMANN. 1982. *Introduction to Insect Pest Management*. 2nd ed. Wiley, New York.

MILES, J. 1985. The pedogenic effects of different species and vegetation types and the implications of succession. *Journal of Soil Science* 36:571–584.

MILES, J. 1987. Vegetation succession: Past and present perceptions. In *Colonization, Succession and Stability*, ed. A. J. Gray, M. J. Crawley, and P. J. Edwards, pp. 1–29. Blackwell, Oxford, England.

MILLER, B., and K. J. MULLETTE. 1985. Rehabilitation of an endangered Australian bird: The Lord Howe Island woodhen *Tricholimnas sylvestris*. *Biological Conservation* 34:55–95.

MILLS, K. H., S. M. CHALANCHUK, L. C. MOHR, and I. J. DAVIES. 1987. Responses of fish populations in Lake 223 to 8 years of experimental acidification. *Canadian Journal of Fisheries and Aquatic Sciences* 44(Suppl. 1):114–125.

MILNE, A. 1943. The comparison of sheep-tick populations (*Ixodes ricinus* L.). *Annals of Applied Biology* 30:240–250.

MILNE, A. 1958. Theories of natural control of insect populations. *Cold Spring Harbor Symposia on Quantitative Biology* 22:253–271.

MILNER, C., and R. E. HUGHES. 1968. *Methods for the Measurement of the Primary Production of Grassland.* International Biolological Program Handbook No. 6, Blackwell, Oxford, England.

MOEN, J. 1989. Diffuse competition—a diffuse concept. *Oikos* 54:260–263.

MONSON, R. K. 1989. On the evolutionary pathways resulting in C_4 photosynthesis and crassulacean acid metabolism (CAM). *Advances in Ecological Research* 19:57–110.

MONTGOMERY, M. E., and W. E. WALLNER. 1988. The gypsy moth: A westward migrant. In *Dynamics of Forest Insect Populations*, ed. A. A. Berryman, pp. 353–375. Plenum Publishing Corporation, New York.

MOOK, L. J. 1963. Birds and the spruce budworm. In *The Dynamics of Epidemic Spruce Budworm Populations*, ed. R. F. Morris, pp. 268–271. *Memoirs of the Entomological Society of Canada* No. 31.

MOORE, J. C., and H. W. HUNT. 1988. Resource compartmentation and the stability of real ecosystems. *Nature* 333:261–263.

MORAN, R. J., and W. L. PALMER. 1963. Ruffed grouse introductions and population trends on Michigan islands. *Journal of Wildlife Management* 27:606–614.

MOREY, H. F. 1936. A comparison of two virgin forests in northwestern Pennsylvania. *Ecology* 17:43–55.

MORGAN, N. C. 1980. Secondary production. In *The Functioning of Freshwater Ecosystems*, ed. E. D. LeCren and R. H. Lowe-McDonnell, pp. 247–340. Cambridge University Press, Cambridge, England.

MORISITA, M. 1965. The fitting of the logistic equation to the rate of increase of population density. *Researches in Population Ecology* 7:52–55.

MORRIS, R. F., ed. 1963. *The Dynamics of Epidemic Spruce Budworm Populations.* Memoirs of the Entomological Society of Canada No. 31.

MORRIS, R. F. 1957. The interpretation of mortality data in studies on population dynamics. *Canadian Entomologist* 89:49–69.

MORRIS, R. F., W. F. CHESHIRE, C. A. MILLER, and D. G. MOTT. 1958. The numerical response of avian and mammalian predators during a gradation of the spruce budworm. *Ecology* 39:487–494.

MORROW, P. A., and L. R. FOX. 1980. Effects of variation in *Eucalyptus* essential oil yield on insect growth and grazing damage. *Oecologia* 45:209–219.

MORSE, D. H. 1980. Habitat selection. In *Behavioral Mechanisms in Ecology*, chap. 4. Harvard University Press, Cambridge, Mass.

MOSS, R., A. WATSON, and J. OLLASON. 1982. *Animal Population Dynamics.* Chapman-Hall, London.

MOUTIA, L. A., and R. MAMET. 1946. A review of twenty-five years of economic entomology in the islands of Mauritius. *Bulletin of Entomological Research* 36: 439–472.

MUELLER-DOMBOIS, D., and H. ELLENBERG. 1974. Measuring species quantities. In *Aims and Methods of Vegetation Ecology*, chap. 6. Wiley, New York.

MUIRHEAD-THOMSON, R. C. 1951. *Mosquito Behaviour in Relation to Malaria Transmission and Control in the Tropics.* Edward Arnold, London.

MULLER, C. H. 1966. The role of chemical inhibition (allelopathy) in vegetational composition. *Bulletin of the Torrey Botanical Club* 93:332–351.

MULLER, C. H. 1970. Phytotoxins as plant habitat variables. *Recent Advances in Phytochemistry* 3:105–121.

MULLER, C. H., R. B. HANAWALT, and J. K. MC PHERSON. 1968. Allelopathic control of herb growth in the fire cycle of California chaparral. *Bulletin of the Torrey Botanical Club* 95:225–231.

MURDOCH, W. W. 1966a. "Community structure, population control, and competition": A critique. *American Naturalist* 100:219–226.

MURDOCH, W. W. 1966b. Population stability and life history phenomena. *American Naturalist* 100:5–11.

MURDOCH, W. W. 1969. Switching in general predators: Experiments on predator specificity and stability of prey populations. *Ecological Monographs* 39:335–354.

MURDOCH, W. W. 1971. The developmental response of predators to changes in prey density. *Ecology* 52:132–137.

MURDOCH, W. W. 1975. Diversity, complexity, stability, and pest control. *Journal of Applied Ecology* 12:795–807.

MURDOCH, W. W. and J. BENCE. 1987. General predators and unstable prey populations. In *Predation: Direct and Indirect Impacts on Aquatic Communities*, ed. W. C. Kerfoot and A. Sih, pp. 17–30. University Press of New England, Hanover, N. H.

MURDOCH, W. W., J. CHESSON, and P. L. CHESSON. 1985. Biological control in theory and practice. *American Naturalist* 125:344–366.

MURDOCH, W. W., R. M. NISBET, S. P. BLYTHE, W. S. C. GURNEY, and J. D. REEVE. 1987. An invulnerable age class and stability in delay-differential parasitoid-host models. *American Naturalist* 129:263–282.

MURDOCH, W. W., and A. OATEN. 1975. Predation and population stability. *Advances in Ecological Research* 9:1–131.

MURDOCH, W. W., and A. SIH. 1978. Age-dependent interference in a predatory insect. *Journal of Animal Ecology* 47:581–592.

MURPHY, E. C. and E. HAUKIOJA. 1986. Clutch size in nidicolous birds. *Current Ornithology* 4:141–180.

MURPHY, G. I. 1966. Population biology of the Pacific sardine (*Sardinops caerulea*). *Proceedings of the California Academy of Science* 34:1–84.

MURPHY, G. I. 1968. Pattern in life history and the environment. *American Naturalist* 102:391–403.

MURRAY, B. G. 1979. *Populations Dynamics: Alternative Models*. Academic Press, New York.

MURRAY, K. G., P. FEINSINGER, W. H. BUSBY, Y. B. LINHART, J. H. BEACH, and S. KINSMAN. 1987. Evaluation of character displacement among plants in two tropical pollination guilds. *Ecology* 68:1283–1293.

MURRAY, N. D., J. A. BISHOP, and M. R. MACNAIR. 1980. Melanism and predation by birds in the moths *Biston betularia* and *Phigalia pilosaria*. *Proceedings of the Royal Society of London, Series B*, 210: 276–284.

MURRAY, S. 1981. A count of gannets on Boreray, St. Kilda. *Scottish Birds* 11:205–211.

MYERS, J. H. 1987. Population outbreaks of introduced insects: Lessons from the biological control of weeds. In *Insect Outbreaks,* ed P. Barbosa and J. C. Schultz, pp. 173–193. Academic Press, San Diego.

MYERS, J. H. and D. BAZELY. 1991. Thorns, spines, prickles, and hairs: Are they stimulated by herbivory and do they deter herbivores? In *Phytochemical Induction by Herbivores,* ed. T. W. Tallamy and M. J. Raupp, pp. 325–344. Wiley, New York.

MYERS, J. H., J. MONRO, and N. MURRAY. 1981. Egg clumping, host plant selection and population regulation in *Cactoblastis cactorum* (Lepidoptera). *Oecologia* 51:7–13.

MYERS, J. H., C. RISLEY, and R. ENG. 1989. The ability of plants to compensate for insect attack: Why biological control of weeds with insects is so difficult. *Proceedings of the VII International Symposium for the Biological Control of Weeds,* pp. 67–73. (Ed. E. S. Delfosse.)

MYERS, K., I. D. MARSHALL, and F. FENNER. 1954. Studies in epidemiology of infectious myxomatosis of rabbits. III. Observations on two succeeding epizootics in Australian wild rabbits on the Riverine Plain of south-eastern Australia, 1951–1953. *Journal of Hygiene* 52:337–360.

MYSAK, L. A. 1986. El Niño, interannual variability and fisheries in the northeast Pacific ocean. *Canadian Journal of Fisheries and Aquatic Sciences* 43:464–497.

NAGY, K. A. 1987. Field metabolic rate and food requirement scaling in mammals and birds. *Ecological Monographs* 57:111–128.

NEDELMAN, J., J. A. THOMPSON, and R. J. TAYLOR. 1987. The statistical demography of whooping cranes. *Ecology* 68:1401–1411.

NEE, S., R. D. GREGORY, and R. M. MAY. 1991. Core and satellite species: Theory and artifacts. *Oikos* 62:83–87.

NEEDHAM, J. 1986. *Science and Civilisation in China, Vol. 6: Part I. Biology and Biological Technology: Botany.* Cambridge University Press, Cambridge, England.

NEILL, W. E. 1990. Induced vertical migration in copepods as a defence against invertebrate predation. *Nature* 345:524–526.

NELSON, J. B. 1964. Factors influencing clutch-size and chick growth in the North Atlantic gannet, *Sula bassana. Ibis* 106:63–77.

NELSON, J. B. 1978. *The Sulidae: Gannets and Boobies.* Oxford University Press, London.

NEUVONEN, S., and E. HAUKIOJA. 1991. The effects of inducible resistance in host foliage on birch-feeding herbivores. In *Phytochemical Induction by Herbivores,* ed D. W. Tallamy and M. J. Raupp, pp. 277–291. Wiley, New York.

NEWBOULD, P. J. 1967. *Methods for Estimating the Primary Production of Forests.* International Biological Program Handbook No. 2, Blackwell, Oxford, England.

NEWMARK, W. D. 1985. Legal and biotic boundaries of western North American national parks: A problem of congruence. *Biological Conservation* 33:197–208.

NEWSOME, A. E. 1975. An ecological comparison of the two arid-zone kangaroos of Australia, and their anomalous prosperity since the introduction of ruminant stock to their environment. *Quarterly Review of Biology* 50:389–424.

NEWTON, I. 1972. *Finches.* Collins, London.

NICHOLLS, A. O., and C. R. MARGULES. 1991. The design of studies to demonstrate the biological importance of corridors. In *Nature Conservation 2: The Role of Corridors,* ed. D. A. Saunders and R. J. Hobbs, pp. 49–61. Surrey Beatty and Sons, Chipping Norton, Australia.

NICHOLS, J. D. 1991. Science, population ecology, and the management of the American black duck. *Journal of Wildlife Management* 55:790–799.

NICHOLS, J. D., M. J. CONROY, D. R. ANDERSON, and K. P. BURNHAM. 1984. Compensatory mortality in waterfowl populations: A review of the evidence and implications for research and management. *Transactions of the North American Wildlife and Natural Resources Conference* 49:535–554.

NICKELSON, T. E., and J. A. LICHATOWICH. 1983. The influence of the marine environment on the interannual variation in coho salmon abundance: An overview. In *The Influence of Ocean Conditions on the Production of Salmonids in the North Pacific.* Oregon State University Sea Grant Publication ORESU-W-83-001.

NICKLAS, K. J., B. H. TIFFNEY, and A. H. KNOLL. 1980. Apparent changes in the diversity of fossil plants. *Evolutionary Biology* 12:1–89.

NIELSEN, E. S., and E. A. JENSEN. 1957. Primary oceanic production. In *Galathea Report* 1:49–136. Copenhagen.

NIERING, W. A. 1987. Vegetation dynamics (succession and climax) in relation to plant community management. *Conservation Biology* 1:287–295.

NOBLE, I. R. 1981. Predicting successional change. In *Fire Regimes and Ecosystem Properties*, ed. H. A. Mooney, pp. 278–300. U. S. Dept. Agriculture Forest Service, General Technical Report WO-26.

NORBY, R. J., C. A. GUNDERSON, S. D. WULLSCHLEGER, E. G. O'NEILL, and M. K. MC CRACKEN. 1992. Productivity and compensatory responses of yellow-poplar trees in elevated CO_2. *Nature* 357:322–324.

NORTON-GRIFFITHS, M. 1978. *Counting Animals*, 2d ed. African Wildlife Leadership Foundation, Nairobi.

NOSS, R. F. 1987. Corridors in real landscapes: A reply to Simberloff and Cox. *Conservation Biology* 1:159–164.

NUR, N. 1984. The consequences of brood size for breeding blue tits II. Nestling weight, offspring survival and optimal brood size. *Journal of Animal Ecology* 53:497–517.

NUR, N. 1990. The cost of reproduction in birds: Evaluating the evidence from manipulative and non-manipulative studies. In *Population Biology of Passerine Birds*, ed. J. Blondel, A. Gosler, J. Lebreton, and R. McCleery, pp. 281–296. Springer-Verlag, Berlin.

ODUM, E. P. 1983. *Basic Ecology.* Saunders, Philadelphia.

OESTERHELD, M., O. E. SALA, and S. J. MC NAUGHTON. 1992. Effect of animal husbandry on herbivore-carrying capacity at a regional scale. *Nature* 356:234–236.

OHMAN, M. D. 1990. The demographic benefits of diel vertical migration by zooplankton. *Ecological Monographs* 60:257–281.

OJIMA, D. S., T. G. F. KITTEL, T. ROSSWALL, and B. H. WALKER. 1991. Critical issues for understanding global change effects on terrestrial ecosystems. *Ecological Applications* 1:316–325.

OKSANEN, L., S. D. FRETWELL, J. ARRUDA, and P. NIEMELÄ. 1981. Exploitation ecosystems in gradients of primary productivity. *American Naturalist* 118:240–261.

OKSANEN, T. 1993. Does predation prevent Norwegian lemmings from establishing permanent populations in lowland forests? In *The Biology of Lemmings*, ed. N. C. Stenseth and R. A. Ims. Academic Press, London.

OLSON, J. S. 1958. Rates of succession and soil changes on southern Lake Michigan sand dunes. *Botanical Gazette* 119:125–170.

OLSON, S. L. 1973. Evolution of the rails of the South Atlantic Islands (Aves: Rallidae). *Smithsonian Contributions to Zoology* 152:1–53.

O'NEILL, R. V., and D. L. DE ANGELIS. 1981. Comparative productivity and biomass relations of forest ecosystems. In *Dynamic Properties of Forest Ecosystems*, ed. D. E. Reichle, pp. 411–449. Cambridge University Press, Cambridge, England.

O'NEILL, R. V., D. L. DE ANGELIS, J. B. WAIDE, and T. F. H. ALLEN. 1986. *A Hierarchical Concept of Ecosystems*. Princeton University Press, Princeton, N.J.

OOSTING, H. J., and W. D. BILLINGS. 1942. Factors affecting vegetational zonation on coastal dunes. *Ecology* 23:131–142.

ORIANS, G. H. 1969. The number of bird species in some tropical forests. *Ecology* 50:783–797.

ORIANS, G. H., and G. COLLIER. 1963. Competition and blackbird social systems. *Evolution* 17:449–459.

ORIANS, G. H., and R. T. PAINE. 1983. Convergent evolution at the community level. In *Coevolution*, ed. D. J. Futuyma and M. Slatkin, pp. 431–458. Sinauer Associates, Sunderland, Mass.

ORIANS, G. H., and M. F. WILLSON. 1964. Interspecific territories of birds. *Ecology* 45:736–745.

ORIANS, G. H. and J. F. WITTENBERGER. 1991. Spatial and temporal scales in habitat selection. *American Naturalist* 137(Suppl.):S29–S49.

OSMOND, C. B., K. WINTER, and H. ZIEGLER. 1982. Functional significance of different pathways of CO_2 fixation in photosynthesis. In *Encyclopedia of Plant Physiology* (new ser. vol. 12B), ed. O. L. Lange, P. S. Nobel, C. B. Osmond, and H. Ziegler, pp. 479–547. Springer-Verlag, New York.

OSMOND, C. H., and J. MONRO. 1981. Prickly pear. In *Plants and Man in Australia*, ed. D. J. Carr and S. G. M. Carr, pp. 194–222. Academic Press, New York.

OTIS, D. L., K. P. BURNHAM, G. C. WHITE, and D. R. ANDERSON. 1978. Statistical inference from capture data on closed animal populations. *Wildlife Monographs* 62:1–135.

OVERLAND, L. 1966. The role of allelopathic substances in the "smother crop" barley. *American Journal of Botany* 53:423–432.

OVINGTON, D. 1978. *Australian Endangered Species: Mammals, Birds and Reptiles*. Cassell Australia, Melbourne.

PACALA, S. W., and J. ROUGHGARDEN. 1985. Population experiments with the *Anolis* lizards of St. Maarten and St. Eustatius. *Ecology* 66:129–141.

PAGEL, M. D., R. M. MAY, and A. R. COLLIE. 1991. Ecological aspects of the geographical distribution and diversity of mammalian species. *American Naturalist* 137:791–815.

PAINE, R. T. 1966. Food web complexity and species diversity. *American Naturalist* 100:65–75.

PAINE, R. T. 1969. A note on trophic complexity and community stability. *American Naturalist* 104:91–93.

PAINE, R. T. 1974. Intertidal community structure: Experimental studies on the relationship between a dominant competitor and its principal predator. *Oecologia* 15:93–120.

PAINE, R. T. 1983. On paleoecology: An attempt to impose order on chaos. *Paleobiology* 9:86–90.

PAINE, R. T. 1984. Ecological determinism in the competition for space. *Ecology* 65:1339–1348.

PAINE, R. T. 1988. Food webs: Road maps of interactions or grist for theoretical development? *Ecology* 69:1648–1654.

PALMER, W. L. 1962. Ruffed grouse flight capability over water. *Journal of Wildlife Management* 26:338–339.

PARK, T., P. H. LESLIE, and D. B. MERTZ. 1964. Genetic strains and competition in populations of *Tribolium. Physiological Zoology* 37:97–162.

PARKER, J. 1969. Further studies of drought resistance in woody plants. *Botanical Review* 35:317–371.

PASTOR, J., and W. M. POST. 1988. Response of northern forests to CO_2-induced climate change. *Nature* 334:55–58.

PATTERSON, B. D. 1987. The principle of nested subsets and its implications for biological conservation. *Conservation Biology* 1:323–334.

PATTERSON, R. S., D. E. WEIDHAAS, H. R. FORD, and C. S. LOFGREN. 1970. Suppression and elimination of an island population of *Culex pipiens quinquefasciatus* with sterile males. *Science* 168:1368–1370.

PEARCY, R. W., 1983. The light environment and growth of C_3 and C_4 tree species in the understory of a Hawaiian forest. *Oecologia* 58:19–25.

PEARCY, R. W., and J. R., EHLERINGER. 1984. Comparative ecophysiology of C_3 and C_4 plants. *Plant, Cell, and Environment* 7:1–13.

PEARL, R. 1922. *The Biology of Death*. Lippincott, Philadelphia.

PEARL, R. 1927. The growth of populations. *Quarterly Review of Biology* 2:532–548.

PEARL, R. 1928. *The Rate of Living*. Knopf, New York.

PEARL. R. 1930. *Introduction of Medical Biometry and Statistics*. Saunders, Philadelphia.

PEARL, R., and L. J. REED. 1920. On the rate of growth of the population of the United States since 1790 and its mathematical representation. *Proceedings of the National Academy of Science (USA)* 6:275–288.

PEARL, R., L. J. REED, and J. F. KISH. 1940. The logistic curve and the census count of 1940. *Science* 92:486–488.

PEARSE, P. H. 1982. *Turning the Tide: A New Policy for Canada's Pacific Fisheries*. Minister of Supply and Services, Ottawa.

PEARSON, G. A. 1936. Why the prairies are treeless. *Journal of Forestry* 34:405–408.

PEARSON, O. P. 1964. Carnivore-mouse predation: An example of its intensity and bioenergetics. *Journal of Mammalogy* 45:177–188.

PERRINS, C. M. 1964. Survival of young swifts in relation to brood-size. *Nature* 201:1147–1148.

PETERS, R. H. 1977. Unpredictable problems with tropho-dynamics. *Environmental Biology of Fishes* 2:97–101.

PETERS, R. H., and K. WASSENBERG. 1983. The effect of body size on animal abundance. *Oecologia* 60:89–96.

PETERSON, R. O. 1992. Ecological studies of wolves on Isle Royale. Annual report, 1991–1992, Michigan Technological University, Houghton, Mich.

PETRIDES, G. A., and W. G. SWANK. 1966. Estimating the productivity and energy relations of an African elephant population. *Proceedings of the Ninth International Grassland Congress, São Paulo, Brazil*, pp. 831–842.

PETRUSEWICZ, K., and W. L. GRODZINSKI. 1975. The role of herbivore consumers in various ecosystems. In *Productivity of World Ecosystems*, pp. 64–70. National Academy of Sciences, Washington, D. C.

PETRUSEWICZ, K., and A. MAC FADYEN. 1970. *Productivity of Terrestrial Animals: Principles and Methods*. International Biological Program Handbook No. 13, Blackwell, Oxford, England.

PHILLIPS, J. 1934–1935. Succession, development, the climax, and the complex organism: An analysis of concepts. *Journal of Ecology* 22:554–571; 23:210–246, 488–508.

PIANKA, E. R. 1988. *Evolutionary Ecology*, 4th ed. Harper & Row, New York.

PIANKA, E. R., and W. S. PARKER. 1975. Age-specific reproductive tactics. *American Naturalist* 109:453–464.

PICKERING, S. 1917. The effect of one plant on another. *Annals of Botany* 31:181–187.

PICKETT, S. T. A., and P. S. WHITE, eds. 1985. *The Ecology of Natural Disturbance and Patch Dynamics*. Academic Press, Orlando, Fl.

PIELOU, E. C. 1966. The measurement of diversity in different types of biological collections. *Journal of Theoretical Biology* 13:131–144.

PIELOU, E. C. 1969. *An Introduction to Mathematical Ecology*. Wiley-Interscience, New York.

PIELOU, E. C. 1979. *Biogeography*. Wiley, New York.

PIELOU, E. C. 1984. *The Interpretation of Ecological Data: A Primer on Classification and Ordination*. Wiley, New York.

PIELOU, E. C. 1991. *After the Ice Age: The Return of Life to Glaciated North America*. University of Chicago Press, Chicago.

PIEROTTI, R. 1982. Habitat selection and its effect on reproductive output in the herring gull in Newfoundland. *Ecology* 63:854–868.

PIJL, L. VAN DER. 1969. *Principles of Dispersal in Higher Plants*. Springer-Verlag, Berlin.

PIJL, L. VAN DER, and C. H. DODSON. 1967. *Orchids and Their Pollinators*. University of Miami Press, Miami.

PIMENTEL, D. 1961. Species diversity and insect population outbreaks. *Annals of the Entomological Society of America* 54:76–86.

PIMENTEL, D. 1963. Introducing parasites and predators to control native pests. *Canadian Entomologist* 95:785–792.

PIMENTEL, D., ed. 1991. *Handbook of Pest Management in Agriculture*. CRC Press, Boca Raton, Fl.

PIMENTEL, D., H. ACQUAY, M. BILTONEN, P. RICE, M. SILVA, J. NELSON, V. LIPNER, S. GIORDANO, A. HOROWITZ, and M. D'AMORE. 1992. Assessment of environmental and economic impacts of pesticide use. In *The Pesticide Question: Environment, Economics, and Ethics*, ed. D. Pimentel and H. Lehman, pp. 47–83. Chapman and Hall, New York.

PIMENTEL, D., and N. GOODMAN. 1978. Ecological basis for the management of insect populations. *Oikos* 30:422–437.

PIMENTEL, D., L. MC LAUGHLIN, A. ZEPP, B. LAKITAN, T. KRAUS, P. KLEINMAN,

F. VANCINI, W. J. ROACH, E. GRAAP, W. S. KEETON, and G. SELIG. 1991. Environmental and economic effects of reducing pesticide use. *Bioscience* 41:402–409.

PIMENTEL, D., W. P. NAGEL, and J. L. MADDEN. 1963. Space-time structure of the environment and the survival of parasite-host systems. *American Naturalist* 97:141–167.

PIMM, S. L. 1982. *Food Webs.* Chapman and Hall, London.

PIMM, S. L. 1984. The complexity and stability of ecosystems. *Nature* 307:321–326.

PIMM, S. L. 1987. Determining the effects of introduced species. *Trends in Ecology and Evolution* 2:106–108.

PIMM, S. L. 1991. *The Balance of Nature?* University of Chicago Press, Chicago.

PIMM, S. L., H. L. JONES, and J. DIAMOND. 1988. On the risk of extinction. *American Naturalist* 132:757–785.

PIMM, S. L., J. H. LAWTON, and J. E. COHEN. 1991. Food web patterns and their consequences. *Nature* 350:669–674.

PITCHER, T. J., and P. J. B. HART. 1982. *Fisheries Ecology.* Croom Helm, London.

PLATT, J. R. 1964. Strong inference. *Science* 146:347–353.

PLATT, T., and S. SATHYENDRANATH. 1988. Oceanic primary production: Estimation by remote sensing at local and regional scales. *Science* 241:1613–1620.

PLATT, T., S. SATHYENDRANATH, O. ULLOA, W. G. HARRISON, N. HOEPFFNER, and J. GOES. 1992. Nutrient control of phytoplankton photosynthesis in the Western North Atlantic. *Nature* 356:229–231.

PLEASANTS, J. M. 1990. Null-model tests for competitive displacement: the fallacy of not focusing on the whole community. *Ecology* 71:1078–1084.

POLIS, G. A. 1991. Complex trophic interactions in deserts: An empirical critique of food-web theory. *American Naturalist* 138:123–155.

PONDER, F., JR. 1987. Allelopathic interference of black walnut trees with nitrogen-fixing plants in mixed plantings. In *Allelochemicals: Role in Agriculture and Forestry*, ed. G. R. Walker, pp. 195–204. American Chemical Society, Washington, D. C.

POOLE, R. W., and B. J. RATHCKE. 1979. Regularity, randomness, and aggregation in flowering phenologies. *Science* 203:470–471.

POPPER, K. R. 1963. *Conjectures and Refutations.* Routledge and Kegan Paul, London.

POTTER, L. D., and D. L. GREEN. 1964. Ecology of ponderosa pine in western North Dakota. *Ecology* 45:10–23.

POUGH, F. H. 1988. Mimicry of vertebrates: Are the rules different? *American Naturalist* 131(Suppl.):S67–S102.

POWER, M. E. 1990. Effects of fish in river food webs. *Science* 250:811–814.

POWER, M. E. 1992. Top-down and bottom-up forces in food webs: Do plants have primacy? *Ecology* 73:733–746.

PRATT, D. M. 1943. Analysis of population development in *Daphnia* at different temperatures. *Biological Bulletin* 85:116–140.

PRESTON, F. W. 1948. The commonness and rarity of species. *Ecology* 29:254–283.

PRESTON, F. W. 1960. Time and space and the variation of species. *Ecology* 41:611–627.

PRESTON, F. W. 1962. The canonical distribution of commonness and rarity. *Ecology* 43:185–215, 410–432.

PRICE, M. V. 1978. The role of microhabitat in structuring desert rodent communities. *Ecology* 59:910–921.

PRICE, T., M. KIRKPATRICK, and S. J. ARNOLD. 1988. Directional selection and the evolution of breeding date in birds. *Science* 240:798–799.

PRIMACK, R. B. 1985. Patterns of flowering phenology in communities, populations, individuals, and single flowers. In *The Population Structure of Vegetation*, ed. J. White, pp. 571–593. Dr. W. Junk Publishers, Dordrecht, Netherlands.

PRIMACK, R. B., and H. KANG. 1989. Measuring fitness and natural selection in wild plant populations. *Annual Review of Ecology and Systematics* 20:367–396.

PRITCHETT, W. L., and N. B. COMERFORD. 1983. Nutrition and fertilization of slash pine. In *The Managed Slash Pine Ecosystem*, ed. E. L. Stone, pp. 69–90. University of Florida, Gainesville.

PROCTOR, J. and S. R. J. WOODELL. 1975. The ecology of serpentine soils. *Advances in Ecological Research* 9:255–366.

PULLIAINEN, E. 1972. Summer nutrition of crossbills (*Loxia pytyopsittacus, L. curvirostra* and *L. leucoptera*) in northeastern Lapland in 1971. *Annales Zoologici Fennici* 9:28–31.

QUAY, P. D., B. TILBROOK, and C. S. WONG. 1992. Oceanic uptake of fossil fuel CO_2: Carbon-13 evidence. *Science* 256:74–79.

QUINN, J. F., and A. E. DUNHAM. 1983. On hypothesis testing in ecology and evolution. *American Naturalist* 122: 602–617.

RABINOWITZ, D. 1981. Seven forms of rarity. In *The Biological Aspects of Rare Plant Conservation,* ed. H. Synge, pp. 205–217. Wiley, New York.

RABINOWITZ, D., J. K. RAPP, V. L. SORK, B. J. RATHCKE, G. A. REESE, and J. C. WEAVER. 1981. Phenological properties of wind- and insect-pollinated prairie plants. *Ecology* 62:49–56.

RAFFA, K. F., and A. A. BERRYMAN. 1983. The role of host plant resistance in the colonization behaviour and ecology of bark beetles (Coleoptera: Scolytidae). *Ecological Monographs* 53:27–49.

RANDALL, J. E. 1961. Overgrazing of algae by herbivorous marine fishes. *Ecology* 42:812.

RASMUSSEN, D. I. 1941. Biotic communities of Kaibab Plateau. Arizona. *Ecological Monographs* 3:229–275.

RECHER, H. F. 1969. Bird species diversity and habitat diversity in Australia and North America. *American Naturalist* 103:75–80.

REDDINGIUS, J., and P. J. DEN BOER. 1970. Simulation experiments illustrating stabilization of animal numbers by spreading of risk. *Oecologia* 5:240–284.

REES, J. C. 1969. Chemoreceptor specificity associated with choice of feeding site by the beetle *Chrysolina brunsvicensis* on its foodplant, *Hypericum hirsutum. Entomologia Experimentalis et Applicata* 12:565–583.

REEVE, J. D., and W. W. MURDOCH. 1985. Aggregation by parasitoids in the successful control of the California red scale: A test of theory. *Journal of Animal Ecology* 54:797–816.

REEVE, J. D., and W. W. MURDOCH. 1986. Biological control by the parasitoid *Aphytis melinus,* and population stability of the California red scale. *Journal of Animal Ecology* 55:1069–1082.

REGAL, P. J. 1982. Pollination by wind and animals: Ecology of geographic patterns. *Annual Review of Ecology and Systematics* 13:497–524.

Registrar General's Statistical Review of England and Wales for the Year 1967, pt. 1. 1968. Tables, Medical. H.M. Stationery Office, London.

REICHLE, D. E., ed. 1981. *Dynamic Properties of Forest Ecosystems*. Cambridge University Press, Cambridge, England.

REICHLE, D. E., R. V. O'NEILL, and W. F. HARRIS. 1975. Principles of energy and material exchange in ecosystems. In *Unifying Concepts in Ecology*, ed. W. H. van Dobben and R. H. Lowe-McConnell, pp. 27–43. Dr. W. Junk Publishers, The Hague.

REYNOLDSON, T. B. 1958. Triclads and lake typology in northern Britain: Qualitative aspects. *Verhandlungen der Internationale Vereinigung Limnologie* 13:320–330.

REZNICK, D. 1985. Costs of reproduction: An evaluation of the empirical evidence. *Oikos* 44:257–267.

RHOADES, D. F. 1979. Evolution of plant chemical defense against herbivores. In *Herbivores*, ed. G. A. Rosenthal and D. H. Janzen, pp. 3–54. Academic Press, New York.

RHOADES, D. F., and R. G. CATES. 1976. Toward a general theory of plant antiherbivore chemistry. *Recent Advances in Phytochemistry* 19:168–213.

RICE, E. L. 1984. *Allelopathy*. Academic Press, New York.

RICHARDS, O. W. 1928. Potentially unlimited multiplication of yeast with constant environment, and the limiting of growth by changing environment. *Journal of General Physiology* 11:525–538.

RICHARDS, O. W. 1939. An American textbook. (Review of A. S. Pearse, Animal Ecology.) *Journal of Animal Ecology* 8:387–388.

RICHARDS, P. W. 1952. *The Tropical Rain Forest*. Cambridge University Press, Cambridge, England.

RICHARDS, P. W. 1969. Speciation in the tropical rain forest and the concept of the niche. *Biological Journal of the Linnean Society* 1:149–153.

RICHARDSON, D. M. and W. J. BOND. 1991. Determinants of plant distribution: Evidence from pine invasions. *American Naturalist* 137: 639–668.

RICHARDSON, R. H., J. J. ELLISON, and W. W. AVERHOFF. 1982. Autocidal control of screwworms in North America. *Science* 215:361–370.

RICKER, W. E. 1973. Two mechanisms that make it impossible to maintain peak-period yields from stocks of Pacific salmon and other fishes. *Journal of the Fisheries Research Board of Canada* 30:1275–1286.

RICKER, W. E. 1975. Computation and interpretation of biological statistics of fish populations. *Fisheries Research Board of Canada, Bulletin 191*.

RICKER, W. E. 1982. Size and age of British Columbia sockeye salmon (*Oncorhynchus nerka*) in relation to environmental factors and the fishery. *Canadian Technical Reports in Fisheries and Aquatic Sciences* 1115:1–117.

RIDLEY, H. N. 1930. *The Dispersal of Plants Throughout the World*. L. Reeve, Ashford, Kent, England.

RIEKIRK, H. 1983. Impacts of silviculture on flatwoods runoff, water quality and nutrient budgets. *Water Resources Bulletin* 19:73–79.

RINEY, T. 1964. The impact of introductions of large herbivores on the tropical environment. *International Union for the Conservation of Nature Publications, New Series* (4) 261–273.

RISSER, P. G. 1990. Landscape processes and the vegetation of the North American grassland. In *Fire in North American Tallgrass Prairies*, ed. S. L. Collins and L. L. Wallace, pp. 133–146. University of Oklahoma Press, Norman.

RIVKIN, R. B., and M. PUTT. 1987. Diel periodicity of photosynthesis in polar phytoplankton: Influence on primary production. *Science* 238:1285–1288.

ROBBINS, C. S., D. BYSTRAK, and P. H. GEISSLER. 1986. *The Breeding Bird Survey: Its First Fifteen Years, 1965–1979.* U.S. Department of the Interior, Fish and Wildlife Service, Resource Publication 157.

ROBERTS, L. 1989. How fast can trees migrate? *Science* 243:735–737.

ROBERTS, L. 1990. Zebra mussel invasion threatens U.S. waters. *Science* 249:1370–1372.

ROBERTSON, J. H. 1947. Responses of range grasses to different intensities of competition with sagebrush (*Artemesia tridentata* Nutt.). *Ecology* 28:1–16.

ROFF, D. A. 1986. The evolution of wing dimorphism in insects. *Evolution* 40:1009–1020.

ROGERS, H. H., J. F. THOMAS, and G. E. BINGHAM. 1983. Response of agronomic and forest species to elevated atmospheric carbon dioxide. *Science* 220:428–429.

ROLSTAD, J. 1991. Consequences of forest fragmentation for the dynamics of bird populations: Conceptual issues and the evidence. *Biological Journal of the Linnean Society* 42:149–163.

ROOM, P. M. 1981. Biogeography, apparency, and exploration for biological control agents in exotic ranges of weeds. *Proceedings of the 5th International Symposium on the Biological Control of Weeds*, Brisbane, 1980, pp. 113–124.

ROOM, P. M. 1990. Ecology of a simple plant-herbivore system: Biological control of *Salvinia. Trends in Ecology and Evolution* 5:74–79.

ROOT, R. B. 1967. The niche exploitation pattern of the blue-gray gnatcatcher. *Ecological Monographs* 37:317–350.

ROOT, R. B. 1973. Organization of a plant-anthropod association in simple and diverse habitats: The fauna of collards *(Brassica oleracea). Ecological Monographs* 43:95–124.

ROOT, T. 1988. Energy constraints on avian distributions and abundances. *Ecology* 69:330–339.

ROSENBERG, A. A., J. R. BEDDINGTON, and M. BASSON. 1986. Growth and longevity of krill during the first decade of pelagic whaling. *Nature* 324:152–154.

ROSENZWEIG, M. L. 1968. Net primary productivity of terrestrial communities: Prediction from climatological data. *American Naturalist* 102: 67–74.

ROSENZWEIG, M. L. 1969. Why the prey curve has a hump. *American Naturalist* 103:81–87.

ROSENZWEIG, M. L. 1985. Some theoretical aspects of habitat selection. In *Habitat Selection in Birds*, ed. M. L. Cody, pp. 517–540. Academic Press, New York.

ROSENZWEIG, M. L. 1991. Habitat selection and population interactions: The search for mechanism. *American Naturalist* 137(Suppl.):S5–S28.

ROSENZWEIG, M. L., and R. H. MAC ARTHUR. 1963. Graphical representation and stability conditions of predator-prey interactions. *American Naturalist* 97:209–223.

ROSS, J., and A. M. TITTENSOR. 1981. Myxomatosis in selected rabbit populations in southern England and Wales, 1971–1977. *Proceedings of the World Lagomorph Conference*, Guelph, Ontario, Aug. 1979, pp. 830–833.

ROSS, R. 1908. *Reports on the Prevention of Malaria in Mauritius.* Waterloo, London.

ROSS, R. 1911. *The Prevention of Malaria*, 2d ed. Waterloo, London.

ROTH, R. R. 1976. Spatial heterogeneity and bird species diversity. *Ecology* 57:773–782.

ROTHSCHILD, B. J. 1986. *Dynamics of Marine Fish Populations*. Harvard University Press, Cambridge, Mass.

ROUGHGARDEN, J. 1974. Niche width: Biogeographic patterns among *Anolis* lizard populations. *American Naturalist* 108:429–442.

ROUGHGARDEN, J. 1986. A comparison of food-limited and space-limited animal competition communities. In *Community Ecology,* ed. J. Diamond and T. J. Case, pp. 492–516. Harper & Row, New York.

ROUGHGARDEN, J. and J. DIAMOND. 1986. Overview: The role of species interactions in community ecology. In *Community Ecology*, ed. J. Diamond and T. J. Case, pp. 333–343. Harper & Row, New York.

ROUGHGARDEN, J., S. D. GAINES, and S. W. PACALA. 1987. Supply side ecology: The role of physical transport processes. In *Organization of Communities: Past and Present,* ed. J. H. R. Gee and P. S. Giller, pp. 491–518. Blackwell, Oxford, England.

ROUNSEFELL, G. A. 1958. Factors causing decline in sockeye salmon of Karluk River, Alaska. *Fisheries Bulletin (U. S.)* 58(130):83–169.

ROWE, J. S. 1961. The level-of-integration concept and ecology. *Ecology* 42:420–427.

ROYAMA, T. 1969. A model for the global variation of clutch size in birds. *Oikos* 20:562–567.

ROYAMA, T. 1970. Factors governing the hunting behaviour and selection of food by the great tit (*Parus major* L.). *Journal of Animal Ecology* 39:619–668.

ROZEMA, J., P. BIJWAARD, G. PRAST, and R. BROEKMAN. 1985. Ecophysiological strategies of coastal halophytes from sand dunes and salt marshes. In *Ecology of Coastal Vegetation*, ed. W. G. Beeftink, J. Rozema, and A. H. L. Huiskes, pp. 499–522. Dr. W. Junk Publishers, Dordrecht, Netherlands.

ROZEMA, J., M. C. T. SCHOLTEN, P. A. BLAAUW, and J. DIGGELEN. 1988. Distribution limits and physiological tolerances with particular reference to the salt marsh environment. In *Plant Population Ecology*, ed. A. J. Davy, M. J. Hutchings, and A. R. Watkinson, pp. 137–164. Blackwell, Oxford, England.

RUNDEL, P. W. 1980. The ecological distribution of C_4 and C_3 grasses in the Hawaiian Islands. *Oecologia* 45:354–359.

RUSSELL, E. S. 1931. Some theoretical considerations on the "overfishing" problem. *Journal du Conseil Permanente International pour l'Exploration de la Mer* 6:3–27.

RUSSELL, P. F., and T. R. RAO. 1942. On the relation of mechanical obstruction and shade to ovipositing of *Anopheles culicifacies. Journal of Experimental Zoology* 91:303–329.

RYEL, L. A. 1981. Population change in the Kirtland's Warbler. *Jack-Pine Warbler* 59:76–91.

RYTHER, J. H. 1963. Geographic variation in productivity. In *The Sea*, vol. 2, ed. M. N. Hill, pp. 347–380. Wiley-Interscience, New York.

RYTHER, J. H., and W. M. DUNSTAN. 1971. Nitrogen, phosphorus, and eutrophication in the coastal marine environment. *Science* 171:1008–1013.

SAKAI, A. 1970. Freezing resistance in willows from different climates. *Ecology* 51:485–491.

SALA, O. E., W. J. PARTON, L. A. JOYCE, and W. K. LAUENROTH. 1988. Primary production of the Central Grassland region of the United States. *Ecology* 69:40–45.

SALE, P. F. 1977. Maintenance of high diversity in coral reef fish communities. *American Naturalist* 111:337–359.

SALE, P. F. 1982. Stock-recruit relationships and regional coexistence in a lottery competitive system: A simulation study. *American Naturalist* 120:139–159.

SALISBURY, E. J. 1942. *The Reproductive Capacity of Plants: Studies in Quantitative Biology.* G. Bell, London.

SALISBURY, E. J. 1961. *Weeds and Aliens.* Collins, London.

SALT, G., and F. S. J. HOLLICK. 1944. Studies of wireworm populations. I. A census of wireworms in pasture. *Annals of Applied Biology* 31:52–64.

SANCHEZ, P. A., D. E. BANDY, J. H. VILLACHICA, and J. J. NICHOLAIDES. 1982. Amazon Basin soils: Management for continuous crop production. *Science* 216:821–827.

SANDERS, H. L. 1968. Marine benthic diversity: A comparative study. *American Naturalist* 102:243–282.

SANG, J. H. 1950. Population growth in *Drosophila* cultures. *Biological Reviews* 25:188–219.

SARMIENTO, J. L., and E. T. SUNDQUIST. 1992. Revised budget for the oceanic uptake of anthropogenic carbon dioxide. *Nature* 356:589–593.

SATHYENDRANATH, S., T. PLATT, E. P. W. HORNE, W. G. HARRISON, O. ULLOA, R. OUTERBRIDGE, and N. HOEPFFNER. 1991. Estimation of new production in the ocean by compound remote sensing. *Nature* 353: 129–133.

SAUER, C. O. 1969. *Agricultural Origins and Dispersals*, 2d ed. MIT Press, Cambridge, Mass.

SAUNDERS, D. A., and R. J. HOBBS. 1991. *Nature Conservation 2: The Role of Corridors.* Surrey Beatty and Sons, Chipping Norton, NSW, Australia.

SCHAEFER, W. B. 1968. Methods of estimating effects of fishing on fish populations. *Transactions of the American Fisheries Society* 97:231–241.

SCHAFFER, W. M., and P. F. ELSON. 1975. The adaptive significance of variations in life history among local populations of Atlantic salmon in North America. *Ecology* 56:577–590.

SCHAFFER, W. M., and M. KOT. 1986. Chaos in ecological systems: The coals that Newcastle forgot. *Trends in Ecology and Evolution* 1:58–63.

SCHEFFER, V. B. 1951. The rise and fall of a reindeer herd. *Scientific Monthly* 73:356–362.

SCHIMPER, A. F. W. 1903. *Plant Geography upon a Physiological Basis.* Clarendon Press, Oxford, England.

SCHIMPER, A. F. W., and F. C. VON FABER. 1935. *Pflanzengeographie auf physiologischer Grundlage*, 3d ed. G. Fisher, Jena, Germany.

SCHINDLER, D. W. 1977. Evolution of phosphorus limitation in lakes. *Science* 195:260–262.

SCHINDLER, D. W. 1988. Effects of acid rain on freshwater ecosystems. *Science* 239:149–157.

SCHINDLER, D. W., and E. J. FEE. 1974. Experimental Lakes area: Whole-lake experiments in eutrophication. *Journal of the Fisheries Research Board of Canada* 31:937–953.

SCHLESINGER, W. H. 1991. *Biogeochemistry: An Analysis of Global Change*. Academic Press, New York.

SCHNEIDER, S. H. 1989. The greenhouse effect: Science and policy. *Science* 243:771–781.

SCHNEIDERHAN, F. J. 1927. The black walnut (*Juglans nigra* L.) as a cause of the death of apple trees. *Phytopathology* 17:529–540.

SCHOENER, A. 1974. Experimental zoogeography: Colonization of marine mini-islands. *American Naturalist* 108:715–738.

SCHOENER, T. W. 1974. Resource partitioning in ecological communities. *Science* 185:27–39.

SCHOENER, T. W. 1983. Field experiments on interspecific competition. *American Naturalist* 122:240–285.

SCHOENER, T. W. 1985. Some comments on Connell's and my reviews of field experiments on interspecific competition. *American Naturalist* 125:730–740.

SCHOENER, T. W. 1986a. Overview: Kinds of ecological communities—ecology becomes pluralistic. In *Community Ecology*, ed. J. Diamond and T. J. Case, pp. 467–479. Harper & Row, New York.

SCHOENER, T. W. 1986b. Resource partitioning. In *Community Ecology: Pattern and Process*, ed. J. Kikkawa and D. J. Anderson, pp. 91–126. Blackwell, Melbourne, Australia.

SCHOENER, T. W. 1987a. The geographical distribution of rarity. *Oecologia* 74:161–173.

SCHOENER, T. W. 1987b. Axes of controversy in community ecology. In *Community and Evolutionary Ecology of North American Stream Fishes*, ed. W. J. Matthews and D. C. Heins, pp. 8–16. University of Oklahoma Press, Norman, Ok.

SCHOENER, T. W., and C. A. TOFT. 1983. Spider populations: Extraordinarily high densities on islands without top predators. *Science* 219:1353–1355.

SCHOENLY, K., R. A. BEAVER, and T. A. HEUMIER. 1991. On the trophic relations of insects: A food-web approach. *American Naturalist* 137:597–638.

SCHOENLY, K., and J. E. COHEN. 1991. Temporal variation in food web structure: 16 empirical cases. *Ecological Monographs* 61:267–298.

SCHOONHOVEN, L. M. 1968. Chemosensory bases of host plant selection. *Annual Review of Entomology* 13:115–136.

SCHULTZ, A. M. 1969. A study of an ecosystem: The arctic tundra. In *The Ecosystem Concept in Natural Resource Management*, ed. G. Van Dyne, pp. 77–93. Academic Press, New York.

SCHULZE, E. D. 1989. Air pollution and forest decline in a spruce (*Picea abies*) forest. *Science* 244:776–783.

SCOTT, J. A., N. R. FRENCH, and J. W. LEETHAM. 1979. Patterns of consumption in grasslands. In *Perspectives in Grassland Ecology*, ed. N. R. French, pp. 89–105. Springer-Verlag, New York.

SEBER, G. A. F. 1982. *The Estimation of Animal Abundance and Related Parameters*. Griffin, London.

SELANDER, R. K. 1965. On mating systems and sexual selection. *American Naturalist* 99:129–141.

SHAFFER, M. L. 1981. Minimum population sizes for species conservation. *BioScience* 31:131–134.

SHARKEY, M. J. 1970. The carrying capacity of natural and improved land in different climatic zones. *Mammalia* 34:564–572.

SHEAIL, J. 1987. *Seventy-Five Years in Ecology: the British Ecological Society.* Blackwell Oxford, England.

SHEAR, C. L., N. E. STEVENS, and R. J. TILLER. 1917. Endothia parasitica *and Related Species.* U. S. Department of Agriculture Bulletin No. 380.

SHEPHERD, J. G., J. G. POPE, and R. D. COUSENS. 1984. Variations in fish stocks and hypotheses concerning their links with climate. *Rapports et Proces-Verbaux des Réunions Conseil International pour l'Exploration de la Mer* 185:255–267.

SHUGART, H. H. 1984. *A Theory of Forest Dynamics: The Ecological Implications of Forest Succession Models.* Springer-Verlag, New York.

SIH, A. 1991. Reflections on the power of a grand paradigm. *Bulletin of the Ecological Society of America* 72:174–178.

SIH, A., P. CROWLEY, M. MC PEEK, J. PETRANKA, K. STROHMEIER. 1985. Predation, competition, and prey communities: A review of field experiments. *Annual Review of Ecology and Systematics* 16:269–311.

SILKWORTH, D. R., and D. F. GRIGAL. 1982. Determining and evaluating nutrient losses following whole-tree harvesting of aspen. *Soil Science Society of America Journal* 46:626–631.

SILLIMAN, R. P., and J. S. GUTSELL. 1958. Experimental exploitation of fish populations. *Fisheries Bulletin (U. S.)* 58(133): 215–252.

SILVERTOWN, J. W. 1987. *Introduction to Plant Population Ecology.* Wiley, New York.

SIMBERLOFF, D. 1976. Experimental zoogeography of islands: Effects of island size. *Ecology* 57:629–648.

SIMBERLOFF, D. 1988. The contribution of population and community biology to conservation science. *Annual Review of Ecology and Systematics* 19:473–511.

SIMBERLOFF, D., and J. COX. 1987. Consequences and costs of conservation corridors. *Conservation Biology* 1:63–71.

SIMBERLOFF, D., and T. DAYAN. 1991. The guild concept and the structure of ecological communities. *Annual Review of Ecology and Systematics* 22:115–143.

SIMBERLOFF, D. S., and W. BOECKLEN. 1981. Santa Rosalia reconsidered: Size ratios and competition. *Evolution* 35:1206–1228.

SIMBERLOFF, D. S., and E. O. WILSON. 1970. Experimental zoogeography of islands: A two-year record of colonization. *Ecology* 51:934–937.

SIMPSON, E. H. 1949. Measurement of diversity. *Nature* 163:688.

SIMPSON, G. G. 1964. Species density of North American recent mammals. *Systematic Zoology* 13:57–73.

SIMPSON, G. G. 1969. The first three billion years of community evolution. *Brookhaven Symposium in Biology* 22:162–177.

SINCLAIR, A. R. E. 1975. The resource limitation of trophic levels in tropical grassland ecosystems. *Journal of Animal Ecology* 44:497–520.

SINCLAIR, A. R. E. 1977. *The African Buffalo.* University of Chicago Press, Chicago.

SINCLAIR, A. R. E. 1981. Environmental carrying capacity and the evidence for overabundance. In *Problems in Management of Locally Abundant Wild Mammals*, ed. P. A. Jewell and S. Holt, pp. 247–257. Academic Press, New York.

SINCLAIR, A. R. E. 1989. Population regulation in animals. In *Ecological Concepts*, ed. J. M. Cherrett, pp. 197–241. Blackwell, Oxford, England.

SINCLAIR, A. R. E., and M. NORTON-GRIFFITHS. 1982. Does competition or facilitation regulate migrant ungulate populations in the Serengeti? A test of hypotheses. *Oecologia* 53:364–369.

SINCLAIR, A. R. E., and J. N. M. SMITH. 1984. Do plant secondary compounds determine feeding preferences of snowshoe hares? *Oecologia* 61:403–410.

SKELLAM, J. G. 1955. The mathematical approach to population dynamics. In *The Numbers of Man and Animals*, ed. J. B. Cragg and N. W. Pirie, pp. 31–46. Oliver and Boyd, Edinburgh.

SKINNER, J. 1972. Eland vs. beef: Eland cannot compete except in hot semiarid regions. *African Wildlife* 26(1):4–9.

SKUTCH, A. F. 1967. Adaptive limitation of the reproductive rate of birds. *Ibis* 109:579–599.

SLAGSVOLD, T. 1989. On the evolution of clutch size and nest size in passerine birds. *Oecologia* 79:300–305.

SLOBODKIN, L. B. 1961. *Growth and Regulation of Animal Populations*. Holt, Rinehart and Winston, New York.

SLOBODKIN, L. B. 1964. Experimental populations of Hydrida. *Journal of Animal Ecology* 33(Suppl.):131–148.

SMITH, A. P. 1973. Stratification of temperate and tropical forests. *American Naturalist* 107:671–683.

SMITH, F. E. 1963. Population dynamics in *Daphnia magna* and a new model for population growth. *Ecology* 44:651–663.

SMITH, F. E. 1970. Analysis of ecosystems. In *Analysis of Temperate Forest Ecosystems*, ed. D. Reichel, pp. 7–18. Springer-Verlag, Berlin.

SMITH, H. S. 1935. The role of biotic factors in the determination of population densities. *Journal of Economic Entomology* 28:873–898.

SMITH, J. N. M. 1988. Determinants of lifetime reproductive success in the song sparrow. In *Reproductive Success*, ed. T. H. Clutton-Brock, pp. 154–172. University of Chicago Press, Chicago.

SMITH, S. M. 1980. Responses of naive temperate birds to warning coloration. *American Midland Naturalist* 103:346–352.

SMITH, T. J., III. 1987. Seed predation in relation to tree dominance and distribution in mangrove forests. *Ecology* 68:266–273.

SMITH, V. H. 1982. The nitrogen and phosphorous dependence of algal biomass in lakes: An empirical and theoretical analysis. *Limnology and Oceanography* 27:1101–1112.

SMITH, V. H. 1983. Low nitrogen to phosphorus ratios favor dominance by blue-green algae in lake phytoplankton. *Science* 221:669–671.

SNAYDON, R. W. 1991. The productivity of C_3 and C_4 plants: A reassessment. *Functional Ecology* 5:321–330.

SOKAL, R. R., and F. J. ROHLF. 1981. *Biometry*, 2d ed. W. H. Freeman and Company, New York.

SOMMER, U. 1990. Phytoplankton nutrient competition—from laboratory to lake. In *Per-

spectives on Plant Competition, ed. J. B. Grace and D. Tilman, pp. 193–213. Academic Press, New York.

SOULÉ, M. E., ed. 1987. *Viable Populations for Conservation.* Cambridge University Press, Cambridge, England.

SOUSA, W. P. 1979. Experimental investigations of disturbance and ecological succession in a rocky intertidal algal community. *Ecological Monographs* 49:227–254.

SOUSA, W. P. 1985. Disturbance and patch dynamics on rocky intertidal shores. In *The Ecology of Natural Disturbance and Patch Dynamics*, ed. S. T. A. Pickett and P. S. White, pp. 101–124. Academic Press, Orlando.

SOUTAR, A., and J. D. ISAACS. 1969. History of fish populations inferred from fish scales in anaerobic sediments off California. *California Cooperative Oceanic Fisheries Investigations Reports* 13:63–70.

SOUTHERN, H. N. 1970. The natural control of a population of tawny owls (*Strix aluco*). *Journal of Zoology, London* 162:197–285.

SOUTHWOOD, T. R. E. 1962. Migration of terrestrial arthropods in relation to habitat. *Biological Reviews* 37:171–214.

SOUTHWOOD, T. R. E. 1978. *Ecological Methods with Particular Reference to the Study of Insect Populations*, 2d ed. Chapman and Hall, London.

SPENCE, D. H. N. 1967. Factors controlling the distribution of freshwater macrophytes with particular reference to the lochs of Scotland. *Journal of Ecology* 55:147–170.

SPENCE, D. H. N. 1982. The zonation of plants in freshwater lakes. *Advances in Ecological Research* 12:37–125.

SPILLER, D. A., and T. W. SCHOENER. 1988. An experimental study of the effect of lizards on web-spider communities. *Ecological Monographs* 58:57–77.

SPILLER, D. A. and T. W. SCHOENER. 1990. Lizards reduce food consumption by spiders: Mechanisms and consequences. *Oecologia* 83:150–161.

STAMP, N. E. 1992. Theory of plant-insect herbivore interactions on the inevitable brink of re-synthesis. *Bulletin of the Ecological Society of America* 73:28–34.

Statistical Abstract of the United States. 1991. 112th ed. U.S. Department of Commerce, Washington, D.C.

Statistics Canada. 1991. Minister of Supply and Services, Ottawa.

STAVN, R. H. 1971. The horizontal-vertical distribution hypothesis: Langmuir circulation and *Daphnia* distributions. *Limnology and Oceanography* 16:453–466.

STEARNS, S. C. 1976. Life-history tactics: A review of the ideas. *Quarterly Review of Biology* 51:3–47.

STEARNS, S. C. 1989. Trade-offs in life history evolution. *Functional Ecology* 3:259–268.

STEEDMAN, R. J., and H. A. REGIER. 1987. Ecosystem science for the Great Lakes: Perspectives on degradative and rehabilitative transformations. *Canadian Journal of Fisheries and Aquatic Sciences* 44(Suppl. 2):95–103.

STEELE, J. H., ed. 1978. *Spatial Patterns in Plankton Communities.* Plenum, New York.

STENSETH, N. C. 1981. How to control pest species: Application of models from the theory of island biogeography in formulating pest control strategies. *Journal of Applied Ecology* 18:773–794.

STEPHENS, D. W., and J. R. KREBS. 1986. *Foraging Theory.* Princeton University Press, Princeton, N.J.

STEPHENSON, N. L. 1990. Climatic control of vegetation distribution: The role of the water balance. *American Naturalist* 135:649–670.

STEVENS, G. C., and J. F. FOX. 1991. The causes of treeline. *Annual Review of Ecology and Systematics* 22: 177–191.

STILES, F. G. 1977. Coadapted competitors: The flowering seasons of hummingbird-pollinated plants in a tropical forest. *Science* 198:1177–1178.

STILES, F. G. 1979. Regularity, randomness, and aggregation in flowering phenologies: A reply. *Science* 203:471.

STOCKNER, J. G., and W. W. BENSON. 1967. The succession of diatom assemblages in the recent sediments of Lake Washington. *Limnology and Oceanography* 12:513–532.

STRATHMANN, R. R. 1990. Testing size-abundance rules in a human exclusion experiment. *Science* 250:1091.

STRAYER, D. 1988. On the limits to secondary production. *Limnology and Oceanography* 33:1217–1220.

STRAYER, D. L. 1991. Projected distribution of the zebra mussel, *Dreissena polymorpha*, in North America. *Canadian Journal of Fisheries and Aquatic Sciences* 48:1389–1395.

STRONG, D. R. 1986. Density-vague population change. *Trends in Ecology and Evolution* 1:39–42.

STRONG, D. R. 1992. Are trophic cascades all wet? Differentiation and donor-control in speciose ecosystems. *Ecology* 73:747–754.

STRONG, D. R., J. H. LAWTON, and T. R. E. SOUTHWOOD. 1984. *Insects on Plants*. Harvard University Press, Cambridge, Mass.

STRONG, D. R., JR. 1984. Density-vague ecology and liberal population regulation in insects. In *A New Ecology*, ed. P. W. Price, C. N. Slobodchikoff, and W. S. Gaud, pp. 313–327. Wiley, New York.

STRONG, D. R. JR., E. D. MC COY, and J. R. REY. 1977. Time and the number of herbivore species: The pests of sugarcane. *Ecology* 58:167–175.

SUGIHARA, G. 1980. Minimal community structure: An explanation of species abundance patterns. *American Naturalist* 116:770–787.

SUTHERLAND, J. P. 1990. Perturbations, resistance, and alternative views of the existence of multiple stable points in nature. *American Naturalist* 136:270–275.

SWAN, L. A., and C. S. PAPP. 1972. *The Common Insects of North America*. Harper & Row, New York.

SWANK, S. E., and W. C. OECHEL. 1991. Interactions among the effects of herbivory, competition, and resource limitation on chaparral herbs. *Ecology* 72:104–115.

SWIFT, M. C. 1976. Energetics of vertical migration in *Chaoborus trivittatus* larvae. *Ecology* 57:900–914.

SYDEMAN, W. J., H. R. HUBER, S. D. EMSLIE, C. A. RIBIC, and N. NUR. 1991. Age-specific weaning success of northern elephant seals in relation to previous breeding experience. *Ecology* 72:2204–2217.

SYMONIDES, E. 1988. Population dynamics of annual plants. In *Plant Population Ecology*, ed. A. J. Davy, M. J. Hutchings, and A. R. Watkinson, pp. 221–248. Blackwell, Oxford, England.

TABASHNIK, B. E., N. L. CUSHING, N. FINSON, and M. W. JOHNSON. 1990. Field development of resistance to *Bacillus thuringiensis* in diamondback moth (Lepidoptera: Plutellidae). *Journal of Economic Entomology* 83:1671–1676.

TADROS, T. M. 1957. Evidence of the presence of an edapho-biotic factor in the problem of serpentine tolerance. *Ecology* 38:14–23.

TALBOT, F. H., B. C. RUSSELL, and G. R. V. ANDERSON. 1978. Coral reef fish communities: Unstable, high-diversity systems. *Ecological Monographs* 48:425–440.

TALLAMY, D. W., and M. J. RAUPP, eds. 1991. *Phytochemical Induction by Herbivores.* Wiley, New York.

TAMARIN, R. H. 1993. *Principles of Genetics,* 4th ed. Wm. C. Brown Publishers, Dubuque, Iowa.

TANSLEY, A. G. 1917. On competition between *Galium saxatile* L. and *G. sylvestre* Poll. on different types of soil. *Journal of Ecology* 5:173–179.

TANSLEY, A. G. 1935. The use and abuse of vegetational concepts and terms. *Ecology* 16:284–307.

TANSLEY, A. G. 1939. *The British Islands and Their Vegetation.* Cambridge University Press, Cambridge, England.

TAYLOR, O. R., JR. 1985. African bees: Potential impact in the United States. *Bulletin of the Entomological Society of America* 31(4):15–28.

TAYLOR, R. J. 1984. *Predation.* Chapman and Hall, New York.

TEERI, J. A., and L. G. STOWE. 1976. Climatic patterns and the distribution of C_4 grasses in North America. *Oecologia* 23:1–12.

TEMPLE, S. A. 1987. Do predators always capture substandard individuals disproportionately from prey populations? *Ecology* 68:669–674.

TEMPLETON, A. R. 1986. Coadaptation and outbreeding depression. In *Conservation Biology,* ed. M. E. Soulé, pp. 105–116. Sinauer Associates, Sunderland, Mass.

TER BRAAK, C. J. F., and I. C. PRENTICE. 1988. *A Theory of Gradient Analysis.* Academic Press, London.

THOMAS, A. S. 1960. Changes in vegetation since the advent of myxomatosis. *Journal of Ecology* 48:287–306.

THOMAS, A. S. 1963. Further changes in vegetation since the advent of myxomatosis. *Journal of Ecology* 51:151–183.

THOMPSON, S. D. 1982. Structure and species composition of desert heteromyid rodent species assemblages: Effects of a simple habitat manipulation. *Ecology* 63:1313–1321.

THORNTHWAITE, C. W. 1948. An approach toward a rational classification of climate. *Geographical Review* 38:55–94.

TILMAN, D. 1977. Resource competition between planktonic algae: An experimental and theoretical approach. *Ecology* 58:338–348.

TILMAN, D. 1982. *Resource Competition and Community Structure.* Princeton University Press, Princeton, N. J.

TILMAN, D. 1985. The resource ratio hypothesis of plant succession. *American Naturalist* 125:827–852.

TILMAN, D. 1986. Resources, competition and the dynamics of plant communities. In *Plant Ecology,* ed. M. J. Crawley, pp. 51–75. Blackwell, Oxford, England.

TILMAN, D. 1987. The importance of the mechanisms of interspecific competition. *American Naturalist* 129:769–774.

TILMAN, D. 1990. Mechanisms of plant competition for nutrients: The elements of a predictive theory of competition. In *Perspectives on Plant Competition*, ed. J. B. Grace and D. Tilman, pp. 117–141. Academic Press, San Diego.

TILMAN, D., S. S. KILHAM, and P. KILHAM. 1982. Phytoplankton community ecology: The role of limiting nutrients. *Annual Review of Ecology and Systematics* 13: 349–372.

TOMBACK, D. F. 1982. Dispersal of whitebark pine seeds by Clark's nutcracker: A mutualism hypothesis. *Journal of Animal Ecology* 51:451–467.

TOMOFF, C. S. 1974. Avian species diversity in desert scrub. *Ecology* 55:396–403.

TRANQUILLINI, W. 1979. *Physiological Ecology of the Alpine Timberline*. Springer-Verlag, Berlin.

TRANSEAU, E. N. 1935. The prairie peninsula. *Ecology* 16:423–437.

TRÉGUER, P., and G. JACQUES. 1992. Dynamics of nutrients and phytoplankton, and fluxes of carbon, nitrogen and silicon in the Antarctic Ocean. *Polar Biology* 12:149–162.

TREWARTHA, G. T. 1954. *An Introduction to Climate*, 3d ed. McGraw-Hill, New York.

TROSTEL, K., A. R. E. SINCLAIR, C. J. WALTERS, and C. J. KREBS. 1987. Can predation cause the 10–year hare cycle? *Oecologia* 74:185–192.

TURESSON, G. 1922. The species and the variety as ecological units. *Hereditas* 3:100–113.

TURESSON, G. 1925. The plant species in relation to habitat and climate. *Hereditas* 6:147–236.

TURESSON. G. 1930. The selective effect of climate upon the plant species. *Hereditas* 14:99–152.

TURNBULL, A. L., and D. A. CHANT. 1961. The practice and theory of biological control of insects in Canada. *Canadian Journal of Zoology* 39:697–753.

TURNER, J. R. G., J. J. LENNON, and J. A. LAWRENSON. 1988. British bird species distributions and the energy theory. *Nature* 335:539–541.

UDVARDY, M. D. F. 1959. Notes on the ecological concepts of habitat, biotope, and niche. *Ecology* 40:725–728.

UENO, O., and T. TAKEDA. 1992. Photosynthetic pathways, ecological characteristics, and the geographical distribution of the Cyperaceae in Japan. *Oecologia* 89:195–203.

UNDERWOOD, A. J., and E. J. DENLEY. 1984. Paradigms, explanations, and generalizations in models for the structure of intertidal communities on rocky shores. In *Ecological Communities*, ed. D. R. Strong, Jr., D. Simberloff, L. G. Abele, and A. B. Thistle, pp. 153–180. Princeton University Press, Princeton, N. J.

UNDERWOOD, A. J., and P. G. FAIRWEATHER. 1989. Supply-side ecology and benthic marine assemblages. *Trends in Ecology and Evolution* 4:16–20.

U.S. DEPARTMENT OF AGRICULTURE. 1990. Gypsy moth research and development program. Northeast Forest Experiment Station, October 1990.

UTIDA, S. 1957. Cyclic fluctuations of population density intrinsic to the host-parasite system. *Ecology* 38:442–449.

UVAROV, B. P. 1931. Insects and climate. *Transactions of the Royal Entomological Society of London* 79:1–247.

VAN DEN BOSCH, R., P. S. MESSENGER, and A. P. GUTIERREZ. 1982. *An Introduction to Biological Control*. Plenum, New York.

VAN EMDEN, H. F. 1991. The role of host plant resistance in insect pest mis-management. *Bulletin of Entomological Research* 81:123–126.

VAN RIPER, III, C., S. G. VAN RIPER, M. L. GOFF, and M. LAIRD. 1986. The epizo-ootiology and ecological significance of malaria in Hawaiian land birds. *Ecological Monographs* 56:327–344.

VARLEY, G. C. 1970. The concept of energy flow applied to a woodland community. In *Animal Populations in Relation to Their Food Resources*, ed. A. Watson, pp. 389–405. Blackwell, Oxford, England.

VARLEY, G. C., and G. R. GRADWELL. 1960. Key factors in population studies. *Journal of Animal Ecology* 29:399–401.

VARLEY, G. C., G. R. GRADWELL, and M. P. HASSELL. 1973. *Insect Population Ecology*. Blackwell, Oxford, England.

VEPSALAINEN, K. 1978. Wing dimorphism and diapause in Gerris: Determination and adaptive significance. In *Evolution of Insect Migration and Diapause*, ed. H. Dingle, pp. 218–253. Springer-Verlag, New York.

VERBOOM, B., and R. VAN APELDOORN. 1990. Effects of habitat fragmentation on the red squirrel, *Sciurus vulgaris* L. *Landscape Ecology* 4:171–176.

VITOUSEK, P. M. 1982. Nutrient cycling and the nutrient use efficiency. *American Naturalist* 119:553–572.

VOLLENWEIDER, R. A. 1974. *A Manual on Methods of Measuring Primary Production in Aquatic Environments*, 2d ed. International Biological Program Handbook No. 12, Blackwell, Oxford, England.

VOLTERRA, V. 1926. Fluctuations in the abundance of a species considered mathematically. *Nature* 118:558–560.

WAAGE, J. 1990. Ecological theory and the selection of biological control agents. In *Critical Issues in Biological Control*, ed. M. MacKauer, L. Ehler, and J. Roland, pp. 135–157. Intercept Ltd., Andover, U.K.

WAGNER, R. H. 1964. The ecology of *Uniola paniculata* L. in the dune-strand habitat of North Carolina. *Ecological Monographs* 34:79–96.

WALKER, B. H. 1976. An assessment of the ecological basis of game ranching in southern African savannas. *Proceedings of the Grasslands Society of South Africa* 1:125–130.

WALKER, B. H. 1992. Biodiversity and ecological redundancy. *Conservation Biology* 6:18–23.

WALKER, L. R., and F. S. CHAPIN III. 1987. Interactions among processes controlling successional change. *Oikos* 50:131–135.

WALKER, R. B. 1954. The ecology of serpentine soils. II. Factors affecting plant growth on serpentine soils. *Ecology* 35:259–266.

WALKINSHAW, L. H. 1983. Kirtland's Warbler: The natural history of an endangered species. Bulletin 58. Cranbrook Institute of Science, Bloomfield Hills, Mich.

WALLACE, A. R. 1878. *Tropical Nature and Other Essays*. Macmillan, New York.

WALLER, D. M. 1986. The dynamics of growth and form. In *Plant Ecology*, ed. M. J. Crawley, pp. 291–320. Blackwell, Oxford, England.

WALTERS, C. 1986. *Adaptive Management of Renewable Resources*. Macmillan, New York.

WALTERS, J. R. 1991. Application of ecological principles to the management of endangered species: The case of the red-cockaded woodpecker. *Annual Review of Ecology and Systematics* 22:505–523.

WAPLES, R. S. 1991. Genetic interactions between hatchery and wild salmonids: Lessons from the Pacific Northwest. *Canadian Journal of Fisheries and Aquatic Sciences* 48(Suppl.)124–133.

WARDLE, P. 1965. A comparison of alpine timberlines in New Zealand and North America. *New Zealand Journal of Botany* 3:113–135.

WARDLE, P. 1981. Is the alpine timberline set by physiological tolerance, reproductive capacity, or biological interactions? *Proceedings of the Ecological Society of Australia* 11: 53–66.

WARING, R. H., and W. H. SCHLESINGER. 1985. *Forest Ecosystems: Concepts and Management.* Academic Press, Orlando.

WARMING, J. E. B. 1895. *Plantesamfundgrundträk af den ökologiska plantegeogrefi.* Copenhagen.

WARMING, J. E. B. 1896. *Lehrbuch der ökologischen Pflanzengeographie.* Berlin.

WARMING, J. E. B. 1909. *Oecology of Plants.* Oxford University Press, New York.

WARNER, R. E. 1968. The role of introduced diseases in the extinction of the endemic Hawaiian avifauna. *Condor* 70:101–120.

WARREN, C. E., and G. E. DAVIS. 1967. Laboratory studies on the feeding, bioenergetics, and growth of fish. In *The Biological Basis of Freshwater Fish Production*, ed. S. D. Gerking, pp. 175–214. Blackwell, Oxford, England.

WASER, N. M. 1978a. Competition for hummingbird pollination and sequential flowering in two Colorado wildflowers. *Ecology* 59:934–944.

WASER, N. M. 1978b. Interspecific pollen transfer and competition between co-occurring plant species. *Oecologia* 36:223–236.

WATERS, T. F. 1957. The effects of lime application to acid bog lakes in northern Michigan. *Transactions of the American Fisheries Society* 86:329–344.

WATSON, A., and R. MOSS. 1970. Dominance, spacing behaviour and aggression in relation to population limitation in vertebrates. In *Animal Populations in Relation to Their Food Resources*, ed. A. Watson, pp. 167–218. Blackwell, Oxford, England.

WATSON, N. H. F. 1974. Zooplankton of the St. Lawrence Great Lakes—species composition, distribution, and abundance. *Journal of the Fisheries Research Board of Canada* 31:783–794.

WATT, A. S. 1940. Contributions to the ecology of bracken *(Pteridium aquilinum)*. I. The rhizome. *New Phytologist* 39:401–422.

WATT, A. S. 1947a. Contributions to the ecology of bracken *(Pteridium aquilinum)*. IV. The structure of the community. *New Phytologist* 46:97–121.

WATT, A. S. 1947b. Pattern and process in the plant community. *Journal of Ecology* 35:1–22.

WATT, A. S. 1955. Bracken versus heather, a study in plant sociology. *Journal of Ecology* 43:490–506.

WEAVER, J. E. 1968. *Prairie Plants and Their Environment.* University of Nebraska Press, Lincoln.

WEBB, D. A. 1954. Is the classification of plant communities either possible or desirable? *Botanisk Tidsskrift* 51:362–370.

WEBB, L. J., J. G. TRACEY, and K. P. HAYDOCK. 1967. A factor toxic to seedlings of the same species associated with living roots of the nongregarious subtropical rain forest tree *Grevillea robusta*. *Journal of Applied Ecology* 4:13–25.

WECKER, S. C. 1963. The role of early experience in habitat selection by the prairie deer mouse, *Peromyscus maniculatus bairdi*. *Ecological Monographs* 33:307–325.

WECKER, S. C. 1964. Habitat selection. *Scientific American* 211(4):109–116.

WEEKS, J. R. 1989. *Population: An Introduction to Concepts and Issues,* 4th ed. Wadsworth Publishing Company, Belmont, California.

WEINER, J. 1992. Physiological limits to sustainable energy budgets in birds and mammals: Ecological implications. *Trends in Ecology and Evolution* 7:384–388.

WELLER, D. E. 1987. A reevalutation of the $-3/2$ power rule of plant self-thinning. *Ecological Monographs* 57:23–43.

WELLER, D. E. 1991. The self-thinning rule: Dead or unsupported?—a reply to Lonsdale. *Ecology* 72:747–750.

WELLS, B. W., and I. V. SHUNK. 1938. Salt spray: An important factor in coastal ecology. *Bulletin of the Torrey Botanical Club* 65:485–492.

WELLS, P. V. 1965. Scarp woodlands, transported grassland soils, and concept of grassland climate in the Great Plains region. *Science* 148:246–249.

WERNER, E. E., J. F. GILLIAM, D. J. HALL, and G. G. MITTELBACH. 1983. An experimental test of the effects of predation risk on habitat use in fish. *Ecology* 64:1540–1548.

WESTERN, D., and M. C. PEARL, eds. 1989. *Conservation for the Twenty-first Century.* Oxford University Press, New York.

WESTOBY, M. 1984. The self-thinning rule. *Advances in Ecological Research* 14:167–225.

WETZEL, R. G., and H. L. ALLEN. 1971. Functions and interactions of dissolved organic matter and the littoral zone in lake metabolism and eutrophication. In *Productivity Problems of Freshwaters*, Proceedings of IBP-UNESCO Symposium, Poland, May 1970, ed. Z. Kajak and A. Hillbricht-Ilkowska, Warsaw.

WHITE, G. G. 1981. Current status of prickly pear control by *Cactoblastis cactorum* in Queensland. *Proceedings of the 5th International Symposium on the Biological Control of Weeds*, Brisbane, 1980, pp. 609–616.

WHITE, H. C. 1939. Bird control to increase the Margaree River salmon. *Fisheries Research Board of Canada Bulletin* 58:1–30.

WHITE, T. C. R. 1974. A hypothesis to explain outbreaks of looper caterpillars, with special reference to populations of *Selidosema suavis* in a plantation of *Pinus radiata* in New Zealand. *Oecologia* 16:279–301.

WHITE, T. C. R. 1984. The abundance of invertebrate herbivory in relation to the availability of nitrogen in stressed food plants. *Oecologia* 63:90–105.

WHITEHEAD, D. R. 1969. Wind pollination in the angiosperms: Evolutionary and environmental considerations. *Evolution* 23:28–35.

WHITEHEAD, D. R. 1973. Late-Wisconsin vegetational changes in unglaciated eastern North America. *Quaternary Research* 3:621–631.

WHITEHEAD, D. R. 1981. Late-Pleistocene vegetational changes in northeastern North Carolina. *Ecological Monographs* 51:451–471.

WHITTAKER, R. H. 1953. A consideration of climax theory: The climax as a population and pattern. *Ecological Monographs* 23:41–78.

WHITTAKER, R. H. 1954. The ecology of serpentine soils. I. Introduction. *Ecology* 35:258–259.

WHITTAKER, R. H. 1956. Vegetation of the Great Smoky Mountains. *Ecological Monographs* 26:1–80.

WHITTAKER, R. H. 1960. Vegetation of the Siskiyou Mountains, Oregon and California. *Ecological Monographs* 30:279–338.

WHITTAKER, R. H. 1961. Experiments with radiophosphorus tracer in aquarium microcosms. *Ecological Monographs* 31:157–188.

WHITTAKER, R. H. 1962. Classification of natural communities. *Botanical Review* 28:1–239.

WHITTAKER, R. H. 1967. Gradient analysis of vegetation. *Biological Reviews* 42:207–264.

WHITTAKER, R. H. 1972. Evolution and measurement of species diversity. *Taxon* 21:213–251.

WHITTAKER, R. H. 1975. *Communities and Ecosystems*, 2d ed. Macmillan, New York.

WHITTAKER, R. H., and P. P. FEENY. 1971. Allelochemics: Chemical interactions between species. *Science* 171:757–770.

WHITTAKER, R. H., S. A. LEVIN, and R. B. ROOT. 1973. Niche, habitat, and ecotope. *American Naturalist* 107:321–338.

WICKLOW, D. T. 1966. Further observations on serpentine response in *Emmenanthe*. *Ecology* 47:864–865.

WIENS, J. A. 1977. On competition and variable environments. *American Scientist* 65:590–597.

WIENS, J. A. 1984a. Resource systems, populations, and communities. In *A New Ecology*, ed. P. W. Price, C. N. Slobodchikoff, and W. S. Gaud, pp. 397–436. Wiley, New York.

WIENS, J. A. 1984b. On understanding a non-equilibrium world: Myth and reality in community patterns and processes. In *Ecological Communities*, ed. D. R. Strong Jr., D. Simberloff, L. G. Abele and A. B. Thistle, pp. 439–457. Princeton University Press, Princeton, N.J.

WIENS, J. A. 1985. Habitat selection in variable environments: Shrub-steppe birds. In *Habitat Selection in Birds*, ed. M. L. Cody, pp. 227–251. Academic Press, New York.

WIENS, J. A. 1986. Spatial scale and temporal variation in studies of shrub–steppe birds. In *Community Ecology*, ed. J. Diamond and T. J. Case, pp. 154–172. Harper & Row, New York.

WIENS, J. A. 1989. *The Ecology of Bird Communities. Volume 2. Processes and variations*. Cambridge University Press, Cambridge, England.

WIENS, J. A., J. F. ADDICOTT, T. J. CASE, and J. DIAMOND. 1986. Overview: The importance of spatial and temporal scale in ecological investigations. In *Community Ecology*, ed. J. Diamond and T. J. Case, pp. 145–153. Harper & Row, New York.

WILCOVE, D. S., and S. K. ROBINSON. 1990. The impact of forest fragmentation on bird communities in eastern North America. In *Biogeography and Ecology of Forest Bird Communities*, ed. A. Keast, pp. 319–331. SPB Academic Publishing bv, The Hague.

WILLIAMS, C. B. 1964. *Patterns in the Balance of Nature and Related Problems in Quantitative Ecology*. Academic Press, New York.

WILLIAMS, G. C. 1966. *Adaptation and Natural Selection*. Princeton University Press, Princeton, N. J.

WILLIAMS, W. E., K. GARBUTT, F. A. BAZZAZ, and P. M. VITOUSEK. 1986. The response of plants to elevated CO_2 IV. Two deciduous-forest tree communities. *Oecologia* 69:454–459.

WILLIAMSON, G. B., and E. M. BLACK. 1981. High temperature of forest fires under pines as a selective advantage over oaks. *Nature* 293:643–644.

WILLIAMSON, M. 1987. Are communities ever stable? In *Colonization, Succession and Stability*, ed. A. J. Bray, M. J. Crawley and P. J. Edwards, pp. 353–371. Blackwell, Oxford, England.

WILLIAMSON, M. 1989. The MacArthur and Wilson theory today: True but trivial. *Journal of Biogeography* 16:3–4.

WILSON, D. S. 1983. The group selection controversy: History and current status. *Annual Review of Ecology and Systematics* 14:159–187.

WILSON, E. O. 1969. The species equilibrium. *Brookhaven Symposia in Biology* 22:38–47.

WILSON, E. O. 1975. *Sociobiology: The New Synthesis.* Harvard University Press (Belknap Press), Cambridge, Mass.

WILSON, J. W., III. 1974. Analytical zoogeography of North American mammals. *Evolution* 28:124–140.

WINKLER, D. W., and J. R. WALTERS. 1983. The determination of clutch size in precocial birds. *Current Ornithology* 1:33–60.

WINTERS, G. H. and J. P. WHEELER. 1985. Interaction between stock area, stock abundance, and catchability coefficient. *Canadian Journal of Fisheries and Aquatic Sciences* 42:989–998.

WOLCOTT, T. G. 1973. Physiological ecology and intertidal zonation in limpets (Acmaea): A critical look at "limiting factors." *Biological Bulletin* 145:389–422.

WOLFSON, A. 1964. Animal photoperiodism. In *Photophysiology*, vol. 2, ed. A. C. Giese, pp. 1–49. Academic Press, New York.

WOOD, D. M., and R. DEL MORAL. 1987. Mechanisms of early primary succession in subalpine habitats on Mount St. Helens. *Ecology* 68:780–790.

WOOD, D. M., and R. DEL MORAL. 1988. Colonizing plants on the pumice plains, Mount St. Helens, Washington. *American Journal of Botany* 75:1228–1237.

WOODS, K. D., and M. B. DAVIS. 1989. Paleoecology of range limits: Beech in the upper peninsula of Michigan. *Ecology* 70:681–696.

WRIGHT, H. E., JR. 1968. The roles of pine and spruce in the forest history of Minnesota and adjacent areas. *Ecology* 49:937–955.

WRIGHT, R. D., and H. A. MOONEY. 1965. Substrate-oriented distribution of bristlecone pine in the White Mountains of California. *American Midland Naturalist* 73:257–284.

WU, L., A. D. BRADSHAW, and D. A. THURMAN. 1975. The potential for evolution of heavy metal tolerance in plants. III. The rapid evolution of copper tolerance in *Agrostis stolonifera. Heredity* 34:165–187.

WYNNE-EDWARDS, V. C. 1962. *Animal Dispersion in Relation to Social Behavior.* Oliver and Boyd, Edinburgh.

WYNNE-EDWARDS, V. C. 1986. *Evolution Through Group Selection.* Blackwell, Oxford, England.

YODA, K., T. KIRA, H. OGAWA, and K. HOZUMI. 1963. Self-thinning in overcrowded pure stands under cultivated and natural conditions. *Journal of the Institute of Polytechnics, Osaka City University,* Series D, 14:107–129.

YODZIS, P. 1986. Competition, mortality, and community structure. In *Community Ecology,* ed. J. Diamond and T. J. Case, pp. 480–491. Harper & Row, New York.

YOM-TOV, Y. 1979. Is air temperature limiting northern breeding distribution of birds? *Ornis Scandinavica* 11:71–72.

ZAR, J. H. 1984. *Biostatistical Analysis,* 2d. ed. Prentice-Hall, Englewood Cliffs, N.J.

ZARET, T. M. 1982. The stability/diversity controversy: A test of hypotheses. *Ecology* 63:721–731.

ZELITCH, I. 1971. *Photosynthesis, Photorespiration, and Plant Productivity.* Academic Press, New York.

FIGURE ACKNOWLEDGMENTS

Chapter 1   Figure 1.2 G. Caughly, N. Shepherd, and J. Short. *Kangaroos: Their Ecology and Management in the Sheep Rangelands of Australia*, p. 12. Copyright © 1987 Cambridge University Press, Cambridge. Reprinted with permission.

Chapter 2   Figure 2.2 Robet H. Tamarin. *Principles of Genetics*, 4th ed., p. 548. Copyright © 1993 Wm. C. Brown Communications Inc., Dubuque, Iowa. All rights reserved. Reprinted by permission. **Figure 2.5** F. Cooke and C. S. Findlay. Polygenic variation and stabilizing selection in a wild population of lesser snow geese (*Anser Caerulescens*). In *American Naturalist*, Vol. 120, p. 544. Copyright © 1982 the University of Chicago Press, Chicago. Reprinted by Permission. **Figure 2.8** N. Nur. Consequences of brood size for breeding blue kits II. Nestling weight, offspring survival and optimal brood size, p. 505. Copyright © 1984, *Journal of Animal Ecology* 53:497–517. Reprinted by permission. **Figure 2.9** Robet H. Tamarin. *Principles of Genetics*, 4th ed., p. 561. Copyright © 1993 Wm. C. Brown Communications, Inc., Dubuque, Iowa. All rights reserved. Reprinted by permission. **Figure 2.10** D. J. Fukuyama. *Evolutionary Biology*, p. 402. Copyright © 1986 Sinauer Associates, Sunderland, MA. Reprinted by permission. **Figure 2.11** B. J. MacFadden. Patterns of phylogeny and rates of evolution in fossil horses: Hipparions from the Miocene and Pliocene of North America, p. 247. Copyright © 1985 *Paleobiology II*: 245–257. Reprinted by permission.

Chapter 3   Figure 3.2 M. R. MacNair. Tolerance of higher plants to toxic materials. In *Genetic Consequences of Man-Made Change*, ed. J. A. Bishop and L. M. Cook, p. 199. Copyright © 1981 Academic Press, London. Reprinted by permission.

Chapter 4   Figure 4.2 William C. Grimm. *Familiar Trees of America*, p. 109. Copyright © 1967 by William C. Grimm. Reprinted by permission of HarperCollins Publishers Inc. **Figures 4.3, 4.4** J. A. Lubena and S. A. Levin. The spread of a reinvading species, range expansion in the California sea otter. *American Naturalist*, Vol. 131, Fig. 1, p. 529; Fig. 2, p. 535. Copyright © 1988 the University of Chicago Press, Chicago. Reprinted by permission. **Figure 4.5** A. Schoener. Experimental zoogeography: Colonization of marine mini-islands. *American Naturalist*, Vol. 108, p. 723. Copyright © 1974 the University of Chicago Press, Chicago. Reprinted by permission. **Figure 4.6** E. C. Pielou. *Biogeography*, p. 247. Copyright © 1979 John Wiley and Sons, Inc. Reprinted by permission. **Figure 4.7** L. A. Swan and C. S. Papp. *The Common Insects of North America*, p. 586. Copyright © 1972 by L. A. Swan and C. S. Papp. Reprinted by permission of HarperCollins Publishers Inc.

Chapter 5   Figure 5.1 R. L. Hutto. Habitat selection by nonbreeding, migratory land birds. *Habitat Selection in Birds*, Chapter 16, ed. M. L. Cody, p. 471. Copyright © 1985 Academic Press, Orlando. Reprinted by permission. **Figure 5.2** J. A. Wiens. Habitat selection in variable environments: Shrub-steppe birds. *Habitat Selection in Birds*, ed. M. L. Cody, pp. 227–251, Fig. 5. Copyright © 1985 Academic Press, Orlando. Reprinted by permission. **Figure 5.3** M. C. Baker, D. B. Thompson, G. L. Sherman, M. A. Cunningham, and D. F. Timback. Allozyne frequencies in a linear series of song dialect populations. *Evolution*, Vol. 36, pp. 1020–1029. Copyright © 1982 Society for the Study of Evolution. Reprinted by permission. **Figure 5.4** M. V. Price. The role of microhabitat in structuring desert rodent communities, p. 913. Copyright © 1978. *Ecology*: 59:910–921. Reprinted by permission. **Figure 5.5** S. D. Thompson. Structure and species composition of desert hetermyid rodent species assemblages: Effects of a simple habitat manipulation, Fig. 1, p. 1314; Fig. 2, p. 1316. Copyright © 1982 *Ecology* 63:1313–1321.

Chapter 6 **Figures 6.1, 6.2** J. A. Kitching and F. J. Ebling. Ecological studies at Lough Inc. *Advances in Ecological Research*, Vol. 4, pp. 259–260. Copyright © 1967 Academic Press, Inc. Reprinted with permission. **Figure 6.4** J. D. Ovington. *Australian Endangered Species: Mammals, Birds and Reptiles*, pp. 76–77. Cassell Australia, Melbourne. Copyright © 1978 Derrick Ovington. Reprinted with permission. **Figure 6.5** C. Van Riper, III, S. G. Van Riper, M. L. Goff, and M. Laird. The epizootiology and ecological significance of malaria in Hawaiian land birds. *Ecological Monographs*, Vol. 56, Fig. 2, p. 331; Fig. 4, p. 333; Fig. 11, p. 338. Copyright © 1986 Ecological Society of America. Reprinted by permission. **Figure 6.8** J. M. Diamond. Assembly of species communities, p. 390. Reprinted by permission of the publishers from *Ecology and Evolution of Communities*, ed. M. L. Cody and J. M. Diamond, Cambridge, MA: The Belknap Press of Harvard University Press. Copyright © 1975 by the President and Fellows of Harvard College. **Figure 6.11** I. Newton. *Finches*, p. 109. Collins London. Copyright © 1972 Ian Newton. Reprinted by permission.

Chapter 7 **Figure 7.1** W. Tranquillini. Physiological ecology of the alpine timberline. *Ecological Studies*, Vol. 31, pp. 110, 118. Copyright © 1979 Springer-Verlag, Inc., Heidelberg. Reprinted with permission from the publisher and author. **Figure 7.2** G. T. Trewartha. *Introduction to Climate*, p. 55. Copyright © 1954 McGraw-Hill Book Company, New York. Used with permission of McGraw-Hill Book Company. **Figure 7.3** H. J. Crichfield. *General Climatology*, 4th ed., p. 69. Copyright © 1983 Prentice Hall, Englewood Cliffs, NJ. Reprinted by permission. **Figure 7.6** T. Root. Energy constraints on avian distributions and abundances. *Ecology*, Vol. 69, p. 331. Copyright © 1988 Ecological Society of America. Reprinted by permission. **Figure 7.10** W. Tranquillini. Physiological ecology of the alpine timberline. *Ecological Studies*, Vol. 31, pp. 110, 118. Copyright © 1979 Springer-Verlag, Berlin. Reprinted with permission of the publisher and the author. **Figure 7.17** M. B. Davis. Quaternary history and the stability of forest communities. *Forest Succession: Concepts and Application*, ed. D. C. West, H. H. Shugart, and D. B. Botkin, Fig. 10.10, p. 146; Fig. 10.11, p. 147. Copyright © 1981 Springer-Verlag, New York. Reprinted by permission.

Chapter 8 **Figure 8.1** I. Zelitch. *Photosynthesis, Photorespiration, and Plant Productivity*, p. 247. Copyright © 1971 Academic Press, New York. Reprinted by permission. **Figure 8.3** M. E. Hay. Functional geometry of seaweeds: Ecological consequences of thallus layering and shape in contrasting light environments. *On the Economy of Plant Form and Function*, ed. T. J. Givnish, p. 648. Copyright © 1986 Cambridge University Press, Cambridge. Reprinted with permission of the publisher and the author. **Figure 8.4** O. Bjorkman. Comparative studies on photosynthesis in higher plants. *Photophysiology*, Vol. 8, ed. A. C. Giese, p. 53. Copyright © 1973 Academic Press, New York. Reprinted with permission of the publisher and the author. **Figure 8.5** O. Bjorkman and J. Berry. High efficiency photosynthesis. *Scientific American*, Vol. 229, p. 86. Copyright © 1973 Scientific American, Inc. All rights reserved. Reprinted by permission **Figure 8.6** P. W. Hattersley. The distribution of C_3 and C_4 grasses in Australia in relation to climate. *Oecologia*, Vol. 57, p. 116. Copyright © 1983 Springer-Verlag, Heidelberg. Reprinted by permission of the publisher and the author. **Figure 8.9** G. E. Hutchinson. The chemical ecology of three species of Myriophyllum (Axgiospermae, Haloragaceae). *Limnology and Oceanography*, Vol. 15, p. 2. Copyright © 1970 American Society of Limnology and Oceanography. Reprinted by permission. **Figure 8.10** R. H. Wagner. The ecology of *Uniola paniculata L.* in the dune-strand habitat of North Carolina. *Ecological Monographs*, Vol. 34, p. 83. Copyright © 1964 Ecological Society of America. Reprinted by permission. **Figure 8.11** J. Rozema, M. C. T. Scholten, P. A. Blaauw, and J. Diggeen. Distribution limits and physiological tolerances with particular reference to the salt marsh environment. *Plant Population Ecology*, ed. A. J. Davy, M. J. Hutchings, and A. R. Watkinson, p. 145. Copyright © 1988 Blackwell Scientific Publications Ltd, Oxford. Reprinted by permission. **Figure 8.13** R. C. Anderson. The historic role of fire in the North American grassland. *Fire in the North American Tallgrass Prairies*, by Scott L. Collins and Linda L. Wallace. Copyright © 1990 University of Oklahoma Press. Reprinted by permission.

Chapter 9   **Figure 9.1** R. T. T. Forman. Growth under controlled conditions to explain the hierarchical distributions of a moss, *Tetraphis pellucida*. *Ecological Monographs*, Vol. 34, p. 21. Copyright © 1964 Ecological Society of America. Reprinted by permission. **Figure 9.3** S. Anderson. The theory of range-size (RS) distributions. *American Museum Novitates*, Vol. 2833, p. 6. Copyright © 1985 American Museum of Natural History. Reprinted by permission. **Figure 9.4** M. D. Pagel, R. M. May, and A. R. Collie. Ecological aspects of the geographical distribution and diversity of mammalian species. *American Naturalist*, Vol. 137, pp. 793, 800. Copyright © 1991 the University of Chicago Press, Chicago. Reprinted by permission. **Figure 9.5** G. Caughley, N. Shepherd, and J. Short. *Kangaroos: Their Ecology and Management in the Sheep Rangelands of Australia*, p. 12. Copyright © 1987 Cambridge University Press, Cambridge. Reprinted with permission. **Figure 9.6** K. J. Gaston. Patterns in the local and regional dynamics of moth populations. *Oikos,* Vol. 53, p. 51. Copyright © 1988 Oikos, Sweden. Reprinted by permission. **Figure 9.8** N. J. Gotelli and D. Simberloff. The distribution and abundance of tallgrass prairie plants: A test of the core-satellite hypothesis. *American Naturalist*, Vol. 130, p. 25. Copyright © 1987 the University of Chicago Press, Chicago. Reprinted by permission.

Chapter 10   **Figure 10.1** R. H. Peters and K. Wassenberg. The effect of body size on animal abundance. *Oecologia*, Vol. 60, p. 91. Copyright © 1983 Springer-Verlag, Heidelberg. Reprinted by permission of the authors and the publisher. **Figure 10.5** Photographs courtesy of Great Barrier Reef Marine Park Authority. Copyright © G.B.R.M.P.A.

Chapter 11   **Figure 11.12** R. C. Lewontin. Selection of colonizing ability. *The Genetics of Colonizing Species*, ed. H. G. Baker and G. L. Stebbins, p. 81. Copyright © 1965 Academic Press, New York. Reprinted by permission. **Figure 11.13** U. Halbach. Introductory remarks. *Population Ecology*, ed. U. Halbach and J. Jacobs, pp. 19, 21. Copyright © 1979 Gustav Fischer Verlag, Stuttgart. Reprinted by permission. **Figures 11.14, 11.19** T. H. Clutton-Brock, F. E. Guinness, and S. D. Albon. *Red Deer: Behavior and Ecology of Two Sines*, pp. 77, 154. Copyright © 1982 University of Chicago, Edinburgh University Press. Reprinted by permission.

Chapter 12   **Figure 12.17** E. C. Pielou. *An Introduction to Mathematical Ecology*, p. 11. Copyright © 1969 Wiley-Interscience, New York. Reprinted by permission. **Figure 12.19** H. Caswell. *Matrix Population Models*, p. 47. Copyright © 1989 Sinauer Associates, Inc., Sunderland, MA. Reprinted by permission. **Figure 12.20** D. T. Crouse, L. B. Crowder, and H. Caswell. A stage-based population model for loggerhead sea turtles and implications for conservation. *Ecology*, Vol. 68, p. 1417. Copyright © 1987 Ecological Society of America. Reprinted by permission.

Chapter 13   **Figure 13.13** D. Tilman. Resource competition between planktonic algae: An experimental and theoretical approach. *Ecology*, Vol. 58, p. 344. Copyright © 1977 Ecological Society of America. Reprinted by permission. **Figure 13.20** D. Lack. *Darwin's Finches*, p. 82. Copyright © 1947 Cambridge University Press, Cambridge. Reprinted by permission.

Chapter 14   **Figure 14.11** G. Caughley, G. C. Grigg, J. Caughley, and G. J. E. Hill. Does dingo predation control the densities of kangaroos and emus? *Australian Wildlife Research*, Vol. 7, No. 1, p. 5. Copyright © 1980 CSIRO. Reprinted with permission from the publisher and the authors. **Figure 14.14** M. P. Hassell. *The Dynamics of Competition and Predation*, p. 37. Copyright © 1976 Edward Arnold, London. Reprinted with permission. **Figure 14.15** W. W. Murdoch and A. Sih. Age-dependent interference in a predatory insect. *Journal of Animal Ecology*, Vol. 47, p. 586. Copyright © 1978 Blackwell Scientific Publications Ltd. Reprinted by permission. **Figure 14.16** M. P.

Hassell and R. M. May. Aggregation in predators and insect parasites and its effect on stability. *Journal of Animal Ecology*, Vol. 43, No. 2, p. 569. Copyright © 1974 Blackwell Scientific Publications, Ltd. Reprinted by permission. **Figure 14.17** M. P. Hassell. Arthropod predator-prey systems. *Theoretical Ecology*, ed. R. M. May, p. 129. Copyright © 1981, Blackwell Scientific Publications, Ltd, Oxford, England. Reprinted with permission. **Figure 14.19** W. W. Murdoch and A. Oaten. Predation and population stability. *Advances in Ecological Research*, Vol. 9, p. 67. Copyright © 1975, Academic Press, Ltd., London, England. Reprinted with permission. **Figure 14.22** D. A. Spiller and T. W. Schoener. An experimental study of the effect of lizards on web-spider communities. *Ecological Monographs*, Vol. 58, p. 63. Copyright © 1988 Ecological Society of America. Reprinted with permission. **Figure 14.23** H. W. Greene and R. W. McDiarmid. Coral snake mimicry: Does it occur? *Science*, Vol. 213, p. 1209. Copyright © 1981 by the AAAS. Reprinted with permission. **Figure 14.24** R. E. Kenward. Hawks and doves: Factors affecting success and selection in goshawk attacks on woodpigeons. *Journal of Animal Ecology*, Vol. 47, p. 455. Copyright © 1978 Blackwell Scientific Publications, Ltd, Oxford, England. Reprinted with permission. **Figure 14.25** H. Kruuk. Predators and anti-predator behaviour of the black-headed gull (*Larus ridibundus L.*). *Behavior*, Supplement, Vol. 11, p. 25. Copyright © 1964 E. J. Brill. Reprinted with permission.

Chapter 15   **Figure 15.1** R. H. Whittaker and P. P. Feeny. Allelochemics: Chemical interactions between species. *Science*, Vol. 17, p. 760. Copyright © 1971 by the AAAS. Reprinted with permission. **Figure 15.2** P. D. Coley, J. P. Bryant, and F. S. Chapin, III. Resource availability and plant antiherbivore defense. *Science*, Vol. 230, F. 1. Copyright © 1985 by the AAAS. Reprinted with permission. **Figure 15.6** C. D. Harvell. The ecology and evolution of inducible defenses in a marine bryozona: Cues, costs, and consequences. *American Naturalist*, Vol. 128, p. 812. Copyright © 1986 the University of Chicago Press. Reprinted by permission; Photo of bryozoan spines from *Ecology*, Vol. 73, p. 1568. Copyright © C. Drew Harvell. Reprinted by permission. **Figure 15.7** J. H. Myers and D. Bazely. Thorns, spines, prickles, and hairs: Are they stimulated by herbivory and do they deter herbivores? *Phytochemical Induction by Herbivores*, ed. T. W. Tallamy and M. J. Raupp, p. 332. Copyright © 1991 John Wiley and Sons, Inc. Reprinted with permission. **Figure 15.9** R. H. V. Bell. A grazing ecosystem in the Serengeti. *Scientific American*, Vol. 255, pp. 92–93. Copyright © 1971 by Scientific American, Inc. All rights reserved. Reprinted with permission from W. H. Freeman and Company. **Figure 15.10** M. D. Gwynne and R. H. V. Bell. Selection of vegetation components by grazing ungulates in the Serengeti National Park. *Nature*, Vol. 220, p. 391. Copyright © 1968 Macmillan Magazines Ltd., London, England. Reprinted with permission. **Figure 15.12** A. R. E. Sinclair and M. Norton-Grifiths. Does competition or facilitation regulate migrant ungulate populations in the Serengeti? A test of hypothesis. *Oecologia*, Vol. 53, p. 366. Copyright © 1982 Springer-Verlag, Heidelberg. Reprinted with permission. **Figure 15.16** G. Caughley. Plant-herbivore systems. *Theoretical Ecology*, ed. R. M. May, p. 103. Copyright © 1976 Blackwell Scientific Publications, Ltd, Oxford, England. Reprinted by permission of the author and the publisher. **Figure 15.17** S. Larsson. Stressful times for the plant stress–insect performance hypothesis. *Oikos*, Vol. 56, p. 281. Copyright © 1989 Oikos, Lund, Sweden. Reprinted with permission. **Figure 15.20** C. E. Bach. Direct and indirect interactions between ants (*Pheidole megacephala*), scales (*Coccus viridis*) and plants (*Pluchea indica*). *Oecologia*, Vol. 87, p. 235. Copyright © 1991 Springer-Verlag, Heidelberg. Reprinted with permission.

Chapter 16   **Figure 16.4** D. L. Deangelis and J. C. Waterhouse. Equilibrium and nonequilibrium concepts in ecological models. *Ecological Monographs*, Vol. 57, pp. 4, 5. Copyright © 1987 Ecological Society of America. Reprinted by permission. **Figure 16.6** D. R. Strong, Jr. Density-vague ecology and liberal population regulation in insects. *A New Ecology*, ed. P. W. Price, C. N. Slobodchikoff, and W. S. Gaud, p. 323. Copyright © 1984 John Wiley and Sons, Inc. Reprinted by permission of John Wiley and Sons, Inc. **Figure 16.7** J. D. Nichols, M. J. Conroy, D. R. Anderson, and K. P. Burnham. Compensatory mortality in waterfowl populations: A review of the evidence and implications for research and management. *Transactions of the North American Wildlife* *and Natural Resources Conference*, Vol. 49, p. 539. Copyright © 1984 Wildlife Management Institute. Reprinted by permission. **Figures 16.8, 16.9** G. C. Varley, G. R. Gradwell, and M. P. Hassell. *Insect Population Ecology*, pp. 121, 123. Copyright © 1973 Blackwell Scientific Publications Ltd, Oxford, England. Reprinted by permission. **Figure 16.10** T. R. E. Southwood. *Ecological Methods with Particular Reference to the Study of Insect Populations*, 2nd ed., p. 381. Copyright © 1978. Chapman and Hall Ltd., London, England. Reprinted by permission. **Figure 16.13** K. Myers, I. D. Marshall, and F. Fenner. Studies in epidemiology of infectious myxomatosis of rabbits. III. Observations on two succeeding epizootics in Australian wild rabbits on the riverine plain of southeastern Australia, 1951–1953. *Journal of Hygiene*, Vol. 52, p. 344. Copyright © 1954 Cambridge University Press. Reprinted with permission. **Figure 16.14** F. Fenner and K. Myers. Myxoma virus and myxomatosis in retrospect: The first quarter century of a new disease. *Viruses and Environment*, ed. E. Kurstak and K. Maramorosch, p. 555. Copyright © 1978 Academic Press, Inc. Reprinted with permission of the publisher and the authors.

Chapter 17   **Figure 17.7** J. A. Gulland. The mathematical models to fish populations. *The Exploitation of Natural Animal Populations*, ed. E. D. Le Cren and M. W. Holdgate, p. 214. Copyright © 1962 Blackwell Scientific Publications Ltd, Oxford, England. **Figure 17.8** R. J. H. Beverton. Long-term dynamics of certain North Sea fish populations. *The Exploitation of Natural Animal Populations*, ed. E. D. LeCren and M. W. Holdgate, p. 248. Copyright © 1962 Blackwell Scientific Publications Ltd, Oxford, England. Reprinted by permission. **Figure 17.11** G. H. Kruse and A. V. Tyler. Exploratory simulation of English sole recruitment mechanisms. *Transactions of the American Fisheries Society*, Vol. 118, p. 104. Copyright © 1989 American Fisheries Society. Reprinted by permission. **Figure 17.20** K. R. Allen. *Conservation and Management of Whales*, p. 13. Copyright © 1980 University of Washington Press, Seattle. Reprinted by permission. **Figure 17.21** D. G. Chapman. Evaluation of marine mammal population models. *Dynamics of Large Mammal Populations*, ed. C. W. Fowler and T. D. Smith, p. 289. Copyright © 1981 John Wiley and Sons, Inc. Reprinted by permission.

Chapter 18   **Figure 18.2** P. Debach. *Biological Control by Natural Enemies*, p. 4. Copyright © 1974 Cambridge University Press, London, England. Reprinted by permission. **Figure 18.3** R. Van Den Bosch, P. S. Messenger, and A. P. Gutierrez. *An Introduction to Biological Control*, p. 2. Copyright © 1982 Plenum Publishing Corp., New York. Reprinted by permission. **Figure 18.4** G. G. White. Current status of prickly pear control by *Cactoblastis cactorum* in Queensland. *Proceedings of the 5th International Symposium on the Biological Control of Weeds, Brisbane, 1980*, p. 610. Copyright 1981 CSIRO. Reprinted by permission. **Figure 18.6** P. M. Room. *Salvinia*. Copyright CSIRO Division of Entomology. Reprinted with permission. **Figure 18.7** J. D. Reeve and W. W. Murdoch. Biological control by the parasitoid *Apytis melinus*, and population stability of the California red scale. *Journal of Animal Ecology*, Vol. 55, p. 1077. Copyright © 1986 Blackwell Scientific Publications Ltd, Oxford, England. Reprinted by permission. **Figure 18.8** J. D. Reeve and W. W. Murdoch. Aggregation by parasitoids in the successful control of the California red scale: A test of theory. *Journal of Animal Ecology*, Vol. 54, p. 807. Copyright © 1985 Blackwell Scientific Publications Ltd, Oxford, England. Reprinted by permission. **Figures 18.9, 18.10** E. J. Armburst and G. G. Gyrisco. Forage crops insect pest management. *Introduction to Insect Pest Management*, 2nd ed., ed. R. L. Metcalf and W. Luckman, pp. 447, 457. Copyright © 1982 John Wiley and Sons, Inc. Reprinted by permission. **Figure 18.11** L. E. Ehler and R. W. Hall. Evidence for competitive exclusion of introduced natural enemies in biological control. *Environmental Entomology*, Vol. 11, p. 2. Copyright © 1982 Entomological Society of America. Reprinted by permission.

Chapter 19   **Figure 19.1** G. Caughley, H. Dublin, and I. Parker. Projected decline of the African elephant. *Biological Conservation*, Vol. 54, p. 160. Copyright © 1990 Elsevier Applied Science Publishers Ltd, Barking, England. Reprinted by permission. **Figure 19.2** J. R. Walters. Application of ecological principles to the management of endangered species: The case of the red-cockaded woodpecker. *Annual Review of Ecology and*

Systematics, Vol. 22, pp. 509, 510. Copyright © 1991 Annual Reviews Inc. Reproduced, with permission, from the *Annual Review of Ecology and Systematics*. **Figure 19.3** M. F. Kinnaird and T. G. O'Brien. Viable populations for an endangered forest primate, the Tana River Crested Mangabey (*Cercocebus galeritus galeritus*). *Conservation Biology*, Vol. 5, p. 210. Copyright © 1991 Blackwell Scientific Publications, Ltd, Cambridge, MA. Reprinted by permission of the Society for Conservation Biology and Blackwell Scientific Publications, Inc. **Figure 19.4** G. E. Belovsky. Extinction models and mammalian persistence. *Viable Populations for Conservation*, ed. M. E. Soulé, p. 46. Copyright © 1987 Cambridge University Press, Cambridge, England. Reprinted by permission. **Figure 19.5** W. D. Newmark. Legal and biotic boundaries of western North American national parks: A problem of congruence. *Biological Conservation*, Vol. 33, p. 199. Copyright © 1985 Elsevier Applied Science Publishers Ltd, Barking, England. Reprinted by permission. **Figures 19.6, 19.7** E. S. Menges. Population viability analysis of an endangered plant. *Conservation Biology*, Vol. 4, pp. 53, 58. Copyright © 1990 Blackwell Scientific Publications, Inc., Cambridge, MA. Reprinted by permission of the Society for Conservation Biology and Blackwell Scientific Publications, Inc. **Figure 19.9** R. Dirzo and M. C. Garcia. Rates of deforestation in Los Tuxtlas, a neotropical area in southeast Mexico. *Conservation Biology*, Vol. 6, p. 88. Copyright © 1992 Blackwell Scientific Publications, Inc., Cambridge, MA. Reprinted by permission of the Society for Conservation Biology and Blackwell Scientific Publications, Inc. **Figure 19.10** J. Rolstad. Consequences of forest fragmentation for the dynamics of bird populations: Conceptual issues and the evidence. *Biological Journal of the Linnean Society*, Vol. 42, p. 152. Copyright © 1991 The Linnean Society of London. Reprinted by permission. **Figure 19.11** B. Verboom and R. Van Apeldoorn. Effects of habitat fragmentation on the red squirrel, *Sciurus vulgaris L. Landscape Ecology*, Vol. 4, p. 174. Copyright © 1990 SPB Academic Publishing bv, the Netherlands. Reprinted by permission. **Figure 19.12** H. Andren and P. Angelstam. Elevated predation rates as an edge effect in habitat islands: Experimental evidence. *Ecology*, Vol. 69, p. 546. Copyright © 1988 Ecological Society of America. Reprinted by permission. **Figure 19.13** A. O. Nicholls and C. R. Margules. The design of studies to demonstrate the biological importance of corridors. *Nature Conservation 2: The Role of Corridors*, ed. D. A. Saunders and R. J. Hobbs, p. 53. Copyright © 1991 Surrey Beatty and Sons Pty. Ltd., Chipping North, Australia. Reprinted by permission. **Figure 19.14** A. Cutler. Nested faunas and extinction in fragmented habitats. *Conservation Biology*, Vol. 5, p. 497. Copyright © 1991 Blackwell Scientific Publications, Inc., Cambridge, MA. Reprinted by permission of the Society for Conservation Biology and Blackwell Scientific Publications, Inc. **Figure 19.17** A. B. Carey, S. P. Harton, and B. L. Biswell. Northern spotted owls: Influence of prey-base and landscape character. *Ecological Monographs*, Vol. 62, p. 229. Copyright © 1992 Ecological Society of America. Reprinted by permission. **Figure 19.18** J. Bart and E. D. Forsman. Dependence of northern spotted owls (*Strix occidentalis caurina*) on old-growth forests in the western USA. *Biological Conservation*, Vol. 62, p. 98. Copyright © 1992 Elsevier Applied Science Publishers Ltd., Barking, England. Reprinted by permission.

Chapter 20   **Figure 20.4** M. P. Austin. Continuum concept, ordination methods, and niche theory. *Annual Review of Ecology and Systematics*, Vol. 16, p. 42. Copyright © 1985 by Annual Reviews Inc. Reprinted, with permission, from the *Annual Review of Ecology and Systematics*. **Figure 20.7** R. Daubenmire. Vegetation: Identification of typal communities. *Science*, Vol. 151, p. 295. Copyright © 1966 by the AAAS. Reprinted with permission. **Figure 20.8** M. P. Austin and T. M. Smith. A new model for the continuum concept. *Vegetatio*, Vol. 83, p. 37. Copyright © 1989 Kluwer Academic Publishers. Reprinted by permission. **Figure 20.9** H. A. Gleason and A. Cronquist. *The Natural Geography of Plants*, pp. 160, 174. Copyright © 1964 Columbia University Press, New York. Reprinted with permission. **Figure 20.11** J. T. Curtis. *The Vegetation of Wisconsin: An Ordination of Plant Communities*. Madison: University of Wisconsin Press, p. 20. Copyright © 1959 by the Regents of the University of Wisconsin. Reprinted by permission. **Figure 20.13** D. R. Whitehead. Late-Pleistocene vegetational changes in northeastern North Carolina. *Ecological Monographs*, Vol. 51, p. 457. Copyright © 1981 Ecological Society of America. Reprinted by permission. **Figure 20.15** R. H.

Whittaker. Vegetation of the Great Smoky Mountains. *Ecological Monographs*, Vol. 26, p. 58. Copyright © 1956 Ecological Society of America. Reprinted by permission.

Chapter 21   **Figures 21.1, 21.2, 21.6** R. H. Whittaker. *Communities and Ecosystems*, 2nd ed., pp. 66, 167. New York: Macmillan. Copyright © 1975 by Robert H. Whittaker. Reprinted with the permission of Macmillan Publishing Company. **Figure 21.4** K. Kikuzawa. A cost–benefit analysis of leaf habit and leaf longevity of trees and their geographical pattern. *American Naturalist*, Vol. 138, p. 1257. Copyright © 1991 the University of Chicago Press. Reprinted by permission. **Figures 21.7, 21.8** T. J. Givnish. On the adaptive significance of leaf height in forest herbs. *American Naturalist*, Vol. 120, pp. 335, 358, 359. Copyright © 1982 the University of Chicago Press. Reprinted by permission. **Figure 21.10** W. M. Lewis, Jr. Surface/volume ratio: Implications for phytoplankton morphology. *Science*, Vol. 192, p. 885. Copyright © 1976 by the AAAS. Reprinted by permission. **Figure 21.11** J. Mauchline and L. R. Fisher. The biology of euphausiids. *Advances in Marine Biology*, Vol. 7, p. 147. Copyright © 1969 Academic Press Ltd., London, England. Reprinted by permission. **Figure 21.13** M. D. Ohman. The demographics benefits of diel vertical migration by zooplankton. *Ecological Monographs*, Vol. 60, p. 277. Copyright © 1990 Ecological Society of America. Reprinted by permission. **Figure 21.15** N. M. Waser. Competition for hummingbird pollination and sequential flowering in two Colorado wildflowers. *Ecology*, Vol. 59, p. 939. Copyright © 1978 Ecological Society of America. Reprinted by permission.

Chapter 22   **Figure 22.3** M. Huston and T. Smith. Plant succession: Life history and competition. *American Naturalist*, Vol. 130, p. 184. Copyright © 1987 the University of Chicago Press. Reprinted by permission. **Figure 22.12** R. Del Moral and D. M. Wood. Dynamics of herbaceous vegetation recovery on Mount St. Helens, Washington, USA, after a volcanic eruption. *Vegetatio*, Vol. 74, p. 12. Copyright © 1988 Kluwer Academic Publishers. Reprinted by permission. **Figure 22.13** Copyright © R. Del Moral. Reprinted by permission. **Figures 22.14, 22.15** W. P. Sousa. Experimental investigations of disturbance and ecological succession in a rocky intertidal algae community. *Ecological Monographs*, Vol. 49, pp. 233, 236. Copyright © 1979 Ecological Society of America. Reprinted with permission from Duke University Press. **Figure 22.21** N. V. Brokaw. Gap-phase regeneration of three pioneer tree species in a tropical forest. *Journal of Ecology*, Vol. 75, p. 14. Copyright © 1987 Blackwell Scientific Publications Ltd, Oxford, England. Reprinted by permission.

Chapter 23   **Figures 23.4, 23.5** G. Sugihara. Minimal community structure: An explanation of species abundance patterns. *American Naturalist*, Vol. 116, pp. 772, 779. Copyright © 1980 the University of Chicago Press. Reprinted by permission. **Figure 23.6** D. J. Currie and V. Paguin. Large-scale biogeographical patterns of species richness of trees. *Nature*, Vol. 329, p. 326. Copyright © 1987 Macmillan Magazines Ltd. Reprinted by permission. **Figure 23.9** C. R. Dickman. Patterns in the structure and diversity of marsupial carnivore communities. *Patterns in the Structure of Mammalian Communities*, ed. D. W. Morris, Z. Abramsky, B. J. Fox, and M. R. Willig, p. 243. Copyright © 1989 Texas Tech University Press, Lubbock. Reprinted by permission. **Figures 23.13, 23.19** D. J. Currie. Energy and large-scale patterns of animal- and plant-species richness. *American Naturalist*, Vol. 137, pp. 35, 38. Copyright © 1991 the University of Chicago Press. Reprinted by permission. **Figure 23.16** J. Roughgarden. Niche width: Biogeographic patterns among *Anolis* lizard populations. *American Naturalist*, Vol. 108, p. 432. Copyright © 1974 the University of Chicago Press. Reprinted by permission. **Figures 23.21, 23.22** M. Huston. A general hypothesis of species diversity. *American Naturalist*, Vol. 113, pp. 86, 87. Copyright © 1979 the University of Chicago Press. Reprinted by permission. **Figure 23.23** J. Lubchenco. Plant species diversity in a marine intertidal community: Importance of herbivore food preference and algae competitive abilities. *American Naturalist*, Vol. 112, p. 31. Copyright © 1978 the University of Chicago Press. Reprinted by permission.

Chapter 24   **Figure 24.2** J. A. Weins. On understanding a nonequilibrium world: Myth and reality in community patterns and processes. *Ecological Communities*, ed. D. R. Strong, Jr., D. Simberloff, L. G. Abele, and A. B. Thistle, p. 451. Copyright © 1984 Princeton

University Press, Princeton, NJ. Reprinted by permission. **Figure 24.5** G. C. Varley. The concept of energy flow applied to a woodland community. *Animal Populations in Relation to Their Food Resources*, ed. A. Watson, p. 389. Copyright © 1970 Blackwell Scientific Publications Ltd, Oxford, England. Reprinted by permission. **Figure 24.6** S. L. Pimm. *Food Webs*, p. 90. Copyright © 1982 Chapman and Hall Ltd, London. Reprinted by permission. **Figure 24.8** R. B. Root. Organization of a plant–anthropod association in simple and diverse habitats: The fauna of collars (*Brassica oleracea*). *Ecological Monographs*, Vol. 43, p. 98. Copyright © 1973 Ecological Society of America. Reprinted by permission. **Figure 24.9** R. W. Elner and R. L. Vadas. Inference in ecology: The sea urchin phenomenon in the northwestern Atlantic. *American Naturalist*, Vol. 136, pp. 108–125. Copyright © 1990 the University of Chicago Press. Reprinted by permission. **Figure 24.11** S. L. Pimm. Determining the effects of introduced species. *Trends in Ecology and Evaluation*, Vol. 2, p. 107. Copyright 1987 Elsevier Trends Journals. Reprinted by permission. **Figure 24.13** L. W. Buss and J. B. C. Jackson. Competitive networks: Nontransitive competitive relationship in cryptic coral reef environments. *American Naturalist*, Vol. 113, p. 232. Copyright © 1979 the University of Chicago Press. Reprinted by permission. **Figure 24.14** T. J. Smith, III. Seed predation in relation to due dominance and distribution in mangrove forests. *Ecology*, Vol. 68, p. 271. Copyright © 1987 Ecological Society of America. Reprinted by permission. **Figure 24.15** J. L. Brooks and S. I. Dodson. Predation, body size, and composition of plankton. *Science*, Vol. 150, p. 33. Copyright © 1965 by the AAAS. Reprinted by permission. **Figure 24.16** Z. M. Gliwicz. Food thresholds and body size in cladocerons. *Nature*, Vol. 343, p. 640. Copyright © 1990 Macmillan Magazines Ltd, London. Reprinted with permission.

Chapter 25   **Figure 25.1** D. L. Deangelis and J. C. Waterhouse. Equilibrium and non equilibrium concepts in ecological models. *Ecological Monographs*, Vol. 57, p. 8. Copyright © 1987 Ecological Society of America. Reprinted by permission. **Figure 25.2** G. P. Jones. The importance of recruitment to the dynamics of a coral reef fish population. *Ecology*, Vol, 71, p. 1693. Copyright © 1990 Ecological Society of America. Reprinted by permission. **Figure 25.3** G. E. Forrester. Factors influencing the juvenile demography of a coral reef fish. *Ecology*, Vol. 71, p. 1675. Copyright © 1990 Ecological Society of America. Reprinted by permission. **Figure 25.4** J. Lubchenko. Relative importance of competition and predation. *Community Ecology*, ed. J. Diamond and T. J. Chase, p. 549. Copyright © 1986 Harper & Row Publishers, Inc. Reprinted by permission of HarperCollins Publishers Inc. **Figure 25.5** R. T. Paine. Ecological determinism in the competition for space. *Ecology*, Vol. 65, p. 1345. Copyright © 1984 Ecological Society of America. Reprinted by permission. **Figure 25.6** B. A. Menge and J. P. Sutherland. Community regulation: Variation in disturbance, competition, and predation in relation to environmental stress and recruitment. *American Naturalist*, Vol. 130, p. 739. Copyright © 1987 the University of Chicago Press. Reprinted by permission. **Figure 25.8** D. J. McQueen, M. R. S. Johannes, J. R. Post, T. J. Stewart, and D. R. S. Lean. Bottom-up and top-down impacts on freshwater pelagic community structure. *Ecological Monographs*, Vol. 59, p. 301. Copyright © 1989 Ecological Society of America. Reprinted by permission. **Figure 25.9** M. E. Power. Effects of fish in river food webs. *Science*, Vol. 250, p. 8121. Copyright © 1990 by the AAAS. Reprinted by permission. **Figure 25.17** J. H. Brown. The theory of insular biogeography and the distribution of boral birds and mammals. *Great Basin Naturalist Memoirs* No. 2, p. 214. Copyright © 1978. Reprinted with permission. **Figures 25.18, 25.19** T. E. Lawlor. Comparative biogeography of mammals on islands. *Biological Journal of the Linnean Society*, Vol. 28, pp. 100, 101. Copyright © 1986. The Linnean Society of London. Reprinted by permission. **Figure 25.20** M. V. Lomolino. Mammalian community structure on islands: The importance of immigration, extinction and interactive effects. *Biological Journal of the Linnean Society*, Vol. 28, p. 10. Copyright © 1986 The Linnean Society of London. Reprinted by permission. **Figure 25.22** Copyright © T. Sinclair. Reprinted with permission.

Chapter 26   **Figure 26.2** D. L. Deangelis. *Dynamics of Nutrient Cycling and Food Webs*, p. 416. Copyright © 1992 Chapman and Hall Ltd., London. Reprinted by permission. **Figure 26.6** T. L. Hayward. Primary production in the North Pacific Central Gyre: A controversy with important implications. *Trends in Ecology and Evolution*, Vol. 6, p. 283. Copyright © 1991 Elsevier Trends Journals, Cambridge. Reprinted by permission. **Figure 26.8** Food and Agricultural Organization of the United Nations. *Atlas of the Living Resources of the Seas*, Maps 1.1 and 1.2. Copyright © 1972 FAR. Reprinted by permission. **Figure 26.9** J. H. Ryther and W. M. Dunstan. Nitrogen, phosphorus, and eutrophication in the coastal marine environment. *Science*, Vol. 171, pp. 1008, 1009. Copyright © 1971 by the AAAS. Reprinted by permission. **Figure 26.11** S. Sathyendranath, T. Platt, E. P. W. Horne, W. G. Harrison, O. Ulloa, R. Outerbridge, and N. Hoepffner. Estimation of new production in the ocean by compound remote sensing. *Nature*, Vol. 353, p. 129. Copyright © 1991 Macmillan Magazines Ltd, London. Reprinted by permission. **Figure 26.13** D. W. Schindler and E. J. Fee. Experimental lakes area: Whole-lake experiments in eutrophication. *Journal of the Fisheries Research Board of Canada*, Vol. 31, p. 940. Copyright © 1974 Department of Fisheries and Oceans, Ontario, Canada. Reprinted by permission. **Figure 26.15** D. W. Schindler. Evolution of phosphorus limitation in lakes. *Science*, Vol. 195, p. 261. Copyright © 1977 by the AAAS. Reprinted by permission. **Figure 26.16** V. H. Smith. Low nitrogen to phosphorus ratios favor dominance by blue-green algae in lake phytoplankton. *Science*, Vol. 221, p. 1983. Copyright © 1983 by the AAAS. Reprinted by permission. **Figure 26.17** R. D. Graez. Copyright © 1988 by R. D. Graez, CSIRO Division of Wildlife and Ecology. Reprinted by permission. **Figures 26.19, 26.21** R. V. O'Neill and D. L. Deangelis. Comparative productivity and biomass relations of forest ecosystems. *Dynamic Properties of Forest Ecosystems*, ed. D. E. Reichele, pp. 416, 432. Copyright © 1981 Cambridge University Press, Cambridge, England. Reprinted with the permission of Cambridge University Press. **Figures 26.20, 26.22** T. Kira. Primary production of forests. *Photosynthesis and Productivity in Different Environments*, ed. J. P. Cooper, pp. 7, 15. Copyright © 1975 Cambridge University Press, Cambridge, England. Reprinted with the permission of Cambridge University Press. **Figure 26.23** S. M. Cargill and R. L. Jeffries. Nutrient limitation of primary production in a subarctic saltmarsh. *Journal of Applied Ecology*, Vol. 21, p. 664. Copyright © 1984 Journal of Applied Ecology. Reprinted by permission. **Figure 26.24** H. Breman and C. T. DeWit. Rangeland productivity and exploitation in the Sahel. *Science*, Vol. 221, p. 1343. Copyright © 1983 by the AAAS. Reprinted by permission. **Figure 26.25** R. H. Stavn. The horizontal-vertical distribution hypothesis: Langmuir circulation and Daphnia distributions. *Limnology and Oceanography*, Vol. 16, p. 454. Copyright © 1971 American Society of Limnology and Oceanography. Reprinted by permission.

Chapter 27 **Figure 27.1** K. A. Nagy. Field metabolic rate and food requirement scaling in mammals and birds. *Ecological Monographs*, Vol. 57, p. 120. Copyright © 1987 Ecological Society of America. Reprinted by permission. **Figure 27.4** W. F. Humphreys. Production and respiration in animal populations. *Journal of Animal Ecology*, Vol. 48, p. 434. Copyright © 1979 Blackwell Scientific Publications Ltd. Reprinted by permission. **Figure 27.8** S. J. McNaughton, M. Oesterheld, D. A. Frank, and K. J. Williams. Ecosystem-level patterns of primary productivity and herbivory in terrestrial habitats. *Nature*, Vol. 341, p. 143. Copyright © 1989 Macmillan Magazines Ltd. Reprinted by permission. **Figure 27.9** N. R. French, ed. *Perspectives in Greenland Ecology*, p. 13. Copyright © 1979 Springer-Verlag, New York. Reprinted by permission. **Figure 27.10** W. K. Lauenroth. Grassland primary-production: North American grasslands in perspective. *Perspectives in Grassland Ecology*, ed. N. R. French, p. 18. Copyright © 1979 Springer-Verlag, New York. Reprinted by permission. **Figure 27.11** N. R. French, R. K. Steinhorst, and D. M. Swift. Grassland biomass trophic pyramids. *Perspectives in Grassland Ecology*, ed. N. R. French, p. 67. Copyright © 1979 Springer-Verlag, New York. Reprinted by permission. **Figure 27.12** J. A. Scott, N. R. French, and J. W. Leetham. Patterns of consumption in grasslands. *Perspectives in Grassland Ecology*, ed. N. R. French, p. 18. Copyright © 1979 Springer-Verlag, New York. Reprinted by permission. **Figure 27.13** J. L. Dodd and W. K. Lauenroth. Analysis of the response of a grassland ecosystem to stress. *Perspectives in Grassland Ecology*, ed. N. R. French, p. 48. Copyright © 1979 Springer-Verlag, New York. Reprinted by permission **Figure 27.14** J. MacNab. Does game cropping serve conservation? A reexamination of the African data. *Canadian Journal of Zoology*, Vol. 69, p. 2284. Copyright © 1991 National Research Council of Canada. Reprinted by permission. **Figure 27.15** M. Oesterheld, O. E. Sala, and S. J. McNaughton. Effect of animal husbandry on herbivore-carrying capacity at a regional scale. *Nature*, Vol. 356, p. 235. Copyright © 1992 Macmillan Magazines Ltd. Reprinted by permission. **Figure 27.16** J. Weiner. Physiological limits to sustainable energy budgets in birds and mammals: Ecological implications. *Trends in Ecology and Evolution*, Vol. 7, p. 384. Copyright © 1992 Elsevier Trends Journals, Cambridge, England. Reprinted by permission.

Chapter 28   Figure 28.1 D. L. Deangelis. *Dynamics of Nutrient Cycling and Food Webs*, p. 5. Copyright © 1992 Chapman and Hall Ltd., London. Reprinted by permission; C. Krebs. Copyright © C. Krebs. Reprinted by permission. **Figure 28.2** F. E. Smith. Analysis of ecosystems. *Analysis of Temperate Forest Ecosystems*, ed. D. Reichle, p. 12. Copyright © 1970 Springer-Verlag, Berlin. Reprinted by permission. **Figure 28.4** R. H. Waring and W. H. Schlesinger. *Forest Ecosystems: Concepts and Management*, p. 123. Copyright © 1985 Academic Press, Orlando. Reprinted by permission. **Figure 28.7** G. E. Likens, F. H. Bormann, N. M. Johnson, D. W. Fisher, and R. S. Pierce. Effects of forest cutting and herbicide treatment on nutrient budgets in the Hubbard Brook watershed-ecosystem. *Ecological Monographs*, Vol. 40, p. 33. Copyright © 1970 Ecological Society of America. Reprinted with permission from Duke University Press. **Figure 28.8** P. M. Vitousek. Nutrient cycling and the nutrient use efficiency. *American Naturalist*, Vol. 119, p. 558. Copyright © 1982 the University of Chicago Press. Reprinted by permission. **Figure 28.9** G. E. Likens, F. H. Bormann, and N. M. Johnson. Interactions between major biogeochemical cycles in terrestrial ecosystems. *Some Perspectives of the Major Biogeochemical Cycles*, ed. G. E. Likens, p. 103. Copyright © 1981 John Wiley and Sons, New York. Reprinted by permission. **Figure 28.10** P. Brimblecombe, C. Hammer, H. Rodhe, A. Ryaboshapko, and C. F. Boutron. Human influence on the sulphur cycle. Scope 39, *Evolution of the Global Biogeochemical Sulphur Cycle*, ed. P. Brimblecombe and A. Y. Lein, p. 82. Copyright © 1989 John Wiley and Sons, Chichester, U.K. Reprinted by permission. **Figure 28.12** E. D. Schulze. Air pollution and forest decline in a spruce (*Picea abies*) forest. *Science*, Vol. 24, p. 777. Copyright © 1989 by the AAAS. Reprinted by permission. **Figure 28.13** W. H. Schlesinger. *Biogeochemistry: An Analysis of Global Change*, p. 309. Copyright © 1991 Academic Press, San Diego. Reprinted by permission. **Figure 28.16** J. M. Barnola, D. Raynaud, Y. S. Korotkevich, and C. Lorius. Vostok icecore provides 160,000 year record of atmospheric CO_2. *Nature*, Vol. 329, p. 410. Copyright © 1987 Macmillan Magazines Ltd. Reprinted by permission. **Figure 28.18** W. D. Billings, J. O. Luken, D. A. Mortensen, and K. M. Peterson. Increasing atmospheric carbon dioxide: Possible effects on arctic tundra. *Oecologia*, Vol. 58, p. 288. Copyright © 1983 Springer-Verlag, Berlin. Reprinted by permission. **Figure 28.19** W. E. Williams, K. Garbutt, F. A. Bazzaz, and P. M. Vitousek. The response of plants to elevated CO_2IV. Two deciduous-forest tree communities. *Oecologia*, Vol. 69, p. 455. Copyright © 1986 Springer-Verlag, Berlin. Reprinted by permission. **Figure 28.20** E. D. Fajer, M. D. Bowers, and F. A. Bazzaz. The effects of enriched carbon dioxide atmospheres on plant-insect herbivore intersections. *Science*, Vol. 243, p. 1199. Copyright © 1989 by the AAAS. Reprinted by permission.

TABLE ACKNOWLEDGMENTS

Table 8.2 C. C. Black. Ecological implications of dividing plants into groups with distinct photosynthetic production capacities. *Advances in Ecological Research*, Vol. 7, p. 100. Copyright © 1971 Academic Press Ltd., London. Reprinted by permission. **Table 12.4** D. T. Crouse, L. B. Crowder, and H. Caswell. A stage-based population model for loggerhead sea turtles and implications for conservation. *Ecology*, Vol. 68, p. 1416. Copyright © 1987 Ecological Society of America. Reprinted by permission. **Table 13.1** J. H. Cornell. On the prevalence and relative importance of interspecific competition: Evidence from field experiments. *American Naturalist*, Vol. 122, p. 671. Copyright © 1983 the University of Chicago Press. Reprinted by permission. **Table 13.2** J. A. Wiens. *The Ecology of Bird Communities, Vol. 2, Processes and Variations*, p. 17. Copyright © 1989 Cambridge University Press, Cambridge. Reprinted by permission. **Table 15.1** P. D. Caley, J. P. Bryant, and F. S. Chapin, III. Resource availability and plant antiherbivore defense. *Science*, Vol. 230, p. 897. Copyright © 1985 by the AAAS. Reprinted by permission. **Table 15.3** C. M. Herrera. Defense of ripe fruit from pests: Its significance in relation to plant–disperser interactions. *American Naturalist*, Vol. 120, p. 230. Copyright © 1982 the University of Chicago Press. Reprinted by permission. **Table 16.1** G. D. Varley, G. R. Gradwell, and M. P. Hassell. *Insect Population Ecology*, p. 117. Copyright © 1973 Blackwell Scientific Publications Ltd, Oxford, England. Reprinted by permission. **Table 16.2** F. Fenner and K. Myers. Myxoma virus and myxomatosis in retrospect: The first quarter century of a new disease. *Viruses and Environment*, ed. E. Kurstak and K. Maramorosch, p. 552. Copyright © 1978 Academic Press, Inc., New York. Reprinted by permission. **Table 18.2** R. S. Patterson, D. E. Weidhaas, H. R. Ford, and C. S. Lofgren. Suppression and elimination of an island population of *Culex pipiens quiquefasciatus* with sterile males. *Science*, Vol. 1369. Copyright © 1970 by the AAAS. Reprinted by permission. **Table 18.3** R. Vanden Bosch, P. S. Messenger, and A. P. Gutierrez. *An Introduction to Biological Control*, p. 181. Copyright © 1982 Plenum Publishing Corp., New York. Reprinted by permission. **Table 19.1** D. Rabinowitz. Seven forms of rarity. *The Biological Aspects of Rare Plant Conservation*, ed. H. Synge, p. 208. Copyright © 1981 John Wiley and Sons, Chichester, Sussex, U.K. Reprinted by permission of John Wiley and Sons, Ltd. **Table 19.2** E. S. Menges. Population viability analysis for an endangered plant. *Conservation Biology*, Vol. 4, p. 60. Copyright © 1990 Blackwell Scientific Publications, Cambridge. Reprinted by permission of the Society for Conservation Biology and Blackwell Scientific Publications, Inc. **Table 19.3** J. Rolstad. Consequences of forest fragmentation for the dynamics of bird populations: Conceptual issues and the evidence. *Biological Journal of the Linnean Society*, Vol. 42, p. 151. Copyright © 1991 The Linnean Society of London. Reprinted by permission of Academic Press Ltd. **Table 19.4** R. F. Noss. Corridors in real landscapes: A reply to Simberloff and Cox. *Conservation Biology*, Vol. 1, p. 160. Copyright © 1987 Blackwell Scientific Publications, Cambridge. Reprinted by permission of the Society for Conservation Biology and Blackwell Scientific Publications, Inc. **Tables 21.1, 26.1** R. H. Whittaker. *Communities and Ecosystems*, 2nd ed., pp. 62, 224. New York: Macmillan. Copyright © 1975 by Robert H. Whittaker. Reprinted with the permission of Macmillan Publishing Company. **Table 22.1** M. Huston and T. Smith. Plant succession: Life history and competition. *American Naturalist*, Vol. 130, p. 170. Copyright © 1987 the University of Chicago Press. Reprinted by permission. **Table 22.2** H. S. Horn. Markovian properties of forest succession. *Ecology and Evolution of Communities*, ed. M. L. Cody and J. M. Diamond, pp. 199, 200. Copyright © 1975 by the President and Fellows of Harvard College. Reprinted by permission of the publishers from *Ecology and Evolution of Communities*. Cambridge, MA: Belknap Press of Harvard University Press. **Table 25.3** G. R. V. Anderson, A. H. Ehrlich, P. R. Ehrlich, J. D. Roughgarden, B. C. Russell, and F. H. Talbot. The community structure of coral reef fishes. *American Naturalist*, Vol. 117, p. 480. Copyright © 1981 the University of Chicago Press. Reprinted by permission. **Table 25.4** A. Sih, P. Crowley, M. McPeek, J. Petranka, and K. Strohmeier. Predation, competition and prey communities: A review of field experiments. *Annual Review of Ecology and Systematics*, Vol. 16, p. 289. Copyright © 1985 by Annual Reviews Inc. Reproduced, with permission, from the *Annual Review of Ecology and Systematics*. **Table 27.2** W. F. Humphreys. Production and respiration in animal populations. *Journal of Animal Ecology*, Vol. 48, p. 434. Copyright © 1979 Blackwell Scientific Publications Ltd. Reprinted by permission. **Table 27.3** J. A. Scott, N. R. French, and J. W. Leetham. Patterns of consumption in grasslands. *Perspectives in Grassland Ecology*, ed. N. R. French, p. 101. Copyright © 1979 Springer-Verlag, New York. Reprinted by permission. **Table 27.4** M. J. Coe, D. H. Cumming, and J. Phillipson. Biomass and production of large African herbivores in relation to rainfall and primary production. *Oecologia*, Vol. 22, p. 344. Copyright © 1976 Springer-Verlag, Heidelberg. Reprinted by permission. **Table 28.1** D. W. Coe and M. Rapp. Elemental cycling in forest ecosystems. *Dynamic Properties of Forest Ecosystems*, ed. D. E. Reichle, pp. 354, 357. Copyright © 1981 Cambridge University Press, London, England. Reprinted with the permission of Cambridge University Press. **Table 28.6** M. C. Feller. Biomass and nutrient distribution in two eucalypt forest ecosystems. *Australian Journal of Ecology*, Vol. 5, pp. 328–330. Copyright © 1980 Blackwell Scientific Publications, Victoria, Australia. Reprinted by permission. **Table 28.7** C. F. Jordan and R. Herrera. Tropical rain forests: Are nutrients really critical? *American Naturalist*, Vol. 117, pp. 170–171. Copyright © 1981 the University of Chicago Press. Reprinted by permission.

Species Index

Subject Index